Fuzzy Techniques for Decision Making

Special Issue Editor

José Carlos R. Alcantud

MDPI • Basel • Beijing • Wuhan • Barcelona • Belgrade

Special Issue Editor
José Carlos R. Alcantud
University of Salamanca
Spain

Editorial Office
MDPI
St. Alban-Anlage 66
Basel, Switzerland

This edition is a reprint of the Special Issue published online in the open access journal *Symmetry* (ISSN 2073-8994) in 2017–2018 (available at: http://www.mdpi.com/journal/symmetry/special_issues/Fuzzy_techniques_decision_making).

For citation purposes, cite each article independently as indicated on the article page online and as indicated below:

Lastname, F.M.; Lastname, F.M. Article title. *Journal Name* **Year**, *Article number*, page range.

First Editon 2018

Cover photo courtesy of https://pxhere.com/

ISBN 978-3-03842-887-9 (Pbk)
ISBN 978-3-03842-888-6 (PDF)

Table of Contents

About the Special Issue Editor

José Carlos R. Alcantud, Professor, received his M.Sc. in Mathematics in 1991 from the University of Valencia, Spain, and his Ph.D. in Mathematics in 1996 from the University of Santiago de Compostela, Spain. Since 2010, he is a Professor of Economic Analysis in the Department of Economics and Economic History at the University of Salamanca, Spain. He has been Head of the Department of Economics and Economic History at the University of Salamanca, Spain, since 2012. Professor Alcantud is a founding member of the Multidisciplinary Institute of Enterprise of the University of Salamanca. He has published well over 50 papers in mathematical economics and mathematics. His current research interests include soft computing models, decision making and its applications, and social choice.

symmetry

Editorial

Fuzzy Techniques for Decision Making

José Carlos R. Alcantud

BORDA Research Unit and IME, University of Salamanca, 37008 Salamanca, Spain; jcr@usal.es

Received: 22 December 2017; Accepted: 25 December 2017; Published: 27 December 2017

This book contains the successful invited submissions [1–21] to a Special Issue of *Symmetry* on the subject area of "Fuzzy Techniques for Decision Making".

We invited contributions addressing novel techniques and tools for decision making (e.g., group or multi-criteria decision making), with notions that overcome the problem of finding the membership degree of each element in Zadeh's original model. We could garner interesting articles in a variety of setups, as well as applications. As a result, this Special Issue includes some novel techniques and tools for decision making, such as:

- Instrumental tools for analysis like correlation coefficients [1,16] or similarity measures [4] and aggregation operators [2,21] in various settings.
- Novel contributions to methodologies, like discrete optimization with fuzzy constraints [3], COMET [5], or fuzzy bi-matrix games [7].
- New methodologies for hybrid models [12,15,18,20] inclusive of theoretical novelties [9].
- Applications to project delivery systems [6], maintenance performance in industry [8], group emergencies [10], pedestrians flows [11], valuation of assets [13], water pollution control [17], or aquaculture enterprise sustainability [19].
- A comparative study of some classes of soft rough sets [14].

Response to our call had the following statistics:

- Submissions (58);
- Publications (21);
- Rejections (37);
- Article types: Research Article (21);

Authors' geographical distribution (published papers) is:

- China (11)
- Spain (4)
- Pakistan (2)
- Poland (1)
- Japan (1)
- Taiwan (1)
- Slovenia (1)

Published submissions are related to various settings like fuzzy soft sets, hesitant fuzzy sets, (fuzzy) soft rough sets, neutrosophic sets, as well as other hybrid models.

I found the edition and selections of papers for this book very inspiring and rewarding. I also thank the editorial staff and reviewers for their efforts and help during the process.

Conflicts of Interest: The authors declare no conflict of interest.

References

1. Ye, J. Multiple Attribute Decision-Making Method Using Correlation Coefficients of Normal Neutrosophic Sets. *Symmetry* **2017**, *9*, 80. [CrossRef]
2. Chen, J.; Ye, J. Some Single-Valued Neutrosophic Dombi Weighted Aggregation Operators for Multiple Attribute Decision-Making. *Symmetry* **2017**, *9*, 82. [CrossRef]
3. Jelušič, P.; Žlender, B. Discrete Optimization with Fuzzy Constraints. *Symmetry* **2017**, *9*, 87. [CrossRef]
4. Jiang, W.; Shou, Y. A Novel Single-Valued Neutrosophic Set Similarity Measure and Its Application in Multicriteria Decision-Making. *Symmetry* **2017**, *9*, 127. [CrossRef]
5. Faizi, S.; Sałabun, W.; Rashid, T.; Wątróbski, J.; Zafar, S. Group Decision-Making for Hesitant Fuzzy Sets Based on Characteristic Objects Method. *Symmetry* **2017**, *9*, 136. [CrossRef]
6. Luo, S.; Cheng, P.; Wang, J.; Huang, Y. Selecting Project Delivery Systems Based on Simplified Neutrosophic Linguistic Preference Relations. *Symmetry* **2017**, *9*, 151. [CrossRef]
7. Zhang, W.; Xing, Y.; Qiu, D. Multi-objective Fuzzy Bi-matrix Game Model: A Multicriteria Non-Linear Programming Approach. *Symmetry* **2017**, *9*, 159. [CrossRef]
8. Carnero, M. Asymmetries in the Maintenance Performance of Spanish Industries before and after the Recession. *Symmetry* **2017**, *9*, 166. [CrossRef]
9. Tang, H. Decomposition and Intersection of Two Fuzzy Numbers for Fuzzy Preference Relations. *Symmetry* **2017**, *9*, 228. [CrossRef]
10. Wang, L.; Labella, Á.; Rodríguez, R.; Wang, Y.; Martínez, L. Managing Non-Homogeneous Information and Experts' Psychological Behavior in Group Emergency Decision Making. *Symmetry* **2017**, *9*, 234. [CrossRef]
11. Xue, Z.; Dong, Q.; Fan, X.; Jin, Q.; Jian, H.; Liu, J. Fuzzy Logic-Based Model That Incorporates Personality Traits for Heterogeneous Pedestrians. *Symmetry* **2017**, *9*, 239. [CrossRef]
12. Liu, Z.; Qin, K.; Pei, Z. A Method for Fuzzy Soft Sets in Decision-Making Based on an Ideal Solution. *Symmetry* **2017**, *9*, 246. [CrossRef]
13. Alcantud, J.; Rambaud, S.; Torrecillas, M. Valuation Fuzzy Soft Sets: A Flexible Fuzzy Soft Set Based Decision Making Procedure for the Valuation of Assets. *Symmetry* **2017**, *9*, 253. [CrossRef]
14. Liu, Y.; Martínez, L.; Qin, K. A Comparative Study of Some Soft Rough Sets. *Symmetry* **2017**, *9*, 252. [CrossRef]
15. Katagiri, H.; Kato, K.; Uno, T. Possibility/Necessity-Based Probabilistic Expectation Models for Linear Programming Problems with Discrete Fuzzy Random Variables. *Symmetry* **2017**, *9*, 254. [CrossRef]
16. Wang, Z.; Li, J. Correlation Coefficients of Probabilistic Hesitant Fuzzy Elements and Their Applications to Evaluation of the Alternatives. *Symmetry* **2017**, *9*, 259. [CrossRef]
17. Liu, J.; Li, Y.; Huang, G.; Chen, L. A Recourse-Based Type-2 Fuzzy Programming Method for Water Pollution Control under Uncertainty. *Symmetry* **2017**, *9*, 265. [CrossRef]
18. Akram, M.; Ali, G.; Alshehri, N. A New Multi-Attribute Decision-Making Method Based on m-Polar Fuzzy Soft Rough Sets. *Symmetry* **2017**, *9*, 271. [CrossRef]
19. Wu, T.; Chen, C.; Mao, N.; Lu, S. Fishmeal Supplier Evaluation and Selection for Aquaculture Enterprise Sustainability with a Fuzzy MCDM Approach. *Symmetry* **2017**, *9*, 286. [CrossRef]
20. Sarwar, M.; Akram, M. New Applications of m-Polar Fuzzy Matroids. *Symmetry* **2017**, *9*, 319. [CrossRef]
21. Kobina, A.; Liang, D.; He, X. Probabilistic Linguistic Power Aggregation Operators for Multi-criteria Group Decision Making. *Symmetry* **2017**, *9*, 320. [CrossRef]

symmetry

MDPI

Article

Some Single-Valued Neutrosophic Dombi Weighted Aggregation Operators for Multiple Attribute Decision-Making

Jiqian Chen [1] and Jun Ye [1,2,*]

[1] Department of Civil engineering, Shaoxing University, Shaoxing 312000, China; chenjiquian@yahoo.com
[2] Department of Electrical and Information Engineering, Shaoxing University, Shaoxing 312000, China
* Correspondence: yejun@usx.edu.cn; Tel.: +86-575-8832-7323

Academic Editor: José Carlos R. Alcantud
Received: 2 May 2017; Accepted: 30 May 2017; Published: 2 June 2017

Abstract: The Dombi operations of T-norm and T-conorm introduced by Dombi can have the advantage of good flexibility with the operational parameter. In existing studies, however, the Dombi operations have so far not yet been used for neutrosophic sets. To propose new aggregation operators for neutrosophic sets by the extension of the Dombi operations, this paper firstly presents the Dombi operations of single-valued neutrosophic numbers (SVNNs) based on the operations of the Dombi T-norm and T-conorm, and then proposes the single-valued neutrosophic Dombi weighted arithmetic average (SVNDWAA) operator and the single-valued neutrosophic Dombi weighted geometric average (SVNDWGA) operator to deal with the aggregation of SVNNs and investigates their properties. Because the SVNDWAA and SVNDWGA operators have the advantage of good flexibility with the operational parameter, we develop a multiple attribute decision-making (MADM) method based on the SVNWAA or SVNWGA operator under a SVNN environment. Finally, an illustrative example about the selection problem of investment alternatives is given to demonstrate the application and feasibility of the developed approach.

Keywords: single-valued neutrosophic number; Dombi operation; single-valued neutrosophic Dombi weighted arithmetic average (SVNDWAA) operator; single-valued neutrosophic Dombi weighted geometric average (SVNDWGA) operator; multiple attribute decision-making

1. Introduction

In 1965, Zadeh [1] introduced a membership function between 0 and 1 instead of traditional crisp value of 0 and 1 and defined the fuzzy set (FS). Fuzzy theory is an important and interesting research topic in decision-making theory and science. However, FS is characterized only by its membership function between 0 and 1, but not a non-membership function. To overcome the insufficient of FS, Atanassov [2] introduced the concept of an intuitionistic fuzzy set (IFS), which is characterized by its membership function and non-membership function between 0 and 1. As a further generalization of an IFS, Atanassov and Gargov [3] further introduced the concept of an interval-valued intuitionistic fuzzy set (IVIFS), which is characterized by its interval membership function and interval non-membership function in the unit interval [0, 1]. Because IFSs and IVIFSs cannot represent indeterminate and inconsistent information, Smarandache [4] introduced a neutrosophic set (NS) from a philosophical point of view to express indeterminate and inconsistent information. In a NS A, its truth, falsity, and indeterminacy membership functions $T_A(x)$, $I_A(x)$ and $F_A(x)$ are represented independently, which lie in real standard or nonstandard subsets of $]^-0, 1^+[$, i.e., $T_A(x)$: $X \to]^-0, 1^+[$, $I_A(x)$: $X \to]^-0, 1^+[$, and $F_A(x)$: $X \to]^-0, 1^+[$. Thus, the nonstandard interval $]^-0, 1^+[$ may result in the difficulty of actual applications. Based on the real standard interval [0, 1], therefore, the concepts of a single-valued neutrosophic set (SVNS) [5] and an interval neutrosophic set (INS) [6] was presented as subclasses of NS to be easily used for actual applications, and then Ye [7] introduced a simplified neutrosophic set (SNS), including the concepts of SVNS and INS, which are the extension of IFS and IVIFS. Obviously, SNS is a subclass of NS, while SVNS and INS are subclasses of SNS. As mentioned in the literature [4–7], NS is the generalization of FS, IFS, and IVIFS. Thereby, Figure 1 shows the flow chart extended from FS to NS (SNS, SVNS, INS).

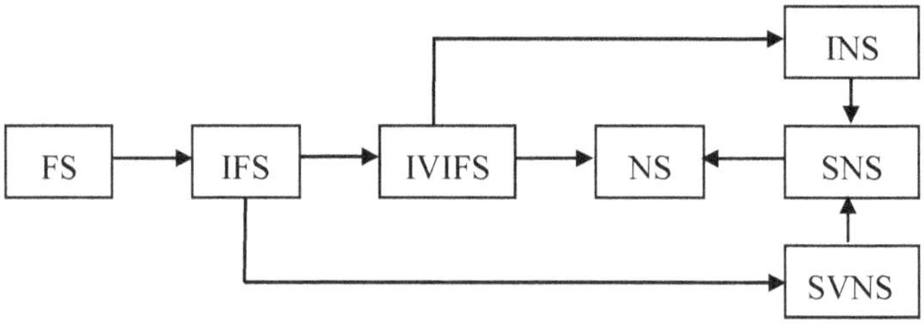

Figure 1. Flow chart extended from fuzzy set (FS) to neutrosophic set (NS) (simplified neutrosophic set (SNS), single-valued neutrosophic set (SVNS), interval neutrosophic set (INS)). IFS: intuitionistic fuzzy set; IVIFS: interval-valued intuitionistic fuzzy set.

On the other hand, some researchers also introduced other fuzzy extensions, such as fuzzy soft sets, hesitant FSs, and hesitant fuzzy soft sets (see [8,9] for detail).

However, SNS (SVNS and INS) is very suitable for the expression of incomplete, indeterminate, and inconsistent information in actual applications. Recently, SNSs (INSs, and SVNSs) have been widely applied in many areas [10–28], such as decision-making, image processing, medical diagnosis, fault diagnosis, and clustering analysis. Especially, many researchers [7,29–36] have developed various aggregation operators, like simplified neutrosophic weighted aggregation operators, simplified neutrosophic prioritized aggregation operators, single-valued neutrosophic normalized weighted Bonferroni mean operators, generalized neutrosophic Hamacher aggregation operators, generalized weighted aggregation operators, interval neutrosophic prioritized ordered weighted average operators, interval neutrosophic Choquet integral operators, interval neutrosophic exponential weighted aggregation operators, and so on, and applied them to decision-making problems with SNS/SVNS/INS information. Obviously, the aggregation operators give us powerful tools to deal with the aggregation of simplified (single-valued and interval) neutrosophic information in the decision making process.

In 1982, Dombi [37] developed the operations of the Dombi T-norm and T-conorm, which show the advantage of good flexibility with the operational parameter. Hence, Liu et al. [38] extended the Dombi operations to IFSs and proposed some intuitionistic fuzzy Dombi Bonferroni mean operators and applied them to multiple attribute group decision-making (MAGDM) problems with intuitionistic fuzzy information. From the existing studies, we can see that the Dombi operations are not extended to neutrosophic sets so far. To develop new aggregation operators for NSs based on the extension of the Dombi operations, the main purposes of this study are (1) to present some Dombi operations of single-valued neutrosophic numbers (SVNNs) (basic elements in SVNS), (2) to propose a single-valued neutrosophic Dombi weighted arithmetic average (SVNDWAA) operator and a single-valued neutrosophic Dombi weighted geometric average (SVNDWGA) operator for the aggregation of SVNN information and to investigate their properties, and (3) to develop a decision-making approach based on the SVNDWAA and SVNDWGA operators for solving multiple attribute decision-making (MADM) problems with SVNN information.

The rest of the paper is organized as follows. Section 2 briefly describes some concepts of SVNSs to be used for the study. Section 3 presents some new Dombi operations of SVNNs. In Section 4, we propose the SVNDWAA and SVNDWGA operators and investigate their properties. Section 5 develops a MADM approach based on the SVNDWAA and SVNDWGA operators. An illustrative example is presented in Section 6. Section 7 gives conclusions and future research directions.

2. Some Concepts of SVNSs

As the extension of IFSs, Wang et al. [5] introduced the definition of a SVNS as a subclass of NS proposed by Smarandache [4] to easily apply in real scientific and engineering areas.

Definition 1. *[5] Let X be a universal set. A SVNS N in X is described by a truth-membership function $t_N(x)$, an indeterminacy-membership function $u_N(x)$, and a falsity-membership function $v_N(x)$. Then, a SVNS N can be denoted as the following form:*

$$N = \{\langle x, t_N(x), u_N(x), v_N(x)\rangle | x \in X\},$$

where the functions $t_N(x)$, $u_N(x)$, $v_N(x) \in [0, 1]$ satisfy the condition $0 \leq t_N(x) + u_N(x) + v_N(x) \leq 3$ for $x \in X$.

For convenient expression, a basic element $<x, t_N(x), u_N(x), v_N(x)>$ in N is denoted by $s = <t, u, v>$, which is called a SVNN.

For any SVNN $s = <t, u, v>$, its score and accuracy functions [29] can be introduced, respectively, as follows:

$$E(s) = (2 + t - u - v)/3, \quad E(s) \in [0, 1], \tag{1}$$

$$H(s) = t - v, \quad H(s) \in [-1, 1]. \tag{2}$$

According to the two functions E(s) and H(s), the comparison and ranking of two SVNNs are introduced by the following definition [29].

Definition 2. *[29] Let $s_1 = <t_1, u_1, v_1>$ and $s_2 = <t_2, u_2, v_2>$ be two SVNNs. Then the ranking method for s_1 and s_2 is defined as follows:*

(1) *If $E(s_1) > E(s_2)$, then $s_1 \succ s_2$,*
(2) *If $E(s_1) = E(s_2)$ and $H(s_1) > H(s_2)$, then $s_1 \succ s_2$,*
(3) *If $E(s_1) = E(s_2)$ and $H(s_1) = H(s_2)$, then $s_1 = s_2$.*

3. Some Single-Valued Neutrosophic Dombi Operations

Definition 3. *[37]. Let p and q be any two real numbers. Then, the Dombi T-norm and T-conorm between p and q are defined as follows:*

$$O_D(p, q) = \frac{1}{1 + \left\{\left(\frac{1-p}{p}\right)^\rho + \left(\frac{1-q}{q}\right)^\rho\right\}^{1/\rho}}, \tag{3}$$

$$O_D^c(p, q) = 1 - \frac{1}{1 + \left\{\left(\frac{p}{1-p}\right)^\rho + \left(\frac{q}{1-q}\right)^\rho\right\}^{1/\rho}}, \tag{4}$$

where $\rho \geq 1$ and $(p, q) \in [0, 1] \times [0, 1]$.

According to the Dombi T-norm and T-conorm, we define the Dombi operations of SVNNs.

Definition 4. *Let* s_1 = <t_1, u_1, v_1> *and* s_2 = <t_2, u_2, v_2> *be two SVNNs,* $\rho \geq 1$, *and* $\lambda > 0$. *Then, the Dombi T-norm and T-conorm operations of SVNNs are defined below:*

$$(1)\, s_1 \oplus s_2 = \left\langle 1 - \frac{1}{1+\left\{\left(\frac{t_1}{1-t_1}\right)^\rho + \left(\frac{t_2}{1-t_2}\right)^\rho\right\}^{1/\rho}}, \frac{1}{1+\left\{\left(\frac{1-u_1}{u_1}\right)^\rho + \left(\frac{1-u_2}{u_2}\right)^\rho\right\}^{1/\rho}}, \frac{1}{1+\left\{\left(\frac{1-v_1}{v_1}\right)^\rho + \left(\frac{1-v_2}{v_2}\right)^\rho\right\}^{1/\rho}} \right\rangle;$$

$$(2)\, s_1 \otimes s_2 = \left\langle \frac{1}{1+\left\{\left(\frac{1-t_1}{t_1}\right)^\rho + \left(\frac{1-t_2}{t_2}\right)^\rho\right\}^{1/\rho}}, 1 - \frac{1}{1+\left\{\left(\frac{u_1}{1-u_1}\right)^\rho + \left(\frac{u_2}{1-u_2}\right)^\rho\right\}^{1/\rho}}, 1 - \frac{1}{1+\left\{\left(\frac{v_1}{1-v_1}\right)^\rho + \left(\frac{v_2}{1-v_2}\right)^\rho\right\}^{1/\rho}} \right\rangle;$$

$$(3)\, \lambda s_1 = \left\langle 1 - \frac{1}{1+\left\{\lambda\left(\frac{t_1}{1-t_1}\right)^\rho\right\}^{1/\rho}}, \frac{1}{1+\left\{\lambda\left(\frac{1-u_1}{u_1}\right)^\rho\right\}^{1/\rho}}, \frac{1}{1+\left\{\lambda\left(\frac{1-v_1}{v_1}\right)^\rho\right\}^{1/\rho}} \right\rangle;$$

$$(4)\, s_1^\lambda = \left\langle \frac{1}{1+\left\{\lambda\left(\frac{1-t_1}{t_1}\right)^\rho\right\}^{1/\rho}}, 1 - \frac{1}{1+\left\{\lambda\left(\frac{u_1}{1-u_1}\right)^\rho\right\}^{1/\rho}}, 1 - \frac{1}{1+\left\{\lambda\left(\frac{v_1}{1-v_1}\right)^\rho\right\}^{1/\rho}} \right\rangle.$$

4. Dombi Weighted Aggregation Operators of SVNNs

Based on the Dombi operations of SVNNs in Definition 4, we propose the two Dombi weighted aggregation operators: the SVNDWAA and SVNDWGA operators, and then investigate their properties.

Definition 5. *Let* s_j = <t_j, u_j, v_j> *(j = 1, 2, ... , n) be a collection of SVNNs and* $\boldsymbol{w}$ = (w_1, w_2, ... , w_n) *be the weight vector for* s_j *with* $w_j \in [0, 1]$ *and* $\sum_{j=1}^{n} w_j = 1$. *Then, the SVNDWAA and SVNDWGA operators are defined, respectively, as follows:*

$$SVNDWAA(s_1, s_2, \ldots, s_n) = \mathop{\oplus}_{j=1}^{n} w_j s_j, \tag{5}$$

$$SVNDWGA(s_1, s_2, \ldots, s_n) = \mathop{\otimes}_{j=1}^{n} s_j^{w_j}. \tag{6}$$

Theorem 1. *Let* s_j = <t_j, u_j, v_j> *(j = 1, 2, ... , n) be a collection of SVNNs and* $\boldsymbol{w}$ = (w_1, w_2, ... , w_n) *be the weight vector for* s_j *with* $w_j \in [0, 1]$ *and* $\sum_{j=1}^{n} w_j = 1$. *Then, the aggregated value of the SVNDWAA operator is still a SVNN, which is calculated by the following formula:*

$$SVNDWAA(s_1, s_2, \ldots, s_n) = \left\langle 1 - \frac{1}{1+\left\{\sum_{j=1}^{n} w_j\left(\frac{t_j}{1-t_j}\right)^\rho\right\}^{1/\rho}}, \frac{1}{1+\left\{\sum_{j=1}^{n} w_j\left(\frac{1-u_j}{u_j}\right)^\rho\right\}^{1/\rho}}, \frac{1}{1+\left\{\sum_{j=1}^{n} w_j\left(\frac{1-v_j}{v_j}\right)^\rho\right\}^{1/\rho}} \right\rangle, \tag{7}$$

By the mathematical induction, we can prove Theorem 1.

Proof. If $n = 2$, based on the Dombi operations of SVNNs in Definition 4 we can obtain the following result:

$$\begin{aligned} SVNDWAA(s_1, s_2) &= s_1 \oplus s_2 \\ &= \left\langle 1 - \frac{1}{1+\left\{w_1\left(\frac{t_1}{1-t_1}\right)^\rho + w_2\left(\frac{t_2}{1-t_2}\right)^\rho\right\}^{1/\rho}}, \frac{1}{1+\left\{w_1\left(\frac{1-u_1}{u_1}\right)^\rho + w_2\left(\frac{1-u_2}{u_2}\right)^\rho\right\}^{1/\rho}}, \frac{1}{1+\left\{w_1\left(\frac{1-v_1}{v_1}\right)^\rho + w_2\left(\frac{1-v_2}{v_2}\right)^\rho\right\}^{1/\rho}} \right\rangle \\ &= \left\langle 1 - \frac{1}{1+\left\{\sum_{j=1}^{2} w_j\left(\frac{t_j}{1-t_j}\right)^\rho\right\}^{1/\rho}}, \frac{1}{1+\left\{\sum_{j=1}^{2} w_j\left(\frac{1-u_j}{u_j}\right)^\rho\right\}^{1/\rho}}, \frac{1}{1+\left\{\sum_{j=1}^{2} w_j\left(\frac{1-v_j}{v_j}\right)^\rho\right\}^{1/\rho}} \right\rangle. \end{aligned}$$

If $n = k$, based on Equation (7), we have the following equation:

$$SVNDWAA(s_1, s_2, \ldots, s_k) = \left\langle 1 - \frac{1}{1+\left\{\sum_{j=1}^{k} w_j \left(\frac{t_j}{1-t_j}\right)^\rho\right\}^{1/\rho}}, \frac{1}{1+\left\{\sum_{j=1}^{k} w_j \left(\frac{1-u_j}{u_j}\right)^\rho\right\}^{1/\rho}}, \frac{1}{1+\left\{\sum_{j=1}^{k} w_j \left(\frac{1-v_j}{v_j}\right)^\rho\right\}^{1/\rho}} \right\rangle.$$

If $n = k + 1$, there is the following result:

$$\begin{aligned} & SVNDWAA(s_1, s_2, \ldots, s_k, s_{k+1}) \\ & = \left\langle 1 - \frac{1}{1+\left\{\sum_{j=1}^{k} w_j \left(\frac{t_j}{1-t_j}\right)^\rho\right\}^{1/\rho}}, \frac{1}{1+\left\{\sum_{j=1}^{k} w_j \left(\frac{1-u_j}{u_j}\right)^\rho\right\}^{1/\rho}}, \frac{1}{1+\left\{\sum_{j=1}^{k} w_j \left(\frac{1-v_j}{v_j}\right)^\rho\right\}^{1/\rho}} \right\rangle \oplus w_{k+1} s_{k+1} \\ & = \left\langle 1 - \frac{1}{1+\left\{\sum_{j=1}^{k+1} w_j \left(\frac{t_j}{1-t_j}\right)^\rho\right\}^{1/\rho}}, \frac{1}{1+\left\{\sum_{j=1}^{k+1} w_j \left(\frac{1-u_j}{u_j}\right)^\rho\right\}^{1/\rho}}, \frac{1}{1+\left\{\sum_{j=1}^{k+1} w_j \left(\frac{1-v_j}{v_j}\right)^\rho\right\}^{1/\rho}} \right\rangle \end{aligned}.$$

Hence, Theorem 1 is true for $n = k + 1$. Thus, Equation (7) holds for all n. □

Then, the SVNDWAA operator contains the following properties:

(1) Reducibility: When $w = (1/n, 1/n, \ldots, 1/n)$, it is obvious that there exists

$$SVNDWAA(s_1, s_2, \ldots, s_n) = \left\langle 1 - \frac{1}{1+\left\{\sum_{j=1}^{n} \frac{1}{n} \left(\frac{t_j}{1-t_j}\right)^\rho\right\}^{1/\rho}}, \frac{1}{1+\left\{\sum_{j=1}^{n} \frac{1}{n} \left(\frac{1-u_j}{u_j}\right)^\rho\right\}^{1/\rho}}, \frac{1}{1+\left\{\sum_{j=1}^{n} \frac{1}{n} \left(\frac{1-v_j}{v_j}\right)^\rho\right\}^{1/\rho}} \right\rangle.$$

(2) Idempotency: Let all the SVNNs be $s_j = <t_j, u_j, v_j> = s$ ($j = 1, 2, \ldots, n$). Then, SVNDWAA($s_1, s_2, \ldots, s_n$) = s.

(3) Commutativity: Let the SVNS ($s_1', s_2', \ldots, s_n'$) be any permutation of ($s_1, s_2, \ldots, s_n$). Then, there is SVNDWAA($s_1', s_2', \ldots, s_n'$) = SVNDWAA($s_1, s_2, \ldots, s_n$).

(4) Boundedness: Let $s_{\min} = \min(s_1, s_2, \ldots, s_n)$ and $s_{\max} = \max(s_1, s_2, \ldots, s_n)$. Then, $s_{\min} \leq$ SVNDWAA($s_1, s_2, \ldots, s_n$) $\leq s_{\max}$.

Proof. (1) Based on Equation (7), the property is obvious.

(2) Since $s_j = <t_j, u_j, v_j> = s$ ($j = 1, 2, \ldots, n$). Then, by using Equation (7) we can obtain the following result:

$$\begin{aligned} SVNDWAA(s_1, s_2, \ldots, s_n) &= \left\langle 1 - \frac{1}{1+\left\{\sum_{j=1}^{n} w_j \left(\frac{t_j}{1-t_j}\right)^\rho\right\}^{1/\rho}}, \frac{1}{1+\left\{\sum_{j=1}^{n} w_j \left(\frac{1-u_j}{u_j}\right)^\rho\right\}^{1/\rho}}, \frac{1}{1+\left\{\sum_{j=1}^{n} w_j \left(\frac{1-v_j}{v_j}\right)^\rho\right\}^{1/\rho}} \right\rangle \\ &= \left\langle 1 - \frac{1}{1+\left\{\left(\frac{t}{1-t}\right)^\rho\right\}^{1/\rho}}, \frac{1}{1+\left\{\left(\frac{1-u}{u}\right)^\rho\right\}^{1/\rho}}, \frac{1}{1+\left\{\left(\frac{1-v}{v}\right)^\rho\right\}^{1/\rho}} \right\rangle = \left\langle 1 - \frac{1}{1+\frac{t}{1-t}}, \frac{1}{1+\frac{1-u}{u}}, \frac{1}{1+\frac{1-v}{v}} \right\rangle = \langle t, u, v \rangle = s. \end{aligned}$$

Hence, SVNDWAA($s_1, s_2, \ldots, s_n$) = s holds.

(3) The property is obvious.

(4) Let $s_{\min} = \min(s_1, s_2, \ldots, s_n) = <t^-, u^-, v^->$ and $s_{\max} = \max(s_1, s_2, \ldots, s_n) = <t^+, u^+, v^+>$. Then, we have $t^- = \min_j(t_j)$, $u^- = \max_j(u_j)$, $v^- = \max_j(v_j)$, $t^+ = \max_j(t_j)$, $u^+ = \min_j(u_j)$, and $v^+ = \min_j(v_j)$. Thus, there are the following inequalities:

$$1 - \frac{1}{1+\left\{\sum_{j=1}^{n} w_j \left(\frac{t^-}{1-t^-}\right)^\rho\right\}^{1/\rho}} \leq 1 - \frac{1}{1+\left\{\sum_{j=1}^{n} w_j \left(\frac{t_j}{1-t^j}\right)^\rho\right\}^{1/\rho}} \leq 1 - \frac{1}{1+\left\{\sum_{j=1}^{n} w_j \left(\frac{t^+}{1-t^+}\right)^\rho\right\}^{1/\rho}},$$

$$\frac{1}{1+\left\{\sum_{j=1}^{n} w_j\left(\frac{1-u^+}{u^+}\right)^{\rho}\right\}^{1/\rho}} \le \frac{1}{1+\left\{\sum_{j=1}^{n} w_j\left(\frac{1-u_j}{u_j}\right)^{\rho}\right\}^{1/\rho}} \le \frac{1}{1+\left\{\sum_{j=1}^{n} w_j\left(\frac{1-u^-}{u^-}\right)^{\rho}\right\}^{1/\rho}},$$

$$\frac{1}{1+\left\{\sum_{j=1}^{n} w_j\left(\frac{1-v^+}{v^+}\right)^{\rho}\right\}^{1/\rho}} \le \frac{1}{1+\left\{\sum_{j=1}^{n} w_j\left(\frac{1-v_j}{v_j}\right)^{\rho}\right\}^{1/\rho}} \le \frac{1}{1+\left\{\sum_{j=1}^{n} w_j\left(\frac{1-v^-}{v^-}\right)^{\rho}\right\}^{1/\rho}}.$$

Hence, $s_{\min} \le \text{SVNDWAA}(s_1, s_2, \ldots, s_n) \le s_{\max}$ holds. □

Theorem 2. *Let* $s_j = \langle t_j, u_j, v_j\rangle$ *(j = 1, 2, ..., n) be a collection of SVNNs and* $w = (w_1, w_2, \ldots, w_n)$ *be the weight vector for* s_j *with* $w_j \in [0, 1]$ *and* $\sum_{j=1}^{n} w_j = 1$. *Then, the aggregated value of the SVNDWGA operator is still a SVNN, which is calculated by the following formula:*

$$SVNDWGA(s_1, s_2, \ldots, s_n) = \left\langle \frac{1}{1+\left\{\sum_{j=1}^{n} w_j\left(\frac{1-t_j}{t_j}\right)^{\rho}\right\}^{1/\rho}}, 1-\frac{1}{1+\left\{\sum_{j=1}^{n} w_j\left(\frac{u_j}{1-u_j}\right)^{\rho}\right\}^{1/\rho}}, 1-\frac{1}{1+\left\{\sum_{j=1}^{n} w_j\left(\frac{v_j}{1-v_j}\right)^{\rho}\right\}^{1/\rho}} \right\rangle. \quad (8)$$

The proof of Theorem 2 is the same as that of Theorem 1. Thus, it is omitted here.

Obviously, the SVNDWGA operator also contains the following properties:

(1) Reducibility: When the weight vector is $w = (1/n, 1/n, \ldots, 1/n)$, it is obvious that there exists the following result:

$$SVNDWGA(s_1, s_2, \ldots, s_n) = \left\langle \frac{1}{1+\left\{\sum_{j=1}^{n} \frac{1}{n}\left(\frac{1-t_j}{t_j}\right)^{\rho}\right\}^{1/\rho}}, 1-\frac{1}{1+\left\{\sum_{j=1}^{n} \frac{1}{n}\left(\frac{u_j}{1-u_j}\right)^{\rho}\right\}^{1/\rho}}, 1-\frac{1}{1+\left\{\sum_{j=1}^{n} \frac{1}{n}\left(\frac{v_j}{1-v_j}\right)^{\rho}\right\}^{1/\rho}} \right\rangle.$$

(2) Idempotency: Let all the SVNNs be $s_j = \langle t_j, u_j, v_j\rangle = s$ $(j = 1, 2, \ldots, n)$. Then, $\text{SVNDWGA}(s_1, s_2, \ldots, s_n) = s$.

(3) Commutativity: Let the SVNS $(s_1', s_2', \ldots, s_n')$ be any permutation of $(s_1, s_2, \ldots, s_n)$. Then, there is $\text{SVNDWGA}(s_1', s_2', \ldots, s_n') = \text{SVNDWGA}(s_1, s_2, \ldots, s_n)$.

(4) Boundedness: Let $s_{\min} = \min(s_1, s_2, \ldots, s_n)$ and $s_{\max} = \max(s_1, s_2, \ldots, s_n)$. Then, $s_{\min} \le \text{SVNDWGA}(s_1, s_2, \ldots, s_n) \le s_{\max}$.

The proof processes of these properties are the same as the ones of the properties for the SVNDWAA operator. Hence, they are not repeated here.

5. MADM Method Using the SVNDWAA Operator or the SVNDWGA Operator

In this section, we propose a MADM method by using the SVNDWAA operator or the SVNDWGA operator to handle MADM problems with SVNN information.

For a MADM problem with SVNN information, let $S = \{S_1, S_2, \ldots, S_m\}$ be a discrete set of alternatives and $G = \{G_1, G_2, \ldots, G_n\}$ be a discrete set of attributes. Assume that the weight vector of the attributes is given as $w = (w_1, w_2, \ldots, w_n)$ such that $w_j \in [0, 1]$ and $\sum_{j=1}^{n} w_j = 1$. If the decision makers are required to provide their suitability evaluation about the alternative S_i $(i = 1, 2, \ldots, m)$ under the attribute G_j $(j = 1, 2, \ldots, n)$ by the SVNN $s_{ij} = \langle t_{ij}, u_{ij}, v_{ij}\rangle$ $(i = 1, 2, \ldots, m; j = 1, 2, \ldots, n)$, then, we can elicit a SVNN decision matrix $D = (s_{ij})_{m \times n}$.

Thus, we utilize the SVNDWAA operator or the SVNDWGA operator to develop a handling approach for MADM problems with SVNN information, which can be described by the following decision steps:

Step 1. Derive the collective SVNN s_i (i = 1, 2, ..., m) for the alternative S_i (i = 1, 2, ..., m) by using the SVNDWAA operator:

$$s_i = SVNDWAA(s_{i1}, s_{i2}, \ldots, s_{in}) = \left\langle 1 - \frac{1}{1+\left\{\sum_{j=1}^{n} w_j \left(\frac{t_{ij}}{1-t_{ij}}\right)^{\rho}\right\}^{1/\rho}}, \frac{1}{1+\left\{\sum_{j=1}^{n} w_j \left(\frac{1-u_{ij}}{u_{ij}}\right)^{\rho}\right\}^{1/\rho}}, \frac{1}{1+\left\{\sum_{j=1}^{n} w_j \left(\frac{1-v_{ij}}{v_{ij}}\right)^{\rho}\right\}^{1/\rho}} \right\rangle, \quad (9)$$

or by using the SVNDWGA operator:

$$s_i = SVNDWGA(s_{i1}, s_{i2}, \ldots, s_{in}) = \left\langle \frac{1}{1+\left\{\sum_{j=1}^{n} w_j \left(\frac{1-t_{ij}}{t_{ij}}\right)^{\rho}\right\}^{1/\rho}}, 1 - \frac{1}{1+\left\{\sum_{j=1}^{n} w_j \left(\frac{u_{ij}}{1-u_{ij}}\right)^{\rho}\right\}^{1/\rho}}, 1 - \frac{1}{1+\left\{\sum_{j=1}^{n} w_j \left(\frac{v_{ij}}{1-v_{ij}}\right)^{\rho}\right\}^{1/\rho}} \right\rangle, \quad (10)$$

where $w = (w_1, w_2, \ldots, w_n)$ is the weight vector such that $w_j \in [0, 1]$ and $\sum_{j=1}^{n} w_j = 1$.

Step 2. Calculate the score values of $E(s_i)$ (the accuracy degrees of $H(s_i)$ if necessary) of the collective SVNN s_i (i = 1, 2, ..., m) by using Equations (1) and (2).

Step 3. Rank the alternatives and select the best one(s).

Step 4. End.

6. Illustrative Example

An illustrative example about investment alternatives for a MADM problem adapted from Ye [10] is used for the applications of the proposed decision-making method under a SVNN environment. An investment company wants to invest a sum of money in the best option. To invest the money, a panel provides four possible alternatives: (1) S_1 is a car company; (2) S_2 is a food company; (3) S_3 is a computer company; (4) S_4 is an arms company. The investment company must take a decision corresponding to the requirements of the three attributes: (1) G_1 is the risk; (2) G_2 is the growth; (3) G_3 is the environmental impact. The suitability evaluations of the alternative S_i (i = 1, 2, 3, 4) corresponding to the three attributes of G_j (j = 1, 2, 3) are given by some decision makers or experts and expressed by the form of SVNNs. Thus, when the four possible alternatives corresponding to the above three attributes are evaluated by the decision makers, we can give the single-valued neutrosophic decision matrix $D(s_{ij})_{m\times n}$, where s_{ij} = <t_{ij}, u_{ij}, v_{ij}> (i = 1, 2, 3, 4; j = 1, 2, 3) is SVNN, as follows:

$$D(s_{ij})_{4\times 3} = \begin{bmatrix} \langle 0.4, 0.2, 0.3\rangle & \langle 0.4, 0.2, 0.3\rangle & \langle 0.8, 0.2, 0.5\rangle \\ \langle 0.6, 0.1, 0.2\rangle & \langle 0.6, 0.1, 0.2\rangle & \langle 0.5, 0.2, 0.8\rangle \\ \langle 0.3, 0.2, 0.3\rangle & \langle 0.5, 0.2, 0.3\rangle & \langle 0.5, 0.3, 0.8\rangle \\ \langle 0.7, 0.0, 0.1\rangle & \langle 0.6, 0.1, 0.2\rangle & \langle 0.6, 0.3, 0.8\rangle \end{bmatrix}.$$

The weight vector of the three attributes is given as w = (0.35, 0.25, 0.4).

Then, we utilize the SVNDWAA operator or the SVNDWGA operator to handle the MADM problem with SVNN information.

In this decision-making problem, the MADM steps based on the SVNDWAA operator can be described as follows:

Step 1. Derive the collective SVNNs of s_i for the alternative S_i (i = 1, 2, 3, 4) by using Equation (9) for ρ = 1 as follows:

s_1 = <0.6667, 0.2000, 0.3571>, s_2 = <0.5652, 0.1250, 0.2857>, s_3 = <0.4444, 0.2308, 0.4000>, and s_4 = <0.6418, 0, 0.1905>.

Step 2. Calculate the score values of $E(s_i)$ of the collective SVNN s_i (i = 1, 2, 3, 4) for the alternatives S_i (i = 1, 2, 3, 4) by using Equation (1) as the following results:

$E(s_1)$ = 0.7032, $E(s_2)$ = 0.7182, $E(s_3)$ = 0.6046, and $E(s_4)$ = 0.8171.

Step 3. Based on the obtained score values, the ranking order of the alternatives is $S_4 \succ S_2 \succ S_1 \succ S_3$ and the best one is S_4.

Or we use the SVNDWGA operator for the MADM problem, which can be described as the following steps:

Step 1′. Derive the collective SVNNs of s_i for the alternative S_i (i = 1, 2, 3, 4) by using Equation (10) for $\rho = 1$ as follows:

s_1 = <0.5000, 0.2000, 0.3966>, s_2 = <0.5556, 0.1429, 0.6364>, s_3 = <0.4054, 0.2432, 0.6500>, and s_4 = <0.6316, 0.1661, 0.6298>.

Step 2′. Calculate the score values of $E(s_i)$ of the collective SVNN s_i (i = 1, 2, 3, 4) for the alternatives S_i (i = 1, 2, 3, 4) by using Equation (1) as the following results:

$E(s_1) = 0.6345$, $E(s_2) = 0.5921$, $E(s_3) = 0.5041$, and $E(s_4) = 0.6119$.

Step 3′. Based on the obtained score values, the ranking order of the alternatives is $S_1 \succ S_4 \succ S_2 \succ S_3$ and the best one is S_1.

In order to ascertain the effects on the ranking alternatives by changing parameters of $\rho \in [1, 10]$ in the SVNDWAA and SVNDWGA operators, all the results are depicted in Tables 1 and 2.

Table 1. Ranking results for different operational parameters of the single-valued neutrosophic Dombi weighted arithmetic average (SVNDWAA) operator.

ρ	$E(s_1)$, $E(s_2)$, $E(s_3)$, $E(s_4)$	Ranking Order
1	0.7032, 0.7182, 0.6046, 0.8171	$S_4 \succ S_2 \succ S_1 \succ S_3$
2	0.7259, 0.7356, 0.6257, 0.8326	$S_4 \succ S_2 \succ S_1 \succ S_3$
3	0.7380, 0.7434, 0.6364, 0.8396	$S_4 \succ S_2 \succ S_1 \succ S_3$
4	0.7449, 0.7480, 0.6429, 0.8441	$S_4 \succ S_2 \succ S_1 \succ S_3$
5	0.7492, 0.7511, 0.6472, 0.8474	$S_4 \succ S_2 \succ S_1 \succ S_3$
6	0.7521, 0.7533, 0.6503, 0.8499	$S_4 \succ S_2 \succ S_1 \succ S_3$
7	0.7542, 0.7550, 0.6525, 0.8520	$S_4 \succ S_2 \succ S_1 \succ S_3$
8	0.7558, 0.7564, 0.6543, 0.8536	$S_4 \succ S_2 \succ S_1 \succ S_3$
9	0.7571, 0.7574, 0.6556, 0.8549	$S_4 \succ S_2 \succ S_1 \succ S_3$
10	0.7580, 0.7583, 0.6567, 0.8560	$S_4 \succ S_2 \succ S_1 \succ S_3$

Table 2. Ranking results for different operational parameters of the single-valued neutrosophic Dombi weighted geometric average (SVNDWGA) operator.

ρ	$E(s_1)$, $E(s_2)$, $E(s_3)$, $E(s_4)$	Ranking Order
1	0.6345, 0.5921, 0.5041, 0.6119	$S_1 \succ S_4 \succ S_2 \succ S_3$
2	0.6145, 0.5602, 0.4722, 0.5645	$S_1 \succ S_4 \succ S_2 \succ S_3$
3	0.6026, 0.5460, 0.4549, 0.5454	$S_1 \succ S_2 \succ S_4 \succ S_3$
4	0.5950, 0.5374, 0.4439, 0.5351	$S_1 \succ S_2 \succ S_4 \succ S_3$
5	0.5898, 0.5316, 0.4363, 0.5286	$S_1 \succ S_2 \succ S_4 \succ S_3$
6	0.5861, 0.5272, 0.4308, 0.5241	$S_1 \succ S_2 \succ S_4 \succ S_3$
7	0.5834, 0.5238, 0.4266, 0.5208	$S_1 \succ S_2 \succ S_4 \succ S_3$
8	0.5813, 0.5211, 0.4234, 0.5183	$S_1 \succ S_2 \succ S_4 \succ S_3$
9	0.5797, 0.5190, 0.4208, 0.5163	$S_1 \succ S_2 \succ S_4 \succ S_3$
10	0.5784, 0.5172, 0.4188, 0.5147	$S_1 \succ S_2 \succ S_4 \succ S_3$

From Tables 1 and 2, we see that the ranking orders based on the SVNDWAA and SVNDWGA operators indicate their obvious difference due to using different aggregation operators. Then, the different operational parameters of ρ can change the ranking orders corresponding to the SVNDWGA operator, which is more sensitive to ρ in this decision-making problem; while the different operation parameters of ρ show the same ranking orders corresponding to the SVNDWAA operator, which is not sensitive to ρ in this decision-making problem.

Compared with existing related method [38], the decision-making method developed in this paper can deal with single-valued neutrosophic or intuitionistic fuzzy MADM problems, while existing method [38] cannot handle single-valued neutrosophic MADM problems.

However, this MADM method based on the SVNDWAA and SVNDWGA operators indicates the advantage of its flexibility in actual applications. Therefore, the developed MADM method provides a new effective way for decision makers to handle single-valued neutrosophic MADM problems.

7. Conclusions

This paper presented some Dombi operations of SVNNs based on the Dombi T-norm and T-conorm operations, and then proposed the SVNDWAA and SVNDWGA operators and investigated their properties. Further, we developed to a MADM method by using the SVNDWAA operator or the SVNDWGA operator to deal with MADM problems under a SVNN environment, in which attribute values with respect to alternatives are evaluated by the form of SVNNs and the attribute weights are known information. We utilized the SVNDWAA operator or the SVNDWGA operator and the score (accuracy) function to rank the alternatives and to determine the best one(s) according to the score (accuracy) values in the different operational parameters. Finally, an illustrative example about the decision-making problem of investment alternatives was provided to demonstrate the application and feasibility of the developed approach. The decision-making results of the illustrative example demonstrated the main highlights of the proposed MADM method: (1) different operational parameters of ρ in the SVNDWGA and SVNDWAA operators can affect the ranking orders; (2) the decision-making process is more flexible corresponding to some operational parameter ρ specified by decision makers' preference and/or actual requirements; (3) the SVNDWGA and SVNDWAA operators provide new aggregation methods of SVNNs to solve MADM problems under an SVNN environment.

In the future work, we shall further develop new Dombi aggregation operators for simplified neutrosophic sets (including SVNSs and INSs) and apply them to solve practical applications in these areas like group decision-making in [39,40], expert system, information fusion system, fault diagnosis, medical diagnosis, and so on.

Acknowledgments: This paper was supported by the National Natural Science Foundation of China (No. 71471172).

Author Contributions: J. Ye proposed the single-valued neutrosophic Dombi weighted arithmetic average (SVNDWAA) and single-valued neutrosophic Dombi weighted geometric average (SVNDWGA) operators, and their decision-making method; J. Chen performed the calculation and analysis of the illustrative example; J. Chen and J. Ye wrote the paper.

Conflicts of Interest: The author declares no conflict of interest.

References

1. Zadeh, L.A. Fuzzy Sets. *Inf. Control* **1965**, *8*, 338–353. [CrossRef]
2. Atanassov, K. Intuitionistic fuzzy sets. *Fuzzy Sets Syst.* **1986**, *20*, 87–96. [CrossRef]
3. Atanassov, K.; Gargov, G. Interval-valued intuitionistic fuzzy sets. *Fuzzy Sets Syst.* **1989**, *31*, 343–349. [CrossRef]
4. Smarandache, F. *Neutrosophy: Neutrosophic Probability, Set, and Logic: Analytic Synthesis & Synthetic Analysis*; American Research Press: Rehoboth, DE, USA, 1998.
5. Wang, H.; Smarandache, F.; Zhang, Y.Q.; Sunderraman, R. Single valued neutrosophic sets. *Multisp. Multistruct.* **2010**, *4*, 410–413.
6. Wang, H.; Smarandache, F.; Zhang, Y.Q.; Sunderraman, R. *Interval Neutrosophic Sets and Logic: Theory and Applications in Computing*; Hexis: Phoenix, AZ, USA, 2005.
7. Ye, J. A multicriteria decision-making method using aggregation operators for simplified neutrosophic sets. *J. Intell. Fuzzy Syst.* **2014**, *26*, 2459–2466. [CrossRef]
8. Bustince, H.; Barrenechea, E.; Fernandez, J.; Pagola, M.; Montero, J. The origin of fuzzy extensions. In *Springer Handbook of Computational Intelligence*; Springer: Heidelberg, Germany, 2015; pp. 89–112.
9. Alcantud, J.C.R. Some formal relationships among soft sets, fuzzy sets, and their extensions. *Int. J. Approx. Reason.* **2016**, *68*, 45–53. [CrossRef]

10. Ye, J. Multicriteria decision-making method using the correlation coefficient under single-value neutrosophic environment. *Int. J. Gen. Syst.* **2013**, *42*, 386–394. [CrossRef]
11. Majumdar, P.; Samant, S.K. On similarity and entropy of neutrosophic sets. *J. Intell. Fuzzy Syst.* **2014**, *26*, 1245–1252. [CrossRef]
12. Peng, J.J.; Wang, J.Q.; Zhang, H.Y.; Chen, X.H. An outranking approach for multi-criteria decision-making problems with simplified neutrosophic sets. *Appl. Soft Comput.* **2014**, *25*, 336–346. [CrossRef]
13. Ye, J. Similarity measures between interval neutrosophic sets and their applications in multicriteria decision making. *J. Intell. Fuzzy Syst.* **2014**, *26*, 165–172. [CrossRef]
14. Ye, J. Single valued neutrosophic cross-entropy for multicriteria decision making problems. *Appl. Math. Model.* **2014**, *38*, 1170–1175. [CrossRef]
15. Guo, Y.; Sengur, A.; Ye, J. A novel image thresholding algorithm based on neutrosophic similarity score. *Measurement* **2014**, *58*, 175–186. [CrossRef]
16. Ye, J. Multiple attribute decision-making method based on the possibility degree ranking method and ordered weighted aggregation operators of interval neutrosophic numbers. *J. Intell. Fuzzy Syst.* **2015**, *28*, 1307–1317. [CrossRef]
17. Ye, J. Improved cosine similarity measures of simplified neutrosophic sets for medical diagnoses. *Artif. Intell. Med.* **2015**, *63*, 171–179. [CrossRef] [PubMed]
18. Sahin, R.; Kucuk, A. Subsethood measure for single valued neutrosophic sets. *J. Intell. Fuzzy Syst.* **2015**, *29*, 525–530. [CrossRef]
19. Zhang, H.Y.; Ji, P.; Wang, J.; Chen, X.H. An improved weighted correlation coefficient based on integrated weight for interval neutrosophic sets and its application in multi-criteria decision-making problems. *Int. J. Comput. Intell. Syst.* **2015**, *8*, 1027–1043. [CrossRef]
20. Ye, J. Interval neutrosophic multiple attribute decision-making method with credibility information. *Int. J. Fuzzy Syst.* **2016**, *18*, 914–923. [CrossRef]
21. Ye, J.; Fu, J. Multi-period medical diagnosis method using a single valued neutrosophic similarity measure based on tangent function. *Comput. Methods Program Biomed.* **2015**, *123*, 142–149. [CrossRef] [PubMed]
22. Biswas, P.; Pramanik, S.; Giri, B.C. TOPSIS method for multi-attribute group decision-making under single-valued neutrosophic environment. *Neural Comput. Appl.* **2016**, *27*, 727–737. [CrossRef]
23. Zhang, H.Y.; Wang, J.Q.; Chen, X.H. An outranking approach for multi-criteria decision-making problems with interval-valued neutrosophic sets. *Neural Comput. Appl.* **2016**, *27*, 615–627. [CrossRef]
24. Peng, J.J.; Wang, J.Q.; Wang, J.; Zhang, H.Y.; Chen, X.H. Simplified neutrosophic sets and their applications in multi-criteria group decision-making problems. *Int. J. Syst. Sci.* **2016**, *47*, 2342–2358. [CrossRef]
25. Tian, Z.P.; Zhang, H.Y.; Wang, J.; Wang, J.Q.; Chen, X.H. Multi-criteria decision-making method based on a cross-entropy with interval neutrosophic sets. *Int. J. Syst. Sci.* **2016**, *47*, 3598–3608. [CrossRef]
26. Ye, J. Projection and bidirectional projection measures of single valued neutrosophic sets and their decision-making method for mechanical design schemes. *J. Exp. Theor. Artif. Intell.* **2016**, 1–10. [CrossRef]
27. Ye, J. Single-valued neutrosophic clustering algorithms based on similarity measures. *J. Classif.* **2017**, *34*, 148–162. [CrossRef]
28. Ye, J. Single valued neutrosophic similarity measures based on cotangent function and their application in the fault diagnosis of steam turbine. *Soft Comput.* **2017**, *21*, 817–825. [CrossRef]
29. Zhang, H.Y.; Wang, J.Q.; Chen, X.H. Interval neutrosophic sets and their application in multicriteria decision making problems. *Sci. World J.* **2014**, *2014*, 645953. [CrossRef] [PubMed]
30. Liu, P.D.; Wang, Y.M. Multiple attribute decision making method based on single-valued neutrosophic normalized weighted Bonferroni mean. *Neural Comput. Appl.* **2014**, *25*, 2001–2010. [CrossRef]
31. Liu, P.D.; Chu, Y.C.; Li, Y.W.; Chen, Y.B. Some generalized neutrosophic number Hamacher aggregation operators and their application to group decision making. *J. Intell. Fuzzy Syst.* **2014**, *16*, 242–255.
32. Zhao, A.W.; Du, J.G.; Guan, H.J. Interval valued neutrosophic sets and multi-attribute decision-making based on generalized weighted aggregation operator. *J. Intell. Fuzzy Syst.* **2015**, *29*, 2697–2706. [CrossRef]
33. Sun, H.X.; Yang, H.X.; Wu, J.Z.; Yao, O.Y. Interval neutrosophic numbers Choquet integral operator for multi-criteria decision making. *J. Intell. Fuzzy Syst.* **2015**, *28*, 2443–2455. [CrossRef]
34. Liu, P.D.; Wang, Y.M. Interval neutrosophic prioritized OWA operator and its application to multiple attribute decision making. *J. Syst. Sci. Complex.* **2016**, *29*, 681–697. [CrossRef]

35. Wu, X.H.; Wang, J.Q.; Peng, J.J.; Chen, X.H. Cross-entropy and prioritized aggregation operator with simplified neutrosophic sets and their application in multi-criteria decision-making problems. *J. Intell. Fuzzy Syst.* **2016**, *18*, 1104–1116. [CrossRef]
36. Ye, J. Exponential operations and aggregation operators of interval neutrosophic sets and their decision making methods. *Springerplus* **2016**, *5*, 1488. [CrossRef] [PubMed]
37. Dombi, J. A general class of fuzzy operators, the demorgan class of fuzzy operators and fuzziness measures induced by fuzzy operators. *Fuzzy Sets Syst.* **1982**, *8*, 149–163. [CrossRef]
38. Liu, P.D.; Liu, J.L.; Chen, S.M. Some intuitionistic fuzzy Dombi Bonferroni mean operators and their application to multi-attribute group decision making. *J. Oper. Res. Soc.* **2017**, 1–26. [CrossRef]
39. Maio, C.D.; Fenza, G.; Loia, V.; Orciuoli, F.; Herrera-Viedma, E. A framework for context-aware heterogeneous group decision making in business processes. *Knowl-Based Syst.* **2016**, *102*, 39–50. [CrossRef]
40. Maio, C.D.; Fenza, G.; Loia, V.; Orciuoli, F.; Herrera-Viedma, E. A context-aware fuzzy linguistic consensus model supporting innovation processes. In Proceedings of the 2016 IEEE International Conference on Fuzzy Systems (FUZZ-IEEE), Vancouver, BC, Canada, 24–29 July 2016; pp. 1685–1692.

symmetry

Article

A Novel Single-Valued Neutrosophic Set Similarity Measure and Its Application in Multicriteria Decision-Making

Wen Jiang * and Yehang Shou

School of Electronics and Information, Northwestern Polytechnical University, Xi'an 710072, Shaanxi, China; shouyehang@mail.nwpu.edu.cn

* Correspondence: jiangwen@nwpu.edu.cn; Tel.: +86-29-8843-1267

Received: 13 June 2017; Accepted: 17 July 2017; Published: 25 July 2017

Abstract: The single-valued neutrosophic set is a subclass of neutrosophic set, and has been proposed in recent years. An important application for single-valued neutrosophic sets is to solve multicriteria decision-making problems. The key to using neutrosophic sets in decision-making applications is to make a similarity measure between single-valued neutrosophic sets. In this paper, a new method to measure the similarity between single-valued neutrosophic sets using Dempster–Shafer evidence theory is proposed, and it is applied in multicriteria decision-making. Finally, some examples are given to show the reasonable and effective use of the proposed method.

Keywords: single-valued neutrosophic set; multicriteria decision-making; neutrosophic set; Dempster–Shafer evidence theory; correlation coefficient; similarity

1. Introduction

The concept of a neutrosophic set, which generalizes the above-mentioned sets from philosophical point of view, was proposed by Smarandache [1] in 1999, and it is defined as a set about the degree of truth, indeterminacy, and falsity. According to the definition of a neutrosophic set, a neutrosophic set A in a universal set X is characterized independently by a truth-membership function $T_A(x)$, an indeterminacy-membership function $I_A(x)$, and a falsity-membership function $F_A(x)$, where in X are real standard or nonstandard subsets of $]^{-}0, 1^{+}[$. However, the neutrosophic set generalizes the above-mentioned sets from a philosophical point of view. It is difficult to apply in practical problems. So, the concept of interval neutrosophic sets (INSs) [2] and single-valued neutrosophic sets (SVNSs) [3] were proposed by Wang et al. Because INS and SVNS are easy to express, and due to the fuzziness of subjective judgment, these two kinds of neutrosophic sets are widely used in reality, such as in multicriteria decision-making [4–9], image processing [10–12], medicine [13], fault diagnosis [14], and personnel selection [15]. Among them, decision-making is a hot issue of research for scholars in various fields [16,17].

This paper introduces how to measure the similarity between SVNSs, which is an important topic in the neutrosophic theory. Similarity measure is key in the SVNS application, and many similarity measures have been proposed by some researchers. In [18], the notion of distance between two SVNSs is introduced and entropy of SVNS is also defined by Majumdar and Samanta. In [6], Ye presented similarity measure of SVNSs using the weighted cosine. Then, in [5], a multicriteria decision-making method is introduced based on the two aggregation operators and cosine similarity measure for SVNSs. In [14], Ye presented two cotangent similarity measures for SVNSs based on cotangent function, and these similarity measures were applied in the fault diagnosis of a steam turbine. However, the cosine similarity measures defined [5,6] have some drawbacks. The results using cosine similarity measures are not consistent with intuition in some situations. Specific analysis will be presented in the following sections. Therefore, similarity measure is still an open problem.

In this paper, a new method of calculating the similarity between SVNSs using Dempster–Shafer (D–S) evidence theory is proposed and is applied in multicriteria decision-making. D–S evidence theory was first proposed by Dempster and then developed by Shafer, and has received recent attention in the field of information fusion. A significant advantage of D–S evidence theory over the traditional probability is that a better fusion result can be obtained via simple reasoning process without knowing the prior probability. In addition, an important property of this theory is that it can easily describe the uncertainty of information [19–21]. In recent years, a number of papers about further research in D–S evidence theory have been published. The main research topics include combination rule [22–24], basic probability assignment(BPA) approximation [25,26], distance of evidence [27,28], and BPA generation [29]. Besides, to solve the problem that frame of discernment is often not incomplete, the generalized evidence theory (GET) is proposed and now there are many studies on GET [30–32]. What is more, D–S evidence theory has been widely used in many fields, such as image fusion [33,34], sensor data fusion [35], gender profiling [36], device fault diagnosis [37,38], and so on [39–43]. In this paper, the method of measuring similarity between SVNSs is that SVNSs are first converted to the basic probability assignment(BPA). Then, the value of similarity measure between SVNSs can be obtained by computing the similarity of BPAs. The correlation between BPAs can be obtained by the correlation coefficient or the evidential distance [44] in D–S evidence theory. The correlation coefficient by Jiang [45] will be applied in our method to measure the similarity between the SVNSs. This correlation coefficient is one of the coefficients which can effectively measure the similarity or relevance of evidence. Finally, the new SVNSs similarity measure method-based D–S evidence theory is used in multicriteria decision-making for single-valued neutrosophic sets, and some examples are given to show its effectiveness.

The remainder of this paper is organized as follows: Section 2 introduces the theoretical background of this research. The method which measures the similarity between SVNSs using D–S evidence theory is introduced in Section 3. Next, compared with existing methods, some tests and analysis are given in Section 4. Finally, D–S evidence theory is applied in multicriteria decision-making under SVNS in Section 4 and we present our conclusions in Section 6.

2. Preliminaries

2.1. Neutrosophic Sets

The definition of a neutrosophic set as proposed by Smarandache [1] is as follows.

Definition 1. *Let X be a space of points (objects), with a generic element in X denoted by x. A neutrosophic set A in X is characterized by a truth-membership function T_A, an indeterminacy membership function I_A, and a falsity membership function F_A. $T_A(x)$, $I_A(x)$, and $F_A(x)$ are real standard or non-standard subsets of $]0^-,1^+[$. That is*

$$\begin{aligned} T_A &: X \rightarrow]0^-,1^+[\\ I_A &: X \rightarrow]0^-,1^+[\\ F_A &: X \rightarrow]0^-,1^+[\end{aligned} \tag{1}$$

There is no restriction on the sum of $T_A(x)$, $I_A(x)$, and $F_A(x)$, so $0^- \leq \sup T_A(x) + \sup I_A(x) + \sup F_A(x) \leq 3^+$.

Definition 2. *The complement of a neutrosophic set A is denoted by c(A), and is defined by*

$$\begin{aligned} T_{c(A)}(x) &= \{1^+\} - T_A(x) \\ I_{c(A)}(x) &= \{1^+\} - I_A(x) \\ F_{c(A)}(x) &= \{1^+\} - F_A(x) \end{aligned} \tag{2}$$

where all x is in X.

Definition 3. *A neutrosophic set A is contained in the other neutrosophic set B,* $A \subseteq B$*, if and only if*

$$\begin{aligned} \inf T_A(x) \le \inf T_B(x), \sup T_A(x) \le \sup T_B(x) \\ \inf F_A(x) \le \inf F_B(x), \sup F_A(x) \le \sup F_B(x) \end{aligned} \quad (3)$$

where all x is in X.

Definition 4. *The union of two neutrosophic sets A and B is a neutrosophic set C, written as* $C = A \cup B$*, whose truth-membership, indeterminacy membership, and falsity membership functions are related to those of A and B by*

$$\begin{aligned} T_C(x) &= T_A(x) + T_B(x) - T_A(x) \times T_B(x) \\ I_C(x) &= I_A(x) + I_B(x) - I_A(x) \times I_B(x) \\ F_C(x) &= F_A(x) + F_B(x) - F_A(x) \times F_B(x) \end{aligned} \quad (4)$$

where all x is in X.

Definition 5. *The intersection of two neutrosophic sets A and B is a neutrosophic set C, written as* $C = A \cap B$*, whose truth-membership, indeterminacy membership, and falsity membership functions are related to those of A and B by*

$$\begin{aligned} T_C(x) &= T_A(x) \times T_B(x) \\ I_C(x) &= I_A(x) \times I_B(x) \\ F_C(x) &= F_A(x) \times F_B(x) \end{aligned} \quad (5)$$

where all x is in X.

2.2. Single-Valued Neutrosophic Set

The definition of SVNS [3] and the weighted aggregation operators of SVNS [5] are introduced as follows:

Definition 6. [3] *Let X be a space of points (objects), with a generic element in X denoted by x. A single-valued neutrosophic set (SVNS) A in X is characterized by truth-membership function* T_A*, indeterminacy-membership function* I_A*, and falsity-membership function* F_A*, For each point x in X,* $T_A(x), I_A(x), F_A(x) \in [0,1]$*. Then, a simplification of the neutrosophic set A is denoted by*

$$A = \{\langle x, T(x), I(x), F(x)\rangle \,|x \in X\} \quad (6)$$

It is a subclass of neutrosophic sets. In this paper, for the sake of simplicity, the SVNS $A = \{\langle x, T(x), I(x), F(x)\rangle \,|x \in X\}$ *is denoted by the simplified symbol* $A = \{\langle T(x), I(x), F(x)\rangle \,|x \in X\}$*.*

Definition 7. [3] *The complement of an SVNS A is denoted by c(A), and is defined:*

$$\begin{aligned} T_{c(A)}(x) &= F_A(x) \\ I_{c(A)}(x) &= 1 - I_A(x) \\ F_{c(A)}(x) &= T_A(x) \end{aligned} \quad (7)$$

where all x is in X.

Definition 8. [3] *An SVNS A is contained in the other SVNS B,* $A \subseteq B$*, if and only if* $T_A(x) \le T_B(x), I_A(x) \ge I_B(x), F_A(x) \ge F_B(x)$*.*

Definition 9. [3] *Two SVNSs A and B are equal, written as* $A = B$ *if and only if* $A \subseteq B$ *and* $B \subseteq A$*.*

Definition 10. [5] *Let A, B be two SVNSs. Operational relations are defined by*

$$\begin{aligned} A+B = \langle & T_A(x) + T_B(x) - T_A(x)T_B(x), \\ & I_A(x) + I_B(x) - I_A(x)I_B(x), \\ & F_A(x) + F_B(x) - F_A(x)F_B(x)\rangle \end{aligned} \quad (8)$$

$$A \cdot B = \langle T_A(x)T_B(x), I_A(x)I_B(x), F_A(x)F_B(x) \rangle \tag{9}$$

$$\lambda A = \left\langle 1-(1-T_A(x))^{\lambda}, 1-(1-I_A(x))^{\lambda}, 1-(1-F_A(x))^{\lambda} \right\rangle, \lambda > 0 \tag{10}$$

$$A^{\lambda} = \left\langle T_A^{\lambda}(x), I_A^{\lambda}(x), F_A^{\lambda}(x) \right\rangle, \lambda > 0 \tag{11}$$

Definition 11. [5] *Let $A_j (j = 1, 2, ..., n)$ be an SVNS. The simplified neutrosophic weighted arithmetic average operator is defined by*

$$F_w(A_1, A_2, ..., A_n) = \sum_{j=1}^{n} w_j A_j \tag{12}$$

where $W = (w_1, w_2, ..., w_n)$ is the weight vector of $A_j (j = 1, 2, ..., n)$, $w_j \in [0, 1]$ and $\sum_{j=1}^{n} w_j = 1$

2.3. Dempster–Shafer Evidence Theory

The Dempster–Shafer (D–S) evidence theory was introduced by Dempster and then developed by Shafer, and has emerged from their works on statistical inference and uncertain reasoning.

Definition 12. *In D–S evidence theory, there is a fixed set of N mutually exclusive and exhaustive elements, called the frame of discernment. Let Θ be a set, indicated by*

$$\Theta = \{\theta_1, \theta_2, \cdots \theta_i, \cdots, \theta_N\}. \tag{13}$$

Let us denote $P(\Theta)$ as the power set composed of 2^N elements of Θ

$$P(\Theta) = \{\emptyset, \{\theta_1\}, \cdots \{\theta_N\}, \{\theta_1, \theta_2\}, \cdots, \{\theta_1, \theta_2, \cdots \theta_i\}, \cdots, \Theta\}. \tag{14}$$

Definition 13. *A mass function is a mapping m from $P(\Theta)$ to [0,1], formally defined by:*

$$m : P(\Theta) \to [0, 1], A \to m(A) \tag{15}$$

which satisfies the following condition:

$$m(\emptyset) = 0 \tag{16}$$

$$\sum_{A \in P(\Theta)} m(A) = 1. \tag{17}$$

A represents any one of the elements in the $P(\Theta)$. The mass $m(A)$ represents how strongly the evidence supports A. When m(A)>0, A, which is a member of the power set, is called a focal element of the mass function.

Definition 14. *In D–S evidence theory, a mass function is also called a basic probability assignment (BPA). Let us assume there are two BPAs, operating on two sets of propositions B and C, respectively, indicated by m_1 and m_2. The Dempster's combination rule is used to combine them as follows:*

$$m(A) = \begin{cases} 0, & A = \emptyset \\ \frac{1}{1-K} \sum_{B \cap C = A} m_1(B) m_2(C) & A \neq \emptyset \end{cases} \tag{18}$$

$$K = \sum_{B \cap C = \emptyset} m_1(B) m_2(C), \tag{19}$$

In Equations (18) and (19), K shows the conflict between the two BPAs m_1 and m_2.

2.4. A Correlation Coefficient

In order to measure the similarity between two BPAs, a correlation coefficient of belief functions is proposed in [45], detailed as follows:

Definition 15. *For a discernment frame Θ with N elements, suppose the mass of two pieces of evidence denoted by m_1, m_2. The correlation coefficient is defined as:*

$$r_{BPA}\left(m_1, m_2\right) = \frac{c\left(m_1, m_2\right)}{\sqrt{c\left(m_1, m_1\right) \cdot c\left(m_2, m_2\right)}} \tag{20}$$

where the correlation coefficient $r_{BPA} \in [0, 1]$ and $c(m_1, m_2)$ is the degree of correlation denoted as:

$$c(m_1, m_2) = \sum_{i=1}^{2^N} \sum_{j=1}^{2^N} m_1(A_i) m_2(A_j) \frac{|A_i \cap A_j|}{|A_i \cup A_j|} \tag{21}$$

where $i, j = 1, \ldots, 2^N$; A_i, A_j are the focal elements of mass, and $|\cdot|$ is the cardinality of a subset.

When the two BPAs are different absolutely, the degree of difference should be the least, which is 0. When the two BPAs are consistent absolutely, the result of the correlation coefficient is 1.

3. A New Similarity Measures for SNVS

The idea of the new similarity measure is that SNVSs can be converted to BPAs and the similarity measure between the two SNVSs be obtained by computing the correlation coefficient between the two BPAs by Equations (20) and (21). The innovation of our method is that a reasonable method for converting SVNS to BPA is proposed and the new idea of using correlation coefficients by Jiang [45] to measure similarity between SNVSs. The key of this method is to find the connection between SNVSs and BPA. In other words, finding a way to generate the BPA reasonably is very important.

Definition 16. *Suppose neutrosophic set A is an SNVS. A simplification of the neutrosophic set A is denoted by $A = \{\langle T_A(x), I_A(x), F_A(x)\rangle \,|x \in X\}$. Each point x is in X and $T_A(x)$,$I_A(x)$, $F_A(x) \in [0, 1]$. According to the meaning of SNVS and D–S evidence theory, the frame of discernment can be defined $\Theta = \{T_x, F_x\}$. T_x and F_x represent the trust for x and the opposition for x respectively. Subset $\{T_x, F_x\}$ of Θ represents supporting for both T_x and F_x. In other words, it means choice between T_x and F_x cannot be made. So, the power set $P(\Theta)$ is $\{\emptyset, \{T_x\}, \{F_x\}, \{T_x, F_x\}\}$. The BPA can be defined $m_A(\emptyset)$, $m_A(T_x)$, $m_A(F_x)$, $m_A(T_x, F_x)$. They respectively indicate the degree of the A support for $\emptyset$, T_x, F_x and T_x, F_x.*

Definition 17. *Suppose neutrosophic set A is an SNVS. A simplification of the neutrosophic set A is denoted by $A = \{\langle T_A(x), I_A(x), F_A(x)\rangle \,|x \in X\}$. Each point x is in X and $T_A(x)$,$I_A(x)$, $F_A(x) \in [0, 1]$. The relationship between SVNS and BPA can be expressed as follows:*

$$\begin{aligned} m_A(T) &= T_A(x)/(T_A(x) + F_A(x) + (1 - I_A(x))) \\ m_A(F) &= F_A(x)/(T_A(x) + F_A(x) + (1 - I_A(x))) \\ m_A(T, X) &= (1 - I_A(x))/(T_A(x) + F_A(x) + (1 - I_A(x))) \end{aligned} \tag{22}$$

$T_A(x)$ represents the degree of the trust for x. Similarly, $m_A(T_x)$ can also express the degree of the trust for x. $T_A(x)$ and $m_A(F_x)$ both represent the degree of the opposition for x. So, $m_A(T_x) = T_A(x)$ and $m_A(F_x) = F_A(x)$ are defined when generating the BPA according to the SNVS. $I_A(x)$ indicates the degree of support for other, with the exception of trust and opposition. So, $1 - I_A(x)$ expresses the degree of trust for x and opposition for x. The meaning of $m_A(T_x, F_x)$ is the same as $1 - I_A(x)$. So, $m_A(T_x, F_x) = 1 - I_A(x)$ is defined in Equation (22). We can see that using basic probability assignment (BPA) can express the SNVS reasonably.

The new method to calculate the similarity between two SVNSs is presented as follows:

Assume that two SVNSs A and B are denoted by $A = \{\langle T_A(x), I_A(x), F_A(x)\rangle \,|x \in X\}$, $B = \{\langle T_B(x), I_B(x), F_B(x)\rangle \,|x \in X\}$.

Step 1: According to A and B, two groups of BPAs m_A and m_B can be obtained by Equation (22);
Step 2: Compute correlation coefficient r_{BPA} between m_A and m_B by Equations (20) and (21);

Step 3: The similarity measure $S_r(A, B)$ can be obtained as:

$$S_r(A, B) = r_{BPA}(m_A, m_B)$$

In the following, one example is used to show the steps of our proposed method.

Example 1. *Suppose two SVNSs A and B in X = x. The value of A is* <0.7 0.8 0.2> *and the value of B is* <0.6 0.8 0.1>*. Calculating similarity using D–S evidence theory is as follows.*

Step 1: According to A and B of SVNS, we can know

$$T_A(x) = 0.7, I_A(x) = 0.8, F_A(x) = 0.2$$

$$T_B(x) = 0.6, I_B(x) = 0.8, F_B(x) = 0.1$$

So, two groups of BPAs m_A and m_B can be obtained by Equation (22).

$$\begin{aligned} m_A(\varnothing) &= 0, \\ m_A(T_x) &= T_A(x)/(T_A(x) + F_A(x) + (1 - I_A(x))) = 0.6364, \\ m_A(F_x) &= T_A(x)/(T_A(x) + F_A(x) + (1 - I_A(x))) = 0.1818, \\ m_A(T_x, F_x) &= T_A(x)/(T_A(x) + F_A(x) + (1 - I_A(x))) = 0.1818 \end{aligned}$$

$$\begin{aligned} m_B(\varnothing) &= 0, \\ m_B(T_x) &= T_B(x)/(T_B(x) + F_B(x) + (1 - I_B(x))) = 0.6667, \\ m_B(F_x) &= T_B(x)/(T_B(x) + F_B(x) + (1 - I_B(x))) = 0.1111, \\ m_B(T_x, F_x) &= T_B(x)/(T_B(x) + F_B(x) + (1 - I_B(x))) = 0.2222 \end{aligned}$$

Step 2: Compute correlation coefficient r_{BPA} between m_A and m_B by Equations (20) and (21). Firstly, we can get the c by Equation (21).

$$c(m_A, m_A) = 0.5500, c(m_B, m_B) = 0.7500, c(m_A, m_B) = 0.6400$$

Then, r_{BPA} can be obtained by Equation (20).

$$r_{BPA}(m_A, m_B) = 0.9965$$

Step 3: The similarity measure of SNVSs can be obtained as

$$S_r(A, B) = 0.9965.$$

4. Test and Analysis

In this section, to compare the similarity measures with existing similarity measures [6], the examples to demonstrate the effectiveness and rationality of similarity measures proposed of SVNSs is provided.

Example 2. *Suppose two SVNSs A and B in X = x. In this example, the value of A remains the unchanged, which is* <1 0 0>*. Meanwhile,* $T_B(x)$ *in B increased gradually. The similarity measures are computed by the method proposed by Ye's method* [6] *and our method, respectively, and the results are shown in Table* 1 *and Figure* 1.

From Table 1 and Figure 1, we can see that our result is gradually increasing with the change of B. However, the similarity measure remains unchanged by Ye's method. Obviously, Ye's results are not consistent with intuition. With the growth of B, A and B are becoming more and more similar. In other words, the trend of similarity should be changed from small to large in this example. Our results are consistent with the intuitive analysis.

Table 1. Similarity measure values.

Group Number	A	B	S_r	S [6]
1	<1 0 0>	<0 0 0>	0.8660	1
2	<1 0 0>	<0.1 0 0>	0.9042	1
3	<1 0 0>	<0.2 0 0>	0.9333	1
4	<1 0 0>	<0.3 0 0>	0.9549	1
5	<1 0 0>	<0.4 0 0>	0.9707	1
6	<1 0 0>	<0.5 0 0>	0.9820	1
7	<1 0 0>	<0.6 0 0>	0.9897	1
8	<1 0 0>	<0.7 0 0>	0.9948	1
9	<1 0 0>	<0.8 0 0>	0.9979	1
10	<1 0 0>	<0.9 0 0>	0.9995	1
11	<1 0 0>	<1 0 0>	1	1

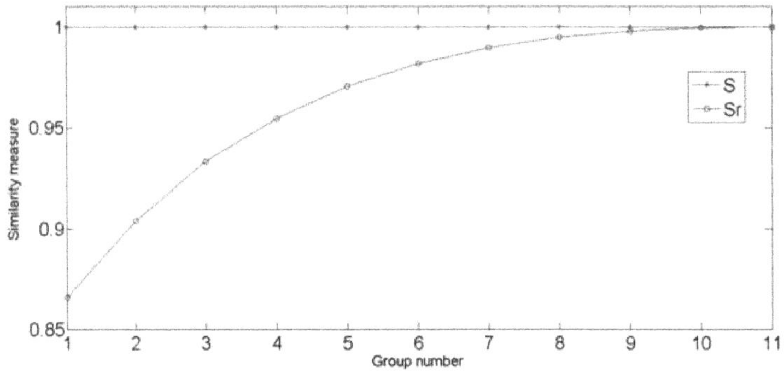

Figure 1. Comparison of correlation degree.

Example 3. *Suppose two SVNSs A and B in* $X = x$ *and we compare similarity measures using D–S evidence theory with similarity measure using* [6]. *The comparison results of similarity measure are shown in* Table 2.

Table 2. Similarity measure values.

Group Number	A	B	S_r	S [6]
1	<0.6 0.2 0.8>	<0.3 0.1 0.4>	0.9862	1
2	<0.7 0.8 0.2>	<0.6 0.8 0.1>	0.9965	0.9935
3	<0.2 0.1 0.5>	<0.2 0.1 0.5>	1	1
4	<0.9 0.8 0.7>	<0.1 0.2 0.1>	0.8440	0.9379

In Example 3, SVNSs of the four groups were calculated and the results are shown in Table 2. From the analysis of data, we can see that the similarity measures of all SVNSs can be obtained reasonably by our method. However, Ye's result is not consistent with intuition in group 1.

5. Multicriteria Decision-Making

In this section, the similarity measure we proposed is applied to solve multicriteria decision-making problems.

Assume that there are multiple groups of alternatives, which can be expressed as $A_1, A_2, \ldots, A_m$. $C_1, C_2, \ldots, C_n$ is expressed as N criteria. In the decision process, the evaluation information of the alternative A_i on the criteria is represented in the form of an SVNS:

$$A_i = \langle C_j, T_{A_i}(C_j), I_{A_i}(C_j), F_{A_i}(C_j) \rangle . \tag{23}$$

where $A_i \in \{A_1, A_2, ..., A_m\}$, $C_j \in \{C_1, C_2, ..., C_n\}$ and $0 \leq T_{A_i}(C_j) \leq 1, 0 \leq I_{A_i}(C_j) \leq 1, 0 \leq F_{A_i}(C_j) \leq 1$. Besides, the importance of each criterion is expressed as $w_1, w_2, ..., w_n$, which is $w_j \in [0,1]$ and $\sum_{j=1}^{n} w_j = 1$. For convenience, Equation (23) is denoted by $\alpha_{ij} = \langle t_{ij}, i_{ij}, f_{ij} \rangle, (i = 1, 2, ..., m; j = 1, 2, ..., n)$. Therefore, simplified neutrosophic decision matrix D can be obtained as follows:

$$D = \begin{bmatrix} \alpha_{11} & \alpha_{12} & \cdots & \alpha_{1n} \\ \alpha_{21} & \alpha_{22} & \cdots & \alpha_{2n} \\ \cdots & \cdots & \cdots & \cdots \\ \alpha_{m1} & \alpha_{m2} & \cdots & \alpha_{mn} \end{bmatrix} = \begin{bmatrix} \langle t_{11}, i_{11}, f_{11} \rangle & \cdots & \langle t_{1n}, i_{1n}, f_{1n} \rangle \\ \cdots & \cdots & \cdots \\ \langle t_{m1}, i_{m1}, f_{m1} \rangle & \cdots & \langle t_{mn}, i_{mn}, f_{mn} \rangle \end{bmatrix} \tag{24}$$

Then, the simplified neutrosophic value α_i for A_i is $\alpha_i = \langle t_i, i_i, f_i \rangle = F_{iw}(\alpha_{i1}, \alpha_{i2}, \cdots, \alpha_{in})$ by Equation (12) according to each row in the simplified neutrosophic decision matrix D.

In order to find the best alternative in all existing alternatives $\alpha_i (i \in 1, 2, \cdots, m)$, the similarity measure needs to be computed between the alternative α_i and ideal alternative α^* by Equation (20) individually. Through the similarity measure between each alternative and the ideal alternative α^*, the ranking order of all alternatives can be determined. The largest value of similarity measures is the best alternative in all existing alternatives. In other words, the greater the similarity measure is, the closer it is to the ideal alternative.

An example for a multicriteria decision-making problem of engineering alternatives is used as a demonstration to show the effectiveness of the similarity measure proposed in this paper.

Example 4. *Let us consider the decision-making problem adapted from* [5]. *There is an investment company which wants to invest a sum of money in the best option. There is a panel with four possible alternatives to invest the money: (1) A_1 is a car company; (2) A_2 is a food company; (3) A_3 is a computer company; (4) A_4 is an arms company. The investment company must take a decision according to the following three criteria: (1) C_1 is the risk analysis; (2) C_2 is the growth analysis; (3) C_3 is the environmental impact analysis. Then, the weight vector of the criteria is given by $W = (0.35, 0.25, 0.4)$, where the value of W is given according to the importance of three criteria. Ideal alternative $\alpha^* = \langle 1, 0, 0 \rangle$.*

Four investment plans A_1, A_2, A_3, A_4 were scored by an expert. For example, when we ask the opinion of an expert about investment in a car company in terms of the risk, they may say that the possibility in which the statement is good is 0.4. The possibility in which the statement is poor is 0.3, and the degree to which they are not sure is 0.2. The opinion of the expert can be expressed as $\alpha_{11} = \langle 0.4, 0.2, 0.3 \rangle$. Thus, all information given by the expert is represented by a simplified neutrosophic decision matrix D:

$$D = \begin{bmatrix} \langle 0.4, 0.2, 0.3 \rangle & \langle 0.4, 0.2, 0.3 \rangle & \langle 0.2, 0.2, 0.5 \rangle \\ \langle 0.6, 0.1, 0.2 \rangle & \langle 0.6, 0.1, 0.2 \rangle & \langle 0.5, 0.2, 0.2 \rangle \\ \langle 0.3, 0.2, 0.3 \rangle & \langle 0.5, 0.2, 0.3 \rangle & \langle 0.5, 0.3, 0.2 \rangle \\ \langle 0.7, 0, 0.1 \rangle & \langle 0.6, 0.1, 0.2 \rangle & \langle 0.4, 0.3, 0.2 \rangle \end{bmatrix}$$

The proposed method in Section 3 is applied to solve this problem separately according to the following computational procedure:

Step 1: $\alpha_i (i = 1, 2, 3, 4)$ can be obtained by Equation (12). The computing results are:

$$\alpha_1 = \langle 0.3268, 0.2000, 0.3881 \rangle$$
$$\alpha_2 = \langle 0.5627, 0.1414, 0.2000 \rangle$$
$$\alpha_3 = \langle 0.4375, 0.2416, 0.2616 \rangle$$
$$\alpha_4 = \langle 0.5746, 0.1555, 0.1663 \rangle$$

Step 2: Correlation measure between each alternative α_i and the ideal alternative α^* were calculated by the method proposed in Section 3. The results are as follows:

$$S_r(\alpha^*, \alpha_1) = 0.8975$$
$$S_r(\alpha^*, \alpha_2) = 0.9745$$
$$S_r(\alpha^*, \alpha_3) = 0.9510$$
$$S_r(\alpha^*, \alpha_4) = 0.9804$$

Step 3: According to the results in the second step, the ranking order of four alternatives is:

$$A_4 > A_2 > A_3 > A_1$$

It can be seen from the results that A_4 is the closest to the ideal replacement. This result is consistent with the result of Ye [5]. This example shows that our method is effective and correct in general.

Example 5. *The application background of this example is the same as in Example 4. In this example, an expert gives different suggestions for investment. So, the value of simplified neutrosophic decision matrix D is changed. A new D matrix is given as follows:*

$$D = \begin{bmatrix} \langle 0.001, 0, 0 \rangle & \langle 0.001, 0, 0 \rangle & \langle 0.001, 0, 0 \rangle \\ \langle 0.9, 0, 0.01 \rangle & \langle 0.95, 0, 0.05 \rangle & \langle 0.96, 0, 0.01 \rangle \\ \langle 0.3, 0.2, 0.3 \rangle & \langle 0.5, 0.2, 0.3 \rangle & \langle 0.5, 0.3, 0.2 \rangle \\ \langle 0.7, 0.0, 0.1 \rangle & \langle 0.6, 0.1, 0.2 \rangle & \langle 0.4, 0.3, 0.2 \rangle \end{bmatrix}$$

The final result of our method is:

$$S_r(\alpha^*, \alpha_1) = 0.8665$$
$$S_r(\alpha^*, \alpha_2) = 0.9997$$
$$S_r(\alpha^*, \alpha_3) = 0.9510$$
$$S_r(\alpha^*, \alpha_4) = 0.9804$$

The result calculated using the Ye's method [5] is:

$$S(\alpha^*, \alpha_1) = 1$$
$$S(\alpha^*, \alpha_2) = 0.9998$$
$$S(\alpha^*, \alpha_3) = 0.9510$$
$$S(\alpha^*, \alpha_4) = 0.9804$$

According to results of our method , the ranking order of four alternatives is $A_4 > A_2 > A_3 > A_1$; this result is reasonable and consistent with intuitive analysis. However, when the results of Ye are sorted, the ranking order of four alternatives is $A_1 > A_2 > A_4 > A_3$. This is obviously not true. The reason is that the similarity measure between α_2 and α^* is greater than the similarity measure between α_1 and α^* by intuitive analysis, where α_1 is $\langle 0.001, 0, 0 \rangle$ and α_2 is $\langle 0.9417, 0, 0.0202 \rangle$.

Example 6. *This example shows steam turbine faults diagnosed using multi-attribute decision-making. In the vibration fault diagnosis of the steam turbine, the relation between the cause and the fault phenomena of the steam turbine has been investigated in [46]. For the vibration fault diagnosis problem of steam turbine, the fault diagnosis of the turbine realized by the frequency features—which are extracted from the vibration signals of the steam turbine—is a simple and effective method. In the fault diagnosis problem of the turbine, we consider a set of ten fault patterns* $A = \{A_1(\text{Unbalance}), A_2(\text{Pneumatic force couple}), A_3(\text{Offset center}), A_4(\text{Oil} - \text{membrane oscillation}), A_5(\text{Radial impact friction of rotor}), A_6(\text{Symbiosis looseness}), A_7(\text{Damage of antithrust bearing}), A_8(\text{Surge}), A_9(\text{Looseness of bearing block}), A_{10}(\text{Nonuniform bearing stiffness})\}$ *as the fault knowledge and a set of nine frequency ranges for different frequency spectrum* $C = \{C_1(0.01 - 0.39f), C_2(0.4 - 0.49f), C_3(0.5f),$

$C_4(0.51 - 0.99f), C_5(f),\ C_6(2f), C_7(3 - 5f),\ C_8(\text{Odd times of} f), C_9\{\text{High frequency} > 5f\}\}$ *under operating frequency f as a characteristic set. Then, the information of the fault knowledge can be introduced from [14], which is shown in Table 3 denoted by* $\langle T_{ij}, I_{ij}, F_{ij}\rangle$.

In the vibration fault diagnosis of steam turbine, the real testing samples are introduced from [], which are represented by the form of single-valued neutrosophic sets: $B = \{\langle 0,0,1\rangle, \langle 0,0,1\rangle, \langle 0.1,0,0.9\rangle, \langle 0.9,0,0.1\rangle, \langle 0,0,1\rangle, \langle 0,0,1\rangle, \langle 0,0,1\rangle, \langle 0,0,1\rangle, \langle 0,0,1\rangle\}$.

Steam turbine faults can be found using multi-attribute decision-making according to the following computational procedure:

Step 1: According to the data in Table 3, the D matrix can be obtained, which is not listed. Then, $\alpha_i(i = 1,2,...,10)$ can be obtained by Equation (12). The computing results are:

$$\begin{aligned}
\alpha_1 &= \langle 0.1974, 0.0234, 1\rangle, \alpha_2 = \langle 0.1349, 0.0301, 1\rangle \\
\alpha_3 &= \langle 0.1003, 0.0674, 1\rangle, \alpha_4 = \langle 0.1714, 0.0101, 1\rangle \\
\alpha_5 &= \langle 0.0950, 0.0367, 0.8713\rangle, \alpha_6 = \langle 0.1088, 0.0260, 1\rangle \\
\alpha_7 &= \langle 0.2037, 0.0125, 1\rangle, \alpha_8 = \langle 0.1224, 0.0193, 1\rangle \\
\alpha_9 &= \langle 0.1975, 0.0137, 1\rangle, \alpha_{10} = \langle 0.1703, 0.0113, 1\rangle
\end{aligned}$$

Step 2: According to SVNS of the real testing samples B, $\beta = \langle 0.2347, 0, 1\rangle$ can be obtained by Equation (12). Additionally, correlation measure between each fault diagnosis problem α_i and the sample β were calculated by the method we proposed in Section 3 and that of Ye [5] individually. Therefore, the two method ranking order of all faults is as follows:

$$Ours: A_7 > A_9 > A_1 > A_4 > A_{10} > A_2 > A_3 > A_5 > A_8 > A_6$$

$$Ye's: A_7 > A_9 > A_1 > A_4 > A_{10} > A_2 > A_8 > A_6 > A_5 > A_3$$

By the comparison between our measure method and the Ye's mehtod in the fault diagnosis of the turbine, the same result can be obtained, which is the main fault of the testing sample B is the damage of antithrust bearing (A7). Besides, the results obtained using our method are also same as Ye's result [14]. This example shows that our method of measuring SVNSs is very effective again, and the expected results can be met in practical applications.

Through the above three examples, it can be seen that solving multicriteria desion-making problems using the new method is very reasonable and effective. Through Examples 4 and 5, compared with other method, we can see that the same correct result can be obtained by our method in general, and more reasonable and logical results can be obtained by our method in some cases. By Example 6, a real-world example is given, which is finding fault using multiple attribute decision making. By comparing the final results, our method has also obtained the right conclusion and it can effectively solve the problems in practical applications.

Table 3. Fault knowledge with single-valued neutrosophic values.

A_i(Fault Knowledge)	C_1(0.01–0.39 f)	C_2(0.4–0.49 f)	C_3(0.5 f)	C_4(0.51–0.99 f)	C_5(f)	C_6(2 f)	C_7(3–5 f)	C_8(Odd Times of f)	C_9(High Frequency > 5 f)
A_1(unbalance)	<0 0 1>	<0 0 1>	<0 0 1>	<0 0 1>	<0.85 0.15 0>	<0.04 0.02 0.94>	<0.04 0.03 0.93>	<0 0 1>	<0 0 1>
A_2(pneumatic force couple)	<0 0 1>	<0.03 0.28 0.69>	<0.9 0.3 0.88>	<0.55 0.15 0.3>	<0 0 1>	<0 0 1>	<0 0 1>	<0 0 1>	<0.08 0.05 0.87>
A_3(offset center)	<0 0 1>	<0 0 1>	<0 0 1>	<0 0 1>	<0.3 0.28 0.42>	<0.40 0.22 0.38>	<0.08 0.05 0.87>	<0 0 1>	<0 0 1>
A_4(oil-membrane oscillation)	<0.09 0.22 0.89>	<0.78 0.04 0.18>	<0 0 1>	<0.08 0.03 0.89>	<0 0 1>	<0 0 1>	<0 0 1>	<0 0 1>	<0 0 1>
A_5(radial impact friction of rotor)	<0.09 0.03 0.88>	<0.09 0.02 0.89>	<0.08 0.04 0.88>	<0.09 0.03 0.88>	<0.18 0.03 0.79>	<0.08 0.05 0.87>	<0.08 0.05 0.87>	<0.08 0.04 0.88>	<0.08 0.04 0.88>
A_6(symbiosis looseness)	<0 0 1>	<0 0 1>	<0 0 1>	<0 0 1>	<0.18 0.04 0.78>	<0.12 0.05 0.83>	<0.37 0.08 0.55>	<0 0 1>	<0.22, 0.06, 0.72>
A_7(damage of antithrust bearing)	<0 0 1>	<0 0 1>	<0.08 0.04 0.88>	<0.86 0.07 0.07>	<0 0 1>	<0 0 1>	<0 0 1>	<0 0 1>	<0 0 1>
A_8(surge)	<0 0 1>	<0.27 0.05 0.68>	<0.08 0.04 0.88>	<0.54 0.08 0.38>	<0 0 1>	<0 0 1>	<0 0 1>	<0 0 1>	<0 0 1>
A_9(looseness of bearing block)	<0.85, 0.08, 0.07>	<0 0 1>	<0 0 1>	<0 0 1>	<0 0 1>	<0 0 1>	<0 0 1>	<0.08 0.04 0.88>	<0 0 1>
A_{10}(non-uniform bearing stiffness)	<0 0 1>	<0 0 1>	<0 0 1>	<0 0 1>	<0 0 1>	<0.77 0.06 0.17>	<0.19 0.04 0.7>	<0 0 1>	<0 0 1>

6. Conclusions

This paper presented a new method to measure the similarity between SVNSs using D–S evidence theory and it is applied in multicriteria decision-making. First of all, the background of neutrosophic sets, single-valued neutrosophic set (SVNS), D–S evidence theory, and correlation coefficient are introduced. Next, the proposed similarity measure method is introduced in detail. Some numerical examples demonstrate that the proposed method can measure similarity more reasonably and effectively compared with the exiting methods. Finally, this method is applied to solve multicriteria decision-making problems. According to the experimental results, it is seen that the proposed method can produce the expected results compared with exiting multicriteria decision-making method for simplified neutrosophic sets. Therefore, it is concluded that the achievement of this paper has a great application prospect and potential in solving multicriteria decision-making problems. Now, our method is limited to measuring the similarity between SVNSs. In the future, we will study its application in interval neutrosophic sets (INSs). At the end of this paper, we hope that the new method can bring some new enlightenments to the related research.

Acknowledgments: The work is partially supported by National Natural Science Foundation of China (Grant No. 61671384), Natural Science Basic Research Plan in Shaanxi Province of China (Program No. 2016JM6018), Aviation Science Foundation (Program No. 20165553036), Fundamental Research Funds for the Central Universities (Program No. 3102017OQD020).

Author Contributions: Wen Jiang conceived and designed the study. Besides, Wen Jiang analyzed the validity of this method. Yehang Shou performed the experiments and wrote the paper.

Conflicts of Interest: The authors declare that they have no competing interests.

References

1. Smarandache, F. A Unifying Field in Logics: Neutrosophic Logic. *Philosophy* **1999**, *8*, 1–141.
2. Wang, H.; Smarandache, F.; Sunderraman, R.; Zhang, Y.Q. *Interval Neutrosophic Sets and Logic: Theory and Applications in Computing*; Infinite Study: Hexis, AZ, USA, 2005.
3. Wang, H.; Smarandache, F.; Zhang, Y.; Sunderraman, R. Single valued neutrosophic sets. *Rev. Air Force Acad.* **2010**, *17*, 10–14.
4. Zhang, H.Y.; Wang, J.Q.; Chen, X.H. Interval neutrosophic sets and their application in multicriteria decision making problems. *Sci. World J.* **2014**, *2014*, 645953.
5. Ye, J. A multicriteria decision-making method using aggregation operators for simplified neutrosophic sets. *J. Intell. Fuzzy Syst.* **2014**, *26*, 2459–2466.
6. Ye, J. Multicriteria decision-making method using the correlation coefficient under single-valued neutrosophic environment. *Int. J. Gen. Syst.* **2013**, *42*, 386–394.
7. Peng, J.J.; Wang, J.Q.; Wang, J.; Zhang, H.Y.; Chen, X.H. Simplified neutrosophic sets and their applications in multi-criteria group decision-making problems. *Int. J. Syst. Sci.* **2016**, *47*, 2342–2358.
8. Zhang, H.; Wang, J.; Chen, X. An outranking approach for multi-criteria decision-making problems with interval-valued neutrosophic sets. *Neural Comput. Appl.* **2016**, *27*, 615–627.
9. Liu, P.; Wang, Y. Multiple attribute decision-making method based on single-valued neutrosophic normalized weighted Bonferroni mean. *Neural Comput. Appl.* **2014**, *25*, 2001–2010.
10. Guo, Y.; Cheng, H.D. New neutrosophic approach to image segmentation. *Pattern Recognit.* **2009**, *42*, 587–595.
11. Broumi, S.; Smarandache, F. Correlation coefficient of interval neutrosophic set. *Appl. Mech. Mater.* **2013**, *436*, 511–517.
12. Guo, Y.; Sengur, A. A novel color image segmentation approach based on neutrosophic set and modified fuzzy c-means. *Circuits Syst. Signal Process.* **2013**, *32*, 1699–1723.
13. Ma, Y.X.; Wang, J.Q.; Wang, J.; Wu, X.H. An interval neutrosophic linguistic multi-criteria group decision-making method and its application in selecting medical treatment options. *Neural Comput. Appl.* **2016**, 1–21, doi:10.1007/s00521-016-2203-1.
14. Ye, J. Single-valued neutrosophic similarity measures based on cotangent function and their application in the fault diagnosis of steam turbine. *Soft Comput.* **2017**, *21*, 1–9.

15. Ji, P.; Zhang, H.Y.; Wang, J.Q. A projection-based TODIM method under multi-valued neutrosophic environments and its application in personnel selection. *Neural Comput. Appl.* **2016**, 1–14, doi:10.1007/s00521-016-2436-z.
16. Fatimah, F.; Rosadi, D.; Hakim, R.F.; Alcantud, J.C.R. Probabilistic soft sets and dual probabilistic soft sets in decision-making. *Neural Comput. Appl.* **2017**, 1–11, doi: 10.1007/s00521-017-3011-y.
17. Alcantud, J.C.R.; Santos-García, G. *A New Criterion for Soft Set Based Decision Making Problems under Incomplete Information*; Technical Report; Mimeo: New York, NY, USA, 2015.
18. Majumdar, P.; Samanta, S.K. On similarity and entropy of neutrosophic sets. *J. Intell. Fuzzy Syst. Appl. Eng. Technol.* **2014**, *26*, 1245–1252.
19. Zadeh, L.A. A simple view of the Dempster-Shafer theory of evidence and its implication for the rule of combination. *AI Mag.* **1986**, *7*, 85–90.
20. Jiang, W.; Xie, C.; Luo, Y.; Tang, Y. Ranking Z-numbers with an improved ranking method for generalized fuzzy numbers. *J. Intell. Fuzzy Syst.* **2017**, *32*, 1931–1943.
21. Jiang, W.; Wei, B.; Tang, Y.; Zhou, D. Ordered visibility graph average aggregation operator: An application in produced water management. *Chaos* **2017**, *27*, 023117.
22. Yang, J.B.; Xu, D.L. Evidential reasoning rule for evidence combination. *Artif. Intell.* **2013**, *205*, 1–29.
23. Chin, K.S.; Fu, C. Weighted cautious conjunctive rule for belief functions combination. *Inf. Sci.* **2015**, *325*, 70–86.
24. Wang, J.; Xiao, F.; Deng, X.; Fei, L.; Deng, Y. Weighted evidence combination based on distance of evidence and entropy function. *Int. J. Distrib. Sens. Netw.* **2016**, *12*, 3218784.
25. Yang, Y.; Liu, Y. Iterative Approximation of Basic Belief Assignment Based on Distance of Evidence. *PLoS ONE* **2016**, *11*, e0147799.
26. Yang, Y.; Han, D.; Han, C.; Cao, F. A novel approximation of basic probability assignment based on rank-level fusion. *Chin. J. Aeronaut.* **2013**, *26*, 993–999.
27. Yang, Y.; Han, D. A new distance-based total uncertainty measure in the theory of belief functions. *Knowl. Based Syst.* **2016**, *94*, 114–123.
28. Deng, X.; Xiao, F.; Deng, Y. An improved distance-based total uncertainty measure in belief function theory. *Appl. Intell.* **2017**, *46*, 898–915.
29. Jiang, W.; Zhuang, M.; Xie, C.; Wu, J. Sensing Attribute Weights: A Novel Basic Belief Assignment Method. *Sensors* **2017**, *17*, 721.
30. He, Y.; Hu, L.; Guan, X.; Han, D.; Deng, Y. New conflict representation model in generalized power space. *J. Syst. Eng. Electron.* **2012**, *23*, 1–9.
31. Jiang, W.; Zhan, J. A modified combination rule in generalized evidence theory. *Appl. Intell.* **2017**, *46*, 630–640.
32. Mo, H.; Lu, X.; Deng, Y. A generalized evidence distance. *J. Syst. Eng. Electron.* **2016**, *27*, 470–476.
33. Dong, J.; Zhuang, D.; Huang, Y.; Fu, J. Advances in multi-sensor data fusion: Algorithms and applications. *Sensors* **2009**, *9*, 7771–7784.
34. Yang, F.; Wei, H. Fusion of infrared polarization and intensity images using support value transform and fuzzy combination rules. *Infrared Phys. Technol.* **2013**, *60*, 235–243.
35. Jiang, W.; Xie, C.; Zhuang, M.; Shou, Y.; Tang, Y. Sensor Data Fusion with Z-Numbers and Its Application in Fault Diagnosis. *Sensors* **2016**, *16*, 1509.
36. Ma, J.; Liu, W.; Miller, P.; Zhou, H. An evidential fusion approach for gender profiling. *Inf. Sci.* **2016**, *333*, 10–20.
37. Islam, M.S.; Sadiq, R.; Rodriguez, M.J.; Najjaran, H.; Hoorfar, M. Integrated Decision Support System for Prognostic and Diagnostic Analyses of Water Distribution System Failures. *Water Resour. Manag.* **2016**, *30*, 2831–2850.
38. Jiang, W.; Xie, C.; Zhuang, M.; Tang, Y. Failure Mode and Effects Analysis based on a novel fuzzy evidential method. *Appl. Soft Comput.* **2017**, *57*, 672–683.
39. Zhang, X.; Deng, Y.; Chan, F.T.S.; Adamatzky, A.; Mahadevan, S. Supplier selection based on evidence theory and analytic network process. *J. Eng. Manuf.* **2016**, *230*, 562–573.
40. Deng, Y. Deng entropy. *Chaos Solitons Fractals* **2016**, *91*, 549–553.
41. Deng, X.; Jiang, W.; Zhang, J. Zero-sum matrix game with payoffs of Dempster-Shafer belief structures and its applications on sensors. *Sensors* **2017**, *17*, 922.

42. Zhang, X.; Mahadevan, S.; Deng, X. Reliability analysis with linguistic data: An evidential network approach. *Reliab. Eng. Syst. Saf.* **2017**, *162*, 111–121.
43. Jiang, W.; Wei, B.; Zhan, J.; Xie, C.; Zhou, D. A visibility graph power averaging aggregation operator: A methodology based on network analysis. *Comput. Ind. Eng.* **2016**, *101*, 260–268.
44. Jousselme, A.L.; Grenier, D.; Bossé, É. A new distance between two bodies of evidence. *Inf. Fusion* **2001**, *2*, 91–101.
45. Jiang, W. A Correlation Coefficient of Belief Functions. Available online: http://arxiv.org/abs/1612.05497 (accessed on 2 February 2017).
46. Ye, J. Fault diagnosis of turbine based on fuzzy cross entropy of vague sets. *Expert Syst. Appl.* **2009**, *36*, 8103–8106.

symmetry

MDPI

Article

Probabilistic Linguistic Power Aggregation Operators for Multi-Criteria Group Decision Making

Agbodah Kobina [1,2], Decui Liang [1,2,*] and Xin He [1]

1 School of Management and Economics, University of Electronic Science and Technology of China, Chengdu 610054, China; kobinaagbodah@yahoo.com (A.K.); iamhexin@163.com (X.H.)

2 Center for West African Studies, University of Electronic Science and Technology of China, Chengdu 610054, China

* Correspondence: decuiliang@126.com

Received: 4 November 2017; Accepted: 13 December 2017; Published: 19 December 2017

Abstract: As an effective aggregation tool, power average (PA) allows the input arguments being aggregated to support and reinforce each other, which provides more versatility in the information aggregation process. Under the probabilistic linguistic term environment, we deeply investigate the new power aggregation (PA) operators for fusing the probabilistic linguistic term sets (PLTSs). In this paper, we firstly develop the probabilistic linguistic power average (PLPA), the weighted probabilistic linguistic power average (WPLPA) operators, the probabilistic linguistic power geometric (PLPG) and the weighted probabilistic linguistic power geometric (WPLPG) operators. At the same time, we carefully analyze the properties of these new aggregation operators. With the aid of the WPLPA and WPLPG operators, we further design the approaches for the application of multi-criteria group decision-making (MCGDM) with PLTSs. Finally, we use an illustrated example to expound our proposed methods and verify their performances.

Keywords: power average operator; probabilistic linguistic term sets; multi-criteria decision making; group decision making

1. Introduction

Yager [1] introduced an operator of power average (PA) to provide more versatility in the information aggregation process. PA is a nonlinear weighted average aggregation tool for which the weight vector depends on the input arguments and that allows the values being aggregated to support and reinforce each other [2]. It has received a large amount of attention in the literature. For instance, Xu and Yager [2] developed power geometric operator on the basis of a geometric mean (GM) and power average (PA). Under the linguistic environment, Xu et al. [3] developed new linguistic aggregation operators based on the power average (PA) to address the relationship of input arguments. Zhou and Chen [4] discussed a generalization of the power aggregation operators for linguistic environment and its application in group decision making (GDM). By extending the PA to the linguistic hesitant fuzzy environment, Zhu et al. [5] established a series of linguistic hesitant fuzzy power aggregation operators. With the above-mentioned literature, PA has successfully been extended to many complex and real situations.

One of the useful theories in dealing with the multi-criteria decision making (MCDM) problems is the theory of probabilistic linguistic term sets (PLTSs). This theory proposed by Pang et al. [6] plays a key role in the decision process where experts express their preferences [7–9]. Nowadays, PLTSs have become a hot topic in the area of hesitant fuzzy linguistic term sets (HFLTSs) [10–12] and hesitant fuzzy sets (HFSs) [13,14]. For example, Pang et al. [6] established a framework for ranking PLTSs and they conducted a comparison method via the score or deviation degree of each PLTS. Bai et al. [7] stated that the existing approaches associated with PLTSs are limited or highly complex in real applications. Thus they established more appropriate comparison method and developed a more efficient way to handle PLTSs. Gou and Xu [15] defined novel operational laws for the probability information. He et al. [16] proposed an algorithm for multi-criteria group decision making (MCGDM) with probabilistic interval preference orderings. Wu and Xu [17] defined the concept of possibility distribution and presented a new framework model to address MCDM. Zhang et al. [18] introduced the concept of probabilistic linguistic preference relations to present the DMs preferences. Under the hesitant probabilistic fuzzy environment, Zhou and Xu [19] studied the consensus building with a group of decision makers. PLTSs generalize the existing models of HFLTSs and HFSs so as to contain hesitations and probabilities. Compared with HFLTSs, the PLTSs have strong ability to express the information vagueness and uncertainty in the hesitant situations under qualitative setting. With respect to the PLTSs, the decision makers (DMs) can not only provide several possible linguistic values over an object (alternative or attribute), but also reflect the probabilistic information of the set of values [6]. In the existing literature, most aggregation operators developed for PLTSs are based on the independence assumption and do not take into account information about the interrelationship between PLTSs being aggregated.

For the PLTSs, it also can encounter the relationship phenomenon between the input arguments. Meanwhile, PA provides a versatility in the aggregation process and has the ability to depict the interrelationship of input arguments, i.e., it allows the input argument being aggregated to support and reinforce each other. However, it rarely discusses in the research works of PLTSs. Hence, we introduce PA into PLTSs and come out with new operators that will improved upon the existing aggregation operators of PLTSs. In this paper, we firstly develop four new aggregation operators based on the Power Average (PA) and the Power Geometric (PG), i.e., probabilistic linguistic power average (PLPA), weighted probabilistic linguistic power average (WPLPA), probabilistic linguistic power geometric (PLPG) and weighted probabilistic linguistic power geometric (WPLPG). These operators take into account all the decision arguments and their relationships. On the basis of probabilistic linguistic GDM, we utilize the WPLPA or WPLPG operator to aggregate the information and design the corresponding approach. In a word, the desirable advantages of our research work are summarized as follows: (1) We involve the probabilistic information. Our proposed methods can allow the collection of a few different linguistic terms evaluated by the DMs and the opinions of the DMs will still remain the same. (2) Our proposed methods also consider the interrelationship of the individual evaluation.

The rest of the paper is structured as follows: Some basic concepts and operations in relation to PLTSs and PA are introduced in Section 2. In Section 3, we develop the PLPA operator, PLPG operator and their own corresponding weighted forms. Meanwhile, we also study several desired properties of these operators. In Section 4, we design the approaches for the application of MCGDM utilizing the WPLPG and WPLPA operators. In Section 5, we give an illustrative example to elaborate and verify our proposed methods. Section 6 concludes the paper and elaborates on future studies.

2. Preliminaries

In this section, we mainly review some basic concepts and operations in relation to PLTSs and PA.

2.1. Probabilistic Linguistic Term Sets (PLTSs)

The concept of PLTSs [6] is an extension of the concepts of HFLTSs. In the following, we review some basic concepts of PLTSs and the corresponding operations.

Definition 1. *[6] Let $S = \{s_t | t = 0, 1, \cdots, \tau\}$ be a linguistic term set. Then a probabilistic linguistic term set (PLTS) is defined as:*

$$L(p) = \{L^{(k)}(p^{(k)}) | L^{(k)} \in S, r^{(k)} \in t, p^{(k)} \geq 0, k = 1, 2, \cdots, \#L(p), \sum_{k=1}^{\#L(p)} p^{(k)} \leq 1\}, \quad (1)$$

where $L^{(k)}(p^{(k)})$ is the linguistic term $L^{(k)}$ associated with the probability $p^{(k)}$, $r^{(k)}$ is the subscript of $L^{(k)}$ and $\#L(p)$ is the number of all linguistic terms in $L(p)$.

Since the positions of elements in a set can be swapped arbitrarily, Pang et al. [6] proposed the ordered PLTSs to ensure that the operational results among PLTSs can be straightforwardly determined. It is described as:

Definition 2. *Given a PLTS $L(p) = \{L^{(k)}(p^{(k)})|k = 1, 2, \cdots, \#L(p)\}$, and $r^{(k)}$ is the subscript of linguistic term $L^{(k)}$. $L(p)$ is called an ordered PLTS, if the linguistic terms $L^{(k)}(p^{(k)})$ are arranged according to the values of $r^{(k)}p^{(k)}$ in descending order.*

Definition 3. *Let $S = \{s_t|t = 0, 1, \cdots, \tau\}$ be a linguistic term set. Given three PLTSs $L(p)$, $L_1(p)$ and $L_2(p)$, their basic operations are summarized as follows [6]:*

(1) $L_1(p) \oplus L_2(p) = \cup_{L_1^{(k)} \in L_1(p), L_2^{(k)} \in L_2(p)} \left\{ p_1^{(k)} L_1^{(k)} \oplus p_2^{(k)} L_2^{(k)} \right\}$;

(2) $L_1(p) \otimes L_2(p) = \cup_{L_1^{(k)} \in L_1(p), L_2^{(k)} \in L_2(p)} \left\{ (L_1^{(k)})^{p_1^{(k)}} \otimes (L_2^{(k)})^{p_2^{(k)}} \right\}$;

(3) $\lambda(L(p)) = \cup_{L^{(k)} \in L(p)} \left\{ \lambda p^{(k)} L^{(k)} \right\}$ *and* $\lambda \geq 0$;

(4) $(L(p))^{\lambda} = \cup_{L^{(k)} \in L(p)} \left\{ (L^{(k)})^{\lambda p^{(k)}} \right\}$ *and* $\lambda \geq 0$.

To compare the PLTSs, Pang et al. [6] defined the score and the deviation degree of a PLTS:

Definition 4. *Let $L(p) = \{L^{(k)}(p^{(k)})|k = 1, 2, \cdots, \#L(p)\}$ be a PLTS, and $r^{(k)}$ is the subscript of linguistic term $L^{(k)}$. Then, the score of $L(p)$ is defined as follows:*

$$E(L(p)) = s_{\bar{\alpha}}, \tag{2}$$

where $\bar{\alpha} = \sum_{k=1}^{\#L(p)} r^{(k)}p^{(k)} / \sum_{k=1}^{\#L(p)} p^{(k)}$. The deviation degree of $L(p)$ is:

$$\sigma(L(p)) = \frac{(\sum_{k=1}^{\#L(p)} (p^{(k)}(r^{(k)} - \bar{\alpha}))^2)^{0.5}}{\sum_{k=1}^{\#L(p)} p^{(k)}}. \tag{3}$$

Based on the score and the deviation degree of a PLTS, Pang et al. [6] further proposed the following laws to compare them.

Definition 5. *Given two PLTSs $L_1(p)$ and $L_2(p)$. $E(L_1(p))$ and $E(L_2(p))$ are the scores of $L_1(p)$ and $L_2(p)$, respectively. $\sigma(L_1(p))$ and $\sigma(L_2(p))$ denote the deviation degrees of $L_1(p)$ and $L_2(p)$. Then, we have:*

(1) *If $E(L_1(p)) > E(L_2(p))$, then $L_1(p)$ is bigger than $L_2(p)$, denoted by $L_1(p) > L_2(p)$;*
(2) *If $E(L_1(p)) < E(L_2(p))$, then $L_1(p)$ is smaller than $L_2(p)$, denoted by $L_1(p) < L_2(p)$;*
(3) *If $E(L_1(p)) = E(L_2(p))$, then we need to compare their deviation degrees:*

    (a) *If $\sigma(L_1(p)) = \sigma(L_2(p))$, then $L_1(p)$ is equal to $L_2(p)$, denoted by $L_1(p) \sim L_2(p)$;*
    (b) *If $\sigma(L_1(p)) > \sigma(L_2(p))$, then $L_1(p)$ is smaller than $L_2(p)$, denoted by $L_1(p) < L_2(p)$;*
    (c) *If $\sigma(L_1(p)) < \sigma(L_2(p))$, then $L_1(p)$ is bigger than $L_2(p)$, denoted by $L_1(p) > L_2(p)$.*

When we analyze and discuss the comparison of PLTSs, we may realise that the number of their corresponding number of the linguistic terms may not be equal. To solve this problem, Pang et al. [6] normalized the PLTSs by increasing the numbers of linguistic terms for PLTSs. The normalized Definition of PLTSs is the following.

Definition 6. *Let* $L_1(p) = \{L_1^{(k)}(p_1^{(k)})|k = 1,2,\cdots,\#L_1(p)\}$ *and* $L_2(p) = \{L_2^{(k)}(p_2^{(k)})|k = 1,2,\cdots,\#L_2(p)\}$ *be any two PLTSs.* $\#L_1(p)$ *and* $\#L_2(p)$ *are the numbers of the linguistic terms in* $L_1(p)$ *and* $L_2(p)$. *If* $\#L_1(p) > \#L_2(p)$, *then we will add* $\#L_1(p) - \#L_2(p)$ *linguistic terms to* $L_2(p)$ *so that the numbers of linguistic terms in* $L_1(p)$ *and* $L_2(p)$ *are identical. The added linguistic terms are the smallest ones in* $L_2(p)$ *and the probabilities of all the linguistic terms are zero. Analogously, if* $\#L_1(p) < \#L_2(p)$, *we can use the similar method.*

Based on the normalized PLTSs, Pang et al. [6] further defined the deviation degree between PLTSs. The result is shown as follows.

Definition 7. *[6] Let* $L_1(p) = \{L_1^{(k)}(p_1^{(k)})|k = 1,2,\cdots,\#L_1(p)\}$ *and* $L_2(p) = \{L_2^{(k)}(p_2^{(k)})|k = 1,2,\cdots,\#L_2(p)\}$ *be any two PLTSs, if* $\#L_1(p) = \#L_2(p)$, *then the deviation degree between PLTSs is defined as:*

$$d(L_1(p), L_2(p)) = \sqrt{\sum_{k=1}^{\#L_1(p)} (r_1^{(k)} p_1^{(k)} - r_2^{(k)} p_2^{(k)})^2 / \#L_1(p)}. \tag{4}$$

2.2. Power Average (PA)

Information fusion is a process of aggregating data operators from different resources by proper aggregating operators. Power average (PA) operator, as a technique of fusing information, was introduced by Yager [1], which allows the arguments to support each other in the aggregation process.

Definition 8. *[1] Let* $A = \{a_1, a_2, \cdots, a_n\}$ *be a collection of non-negative numbers. The power aggregation is defined as follows:*

$$PA(a_1, a_2, \cdots, a_n) = \frac{\sum_{i=1}^{n}(1 + T(a_i))a_i}{\sum_{i=1}^{n}(1 + T(a_i))}, \tag{5}$$

where

$$T(a_i) = \sum_{j=1, j \neq i}^{n} sup(a_i, a_j). \tag{6}$$

In this case, $sup(a_i, a_j)$ *is denoted as the support for* a_i *from* a_j, *which satisfies the following three properties:*

(1) $sup(a_i, a_j) \in [0,1]$;
(2) $sup(a_i, a_j) = sup(a_j, a_i)$;
(3) $sup(a_i, a_j) \geq sup(a_i, a_k)$, *if* $|a_i - a_j| < |a_i - a_k|$.

From the result of Definition 7, the supports among the input arguments are involved in the PA. In general, $sup(a_i, a_j)$ can be measured by the distance between the arguments, e.g., $d(a_i, a_j)$. By introducing geometric mean (GM), Xu and Yager [2] defined a power geometric (PG) operator as follows:

$$PG(a_1, a_2, \cdots, a_n) = \prod_{i=1}^{n} a_i^{\frac{(1+T(a_i))}{\sum_{i=1}^{n}(1+T(a_i))}}, \tag{7}$$

where a_i $(i = 1, 2, \cdots, n)$ are a collection of arguments, and $T(a_i)$ satisfies the condition above.

3. Probabilistic Linguistic Power Aggregation Operators

Under the probabilistic linguistic environment, we assume that the input arguments are PLTSs and we mainly study the extension of power average (PA) and power geometric (PG) aggregation operators.

3.1. Probabilistic Linguistic Power Average (PLPA) Aggregation Operators

In this section, we discuss the extension of power average (PA) aggregation operators to accommodate the probabilistic linguistic environment. In the following, some probabilistic linguistic power average aggregation operators should be developed, which allow the input data to support each other in the aggregation process, i.e., Probabilistic Linguistic Power Average (PLPA) and Weighted Probabilistic Linguistic Power Average (WPLPA).

3.1.1. PLPA

Based on the results of Definitions 1 and 7, we present the Definition of the PLPA aggregation operator as follows:

Definition 9. *Let* $L(p) = \left\{ L_i^{(k)}(p_i^{(k)}) \mid k = 1, 2, \cdots, \#L_i(p) \right\}$ $(i = 1, 2, ..., n)$ *be a collection of PLTSs. A probabilistic linguistic power average (PLPA) is a mapping* $L^n(p) \to L(p)$ *such that:*

$$PLPA(L_1(p), L_2(p), \cdots, L_n(p)) = \bigoplus_{i=1}^{n} \frac{(1 + T(L_i(p)))L_i(p)}{\sum_{i=1}^{n}(1 + T(L_i(p)))}, \tag{8}$$

where:

$$T(L_i(p)) = \sum_{j=1, j \neq i}^{n} sup(L_i(p), L_j(p)). \tag{9}$$

and $sup(L_i(p), L_j(p))$ *is considered to be the support for* $L_i(p)$ *from* $L_j(p)$ *which satisfies the following properties:*

(1) $sup(L_i(p), L_j(p)) \in [0, 1]$;
(2) $sup(L_i(p), L_j(p)) = sup(L_j(p), L_i(p))$;
(3) $sup(L_i(p), L_j(p)) \geq sup(L_i(p), L_k(p))$ *if* $d(L_i(p), L_j(p)) < d(L_i(p), L_k(p))$.

In light of the operations law (1) of Definition 3, Definition 8 can be transformed into the following form:

$$\begin{aligned} & PLPA(L_1(p), L_2(p), \cdots, L_n(p)) \\ = & \frac{(1 + T(L_1(p)))}{\sum_{i=1}^{n}(1 + T(L_i(p)))} L_1(p) \oplus \frac{(1 + T(L_2(p)))}{\sum_{i=1}^{n}(1 + T(L_i(p)))} L_2(p) \oplus \cdots \oplus \frac{(1 + T(L_n(p)))}{\sum_{i=1}^{n}(1 + T(L_i(p)))} L_n(p). \end{aligned}$$

Hence, we can deduce the following result from Definition 8.

Proposition 1. *Let* $L(p) = \left\{ L_i^{(k)}(p_i^{(k)}) \mid k = 1, 2, \cdots, \#L_i(p) \right\}$ $(i = 1, 2, ..., n)$ *be a collection of PLTSs. A probabilistic linguistic power average (PLPA) is calculated as:*

$$\begin{aligned} & PLPA(L_1(p), L_2(p), \cdots, L_n(p)) = \bigoplus_{i=1}^{n} v_i L_i(p) \\ = & \bigcup_{L_1^{(k)} \in L_1(p)} \left\{ v_1 p_1^{(k)} L_1^{(k)} \right\} \oplus \bigcup_{L_2^{(k)} \in L_2(p)} \left\{ v_2 p_2^{(k)} L_2^{(k)} \right\} \oplus \cdots \oplus \bigcup_{L_n^{(k)} \in L_n(p)} \left\{ v_n p_n^{(k)} L_n^{(k)} \right\}, \end{aligned} \tag{10}$$

where $v_i = \frac{(1+T(L_i(p)))}{\sum_{j=1}^{n}(1+T(L_j(p)))}$ $(i = 1, 2, ..., n)$.

On the basis of Definition 8 and Proposition 1, it can easily be proven that the PLPA aggregation operator has the following desirable properties.

Theorem 1. *(Commutativity) Let* $(L_1(p)^*, L_2(p)^*, \cdots, L_n(p)^*)$ *be any permutation of* $(L_1(p), L_2(p), \cdots, L_n(p))$*, then* $PLPA(L_1(p), L_2(p), \cdots, L_n(p)) = PLPA(L_1(p)^*, L_2(p)^*, \cdots, L_n(p)^*)$.

Proof. According to the result of Definition 2, $L_i(p)$ is called an ordered PLTS $(i = 1, 2, ..., n)$. By Proposition 1 and the operations law (1) of Definition 3, we can conclude that:

$$PLPA(L_1(p), L_2(p), \cdots, L_n(p)) = PLPA(L_1(p)^*, L_2(p)^*, \cdots, L_n(p)^*).$$

Therefore, we complete the proof of Theorem 1. □

Theorem 2. *(Idempotency) Let* $L_i(p)$ $(i = 1, 2, \cdots, n)$ *be a collection of PLTSs. If all* $L_i(p)$ $(i = 1, 2, \cdots, n)$ *are equal, i.e.,* $L_i(p) = L(p)$*, then* $PLPA(L_1(p), L_2(p), \cdots, L_n(p)) = L(p)$.

Proof. If $L_i(p) = L(p)$ for all i, then $PLPA(L_1(p), L_2(p), \cdots, L_n(p))$ is computed as:

$$\begin{aligned} PLPA(L_1(p), L_2(p), \cdots, L_n(p)) &= \bigoplus_{i=1}^{n} \frac{(1+T(L_i(p)))}{\sum_{i=1}^{n}(1+T(L_i(p)))} L_i(p) \\ &= \bigoplus_{i=1}^{n} \frac{1}{n} L_i(p) = L_i(p). \end{aligned}$$

Hence, the statement of Theorem 2 holds. □

Theorem 3. *(Boundedness) Let* $L_i(p)$ $(i = 1, 2, \cdots, n)$ *be a collection of PLTSs, then we have:*

$$\min_{i=1}^{n} \min_{k=1}^{\#L_i(p)} p_i^{(k)} L_i^{(k)} \leq L \leq \max_{i=1}^{n} \max_{k=1}^{\#L_i(p)} p_i^{(k)} L_i^{(k)},$$

where $L \in PLPA(L_1(p), L_2(p), \cdots, L_n(p))$.

Proof. According to the result of Proposition 1, $PLPA(L_1(p), L_2(p), \cdots, L_n(p))$ is computed as:

$$PLPA(L_1(p), L_2(p), \cdots, L_n(p)) = \bigoplus_{i=1}^{n} v_i L_i(p)$$
$$= \bigcup_{L_1^{(k)} \in L_1(p)} \left\{ v_1 p_1^{(k)} L_1^{(k)} \right\} \oplus \bigcup_{L_2^{(k)} \in L_2(p)} \left\{ v_2 p_2^{(k)} L_2^{(k)} \right\} \oplus \cdots \oplus \bigcup_{L_n^{(k)} \in L_n(p)} \left\{ v_n p_n^{(k)} L_n^{(k)} \right\}.$$

Then, we can deduce the following relationship:

$$\min_{i=1}^{n} \min_{k=1}^{\#L_i(p)} p_i^{(k)} L_i^{(k)} \leq p_i^{(k)} L_i^{(k)} \leq \max_{i=1}^{n} \max_{k=1}^{\#L_i(p)} p_i^{(k)} L_i^{(k)}.$$

By utilizing the result of Theorem 2, we can easily finish the proof of Theorem 1. □

Theorem 4. *(Monotonicity) Let* $L_i(p)$ *and* $L_i(p)^*$ *be two sets of PLTSs and the numbers of linguistic terms in* $L_i(p)$ *and* $L_i(p)^*$ *are identical* $(i = 1, 2, \cdots, n)$*. If* $L_i^{(k)}(p_i^{(k)}) \leq L_i^{(k)}(p_i^{(k)})^*$ *for all i, i.e.,* $L_i(p) \leq L_i(p)^*$*, then* $PLPA(L_1(p), L_2(p), \cdots, L_n(p)) \leq PLPA(L_1(p)^*, L_2(p)^*, \cdots, L_n(p)^*)$.

Theorem 5. *Let $sup(L_i(p), L_j(p)) = k$ for all $i \neq j$, then $PLPA(L_1(p), L_2(p), \cdots, L_n(p)) = \bigoplus \frac{1}{n} L_i(p)$.*

Proof. If $sup(L_i(p), L_j(p)) = k$ for all $i \neq j$, it indicates that all the supports are the same. In this situation, the PLPA operator is computed as follows:

$$\begin{aligned} PLPA(L_1(p), L_2(p), \cdots, L_n(p)) &= \bigoplus_{i=1}^{n} \frac{(1+T(L_i(p)))}{\sum_{i=1}^{n}(1+T(L_i(p)))} L_i(p) \\ &= \bigoplus_{i=1}^{n} \frac{(1+(n-1)k)}{\sum_{i=1}^{n}(1+(n-1)k)} L_i(p) = \bigoplus_{i=1}^{n} \frac{1}{n} L_i(p). \end{aligned}$$

It is a simple probabilistic linguistic averaging operator. Hence, the statement of Theorem 5 holds. □

3.1.2. WPLPA

With respect to the PLPA operator, the weights of the arguments should be considered, because each argument that is being aggregated has a weight indicating its importance [3]. Based on this idea, we extend the PLPA and give the Definition of the weighted probabilistic linguistic power average (WPLPA) operator as follows:

Definition 10. *Let $L_i(p)$ be a collection of PLTSs. $w = (w_1, w_2, ..., w_n)^T$ denotes the weighting vector of $L_i(p)$ and $w_i \in [0,1]$, $\sum_{i=1}^{n} w_i = 1$. Given the value of the weight vector $w = (w_1, w_2, ..., w_n)^T$, we define weighted probabilistic linguistic power average (WPLA) operator as follows:*

$$WPLPA(L_1(p), L_2(p), \cdots, L_n(p)) = \bigoplus_{i=1}^{n} \frac{w_i(1+T'(L_i(p)))L_i(p)}{\sum_{i=1}^{n} w_i(1+T'(L_i(p)))}. \tag{11}$$

In this case, $T'(L_i(p)) = \sum_{j=1, j\neq i}^{n} w_j sup(L_i(p), L_j(p))$.

Based on the operations of the PLTSs described in Definition 3, we can derive the following Proposition 2.

Proposition 2. *Let $L_i(p)$ $(i = 1, 2, \cdots, n)$ be a collection of PLTSs, then their aggregated values by using the WPLPA operator is also a PLTS, and:*

$$\begin{aligned} WPLPA(L_1(p), L_2(p), \cdots, L_n(p)) &= \bigcup_{L_1^{(k)} \in L_1(p)} \left\{ v'_1 p_1^{(k)} L_1^{(k)} \right\} \oplus \bigcup_{L_2^{(k)} \in L_2(p)} \left\{ v'_2 p_2^{(k)} L_2^{(k)} \right\} \\ &\oplus \cdots \oplus \bigcup_{L_n^{(k)} \in L_n(p)} \left\{ v'_n p_n^{(k)} L_n^{(k)} \right\}. \end{aligned} \tag{12}$$

where $v'_i = \frac{w_i(1+T'(L_i(p)))}{\sum_{j=1}^{n} w_j(1+T'(L_j(p)))}$ $(i = 1, 2, \cdots, n)$.

Especially, if $sup(L_i(p), L_j(p)) = 0$ for all $i \neq j$, then $T(L_i(p) = 0$. Thus, $WPLPA(L_1(p), L_2(p), \cdots, L_n(p)) = \bigoplus_{i=1}^{n} w_i L_i(p)$. Under this situation, the WPLPA operator reduces to PLWA proposed by Ref. [6]. If the weight vector $w = (w_1, w_2, ..., w_n)^T = (\frac{1}{n}, \frac{1}{n}, \cdots, \frac{1}{n})^T$, v'_i of Proposition 2 is computed as:

$$\begin{aligned} v'_i &= \frac{w_i(1+T'(L_i(p)))}{\sum_{j=1}^{n} w_j(1+T'(L_j(p)))} \\ &= \frac{(1+T'(L_i(p)))}{\sum_{i=1}^{n}(1+T'(L_i(p)))} = v_i. \end{aligned}$$

Thus, the WPLPA operator is computed as:

$$\begin{aligned}
&WPLPA(L_1(p), L_2(p), \cdots, L_n(p)) \\
&= \bigcup_{L_1^{(k)} \in L_1(p)} \left\{ v_1 p_1^{(k)} L_1^{(k)} \right\} \oplus \bigcup_{L_2^{(k)} \in L_2(p)} \left\{ v_2 p_2^{(k)} L_2^{(k)} \right\} \oplus \cdots \oplus \bigcup_{L_n^{(k)} \in L_n(p)} \left\{ v_n p_n^{(k)} L_n^{(k)} \right\} \\
&= PLPA(L_1(p), L_2(p), \cdots, L_n(p)).
\end{aligned}$$

It indicates that the WPLPA reduces to the PLPA operator. According to the results of Definitions 3 and 10, it can easily prove that the WPLPA operator has the following properties.

Theorem 6. *(Idempotency) Let $L_i(p)$ $(i = 1, 2, \cdots, n)$ be a collection of PLTSs, if all $L_i(p)$ $(i = 1, 2, \cdots, n)$ are equal, i.e., $L_i(p) = L(p)$, then $WPLPA(L_1(p), L_2(p), \cdots, L_n(p)) = L(p)$.*

Proof. If $L_i(p) = L(p)$ for all i, then $WPLPA(L_1(p), L_2(p), \cdots, L_n(p))$ is computed as:

$$\begin{aligned}
WPLPA(L_1(p), L_2(p), \cdots, L_n(p)) &= \bigoplus_{i=1}^{n} \frac{w_i(1 + T'(L_i(p)))L_i(p)}{\sum_{i=1}^{n} w_i(1 + T'(L_i(p)))} \\
&= \bigoplus_{i=1}^{n} \frac{1}{n} L_i(p) = L_i(p).
\end{aligned}$$

Thus, the statement of Theorem 6 holds. □

Theorem 7. *(Boundedness) Let $L_i(p)$ $(i = 1, 2, \cdots, n)$ be a collection of PLTSs, then we have:*

$$\min_{i=1}^{n} \min_{k=1}^{\#L_i(p)} p_i^{(k)} L_i^{(k)} \leq L \leq \max_{i=1}^{n} \max_{k=1}^{\#L_i(p)} p_i^{(k)} L_i^{(k)}.$$

where $L \in WPLPA(L_1(p), L_2(p), \cdots, L_n(p))$.

If we let $sup(L_i(p), L_j(p)) = k$ for all $i \neq j$, we have: $T'(L_i(p)) = \sum_{j=1, j\neq i}^{n} w_j sup(L_i(p), L_j(p)) = k\sum_{j=1, j\neq i}^{n} w_j$ $(i = 1, 2, \cdots, n)$. Based on the result of Definition 10, we have:

$$WPLPA(L_1(p), L_2(p), \cdots, L_n(p)) = \bigoplus_{i=1}^{n} \frac{w_i(1 + k\sum_{j=1, j\neq i}^{n} w_j)}{\sum_{i=1}^{n} w_i(1 + k\sum_{j=1, j\neq i}^{n} w_j)} L_i(p).$$

In this case, $WPLPA(L_1(p), L_2(p), \cdots, L_n(p))$ is not equivalent to $PLPA(L_1(p), L_2(p), \cdots, L_n(p)) = \frac{1}{n} \bigoplus_{i=1}^{n} L_i(p)$.

Theorem 8. *Let $(L_1(p)^*, L_2(p)^*, \cdots, L_n(p)^*)$ be any permutation of $(L_1(p), L_2(p), \cdots, L_n(p))$, then we can deduce the following relationship:*

$$WPLPA(L_1(p)^*, L_2(p)^*, \cdots, L_n(p)^*) \neq WPLPA(L_1(p), L_2(p), \cdots, L_n(p)).$$

Proof. According to the result of Definition 10, we can obtain:

$$T'(L_p^*) = \sum_{j=1, j\neq i}^{n} w_j sup(L_i(p)^*, L_j(p)^*).$$

Then, we can deduce:

$$WPLPA(L_1(p), L_2(p), \cdots, L_n(p)) = \bigoplus_{i=1}^{n} \frac{w_i(1 + T'(L_i(p)^*))}{\sum_{i=1}^{n} w_i(1 + T'(L_i(p)^*))} L_i(p)^*.$$

Since $(T'(L_1(p)^*), T'(L_2(p)^*), \cdots, T'(L_2(p)^*))$ may not be the permutation of $(T'(L_1(p)), T'(L_2(p)), \cdots, T'(L_n(p)))$, we can judge that the WPLPA operator is not commutative. Therefore, we complete the proof of Theorem 8. □

3.2. Probabilistic Linguistic Power Geometric (PLPG) Aggregation Operators

In this section, we investigate the extension of power geometric (PG) aggregation operators under the probabilistic linguistic environment, i.e., the probabilistic linguistic power geometric (PLPG) and weighted probabilistic linguistic power geometric (WPLPG).

3.2.1. PLPG

By utilizing the results of Definition 1 and Equation (7), we present the Definition of the PLPG operator as follows.

Definition 11. *Let* $L(p) = \left\{ L_i^{(k)}(p_i^k) | k = 1, 2, \cdots, \#L_i(p) \right\}$ $(i = 1, 2, \cdots, n)$ *be a collection of PLTSs. A probabilistic linguistic power geometric (PLPG) operator is a mapping* $L^n(p) \to L(p)$ *such that:*

$$PLPG(L_1(p), L_2(p), \cdots, L_n(p)) = \bigotimes_{i=1}^{n} (L_i(p))^{\frac{(1+T(L_i(p)))}{\sum_{i=1}^{n}(1+T(L_i(p)))}}, \tag{13}$$

where $T(L_i(p)) = \sum_{j=1, j\neq i}^{n} sup(L_i(p), L_j(p))$*.* $sup(L_i(p), L_j(p))$ *is considered to be the support of* $L_i(p)$ *from* $L_j(p)$ *which also satisfies the following properties:*

(1) $sup(L_i(p), L_j(p)) \in [0, 1]$*;*
(2) $sup(L_i(p), L_j(p)) = sup(L_j(p), L_i(p))$*;*
(3) $sup(L_i(p), L_j(p)) \geq sup(L_j(p), L_i(p))$ *if* $d(L_i(p), L_j(p)) < d(L_i(p), L_k(p))$*.*

By the operations law (2) of Definition 3, Definition 11 can be transformed into the following form:

$$\begin{aligned} & PLPG(L_1(p), L_2(p), \cdots, L_n(p)) \\ &= (L_1(p))^{\frac{(1+T(L_1(p)))}{\sum_{i=1}^{n}(1+T(L_i(p)))}} \otimes (L_2(p))^{\frac{(1+T(L_2(p)))}{\sum_{i=1}^{n}(1+T(L_i(p)))}} \otimes \cdots \otimes (L_n(p))^{\frac{(1+T(L_n(p)))}{\sum_{i=1}^{n}(1+T(L_i(p)))}}. \end{aligned}$$

Therefore, we can deduce the following based on the results of Definition 9:

Proposition 3. *Let* $L(p) = \left\{ L_i^{(k)}(p_i^k) | k = 1, 2, \cdots, \#L_i(p) \right\}$ $(i = 1, 2, \cdots, n)$ *be a collection of PLTSs. A probabilistic linguistic power geometric (PLPG) operator is calculated as:*

$$\begin{aligned} & PLPG(L_1(p), L_2(p), \cdots, L_n(p)) = \bigotimes_{i=1}^{n} (L_i(p))^{v_i} \\ &= \bigcup_{L_1^{(k)} \in L_1(p)} \left\{ (L_1^{(k)})^{v_1 p_1^{(k)}} \right\} \otimes \bigcup_{L_2^{(k)} \in L_2(p)} \left\{ (L_2^{(k)})^{v_2 p_2^{(k)}} \right\} \otimes \cdots \otimes \bigcup_{L_n^{(k)} \in L_n(p)} \left\{ (L_n^{(k)})^{v_n p_n^{(k)}} \right\}, \end{aligned} \tag{14}$$

where $v_i = \frac{(1+T(L_i(p)))}{\sum_{j=1}^{n}(1+T(L_j(p)))}$ $(i = 1, 2, ..., n)$*.*

On the basis of Definition 11 and Proposition 3, it can be proved that the PLPG operator has the following desirable properties:

Theorem 9. *(Commutativity) Let* $(L_1(p)^*, L_2(p)^*, \cdots, L_n(p)^*)$ *be any permutation of* $(L_1(p), L_2(p), \cdots, L_n(p))$ *then* $PLPG(L_1(p), L_2(p), \cdots, L_n(p))=PLPG(L_1(p)^*, L_2(p)^*, \cdots, L_n(p)^*)$.

Proof. According to the result of Definition 2, $L_i(p)$ is called an ordered PLTS $(i = 1, 2, \cdots, n)$. By the results of Proposition 3 and the operations laws (2) of Definition 3, we can conclude that:

$$PLPG(L_1(p), L_2(p), \cdots, L_n(p)) = PLPG(L_1(p)^*, L_2(p)^*, \cdots, L_n(p)^*).$$

Therefore, we complete the proof of Theorem 9. □

Theorem 10. *(Idempotency) Let* $L_i(p)$ $(i = 1, 2, \cdots, n)$ *be a collection of PLTSs. If all* $L_i(p)$ $(i = 1, 2, \cdots, n)$ *are equal, i.e.,* $L_i(p) = L(p)$*, then* $PLPG(L_1(p), L_2(p), \cdots, L_n(p)) = L(p)$.

Proof. If $L_i(p) = L(p)$ for all i, then $PLPG(L_1(p), L_2(p), \cdots, L_n(p))$ is computed as:

$$\begin{aligned} PLPG(L_1(p), L_2(p), \cdots, L_n(p)) &= \bigotimes_{i=1}^{n} (L_i(p))^{\frac{(1+T(L_i(p)))}{\sum_{i=1}^{n}(1+T(L_i(p)))}} \\ &= \bigotimes_{i=1}^{n} (L_i(p))^{\frac{1}{n}} = L(p). \end{aligned}$$

Hence, the statement of Theorem 10 holds. □

Theorem 11. *(boundedness) Let* $L_i(p)$ $(i = 1, 2, \cdots, n)$ *be a collection of PLTSs, then we have:*

$$\min_{i=1}^{n} \min_{k=1}^{\#L_i(p)} (L_i^{(k)})^{p_i^{(k)}} \leq L \leq \max_{i=1}^{n} \max_{k=1}^{\#L_i(p)} (L_i^{(k)})^{p_i^{(k)}},$$

where $L \in PLPG(L_1(p), L_2(p), \cdots, L_n(p))$.

Proof. According to the result of Proposition 3, $PLPG(L_1(p), L_2(p), \cdots, L_n(p))$ is computed as:

$$PLPG(L_1(p), L_2(p), \cdots, L_n(p)) = \bigotimes_{i=1}^{n} (L_i(p))^{v_i}$$

$$= \bigcup_{L_1^{(k)} \in L_1(p)} \left\{(L_1^{(k)})^{v_1 p_1^{(k)}}\right\} \otimes \bigcup_{L_2^{(k)} \in L_2(p)} \left\{(L_2^{(k)})^{v_2 p_2^{(k)}}\right\} \otimes \cdots \otimes \bigcup_{L_n^{(k)} \in L_n(p)} \left\{(L_n^{(k)})^{v_n p_n^{(k)}}\right\}.$$

Then, we can deduce the following relationship:

$$\min_{i=1}^{n} \min_{k=1}^{\#L_i(p)} (L_i^{(k)})^{p_i^{(k)}} \leq (L_i^{(k)})^{p_i^{(k)}} \leq \max_{i=1}^{n} \max_{k=1}^{\#L_i(p)} (L_i^{(k)})^{p_i^{(k)}}.$$

In light of the results of Theorem 10, we can easily finish the proof of Theorem 11. □

3.2.2. WPLPG

Considering the importance of the aggregated arguments, we extend the PLPG and give the Definition of the weighted probabilistic linguistic power geometric (WPLPG) operator as following.

Definition 12. *Let $L_i(p)$ be a collection of PLTSs. $w = (w_1, w_2, \cdots, w_n)^T$ denotes the weighting vector of $L_i(p)$, $w_i \in [0,1]$ and $\sum_{i=1}^{n} w_i = 1$. Given the value of the weight vector $w = (w_1, w_2, \cdots, w_n)^T$, we define weighted probabilistic linguistic power geometric (WPLPG) operator as follows:*

$$WPLPG(L_1(p), L_2(p), \cdots, L_n(p)) = \bigotimes_{i=1}^{n} (L_i(p))^{\frac{w_i(1+T'(L_i(p)))}{\sum_{i=1}^{n} w_i(1+T'(L_i(p)))}}. \tag{15}$$

In this case, $T'(L_i(p)) = \sum_{j=1, j \neq i}^{n} w_j sup(L_i(p), L_j(p))$.

Based on the operations of the PLPTs described in Definition 3, we can derive the following Proposition:

Proposition 4. *Let $L_i(p)$ $(i = 1, 2, \cdots, n)$ be a collection of PLTSs, then their aggregated values by using the WPLPG operator is also a PLTS, and:*

$$\begin{aligned} WPLPG(L_1(p), L_2(p), \cdots, L_n(p)) &= \bigcup_{L_1^{(k)} \in L_1(p)} \left\{ (L_1^{(k)})^{v_1' p_1^{(k)}} \right\} \otimes \bigcup_{L_2^{(k)} \in L_2(p)} \left\{ (L_2^{(k)})^{v_2' p_2^{(k)}} \right\} \\ &\otimes \cdots \otimes \bigcup_{L_n^{(k)} \in L_n(p)} \left\{ (L_n^{(k)})^{v_n' p_n^{(k)}} \right\}. \end{aligned} \tag{16}$$

where $v_i' = \frac{w_i(1+T'(L_i(p)))}{\sum_{j=1}^{n} w_j(1+T'(L_j(p)))}$ $(i = 1, 2, \cdots, n)$.

For the result of Proposition 4, if $sup(L_i(p), L_j(p)) = 0$ for all $i \neq j$, then $T(L_i(p)) = 0$. Thus, we have:

$$WPLPG(L_1(p), L_2(p), \cdots, L_n(p)) = \bigotimes_{i=1}^{n} (L_i(p))^{w_i}.$$

Under this situation, the WPLPG operator reduces to PLWG proposed by Ref. [6]. If the weight vector $w = (w_1, w_2, ..., w_n)^T = (\frac{1}{n}, \frac{1}{n}, \cdots, \frac{1}{n})^T$, v_i' of Proposition 4 is computed as:

$$\begin{aligned} v_i' &= \frac{w_i(1 + T'(L_i(p)))}{\sum_{j=1}^{n} w_j(1 + T'(L_j(p)))} \\ &= \frac{(1 + T'(L_i(p)))}{\sum_{i=1}^{n}(1 + T'(L_i(p)))} = v_i. \end{aligned}$$

Hence, the WPLPG operator is computed as:

$$\begin{aligned} WPLPG(L_1(p), L_2(p), \cdots, L_n(p)) &= \bigcup_{L_1^{(k)} \in L_1(p)} \left\{ (L_1^{(k)})^{v_1' p_1^{(k)}} \right\} \otimes \bigcup_{L_2^{(k)} \in L_2(p)} \left\{ (L_2^{(k)})^{v_2' p_2^{(k)}} \right\} \\ &\otimes \cdots \otimes \bigcup_{L_n^{(k)} \in L_n(p)} \left\{ (L_n^{(k)})^{v_n' p_n^{(k)}} \right\} \\ &= PLPG(L_1(p), L_2(p), \cdots, L_n(p)). \end{aligned}$$

Thus, it indicates that the WPLPG can be reduced to the PLPG operator. According to the results of Definitions 3 and 12, it can easily prove that the WPLPG operator has the following properties.

Theorem 12. *(Idempotency) Let $L_i(p)$ $(i = 1, 2, \cdots, n)$ be a collection of PLTSs, if all $L_i(p)$ $(i = 1, 2, \cdots, n)$ are equal, i.e., $L_i(p) = L(p)$, then $WPLPG(L_1(p), L_2(p), \cdots, L_n(p)) = L(p)$.*

Proof. If $L_i(p) = L(p)$ for all i, then $WPLPG(L_1(p), L_2(p), \cdots, L_n(p))$ is computed as:

$$\begin{aligned} WPLPG(L_1(p), L_2(p), \cdots, L_n(p)) &= \bigotimes_{i=1}^{n} (L_i(p))^{\frac{w_i(1+T'(L_i(p)))}{\sum_{i=1}^{n} w_i(1+T'(L_i(p)))}} \\ &= \bigotimes_{i=1}^{n} (L_i(p))^{\frac{1}{n}} = L(p). \end{aligned}$$

Thus, the statement of Theorem 12 holds. □

Theorem 13. *(Boundedness) Let $L_i(p)$ $(i = 1, 2, \cdots, n)$ be a collection of PLTSs, then we have:*

$$\min_{i=1}^{n} \min_{k=1}^{\#L_i(p)} p_i^{(k)} L_i^{(k)} \leq L \leq \max_{i=1}^{n} \max_{k=1}^{\#L_i(p)} p_i^{(k)} L_i^{(k)},$$

where $L \in WPLPG(L_1(p), L_2(p), \cdots, L_n(p))$.

Theorem 14. *Let $(L_1(p)^*, L_2(p)^*, \cdots, L_n(p)^*)$ is any permutation of $(L_1(p), L_2(p), \cdots, L_n(p))$, then we can deduce the following relationship:*

$$WPLPG(L_1(p)^*, L_2(p)^*, \cdots, L_n(p)^*) \neq WPLPG(L_1(p), L_2(p), \cdots, L_n(p)).$$

Proof. According to the result of Definition 12, we can obtain:

$$T'(L_p^*) = \sum_{j=1, j\neq i}^{n} w_j sup(L_i(p)^*, L_j(p)^*).$$

Then, we can deduce:

$$WPLPG(L_1(p), L_2(p), \cdots, L_n(p)) = \bigotimes_{i=1}^{n} (L_i(p)^*)^{\frac{w_i(1+T'(L_i(p)^*))}{\sum_{i=1}^{n} w_i(1+T'(L_i(p)^*))}}.$$

Since $(T'(L_1(p)^*), T'(L_2(p)^*), \cdots, T'(L_2(p)^*))$ may not be the permutation of $(T'(L_1(p)), T'(L_2(p)), \cdots, T'(L_n(p)))$, we can judge that the WPLPG operator is not commutative. Hence, we complete the proof of Theorem 14. □

4. Approaches to Multi-Criteria Group Decision Making with Probabilistic Linguistic Power Aggregation Operators

In this section, we firstly present a MCGDM problem in which the evaluation information may be expressed by PLTSs. Then, we utilize the WPLPA or WPLPG operator to support our decision. Let $X = \{x_1, x_2, \cdots, x_m\}$ be a finite set of m alternatives and $C = \{c_1, c_2, \cdots, c_n\}$ be a set of n attributes. Suppose that $D = \{d_1, d_2, \cdots, d_e\}$ denotes the set of DMs. By using the linguistic scale $S = \{s_\alpha | \alpha = 0, 1, \cdots, \tau\}$, each DM d_q provides his or her linguistic evaluations over the alternative x_i with respect to the attribute a_j, i.e., $A^q = (L_{ij}^q)_{m\times n}$ $(i = 1, 2, \cdots, m; j = 1, 2, \cdots, n; q = 1, 2, \cdots, e)$. Then, we determine the collective evaluations of DMs for each alternative in terms of PLTSs. In the context of GDM, the linguistic evaluation values $L_{ij}^{(k)}$ $(k = 1, 2, \cdots, \#L_{ij}(p))$ with the corresponding probability $p_{ij}^{(k)}$ are described as the PLTS $L_{ij}(p) = \{L_{ij}^{(k)}(p_{ij}^{(k)}) | k = 1, 2, \cdots, \#L_{ij}(p)\}$ and $\#L_{ij}(p)$ is the number of linguistic terms in $L_{ij}(p)$. The PLTS $L_{ij}(p)$ denotes the evaluation values over the alternative x_i $(i = 1, 2, \cdots, m)$ with respect to the attributes c_j $(j = 1, 2, \cdots, n)$, where $L_{ij}^{(k)}$ is the kth value of $L_{ij}(p)$, and $p_{ij}^{(k)}$ is the probability of $L_{ij}^{(k)}$ $(k = 1, 2, \cdots, \#L_{ij}(p))$. In the case, $p_{ij}^{(k)} \geq 0$ and $\sum_{k=1}^{\#L_{ij}(p)} p_{ij}^{(k)} = 1$.

All the PLTSs are contained in the probabilistic linguistic decision matrix R. Hence, the result is shown as follows:

$$R = (L_{ij}(p))_{m\times n} = \begin{pmatrix} L_{11}(p) & L_{12}(p) & \cdots & L_{1n}(p) \\ L_{21}(p) & L_{22}(p) & \cdots & L_{2n}(p) \\ \vdots & \vdots & \vdots & \vdots \\ L_{m1}(p) & L_{m2}(p) & \cdots & L_{mn}(p) \end{pmatrix}. \quad (17)$$

Without loss of generality, we assume that each PLTS $L_{ij}(p)$ is an ordered PLTS. $w = (w_1, w_2, \cdots, w_n)^T$ denotes the weighting vector of the attributes C and $w_j \in [0,1]$, $\sum_{j=1}^{n} w_j = 1$. Based on the above results, we will use the WPLPA or WPLPG aggregation operator to develop the corresponding approach for MCGDM with probabilistic linguistic information. This approach is designed as follows:

Step 1: According to the practical decision-making problem, we determine the alternatives $X = \{x_1, x_2, \cdots, x_m\}$ and a set of the attributes $C = \{c_1, c_2, \cdots, c_n\}$. Then, we can obtain the decision matrix $A^q = (L_{ij}^q)_{m\times n}$ provided by the DM d_q. By using the PLTSs, we construct the collective matrix $R = (L_{ij}(p))_{m\times n}$.

Step 2: With respect to the collective matrix $R = (L_{ij}(p))_{m\times n}$, we can normalize the entries of R as stated in Definition 6.

Step 3: Based on the matrix R and the result of Definition 7, the deviation degree between PLTSs $L_{ij}(p)$ and $L_{it}(p)$ is calculated below $(i = 1, 2, \cdots, m; j, t = 1, 2, \cdots, n)$:

$$d(L_{ij}(p), L_{it}(p)) = \sqrt{\frac{\sum_{k=1}^{\#L_{ij}(p)} (p_{ij}^{(k)} r_{ij}^{(k)} - p_{it}^{(k)} r_{it}^{(k)})^2}{\#L_{ij}(p)}}.$$

Step 4: By using the results of Definitions 7 and 8, we calculate the support of the alternative x_i as follows:

$$sup(L_{ij}(p), L_{it}(p)) = 1 - \frac{d(L_{ij}(p), L_{it}(p))}{\sum_{g=1, g\neq j}^{n} d(L_{ij}(p), L_{ig}(p))}, \quad (18)$$

which satisfies the support conditions (1)–(3) of Definition 9.

Step 5: According to the result of Definition 10, we can calculate the support $T'(L_{ij}(p))$ of $L_{ij}(p)$ by all of other $L_{it}(p)$ $(j, t = 1, 2, \cdots, n; t \neq j)$:

$$T'(L_{ij}(p)) = \sum_{t=1, t\neq j}^{n} w_t sup(L_{ij}(p), L_{it}(p)).$$

Step 6: With the aid of Proposition 2, we further compute the weight v_{ij}' associated with the PLTS $L_{ij}(p)$:

$$v_{ij}' = \frac{w_j(1 + T'(L_{ij}(p)))}{\sum_{j=1}^{n} w_j(1 + T'(L_{ij}(p)))}.$$

Step 7: If the DM prefers the WPLPA operator, then the aggregated value of the alternative x_i is determined based on Equation (12). The result is:

$$\begin{aligned} WPLPA(L_{i1}(p), L_{i2}(p), \cdots, L_{in}(p)) &= \bigcup_{L_{i1}^{(k)} \in L_{i1}(p)} \left\{ v'_{i1} p_{i1}^{(k)} L_{i1}^{(k)} \right\} \oplus \bigcup_{L_{i2}^{(k)} \in L_{i2}(p)} \left\{ v'_{i2} p_{i2}^{(k)} L_{i2}^{(k)} \right\} \\ &\oplus \cdots \oplus \bigcup_{L_{in}^{(k)} \in L_{in}(p)} \left\{ v'_{in} p_{in}^{(k)} L_{in}^{(k)} \right\}. \end{aligned}$$

If the DM uses the WPLPG operator, then the aggregated value of the alternative x_i is determined based on Equation (16). The result is:

$$\begin{aligned} WPLPG(L_{i1}(p), L_{i2}(p), \cdots, L_{in}(p)) &= \bigcup_{L_{i1}^{(k)} \in L_{i1}(p)} \left\{ (L_{i1}^{(k)})^{v'_{i1} p_{i1}^{(k)}} \right\} \otimes \bigcup_{L_{i2}^{(k)} \in L_{i2}(p)} \left\{ (L_{i2}^{(k)})^{v'_{i2} p_{i2}^{(k)}} \right\} \\ &\otimes \cdots \otimes \bigcup_{L_{in}^{(k)} \in L_{in}(p)} \left\{ (L_{in}^{(k)})^{v'_{in} p_{in}^{(k)}} \right\}. \end{aligned}$$

In this case, we denote the aggregated value of the alternative x_i as Z_i.

Step 8: Based on the results of Definition 4, the score and the deviation degree of Z_i of the alternative x_i are computed, i.e., $E(Z_i)$ and $\sigma(Z_i)$ $(i = 1, 2, \cdots, m)$.

Step 9: Rank all of the alternatives in accordance with the ranking results of Definition 5.

5. An Illustrative Example

In recent years, there has been considerable concern regarding problems associated with undergraduate school rankings, graduate school rankings, evaluating and rewarding university professors in China and other countries of the world. Katz et al. [20] mentioned that these problems always existed and political activism together with various economic recession have worsen them. Katz and his partners were concerned with the criteria for evaluating them. They came out with multiple regression analysis to determine the factors important in salary and promotion decision-making at the university level and developed a more rational means of evaluating and rewarding university professors. They were motivated by the fact that there is a discriminatory policy in rank and reward in the universities which is not necessarily justifiable. They went further to state that rewarding professors goes through an arbitrary and chaotic process and a more equitable system could be instituted to enhance decision-making process. Another concern raised was that decisions on salaries and promotions were made in an intuitive manner in such a way that the weights attached to the various criteria for classification lack clear understanding. In this section, we illustrate our proposed approach by evaluating some university faculty for tenure and promotion in China adapted from Bryson et al. [21]. Hence, we firstly present a MCGDM problem in which the evaluation information may be expressed by PLTSs. Then, we utilize the WPLPA and WPLPG operator to support our decision. In light of the results of Ref. [21], the criteria considered for the assessment of the decision problem are summarized as follows: (1) teaching (c_1); (2) research (c_2); (3) service (c_3). Let $X = \{x_1, x_2, x_3, x_4, x_5\}$ be the set of five alternatives and $C = \{c_1, c_2, c_3\}$ be the set of three attributes. The linguistic scale is $S = \{s_\alpha | \alpha = 0, 1, \cdots, 8\}$. Suppose that $D = \{d_1, d_2, d_3, d_4\}$ denotes the set of DMs. Based on the results of Ref. [22], their evaluations are shown in Tables 1–4.

Table 1. Decision matrix A^1 provided by d_1.

	c_1	c_2	c_3
x_1	s_8	s_6	s_6
x_2	s_6	s_7	s_7
x_3	s_5	s_8	s_7
x_4	s_7	s_4	s_6
x_5	s_8	s_6	s_7

Table 2. Decision matrix A^2 provided by d_2.

	c_1	c_2	c_3
x_1	s_6	s_8	s_5
x_2	s_5	s_6	s_7
x_3	s_7	s_6	s_7
x_4	s_8	s_6	s_7
x_5	s_8	s_7	s_6

Table 3. Decision matrix A^3 provided by d_3.

	c_1	c_2	c_3
x_1	s_7	s_8	s_6
x_2	s_4	s_5	s_6
x_3	s_8	s_7	s_6
x_4	s_7	s_5	s_8
x_5	s_6	s_7	s_6

Table 4. Decision matrix A^4 provided by d_4.

	c_1	c_2	c_3
x_1	s_6	s_7	s_6
x_2	s_8	s_7	s_7
x_3	s_7	s_6	s_8
x_4	s_5	s_7	s_6
x_5	s_5	s_6	s_5

5.1. Decision Analysis with Our Proposed Approaches

Based on the proposed approaches of Section 4, we need to fuse the information presented in the decision matrices $A^1 - A^4$ by (17). In the context of GDM, all the PLTSs are contained in the probabilistic linguistic decision matrix R. Hence, the result is shown in Table 5.

Table 5. The probabilistic linguistic decision matrix R.

	c_1	c_2	c_3
x_1	$\left\{s_8(0.25), s_6(0.5), s_7(0.25)\right\}$	$\left\{s_6(0.25), s_8(0.5), s_7(0.25)\right\}$	$\left\{s_6(0.75), s_5(0.25)\right\}$
x_2	$\left\{s_6(0.25), s_5(0.25), s_4(0.25), s_8(0.25)\right\}$	$\left\{s_7(0.5), s_6(0.25), s_5(0.25)\right\}$	$\left\{s_7(0.75), s_6(0.25)\right\}$
x_3	$\left\{s_5(0.25), s_7(0.5), s_8(0.25)\right\}$	$\left\{s_8(0.25), s_6(0.5), s_7(0.25)\right\}$	$\left\{s_7(0.5), s_6(0.25), s_8(0.25)\right\}$
x_4	$\left\{s_7(0.5), s_8(0.25), s_5(0.25)\right\}$	$\left\{s_4(0.25), s_6(0.25), s_5(0.25), s_7(0.25)\right\}$	$\left\{s_6(0.5), s_7(0.25), s_8(0.25)\right\}$
x_5	$\left\{s_8(0.5), s_6(0.25), s_5(0.25)\right\}$	$\left\{s_6(0.5), s_7(0.5)\right\}$	$\left\{s_7(0.25), s_6(0.5), s_5(0.25)\right\}$

For Table 5, each PLTS $L_{ij}(p)$ is assumed to be an ordered PLTS ($i = 1,2,3,4,5; j = 1,2,3$). In this case, the weighting vector of the attributes C is $w = (w_1, w_2, w_3)^T = (0.3, 0.4, 0.3)^T$. We use the WPLPA or WPLPG aggregation operator to analyze the results of Table 5. Based on the above results and the proposed methods of Section 4, the detailed steps are shown as follows:

Step 2: With respect to the collective matrix $R = (L_{ij}(p))_{5\times 3}$, we can find that the number of their corresponding number of the linguistic terms is not equal. Thus, we normalize the entries of R as stated in Definition 6. The normalized probabilistic linguistic decision matrix is shown in Table 6.

Table 6. The normalized probabilistic linguistic decision matrix.

	c_1	c_2	c_3
x_1	$\{s_6(0.5), s_8(0.25), s_7(0.25), s_6(0)\}$	$\{s_8(0.5), s_7(0.25), s_6(0.25), s_6(0)\}$	$\{s_6(0.75), s_5(0.25), s_5(0), s_5(0)\}$
x_2	$\{s_8(0.25), s_6(0.25), s_5(0.25), s_4(0.25)\}$	$\{s_7(0.5), s_6(0.25), s_5(0.25), s_5(0)\}$	$\{s_7(0.75), s_6(0.25), s_6(0), s_6(0)\}$
x_3	$\{s_7(0.5), s_8(0.25), s_5(0.25), s_5(0)\}$	$\{s_6(0.5), s_8(0.25), s_7(0.25), s_6(0)\}$	$\{s_7(0.5), s_8(0.25), s_6(0.25), s_6(0)\}$
x_4	$\{s_7(0.5), s_8(0.25), s_5(0.25), s_5(0)\}$	$\{s_7(0.25), s_6(0.25), s_5(0.25), s_4(0.25)\}$	$\{s_6(0.5), s_8(0.25), s_7(0.25), s_6(0)\}$
x_5	$\{s_8(0.5), s_6(0.25), s_5(0.25), s_5(0)\}$	$\{s_7(0.5), s_6(0.5), s_6(0), s_6(0)\}$	$\{s_6(0.5), s_7(0.25), s_5(0.25), s_5(0)\}$

Step 3: According to the results of Definition 7 and Table 6, the deviation degree between PLTSs $L_{ij}(p)$ and $L_{it}(p)$ ($i = 1,2,3,4,5; j,t = 1,2,3$) can be calculated by the following equation:

$$d(L_{ij}(p), L_{it}(p)) = \sqrt{\frac{\sum_{k=1}^{\#L_{ij}(p)} (p_{ij}^{(k)} r_{ij}^{(k)} - p_{it}^{(k)} r_{it}^{(k)})^2}{\#L_{ij}(p)}}.$$

Then, we can calculate the deviation degree of any two $L_{ij}(p)$, respectively. For alternative x_1, the deviation degrees are shown as follows:

$$d(L_{11}, L_{12}) = 0.5303;\ \ d(L_{12}, L_{13}) = 0.8292;\ \ d(L_{11}, L_{13}) = 1.2119.$$

For alternative x_2, the deviation degrees are shown as follows:

$$d(L_{21}, L_{22}) = 0.9014;\ \ d(L_{22}, L_{23}) = 1.0752;\ \ d(L_{21}, L_{23}) = 1.8114.$$

For alternative x_3, the deviation degrees are shown as follows:

$$d(L_{31}, L_{32}) = 0.3536;\ \ d(L_{32}, L_{33}) = 0.2795;\ \ d(L_{31}, L_{33}) = 0.125.$$

For alternative x_4, the deviation degrees are shown as follows:

$$d(L_{41}, L_{42}) = 1.0383;\ \ d(L_{42}, L_{43}) = 0.875;\ \ d(L_{41}, L_{43}) = 0.3536.$$

For alternative x_5, the deviation degrees are shown as follows:

$$d(L_{51}, L_{52}) = 1.00778;\ d(L_{52}, L_{53}) = 0.9185;\ d(L_{51}, L_{53}) = 0.5154.$$

Step 4: Based on the results of Definitions 7 and 8, we can calculate the support of the alternative x_i by using (18) ($i = 1,2,3,4,5$). The results are summarized as follows:

$$sup(L_{11}, L_{12}) = 0.7938;\ \ sup(L_{12}, L_{13}) = 0.6775;\ \ sup(L_{11}, L_{13}) = 0.5287.$$

$$sup(L_{21}, L_{22}) = 0.7620; \; sup(L_{22}, L_{23}) = 0.7161; \; sup(L_{21}, L_{23}) = 0.5218.$$

$$sup(L_{31}, L_{32}) = 0.5335; \; sup(L_{32}, L_{33}) = 0.6313; \; sup(L_{31}, L_{33}) = 0.8351.$$

$$sup(L_{41}, L_{42}) = 0.5419; \; sup(L_{42}, L_{43}) = 0.6140; \; sup(L_{41}, L_{43}) = 0.8440.$$

$$sup(L_{51}, L_{52}) = 0.5873; \; sup(L_{52}, L_{53}) = 0.6238; \; sup(L_{51}, L_{53}) = 0.7889.$$

Step 5: In light of the result of Definition 10, we can calculate the support $T^{'}(L_{ij}(p))$ of $L_{ij}(p)$ by all of other $L_{it}(p)$ $(j, t = 1, 2, 3; t \neq j)$ by the following equation:

$$T^{'}(L_{ij}(p)) = \sum_{t=1, t\neq j}^{3} w_t sup(L_{ij}(p), L_{it}(p)).$$

These results are shown as the following matrix:

$$T^{'}(L_{ij}(p)) = \begin{pmatrix} 0.47613 & 0.44139 & 0.42961 \\ 0.46134 & 0.44343 & 0.44298 \\ 0.46393 & 0.34944 & 0.50305 \\ 0.46996 & 0.34677 & 0.49880 \\ 0.47159 & 0.36333 & 0.48619 \end{pmatrix}.$$

Step 6: With the aid of Proposition 2, we further compute the weight $v^{'}_{ij}$ associated with the PLTS $L_{ij}(p)$ by the following equation $(i = 1, 2, 3, 4, 5; j = 1, 2, 3)$:

$$v^{'}_{ij} = \frac{w_j(1 + T^{'}(L_{ij}(p)))}{\sum_{j=1}^{3} w_j(1 + T^{'}(L_{ij}(p)))}.$$

These results are shown as the following matrix:

$$v^{'}_{ij} = \begin{pmatrix} 0.3057 & 0.3981 & 0.2961 \\ 0.3026 & 0.3986 & 0.2988 \\ 0.3071 & 0.3775 & 0.3154 \\ 0.3085 & 0.3769 & 0.3146 \\ 0.3082 & 0.3806 & 0.3112 \end{pmatrix}$$

Step 7: If the DM prefers the WPLPA operator, then the aggregated value of the alternative x_i is determined based on Equation (12) $(i = 1, 2, 3, 4, 5)$. We denote the aggregated value of the alternative x_i as Z_i. The results are:

$$\begin{aligned} Z_1 &= WPLPA(L_{11}(p), L_{12}(p), L_{13}(p)) \\ &= ((0.9171, 0.6114, 0.5349, 0); (1.5924, 0.6967, 0.5972, 0); (1.3325, 0.3701, 0, 0)) \\ &= (3.8420, 1.6782, 1.1321, 0), \end{aligned}$$

$$\begin{aligned} Z_2 &= WPLPA(L_{21}(p), L_{22}(p), L_{23}(p)) \\ &= ((0.6052, 0.4539, 0.3738, 0.3026); (1.3951, 0.5979, 0.4983, 0); (1.5687, 0.4482, 0, 0)) \\ &= (3.569, 1.5, 0.8766, 0.3026), \end{aligned}$$

$$\begin{aligned} Z_3 &= WPLPA(L_{31}(p), L_{32}(p), L_{33}(p)) \\ &= ((0.6052, 0.4539, 0.3738, 0.3026); (1.3951, 0.5979, 0.4983, 0); (1.5687, 0.4482, 0, 0)) \\ &= (3.3113, 2, 1.5176, 0), \end{aligned}$$

$$\begin{aligned} Z_4 &= WPLPA(L_{41}(p), L_{42}(p), L_{43}(p)) \\ &= ((0.6052, 0.4539, 0.3738, 0.3026); (1.3951, 0.5979, 0.4983, 0); (1.5687, 0.4482, 0, 0)) \\ &= (2.6824, 1.8116, 1.4073, 0.3769), \end{aligned}$$

$$\begin{aligned} Z_5 &= WPLPA(L_{51}(p), L_{52}(p), L_{53}(p)) \\ &= ((1.2326, 0.4622, 0.3852, 0); (1.3322, 1.1419, 0, 0); (0.9336, 0.5446, 0.3890, 0)) \\ &= (3.4985, 2.1488, 0.7742, 0). \end{aligned}$$

If the DM uses the WPLPG operator, then the aggregated value of the alternative x_i is determined based on Equation (16) ($i = 1, 2, 3, 4, 5$). In the same way, we denote the aggregated value of the alternative x_i as Z_i. The results are:

$$\begin{aligned} Z_1 &= WPLPG(L_{11}(p), L_{12}(p), L_{13}(p)) \\ &= ((1.3150, 1.1722, 1.1603, 1); (1.5127, 1.2137, 1.1952, 1); (1.4887, 1.1265, 1, 1)) \\ &= (2.9613, 1.6026, 1.3868, 1), \end{aligned}$$

$$\begin{aligned} Z_2 &= WPLPG(L_{21}(p), L_{22}(p), L_{23}(p)) \\ &= ((1.1703, 1.1452, 1.1295, 1.1106); (1.4738, 1.1955, 1.1739, 1); (1.5466, 1.132, 1, 1)) \\ &= (2.6676, 1.5651, 1.3259, 1.1106), \end{aligned}$$

$$\begin{aligned} Z_3 &= WPLPG(L_{31}(p), L_{32}(p), L_{33}(p)) \\ &= ((1.3482, 1.1731, 1.1315, 1); (1.4024, 1.2168, 1.2016, 1); (1.3592, 1.1782, 1.1517, 1)) \\ &= (2.5698, 1.6818, 1.5658, 1), \end{aligned}$$

$$\begin{aligned} Z_4 &= WPLPG(L_{41}(p), L_{42}(p), L_{43}(p)) \\ &= ((1.3500, 1.1739, 1.1322, 1); (1.2012, 1.1839, 1.1637, 1.1395); (1.3256, 1.1777, 1.1654, 1)) \\ &= (2.1496, 1.6367, 1.5355, 1.1395), \end{aligned}$$

$$\begin{aligned} Z_5 &= WPLPG(L_{51}(p), L_{52}(p), L_{53}(p)) \\ &= ((1.3777, 1.1480, 1.1320, 1); (1.4482, 1.4064, 1, 1); (1.3216, 1.1635, 1.1334, 1)) \\ &= (2.6367, 1.8784, 1.2830, 1). \end{aligned}$$

Step 8: Based on the results of Definition 4, the scores of the alternative x_i can be computed, i.e., $E(Z_i)$. If the DM uses WPLPA operator to aggregate the decision formation, the scores are determined as follows:

$$E(Z_1) = 1.6632; E(Z_2) = 1.5620; E(Z_3) = 1.7072; E(Z_4) = 1.5697; E(Z_5) = 1.6054.$$

If the the DM uses WPLPG operator to aggregate the decision formation, the scores are determined as follows:

$$E(Z_1) = 1.7379; E(Z_2) = 1.6673; E(Z_3) = 1.7044; E(Z_4) = 1.6154; E(Z_5) = 1.6995.$$

Step 9: If the DM uses WPLPA operator, we can determine the ranking of the scores of the alternatives based on the results of the Step 8. It is shown as follows:

$$E(Z_3) > E(Z_1) > E(Z_5) > E(Z_4) > E(Z_2).$$

That is to say, the ordering of the alternatives is:

$$x_3 > x_1 > x_5 > x_4 > x_2.$$

If the DM uses WPLPG operator, we can obtain the ranking of the scores of the alternatives as follows:

$$E(Z_1) > E(Z_3) > E(Z_5) > E(Z_2) > E(Z_4).$$

In this situation, the ordering of the alternatives is:

$$x_1 > x_3 > x_5 > x_2 > x_4.$$

5.2. Comparison Analysis

Under the probabilistic linguistic information, Pang et al. [6] have developed an aggregation-based method for MAGDM. In order to verify the performance of our proposed methods, we compare our decision results with Pang et al. [6] based on our illustrative example. Torra [13], Merigó et al. [22] and Zhang et al. [23] also developed some methods for the lingusitic information and GDM. Thus, we also compare our results with the methods of Refs. [12,22,23]. The decision results are shown in Table 7.

Table 7. The decision results of different methods.

Method	Rank
Aggregation-based method of Ref. [6]	$x_3 > x_1 > x_5 > x_2 > x_4$
The method with HFLWA of Ref. [23]	$x_3 > x_1 > x_5 > x_2 = x_4$
The method with HFLWG of Ref. [23]	$x_1 > x_2 > x_5 > x_3 > x_4$
Max lower operator of Ref. [12]	$x_3 > x_2 = x_5 = x_4 > x_1$
ILGCIA with group decision making of Ref. [22]	$x_3 > x_2 > x_1 > x_4 > x_5$
Our proposed method with WPLPA	$x_3 > x_1 > x_5 > x_4 > x_2$
Our proposed method with WPLPG	$x_1 > x_3 > x_5 > x_2 > x_4$

In Table 7, we can find the rank result of the method proposed in Ref. [6] is: $x_3 > x_1 > x_5 > x_2 > x_4$. Compared with the decision results of our proposed method with WPLPA, the aggregation-based method with PLTSs can select the same best candidate, i.e., x_3. Meanwhile, for the WPLPG, the best candidate is x_1. Under the result of Ref. [23], HFLWA has the rank: $x_3 > x_1 > x_5 > x_2 = x_4$. Meanwhile, the ranking of HFLWG is $x_1 > x_2 > x_5 > x_3 > x_4$. By using the max lower operator of Ref. [12], we can find the rank is: $x_3 > x_2 = x_5 = x_4 > x_1$. For ILGCIA with group decision making of Ref. [22], the result is $x_3 > x_2 > x_1 > x_4 > x_5$. On the MCGDM problems under linguistic environment, we introduced our model to achieve the same acceptable performance with the existing techniques or to improve upon them. Unlike the existing models considered in this paper, our model contains probabilities which normally help in getting a comprehensive and accurate preference information of the DMs [6]. In Ref. [3], for instance, the developed approaches take all the decisions and their relationships into account, and the decision arguments reinforce and support each other, but since probabilities were not considered, the accuracy of preference information of the DMs might be questionable. In addition, without the PLTS, it might not be easy for the DMs to provide several possible linguistic values over an alternative or an attribute. This situation translates into some kind of limitation of the model proposed in Ref. [3] inspite of the power average (PA) involvement in the aggregation process.The PLTSs itself as a theory has some limitations . In general, WPLPG applies to the average of the ratio data and is mainly used to calculate the average growth (or change) rate of the data. From the trait of Table 6, the WPLPA is much better than WPLPG.

6. Conclusions

With respect to the support and reinforcement among input arguments with PLTSs, we introduce PA into the probabilistic linguistic environment. Meanwhile, we develop the corresponding new operators, i.e., the PLPA, PLPG, WPLPA and WPLPG operators. In light of the PLMCGDM, we describe the decision-making problem and design corresponding approaches by employing the WPLPA and WPLPG. In this paper, we expanded the applied field of the original PA and enrich the research work of PLTSs. Future research work may focus on exploring the decision-making mechanisms when the weight information is unknown or incomplete and developing some new generalized aggregation operators of PLTSs. In addition, we also deeply investigate a more complex case study with more alternatives and criteria.

Acknowledgments: This work is partially supported by the National Science Foundation of China (Nos. 71401026, 71432003, 71571148), the Fundamental Research Funds for the Central Universities of China (No. ZYGX2014J100), the Social Science Planning Project of the Sichuan Province (No. SC15C009) and the Sichuan Youth Science and Technology Innovation Team (2016TD0013).

Author Contributions: Decui Liang designed the reaserach work and the basic idea. Agbodah Kobina analyzed the data and finished the deduction procedure. Xin He also analyzed the data and modified the expression.

Conflicts of Interest: The authors declare no conflict of interest.

References

1. Yager, R.R. The power average operator. *IEEE Trans. Syst. Man Cybern. Part A Syst. Hum.* **2001**, *31*, 724–731.
2. Xu, Z.S.; Yager, R.R. Power-Geometric operators and their use in group decision making. *IEEE Trans. Fuzzy Syst.* **2010**, *18*, 94–105.
3. Xu, Y.J.; Merigó, J.M.; Wang, H.M. Linguistic power aggregation operators and their application to multiple attribute group decision making. *Appl. Math. Model.* **2012**, *36*, 5427–5444.
4. Zhou, L.G.; Chen, H.Y. A generalization of the power aggregation operators for linguistic environment and its application in group decision making. *Knowl. Based Syst.* **2012**, *26*, 216–224.
5. Zhu, C.; Zhu, L.; Zhang, X. Linguistic hesitant fuzzy power aggregation operators and their applications in multiple attribute decision-making. *Inf. Sci.* **2016**, *367–368*, 809–826.
6. Pang, Q.; Wang, H.; Xu, Z.S. Probabilistic linguistic term sets in multi-attribute group decision making. *Inf. Sci.* **2016**, *369*, 128–143.
7. Bai, C.Z.; Zhang, R.; Qian, L.X.; Wu, Y.N. Comparisons of probabilistic linguistic term sets for multi-criteria decision making. *Knowl. Based Syst.* **2017**, *119*, 284–291.
8. Merigó, J.M.; Casanovas, M.; Martínez, L. Linguistic aggregation operators for linguistic decision making based on the Dempster-Shafer theory of evidence. *Int. J. Uncertain. Fuzziness Knowl. Based Syst.* **2010**, *18*, 287–304.
9. Zhai, Y.L.; Xu, Z.S.; Liao, H.C. Probabilistic linguistic vector-term set and its application in group decision making with multi-granular linguistic information. *Appl. Soft Comput.* **2016**, *49*, 801–816.
10. Liao, H.C.; Xu, Z.S.; Zeng, X.J.; Merigó, J.M. Qualitative decision making with correlation coefficients of hesitant fuzzy linguistic term sets. *Knowl. Based Syst.* **2015**, *76*, 127–138.
11. Liao, H.C.; Xu, Z.S.; Zeng, X.J. Hesitant fuzzy linguistic vikor method and its application in qualitative multiple criteria decision making. *IEEE Trans. Fuzzy Syst.* **2015**, *23*, 1343–1355.
12. Rodriguez, R.M.; Martinez, L.; Herrera, F. Hesitant fuzzy linguistic term sets for decision making. *IEEE Trans. Fuzzy Syst.* **2012**, *20*, 109–119.

13. Torra, V. Hesitant fuzzy sets. *Int. J. Intell. Syst.* **2010**, *25*, 529–539.
14. Liang, D.C.; Liu, D. A novel risk decision making based on decision-theoretic rough sets under hesitant fuzzy information. *IEEE Trans. Fuzzy Syst.* **2015**, *23*, 237–247.
15. Gou, X.J.; Xu, Z.S. Novel basic operational laws for linguistic terms, hesitant fuzzy linguistic term sets and probabilistic linguistic term sets. *Inf. Sci.* **2016**, *372*, 407–427.
16. He, Y.; Xu, Z.S.; Jiang, W.L. Probabilistic interval reference ordering sets in multi-criteria group decision making. *Int. J. Uncertain. Fuzziness Knowl. Based Syst.* **2017**, *25*, 189–212.
17. Wu, Z.B.; Xu, J.C. Possibility distribution-based approach for MAGDM with hesitant fuzzy linguistic information. *IEEE Trans. Cybern.* **2016**, *46*, 694–705.
18. Zhang, Y.X.; Xu, Z.S.; Wang, H.; Liao, H.C. Consistency-based risk assessment with probabilistic linguistic preference relation. *Appl. Soft Comput.* **2016**, *49*, 817–833.
19. Zhou, W.; Xu, Z.S. Consensus building with a group of decision makers under the hesitant probabilistic fuzzy environment. *Fuzzy Optim. Decis. Mak.* **2016**, doi:10.1007/s10700-016-9257-5.
20. Katz, D.A. Faculty salaries, promotions and productivity at a large University. *Am. Econ. Rev.* **1973**, *63*, 469–477.
21. Bryson, N.; Mobolurin, A. An action learning evaluation procedure for multiple criteria decision making problems. *Eur. J. Oper. Res.* **1995**, *96*, 379–386.
22. Merigó, J.M.; Gil-Lafuente, A.M.; Zhou, L.G.; Chen, H.Y. Induced and linguistic generalized aggregation operators and their application in linguistic group decision making. *Group Decis. Negot.* **2012**, *21*, 531–549.
23. Zhang, Z.M.; Wu, C. Hesitant fuzzy linguistic aggregation operators and their applications to multiple attribute group decision making. *J. Intell. Fuzzy Syst.* **2014**, *26*, 2185–2202.

symmetry

MDPI

Article

A Comparative Study of Some Soft Rough Sets

Yaya Liu [1,2], Luis Martínez [2] and Keyun Qin [1,*]

[1] College of Mathematics, Southwest Jiaotong University, Chengdu 610031, China; yayaliu@my.swjtu.edu.cn
[2] Department of Computer Science, University of Jaén, E-23071 Jaén, Spain; martin@ujaen.es
* Correspondence: keyunqin@263.net

Received: 22 September 2017; Accepted: 24 October 2017; Published: 27 October 2017

Abstract: Through the combination of different types of sets such as fuzzy sets, soft sets and rough sets, abundant hybrid models have been presented in order to take advantage of each other and handle uncertainties. A comparative study of relationships and interconnections of some existing hybrid models has been carried out. Some foundational properties of modified soft rough sets (MSR sets) are analyzed. It is pointed out that MSR approximation operators are some kinds of Pawlak approximation operators, whereas approximation operators of Z-soft rough fuzzy sets are equivalent to approximation operators of rough fuzzy sets. The relationships among F-soft rough fuzzy sets, M-soft rough fuzzy sets and Z-soft rough fuzzy sets are surveyed. A new model called soft rough soft sets has been provided as the generalization of F-soft rough sets, and its application in group decision-making has been studied. Various soft rough sets models show great potential as a tool to solve decision-making problems, and a depth study of the connections among these models contributes to the flexible application of soft rough sets based decision-making approaches.

Keywords: rough set; soft set; soft rough set; soft rough fuzzy set

1. Introduction

Various types of uncertainties exist in real life situations, which calls for useful mathematic tools to meet various information process demands. Usually complicated problems take place with uncertainties, and most of these complex situations can not be handled by adopting classical mathematic methods, considering the fact that with classical mathematic tools all notions are requested to be strict. Up to now, abundant mathematic tools such as fuzzy set theory [1] and rough set theory [2,3] have already been developed and proved to be useful in handling several kinds of the problems that contain uncertainties, and all of these theories share a common inherent difficulty, which is mainly the inadequacy of the parametrization tool [4,5]. However, it is noticed that, without proper parametrization tools, sometimes a practical problem can not be described in a way as much as information collected from different aspects could be taken into account. To handle this issue and to enrich mathematical methodologies for coping with uncertainties, soft set theory was initially proposed by Molodtsov [4] in 1999, which considers every specific object from different attributes' aspects, in this way, this new model goes beyond all other existing mathematical tools to avoid the above-mentioned difficulties. After soft set theory comes out, in the past few years, there appears a continuous growth of interest in studying theoretical aspects of soft set theory, as well as the practical applications of soft sets.

Abundant mathematical models have already been designed in order to model and process vague concepts, among which it is noteworthy that fuzzy set theory and rough set theory have already drawn worldwide attention from researchers. The development of these two theories makes contributions to handle lots of complicated problems in engineering, economics, social science, et al. The main character of fuzzy set theory is that it describes a vague concept by using a membership function, and the allowance of partial memberships contributes to providing an appropriate framework to represent and process vague concepts. The character of rough set theory relies on handling vagueness and granularity in information systems by indirectly describing a vague concept through two exact concepts called its lower and upper approximations. In Pawlak's rough set model, the equivalence relation is a vital concept, by replacing the equivalence relation with a fuzzy similarity relation, fuzzy rough sets and rough fuzzy sets have been proposed [6].

The combinations of soft sets, rough sets and fuzzy sets have been extensively studied to benefit each other and to take the best advantage of them. Research on generalization models of soft sets is promising since usually the generalized models are not short of parameter tools, that is, all of the generalized soft set models usually keep the most important feature of soft set theory in considering issues from various aspects. The history of research on extending soft sets applying fuzzy set theory goes beyond fifteen years already since Maji et al. introduced fuzzy soft sets in [7]. Therefore far, the soft sets have been extended to intuitionistic fuzzy soft sets [8], interval-valued intuitionistic fuzzy soft sets [5,9], vague soft sets [10], soft interval sets [11] and many other hybrid soft sets models. The history of research on the generalization of soft sets by using rough set theory is relatively short. To introduce parametrization tools to rough set theory, Feng et al. [12,13] initially put forward the concept of soft rough sets and soft rough fuzzy sets, in which a soft set looks for the lower and upper approximations of a subset of the universe. Afterwards, Meng et al. [14] proposed soft fuzzy rough set, in which model the fuzzy soft set has been adopted into granulate the universe. Benefitting from similarity measures induced by soft sets and soft fuzzy sets, Qin et al. [15] provided several soft fuzzy rough set models through introducing confidence threshold values. Recently, Shabir et al. [16] noticed that Feng et al.'s soft rough sets [12] suffer from some unexpected properties such as the upper approximation of a non-empty set might be empty and a subset set X might not be contained in its upper approximation. To resolve this problem, Shabir et al. [16] modified their soft rough sets and introduced the modified soft rough set (MSR set), which has already been extended to fuzzy soft sets [17], and Z-soft rough fuzzy sets was proposed, and its application in decision-making problems was analyzed.

The exploitation of soft sets and hybrid soft sets models in decision-making shows a great development in the recent years [18–22]. The utilization of soft rough sets models in decision-making shows a promising prospect. Different decision-making approaches have been put forth based on MSR set [20], Z-soft rough fuzzy sets [17], Z-soft fuzzy rough set [21], and other soft rough sets models [23,24]. If the researchers could have a thorough knowledge of the connections among various soft rough sets, we believe that decision-making approaches under framework of soft rough sets could be applied in a more flexible and reliable way. However, the relationships among these hybrid sets have not been systematically studied so far. Furthermore, we notice that a soft set S can be looked upon an information system I_S. Based on this information system, we can establish Pawlak rough approximations and rough fuzzy approximations. What is the relationship between soft rough approximations (soft rough fuzzy approximations) in S and Pawlak rough approximations (rough fuzzy approximations) in I_S? Additionally, soft set and formal context are mathematically equivalent. The relationships among soft rough approximation operators and derivation operators used in formal concept analysis (FCA) are also interesting issues to be addressed. In this paper, we will concentrate on the discussion of these problems. The paper is structured as follows: Section 2 revises several basic concepts of soft sets, fuzzy sets and rough set. Section 3 studies relationships among several soft rough sets. The properties of MSR approximation operators and different connections between MSR approximation operators and F-soft rough approximation operators are analyzed. It is shown that MSR approximation operators and a kind of Pawlak approximation operators are equivalent, while Z-soft rough fuzzy approximation operators and a kind of rough fuzzy approximation operators are equivalent. The relationships among F-soft rough fuzzy sets, M-soft rough fuzzy sets and Z-soft rough fuzzy sets have also been investigated. Section 4 discusses the relationship between F-soft rough sets and modal-style operators in formal concept analysis. Section 5 proposes a new generalization of F-soft rough set, which is called a soft rough soft set, and a simple application of soft rough soft sets in group decision-making has been studied. Eventually, Section 6 concludes the paper by presenting some remarks and future works.

2. Preliminaries

Here, several concepts of fuzzy sets, soft sets and rough sets are briefly reviewed. Please refer to [1,2,4,7] for details.

An advantageous framework has been offered by fuzzy set theory [1] to handle vague concepts through the allowance for partial memberships. Let U be the universe set. Define a fuzzy set μ on U by its membership function $\mu : U \to [0,1]$. $\mu(x)$ indicates the degree to which x belongs to the fuzzy set μ for all $x \in U$. In what follows, we denote the family of all subsets of U by $P(U)$ and the family of all fuzzy sets on U by $F(U)$. The operations of fuzzy sets can be found in [1].

Molodtsov [4] introduced the concept of soft set. Let U be the universe set and E the set consisted of all parameters that is related to U. Hence, a soft set is defined as below:

Definition 1. *A pair (F, A) is called a soft set over U, where $A \subseteq E$ and F is a mapping given by $F : A \to P(U)$ [4].*

The soft set is characterized by a parameter set and a function defined on the parameter set. For every parameter $e \in A$, $F(e)$ is said to be the e-approximate elements and, correspondingly, the soft set can be viewed as a parameterized family of subsets of U.

A soft set (F, A) is called a full soft set if $\cup_{e \in A} F(e) = U$ [12]; $\tilde{N}_{(U,A)} = (N, A)$ is called a relative null soft set (with respect to the parameter set A), if $N(e) = \emptyset$ for all $e \in A$; $\tilde{W}_{(U,B)} = (W, B)$ is called a relative whole soft set (with respect to the parameter set B) if $W(e) = U$ for all $e \in B$ [25]. Maji et al. in [7] introduced the concept of fuzzy soft set.

Definition 2. *Let (U, E) be a soft space. A pair (F, A) is called a fuzzy soft set over U, where $A \subseteq E$ and F is a mapping defined as $F : A \to F(U)$ [7].*

The fuzzy soft set is also characterized by a parameter set and a function on the parameter set, whereas a fuzzy set on U takes place of a crisp subset of U corresponds to each parameter. It follows that, to a certain degree, a soft set can also be viewed as a special kind of fuzzy soft set.

Pawlak introduced rough set theory in [2], the application of which is based on a structure called information system.

Definition 3. *An information system is a pair $I = (U, A)$ of non-empty finite sets U and A, where U is a set of objects and A is a set of attributes; each attribute $a \in A$ is a function $a : U \to V_a$, where V_a is the set of all values (called domain) of attribute a [3].*

Soft sets and information systems are closely related [13,26,27]. $S = (F, A)$ is assumed to be a soft set over U and $I_S = (U, A)$ an information system induced by S. For any attribute $a \in A$, a function $a : U \to V_a = \{0, 1\}$ is defined by $a(x) = 1$ if $x \in F(a)$; or else $a(x) = 0$. In this way, every soft set could be viewed as an information system. In what follows, I_S is called the information system induced by soft set S.

By contrast, suppose the information system, $I = (U, A)$. It uses a parameter set as

$$B = \{(a, v_a); a \in A \wedge v_a \in V_a\},$$

and it follows that through setting $F(a, v_a) = \{x \in U; a(x) = v_a\}$ for each $a \in A$ and $v_a \in V_a$, a soft set (F, B) can be defined, which is the soft set induced by I.

Let U be the universe of discourse and R be an equivalence relation on U. (U,R) is called Pawlak approximation space. For each $X \subseteq U$, the upper approximation $\overline{R}(X)$ and lower approximation $\underline{R}(X)$ of X with respect to (U,R) are defined as [2]:

$$\overline{R}(X) = \{x \in U; [x]_R \cap X \neq \emptyset\}, \tag{1}$$

$$\underline{R}(X) = \{x \in U; [x]_R \subseteq X\}. \tag{2}$$

X is so-called definable in (U,R) if $\underline{R}(X) = \overline{R}(X)$, or else X is a rough set. Thus, in rough set theory, a rough concept is characterized by a couple of exact concepts, namely, its lower approximation and upper approximation. $Pos_R(X) = \underline{R}(X)$ and $Neg_R(X) = U - \overline{R}(X)$ are the R-positive region and R-negative region of X, respectively. Furthermore, $Bnd_R(X) = \overline{R}(X) - \underline{R}(X)$ is called the R-boundary region.

Up to now, various types of extension models of the Pawlak rough set have been proposed to enrich the theory and to meet different application demands [28,29]. In [12], by the combination of soft set, rough set and fuzzy set theory, soft rough sets and soft rough fuzzy sets were introduced. To make them easy to be distinguished from other models mentioned in the current work and also to facilitate the discussion, these two notions are called F-soft rough sets and F-soft rough fuzzy sets.

Definition 4. *Let $S = (f,A)$ be a soft set over U. $P = (U,S)$ is called a soft approximation space. Two operations can be defined based on P as follows [12]:*

$$\underline{apr}_P(X) = \{u \in U; \exists a \in A(u \in f(a) \subseteq X)\}, \tag{3}$$

$$\overline{apr}_P(X) = \{u \in U; \exists a \in A(u \in f(a), f(a) \cap X \neq \emptyset)\}. \tag{4}$$

For all $X \subseteq U$, $\underline{apr}_P(X)$ and $\overline{apr}_P(X)$ are respectively called the F-lower and F-upper soft rough approximations of X in S. X is F-soft definable in P if $\underline{apr}_P(X) = \overline{apr}_P(X)$, or else X is a F-soft rough set.

It is noted that we can present $\underline{apr}_P(X)$ and $\overline{apr}_P(X)$ in a more concise manner [13]:

$$\underline{apr}_P(X) = \cup\{f(a); a \in A \wedge f(a) \subseteq X\}, \tag{5}$$

$$\overline{apr}_P(X) = \cup\{f(a); a \in A \wedge f(a) \cap X \neq \emptyset\}. \tag{6}$$

In this definition, the soft set S is regarded as the elementary knowledge on the universe. F-lower and F-upper soft rough approximation operators are not dual to each other, that is, $\underline{apr}_P(X^c) = (\overline{apr}_P(X))^c$ usually does not hold, where the complement of set X is computed by $X^c = U - X$. If the condition $\cup_{a\in A} f(a) = U$ holds in a soft set $S = (f,A)$ over U, this soft set is a full soft set [12]. In this case, $\{f(a); a \in A\}$ comes into being a cover of the universe U. It is pointed out that $\underline{apr}_P$, $\overline{apr}_P$ and covering rough approximations [30] are closely related but fundamentally different [13]. Additionally, if $\{f(a); a \in A\}$ forms a partition of U, we will call $S = (f,A)$ a partition soft set [13,31].

It is pointed out by Shabir et al. [16] that $\exists\, x \in U$ s.t. $x \in Neg_P(X) = U - \overline{apr}_P(X)$ for all $X \subseteq U$, if $S = (f,A)$ is not a full soft set. In other words, $x \notin \overline{apr}_P(X)$ for all $X \subseteq U$. Thus, $X \subseteq \overline{apr}_P(X)$ and some basic properties of rough set do not hold in general. Based on these observations, modified soft rough sets (MSR sets) was defined as follows.

Definition 5. *Let (f,A) be a soft set over U and $\varphi : U \to P(A)$ be a map defined as $\varphi(x) = \{a \in A; x \in f(a)\}$. Then, (U,φ) is called MSR-approximation space and for any $X \subseteq U$, its lower MSR approximation $\underline{X}_\varphi$ and upper MSR approximation $\overline{X}_\varphi$ are defined as [16]:*

$$\underline{X}_\varphi = \{x \in U; \forall y \in X^c(\varphi(x) \neq \varphi(y))\}, \tag{7}$$

$$\overline{X}_\varphi = \{x \in U; \exists y \in X(\varphi(x) = \varphi(y)).\tag{8}$$

X is MSR definable if the condition $\underline{X}_\varphi = \overline{X}_\varphi$ *holds, or else X is a MSR set.*

Mathematically speaking, (U, φ) can be looked upon a soft set over A. In [32], (U, φ) was considered as a pseudo soft set that is induced by (f, A), afterwards a decision-making method related to pseudo soft set was provided.

3. Relationships among Several Soft Rough Sets

3.1. Relationships between F-Soft Rough Approximations and MSR Approximations

The notion of MSR set is the modification of a F-soft rough set, and some inherent connections between these two models should exist, which have not drawn enough attention from scholars yet. In this subsection, a theoretical analysis of F-soft rough sets and MSR sets will be provided, and some connections between F-soft rough approximations and MSR approximations will be pointed out.

It is noted that Ref. [16] $\underline{apr}_P(X) \subseteq \underline{X}_\varphi$ for any $X \subseteq U$ and the containment may be proper. Furthermore, in general, $\overline{X}_\varphi \subseteq \overline{apr}_P(X)$ or $\overline{apr}_P(X) \subseteq \overline{X}_\varphi$ does not hold. Now, we provide an example:

Example 1. *Let* $A = \{a, b, c, d\}$ *be a parameter set and* $U = \{x_1, x_2, x_3, x_4, x_5, x_6\}$ *the universe. Suppose that* $S = (f, A)$ *is a soft set over* U*, in which* $F(a) = \{x_1, x_6\}$, $F(b) = \{x_3\}$, $F(c) = \emptyset$, $F(d) = \{x_1, x_2, x_5\}$.

(1) By the definition, $\overline{apr}_P(U) = \cup_{a \in A} f(a) = \{x_1, x_2, x_3, x_5, x_6\}$. *It follows that* $x_4 \notin \overline{apr}_P(U)$ *and hence* $x_4 \notin \overline{apr}_P(X)$ *for any* $X \subseteq U$.

(2) Let $X = \{x_3, x_4, x_5\}$. *By direct computation, we know that* $\overline{apr}_P(X) = \{x_1, x_2, x_3, x_5\}$, $\overline{X}_\varphi = \{x_2, x_3, x_4, x_5\}$. *Thus,* $\overline{apr}_P(X) \subseteq \overline{X}_\varphi$*, or* $\overline{X}_\varphi \subseteq \overline{apr}_P(X)$ *does not hold.*

However, only a shallow impression can be obtained noticing the above-mentioned conclusions in [16], and no details have been provided discussing the properties of and connections among $\overline{apr}_P(X)$, $\overline{X}_\varphi$, $\underline{apr}_P(X)$ and $\underline{X}_\varphi$. The questions still remain: is there any possibility $\overline{X}_\varphi \subseteq \overline{apr}_P(X)$ or $\overline{apr}_P(X) \subseteq \overline{X}_\varphi$ that holds? Which features will be requested if these conditions need to be established? Now, we will pay attention to these questions and provide answers.

A general assumption for Theorems 1–3 and Corollaries 1 and 2 is presented as below:

Let $S = (f, A)$ be a soft set over U and $P = (U, S)$ a soft approximation space.

Theorem 1. *S is a full soft set iff* $\overline{X}_\varphi \subseteq \overline{apr}_P(X)$ *for any* $X \subseteq U$.

Proof. ($\Rightarrow$). It is assumed that S is a full soft set and $X \subseteq U$. For all $x \in \overline{X}_\varphi$, $\exists\, y \in X$ s.t. $\varphi(x) = \varphi(y)$. By $y \in U = \cup_{a \in A} f(a)$, $\exists\, a \in A$ s.t. $y \in f(a)$. Then, $y \in X \cap f(a)$ and $X \cap f(a) \neq \emptyset$. By $y \in f(a)$ we obtain $a \in \varphi(y) = \varphi(x)$ and hence $x \in f(a)$. Consequently, $x \in \overline{apr}_P(X)$. Thus, $\overline{X}_\varphi \subseteq \overline{apr}_P(X)$.

($\Leftarrow$). Suppose that, for all $X \subseteq U$, the condition $\overline{X}_\varphi \subseteq \overline{apr}_P(X)$ holds. It can be observed that $x \in \overline{\{x\}}_\varphi \subseteq \overline{apr}_P(\{x\}) = \cup\{f(a); f(a) \cap \{x\} \neq \emptyset\} = \cup\{f(a); x \in f(a)\}$, for any x in U.

Thus, $\exists\, a \in A$ s.t. $x \in f(a)$. S is a full soft set by the arbitrary of x. □

Theorem 2. $\overline{apr}_P(X) \subseteq \overline{X}_\varphi$ *for any* $X \subseteq U$ *iff for any* $a, b \in A$, $f(a) \cap f(b) = \emptyset$ *whenever* $f(a) \neq f(b)$.

Proof. ($\Leftarrow$). Assume that for any $a, b \in A$, $f(a) \cap f(b) = \emptyset$ whenever $f(a) \neq f(b)$. Let $X \subseteq U$. For any $x \in \overline{apr}_P(X)$, attribute $a \in A$ exists s.t. $x \in f(a)$ and $f(a) \cap X \neq \emptyset$. Thus, we know that there exists $y \in U$ s.t. $y \in f(a) \cap X$. For any $b \in A$, if $f(a) \neq f(b)$, then $f(a) \cap f(b) = \emptyset$ and hence $x \notin f(b)$ by $x \in f(a)$. Thus, $\varphi(x) = \{b \in A; f(b) = f(a)\}$. Similarly, we have $\varphi(y) = \{b \in A; f(b) = f(a)\}$ and hence $\varphi(x) = \varphi(y)$. By $y \in X$, we know that $x \in \overline{X}_\varphi$ and consequently $\overline{apr}_P(X) \subseteq \overline{X}_\varphi$.

($\Rightarrow$). Assume that $\overline{apr}_P(X) \subseteq \overline{X}_\varphi$ for any $X \subseteq U$. For any $a, b \in A$, if $f(a) \cap f(b) \neq \emptyset$, $\exists\, x \in U$ s.t. $x \in f(a) \cap f(b)$. By $x \in f(a)$, we conclude that

$$\begin{aligned} f(a) &\subseteq \cup\{f(c); x \in f(c)\} = \cup\{f(c); \{x\} \cap f(c) \neq \emptyset\} = \overline{apr}_P(\{x\}) \\ &\subseteq \overline{\{x\}}_\varphi = \{y \in U; \varphi(y) = \varphi(x)\}. \end{aligned}$$

Meanwhile, if $\varphi(y) = \varphi(x)$, then $a \in \varphi(x) = \varphi(y)$ and hence $y \in f(a)$. Therefore, $f(a) = \{y \in U; \varphi(y) = \varphi(x)\}$. Similarly, by $x \in f(b)$, we have $f(b) = \{y \in U; \varphi(y) = \varphi(x)\}$ and hence $f(a) = f(b)$. □

Theorems 1 and 2 shows that $\exists$ containment relationships between $\overline{X}_\varphi$ and $\overline{apr}_P(X)$ if some specific conditions hold. Based on these two theorems, we can have a clear idea about under which conditions the containment relationships can be held. Furthermore, by Theorems 1 and 2, we obtain

Corollary 1. *Let $f(e) \neq \emptyset$ for each $e \in A$. S is a partition soft set iff $\overline{apr}_P(X) = \overline{X}_\varphi$ for any $X \subseteq U$.*

Corollary 2. *S is a full soft set iff $X \subseteq \overline{apr}_P(X)$ for any $X \subseteq U$.*

Proof. It is assumed that S is a full soft set. For all $X \subseteq U$, it is obvious that $X \subseteq \overline{X}_\varphi \subseteq \overline{apr}_P(X)$ by Theorem 1. On the contrary, assume that $X \subseteq \overline{apr}_P(X)$ for any $X \subseteq U$. For each $x \in U$,

$$x \in \{x\} \subseteq \overline{apr}_P(\{x\}) = \cup\{f(a); f(a) \cap \{x\} \neq \emptyset\} = \cup\{f(a); x \in f(a)\}.$$

Thus, $\exists\, a \in A$ s.t. $x \in f(a)$. Consequently, S is a full soft set as required. □

Theorem 3. *$\underline{X}_\varphi \subseteq \underline{apr}_P(X)$ for any $X \subseteq U$ iff for any $x \in U$, $\exists\, a \in A$ s.t. $f(a) = \{y \in U; \varphi(y) = \varphi(x)\}$.*

Proof. ($\Rightarrow$). Suppose that $\underline{X}_\varphi \subseteq \underline{apr}_P(X)$ for all $X \subseteq U$. For any $x \in U$, let $X = \{y \in U; \varphi(y) = \varphi(x)\}$. It follows that

$$\underline{X}_\varphi = \{u \in U; \exists y \in X(\varphi(u) = \varphi(y))\} = \{u \in U; \varphi(u) = \varphi(x)\} = X.$$

By $x \in X$ and $\underline{X}_\varphi \subseteq \underline{apr}_P(X)$, then $x \in \underline{apr}_P(X)$ and hence $\exists\, a \in A$ s.t. $x \in f(a)$ and $f(a) \subseteq X$.

On the other hand, for any $y \in X$, we have $\varphi(y) = \varphi(x)$, therefore $a \in \varphi(x) = \varphi(y)$. Then, $y \in f(a)$ and hence $X \subseteq f(a)$. Thus, $f(a) = X = \{y \in U; \varphi(y) = \varphi(x)\}$.

($\Leftarrow$). Assume that $X \subseteq U$ and $x \in \underline{X}_\varphi$. For each $y \in U$, if $\varphi(x) = \varphi(y)$, we have $y \in X$ by $x \in \underline{X}_\varphi$. It follows that $\{y \in U; \varphi(y) = \varphi(x)\} \subseteq X$ and $\exists\, a \in A$ such that $f(a) = \{y \in U; \varphi(y) = \varphi(x)\}$. Thus, $x \in f(a)$ and $f(a) \subseteq X$. It follows that $x \in \underline{apr}_P(X)$ and consequently $\underline{X}_\varphi \subseteq \underline{apr}_P(X)$. □

By Theorem 3, we obtain a clear mind about the necessary conditions for $\underline{X}_\varphi \subseteq \underline{apr}_P(X)$ to be held, which has not been discussed in other literature yet. The connections between F-soft rough approximations and MSR approximations have been discussed in detail through the theorems presented above.

Keeping in mind that all of the theoretical research should serve practical applications. It is noted that F-soft rough sets and MSR sets group decision-making approaches have been put forward in [20,31], respectively. Based on the analysis about the connections of F-soft rough approximations and MSR approximations, the relationships between decision schemes by using these two different hybrid models could be further discussed in the future, and the decision results obtained by the two decision schemes may have some inherent relationship.

3.2. The Relationships between MSR Approximations and Pawlak's Rough Approximations

After the notion of MSR sets was put forward, it was applied to different circumstances to cope with practical problems. However, since there is systematic research on its relationship with Pawlak's rough sets up to now, the rationality of MSR sets may be questioned by scholars from a theoretical point of view.

Let $S = (f, A)$ be a soft set. S induces an information system $I_S = (U, A)$. According to Pawlak [2], A determines an indiscernibility relation R_S on U given by

$$R_S = \{(x, y) \in U \times U; \forall a \in A(a(x) = a(y))\}. \quad (9)$$

Clearly, (U, R_S) is a Pawlak approximation space. The equivalence class determined by the equivalence relation R_S that contains x is denoted by $[x]_{R_S}$. What is the relationship between Pawlak's rough approximations in (U, R_S) and F-soft rough approximations (MSR approximations) induced by soft set S? This section offers the discussion of this problem.

Theorem 4. *Let $S = (f, A)$ be a partition soft set over U and $P = (U, S)$ a soft approximation space. Define an equivalence relation R on U by*

$$R = \{(x, y) \in U \times U; \exists a \in A(\{x, y\} \subseteq f(a)\}. \quad (10)$$

Then, for all $X \subseteq U$, $\underline{apr}_P(X) = \underline{R}(X)$ and $\overline{apr}_P(X) = \overline{R}(X)$ [13,31].

Theorem 5. *Let $S = (f, A)$ be a partition soft set over U and $I_S = (U, A)$ the information system induced by soft set $S = (f, A)$. Then, $R_S = R$, where R is determined by Equation (10).*

Proof. Let $x, y \in U$ and $(x, y) \in R$. By the definition, $\exists\ a \in A$ s.t. $\{x, y\} \subseteq f(a)$. It follows that $a(x) = 1 = a(y)$. For any $b \in A - \{a\}$, if $f(b) = f(a)$, then $\{x, y\} \subseteq f(a) = f(b)$ and hence $b(x) = 1 = b(y)$; if $f(b) \neq f(a)$, then $f(b) \cap f(a) = \emptyset$ and hence $x \notin f(b)$, $y \notin f(b)$. Then, $b(x) = 0 = b(y)$. Thus, $c(x) = c(y)$ for each $c \in A$. Consequently, $(x, y) \in R_S$.

Conversely, let $x, y \in U$ and $(x, y) \in R_S$. By $x \in U = \cup_{a \in A} f(a)$, $\exists\ a \in A$ s.t. $x \in f(a)$. It follows that $a(y) = a(x) = 1$ and hence $y \in f(a)$. Consequently, $\{x, y\} \subseteq f(a)$ and thus $(x, y) \in R$. □

By Theorems 4 and 5, in cases when a partition soft set is used as the underlying soft set, F-soft rough sets in (U, S) could be identified with Pawlak's rough sets in (U, R_S). For MSR sets, we have the following results.

Theorem 6. *Let $S = (F, A)$ be a soft set over U and $I_S = (U, A)$ be the information system induced by soft set $S = (F, A)$.*

(1) For any $x \in U$, $[x]_{R_S} = \{y \in U; \varphi(x) = \varphi(y)\}$.
(2) For any $X \subseteq U$, $\underline{X}_\varphi = \underline{R_S}(X)$.
(3) For any $X \subseteq U$, $\overline{X}_\varphi = \overline{R_S}(X)$.

Proof. (1) Let $x, y \in U$ and $y \in [x]_{R_S}$. Then, $a(x) = a(y)$ for each $a \in A$. For any $b \in \varphi(x)$, we have $x \in f(b)$ and hence $b(x) = 1$. We can observe that $b(y) = b(x) = 1$ and $y \in f(b)$. Thus, $b \in \varphi(y)$ and hence $\varphi(x) \subseteq \varphi(y)$. Similarly, we have $\varphi(y) \subseteq \varphi(x)$ and consequently $\varphi(x) = \varphi(y)$.

On the contrary, suppose that $\varphi(x) = \varphi(y)$. For any $a \in A$, if $a(x) = 1$, then $x \in f(a)$ and hence $a \in \varphi(x) = \varphi(y)$. Thus, $y \in f(a)$ and $a(y) = 1$; if $a(x) = 0$, then $x \notin f(a)$ and hence $a \notin \varphi(x) = \varphi(y)$. Thus, $y \notin f(a)$ and $a(y) = 0$. Then, $a(x) = a(y)$ for any $a \in A$ and hence $y \in [x]_{R_S}$.

(2) Let $X \subseteq U$ and $x \in \underline{X}_\varphi$. For any $y \in [x]_{R_S}$, we have $\varphi(x) = \varphi(y)$ by (1). By $x \in \underline{X}_\varphi$, we have $\varphi(x) \neq \varphi(z)$ whenever $z \in X^c$. Thus, $y \in X$ by $\varphi(x) = \varphi(y)$. Then, $[x]_{R_S} \subseteq X$ and hence $x \in \underline{R_S}(X)$. We conclude that $\underline{X}_\varphi \subseteq \underline{R_S}(X)$.

On the contrary, assume that $x \in \underline{R_S}(X)$. It follows that $[x]_{R_S} \subseteq X$. For any $y \in X^c$, we have $y \notin X$ and hence $y \notin [x]_{R_S}$. Thus, $\varphi(x) \neq \varphi(y)$ by (1). Consequently, $x \in \underline{X}_\varphi$ and hence $\underline{R_S}(X) \subseteq \underline{X}_\varphi$.

(3) Let $X \subseteq U$ and $x \in \overline{X}_\varphi$. It follows that $\exists\ y \in X$ s.t. $\varphi(x) = \varphi(y)$. Thus, $y \in [x]_{R_S}$. Consequently, $[x]_{R_S} \cap X \neq \emptyset$ and hence $x \in \overline{R_S}(X)$.

Conversely, suppose that $x \in \overline{R_S}(X)$. Thus, $[x]_{R_S} \cap X \neq \emptyset$. It follows that there exists $y \in X$ s.t. $y \in [x]_{R_S}$. Consequently, $\varphi(x) = \varphi(y)$ and hence $x \in \overline{X}_\varphi$. □

Theorem 6 shows that MSR approximation operator is a kind of Pawlak rough approximation operator. The two mathematic models that correspond with these approximation operators have been interconnected by this theorem, which could be regarded as a theoretical proof for the rationality of MSR sets. Benefitting from the notion of MSR set, Zhan et al. provided the definition of Z-soft rough fuzzy set in a recent work [17] .

Definition 6. *Let (f, A) be a soft set over U and (U, φ) the MSR approximation space. For any fuzzy set $\mu \in F(U)$, the Z-lower and Z-upper soft rough approximations of μ are denoted by $\underline{\mu}_\varphi$ and $\overline{\mu}_\varphi$, respectively, which are fuzzy sets on U given by [17]:*

$$\underline{\mu}_\varphi(x) = \wedge\{\mu(y); y \in U \wedge \varphi(x) = \varphi(y)\}, \tag{11}$$

$$\overline{\mu}_\varphi(x) = \vee\{\mu(y); y \in U \wedge \varphi(x) = \varphi(y)\}, \tag{12}$$

for each $x \in U$, and the operators $\underline{\mu}_\varphi$ and $\overline{\mu}_\varphi$ are the Z-lower and Z-upper soft rough approximation operators on a fuzzy set, respectively. Specifically, if $\underline{\mu}_\varphi = \overline{\mu}_\varphi$, μ is a Z-soft definable; or else μ is a Z-soft rough fuzzy set.

By Theorem 6 (1), the following corollary could easily be achieved:

Corollary 3. *Let $S = (F, A)$ be a soft set over U and $I_S = (U, A)$ the information system induced by soft set $S = (F, A)$. Then,*

(1) $\underline{\mu}_\varphi(x) = \wedge\{\mu(y); y \in [x]_{R_S}\}, and$
(2) $\overline{\mu}_\varphi(x) = \vee\{\mu(y); y \in [x]_{R_S}\}$

for any $\mu \in F(U)$, $x \in U$.

By Corollary 3, Z-lower and Z-upper soft rough approximation operators are equivalent to Dubois and Prade's lower and upper rough fuzzy approximation operators in [6]. Benefitting from this corollary, the researchers may refer to both of the theories' aspects and the applications of rough fuzzy sets to better study the development of Z-soft rough sets. Furthermore, the utilization of rough set theory in decision system has been extensively studied during the past few decades. Through discussing the connections between F-soft rough set and and Pawlak rough set, as well as the connections between MSR approximation operators and Pawlak rough approximation operators, the exploitation of various soft rough sets models in decision-making may be studied in a more logic and systematic way in the future.

3.3. The Relationships among Several Soft Rough Fuzzy Sets

A soft rough fuzzy set can be viewed as an extension model of a soft rough set, where the approximations of a fuzzy set in a soft approximation space are characterized. There are several distinct soft rough fuzzy set models in the literature. In the current part, the connections between soft rough fuzzy set and rough fuzzy set will be discussed, as well as the relationships among several soft rough fuzzy sets.

Soft rough approximation operators on fuzzy sets were initially proposed by Feng et al. in [12].

Definition 7. *Let $S = (f, A)$ be a full soft set over U and $P = (U, S)$ a soft approximation space. The lower and upper soft rough approximations of a fuzzy set, $\mu \in F(U)$, with respect to P are noted as $\underline{sap}_P(\mu)$ and $\overline{sap}_P(\mu)$, respectively, which are defined by [12]:*

$$\underline{sap}_P(\mu)(x) = \wedge\{\mu(y); \exists a \in A(\{x,y\} \subseteq f(a))\}, \tag{13}$$

$$\overline{sap}_P(\mu)(x) = \vee\{\mu(y); \exists a \in A(\{x,y\} \subseteq f(a))\}, \tag{14}$$

for all $x \in U$. The operators $\underline{sap}_P$ and $\overline{sap}_P$ are the F-lower and F-upper soft rough approximation operators on fuzzy sets. If $\underline{sap}_P(\mu) = \overline{sap}_P(\mu)$, μ is said to be F-soft definable, or else μ is called a F-soft rough fuzzy set.

Note that $\underline{sap}_P$ and $\overline{sap}_P$ are dual to each other, i.e., $\overline{sap}_P(\mu^c) = (\underline{sap}_P(\mu))^c$ for every $\mu \in F(U)$. It has already been figured out that rough fuzzy sets in Pawlak approximation space (U, R) can be identified with F-soft rough fuzzy sets in soft approximation space (U, S) when the underlying soft set S is a partition soft set [13].

Meng et al. [14] noted that $\overline{sap}_P$ is a generalization of $\overline{apr}_P$, i.e., $\overline{sap}_P(X) = \overline{apr}_P(X)$ if $X \in P(U)$. On the contrary, $\underline{sap}_P$ is a not a generalization of $\underline{apr}_P$. Considering this issue, Meng et al. presented another soft rough fuzzy set model in [14].

Definition 8. *Let $S = (f, A)$ be a full soft set over U and $P = (U, S)$ a soft approximation space. The lower soft rough approximation $\underline{sap}'_P(\mu)$ and upper soft rough approximation $\overline{sap}'_P(\mu)$ of the fuzzy set $\mu \in F(U)$ are fuzzy sets in U defined as [14]:*

$$\underline{sap}'_P(\mu)(x) = \vee_{x \in f(a)} \wedge_{y \in f(a)} \mu(y), \tag{15}$$

$$\overline{sap}'_P(\mu)(x) = \wedge_{x \in f(a)} \vee_{y \in, f(a)} \mu(y) \tag{16}$$

for all $x \in U$. μ is called soft definable if the condition $\underline{sap}'_P(\mu) = \overline{sap}'_P(\mu)$ holds; or else μ is a soft rough fuzzy set. For avoiding confusion with other soft rough fuzzy set models, it will be called M-soft rough fuzzy set in the following parts.

It is proved that [14] $\underline{sap}'_P$ and $\overline{sap}'_P$ are dual to each other, and $\underline{sap}'_P$ is a generalization of $\underline{apr}_P$, i.e., $\underline{sap}'_P(X) = \underline{apr}_P(X)$ for any $X \subseteq U$.

Theorem 7. *Let $S = (f, A)$ be a partition soft set over U, $P = (U, S)$ a soft approximation space, and (U, R) a Pawlak approximation space, where R is given by Equation (10). For each $\mu \in F(U)$, $\underline{sap}'_P(\mu) = \underline{R}(\mu)$ and $\overline{sap}'_P(\mu) = \overline{R}(\mu)$.*

Proof. Assume that $\mu \in F(U)$ and $x \in U$. For each $y \in [x]_R$, $\exists\, a \in A$ s.t. $\{x, y\} \subseteq f(a)$. Suppose that $b \in A$ and $x \in f(b)$. We note that (f, A) is a partition soft set. By $x \in f(a) \cap f(b)$, it follows that $f(a) \cap f(b) \neq \emptyset$ and hence $f(a) = f(b)$. Hence,

$$\underline{sap}'_P(\mu)(x) = \vee_{x \in f(a)} \wedge_{z \in f(a)} \mu(z) = \wedge_{z \in f(a)} \mu(z) \leq \mu(y).$$

Consequently, $\underline{sap}'_P(\mu)(x) \leq \wedge\{\mu(y); y \in [x]_R\} = \underline{R}(\mu)(x)$.

Conversely, suppose that $x \in f(a)$. For each $y \in f(a)$, since $\{x, y\} \subseteq f(a)$, we get $y \in [x]_R$.

$$\mu(y) \geq \wedge\{\mu(z); z \in [x]_R\} = \underline{R}(\mu)(x),$$

hence $\wedge_{y \in f(a)} \mu(y) \geq \underline{R}(\mu)(x)$. Consequently,

$$\underline{sap}'_P(\mu)(x) = \vee_{x \in f(a)} \wedge_{z \in f(a)} \mu(z) \geq \underline{R}(\mu)(x),$$

and $\overline{sap}'_P(\mu) = \overline{R}(\mu)$ can be proved similarly. □

By this theorem, the (classical) rough fuzzy sets in Pawlak approximation space (U, R) and M-soft rough fuzzy sets in soft approximation space (U, S) are equivalent when the underlying soft set S is a partition soft set. It is shown by Corollary 3 that Z-soft rough fuzzy sets could be regarded as a kind of rough fuzzy set, which indicates that there also exist some fantastic relationships between these two distinct models. The following theorem demonstrates the correlation between Z-soft rough approximation operators and M-soft rough approximation operators.

Theorem 8. *Let $S = (f, A)$ be a full soft set over U, $P = (U, S)$ a soft approximation space and $\mu \in F(U)$:*

(1) $\underline{sap}'_P(\mu) \subseteq \underline{\mu}_{\varphi}$,
(2) $\overline{\mu}_{\varphi} \subseteq \overline{sap}'_P(\mu)$.

Proof. (1) Let $x \in U$, $a \in A$, $x \in f(a)$. For any $y \in U$, if $y \in [x]_R$, then $\varphi(x) = \varphi(y)$. It follows that $a \in \varphi(x) = \varphi(y)$ and hence $y \in f(a)$. Then, $[x]_R \subseteq f(a)$ and hence $\wedge_{y \in f(a)}\mu(y) \le \wedge\{\mu(y); y \in [x]_R\} = \underline{\mu}_{\varphi}(x)$. Consequently, we conclude that

$$\underline{sap}'_P(\mu)(x) = \vee_{x \in f(a)} \wedge_{y \in f(a)} \mu(y) \le \underline{\mu}_{\varphi}(x)$$

and hence $\underline{sap}'_P(\mu) \subseteq \underline{\mu}_{\varphi}$.

(2) Let $x \in U$, $a \in A$ and $x \in f(a)$. By (1), we have $[x]_R \subseteq f(a)$ and hence $\overline{\mu}_{\varphi}(x) = \vee\{\mu(y); y \in [x]_R\} \le \vee_{y \in f(a)}\mu(y)$. It follows that

$$\overline{\mu}_{\varphi}(x) \le \wedge_{x \in f(a)} \vee_{y \in f(a)} \mu(y) = \overline{sap}'_P(\mu)(x)$$

and hence $\overline{\mu}_{\varphi} \subseteq \overline{sap}'_P(\mu)$. □

It is noted that F-soft rough approximation operators $\underline{apr}_P(\mu)$, $\overline{apr}_P(\mu)$ can be expressed equivalently as [15]:

$$\underline{sap}_P(\mu)(x) = \wedge\{\mu(y); \exists a \in A(\{x, y\} \subseteq f(a))\} = \wedge_{x \in f(a)} \wedge_{y \in f(a)} \mu(y),$$
$$\overline{sap}_P(\mu)(x) = \vee\{\mu(y); \exists a \in A(\{x, y\} \subseteq f(a))\} = \vee_{x \in f(a)} \vee_{y \in f(a)} \mu(y).$$

Therefore, we have the following corollary:

Corollary 4. *Let $S = (f, A)$ be a full soft set over U and $P = (U, S)$ a soft approximation space. For any $\mu \in F(U)$,*

$$\underline{sap}_P(\mu) \subseteq \underline{sap}'_P(\mu) \subseteq \underline{\mu}_{\varphi} \subseteq \mu \subseteq \overline{\mu}_{\varphi} \subseteq \overline{sap}'_P(\mu) \subseteq \overline{sap}_P(\mu).$$

Meng et al. [14] presented a kind of soft fuzzy approximation space, where a fuzzy soft set is regarded as the elementary knowledge on the universe and used to granulate the universe.

Definition 9. *Let $\mathcal{F} = (f, A)$ be a fuzzy soft set over U. The pair $SF = (U, \mathcal{F})$ is called a soft fuzzy approximation space. For a fuzzy set $\mu \in F(U)$, the lower and upper soft fuzzy rough approximations of μ with respect to SF are denoted by $\underline{Apr}_{SF}(\mu)$ and $\overline{Apr}_{SF}(\mu)$, respectively, which are given by [14]:*

$$\underline{Apr}_{SF}(\mu)(x) = \wedge_{a \in A}((1 - f(a)(x)) \vee (\wedge_{y \in U}((1 - f(a)(y)) \vee \mu(y)))), \tag{17}$$

$$\overline{Apr}_{SF}(\mu)(x) = \vee_{a \in A}(f(a)(x) \wedge (\vee_{y \in U}(f(a)(y) \wedge \mu(y)))), \tag{18}$$

for all $x \in U$. The operators $\underline{Apr}_{SF}$ and $\overline{Apr}_{SF}$ are called the lower and upper soft fuzzy rough approximation operators on fuzzy sets.

It is proved that [14] $\underline{Apr}_{SF}$ and $\overline{Apr}_{SF}$ are extensions of $\underline{sap}_{SF}$ and $\overline{sap}_{SF}$, respectively, i.e., if $\mathcal{F} = (f, A)$ is a soft set, then $\underline{Apr}_{SF}(\mu) = \underline{sap}_{SF}(\mu)$ and $\overline{Apr}_{SF}(\mu) = \overline{sap}_{SF}(\mu)$ for any $\mu \in F(U)$.

Theorem 9. *Suppose that $\mathcal{F} = (f, A)$ is a fuzzy soft set over U and $SF = (U, \mathcal{F})$. Let $R_{\mathcal{F}}$ be the fuzzy relation on U given by $R_{\mathcal{F}}(x, y) = \vee_{a \in A}(f(a)(x) \wedge f(a)(y))$. For each $\mu \in F(U)$,*

(1) $\underline{Apr}_{SF}(\mu) = \underline{R_{\mathcal{F}}}(\mu)$,
(2) $\overline{Apr}_{SF}(\mu) = \overline{R_{\mathcal{F}}}(\mu)$.

By this theorem, the soft fuzzy rough approximation presented in Definition 9 is a kind of Dubois and Prade's fuzzy rough approximation in [6]. We note that $R_{\mathcal{F}}(x, y)$ describes a kind of similarity between x and y, and $R_{\mathcal{F}}$ is symmetric but $R_{\mathcal{F}}(x, x) \neq 1$ in general.

The utilization of Z-soft rough fuzzy set in decision-making has already been studied in [17]. Through discussing the connections among different soft rough fuzzy set models, we can further explore the applications of the other two kinds of soft rough fuzzy sets models in decision-making, enrich the decision mechanisms and pay attention to the selection of the most suitable mechanism according to environments. The soft fuzzy rough approximation operators on fuzzy sets proposed by Meng et al. [14] have the potential to be utilized to handle decision-making problems, discussion on the connections between which and fuzzy rough approximation operators confirm the rationality of this model from the theoretical perspective and lays the foundation for subsequent practical applications.

4. *F*-Soft Rough Sets and Modal-Style Operators in FCA

FCA [22,33,34] provides a methodology for knowledge description and summarization. In this section, several absorbing connections between *F*-soft rough sets and modal-style operators in FCA will be discussed. Formal concept analysis is carried out based on a formal context specifying which objects posses what properties or attributes. A formal concept is formulated as a pair of two sets, one is consists of objects and another consists of properties or attributes, and these two sets are connected by two set-theoretic operators. A complete lattice called concept lattice is constituted by the set of all formal concepts, which reflects the correlation of generalization and specialization for formal concepts.

Definition 10. *A formal context (G, M, I) consists of two sets G and M and a relation I between G and M. The elements of G are called the objects and the elements of M are called the attributes of the context. $(g, m) \in I$ indicate that the object g has the attribute m, or the attribute m is possessed by the object g [33].*

Let (G, M, I) be a formal context. For $A \subseteq G$, $B \subseteq M$, Duntsch and Gediga [6] defined a pair of modal-style operators $\triangle, \bigtriangledown$ as follows:

$$A^{\triangle} = \{m \in M; \exists g \in A((g, m) \in I)\}, \tag{19}$$

$$A^{\bigtriangledown} = \{m \in M; \forall g \in G((g, m) \in I \rightarrow g \in A)\}, \tag{20}$$

$$B^{\triangle} = \{g \in G; \exists m \in B((g, m) \in I)\}, \tag{21}$$

$$B^{\bigtriangledown} = \{g \in G; \forall m \in M((g, m) \in I \rightarrow m \in B)\}. \tag{22}$$

Recently, the granular computing based concept lattice theory has received much attention [35].

Rough set theory, soft set theory and concept lattices have similar basis data description. Mathematically speaking, the notions of soft set and formal context are equivalent. Furthermore, both a formal context and a soft set can be considered as a two-valued information system.

Theorem 10. *Let $S = (F, A)$ be a soft set over U. A formal context $C_S = (U, A, I_S)$ is induced by S, where I_S is provided as*

$$I_S = \{(x,a) \in U \times A; x \in F(a)\}.$$

Conversely, let $C = (U, A, I)$ be a formal context. A set-valued mapping $F_C : A \to P(U)$ is defined by

$$F_C(a) = \{x \in U; (x,a) \in I\}$$

for all $a \in A$, and $S_C = (F_C, A)$ is a soft set. Moreover, we have $S_{C_S} = S$ and $C_{S_C} = C$.

Proof. Only the proof for $S_{C_S} = S$ and $C_{S_C} = C$ will be provided here. Suppose that $S = (F, A)$ is a soft set over U and $a \in A$. For any $x \in U$, from the definition, we obtain that

$$x \in F_{C_S}(a) \Leftrightarrow (x,a) \in I_S \Leftrightarrow x \in F(a).$$

That is, $F_{C_S}(a) = F(a)$ for all $a \in A$. Thus, $F_{C_S} = F$, whence $S_{C_S} = S$.

Next, assume that $C = (U, A, I)$ is a formal context, $x \in U$ and $a \in A$. Then, by definition,

$$(x,a) \in I_{S_C} \Leftrightarrow x \in F_C(a) \Leftrightarrow (x,a) \in I.$$

Therefore, we conclude that $C_{S_C} = C$ as required. □

Theorem 11 shows the relationship among operators $\triangle$, $\bigtriangledown$ and soft rough approximation operators.

Theorem 11. *Let $S = (F, A)$ be a soft set over U. For any $X \subseteq U$, $\underline{apr}_P(X) = X^{\bigtriangledown\triangle}$, $\overline{apr}_P(X) = X^{\triangle\triangle}$.*

Proof. (1) For any $x \in \underline{apr}_P(X)$, $\exists\, a \in A$ s.t. $x \in f(a) \subseteq X$. Then, $x \in a^{\triangle}$ and $a^{\triangle} \subseteq X$. Therefore, $a \in X^{\bigtriangledown}$ and consequently $x \in a^{\triangle} \subseteq X^{\bigtriangledown\triangle}$. We conclude that $\underline{apr}_P(X) \subseteq X^{\bigtriangledown\triangle}$.

Conversely, if $x \in X^{\bigtriangledown\triangle}$, then $\exists\, a \in X^{\bigtriangledown}$ s.t. $x \in a^{\triangle}$. Then, $x \in f(a)$ and $f(a) \subseteq X$. Thus, $x \in \{f(c); f(c) \subseteq X\} = \underline{apr}_P(X)$ and hence $X^{\bigtriangledown\triangle} \subseteq \underline{apr}_P(X)$.

(2) For any $x \in \overline{apr}_P(X)$, $\exists\, a \in A$ satisfying $x \in f(a)$ and $f(a) \cap X \neq \emptyset$. It follows that $x \in a^{\triangle}$ and $a^{\triangle} \cap X \neq \emptyset$. Thus, $a \in X^{\triangle}$ and consequently $x \in a^{\triangle} \subseteq X^{\triangle\triangle}$.

Conversely, $\exists\, a \in X^{\triangle}$ s.t. $x \in a^{\triangle}$ if $x \in X^{\triangle\triangle}$. Then, $x \in f(a)$ and $f(a) \cap X \neq \emptyset$. Consequently, $x \in \{f(c); f(c) \cap X \neq \emptyset\} = \overline{apr}_P(X)$. □

FCA has become increasingly popular among various methods of conceptual data analysis, knowledge representation and decision-making. Depth study on the connections of soft rough sets theory and FCA contributes to the reference and fusion for decision-making approaches in these two different fields.

5. A New Generalization of F-Soft Rough Set: Soft Rough Soft Sets

In this section, by extending the notion of F-soft rough set, a new generalization model called soft rough soft set will be proposed. In this new model, we use a soft set is as the elementary knowledge to compute the approximations of soft set. In this way, parameterized tools can be used to the greatest extent. Some basic properties of the new proposed model are discussed. A multi-group decision-making approach based on soft rough soft sets has been provided.

Definition 11. *Let U be the universe set and A, A_1 be parameter sets. Let $\mathcal{S}_1 = (f_1, A_1)$ be a full soft set over U and $(U, \mathcal{S}_1)$ be a soft approximation space. Let $\mathcal{S} = (f, A)$ be a soft set over U. The lower and upper soft rough approximations of $\mathcal{S}$ in $(U, \mathcal{S}_1)$ are denoted by $\underline{sapr}_{\mathcal{S}_1}(\mathcal{S}) = (f_{\mathcal{S}_1}, A)$ and $\overline{sapr}_{\mathcal{S}_1}(\mathcal{S}) = (f^{\mathcal{S}_1}, A)$, which are soft sets over U defined by:*

$$f_{S_1}(e) = \{x \in U : \exists e' \in A_1[x \in f_1(e') \subseteq f(e)]\},$$
$$f^{S_1}(e) = \{x \in U : \exists e' \in A_1[x \in f_1(e'), f_1(e') \cap f(e) \neq \emptyset]\},$$

for all $e \in A$. $\underline{sapr}_{S_1}$, $\overline{sapr}_{S_1}$ *are the lower and the upper soft rough approximation operators on soft set* $\mathcal{S}$, *respectively. If* $\underline{sapr}_{S_1}(\mathcal{S}) = \overline{sapr}_{S_1}(\mathcal{S})$, *the soft set* $\mathcal{S}$ *is soft definable, or else* $\mathcal{S}$ *is so-called a soft rough soft set.*

Example 2. *Suppose that the universe set* $U = \{x_1, x_2, x_3, x_4, x_5, x_6\}$ *and the parameters set* $E = \{e_1, e_2, e_3, e_4, e_5, e_6, e_7\}$. *Let* $A = \{e_1, e_2, e_3, e_4\} \subseteq E$ *and* $A_1 = \{e_3, e_4, e_5, e_6, e_7\} \subseteq E$. *Let* $\mathcal{S}_1 = (f_1, A_1)$ *be a full soft set and* $\mathcal{S} = (f, A)$ *be a soft set over* U *as shown by Tables 1 and 2, respectively. In the soft approximation space* $(U, \mathcal{S}_1)$, *by Definition 11, we get the lower soft rough approximation* $\underline{sapr}_{S_1}(\mathcal{S}) = (f_{S_1}, A)$ *and the upper soft rough approximation* $\overline{sapr}_{S_1}(\mathcal{S}) = (f^{S_1}, A)$ *of soft set* $\mathcal{S} = (f, A)$, *as shown by Tables 3 and 4, respectively. In order to facilitate the readers to understand, Figure 1 is given to show the process of computing* $f_{S_1}(e_4)$ *and* $f^{S_1}(e_4)$ *from* $f(e_4)$.

Table 1. Soft set (f_1, A_1).

A \ U	x_1	x_2	x_3	x_4	x_5	x_6
e_3	1	0	0	0	0	1
e_4	0	1	1	0	0	0
e_5	0	0	0	0	0	0
e_6	0	0	0	0	1	0
e_7	0	0	0	1	1	1

Table 2. Soft set (f, A).

A \ U	x_1	x_2	x_3	x_4	x_5	x_6
e_1	1	1	0	1	0	1
e_2	0	1	1	0	0	0
e_3	0	0	0	1	1	1
e_4	1	1	1	1	0	1

Table 3. Soft set (f_{S_1}, A).

A \ U	x_1	x_2	x_3	x_4	x_5	x_6
e_1	1	0	0	0	0	1
e_2	0	1	1	0	0	0
e_3	0	0	0	1	1	1
e_4	1	1	1	0	0	1

Table 4. Soft set (f^{S_1}, A).

A \ U	x_1	x_2	x_3	x_4	x_5	x_6
e_1	1	1	1	1	1	1
e_2	0	1	1	0	0	1
e_3	1	0	0	1	1	1
e_4	1	1	1	1	1	1

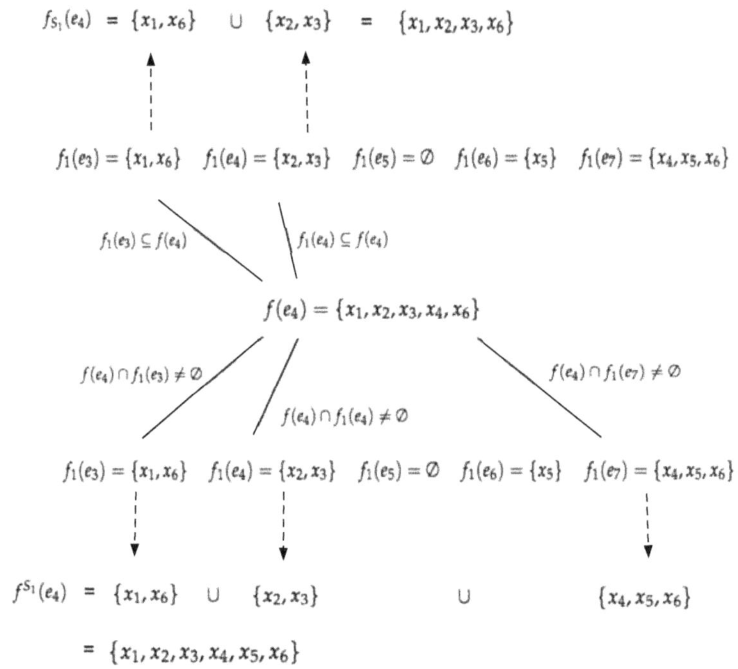

Figure 1. The process of computing $f_{S_1}(e_4)$ and $f^{S_1}(e_4)$ from $f(e_4)$ in Example 2.

Proposition 1. *Let $\mathcal{S}_1 = (f_1, A_1)$ be a full soft set over U and $(U, \mathcal{S}_1)$ be a soft approximation space. Let $\mathcal{S} = (f, A)$ be a soft set over U. The following properties hold:*

(1) $\underline{sapr}_{S_1}(\mathcal{S}) \subseteq \mathcal{S} \subseteq \overline{sapr}_{S_1}(\mathcal{S})$,
(2) $\underline{sapr}_{S_1}(\tilde{N}_{(U,A)}) = \tilde{N}_{(U,A)} = \overline{sapr}_{S_1}(\tilde{N}_{(U,A)})$,
(3) $\overline{sapr}_{S_1}(\tilde{W}_{(U,A)}) = \tilde{W}_{(U,A)} = \underline{sapr}_{S_1}(\tilde{W}_{(U,A)})$.

Proof. The lower and upper soft rough approximations of $\tilde{N}_{(U,A)} = (N, A)$ in $(U, \mathcal{S}_1)$ are denoted by $\underline{sapr}_{S_1}(\tilde{N}_{(U,A)}) = (N_{S_1}, A)$ and $\overline{sapr}_{S_1}(\tilde{N}_{(U,A)}) = (N^{S_1}, A)$; the lower and upper soft rough approximations of $\tilde{W}_{(U,A)} = (W, A)$ in $(U, \mathcal{S}_1)$ are denoted by $\underline{sapr}_{S_1}(\tilde{W}_{(U,A)}) = (W_{S_1}, A)$ and $\overline{sapr}_{S_1}(\tilde{W}_{(U,A)}) = (W^{S_1}, A)$.

(1a) For all $x \in U, e \in A$, if $x \in f_{S_1}(e) = \{x \in U : \exists e' \in A_1[x \in f_1(e') \subseteq f(e)]\}$, then we obtain $x \in f(e)$, so $f_{S_1}(e) \subseteq f(e)$;

(1b) For all $e \in A$, if $x \in f(e)$, since (f_1, A_1) is a full soft set, we obtain that $\exists e' \in A_1$, s.t. $x \in f_1(e')$, then $x \in f_1(e') \cap f(e) \neq \emptyset$, then $x \in \{x \in U : \exists e' \in A_1[x \in f_1(e'), f_1(e') \cap f(e) \neq \emptyset]\}$, that is, $x \in f^{S_1}(e)$ and $f(e) \subseteq f^{S_1}(e)$ for all $e \in A$.

Hence, we know that $f_{S_1}(e) \subseteq f(e) \subseteq f^{S_1}(e)$ for all $e \in A$, that is, $\underline{sapr}_{S_1}(\mathcal{S}) \subseteq \mathcal{S} \subseteq \overline{sapr}_{S_1}(\mathcal{S})$.

(2a) By the definition of relative null soft set, we know $N(e) = \emptyset$ for all $e \in A$. For all $e \in A$, we have $N_{S_1}(e) = \{x \in U : \exists e' \in A_1[x \in f_1(e') \subseteq N(e)]\} = \{x \in U : \exists e' \in A_1[x \in f_1(e') \subseteq \emptyset]\} = \emptyset = N(e)$, that is, $\underline{sapr}_{S_1}(\tilde{N}_{(U,A)}) = \tilde{N}_{(U,A)}$;

(2b) By the definition of relative null soft set, we know $N(e) = \emptyset$ for all $e \in A$. For all $e \in A$, we have $N^{S_1}(e) = \{x \in U : \exists e' \in A_1[x \in f_1(e'), f_1(e') \cap N(e) \neq \emptyset]\} = \emptyset = N(e)$, that is, $\overline{sapr}_{S_1}(\tilde{N}_{(U,A)}) = \tilde{N}_{(U,A)}$.

(3a) By the definition of relative whole soft set, we know $W(e) = U$ for all $e \in A$. For all $e \in A$, we have $W^{S_1}(e) = \{x \in U : \exists e' \in A_1[x \in f_1(e'), f_1(e') \cap W(e) \neq \emptyset]\} = U = W(e)$, that is, $\overline{sapr}_{S_1}(\tilde{W}_{(U,A)}) = \tilde{W}_{(U,A)}$.

(3b) By the definition of relative whole soft set, we know $W(e) = U$ for all $e \in A$. Since (f_1, A_1) is a full soft set over U, for all $e \in A$, we have $W_{S_1}(e) = \{x \in U : \exists e' \in A_1[x \in f_1(e') \subseteq W(e)]\} = \{x \in U : \exists e' \in A_1[x \in f_1(e') \subseteq U]\} = U = W(e)$, that is, $\underline{sapr}_{S_1}(\tilde{W}_{(U,A)}) = \tilde{W}_{(U,A)}$. □

Proposition 2. *Suppose that $\mathcal{S}_1 = (f_1, A_1)$ is a full soft set over U and $(U, \mathcal{S}_1)$ is a soft approximation space. Let $\mathcal{S} = (f, A)$, $\mathcal{T} = (g, A)$ be two soft sets over U. The following properties hold:*

(1) $\mathcal{S} \subseteq \mathcal{T} \Rightarrow \underline{sapr}_{S_1}(\mathcal{S}) \subseteq \underline{sapr}_{S_1}(\mathcal{T})$,
(2) $\mathcal{S} \subseteq \mathcal{T} \Rightarrow \overline{sapr}_{S_1}(\mathcal{S}) \subseteq \overline{sapr}_{S_1}(\mathcal{T})$,
(3) $\underline{sapr}_{S_1}(\mathcal{S} \cap \mathcal{T}) \subseteq \underline{sapr}_{S_1}(\mathcal{S}) \cap \underline{sapr}_{S_1}(\mathcal{T})$,
(4) $\underline{sapr}_{S_1}(\mathcal{S} \cup \mathcal{T}) \supseteq \underline{sapr}_{S_1}(\mathcal{S}) \cup \underline{sapr}_{S_1}(\mathcal{T})$,
(5) $\overline{sapr}_{S_1}(\mathcal{S} \cup \mathcal{T}) \supseteq \overline{sapr}_{S_1}(\mathcal{S}) \cup \overline{sapr}_{S_1}(\mathcal{T})$,
(6) $\overline{sapr}_{S_1}(\mathcal{S} \cap \mathcal{T}) \subseteq \overline{sapr}_{S_1}(\mathcal{S}) \cap \overline{sapr}_{S_1}(\mathcal{T})$.

Proof. The lower and upper soft rough approximations of $\mathcal{S}$ in $(U, \mathcal{S}_1)$ are denoted by $\underline{sapr}_{S_1}(\mathcal{S}) = (f_{S_1}, A)$ and $\overline{sapr}_{S_1}(\mathcal{S}) = (f^{S_1}, A)$; the lower and upper soft rough approximations of $\mathcal{T}$ in $(U, \mathcal{S}_1)$ are denoted by $\underline{sapr}_{S_1}(\mathcal{T}) = (g_{S_1}, A)$ and $\overline{sapr}_{S_1}(\mathcal{T}) = (g^{S_1}, A)$.

(1) If $\mathcal{S} \subseteq \mathcal{T}$, then for all $e \in A$, we have $f(e) \subseteq g(e)$. Assume that $x \in f_{S_1}(e) = \{x \in U : \exists e' \in A_1[x \in f_1(e') \subseteq f(e)]\}$. From $f(e) \subseteq g(e)$, we obtain $x \in \{x \in U : \exists e' \in A_1[x \in f_1(e') \subseteq g(e)]\} = g_{S_1}(e)$. Therefore, we get $f_{S_1}(e) \subseteq g_{S_1}(e)$ for all $e \in A$, i.e., $\underline{sapr}_{S_1}(\mathcal{S}) \subseteq \underline{sapr}_{S_1}(\mathcal{T})$;

(2) If $\mathcal{S} \subseteq \mathcal{T}$, then for all $e \in A$, we have $f(e) \subseteq g(e)$. Assume that $x \in f^{S_1}(e) = \{x \in U : \exists e' \in A_1[x \in f_1(e'), f_1(e') \cap f(e) \neq \emptyset]\}$, from $f(e) \subseteq g(e)$, we obtain $\exists e' \in A_1$, s.t. $x \in f_1(e'), f_1(e') \cap g(e) \neq \emptyset$, so $x \in g^{S_1}(e) = \{x \in U : \exists e' \in A_1[x \in f_1(e'), f_1(e') \cap g(e) \neq \emptyset]\}$, it follows that $f^{S_1}(e) \subseteq g^{S_1}(e)$ for all $e \in A$, i.e., $\overline{sapr}_{S_1}(\mathcal{S}) \subseteq \overline{sapr}_{S_1}(\mathcal{T})$;

(3) It is obvious that $\mathcal{S} \cap \mathcal{T} \subseteq \mathcal{S}$ and $\mathcal{S} \cap \mathcal{T} \subseteq \mathcal{T}$. From property (1), we obtain $\underline{sapr}_{S_1}(\mathcal{S} \cap \mathcal{T}) \subseteq \underline{sapr}_{S_1}(\mathcal{S})$ and $\underline{sapr}_{S_1}(\mathcal{S} \cap \mathcal{T}) \subseteq \underline{sapr}_{S_1}(\mathcal{T})$. Thus, $\underline{sapr}_{S_1}(\mathcal{S} \cap \mathcal{T}) \subseteq \underline{sapr}_{S_1}(\mathcal{S}) \cap \underline{sapr}_{S_1}(\mathcal{T})$.

(4) It is obvious that $\mathcal{S} \cup \mathcal{T} \supseteq \mathcal{S}$ and $\mathcal{S} \cup \mathcal{T} \supseteq \mathcal{T}$. From property (1), we obtain $\underline{sapr}_{S_1}(\mathcal{S} \cup \mathcal{T}) \supseteq \underline{sapr}_{S_1}(\mathcal{S})$ and $\underline{sapr}_{S_1}(\mathcal{S} \cup \mathcal{T}) \supseteq \underline{sapr}_{S_1}(\mathcal{T})$. Thus, $\underline{sapr}_{S_1}(\mathcal{S} \cup \mathcal{T}) \supseteq \underline{sapr}_{S_1}(\mathcal{S}) \cup \underline{sapr}_{S_1}(\mathcal{T})$.

(5) It is obvious that $\mathcal{S} \cup \mathcal{T} \supseteq \mathcal{S}$ and $\mathcal{S} \cup \mathcal{T} \supseteq \mathcal{T}$. From property (2), we obtain $\overline{sapr}_{S_1}(\mathcal{S} \cup \mathcal{T}) \supseteq \overline{sapr}_{S_1}(\mathcal{S})$ and $\overline{sapr}_{S_1}(\mathcal{S} \cup \mathcal{T}) \supseteq \overline{sapr}_{S_1}(\mathcal{T})$. Thus, $\overline{sapr}_{S_1}(\mathcal{S} \cup \mathcal{T}) \supseteq \overline{sapr}_{S_1}(\mathcal{S}) \cup \overline{sapr}_{S_1}(\mathcal{T})$.

(6) It is obvious that $\mathcal{S} \cap \mathcal{T} \subseteq \mathcal{S}$ and $\mathcal{S} \cap \mathcal{T} \subseteq \mathcal{T}$. From property (2), we obtain $\overline{sapr}_{S_1}(\mathcal{S} \cap \mathcal{T}) \subseteq \overline{sapr}_{S_1}(\mathcal{S})$ and $\overline{sapr}_{S_1}(\mathcal{S} \cap \mathcal{T}) \subseteq \overline{sapr}_{S_1}(\mathcal{T})$. Thus, $\overline{sapr}_{S_1}(\mathcal{S} \cap \mathcal{T}) \subseteq \overline{sapr}_{S_1}(\mathcal{S}) \cap \overline{sapr}_{S_1}(\mathcal{T})$. □

Proposition 3. *Let $\mathcal{S}_1 = (f_1, A_1)$ be a full soft set over U and $(U, \mathcal{S}_1)$ be a soft approximation space. Let $\mathcal{S} = (f, A)$ be a soft set over U. The following properties hold:*

(1) $\underline{sapr}_{S_1}(\mathcal{S}) \subseteq \underline{sapr}_{S_1}(\overline{sapr}_{S_1}(\mathcal{S}))$,
(2) $\overline{sapr}_{S_1}(\mathcal{S}) \supseteq \overline{sapr}_{S_1}(\underline{sapr}_{S_1}(\mathcal{S}))$.

Proof. From property (1) in Proposition 1, it is obvious that $\underline{sapr}_{S_1}(\mathcal{S}) \subseteq \mathcal{S} \subseteq \overline{sapr}_{S_1}(\mathcal{S})$. From property (1) and (2) in Proposition 2, we get $\underline{sapr}_{S_1}(\mathcal{S}) \subseteq \underline{sapr}_{S_1}(\overline{sapr}_{S_1}(\mathcal{S}))$ and $\overline{sapr}_{S_1}(\mathcal{S}) \supseteq \overline{sapr}_{S_1}(\underline{sapr}_{S_1}(\mathcal{S}))$, respectively. □

In [12], a group decision-making approach based on F-soft rough sets was proposed; however, if we carefully check their decision scheme, it is not hard to find that they actually use the tool of a soft rough soft set since the best alternatives provided by each specialist gather together to form a soft set and they compute the upper and lower soft rough approximations (soft sets) on the preliminary evaluation soft set during the decision process. That is, although the concept has not been formally proposed, the application of soft rough soft sets has already appeared in literature. From another perspective, the decision-making problem that can be solved by F-soft rough sets in [12] can also be solved by using soft rough soft sets. It is necessary to propose the concept for soft rough soft sets as well as its application to introduce parameter tools to the universe description, that is, make it feasible to describe objects in the universe from different aspects at the same time, information obtained from different aspects be able to be handled as a whole before the approximations of a soft set are computed, and allow the flexibility to make operations such as the restricted intersection "$\cap$" [25] on soft sets whose soft rough approximations need to be computed; in this way, soft rough soft sets have the potential to be applied in more complex decision-making situations to meet demands of applications in real life cases. As follows, we provide a simple application of soft rough soft sets in decision-making.

Let $G = \{T_1, T_2, ..., T_p\}$ and $A_1 = \{e'_1, e'_2, ..., e'_q\}$ be two groups of specialists to evaluate all the candidates $U = \{x_1, x_2, ..., x_m\}$. In group G, each specialist is asked to point out if the candidates satisfy benefit properties in $A = \{e_1, e_2, ..., e_n\}$ or not. In this way, a serious of evaluations provided by specialists are obtained as (g_1, A), (g_2, A), (g_3, A), $\ldots$, (g_p, A). Afterwards, the evaluation made by group G could be obtained by $S = (f, A) = (g_1, A) \cap (g_2, A) \cap (g_3, A) \cap \ldots (g_p, A)$. Meanwhile, in another group $A_1 = \{e'_1, e'_2, ..., e'_q\}$, the specialists are under time pressure, and a lack of patience, or, because of some other issues, each specialist only points out the best alternatives; however, we have no clear idea about which properties are under their consideration. The best alternatives chosen by specialists in group A_1 form another soft set $S_1 = (f_1, A_1)$. We say the assessments provided by group G are more reliable since the assessments provided by them are more specific than group A_1. However, in order to make full use of information provided by the two independent groups, we can compute the lower soft rough approximation on (f, A) in soft approximation space (U, S_1). If $x_i \in f_{S_1}(e_j)$, from the axiomatic definition of soft rough soft sets, we know that the best alternatives of one or more specialists in A_1 are totally contained in $f(e_j)$, that is, the best alternatives chosen by some specialists in A_1 certainly occupy property e_j, which indicates that this benefit property e_j considered by group G may also be very important to group A_1. The final decision is to select the alternative that occupies the most number of beneficial properties that may be important for both groups.

The steps of this soft rough soft sets based multi-group decision-making approach can be listed as:

Step 1. Input the evaluations on alternatives $U = \{x_1, x_2, ..., x_m\}$ provided by specialists group $G = \{T_1, T_2, ..., T_p\}$ as $(g_1, A), (g_2, A), (g_3, A), \ldots, (g_p, A)$.

Step 2. Input the best alternatives selected by specialists group A_1 as $S_1 = (f_1, A_1)$.

Step 3. Compute the group evaluation made by the specialists in G as $S = (f, A) = (g_1, A) \cap (g_2, A) \cap (g_3, A) \cap \ldots (g_p, A)$.

Step 4. Compute the lower soft rough approximation of (f, A) in (U, S_1), i.e. (f_{S_1}, A).

Step 5. Compute the score of alternatives of each x_j $(j = 1, 2, \ldots, m)$ as $s(x_j) = \sum_{i=1}^{n} f_{S_1}(e_i)(x_j)$, and the decision result is x_k if it satisfies $s(x_k) = max_{j=1,2,\ldots,m}\, s(x_j)$.

Example 3. *Suppose that a factory needs to purchase the best machine from $U = \{x_1, x_2, ..., x_6\}$ according to evaluations provided by two specialists groups G and A_1, which form a multi-group decision-making problem. $G = \{T_1, T_2, T_3, T_4\}$ consists of four specialists and each of them provides assessments on machines in U with respect to beneficial properties $A = \{e_1 =$ low price, $e_2 =$ high endurance, $e_3 =$ advanced technology, $e_4 =$ good compatibility$\}$. Each specialist in G points out if the machines satisfy properties in A or not. In this way, a serious of evaluation soft sets provided by specialists are obtained as (g_1, A), (g_2, A), (g_3, A), (g_4, A) (see Tables 5–8 as their tabular representations) and the group evaluation of G can be computed by $(f, A) = (g_1, A) \cap (g_2, A) \cap (g_3, A) \cap (g_4, A)$ (see also Table 2 as the tabular representation for (f, A)). Meanwhile, each specialist in another specialist group $A_1 = \{e'_3, e'_4, e'_5, e'_6, e'_7\}$ only points out the best machines according to his/her own cognition, which form soft set (f_1, A_1) (replace $e'_3 - e'_7$ by $e_3 - e_7$ and see also Table 1 for its tabular representation). The lower soft rough approximation of (f, A) in (U, S_1) can be easily computed as (f_{S_1}, A) (see also Table 3 for its tabular representation). It is easy to obtain that $s(x_1) = s(x_2) = s(x_3) = 2$,*

$s(x_4) = s(x_5) = 1$ and $s(x_6) = 3$, hence x_6 should be the machine purchased by the factory since it satisfies largest number of beneficial properties that are important to two groups.

Table 5. Soft set (g_1, A).

A \ U	x_1	x_2	x_3	x_4	x_5	x_6
e_1	1	1	1	1	0	1
e_2	0	1	1	0	0	0
e_3	0	0	0	1	1	1
e_4	1	1	1	1	0	1

Table 6. Soft set (g_2, A).

A \ U	x_1	x_2	x_3	x_4	x_5	x_6
e_1	1	1	0	1	0	1
e_2	1	1	1	0	0	0
e_3	0	0	0	1	1	1
e_4	1	1	1	1	0	1

Table 7. Soft set (g_3, A).

A \ U	x_1	x_2	x_3	x_4	x_5	x_6
e_1	1	1	0	1	0	1
e_2	0	1	1	0	0	0
e_3	1	0	0	1	1	1
e_4	1	1	1	1	0	1

Table 8. Soft set (g_4, A).

A \ U	x_1	x_2	x_3	x_4	x_5	x_6
e_1	1	1	0	1	0	1
e_2	0	1	1	0	0	0
e_3	0	0	1	1	1	1
e_4	1	1	1	1	0	1

As is mentioned at the beginning of this section, soft rough soft set is an extension model of F-soft rough set. Sometimes, in a practical situation, the universe set that needs to be granulated is presented from different attributes' aspects simultaneously. In other words, the parameter tools are necessary not only for the knowledge presentation, but also for the universe description. The new model provides a framework for dealing with these kinds of problems and the exploration of its potential use in decision-making is promising. Compared to F-soft rough sets, soft rough soft sets introduce parameter tools to the universe description and a soft set (instead of a subset of the universe) is approximated. Compared to rough soft set [12], a soft set instead of an equivalence relation has been adopted in soft rough soft sets to compute the approximations of soft sets [36,37]. In this section, only a small application attempt of soft rough soft sets in decision-making has been provided, which is far from enough to meet various demands in real life situations. More flexible and effective approaches need to be developed in the future.

6. Conclusions

This paper has presented a comparative study of some existing soft rough set models, and new discoveries on the relationships among various hybrid sets have been summarized in Table 9. It has been shown that the Z-soft rough fuzzy set is a kind of rough fuzzy set. Therefore, decision-making approaches based on rough fuzzy sets have the potential to be addicted to more specific situations in which Z-soft rough fuzzy sets should be applied to solve the problem. Various soft rough set models have shown great potential in coping with decision-making problems. Some potential applications of connections among various soft rough set models in decision-making have been briefly discussed in the current work. For instance, benefitting from the connections between F-soft rough approximations and MSR approximations that have been discussed, it is possible to further study the relationships between the decision results made by using soft rough sets and MSR sets. In future works, deeper and more specific research on the applications of these connections in decision-making will be conducted.

Table 9. Summary on relationships among various hybrid models.

Various Hybrid Models	Relationships
F-soft rough approximations and modified soft rough approximations (MSR approximations)	$\overline{X}_{\varphi} \subseteq \overline{apr}_P(X)$, $\overline{apr}_P(X) \subseteq \overline{X}_{\varphi}$, $\underline{X}_{\varphi} \subseteq \underline{apr}_P(X)$, if some specific conditions hold, respectively (see Theorems 1–3)
F-soft rough sets in (U, S) and Pawlak's rough sets in (U, R_S)	F-soft rough sets in (U, S) could be identified with Pawlak's rough sets in (U, R_S), when the underlying soft set is a partition soft set (see Theorems 4 and 5)
MSR approximations and Pawlak's rough approximations	MSR approximation operator is a kind of Pawlak rough approximation operator (see Theorem 6)
Z-lower, Z-upper soft rough approximation operators and Dubois and Prade's lower and upper rough fuzzy approximation operators in [6]	Z-lower and Z-upper soft rough approximation operators are equivalent to Dubois and Prade's lower and upper rough fuzzy approximation operators in [6] (see Corollary 3)
The (classical) rough fuzzy sets and M-soft rough fuzzy sets	The (classical) rough fuzzy sets in Pawlak approximation space (U, R) and M-soft rough fuzzy sets in soft approximation space (U, S) are equivalent when the underlying soft set S is a partition soft set (see Theorem 7)
Z-soft rough approximation operators and M-soft Rough approximation operators and F-soft rough approximation operators	$\underline{sap}_P(\mu) \subseteq \underline{sap}'_P(\mu) \subseteq \underline{\mu}_{\varphi} \subseteq \mu \subseteq \overline{\mu}_{\varphi} \subseteq \overline{sap}'_P(\mu) \subseteq \overline{sap}_P(\mu)$ (see Theorem 8 and Corollary 4)
The soft fuzzy rough approximation in Definition 9 and Dubois and Prade's fuzzy rough approximation in [6]	The soft fuzzy rough approximation is a kind of Dubois and Prade's fuzzy rough approximation in [6] (see Theorem 9)
F-soft rough set and soft rough soft set	Soft rough soft set is an extension of F-soft rough set

Acknowledgments: This work has been supported by the National Natural Science Foundation of China (Grant Nos. 61473239, 61372187, and 61673320) and the Spanish National Research Project TIN2015-66524-P.

Author Contributions: All authors have contributed equally to this paper.

Conflicts of Interest: The authors declare no conflict of interest.

References

1. Zadeh, L.A. Fuzzy sets. *Inf. Control* **1965**, *8*, 338–353.
2. Pawlak, Z. Rough sets. *Int. J. Comput. Inf. Sci.* **1982**, *11*, 341–356.
3. Pawlak, Z.; Skowron, A. Rudiments of rough sets. *Inf. Sci.* **2007**, *177*, 3–27.
4. Molodtsov, D. Soft set theory-First results. *Comput. Math. Appl.* **1999**, *37*, 19–31.

5. Jiang, Y.; Tang, Y.; Chen, Q.; Liu, H.; Tang, J. Interval-valued intuitionistic fuzzy soft sets and their properties. *Comput. Math. Appl.* **2010**, *60*, 906–918.
6. Dubois, D.; Prade, H. Rough fuzzy set and fuzzy rough sets. *Int. J. Gen. Syst.* **1990**, *17*, 191–209.
7. Maji, P.K.; Biswas, R.; Roy, A.R. Fuzzy soft sets. *J. Fuzzy Math.* **2001**, *9*, 589–602.
8. Maji, P.K.; Roy, A.R.; Biswas, R. On intuitionistic fuzzy soft sets. *J. Fuzzy Math.* **2004**, *12*, 669–683.
9. Liu, Y.; Luo, J.; Wang, B.; Qin, K.Y. A theoretical development on the entropy of interval-valued intuitionistic fuzzy soft sets based on the distance measure. *Int. J. Comput. Intell. Syst.* **2017**, *10*, 569–592.
10. Xu, W.; Ma, J.; Wang, S.; Hao, G. Vague soft sets and their properties. *Comput. Math. Appl.* **2010**, *59*, 787–794.
11. Qin, K.Y.; Meng, D.; Pei, Z.; Xun, Y. Combination of interval set and soft set. *Int. J. Comput. Intell. Syst.* **2013**, *2*, 370–380.
12. Feng, F.; Li, C.X.; Davvaz, B.; Ali, M.I. Soft sets combined with fuzzy sets and rough sets: A tentative approach. *Soft Comput.* **2010**, *14*, 899–911.
13. Feng, F.; Liu, X.; Leoreanu-Fotea, V.; Jun,Y.B. Soft sets and soft rough sets. *Inf. Sci.* **2011**, *181*, 1125–1137.
14. Meng, D.; Zhang, X.H.; Qin, K.Y. Soft rough fuzzy sets and soft fuzzy rough sets. *Comput. Math. Appl.* **2011**, *62*, 4635–4645.
15. Qin, K.Y.; Thereforeng, Z.M.; Xu, Y. Soft rough sets based on similarity measures. In *Rough Sets and Knowledge Technology*; RSKT 2012; Li, T., Nguyen, H.S., Wang, G., Grzymala-Busse, J.W., Janicki, R., Hassanien, A.-E., Yu, H., Eds.; Springer: Berlin, Germany, 2012; pp. 40–48.
16. Shabir, M.; Ali, M.I.; Shaheen, T. Another approach to soft rough sets. *Knowl.-Based Syst.* **2013**, *40*, 72–80.
17. Zhan, J.; Zhu, K. A novel soft rough fuzzy sets: Z-soft rough fuzzy ideals of hemirings and corresponding decision-making. *Soft Comput.* **2017**, *21*, 1923–1936.
18. Alcantud, J.C.R.; Santos-Garcia, G. A New Criterion for Soft Set Based Decision Making Problems under Incomplete Information. *Int. J. Comput. Intell. Syst.* **2017**, *10*, 394–404.
19. Alcantud, J.C.R.; Mathew, T.J. Separable fuzzy soft sets and decision-making with positive and negative attributes. *Appl. Soft Comput.* **2017**, *59*, 586–595.
20. Zhan, J.; Liu, Q.; Herawan, T. A novel soft rough set: Soft rough hemirings and its multicriteria group decision-making. *Appl. Soft Comput.* **2017**, *54*, 393–402.
21. Zhan, J.; Ali, M.I.; Mehmood, N. On a novel uncertain soft set model: Z-soft fuzzy rough set model and corresponding decision making methods. *Appl. Soft Comput.* **2017**, *56*, 446–457.
22. Khalil, A.M.; Hassan, N. A novel approach to multi attribute group decision-making based on trapezoidal interval type-2 fuzzy soft sets. *Appl. Math. Model.* **2016**, *41*, 684–690.
23. Zhang, G.; Li, Z.; Qin, B. A method for multi-attribute decision-making applying soft rough sets. *J. Intell. Fuzzy Syst.* **2016**, *30*, 1803–1815.
24. Yu, G. An algorithm for multi-attribute decision-making based on soft rough sets. *J. Comput. Anal. Appl.* **2016**, *20*, 1248–1258.
25. Ali, M.I.; Feng, F.; Liu, X.Y.; Min, W.K.; Shabir, M. On some new operations in soft set theory. *Comput. Math. Appl.* **2009**, *57*, 1547–1553.
26. Pei, D.; Miao, D. From soft sets to information systems. In Proceedings of the 2005 IEEE International Conference on Granular Computing, Beijing, China, 25–27 July 2005; pp. 617–621.
27. Ali, M.I. A note on soft sets, rough soft sets and fuzzy soft sets. *Appl. Soft Comput.* **2011**, *11*, 3329–3332.
28. Wu, W.Z.; Mi, J.S.; Zhang, W.X. Generalized fuzzy rough sets. *Inf. Sci.* **2003**, *151*, 263–282.
29. Yao, Y.Y. Relational interpretations of neighborhood operators and rough set approximation operators. *Inf. Sci.* **1998**, *111*, 239–259.
30. Zhu, W. Relationship between generalized rough sets based on binary relation and covering. *Inf. Sci.* **2009**, *179*, 210–225.
31. Feng, F. Soft rough sets applied to multicriteria group decision-making. *Ann. Fuzzy Math. Inf.* **2011**, *2*, 69–80.
32. Sun, B.; Ma, W. Soft fuzzy rough sets and its application in decision-making. *Artif. Intell. Rev.* **2014**, *41*, 67–80.
33. Wille, R. Restructuring lattice theory: An approach based on hierarchies of concepts. In *Ordered Sets*; Rival, I., Ed.; Springer: Dordrecht, The Netherlands; Reidel Dordrecht: Boston, MA, USA, 1982; pp. 445–470.
34. Benitez-Caballero, M.J.; Medina, J.; Ramirez-Poussa, E. Attribute Reduction in Rough Set Theory and Formal Concept Analysis. In *Rough Sets*; IJCRS 2017; Polkowski, L., Yao, Y., Artiermjew, P., Ciucci, D., Liu, D., Ślęzak, D., Zielo sko, B., Eds.; Springer: Cham, Switzerland; Olsztyn, Poland, 2017; pp. 513–525.

35. Li, J.; Mei, C.; Xu, W.; Qian, Y. Concept learning via granular computing: A cognitive viewpoint. *Inf. Sci.* **2015**, *298*, 447–467.
36. Li, Z.; Xie, T. Roughness of fuzzy soft sets and related results. *Int. J. Comput. Intell. Syst.* **2015**, *8*, 278–296.
37. Basu, K.; Deb, R.; Pattanaik, P.K. Soft sets: An ordinal formulation of vagueness with some applications to the theory of choice. *Fuzzy Sets Syst.* **1992**, *45*, 45–58.

symmetry

MDPI

Article

Decomposition and Intersection of Two Fuzzy Numbers for Fuzzy Preference Relations

Hui-Chin Tang

Department of Industrial Engineering and Management, National Kaohsiung University of Applied Sciences, Kaohsiung 80778, Taiwan; tang@kuas.edu.tw

Received: 11 September 2017; Accepted: 9 October 2017; Published: 14 October 2017

Abstract: In fuzzy decision problems, the ordering of fuzzy numbers is the basic problem. The fuzzy preference relation is the reasonable representation of preference relations by a fuzzy membership function. This paper studies Nakamura's and Kołodziejczyk's preference relations. Eight cases, each representing different levels of overlap between two triangular fuzzy numbers are considered. We analyze the ranking behaviors of all possible combinations of the decomposition and intersection of two fuzzy numbers through eight extensive test cases. The results indicate that decomposition and intersection can affect the fuzzy preference relations, and thereby the final ranking of fuzzy numbers.

Keywords: fuzzy number; ranking; preference relations

1. Introduction

For solving decision-making problems in a fuzzy environment, the overall utilities of a set of alternatives are represented by fuzzy sets or fuzzy numbers. A fundamental problem of a decision-making procedure involves ranking a set of fuzzy sets or fuzzy numbers. Ranking functions, reference sets and preference relations are three categories with which to rank a set of fuzzy numbers. For a detailed discussion, we refer the reader to surveys by Chen and Hwang [1] and Wang and Kerre [2,3]. For ranking a set of fuzzy numbers, this paper concentrates on those fuzzy preference relations that are able to represent preference relations in linguistic or fuzzy terms and to make pairwise comparisons. To propose the fuzzy preference relation, Nakamura [4] employed a fuzzy minimum operation followed by the Hamming distance. Kołodziejczyk [5] considered the common part of two membership functions and used the fuzzy maximum and Hamming distance. Yuan [6] compared the fuzzy subtraction of two fuzzy numbers with real number zero and indicated that the desirable properties of a fuzzy ranking method are the fuzzy preference presentation, rationality of fuzzy ordering, distinguishability and robustness. Li [7] included the influence of levels of possibility of dominance. Lee [8] presented a counterexample to Li's method [7] and proposed an additional comparable property. The methods of Wang et al. [9] and Asady [10] were based on deviation degree. Zhang et al. [11] presented a fuzzy probabilistic preference relation. Zhu et al. [12] proposed hesitant fuzzy preference relations. Wang [13] adopted the relative preference degrees of the fuzzy numbers over average.

This paper evaluates and compares two fundamental fuzzy preference relations—one is proposed by Nakamura [4] and the other by Kołodziejczyk [5]. The intersection of two membership functions and the decomposition of two fuzzy numbers are main differences between these two preference relations. Since the desirable criteria cannot easily be represented in mathematical forms, their performance measures are often tested by using test examples and judged intuitively. To this end, we consider eight complex cases that represent all the possible cases the way two fuzzy numbers can overlap with each other. For Nakamura's and Kołodziejczyk's fuzzy preference relations, this paper analyzes and compares the ordering behaviors of the decomposition and intersection through a group of extensive cases.

The organization of this paper is as follows—Section 2 briefly reviews the fuzzy sets and fuzzy preference relations and presents the eight test cases. Section 3 analyzes Nakamura's fuzzy preference relation and presents an algorithm. Section 4 presents the behaviors of Kołodziejczyk's fuzzy preference relation. Section 5 analyzes the effect of the decomposition and intersection on fuzzy preference relations. Finally, some concluding remarks and suggestions for future research are presented.

2. Fuzzy Sets and Test Problems

We first review the basic notations of fuzzy sets and fuzzy preference relations. Consider a fuzzy set A defined by a universal set of real numbers $\mathcal{R}$ by the membership function $A(x)$, where $A(x): \quad \Re \rightarrow [0, \quad 1]$.

Definition 1. *Let A be a fuzzy set. The support of A is the crisp set* $S_A = \{x \in \Re | \, A(x) > 0\}$. *A is called normal when* $\sup_{x \in S_A} A(x) = 1$. *An* α*-cut of A is a crisp set* $A_\alpha = \{x \in \Re | A(x) \geq \alpha\}$. *A is convex if, and only if, each of its* α*-cut is a convex set.*

Definition 2. *A normal and convex fuzzy set whose membership function is piecewise continuous is called a fuzzy number.*

Definition 3. *A triangular fuzzy number A, denoted* $A = (a, \quad b, \quad c)$, *is a fuzzy number with membership function given by:*

$$A(x) = \begin{cases} \frac{x-a}{b-a} & \text{if } a \leq x \leq b \\ \frac{c-x}{c-b} & \text{if } b \leq x \leq c \\ 0 & \text{otherwise} \end{cases}$$

where $-\infty < a \leq b \leq c < \infty$. *The set of all triangular fuzzy numbers on* $\mathcal{R}$ *is denoted by* $\mathrm{TF}(\Re)$.

Definition 4. *For a fuzzy number A, the upper boundary set* $\overline{A}$ *of A and the lower boundary set* $\underline{A}$ *of A are respectively defined as:*

$$\overline{A}(x) = \sup_{y \geq x} A(y)$$

and:

$$\underline{A}(x) = \sup_{y \leq x} A(y).$$

Definition 5. *The Hamming distance between two fuzzy numbers A and B is defined by:*

$$\mathrm{d}(A, B) = \int_R |A(x) - B(x)| dx$$

$$= \int_{A(x) \geq B(x)} A(x) - B(x) dx + \int_{B(x) \geq A(x)} B(x) - A(x) dx.$$

Definition 6. *Let A and B be two fuzzy numbers and* $\times$ *be an operation on* $\mathcal{R}$, *such as* $+, -, *, \div \ldots$. *By extension principle, the extended operation* $\otimes$ *on fuzzy numbers can be defined by:*

$$\mu_{A \otimes B}(z) = \sup_{x,y: z = x \times y} \min\{A(x), \quad B(y)\}.$$

Definition 7. *A fuzzy preference relation R is a fuzzy binary relation with membership function* $R(A, B)$ *indicating the degree of preference of fuzzy number A over fuzzy number B.*

1. R is reciprocal if, and only if, $R(A,B) = 1 - R(B,A)$ for all fuzzy numbers A and B.
2. R is transitive if, and only if, $R(A,B) \geq 0.5$ and $R(B,C) \geq 0.5$ implies $R(A,C) \geq 0.5$ for all fuzzy numbers A, B and C.
3. R is a fuzzy total ordering if, and only if, R is both reciprocal and transitive.
4. R is robust if, and only if, for any given fuzzy numbers A, B and $\varepsilon > 0$, there exists $\delta > 0$ for which $|R(A, B) - R(A', B)| < \varepsilon$, for all fuzzy number A' and $\max_{\alpha>0}(|\inf A_\alpha - \inf B_\alpha|, |\sup A_\alpha - \sup B_\alpha|) < \delta$.

For simplicity, we denote $R'(A,B)$ for the degree of preference of fuzzy number B over fuzzy number A.

The evaluation criteria for the comparison of two fuzzy numbers cannot easily be represented in mathematical forms therefore it is often tested on a group of selected examples. The membership functions of two fuzzy numbers can be overlapping/nonovelapping, convex/nonconvex, and normal/non-normal. All the approaches proposed in the literature seem to suffer from some questionable examples, especially for the portion of overlap between two membership functions.

Let $A(a_1, b_1, c_1)$ and $B(a_2, b_2, c_2)$ be two triangular fuzzy numbers. Figure 1 displays eight test cases of representing all the possible cases the way two fuzzy numbers A and B can overlap with each other. Table 1 shows the area Q_i of i-th region in each case. More precisely, the eight extensive test cases are as follows:

Case 1. $a_1 \leq a_2, b_1 \leq b_2, c_1 \leq c_2$.
Case 2. $a_1 \leq a_2, b_1 \geq b_2, c_1 \leq c_2$.
Case 3. $a_1 \leq a_2, b_1 \leq b_2, c_1 \geq c_2$.
Case 4. $a_1 \leq a_2, b_1 \geq b_2, c_1 \geq c_2$.
Case 5. $a_1 \geq a_2, b_1 \leq b_2, c_1 \leq c_2$.
Case 6. $a_1 \geq a_2, b_1 \geq b_2, c_1 \leq c_2$.
Case 7. $a_1 \geq a_2, b_1 \leq b_2, c_1 \geq c_2$.
Case 8. $a_1 \geq a_2, b_1 \geq b_2, c_1 \geq c_2$.

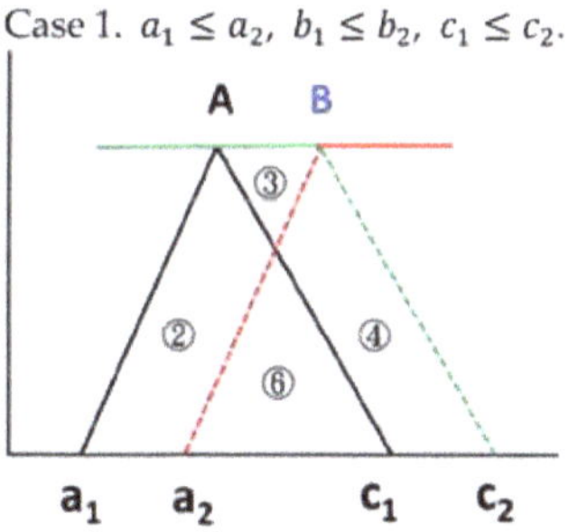

Figure 1. *Cont.*

Case 2. $a_1 \le a_2,\ b_1 \ge b_2,\ c_1 \le c_2$

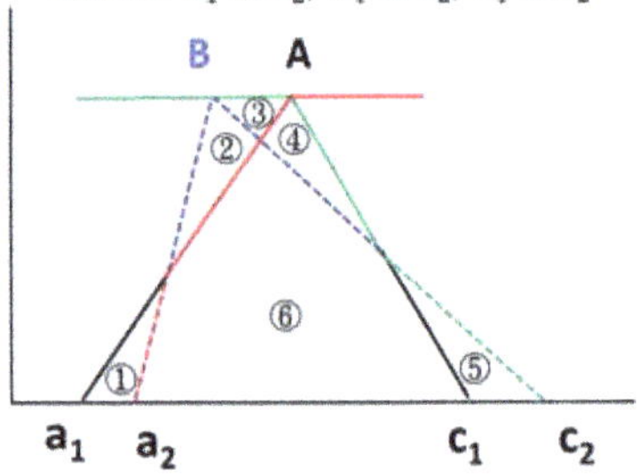

Case 3. $a_1 \le a_2,\ b_1 \le b_2,\ c_1 \ge c_2$.

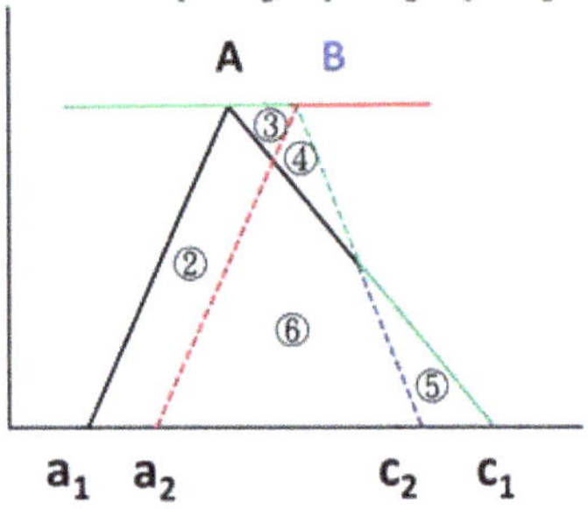

Case 4. $a_1 \le a_2,\ b_1 \ge b_2,\ c_1 \ge c_2$.

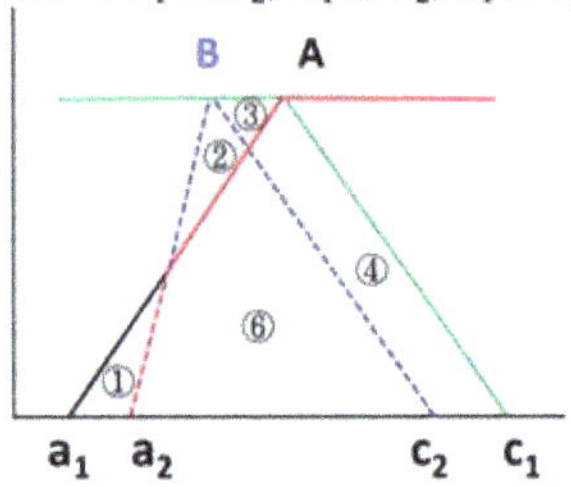

Case 5. $a_1 \ge a_2,\ b_1 \le b_2,\ c_1 \le c_2$.

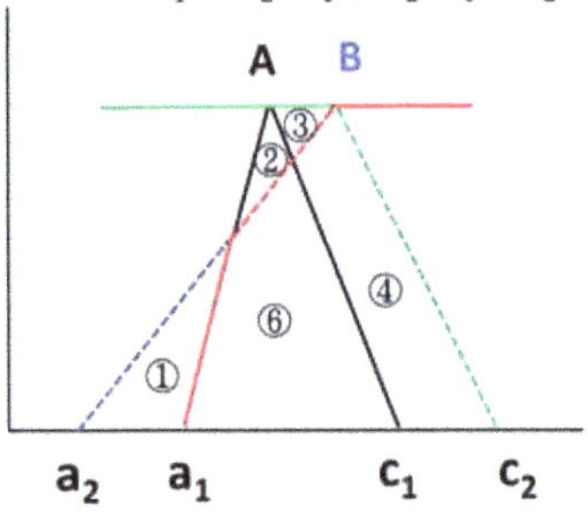

Figure 1. *Cont.*

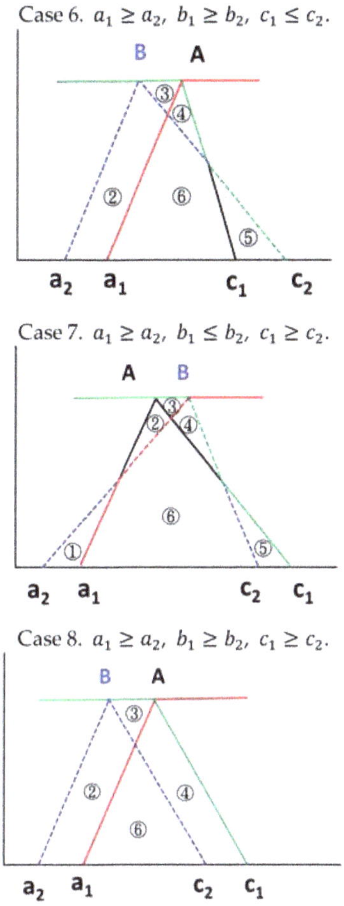

Figure 1. Eight test cases for two fuzzy numbers $A(a_1, b_1, c_1)$ and $B(a_2, b_2, c_2)$.

Table 1. The area Q_i of i-th region for eight cases.

Case	Area
1	$Q_3 = \int_{(c_1-a_2)/(c_1-b_1+b_2-a_2)}^{1} -(c_1 - a_2 - (c_1 - b_1 + b_2 - a_2)\alpha)d\alpha$ $Q_2 = \int_0^1 a_2 - a_1 + (b_2 - a_2 - b_1 + a_1)\alpha d\alpha - Q_3$ $Q_4 = \int_0^1 c_2 - c_1 - (c_2 - b_2 - c_1 + b_1)\alpha d\alpha - Q_3$ $Q_6 = \int_0^{(c_1-a_2)/(c_1-b_1+b_2-a_2)} (c_1 - a_2 - (c_1 - b_1 + b_2 - a_2)\alpha d\alpha$
2	$Q_1 = \int_0^{(-a_2+a_1)/(b_2-a_2-b_1+a_1)} a_2 - a_1 + (b_2 - a_2 - b_1 + a_1)\alpha d\alpha$ $Q_3 = \int_{(c_2-a_1)/(c_2-b_2+b_1-a_1)}^{1} -(c_2 - a_1 - (c_2 - b_2 + b_1 - a_1)\alpha)d\alpha$ $Q_2 = \int_{(-a_1+a_2)/(b_1-a_1-b_2+a_2)}^{1} a_1 - a_2 + (b_1 - a_1 - b_2 + a_2)\alpha d\alpha - Q_3$ $Q_4 = \int_{(c_1-c_2)/(c_1-b_1-c_2+b_2)}^{1} c_1 - c_2 - (c_1 - b_1 - c_2 + b_2)\alpha d\alpha - Q_3$ $Q_5 = \int_0^{(c_2-c_1)/(c_2-b_2-c_1+b_1)} c_2 - c_1 - (c_2 - b_2 - c_1 + b_1)\alpha d\alpha$ $Q_6 = \int_0^{(c_2-a_1)/(c_2-b_2+b_1-a_1)} (c_2 - a_1 - (c_2 - b_2 + b_1 - a_1)\alpha)d\alpha - Q_1 - Q_5$

Table 1. *Cont.*

Case	Area
3	$Q_3 = \int_{(c_1-a_2)/(c_1-b_1+b_2-a_2)}^{1} -(c_1 - a_2 - (c_1 - b_1 + b_2 - a_2)\alpha)d\alpha$ $Q_2 = \int_0^1 a_2 - a_1 + (b_2 - a_2 - b_1 + a_1)\alpha d\alpha - Q_3$ $Q_4 = \int_{(c_2-c_1)/(c_2-b_2-c_1+b_1)}^{1} c_2 - c_1 - (c_2 - b_2 - c_1 + b_1)\alpha d\alpha - Q_3$ $Q_5 = \int_0^{(c_1-c_2)/(c_1-b_1-c_2+b_2)} c_1 - c_2 - (c_1 - b_1 - c_2 + b_2)\alpha d\alpha$ $Q_6 = \int_0^{(c_1-a_2)/(c_1-b_1+b_2-a_2)} (c_1 - a_2 - (c_1 - b_1 + b_2 - a_2)\alpha d\alpha - Q_5$
4	$Q_1 = \int_0^{(-a_2+a_1)/(b_2-a_2-b_1+a_1)} a_2 - a_1 + (b_2 - a_2 - b_1 + a_1)\alpha d\alpha$ $Q_3 = \int_{(c_2-a_1)/(c_2-b_2+b_1-a_1)}^{1} -(c_2 - a_1 - (c_2 - b_2 + b_1 - a_1)\alpha)d\alpha$ $Q_2 = \int_{(-a_1+a_2)/(b_1-a_1-b_2+a_2)}^{1} a_1 - a_2 + (b_1 - a_1 - b_2 + a_2)\alpha d\alpha - Q_3$ $Q_4 = \int_0^1 c_1 - c_2 - (c_1 - b_1 - c_2 + b_2)\alpha d\alpha - Q_3$ $Q_6 = \int_0^{(c_2-a_1)/(c_2-b_2+b_1-a_1)} (c_2 - a_1 - (c_2 - b_2 + b_1 - a_1)\alpha)d\alpha - Q_1$
5	$Q_1 = \int_0^{(-a_1+a_2)/(b_1-a_1-b_2+a_2)} a_1 - a_2 + (b_1 - a_1 - b_2 + a_2)\alpha d\alpha$ $Q_3 = \int_{(c_1-a_2)/(c_1-b_1+b_2-a_2)}^{1} -(c_1 - a_2 - (c_1 - b_1 + b_2 - a_2)\alpha)d\alpha$ $Q_2 = \int_{(-a_2+a_1)/(b_2-a_2-b_1+a_1)}^{1} a_2 - a_1 + (b_2 - a_2 - b_1 + a_1)\alpha d\alpha - Q_3$ $Q_4 = \int_0^1 c_2 - c_1 - (c_2 - b_2 - c_1 + b_1)\alpha d\alpha - Q_3$ $Q_6 = \int_0^{(c_1-a_2)/(c_1-b_1+b_2-a_2)} (c_1 - a_2 - (c_1 - b_1 + b_2 - a_2)\alpha d\alpha - Q_1$
6	$Q_3 = \int_{(c_2-a_1)/(c_2-b_2+b_1-a_1)}^{1} -(c_2 - a_1 - (c_2 - b_2 + b_1 - a_1)\alpha)d\alpha$ $Q_2 = \int_0^1 a_1 - a_2 + (b_1 - a_1 - b_2 + a_2)\alpha d\alpha - Q_3$ $Q_4 = \int_{(c_1-c_2)/(c_1-b_1-c_2+b_2)}^{1} c_1 - c_2 - (c_1 - b_1 - c_2 + b_2)\alpha d\alpha - Q_3$ $Q_5 = \int_0^{(c_2-c_1)/(c_2-b_2-c_1+b_1)} c_2 - c_1 - (c_2 - b_2 - c_1 + b_1)\alpha d\alpha$ $Q_6 = \int_0^{(c_2-a_1)/(c_2-b_2+b_1-a_1)} (c_2 - a_1 - (c_2 - b_2 + b_1 - a_1)\alpha)d\alpha - Q_5$
7	$Q_1 = \int_0^{(-a_1+a_2)/(b_1-a_1-b_2+a_2)} a_1 - a_2 + (b_1 - a_1 - b_2 + a_2)\alpha d\alpha$ $Q_3 = \int_{(c_1-a_2)/(c_1-b_1+b_2-a_2)}^{1} -(c_1 - a_2 - (c_1 - b_1 + b_2 - a_2)\alpha)d\alpha$ $Q_2 = \int_{(-a_2+a_1)/(b_2-a_2-b_1+a_1)}^{1} a_2 - a_1 + (b_2 - a_2 - b_1 + a_1)\alpha d\alpha - Q_3$ $Q_4 = \int_{(c_2-c_1)/(c_2-b_2-c_1+b_1)}^{1} c_2 - c_1 - (c_2 - b_2 - c_1 + b_1)\alpha d\alpha - Q_3$ $Q_5 = \int_0^{(c_1-c_2)/(c_1-b_1-c_2+b_2)} c_1 - c_2 - (c_1 - b_1 - c_2 + b_2)\alpha d\alpha$ $Q_6 = \int_0^{(c_1-a_2)/(c_1-b_1+b_2-a_2)} (c_1 - a_2 - (c_1 - b_1 + b_2 - a_2)\alpha d\alpha - Q_1 - Q_5$
8	$Q_3 = \int_{(c_2-a_1)/(c_2-b_2+b_1-a_1)}^{1} -(c_2 - a_1 - (c_2 - b_2 + b_1 - a_1)\alpha)d\alpha$ $Q_2 = \int_0^1 a_1 - a_2 + (b_1 - a_1 - b_2 + a_2)\alpha d\alpha - Q_3$ $Q_4 = \int_0^1 c_1 - c_2 - (c_1 - b_1 - c_2 + b_2)\alpha d\alpha - Q_3$ $Q_6 = \int_0^{(c_2-a_1)/(c_2-b_2+b_1-a_1)} (c_2 - a_1 - (c_2 - b_2 + b_1 - a_1)\alpha)d\alpha$

3. Nakamura's Fuzzy Preference Relation

Using fuzzy minimum, fuzzy maximum, and Hamming distance, Nakamura's fuzzy preference relations [4] are defined as follows:

Definition 8. *For two fuzzy numbers A and B, Nakamura* [4] *defines* $N(A,B)$ *and* $N'(A,B)$ *as fuzzy preference relations by the following membership functions:*

$$N(A,B) = \frac{d\left(\underline{A}, \widetilde{min}(\underline{A}, \underline{B})\right) + d\left(\overline{A}, \widetilde{min}(\overline{A}, \overline{B})\right)}{d(\underline{A}, \underline{B}) + d(\overline{A}, \overline{B})}$$

and:

$$N'(A,B) = \frac{d(A \cap B, 0) + d(A, \widetilde{max}(A,B))}{d(A,0) + d(B,0)}$$

respectively. Yuan [6] *showed that $N(A,B)$ is reciprocal and transitive, but not robust. Wang and Kerre* [3] *derived that:*

$$d\left(\underline{A}, \widetilde{min}(\underline{A}, \underline{B})\right) = d(\underline{B}, \widetilde{max}(\underline{A}, \underline{B}))$$
$$d(\overline{A}, \widetilde{max}(\overline{A}, \overline{B})) = d\left(\overline{B}, \widetilde{min}(\overline{A}, \overline{B})\right)$$
$$d\left(\underline{A}, \widetilde{min}(\underline{A}, \underline{B})\right) + d(\underline{A}, \widetilde{max}(\underline{A}, \underline{B})) = d(\underline{A}, \underline{B})$$
$$d\left(\overline{A}, \widetilde{min}(\overline{A}, \overline{B})\right) + d(\overline{A}, \widetilde{max}(\overline{A}, \overline{B})) = d(\overline{A}, \overline{B})$$

and:

$$2d(A \cap B, 0) + d(A, \widetilde{max}(A,B)) + d(B, \widetilde{max}(A,B)) = d(A,0) + d(B,0).$$

It follows that:

$$N(A,B) + N(B,A) = 1$$

and:

$$N'(A,B) + N'(B,A) = 1.$$

For two triangular fuzzy numbers $A(a_1, b_1, c_1)$ and $B(a_2, b_2, c_2)$, then:

$$A_\alpha = [L_1, U_1] = [a_1 + (b_1 - a_1)\alpha,\ c_1 - (c_1 - b_1)\alpha]$$
$$B_\alpha = [L_2,\ U_2] = [a_2 + (b_2 - a_2)\alpha,\ c_2 - (c_2 - b_2)\alpha]$$

so:

$$N(A,B) = \frac{d\left(\underline{A}, \widetilde{min}(\underline{A}, \underline{B})\right) + d\left(\overline{A}, \widetilde{min}\left(\overline{A}, \overline{B}\right)\right)}{d(\underline{A}, \underline{B}) + d\left(\overline{A}, \overline{B}\right)}$$
$$= \frac{\int_{L_1 \geq L_2} L_1 - L_2 d\alpha + \int_{U_1 \geq U_2} U_1 - U_2 d\alpha}{\int_{L_1 \geq L_2} L_1 - L_2 d\alpha + \int_{L_2 \geq L_1} L_2 - L_1 d\alpha + \int_{U_1 \geq U_2} U_1 - U_2 d\alpha + \int_{U_2 \geq U_1} U_2 - U_1 d\alpha}.$$

Define:

$$S_1 = \int_{L_1 \geq L_2} L_1 - L_2 d\alpha = \int_{a_1 - a_2 + (b_1 - a_1 - b_2 + a_2)\alpha \geq 0} a_1 - a_2 + (b_1 - a_1 - b_2 + a_2)\alpha d\alpha$$
$$S_2 = \int_{L_2 \geq L_1} L_2 - L_1 d\alpha = \int_{a_2 - a_1 + (b_2 - a_2 - b_1 + a_1)\alpha \geq 0} a_2 - a_1 + (b_2 - a_2 - b_1 + a_1)\alpha d\alpha$$
$$S_3 = \int_{U_1 \geq U_2} U_1 - U_2 d\alpha = \int_{c_1 - c_2 - (c_1 - b_1 - c_2 + b_2)\alpha \geq 0} c_1 - c_2 - (c_1 - b_1 - c_2 + b_2)\alpha d\alpha$$
$$S_4 = \int_{U_2 \geq U_1} U_2 - U_1 d\alpha = \int_{c_2 - c_1 - (c_2 - b_2 - c_1 + b_1)\alpha \geq 0} c_2 - c_1 - (c_2 - b_2 - c_1 + b_1)\alpha d\alpha,$$

then:

$$N(A,B) = \frac{S_1 + S_3}{S_1 + S_2 + S_3 + S_4}.$$

Let $\mathcal{A} = a_2 - a_1$, $\mathcal{B} = b_2 - b_1$ and $\mathcal{C} = c_2 - c_1$. The steps for implementing the Nakamura's fuzzy preference relation $N(A,B)$ are as in Algorithm 1:

Algorithm 1. Nakamura's fuzzy preference relation

If $\mathcal{A} \geq 0$

  If $\mathcal{C} \geq 0$

    If $\mathcal{B} \geq 0$, then $N(A,B) = 0$ else $N(A,B) = \frac{\mathcal{B}^2(\mathcal{A}-2\mathcal{B}+\mathcal{C})}{(\mathcal{A}^2+\mathcal{B}^2)(-\mathcal{B}+\mathcal{C})+(\mathcal{B}^2+\mathcal{C}^2)(\mathcal{A}-\mathcal{B})}$.

  else if $\mathcal{B} \geq 0$, then $N(A,B) = \frac{\mathcal{C}^2}{(\mathcal{A}+\mathcal{B})(\mathcal{B}-\mathcal{C})+\mathcal{B}^2+\mathcal{C}^2}$ else

    $N(A,B) = 1 - \frac{\mathcal{A}^2}{(\mathcal{A}-\mathcal{B})(-\mathcal{B}-\mathcal{C})+\mathcal{A}^2+\mathcal{B}^2}$.

else if $\mathcal{C} \geq 0$

  If $\mathcal{B} \geq 0$, then $N(A,B) = \frac{\mathcal{A}^2}{(-\mathcal{A}+\mathcal{B})(\mathcal{B}+\mathcal{C})+\mathcal{A}^2+\mathcal{B}^2}$ else

    $N(A,B) = 1 - \frac{\mathcal{C}^2}{(\mathcal{A}+\mathcal{B})(\mathcal{B}-\mathcal{C})+\mathcal{B}^2+\mathcal{C}^2}$.

  else if $\mathcal{B} \geq 0$, then $N(A,B) = 1 - \frac{\mathcal{B}^2(\mathcal{A}-2\mathcal{B}+\mathcal{C})}{(\mathcal{A}^2+\mathcal{B}^2)(-\mathcal{B}+\mathcal{C})+(\mathcal{B}^2+\mathcal{C}^2)(\mathcal{A}-\mathcal{B})}$ else $N(A,B) = 1$.

Table 2 shows the values of $N(A,B)$ and $N'(A,B)$ for each test case. The first observation of this table is that:

$$N_1(A,B) + N_8(A,B) = 1$$
$$N_2(A,B) + N_7(A,B) = 1$$
$$N_3(A,B) + N_6(A,B) = 1$$
$$N_4(A,B) + N_5(A,B) = 1.$$

Secondly, comparing the values of $N(A,B)$ with that of $N'(A,B)$, we have that $1 - N_1'(A,B) \geq N_1(A,B)$ and $1 - N_8'(A,B) \leq N_8(A,B)$. If $a_2 + 2b_2 + c_2 \geq a_1 - 2b_1 - c_1$, we obtain that $1 - N_2'(A,B) \leq N_2(A,B)$, $1 - N_3'(A,B) \geq N_3(A,B)$, $1 - N_4'(A,B) \leq N_4(A,B)$, $1 - N_5'(A,B) \geq N_5(A,B)$, $1 - N_6'(A,B) \leq N_6(A,B)$ and $1 - N_7'(A,B) \geq N_7(A,B)$.

Table 2. $N(A,B)$ and $N'(A,B)$ for eight cases.

Case	$N(A,B)$	$N'(A,B)$
1	0	$1+\frac{(a_2-c_1)^2}{(a_2+b_1-b_2-c_1)(-a_1-a_2+c_1+c_2)}$
2	$\frac{(b_2-b_1)^2(a_2-a_1-2(b_2-b_1)+(c_2-c_1))}{((a_2-a_1)^2+(b_2-b_1)^2)(b_1-b_2+c_2-c_1)+((b_2-b_1)^2+(c_2-c_1)^2)(a_2-a_1-b_2+b_1)}$	$\frac{(a_1-c_2)^2}{(a_1-b_1+b_2-c_2)(a_1+a_2-c_1-c_2)}$
3	$\frac{(c_2-c_1)^2}{(a_2-a_1+b_2-b_1)(b_2-b_1-c_2+c_1)+(b_2-b_1)^2+(c_2-c_1)^2}$	$1+\frac{(a_2-c_1)^2}{(a_2+b_1-b_2-c_1)(-a_1-a_2+c_1+c_2)}$
4	$1-\frac{(a_2-a_1)^2}{(a_2-a_1-b_2+b_1)(-b_2+b_1-c_2+c_1)+(a_2-a_1)^2+(b_2-b_1)^2}$	$\frac{(a_1-c_2)^2}{(a_1-b_1+b_2-c_2)(a_1+a_2-c_1-c_2)}$
5	$\frac{(a_2-a_1)^2}{(a_2-a_1-b_2+b_1)(-b_2+b_1-c_2+c_1)+(a_2-a_1)^2+(b_2-b_1)^2}$	$1+\frac{(a_2-c_1)^2}{(a_2+b_1-b_2-c_1)(-a_1-a_2+c_1+c_2)}$
6	$1-\frac{(c_2-c_1)^2}{(a_2-a_1+b_2-b_1)(b_2-b_1-c_2+c_1)+(b_2-b_1)^2+(c_2-c_1)^2}$	$\frac{(a_1-c_2)^2}{(a_1-b_1+b_2-c_2)(a_1+a_2-c_1-c_2)}$
7	$1-\frac{(b_2-b_1)^2(a_2-a_1-2b_2+2b_1+c_2-c_1)}{((a_2-a_1)^2+(b_2-b_1)^2)(b_1-b_2+c_2-c_1)+((b_2-b_1)^2+(c_2-c_1)^2)(a_2-a_1-b_2+b_1)}$	$1+\frac{(a_2-c_1)^2}{(a_2+b_1-b_2-c_1)(-a_1-a_2+c_1+c_2)}$
8	1	$\frac{(a_1-c_2)^2}{(a_1-b_1+b_2-c_2)(a_1+a_2-c_1-c_2)}$

4. Kołodziejczyk's Fuzzy Preference Relation

By considering the common part of two membership functions, Kołodziejczyk's method [5] is based on fuzzy maximum and Hamming distance to propose the following fuzzy preference relations:

Definition 9. *For two fuzzy numbers A and B, Kołodziejczyk [5] defines $K1'(A,B)$ and $K2'(A,B)$ as fuzzy preference relations by the following membership functions:*

$$K1'(A,B) = \frac{d(\underline{A}, \widetilde{max}(\underline{A},\underline{B})) + d(\overline{A}, \widetilde{max}(\overline{A},\overline{B})) + d(A \cap B, 0)}{d(\underline{A},\underline{B}) + d(\overline{A},\overline{B}) + 2d(A \cap B, 0)}$$

and:

$$K2'(A,B) = \frac{d(\underline{A},\widetilde{\max}(\underline{A},\underline{B})) + d(\overline{A},\widetilde{\max}(\overline{A},\overline{B}))}{d(\underline{A},\underline{B}) + d(\overline{A},\overline{B})}$$

respectively. $K1'(A,B)$ *is reciprocal, transitive and robust* [3,5]. *Since:*

$$K2'(A,B) = 1 - N(A,B)$$

the results of $K2'(A,B)$ *can be obtained from those of* $N(A,B)$.

For two triangular fuzzy numbers $A(a_1,b_1,c_1)$ and $B(a_2,b_2,c_2)$, then:

$$A_\alpha = [L_1,U_1] = [a_1 + (b_1 - a_1)\alpha,\ c_1 - (c_1 - b_1)\alpha]$$
$$B_\alpha = [L_2,\ U_2] = [a_2 + (b_2 - a_2)\alpha,\ c_2 - (c_2 - b_2)\alpha].$$

Define:

$$S_1 = d(\underline{A},\widetilde{max}(\underline{A},\underline{B})) = \int_{L_2 \geq L_1} L_2 - L_1 d\alpha$$
$$S_2 = \int_{L_1 \geq L_2} L_1 - L_2 d\alpha$$
$$d(\underline{A},\underline{B}) = S_1 + S_2$$
$$S_3 = d(\overline{A},\widetilde{max}(\overline{A},\overline{B})) = \int_{U_2 \geq U_1} U_2 - U_1 d\alpha$$
$$S_4 = \int_{U_1 \geq U_2} U_1 - U_2 d\alpha$$
$$d(\overline{A},\overline{B}) = S_3 + S_4$$

and:

$$S_5 = d(A \cap B, 0) = \int_{U_1 \geq L_2} U_1 - L_2 d\alpha - \int_{U_1 \geq U_2} U_1 - U_2 d\alpha - \int_{L_1 \geq L_2} L_1 - L_2 d\alpha.$$

Then:

$$K1'(A,B) = \frac{S_1 + S_3 + S_5}{S_1 + S_2 + S_3 + S_4 + 2S_5}$$

and:

$$K2'(A,B) = \frac{S_1 + S_3}{S_1 + S_2 + S_3 + S_4}.$$

In Table 3, we display the values of $K1'(A,B)$ and $K2'(A,B)$ for each test case. An examination of the table reveals that:

$$\begin{aligned} K1'_1(A,B) = K1'_3(A,B) &= K1'_5(A,B) = K1'_7(A,B) \\ &= 1 - \frac{(c_1-a_2)^2}{(c_1-a_2+c_2-a_1)(c_1-a_2-b_1+b_2)+2(b_2-b_1)^2} \end{aligned}$$

and:

$$K1'_2(A,B) = K1'_4(A,B) = K1'_6(A,B) = K1'_8(A,B) = \frac{(c_2-a_1)^2}{(c_2-a_1)^2+(c_2-a_1-b_2+b_1)(c_1-a_2-b_2+b_1)+(b_2-b_1)^2}.$$

If $b_1 = b_2$, we have:

$$K1'_1(A,B) = 1 - \frac{c_1 - a_2}{(c_1 - a_2 + c_2 - a_1)} = \frac{c_2 - a_1}{(c_1 - a_2 + c_2 - a_1)}$$

and:

$$K1'_2(A,B) = \frac{c_2 - a_1}{(c_1 - a_2 + c_2 - a_1)}$$

so:

$$K1'_1(A,B) = K1'_2(A,B)$$

and:

$$K1'_1(A,B) + K1'_2(A,B) = \frac{2(c_2 - a_1)}{(c_1 - a_2 + c_2 - a_1)}.$$

It follows that:

$$K1'_1(A,B) + K1'_2(A,B) = 0 \text{ for } b_1 = b_2 \text{ and } c_2 = a_1.$$

and:

$$K1'_1(A,B) + K1'_2(A,B) = 1 \text{ for } b_1 = b_2 \text{ and } c_1 - a_2 = c_2 - a_1.$$

Table 3. $K1'(A,B)$ and $K2'(A,B)$ for eight cases.

Case	$K1'(A,B)$	$K2'(A,B)$
1	$1 - \frac{(c_1-a_2)^2}{(c_1-a_2+c_2-a_1)(c_1-a_2-b_1+b_2)+2(b_2-b_1)^2}$	1
2	$\frac{(c_2-a_1)^2}{(c_2-a_1)^2+(c_2-a_1-b_2+b_1)(c_1-a_2-b_2+b_1)+(b_2-b_1)^2}$	$1 - \frac{(b_2-b_1)^2(a_2-a_1-2(b_2-b_1)+(c_2-c_1))}{((a_2-a_1)^2+(b_2-b_1)^2)(b_1-b_2+c_2-c_1)+((b_2-b_1)^2+(c_2-c_1)^2)(a_2-a_1-b_2+b_1)}$
3	$1 - \frac{(c_1-a_2)^2}{(c_1-a_2+c_2-a_1)(c_1-a_2-b_1+b_2)+2(b_2-b_1)^2}$	$1 - \frac{(c_2-c_1)^2}{(a_2-a_1+b_2-b_1)(b_2-b_1-c_2+c_1)+(b_2-b_1)^2+(c_2-c_1)^2}$
4	$\frac{(c_2-a_1)^2}{(c_2-a_1)^2+(c_2-a_1-b_2+b_1)(c_1-a_2-b_2+b_1)+(b_2-b_1)^2}$	$\frac{(a_2-a_1)^2}{(a_2-a_1-b_2+b_1)(-b_2+b_1-c_2+c_1)+(a_2-a_1)^2+(b_2-b_1)^2}$
5	$1 - \frac{(c_1-a_2)^2}{(c_1-a_2+c_2-a_1)(c_1-a_2-b_1+b_2)+2(b_2-b_1)^2}$	$1 - \frac{(a_2-a_1)^2}{(a_2-a_1-b_2+b_1)(-b_2+b_1-c_2+c_1)+(a_2-a_1)^2+(b_2-b_1)^2}$
6	$\frac{(c_2-a_1)^2}{(c_2-a_1)^2+(c_2-a_1-b_2+b_1)(c_1-a_2-b_2+b_1)+(b_2-b_1)^2}$	$\frac{(c_2-c_1)^2}{(a_2-a_1+b_2-b_1)(b_2-b_1-c_2+c_1)+(b_2-b_1)^2+(c_2-c_1)^2}$
7	$1 - \frac{(c_1-a_2)^2}{(c_1-a_2+c_2-a_1)(c_1-a_2-b_1+b_2)+2(b_2-b_1)^2}$	$\frac{(b_2-b_1)^2(a_2-a_1-2b_2+2b_1+c_2-c_1)}{((a_2-a_1)^2+(b_2-b_1)^2)(b_1-b_2+c_2-c_1)+((b_2-b_1)^2+(c_2-c_1)^2)(a_2-a_1-b_2+b_1)}$
8	$\frac{(c_2-a_1)^2}{(c_2-a_1)^2+(c_2-a_1-b_2+b_1)(c_1-a_2-b_2+b_1)+(b_2-b_1)^2}$	0

5. Two Comparative Studies of Decomposition and Intersection of Two Fuzzy Numbers

If the fuzzy number A is less than the fuzzy number B, then the Hamming distance between A and $\widetilde{max}(A,B)$ is large. Two representations are adopted. One is $d(A, \widetilde{max}(A,B))$. The other is $d(\underline{A}, \widetilde{max}(\underline{A}, \underline{B})) + d(\overline{A}, \widetilde{max}(\overline{A}, \overline{B}))$ which decomposes A into $\overline{A}$ and $\underline{A}$. To analyze the effect of decomposition, we consider the following preference relations without decomposition:

$$T1'(A,B) = \frac{d(A, \widetilde{\max}(A,B)) + d(A \cap B, 0)}{d(A,0) + d(B,0)}$$

and:

$$T2'(A,B) = \frac{d(A, \widetilde{\max}(A,B))}{d(A,B)}$$

which are the counterparts of the Kołodziejczyk's preference relations $K1'(A,B)$ and $K2'(A,B)$. Therefore, the preference relations $K1'(A,B)$ and $K2'(A,B)$ consider the decomposition of fuzzy numbers, while $T1'(A,B)$ and $T2'(A,B)$ do not. The preference relations $K1'(A,B)$ and $T1'(A,B)$ consider the intersection of two membership functions, while $K2'(A,B)$ and $T2'(A,B)$ do not. For completeness, Table 4 displays the values of $N(A,B)$, $N'(A,B)$, $K1'(A,B)$, $K2'(A,B)$, $T1'(A,B)$ and $T2'(A,B)$ of each test case in terms of the values of Q_i. The $K1'(A,B)$ considers both decomposition and intersection of two fuzzy numbers, while $T2'(A,B)$ do not. From $K1'(A,B)$ to $T2'(A,B)$, two representations are:

$$K1'(A,B) \to K2'(A,B) \to T2'(A,B)$$

and:

$$K1'(A,B) \to T1'(A,B) \to T2'(A,B).$$

Table 4. $N(A,B)$, $N'(A,B)$, $K1'(A,B)$, $K2'(A,B)$, $T1'(A,B)$ and $T2'(A,B)$ for eight cases.

Case	$N(A,B)$	$N'(A,B)$	$K1'(A,B)$	$K2'(A,B)$	$T1'(A,B)$	$T2'(A,B)$
1	0	$\frac{Q_2+Q_4+Q_6}{Q_2+Q_4+2Q_6}$	$\frac{Q_2+2Q_3+Q_4+Q_6}{Q_2+2Q_3+Q_4+2Q_6}$	1	$\frac{Q_2+Q_4+Q_6}{Q_2+Q_4+2Q_6}$	1
2	$\frac{Q_2+2Q_3+Q_4}{Q_1+Q_2+2Q_3+Q_4+Q_5}$	$\frac{Q_1+Q_5+Q_6}{Q_1+Q_2+Q_4+Q_5+2Q_6}$	$\frac{Q_1+Q_5+Q_6}{Q_1+Q_2+2Q_3+Q_4+Q_5+2Q_6}$	$\frac{Q_1+Q_5}{Q_1+Q_2+2Q_3+Q_4+Q_5}$	$\frac{Q_1+Q_5+Q_6}{Q_1+Q_2+Q_4+Q_5+2Q_6}$	$\frac{Q_1+Q_5}{Q_1+Q_2+Q_4+Q_5}$
3	$\frac{Q_5}{Q_2+2Q_3+Q_4+Q_5}$	$\frac{Q_2+Q_4+Q_6}{Q_2+Q_4+Q_5+2Q_6}$	$\frac{Q_2+2Q_3+Q_4+Q_6}{Q_2+2Q_3+Q_4+Q_5+2Q_6}$	$\frac{Q_2+2Q_3+Q_4}{Q_2+2Q_3+Q_4+Q_5}$	$\frac{Q_2+Q_4+Q_6}{Q_2+Q_4+Q_5+2Q_6}$	$\frac{Q_2+Q_4}{Q_2+Q_4+Q_5}$
4	$\frac{Q_2+2Q_3+Q_4}{Q_1+Q_2+2Q_3+Q_4}$	$\frac{Q_1+Q_6}{Q_1+Q_2+Q_4+2Q_6}$	$\frac{Q_1+Q_6}{Q_1+Q_2+2Q_3+Q_4+2Q_6}$	$\frac{Q_1}{Q_1+Q_2+2Q_3+Q_4}$	$\frac{Q_1+Q_6}{Q_1+Q_2+Q_4+2Q_6}$	$\frac{Q_1}{Q_1+Q_2+Q_4}$
5	$\frac{Q_1}{Q_1+Q_2+2Q_3+Q_4}$	$\frac{Q_2+Q_4+Q_6}{Q_1+Q_2+Q_4+2Q_6}$	$\frac{Q_2+2Q_3+Q_4+Q_6}{Q_1+Q_2+2Q_3+Q_4+2Q_6}$	$\frac{Q_2+2Q_3+Q_4}{Q_1+Q_2+2Q_3+Q_4}$	$\frac{Q_2+Q_4+Q_6}{Q_1+Q_2+Q_4+2Q_6}$	$\frac{Q_2+Q_4}{Q_1+Q_2+Q_4}$
6	$\frac{Q_2+2Q_3+Q_4}{Q_2+2Q_3+Q_4+Q_5}$	$\frac{Q_5+Q_6}{Q_2+Q_4+Q_5+2Q_6}$	$\frac{Q_5+Q_6}{Q_2+2Q_3+Q_4+Q_5+2Q_6}$	$\frac{Q_5}{Q_2+2Q_3+Q_4+Q_5}$	$\frac{Q_5+Q_6}{Q_2+Q_4+Q_5+2Q_6}$	$\frac{Q_5}{Q_2+Q_4+Q_5}$
7	$\frac{Q_1+Q_5}{Q_1+Q_2+2Q_3+Q_4+Q_5}$	$\frac{Q_2+Q_4+Q_6}{Q_1+Q_2+Q_4+Q_5+2Q_6}$	$\frac{Q_2+2Q_3+Q_4+Q_6}{Q_1+Q_2+2Q_3+Q_4+Q_5+2Q_6}$	$\frac{Q_2+2Q_3+Q_4}{Q_1+Q_2+2Q_3+Q_4+Q_5}$	$\frac{Q_2+Q_4+Q_6}{Q_1+Q_2+Q_4+Q_5+2Q_6}$	$\frac{Q_2+Q_4}{Q_1+Q_2+Q_4+Q_5}$
8	1	$\frac{Q_6}{Q_2+Q_4+2Q_6}$	$\frac{Q_6}{Q_2+2Q_3+Q_4+2Q_6}$	0	$\frac{Q_6}{Q_2+Q_4+2Q_6}$	0

The first feature of Table 4 is that the differences between $K1'(A,B)$ and $T1'(A,B)$ and between $K2'(A,B)$ and $T2'(A,B)$ are Q_3. More precisely, the numerators and denominators of both $K1'(A,B)$ and $K2'(A,B)$ include $2Q_3$ for cases 1, 3, 5 and 7, the denominators of both $K1'(A,B)$ and $K2'(A,B)$ include $2Q_3$ for cases 2, 4, 6 and 8. Therefore, $2Q_3$ represents the effect of the decomposition of fuzzy numbers. The differences between $K1'(A,B)$ and $K2'(A,B)$ and between $T1'(A,B)$ and $T2'(A,B)$ are Q_6. More precisely, the numerators and denominators of both $K1'(A,B)$ and $T1'(A,B)$ include Q_6 and $2Q_6$, respectively. Therefore, Q_6 represents the effect of the intersection of two membership functions. After some computations, the characteristics of $K1'(A,B)$, $K2'(A,B)$, $T1'(A,B)$ and $T2'(A,B)$ are described as follows:

Theorem 1. *Let* $T2'(A,B) = \frac{\alpha}{\alpha+\beta}$.

(1) If $b_1 \le b_2$, $\beta \le 2Q_3 + \alpha$ or $b_1 \ge b_2$, $\beta + 2Q_3 \le \alpha$, then $K1'(A,B) \le K2'(A,B)$. If $b_1 \le b_2$, $\beta \ge 2Q_3 + \alpha$ or $b_1 \ge b_2$, $\beta + 2Q_3 \ge \alpha$, then $K1'(A,B) \ge K2'(A,B)$.
(2) If $b_1 \le b_2$, then $K2'(A,B) \ge T2'(A,B)$. If $b_1 \ge b_2$, then $K2'(A,B) \le T2'(A,B)$.
(3) If $\alpha \ge \beta$, then $T1'(A,B) \le T2'(A,B)$. If $\alpha \le \beta$, then $T1'(A,B) \ge T2'(A,B)$.
(4) If $b_1 \le b_2$, then $K1'(A,B) \ge T1'(A,B)$. If $b_1 \ge b_2$, then $K1'(A,B) \le T1'(A,B)$.
(5) If $b_1 \le b_2$, $\beta(2Q_3 + Q_6) \le \alpha Q_6$ or $b_1 \ge b_2$, $\beta Q_6 \le \alpha(Q_3 + 2Q_6)$, then $K1'(A,B) \le T2'(A,B)$. If $b_1 \le b_2$, $\beta(2Q_3 + Q_6) \ge \alpha Q_6$ or $b_1 \ge b_2$, $\beta Q_6 \ge \alpha(Q_3 + 2Q_6)$, then $K1'(A,B) \ge T2'(A,B)$.

For each test case of two triangular fuzzy numbers $A(a_1, b_1, c_1)$ and $B(a_2, b_2, c_2)$, we analyze the behaviors of $K1'(A,B)$, $K2'(A,B)$, $T1'(A,B)$ and $T2'(A,B)$ by applying Theorem 1 as follows: Firstly, for $b_1 \le b_2$, we have:

$$T1'(A,B) \le K1'(A,B) \le K2'(A,B) = T2'(A,B) = 1$$

for case 1. For cases 3, 5 and 7, we have the following results.

(1) From $2Q_3 + \alpha - \beta = \frac{1}{2}(a_2 + 2b_2 + c_2 - a_1 - 2b_1 - c_1)$, we have that if $a_2 + 2b_2 + c_2 \ge a_1 + 2b_1 + c_1$, then $K1'(A,B) \le K2'(A,B)$; if $a_2 + 2b_2 + c_2 \le a_1 + 2b_1 + c_1$, then $K1'(A,B) \ge K2'(A,B)$.
(2) $K2'(A,B) \ge T2'(A,B)$.
(3) From $\alpha - \beta = \frac{(a_2-c_1)(a_2-b_1+b_2-c_1+c_2-a_1)+(b_1-b_2)(c_2-a_1)}{2(a_2+b_1-b_2-c_1)}$, it follows that if $a_2 + b_2 + c_2 \ge a_1 + b_1 + c_1$, then $T1'(A,B) \le T2'(A,B)$; if $a_2 + b_2 + c_2 \le a_1 + b_1 + c_1$, then $T1'(A,B) \ge T2'(A,B)$.
(4) $K1'(A,B) \ge T1'(A,B)$.
(5) If $a_2 + 2b_2 + c_2 \ge a_1 + 2b_1 + c_1$, then $K1'(A,B) \ge T2'(A,B)$. If $a_2 + 2b_2 + c_2 \le a_1 + 2b_1 + c_1$, then $K1'(A,B) \le T2'(A,B)$.

Therefore, for the cases 3, 5 and 7, if $a_2 + 2b_2 + c_2 \leq a_1 + 2b_1 + c_1$, then:

$$K1'(A,B) \geq K2'(A,B) \geq T2'(A,B)$$

and:

$$K1'(A,B) \geq T1'(A,B) \geq T2'(A,B).$$

Secondly, for $b_1 \geq b_2$, we have:

$$K2'(A,B) = T2'(A,B) = 0 \leq K1'(A,B) \leq T1'(A,B)$$

for case 8. For cases 2, 4 and 6, we have the following results.

(1) From $\alpha - 2Q_3 - \beta = \frac{1}{2}(a_2 + 2b_2 + c_2 - a_1 - 2b_1 - c_1)$, we obtain if $a_2 + 2b_2 + c_2 \geq a_1 + 2b_1 + c_1$, then $K1'(A,B) \leq K2'(A,B)$; if $a_2 + 2b_2 + c_2 \leq a_1 + 2b_1 + c_1$, then $K1'(A,B) \geq K2'(A,B)$.

(2) $K2'(A,B) \leq T2'(A,B)$.

(3) From $\alpha - \beta = \frac{1}{2}\left(-a_1 + a_2 - 2b_1 + 2b_2 - c_1 + c_2 + \frac{2(b_1-b_2)^2}{-a_1+b_1-b_2+c_2}\right)$, it follows thatif $a_2 + 2b_2 + c_2 \geq a_1 + 2b_1 + c_1$, then $T1'(A,B) \leq T2'(A,B)$; if $a_2 + 2b_2 + c_2 \leq a_1 + 2b_1 + c_1$, then $T1'(A,B) \geq T2'(A,B)$.

(4) $K1'(A,B) \leq T1'(A,B)$.

(5) If $a_2 + 2b_2 + c_2 \geq a_1 + 2b_1 + c_1$, then $K1'(A,B) \leq T2'(A,B)$. If $a_2 + 2b_2 + c_2 \leq a_1 + 2b_1 + c_1$, then $K1'(A,B) \geq T2'(A,B)$.

Therefore, for the cases 2, 4 and 6, if $a_2 + 2b_2 + c_2 \geq a_1 + 2b_1 + c_1$, then:

$$K1'(A,B) \leq K2'(A,B) \leq T2'(A,B)$$

and:

$$K1'(A,B) \leq T1'(A,B) \leq T2'(A,B).$$

For the two triangular fuzzy numbers $A(a_1, b_1, c_1)$ and $B(a_2, b_2, c_2)$, the second comparative study is comprised of the five case studies shown in Figure 2, which compares the fuzzy preference relations $K1'(A,B)$, $K2'(A,B)$, $T1'(A,B)$ and $T2'(A,B)$.

Case (a) $A(a_1, b_1, c_1)$ and $B(a_2, b_2, c_2)$ with $a_2 \geq c_1$.

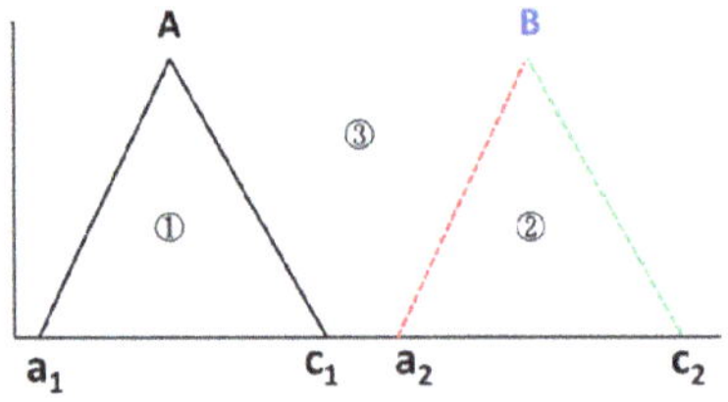

Case (b) $A(c-a, c, c+a)$ and $B(c-b, c, c+b)$.

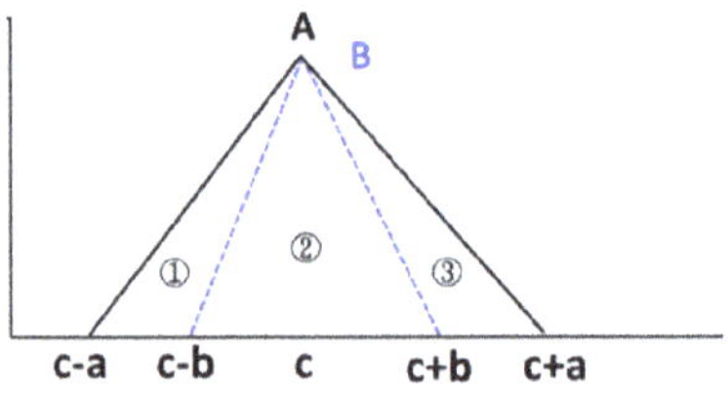

Figure 2. *Cont.*

Case (c) $A(a, a+b, a+2b)$ and $B(a+\alpha, a+b+\alpha+\beta, a+\alpha+2b+2\beta)$.

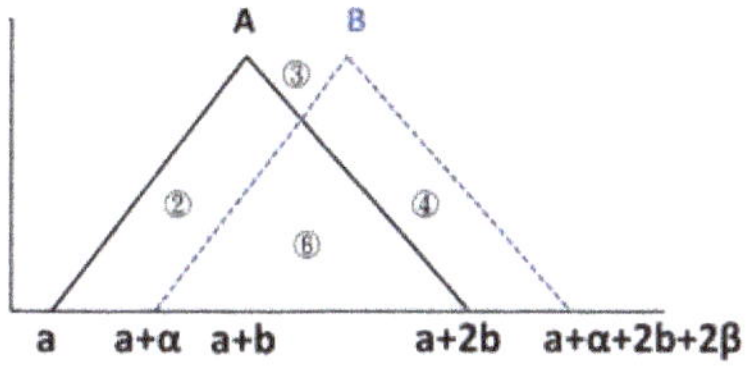

Case (d) $A(a, a, a+b)$ and $B(c, c+b, c+b)$.

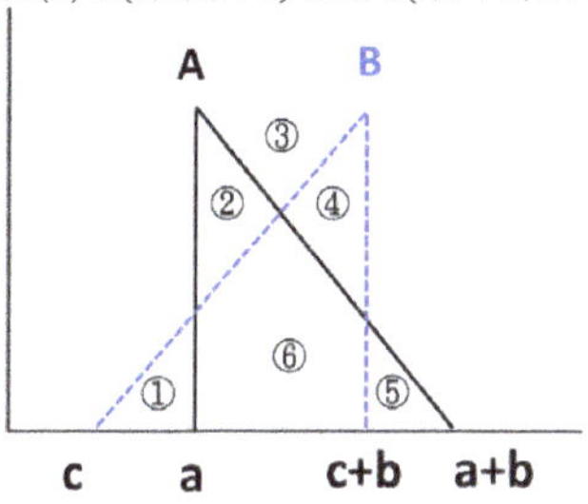

Case (e) $A(c+a, \mathrm{b}, 1-c+a)$ and $B(c, 0.5, 1-c)$.

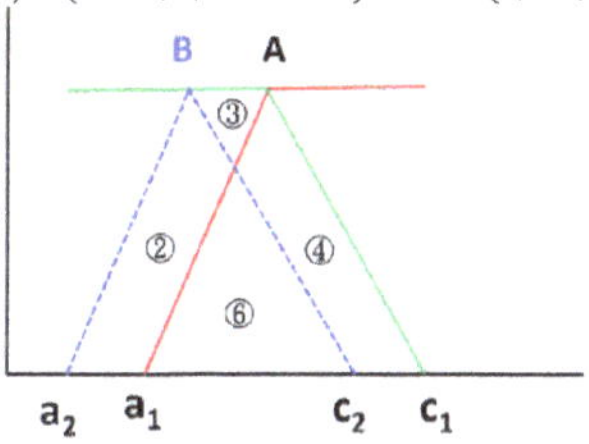

Figure 2. Five case studies of A and B for $K1'(A,B)$, $K2'(A,B)$, $T1'(A,B)$ and $T2'(A,B)$.

Case (a) $A(a_1, b_1, c_1)$ and $B(a_2, b_2, c_2)$ with $a_2 \geq c_1$.

It follows that:

$$K1'(A,B) = \frac{(Q_1+Q_3)+(Q_2+Q_3)+0}{(Q_1+Q_3)+(Q_2+Q_3)+0} = 1$$

$$K2'(A,B) = \frac{(Q_1+Q_3)+(Q_2+Q_3)}{(Q_1+Q_3)+(Q_2+Q_3)} = 1$$

$$T1'(A,B) = \frac{0+(Q_1+Q_2)}{(Q_1+Q_2)} = 1$$

and:

$$T2'(A,B) = \frac{(Q_1+Q_2)}{(Q_1+Q_2)} = 1. \tag{1}$$

For this simple case, all the preference relations give the same degree of preference of B over A.

Case (b) $A(c-a, c, c+a)$ and $B(c-b, c, c+b)$.

We have:

$$K1'(A,B) = \frac{Q_1+0+Q_2}{Q_1+Q_3+2Q_2} = 1/2$$

$$K2'(A,B) = \frac{Q_1+0}{Q_1+Q_3} = 1/2$$

$$T1'(A,B) = \frac{Q_2+Q_1}{(Q_1+Q_2+Q_3)+Q_2} = 1/2$$

and:

$$T2'(A,B) = \frac{Q_1}{Q_1 + Q_3} = 1/2.$$

From the viewpoint of probability, the fuzzy numbers A and B have the same mean, but B has a smaller standard deviation. The results indicate that the differences between the decomposition and intersection of A and B cannot affect the degree of preference for B over A.

Case (c) $A(a, a+b, a+2b)$ and $B(a+\alpha,\ a+b+\alpha+\beta,\ a+\alpha+2b+2\beta)$.

For this case, the fuzzy number B is a right shift of A. Therefore, B should have a higher ranking than A based on the intuition criterion. We obtain:

$$K1'(A,B) = \frac{(Q_2+Q_3)+(Q_3+Q_4)+Q_6}{(Q_2+Q_3)+(Q_3+Q_4)+2Q_6} = \frac{(2b+\alpha+2\beta)^2}{2(\alpha^2+4b^2+4b\beta+2b\alpha+2\beta^2)} > 1/2$$

$$K2'(A,B) = \frac{(Q_2+Q_3)+(Q_3+Q_4)}{(Q_2+Q_3)+(Q_3+Q_4)} = 1$$

$$T1'(A,B) = \frac{Q_6+(Q_2+Q_4)}{(Q_2+Q_6)+(Q_6+Q_4)} = 1 - \frac{(-2b+\alpha)^2}{2(2b+\beta)^2}$$

and:

$$T2'(A,B) = \frac{Q_2+Q_4}{Q_2+Q_4} = 1.$$

All methods prefer B, but $T1'(A,B)$ is indecisive. More precisely,

If $2b+\beta < \alpha$, then $T1'(A,B) < 1/2$, so $A > B$
If $2b+\beta = \alpha$, then $T1'(A,B) = 1/2$, so $A = B$
If $2b+\beta > \alpha$, then $T1'(A,B) > 1/2$, so $A < B$.

Hence, a conflicting ranking order of $T1'(A,B)$ exists in this case.

Case (d) $A(a, a, a+b)$ and $B(c, c+b,\ c+b)$ with $a \geq c$.

This case is more complex for the partial overlap of A and B. The membership function of B has the right peak, B expands to the left of A for the left membership function, and A expands to the right of B for the right membership function. We have:

$$K1'(A,B) = \frac{(Q_2+Q_3)+(Q_3+Q_4)+Q_6}{(Q_1+Q_2+Q_3)+(Q_3+Q_4+Q_5)+2Q_6} = 0.5 + \frac{b(-2a+b+2c)}{a^2+3b^2+2bc+c^2-2ab-2ac}$$

$$K2'(A,B) = \frac{(Q_2+Q_3)+(Q_3+Q_4)}{(Q_1+Q_2+Q_3)+(Q_3+Q_4+Q_5)} = \frac{(-a+b+c)^2}{2a^2+b^2+2bc+2c^2-2ab-4ac}$$

$$T1'(A,B) = \frac{Q_6+(Q_2+Q_4)}{(Q_2+Q_5+Q_6)+(Q_1+Q_6+Q_4)} = \frac{(a+3b-c)(-a+b+c)}{4b^2}$$

and:

$$T2'(A,B) = \frac{Q_2+Q_4}{Q_1+Q_2+Q_4+Q_5} = \frac{(-a+b+c)^2}{3a^2+b^2+2bc+3c^2-2ab-6ac}.$$

It follows that:

If $-2a+b+2c < 0$, then $K1'(A,B) < 1/2$ and $K2'(A,B) < 1/2$, so $A > B$;
If $-2a+b+2c = 0$, then $K1'(A,B) = 1/2$ and $K2'(A,B) = 1/2$, so $A = B$;
If $-2a+b+2c > 0$, then $K1'(A,B) > 1/2$ and $K2'(A,B) > 1/2$, so $A < B$;
If $b < \left(1+\sqrt{2}\right)(a-c)$, then $T1'(A,B) < 1/2$ and $T2'(A,B) < 1/2$, so $A > B$;
If $b = \left(1+\sqrt{2}\right)(a-c)$, then $T1'(A,B) = 1/2$ and $T2'(A,B) = 1/2$, so $A = B$;

If $b > \left(1+\sqrt{2}\right)(a-c)$, then $T1'(A,B) > 1/2$ and $T2'(A,B) > 1/2$, so $A < B$.

Three special subcases are considered as follows:

(1) Subcase (d1) If $b = \left(1+\sqrt{2}\right)(a-c)$, then $A\left(a, a, \left(2+\sqrt{2}\right)a - \left(1+\sqrt{2}\right)c\right)$ and $B\left(c, \left(1+\sqrt{2}\right)a - \sqrt{2}c, \left(1+\sqrt{2}\right)a - \sqrt{2}c\right)$, therefore $T1'(A,B) = T2'(A,B) = 0.5$, so $A = B$. However, $K1'(A,B) = \frac{6-\sqrt{2}}{8}$, $K2'(A,B) = 2/3$ and $A < B$.

(2) Subcase (d2) If $b = 2(a-c)$, then $A(a, a, 3a-2c)$ and $B(c, 2a-c,\ 2a-c)$, therefore $T1'(A,B) = 7/16$, $T2'(A,B) = 1/3$, so $A > B$. However, $K1'(A,B) = K2'(A,B) = 0.5$ and $A = B$.

(3) Subcase (d3) If $A(0.3, 0.3, 0.9)$ and $B(0.1, 0.7,\ 0.7)$, then $K1'(A,B) = 0.5556$, $K2'(A,B) = 0.6667$, $T1'(A,B) = 0.5556$, $T2'(A,B) = 0.6667$, so $A < B$.

Therefore, if $b < 2(a-c)$, then $K1'(A,B) < 1/2$, $K2'(A,B) < 1/2$, $T1'(A,B) < 1/2$ and $T2'(A,B) < 1/2$, so $A > B$; if $b > \left(1+\sqrt{2}\right)(a-c)$, then $K1'(A,B) > 1/2$, $K2'(A,B) > 1/2$, $T1'(A,B) > 1/2$ and $T2'(A,B) > 1/2$, so $A < B$.

Case (e) $A(c+a, \mathrm{b}, 1-c+a)$ and $B(c, 0.5,\ 1-c)$.

For this case, the membership function B is symmetric with respect to $x = 0.5$. The membership function of A is parallel translation of that of B except its peak. We have the following results:

(1) $K1'(A,B) = \frac{(Q_2+Q_3)+(Q_3+Q_4)+Q_6}{(Q_1+Q_2+Q_3)+(Q_3+Q_4+Q_5)+2Q_6} = \frac{5-8b-12c+8bc-2a^2+4b^2+8c^2}{7+4a-8b-20c-8ac+8bc+4b^2+16c^2}$. If $(1-2a-2b)(3+2a-2b-4c) < 0$, then $K1'(A,B) < 1/2$, so $A > B$. For simplicity, the other two conditions are omitted.

(2) $K2'(A,B) = \frac{(Q_2+Q_3)+(Q_3+Q_4)}{(Q_1+Q_2+Q_3)+(Q_3+Q_4+Q_5)} = \frac{(1-2b)^2}{4a^2+(1-2b)^2}$. If $2a+2b-1 > 0$, then $K2'(A,B) < 1/2$, so $A > B$.

(3) $T1'(A,B) = \frac{Q_6+(Q_2+Q_4)}{(Q_2+Q_5+Q_6)+(Q_1+Q_6+Q_4)} = \frac{2(1-b-c)(1-2c)-a^2}{2(1+2a-2b)(3+2a-2b-4c)(1-2c)}$ and $T2'(A,B) = \frac{Q_2+Q_4}{Q_1+Q_2+Q_4+Q_5} = \frac{(1-2b)^2(1-2c)}{4a^3+(1-2b)^2(1-2c)+a^2(6-4b-8c)}$. If $> -0.5+c+\frac{1}{2}\sqrt{(1-2c)(3-4b-2c)}$, then $T1'(A,B) < 1/2$ and $T2'(A,B) < 1/2$, so $A > B$.

Four special subcases are considered as follows:

(1) Subcase (e1) If $a = -0.5+c+\frac{1}{2}\sqrt{(1-2c)(3-4b-2c)}$, then $T1'(A,B) = T2'(A,B) = 0.5$, so $A = B$. However, $K1'(A,B) > 0.5$, $K2'(A,B) > 0.5$ and $A < B$.

(2) Subcase (e2) If $2a+2b-1 = 0$, then $T1'(A,B) = 0.5 - \frac{(1-2b)^2}{16(1-b-c)(1-2c)} < 0.5$, $T2'(A,B) = \frac{(1-2c)}{(3-2b-4c)} < 0.5$, so $A > B$. However, $K1'(A,B) = K2'(A,B) = 0.5$ and $A = B$.

(3) Subcase (e3) If A$(0.3, 0.4, 0.9)$ and B$(0.2, 0.5,\ 0.8)$, then $K1'(A,B) = 0.4896$, $K2'(A,B) = 0.4286$, $T1'(A,B) = 0.4896$, $T2'(A,B) = 0.4286$, so $A > B$.

(4) Subcase (e4) If $b \geq 0.5$, then $(1-2a-2b)(3+2a-2b-4c) < 0$, $2a+2b-1 > 0$ and $a > -0.5+c+\frac{1}{2}\sqrt{(1-2c)(3-4b-2c)}$, so $K1'(A,B) < 1/2$, $K2'(A,B) < 1/2$, $T1'(A,B) < 1/2$ and $T2'(A,B) < 1/2$, hence $A > B$.

6. Conclusions

This paper analyzes and compares two types of Nakamura's fuzzy preference relations—($N(A,B)$ and $N'(A,B)$)—two types of Kołodziejczyk's fuzzy preference relations—($K1'(A,B)$ and $K2'(A,B)$)—and the counterparts of the Kołodziejczyk's fuzzy preference relations—($T1'(A,B)$ and $T2'(A,B)$)—on a group of eight selected cases, with all the possible levels of overlap between two triangular fuzzy numbers $A(a_1, b_1, c_1)$ and $B(a_2, b_2, c_2)$. First, for $N(A,B)$ and $N'(A,B)$ we obtain that $N_j(A,B) + N_{8-j}(A,B) = 1$, $j = 1,2,3,4$. If $a_2 + 2b_2 + c_2 \geq a_1 - 2b_1 - c_1$, we have that $1 - N'_j(A,B) \geq N_j(A,B)$ for $j = 1,3,5,7$ and $1 - N'_j(A,B) \leq N_j(A,B)$ for $j = 2,4,6,8$. Secondly, for $K1'(A,B)$ and $K2'(A,B)$, we have that $K1'_1(A,B) = K1'_j(A,B)$ for $j = 3,5,7$ and

$K1'_2(A,B) = K1'_j(A,B)$ for $j = 4,6,8$. Furthermore, $K1'_1(A,B) + K1'_2(A,B) = 0$ for $b_1 = b_2$ and $c_2 = a_1$ and $K1'_1(A,B) + K1'_2(A,B) = 1$ for $b_1 = b_2$ and $c_1 - a_2 = c_2 - a_1$. Thirdly, for test case 1, $T1'(A,B) \leq K1'(A,B) \leq K2'(A,B) = T2'(A,B) = 1$. For the test cases 3, 5 and 7, if $a_2 + 2b_2 + c_2 \leq a_1 + 2b_1 + c_1$, then $K1'(A,B) \geq K2'(A,B) \geq T2'(A,B)$ and $K1'(A,B) \geq T1'(A,B) \geq T2'(A,B)$. For the test case 8, we have $K2'(A,B) = T2'(A,B) = 0 \leq K1'(A,B) \leq T1'(A,B)$. For the test cases 2, 4 and 6, if $a_2 + 2b_2 + c_2 \geq a_1 + 2b_1 + c_1$, then $K1'(A,B) \leq K2'(A,B) \leq T2'(A,B)$ and $K1'(A,B) \leq T1'(A,B) \leq T2'(A,B)$. These results provide insights into the decomposition and intersection of fuzzy numbers. Among the six fuzzy preference relations, the appropriate fuzzy preference relation can be chosen from the decision-maker's perspective. Given this fuzzy preference relation, the final ranking of a set of alternatives is derived.

Worthy of future research is extending the analysis to other types of fuzzy numbers. First, the analysis can be easily extended to the trapezoidal fuzzy numbers. Second, for the hesitant fuzzy set lexicographical ordering method, Liu et al. [14] modified the method of Farhadinia [15] and this was more reasonable in more general cases. Recently, Alcantud and Torra [16] provided the necessary tools for the hesitant fuzzy preference relations. Thus, the analysis of hesitant fuzzy preference relations is a subject of considerable ongoing research.

Conflicts of Interest: The author declares no conflict of interest.

References

1. Chen, S.J.; Hwang, C.L. *Fuzzy Multiple Attribute Decision-Making*; Springer: Berlin, Germany, 1992.
2. Wang, X.; Kerre, E.E. Reasonable properties for the ordering of fuzzy quantities (I). *Fuzzy Sets Syst.* **2001**, *118*, 375–385. [CrossRef]
3. Wang, X.; Kerre, E.E. Reasonable properties for the ordering of fuzzy quantities (II). *Fuzzy Sets Syst.* **2001**, *118*, 387–405. [CrossRef]
4. Nakamura, K. Preference relations on a set of fuzzy utilities as a basis for decision-making. *Fuzzy Sets Syst.* **1986**, *20*, 147–162. [CrossRef]
5. Kolodziejczyk, W. Orlovsky's concept of decision-making with fuzzy preference relation-further results. *Fuzzy Sets Syst.* **1990**, *19*, 197–212. [CrossRef]
6. Yuan, Y. Criteria for evaluating fuzzy ranking methods. *Fuzzy Sets Syst.* **1991**, *44*, 139–157. [CrossRef]
7. Li, R.J. Fuzzy method in group decision-making. *Comput. Math. Appl.* **1999**, *38*, 91–101. [CrossRef]
8. Lee, H.S. On fuzzy preference relation in group decision-making. *Int. J. Comput. Math.* **2005**, *82*, 133–140. [CrossRef]
9. Wang, Z.X.; Liu, Y.J.; Fan, Z.P.; Feng, B. Ranking LR fuzzy number based on deviation degree. *Inf. Sci.* **2009**, *179*, 2070–2077. [CrossRef]
10. Asady, B. The revised method of ranking LR fuzzy number based on deviation degree. *Expert Syst. Appl.* **2010**, *37*, 5056–5060. [CrossRef]
11. Zhang, F.; Ignatius, J.; Lim, C.P.; Zhao, Y. A new method for ranking fuzzy numbers and its application to group decision-making. *Appl. Math. Model.* **2014**, *38*, 1563–1582. [CrossRef]
12. Zhu, B.; Xu, Z.; Xu, J. Deriving a ranking from hesitant fuzzy preference relations under group decision-making. *IEEE Tran. Cyber.* **2014**, *44*, 1328–1337. [CrossRef] [PubMed]
13. Wang, Y.J. Ranking triangle and trapezoidal fuzzy numbers based on the relative preference relation. *Appl. Math. Model.* **2015**, *39*, 586–599. [CrossRef]
14. Liu, X.; Wang, Z.; Zhang, S. A modification on the hesitant fuzzy set lexicographical ranking method. *Symmetry* **2016**, *8*, 153. [CrossRef]
15. Farhadinia, B. Hesitant fuzzy set lexicographical ordering and its application to multi-attribute decision-making. *Inf. Sci.* **2016**, *327*, 233–245. [CrossRef]
16. Alcantud, J.C.R.; Torra, V. Decomposition theorems and extension principles for hesitant fuzzy sets. *Inf. Fus.* **2018**, *41*, 48–56. [CrossRef]

symmetry

Article

Multiple Attribute Decision-Making Method Using Correlation Coefficients of Normal Neutrosophic Sets

Jun Ye

Department of Electrical and Information Engineering, Shaoxing University, Shaoxing 312000, China; yehjun@aliyun.com; Tel.: +86-575-88327323

Academic Editor: José Carlos R. Alcantud
Received: 29 April 2017; Accepted: 25 May 2017; Published: 26 May 2017

Abstract: The normal distribution is a usual one of various distributions in the real world. A normal neutrosophic set (NNS) is composed of both a normal fuzzy number and a neutrosophic number, which a significant tool for describing the incompleteness, indeterminacy, and inconsistency of the decision-making information. In this paper, we propose two correlation coefficients between NNSs based on the score functions of normal neutrosophic numbers (NNNs) (basic elements in NNSs) and investigate their properties. Then, we develop a multiple attribute decision-making (MADM) method with NNSs under normal neutrosophic environments, where, by correlation coefficient values between each alternative (each evaluated NNS) and the ideal alternative (the ideal NNS), the ranking order of alternatives and the best one are given in the normal neutrosophic decision-making process. Finally, an illustrative example about the selection problem of investment alternatives is provided to demonstrate the application and feasibility of the developed decision-making method. Compared to the existing MADM approaches based on aggregation operators of NNNs, the proposed MADM method based on the correlation coefficients of NNSs shows the advantage of its simple decision-making process.

Keywords: multiple attribute decision-making; normal neutrosophic set; normal neutrosophic number; correlation coefficient

1. Introduction

In probability theory [1], the normal (or Gaussian) distribution is a very common continuous probability distribution. Normal distribution is an important distribution form in statistics and is very useful in the natural and social sciences to express real-valued random variables whose distributions are not known. Hence, it has been widely applied to various fields. Then, the fuzziness and uncertainty of the real decision-making information are a common phenomenon because some numerical values may be inadequate or insufficient to complex decision-making problems. In some occasions, it can be more reasonable to describe the attribute values by the fuzzy numbers in a fuzzy environment. Thus, Zadeh [2] firstly introduced the fuzzy set, which is described by the membership function. After that, Yang and Ko [3] defined a normal fuzzy number (NFN) to express the normal fuzzy information in random fuzzy situations. It is obvious that its main advantage is reasonable and realistic to normal distribution environments. As an extension of the fuzzy set, Atanassov [4] proposed the intuitionistic fuzzy set (IFS) by adding the non-membership function to the fuzzy set. However, because NFN only contains its normal fuzzy membership degree, Wang et al. [5] presented an intuitionistic normal fuzzy number (INFN) based on the combination of both an NFN and an intuitionistic fuzzy number (IFN) (a basic element in IFS), defined the score function and operational laws of INFNs, and presented some aggregation operators of INFNs, including an ordered intuitionistic normal ordered fuzzy weighted averaging operator, an INFN ordered weighted geometric averaging operator, two INFN-related ordered weighted arithmetic and geometric averaging operators, two induced INFN-related ordered weighted arithmetic and geometric averaging operators, and they then applied them to multiple criteria decision-making (MCDM) problems, where the criteria are interactive and the criteria values are the INFNs. Then, Wang and Li [6] proposed a score function of INFN based on relative entropy and an INFN weighted arithmetic averaging operator, and then applied them to normal intuitionistic fuzzy MCAD problems. Wang and Li [7] also introduced Euclidean distance between INFNs and an INFN weighted arithmetic averaging operator and an INFN weighted geometric averaging operator for MCDM problems with INFNs. Wang et al. [8] further introduced a normal intuitionistic fuzzy number (NIFN) weighted arithmetic averaging operator, an NIFN weighted geometric averaging operator, an NIFN-induced ordered weighted averaging operator, an NIFN-induced ordered weighted geometric averaging operator, and an NIFN-induced generalized ordered weighted averaging (NIFN-IGOWA) operator, and then applied the NIFN-IGOWA operator to MCDM problems with NIFN information. To express the truth, indeterminacy, and falsity information in real world, Smarandache [9] proposed a concept of a neutrosophic set from a philosophical point of view. As a subclass of the neutrosophic set, Smarandache [9] and Wang et al. [10] introduced the concept of a single-valued neutrosophic set (SVNS). Obviously, SVNS is a generalization of IFS and represents incomplete, indeterminate, and inconsistent information, which cannot be expressed by IFS. For example, assume that an investment company wants to invest a sum of money to some investment alternative. Then, there are 10 voters in the voting process of the investment alternative. Five vote "aye", four vote 'blackball', and one votes 'indeterminacy/neutrality'. From neutrosophic notation, it can be represented as $(x, 0.5, 0.4, 0.1)$. It is obvious that this expression is beyond the scope of IFS. Hence, SVNS is suitable for the expression of indeterminate and inconsistent information. Recently, the neutrosophic sets have been applied in many decision-making problems [11–17]. Liu and Teng [18] presented a normal neutrosophic number (NNN) as an extension of NIFN and its generalized weighted power averaging operator, and then applied it to multiple attribute decision-making (MADM) problems with normal neutrosophic information. Liu and Li [19] further introduced some normal neutrosophic Bonferroni mean operators for decision-making problems with normal neutrosophic information. After that, Sahin [20] proposed some normal neutrosophic generalized prioritized aggregation operators for MADM problems under normal neutrosophic environments.

However, the aforementioned decision-making methods depend on aggregation operators of NNNs in the normal neutrosophic decision-making process. Then, the correlation coefficient is an important mathematical tool in decision-making problems [11–13]. Compared with the decision-making methods using aggregation operators [18–20], the decision-making methods based on correlation coefficients imply relatively simple decision-making processes. However, there is no research on correlation coefficients of NNSs in existing normal neutrosophic decision-making methods. On the other hand, the applications of NNNs (basic elements in NNSs) in science and engineering fields are necessary and significant because the normal distribution is a typical and common distribution in the real world [18–20]. Additionally, NNN contain much more information than the general neutrosophic number because NNN is expressed by the combination information of both an NFN and a single-valued neutrosophic number (SVNN) (a basic element in SVNS). Hence, NNN used in decision-making can show its rationality and reality. Motivated by the decision-making methods [18–20], this study firstly proposes two correlation coefficients of normal neutrosophic sets (NNSs) based on the score functions of NNNs and then develops an MADM method using the correlation coefficients of NNSs to simplify the decision-making process under normal neutrosophic environments.

The rest of this paper is organized as follows. In Section 2, we review some basic concepts of NIFNs and NNSs. In Section 3, two correlation coefficients between NNSs are presented based on the score functions of NNNs. Section 4 develops an MADM method using the correlation coefficients of NNSs under normal neutrosophic environments. In Section 5, an illustrative example about the selection problem of investment alternatives is provided to demonstrate the applications and effectiveness of the proposed MADM method with normal neutrosophic information. Conclusions and future work are contained in Section 6.

2. Some Basic Concepts of NIFNs and NNSs

Yang and Ko [4] defined an NFN to express the normal fuzzy information in random fuzzy situations.

For a real number set X, if the membership function satisfies the form

$$N(x) = \mathrm{e}^{-(\frac{x-\mu}{\sigma})^2} \tag{1}$$

then $N(x)$ is called NFN, where μ is the mean or expectation of the distribution (and its median and mode) and σ is standard deviation. Then, this NFN is symmetric around $x = \mu$, denoted by $N(\mu, \sigma)$.

Based on the combination of an IFN and an NFN, Wang et al. [8] defined an NIFN $A = \langle x \mid N(\mu, \sigma), t_A(x), v_A(x) \rangle$, where its membership function is expressed as

$$t_A(x) = t_A \mathrm{e}^{-(\frac{x-\mu}{\sigma})^2}, x \in X$$

and its non-membership function is expressed as

$$v_A(x) = 1 - (1 - v_A)\mathrm{e}^{-(\frac{x-\mu}{\sigma})^2}, x \in X$$

where t_A and v_A are a membership degree and a non-membership degree in an IFN and satisfy t_A, $v_A \in [0,1]$, and $0 \leq t_A + v_A \leq 1$.

To express indeterminate and inconsistent information in the real world, Smarandache [9] introduced a concept of a neutrosophic set from a philosophical point of view. A neutrosophic set B in a universe of discourse X can be described independently by its truth, indeterminacy, and falsity membership functions $t_B(x)$, $u_B(x)$, and $v_B(x)$ in real standard interval [0,1] or nonstandard interval $]^-0, 1^+[$, such that $t_B(x)$: $X \rightarrow]^-0, 1^+[$, $u_B(x)$: $X \rightarrow]^-0, 1^+[$, $v_B(x)$: $U \rightarrow]^-0, 1^+[$, and $^-0 \leq \sup t_B(x) + \sup u_B(x) + \sup v_B(x) \leq 3^+$ for $x \in X$.

However, when the three membership functions in the neutrosophic set lie in the nonstandard interval $]^-0, 1^+[$, the neutrosophic set shows the difficulty of its actual applications. Thus, Smarandache [9] and Wang et al. [10] introduced the concept of an SVNS as a subclass of the neutrosophic set when the three membership functions in the neutrosophic set are constrained in the real standard interval [0,1].

Definition 1. [9,10]. *Let X be a universe of discourse. An SVNS S in X is described independently by its truth, indeterminacy, and falsity membership functions* $t_S(x)$, $u_S(x)$, *and* $v_S(x)$, *where* $t_S(x)$, $u_S(x)$, $v_S(x) \in [0,1]$, *and* $0 \leq t_S(x) + u_S(x) + v_S(x) \leq 3$ *for* $x \in X$. *Then, the SVNS S can be denoted as.* $S = \{\langle x, t_S(x), u_S(x), v_S(x)\rangle : x \in X\}$.

For convenience, a basic element $\langle x, t_S(x), u_S(x), v_S(x)\rangle$ in S is denoted by $s = \langle t, u, v \rangle$ for short, which is called an SVNN.

As an extension of NIFN, Liu and Teng [11] and Liu and Li [12] presented a concept of NNS based on the combination of NFN and SVNN.

Definition 2. [11,12]. *Let X be a finite non-empty set and* $N(\mu, \sigma)$ *be a normal distribution function. An NNS is defined as*

$$P = \{\langle x | N(\mu_P, \sigma_P), (t_P(x), u_P(x), v_P(x))\rangle : x \in X\} \tag{2}$$

where the three functions $t_P(x)$, $u_P(x)$, *and* $v_P(x)$ *for* $x \in X$ *satisfy the following properties:*

$$t_P(x) = t_P \mathrm{e}^{-(\frac{x-\mu}{\sigma})^2}$$

$$u_P(x) = 1 - (1 - u_P)\mathrm{e}^{-(\frac{x-\mu}{\sigma})^2}$$

$$v_P(x) = 1 - (1 - v_P)\mathrm{e}^{-(\frac{x-\mu}{\sigma})^2}$$

$$0 \leq t_P(x) + u_P(x) + v_P(x) \leq 1$$

and t_p, u_p, *and* v_p *are the truth, indeterminacy, and falsity degrees in the SVNN, respectively, and satisfy* t_p, u_p, *and* $v_p \in [0,1]$ *and* $0 \leq t_p + u_p + v_p \leq 3$.

Then, an NNN (a basic element) in the NNS P is denoted by $p = <N(\mu, \sigma), (t, u, v)>$ for convenience, where t, u, and v are the truth, indeterminacy, and falsity degrees, respectively, in the SVNN (t, u, v) and satisfy $t, u, v \in [0,1]$ and $0 \leq t + u + v \leq 3$.

Definition 3. [12]. *Let* $p = <N(\mu, \sigma), (t, u, v)>$ *be an NNN. Then, its score functions are defined as*

$$\begin{aligned} S_1(p) &= \mu(2 + t - u - v), \\ S_2(p) &= \sigma(2 + t - u - v). \end{aligned} \tag{3}$$

3. Correlation Coefficients between NNSs

Based on the score functions of NNNs in Definition 3, we can give the definitions of the correlation and correlation coefficients between NNSs under normal neutrosophic environments.

Definition 4. *Let two NNSs be* $P = \{p_1, p_2, \ldots, p_n\}$ *and* $Q = \{q_1, q_2, \ldots, q_n\}$, *where* $p_j = <N(\mu_{pj}, \sigma_{pj}), (t_{pj}, u_{pj}, v_{pj})>$ *and* $q_j = <N(\mu_{qj}, \sigma_{qj}), (t_{qj}, u_{qj}, v_{qj})>$ *for* $j = 1, 2, \ldots, n$ *are NNNs in P and Q. The correlation between two NNSs P and Q is defined as*

$$C(P,Q) = \sum_{j=1}^{n} \left[(2 + t_{Pj} - u_{Pj} - v_{Pj})(2 + t_{qj} - u_{qj} - v_{qj})(\mu_{pj}\mu_{qj} + \sigma_{pj}\sigma_{qj}) \right] \tag{4}$$

Thus, based on the correlation between two NNSs P and Q, we can introduce the definition of the following correlation coefficients between two NNSs P and Q.

Definition 5. *Let two NNSs be* $P = \{p_1, p_2, \ldots, p_n\}$ *and* $Q = \{q_1, q_2, \ldots, q_n\}$, *where* $p_j = <N(\mu_{pj}, \sigma_{pj}), (t_{pj}, u_{pj}, v_{pj})>$ *and* $q_j = <N(\mu_{qj}, \sigma_{qj}), (t_{qj}, u_{qj}, v_{qj})>$ *for* $j = 1, 2, \ldots, n$ *are NNNs in P and Q. The correlation coefficients between two NNSs P and Q are defined as*

$$\begin{aligned} \rho_1(P,Q) &= \frac{C(P,Q)}{[C(P,P)C(Q,Q)]^{1/2}} \\ &= \frac{\sum_{j=1}^{n} \left[(2+t_{Pj}-u_{Pj}-v_{Pj})(2+t_{qj}-u_{qj}-v_{qj})(\mu_{pj}\mu_{qj}+\sigma_{pj}\sigma_{qj}) \right]}{\left\{ \begin{array}{l} \sqrt{\sum_{j=1}^{n} \left\{ \left[(2 + t_{Pj} - u_{Pj} - v_{pj})\mu_{pj} \right]^2 + \left[(2 + t_{pj} - u_{pj} - v_{pj})\sigma_{pj} \right]^2 \right\}} \\ \times \sqrt{\sum_{j=1}^{n} \left\{ \left[(2 + t_{qj} - u_{qj} - v_{qj})\mu_{qj} \right]^2 + \left[(2 + t_{qj} - u_{qj} - v_{qj})\sigma_{qj} \right]^2 \right\}} \end{array} \right\}} \end{aligned} \tag{5}$$

$$\rho_2(P,Q) = \frac{C(P,Q)}{\max[C(P,P),C(Q,Q)]}$$
$$= \frac{\sum_{j=1}^{n}\left[(2+t_{Pj}-u_{Pj}-v_{Pj})(2+t_{qj}-u_{qj}-v_{qj})(\mu_{pj}\mu_{qj}+\sigma_{pj}\sigma_{qj})\right]}{\max\left\{\begin{array}{l}\sum_{j=1}^{n}\left\{\left[(2+t_{Pj}-u_{Pj}-v_{Pj})\mu_{Pj}\right]^2+\left[(2+t_{pj}-u_{pj}-v_{pj})\sigma_{pj}\right]^2\right\}, \\ \sum_{j=1}^{n}\left\{\left[(2+t_{qj}-u_{qj}-v_{qj})\mu_{qj}\right]^2+\left[(2+t_{qj}-u_{qj}-v_{qj})\sigma_{qj}\right]^2\right\}\end{array}\right\}} \tag{6}$$

Proposition 1. *The correlation coefficients of $\rho_k(P, Q)$ $(k = 1, 2)$ satisfy the following properties:*

1. $0 \le \rho_k(P, Q) \le 1$;
2. $\rho_k(P, Q) = 1$ *if* $P = Q$, *i.e.*, $N(\mu_{pj}, \sigma_{pj}) = N(\mu_{qj}, \sigma_{qj})$ *and* $(t_{pj}, u_{pj}, v_{pj}) = (t_{qj}, u_{qj}, v_{qj})$;
3. $\rho_k(P, Q) = \rho_k(Q, P)$.

Proof.
Firstly, we prove that the correlation coefficient of $\rho_1(P, Q)$ satisfies the properties (1)–(3).
The inequality $\rho_1(P, Q) \ge 0$ is obvious. Then, we only prove $\rho_1(P, Q) \le 1$.
Based on the Cauchy–Schwarz inequality:

$$(x_1y_1 + x_2y_2 + \cdots + x_ny_n)^2 \le \left(x_1^2 + x_2^2 + \cdots x_n^2\right) \times \left(y_1^2 + y_2^2 + \cdots y_n^2\right)$$

where $(x_1, x_2, \ldots, x_n) \in R^n$ and $(y_1, y_2, \ldots, y_n) \in R^n$, we can yield the following inequality:

$$(x_1y_1 + x_2y_2 + \cdots + x_ny_n) \le \sqrt{(x_1^2 + x_2^2 + \cdots x_n^2)} \times \sqrt{(y_1^2 + y_2^2 + \cdots y_n^2)}$$

Corresponding to the above inequality and the definition of correlations coefficients in Definition 3, we have the following inequality:

$$\sum_{j=1}^{n}\left[(2+t_{pj}-u_{pj}-v_{pj})\mu_{pj} \times (2+t_{qj}-u_{qj}-v_{qj})\mu_{qj}\right] + \sum_{j=1}^{n}\left[(2+t_{pj}-u_{pj}-v_{pj})\sigma_{pj} \times (2+t_{qj}-u_{qj}-v_{qj})\sigma_{qj})\right] \le$$
$$\sqrt{\sum_{j=1}^{n}\left[(2+t_{pj}-u_{pj}-v_{pj})\mu_{pj}\right]^2 + \sum_{j=1}^{n}\left[(2+t_{pj}-u_{pj}-v_{pj})\sigma_{pj}\right]^2} \times \sqrt{\sum_{j=1}^{n}\left[(2+t_{qj}-u_{qj}-v_{qj})\mu_{qj}\right]^2 + \sum_{j=1}^{n}\left[(2+t_{qj}-u_{qj}-v_{qj})\sigma_{qj}\right]^2}$$

Hence, there is the following result:

$$\sum_{j=1}^{n}\left[(2+t_{pj}-u_{pj}-v_{pj})(2+t_{qj}-u_{qj}-v_{qj})(\mu_{pj}\mu_{qj}+\sigma_{pj}\sigma_{qj})\right] \le$$
$$\sqrt{\sum_{j=1}^{n}\left\{\left[(2+t_{pj}-u_{pj}-v_{pj})\mu_{pj}\right]^2 + \left[(2+t_{pj}-u_{pj}-v_{pj})\sigma_{pj}\right]^2\right\}} \times \sqrt{\sum_{j=1}^{n}\left\{\left[(2+t_{qj}-u_{qj}-v_{qj})\mu_{qj}\right]^2 + \left[(2+t_{qj}-u_{qj}-v_{qj})\sigma_{qj}\right]^2\right\}}$$

Based on Equation (5), we have $\rho_1(P, Q) \le 1$. Hence, $0 \le \rho_1(P, Q) \le 1$ holds.

(2) $P = Q \Rightarrow N(\mu_{pj}, \sigma_{pj}) = N(\mu_{qj}, \sigma_{qj})$ and $(t_{pj}, u_{pj}, v_{pj}) = (t_{qj}, u_{qj}, v_{qj}) \Rightarrow \mu_{pj} = \mu_{qj}, \sigma_{pj} = \sigma_{qj}, t_{pj} = t_{qj}$, $u_{pj} = u_{qj}$, and $v_{pj} = v_{qj}$ for $j = 1, 2, \ldots, n \Rightarrow \rho_1(P, Q) = 1$.

(3) It is straightforward.

Secondly, we prove that the correlation coefficient of $\rho_2(P, Q)$ satisfies the properties (1)–(3).

By the similar proof manner of the properties (1)–(3) of $\rho_1(P, Q)$, we can prove the properties (1)–(3) of $\rho_2(P, Q)$. It is not repeated here.

Therefore, we complete these proofs. □

When the weight of the elements p_j and q_j (j = 1, 2, ..., n) is taken into account, $\boldsymbol{w} = \{w_1, w_2, \ldots, w_n\}$ is given as the weight vector of the elements p_j and q_j (j = 1, 2, ..., n) with $w_j \in [0,1]$ and $\sum_{j=1}^{n} w_j = 1$. Then, we have the following weighted correlation coefficients of NNSs:

$$\rho_{1w}(P,Q) = \frac{\sum_{j=1}^{n} w_j\left[(2+t_{pj}-u_{pj}-v_{pj})(2+t_{qj}-u_{qj}-v_{qj})(\mu_{pj}\mu_{qj}+\sigma_{pj}\sigma_{qj})\right]}{\left\{\begin{array}{l}\sqrt{\sum_{j=1}^{n} w_j\left\{\left[(2+t_{pj}-u_{pj}-v_{pj})\mu_{pj}\right]^2+\left[(2+t_{pj}-u_{pj}-v_{pj})\sigma_{pj}\right]^2\right\}} \\ \times\sqrt{\sum_{j=1}^{n} w_j\left\{\left[(2+t_{qj}-u_{qj}-v_{qj})\mu_{qj}\right]^2+\left[(2+t_{qj}-u_{qj}-v_{qj})\sigma_{qj}\right]^2\right\}}\end{array}\right\}} \tag{7}$$

$$\rho_{2w}(P,Q) = \frac{\sum_{j=1}^{n} w_j\left[(2+t_{pj}-u_{pj}-v_{pj})(2+t_{qj}-u_{qj}-v_{qj})(\mu_{pj}\mu_{qj}+\sigma_{pj}\sigma_{qj})\right]}{\max\left\{\begin{array}{l}\sum_{j=1}^{n} w_j\left\{\left[(2+t_{pj}-u_{pj}-v_{pj})\mu_{pj}\right]^2+\left[(2+t_{pj}-u_{pj}-v_{pj})\sigma_{pj}\right]^2\right\}, \\ \sum_{j=1}^{n} w_j\left\{\left[(2+t_{qj}-u_{qj}-v_{qj})\mu_{qj}\right]^2+\left[(2+t_{qj}-u_{qj}-v_{qj})\sigma_{qj}\right]^2\right\}\end{array}\right\}} \tag{8}$$

Proposition 2. *The weighted correlation coefficients of $\rho_{kw}(P, Q)$ (k = 1, 2) also satisfy the following properties:*

1. $0 \le \rho_{kw}(P, Q) \le 1$;
2. $\rho_{kw}(P, Q) = 1$ *if and only if* $P = Q$, *i.e.*, $N(\mu_{pj}, \sigma_{pj}) = N(\mu_{qj}, \sigma_{qj})$ *and* $(t_{pj}, u_{pj}, v_{pj}) = (t_{qj}, u_{qj}, v_{qj})$;
3. $\rho_{kw}(P, Q) = \rho_{kw}(Q, P)$.

By the similar proofs of the properties in Proposition 1, we can prove the ones in Proposition 2. They are not repeated here.

Especially when $w = \{1/n, 1/n, \ldots, 1/n\}$, Equations (7) and (8) are reduced to Equations (5) and (6).

4. The MADM Method Using the Correlation Coefficients of NNSs

In this section, we present a handling method for the MADM problems with normal neutrosophic information by means of the weighted correlation coefficients between NNSs.

In an MADM problem with normal neutrosophic information, assume that $P = \{P_1, P_2, \ldots, P_m\}$ is a set of m alternatives and $R = \{R_1, R_2, \ldots, R_n\}$ is a set of n attributes. The weight vector of the attributes is given as $\boldsymbol{w} = (w_1, w_2, \ldots, w_n)$ satisfying $w_j \in [0,1]$ and $\sum_{j=1}^{n} w_j = 1$. Then, the average value μ_{ij} and standard derivation σ_{ij} in the normal distribution $N(\mu_{ij}, \sigma_{ij})$ are obtained by the statistical analysis of data corresponding to the alternative P_i (i = 1, 2, ..., m) over the attribute R_j (j = 1, 2, ..., n), while the evaluation values of SVNNs corresponding to the alternative P_i (i = 1, 2, ..., m) over the attribute R_j (j = 1, 2, ..., n) are given by decision-makers. Based on the obtained NNNs $p_{ij} = \langle N(\mu_{ij}, \sigma_{ij}), (t_{ij}, u_{ij}, v_{ij})\rangle$ (i = 1, 2, ..., m; j = 1, 2, ..., n), we can yield the normal neutrosophic decision matrix $M(p_{ij})_{m\times n}$:

$$M(p_{ij})_{m\times n} = \begin{bmatrix} \langle N(\mu_{11},\sigma_{11}),(t_{11},u_{11},v_{11})\rangle & \langle N(\mu_{12},\sigma_{12}),(t_{12},u_{12},v_{12})\rangle & \cdots & \langle N(\mu_{1n},\sigma_{1n}),(t_{1n},u_{1n},v_{1n})\rangle \\ \langle N(\mu_{21},\sigma_{21}),(t_{21},u_{21},v_{21}\rangle & \langle N(\mu_{22},\sigma_{22}),(t_{22},u_{22},v_{22})\rangle & \cdots & \langle N(\mu_{2n},\sigma_{2n}),(t_{2n},u_{2n},v_{2n})\rangle \\ \vdots & \vdots & \vdots & \vdots \\ \langle N(\mu_{m1},\sigma_{m1}),(t_{m1},u_{m1},v_{m1})\rangle & \langle N(\mu_{m2},\sigma_{m2}),(t_{m2},u_{m2},v_{m2})\rangle & \cdots & \langle N(\mu_{mn},\sigma_{mn}),(t_{mn},u_{mn},v_{mn})\rangle \end{bmatrix}$$

In MADM problems, the concept of the ideal point has been used to help the identification of the best alternative in the decision set. It does provide a useful method to evaluate alternatives [13]. However, there are two types of attributes, i.e., benefit type and cost type, in decision-making problems. Hence, we firstly need to determinate an ideal solution/alternative (an ideal NNS) corresponding to the benefit type and cost type of attributes. Then, by correlation coefficient values between each alternative (each evaluated NNS) and the ideal alternative (the ideal NNS), the ranking order of alternatives and the best one are given in the normal neutrosophic decision-making process.

Thus, we use the developed method to deal with the MADM problem with normal neutrosophic information, which is described by the following procedures:

Step 1: Establish an ideal solution (an ideal alternative) $P^* = \{p_1^*, p_2^*, \ldots, p_n^*\}$ by the ideal NNN $p_j^* = \left\langle N\left(\max\limits_i(m_{ij}), \min\limits_i(\sigma_{ij})\right), \left(\max\limits_i(t_{ij}), \min\limits_i(u_{ij}), \min\limits_i(v_{ij})\right)\right\rangle$ corresponding to the benefit type of attributes and $p_j^* = \left\langle N\left(\min\limits_i(m_{ij}), \min\limits_i(\sigma_{ij})\right), \left(\min\limits_i(t_{ij}), \max\limits_i(u_{ij}), \max\limits_i(v_{ij})\right)\right\rangle$ corresponding to the cost type of attributes.

Step 2: Calculate the weighted correlation coefficients between an alternative P_i (i = 1, 2, ... , m) and the ideal solution P^* by using Equation (7) or Equation (8) and obtain the values of $\rho_{1w}(P_i, P^*)$ or $\rho_{2w}(P_i, P^*)$ (i = 1, 2, ... , m).

Step 3: Rank the alternatives in a descending order corresponding to the weighted correlation coefficient values and select the best one(s) according to the bigger value of $\rho_{1w}(P_i, P^*)$ or $\rho_{2w}(P_i, P^*)$.

Step 4: End.

5. Illustrative Example

For convenient comparison, an illustrative example about the selection problem of investment alternatives adopted from [18] is provided to demonstrate the applications and effectiveness of the proposed MADM method with normal neutrosophic information.

An investment company wants to invest a sum of money to the best industry. Then, four possible alternatives are considered as four potential industries: (1) P_1 is a car company; (2) P_2 is a food company; (3) P_3 is a computer company; (4) P_4 is an arms company. In the decision-making process, the four possible alternatives must satisfy the requirements of the three attributes: (1) R_1 is the risk; (2) R_2 is the growth; (3) R_3 is the environment, where the attributes R_1 and R_2 are benefit types and the attribute R_3 is a cost type. Assume that the weighting vector of the attributes is given by w = (0.35, 0.25, 0.4). By the statistical analysis and the evaluation of investment data regarding the four possible alternatives of P_i (i = 1, 2, 3, 4) over the three attributes of R_j (j = 1, 2, 3), we can establish the following NNN decision matrix [18]:

$$M(p_{ij})_{4\times 3} = \begin{bmatrix} \langle N(3,0.4),(0.4,0.2,0.3)\rangle & \langle N(7,0.6),(0.4,0.1,0.2)\rangle & \langle N(5,0.4),(0.7,0.2,0.4)\rangle \\ \langle N(4,0.2),(0.6,0.1,0.2)\rangle & \langle N(8,0.4),(0.6,0.1,0.2)\rangle & \langle N(6,0.7),(0.3,0.5,0.8)\rangle \\ \langle N(3.5,0.3),(0.3,0.2,0.3)\rangle & \langle N(6,0.2),(0.5,0.2,0.3)\rangle & \langle N(5.5,0.6),(0.4,0.2,0.7)\rangle \\ \langle N(5,0.5),(0.7,0.1,0.2)\rangle & \langle N(7,0.5),(0.6,0.1,0.1)\rangle & \langle N(4.5,0.5),(0.6,0.3,0.8)\rangle \end{bmatrix}$$

Then, we use Equation (7) to deal with the MADM problem with normal neutrosophic information, which is described by the following procedures:

Step 1: Establish an ideal solution (an ideal alternative) $P^* = \{p_1^*, p_2^*, \ldots, p_n^*\}$ expressed by the ideal NNS $P^* = \{\langle N(5,0.2),(0.7,0.1,0.2)\rangle, \langle N(8,0.2),(0.6,0.1,0.1)\rangle, \langle N(4.5,0.4),(0.3,0.5,0.8)\rangle\}$ corresponding to the benefit types and cost types of attributes.

Step 2: Calculate the weighted correlation coefficient between the alternative P_1 and the ideal solution P^* by using Equation (7) as follows:

$$\rho_{1w}(P_1,P^*) = \frac{\sum_{j=1}^{3} w_j\left[(2+t_{p_{1j}}-u_{p_{1j}}-v_{p_{1j}})(2+t_{p_j^*}-u_{p_j^*}-v_{p_j^*})(\mu_{p_{1j}}\mu_{p_j^*}+\sigma_{p_{1j}}\sigma_{p_j^*})\right]}{\left\{\begin{array}{l}\sqrt{\sum_{j=1}^{3} w_j\left\{\left[(2+t_{p_{1j}}-u_{P_{1j}}-v_{p_{1j}})\mu_{p_{1j}}\right]^2+\left[(2+t_{p_{1j}}-u_{p_{1j}}-v_{p_{1j}})\sigma_{p_{1j}}\right]^2\right\}}\\ \times\sqrt{\sum_{j=1}^{3} w_j\left\{\left[(2+t_{p_j^*}-u_{p_j^*}-v_{p_j^*})\mu_{p_j^*}\right]^2+\left[(2+t_{p_j^*}-u_{p_j^*}-v_{p_j^*})\sigma_{p_j^*}\right]^2\right\}}\end{array}\right\}}$$

$$= \frac{\left\{\begin{array}{l}0.35\times[(2+0.4-0.2-0.3)\times(2+0.7-0.1-0.2)\times(3\times5+0.4\times0.2)]\\ +0.25\times[(2+0.4-0.1-0.2)\times(2+0.6-0.1-0.1)\times(7\times8+0.6\times0.2)]\\ +0.4\times[(2+0.7-0.2-0.4)\times(2+0.3-0.5-0.8)\times(5\times4.5+0.4\times0.4)]\end{array}\right\}}{\left\{\begin{array}{l}\sqrt{\begin{array}{l}0.35\times\left\{[(2+0.4-0.2-0.3)\times3]^2+[(2+0.4-0.2-0.3)\times0.4]^2\right\}\\ +0.25\times\left\{[(2+0.4-0.1-0.2)\times7]^2+[(2+0.4-0.1-0.2)\times0.6]^2\right\}\\ +0.4\times\left\{[(2+0.7-0.2-0.4)\times5]^2+[(2+0.7-0.2-0.4)\times0.4]^2\right\}\end{array}}\\ \times\sqrt{\begin{array}{l}0.35\times\left\{[(2+0.7-0.1-0.2)\times5]^2+[(2+0.7-0.1-0.2)\times0.2]^2\right\}\\ +0.25\times\left\{[(2+0.6-0.1-0.1)\times8]^2+[(2+0.6-0.1-0.1)\times0.2]^2\right\}\\ +0.4\times\left\{[(2+0.3-0.5-0.8)\times4.5]^2+[(2+0.3-0.5-0.8)\times0.4]^2\right\}\end{array}}\end{array}\right\}}$$

$$= 0.8820.$$

By similar calculations, the weighted correlation coefficients between each alternative P_i (i = 2, 3, 4) and the ideal solution P^* can be given as the following values of $\rho_{1w}(P_i, P^*)$ (i = 2, 3, 4):

$\rho_{1w}(P_2, P^*) = 0.9891$, $\rho_{1w}(P_3, P^*) = 0.9169$, and $\rho_{1w}(P_4, P^*) = 0.9875$.

Step 3: According to the values of $\rho_{1w}(P_i, P^*)$ (i = 1, 2, 3, 4), the ranking order of the alternatives is $P_2 > P_4 > P_3 > P_1$ and the best one is P_2. These results are the same as in [18].

We could also use Equation (8) to deal with the MADM problem with normal neutrosophic information, which is described by the following steps:

Step 1′: The same as Step 1.

Step 2′: Calculate the weighted correlation coefficient between the alternative P_1 and the ideal solution P^* by using Equation (8) as follows:

$$\rho_{2w}(P_1,P^*) = \frac{\sum_{j=1}^{3} w_j\left[(2+t_{p_{1j}}-u_{p_{1j}}-v_{p_{1j}})(2+t_{p_j^*}-u_{p_j^*}-v_{p_j^*})(\mu_{p_{1j}}\mu_{p_j^*}+\sigma_{p_{1j}}\sigma_{p_j^*})\right]}{\max\left\{\begin{array}{l}\sum_{j=1}^{3} w_j\left\{\left[(2+t_{p_{1j}}-u_{P_{1j}}-v_{p_{1j}})\mu_{p_{1j}}\right]^2+\left[(2+t_{p_{1j}}-u_{p_{1j}}-v_{p_{1j}})\sigma_{p_{1j}}\right]^2\right\},\\ \sum_{j=1}^{3} w_j\left\{\left[(2+t_{p_j^*}-u_{p_j^*}-v_{p_j^*})\mu_{p_j^*}\right]^2+\left[(2+t_{p_j^*}-u_{p_j^*}-v_{p_j^*})\sigma_{p_j^*}\right]^2\right\}\end{array}\right\}}$$

$$= \frac{\left\{\begin{array}{l}0.35\times[(2+0.4-0.2-0.3)\times(2+0.7-0.1-0.2)\times(3\times5+0.4\times0.2)]\\ +0.25\times[(2+0.4-0.1-0.2)\times(2+0.6-0.1-0.1)\times(7\times8+0.6\times0.2)]\\ +0.4\times[(2+0.7-0.2-0.4)\times(2+0.3-0.5-0.8)\times(5\times4.5+0.4\times0.4)]\end{array}\right\}}{\max\left\{\begin{array}{l}\left(\begin{array}{l}0.35\times\left\{[(2+0.4-0.2-0.3)\times3]^2+[(2+0.4-0.2-0.3)\times0.4]^2\right\}\\ +0.25\times\left\{[(2+0.4-0.1-0.2)\times7]^2+[(2+0.4-0.1-0.2)\times0.6]^2\right\}\\ +0.4\times\left\{[(2+0.7-0.2-0.4)\times5]^2+[(2+0.7-0.2-0.4)\times0.4]^2\right\}\end{array}\right),\\ \left(\begin{array}{l}0.35\times\left\{[(2+0.7-0.1-0.2)\times5]^2+[(2+0.7-0.1-0.2)\times0.2]^2\right\}\\ +0.25\times\left\{[(2+0.6-0.1-0.1)\times8]^2+[(2+0.6-0.1-0.1)\times0.2]^2\right\}\\ +0.4\times\left\{[(2+0.3-0.5-0.8)\times4.5]^2+[(2+0.3-0.5-0.8)\times0.4]^2\right\}\end{array}\right)\end{array}\right\}}$$

$$= 0.7544.$$

By similar calculations, the weighted correlation coefficients between each alternative P_i (i = 2, 3, 4) and the ideal solution P^* can be given as the following values of $\rho_{2w}(P_i, P^*)$ (i = 2, 3, 4):

$\rho_{2w}(P_2, P^*) = 0.9151$, $\rho_{2w}(P_3, P^*) = 0.6575$, and $\rho_{2w}(P_4, P^*) = 0.9522$.

Step 3′: According to the values of $\rho_{2w}(P_i, P^*)$ ($i = 1, 2, 3, 4$), the ranking order of the alternatives is $P_4 > P_2 > P_1 > P_3$, and the best one is P_4. These results also are the same as in [18].

Obviously, the above two ranking orders are different corresponding to different correlation coefficients for this decision-making problem; these results are thus in accordance with the ones in [18]. Hence, the proposed normal neutrosophic decision-making method based on the correlation coefficients illustrates its feasibility and effectiveness. Compared with existing decision-making methods based on aggregation operators of NNNs, the proposed decision-making method based on the correlation coefficients of NNSs shows that it is simpler to employ than existing normal neutrosophic decision-making methods in [18–20] under normal neutrosophic environments because the decision-making method proposed in this paper implies its simple algorithms and decision steps in the normal neutrosophic decision-making problems.

From the decision results of the illustrative example, we see that different correlation coefficients used in the decision-making problem can result in different ranking orders and selecting alternatives. Hence, the decision-maker can select one of both corresponding to his/her preference or actual requirements.

6. Conclusions

To simplify the complex decision-making process/steps and algorithms of existing normal neutrosophic decision-making methods in [18–20], this paper proposed two correlation coefficients between NNSs based on the score functions of NNNs under normal neutrosophic environments. Then, we developed an MADM method with normal neutrosophic information by using the correlation coefficients of NNSs under normal neutrosophic environments. An illustrative example about the selection problem of investment alternatives was provided to demonstrate the applications and effectiveness of the proposed MADM method under normal neutrosophic environments.

The main advantages of this study are (1) the evaluation information expressed by NNNs is relatively more reasonable and more realistic than the evaluation information expressed by general neutrosophic numbers in the decision-making process; (2) the proposed decision-making method based on the correlation coefficients of NNSs is simpler to employ than existing ones based on aggregation operators of NNNs in the normal neutrosophic decision-making algorithms; (3) the proposed decision-making method with NNNs contains much more information and shows its rationality and reality, while the existing decision-making methods with single neutrosophic information may lose some useful evaluation information of attributes in the decision-making process.

In future work, the study about new similarity measures of NNSs and applications in science and engineering fields are necessary and significant because the applications of the normal distribution widely exist in many domains.

Acknowledgments: This paper was supported by the National Natural Science Foundation of China (No. 71471172).

Author Contributions: The author proposed two correlation coefficients between NNSs and their decision-making method.

Conflicts of Interest: The author declares no conflict of interest.

References

1. Hazewinkel, M. Normal Distribution, Encyclopedia of Mathematics. Springer: Heidelberg/Berlin, Germany, 2001.
2. Zadeh, L.A. Fuzzy sets. *Inf. Control* **1965**, *8*, 338–356. [CrossRef]
3. Yang, M.S.; Ko, C.H. On a class of fuzzy c-numbers clustering procedures for fuzzy data. *Fuzzy Sets Syst.* **1996**, *84*, 49–60. [CrossRef]
4. Atanassov, K.T. Intuitionistic fuzzy sets. *Fuzzy Sets Syst.* **1986**, *20*, 87–96. [CrossRef]

5. Wang, J.Q.; Li, K.J. Multi-criteria decision-making method based on induced intuitionistic normal fuzzy related aggregation operators. *Int. J. Uncertain Fuzziness Knowl. Based Syst.* **2012**, *20*, 559–578. [CrossRef]
6. Wang, J.Q.; Li, K.J.; Zhang, H.Y.; Chen, X.H. A score function based on relative entropy and its application in intuitionistic normal fuzzy multiple criteria decision-making. *J. Intell. Fuzzy Syst.* **2013**, *25*, 567–576.
7. Wang, J.Q.; Li, K.J. Multi-criteria decision-making method based on intuitionistic normal fuzzy aggregation operators. *Syst. Eng. Theory Pract.* **2013**, *33*, 1501–1508. [CrossRef]
8. Wang, J.Q.; Zhou, P.; Li, K.J.; Zhang, H.Y.; Chen, X.H. Multicriteria decision-making method based on normal intuitionistic fuzzy induced generalized aggregation operator. *Top* **2014**, *22*, 1103–1122. [CrossRef]
9. Smarandache, F. *Neutrosophy: Neutrosophic Probability, Set, and Logic: Analytic Synthesis & Synthetic Analysis*; American Research Press: Rehoboth, DE, USA, 1998.
10. Wang, H.; Smarandache, F.; Zhang, Y.Q.; Sunderraman, R. Single valued neutrosophic sets. *Multisp. Multistruct.* **2010**, *4*, 410–413.
11. Ye, J. Fuzzy decision-making method based on the weighted correlation coefficient under intuitionistic fuzzy environment. *Eur. J. Oper. Res.* **2010**, *205*, 202–204. [CrossRef]
12. Ye, J. Another form of correlation coefficient between single valued neutrosophic sets and its multiple attribute decision-making method. *Neutrosophic Sets Syst.* **2013**, *1*, 8–12.
13. Ye, J. Multicriteria decision-making method using the correlation coefficient under single-value neutrosophic environment. *Int. J. Gen. Syst.* **2013**, *42*, 386–394. [CrossRef]
14. Bausys, R.; Zavadskas, E.K.; Kaklauskas, A. Application of neutrosophic set to multicriteria decision-making by COPRAS. *Econ. Comput. Econ. cybern. Stud. Res.* **2015**, *49*, 91–106.
15. Tian, Z.P.; Zhang, H.Y.; Wang, J.; Wang, J.Q.; Chen, X.H. Multi-criteria decision-making method based on a cross-entropy with interval neutrosophic sets. *Int. J. Syst. Sci.* **2016**, *47*, 3598–3608. [CrossRef]
16. Peng, X.D.; Liu, C. Algorithms for neutrosophic soft decision-making based on EDAS, new similarity measure and level soft set. *J. Intell. Fuzzy Syst.* **2017**, *32*, 955–968. [CrossRef]
17. Pouresmaeil, H.; Shivanian, E.; Khorram, E.; Fathabadi, H.S. An extended method using TOPSIS and VIKOR for multiple attribute decision-making with multiple decision-makers and single valued neutrosophic numbers. *Adv. Appl. Stat.* **2017**, *50*, 261–292. [CrossRef]
18. Liu, P.D.; Teng, F. Multiple attribute decision-making method based on normal neutrosophic generalized weighted power averaging operator. *Int. J. Mach. Learn. Cyber.* **2015**, 1–13. [CrossRef]
19. Liu, P.D.; Li, H.G. Multiple attribute decision-making method based on some normal neutrosophic Bonferroni mean operators. *Neural Comput. Applic.* **2017**, *28*, 179–194. [CrossRef]
20. Sahin, R. Normal neutrosophic multiple attribute decision-making based on generalized prioritized aggregation operators. *Neural Comput. Applic.* **2017**, 1–21. [CrossRef]

symmetry

MDPI

Article

Correlation Coefficients of Probabilistic Hesitant Fuzzy Elements and Their Applications to Evaluation of the Alternatives

Zhong-xing Wang [1] **and Jian Li** [2,*]

[1] School of Mathematics and Information Science, Guangxi University, Nanning 530004, China; zxwgx@126.com

[2] School of XingJian College of Science and Liberal Arts, Guangxi University, Nanning 530005, China

* Correspondence: jian2016@csu.edu.cn

Received: 25 September 2017; Accepted: 29 October 2017; Published: 2 November 2017

Abstract: Correlation coefficient is one of the broadly use indexes in multi-criteria decision-making (MCDM) processes. However, some important issues related to correlation coefficient utilization within probabilistic hesitant fuzzy environments remain to be addressed. The purpose of this study is introduced a MCDM method based on correlation coefficients utilize probabilistic hesitant fuzzy information. First, the covariance and correlation coefficient between two PHFEs is introduced, the properties of the proposed covariance and correlation coefficient are discussed. In addition, the northwest corner rule to obtain the expected mean related to the multiply of two PHFEs is introduced. Second, the weighted correlation coefficient is proposed to make the proposed MCDM method more applicable. And the properties of the proposed weighted correlation coefficient are also discussed. Finally, an illustrative example is demonstrated the practicality and effectiveness of the proposed method. An illustrative example is presented to demonstrate the correlation coefficient propose in this paper lies in the interval $[-1, 1]$, which not only consider the strength of relationship between the PHFEs but also whether the PHFEs are positively or negatively related. The advantage of this method is it can avoid the inconsistency of the decision-making result due to the loss of information.

Keywords: probabilistic hesitant fuzzy element; covariance; correlation coefficient; northwest corner rule; multi-criteria decision-making

---

1. Introduction

With the rapid development of economic and the progress of modern society, people are facing more and more complicated decision-making problems, group decision-making plays an increasingly important role when dealing with multi-criteria decision-making (MCDM) problems [1–3]. In our daily life, group decision-making has turned out to be a commonly used tool in human activities, whose purpose is to determine the most preferred alternative among several alternatives (or a series of alternatives) using the evaluation values provided from a group of decision makers. In group decision-making processes, the information provided by the experts has different forms. Because of this, many scholars have investigated the techniques based on various kinds of decision information, including intuitionistic fuzzy sets [4,5], hesitant fuzzy sets (HFSs) [6–8], probabilistic hesitant fuzzy sets (PHFSs) [9,10] and probabilistic linguistic term sets [11] and so on.

Correlation coefficient is one of the broadly use indexes in MCDM processes [12–14]. Since many data may be fuzzy and uncertain, the utilization of correlation coefficient has been extended to fuzzy environments [15–17] and intuitionistic fuzzy environments [18–21]. For example, Huang et al. [18] proposed a correlation coefficient formula utilizing the centroid of intuitionistic fuzzy numbers. Ye [19] utilizing entropy weights of intuitionistic fuzzy numbers proposed the weighted correlation coefficient. In a sequent, Dong et al. [21] proposed weighted correlation coefficient based on the relationship of an arbitrary alternative and the ideal alternative. Afterwards, correlation coefficient has been extended to hesitant fuzzy environments [22–25]. At the same time, some correlation coefficients formulas have been proposed. Such as, Chen et al. [22] derived some correlation coefficients based on the membership degree of the HFSs and applied them in clustering analysis. Liao et al. [23] pointed out there are some shortcoming in the correlation coefficients were introduced in [22] and then proposed a novel correlation coefficient. The significant characteristic of the proposed formula is that it lies in the interval $[-1, 1]$. Based on the same idea, Liao et al. [24] proposed several types of correlation coefficients for hesitant fuzzy linguistic term sets and then applied them to traditional medical diagnosis problems. Because of the potential application of the correlation coefficient, some other extensions are still going on—for example, dual hesitant fuzzy environments [26] and neutrosophic fuzzy environments [27–29] and so on.

Although the concept of correlation coefficient has been extended to various kinds of fuzzy environments and has been applied in many fields. There are still some disputes in the utilization of it. Some decision makers noticed that the correlation coefficients were proposed in above mention papers mainly lies on the statistics formula, that is, the correlation coefficient between two random variables X and Y is $\rho(X,Y) = \frac{E(X-E(X))(Y-E(Y))}{\sqrt{D(X)}\sqrt{D(Y)}}$. In addition, the correlation coefficient has lots of important properties, such as lies in the range of $[-1, 1]$. Unfortunately, most of the correlation coefficients in above mention papers always positive, which lies in the range of [0, 1] and ignored the negative correlation information. This shortcoming has been pointed out by some scholars [18,23,25] and the ignored information may result in unreasonable decision-making results. Based on this consideration, the correlation coefficient be applied in fuzzy environment should be further discussion.

Since Zhu [30] first proposed the concept of PHFSs, it has been attracted some scholars' attention and many achievements have been made. For example, Zhang et al. [31] pointed out that there are some shortcoming in the concept of PHFSs that was proposed in [30], they asserted that there were maybe some incomplete information in the decision-making processes and then they proposed the improvement PHFSs. He et al. [32] extended the PHFSs to the probabilistic interval preference ordering sets and Hao et al. [33] extended it to the probabilistic dual hesitant fuzzy sets. In addition, PHFSs have been extended to probabilistic linguistic term sets [11,34,35]. Although the concept of PHFSs has been extended to various kinds of fuzzy environments and some decision-making methods have been proposed. For example, Zhou et al. [36] discussed group consensus based on additive consistency and Li et al. [9] introduced a MCDM process based on Hausdorff distance. However, some important issues in PHFSs utilization remain to be addressed. For example, the probability part does not pay enough attention, the existing decision-making methods mainly directly integrated the probability part into the membership degree part [9,36], this make cause a lot of information loss.

Considering that PHFSs consists of two parts, that is, the membership degree of the elements to the set and the corresponding probabilities of the membership degree, this information can be interpreted as a probability distribution function. Inspired by statistics knowledge, each probabilistic hesitant fuzzy element (PHFE) can be treated as a discrete random variable. Since every PHFE has two parts, that is: γ_i and p_i, where γ_i can be regarded as the condition of a random variable, p_i can be regarded as the corresponding probability with γ_i, the similar opinion has been proposed by Hung [37]. Based on this consideration, we can apply some concepts in statistics such as expected value, variance, covariance and correlation coefficient, to construct a novel MCDM method within the background of a probabilistic hesitant environment. Considering sometimes two random variables maybe do not mutual independent. In this paper, the expected mean related to the multiply of two PHFEs can be obtained through using the northwest corner rule, from the course of operations research, balance problems of transport model [38].

To overcome the above mention limitations, this paper focuses on the correlation coefficient between two PHFSs and based on the northwest corner rule to obtain the expected mean related to the multiply of two PHFEs when the PHFEs are not mutual independent. Finally, a novel MCDM method with the probabilistic hesitant fuzzy environment is introduced based on the proposed weighted correlation coefficient. The primary motivations and contributions of this paper are summarized as follows.

(1) A novel formula to calculate the correlation coefficient between two PHFSs is proposed. The correlation coefficient is proposed in this paper utilize the knowledge of statistics, the significant characteristic of the proposed formula is that it lies in the interval $[-1, 1]$. The proposed formula not only consider the strength of the PHFSs but also whether the PHFSs are positively or negatively related, it avoids the inconsistency of the decision-making result due to the loss of information.
(2) The existing decision-making methods within probabilistic hesitant fuzzy environments, few papers discussed the condition when two PHFEs are not mutual independent. In this paper, the northwest corner rule to obtain the expected mean related to the multiply of two PHFEs is introduced.
(3) A novel MCDM method within the probabilistic hesitant fuzzy environment is introduced based on the proposed weighted correlation coefficient and this proposed method is applied to practical decision-making problems, that is, the evaluation of the alternatives.

The rest of this paper is organized as follows. Section 2 reviews some basic concepts related to HFSs and PHFSs and some correlation coefficient formulas related to HFSs. In Section 3, we introduce a novel correlation coefficient formula for PHFSs, the properties of the proposed covariance and correlation coefficient are discussed. And the northwest corner rule to obtain the expected mean related to the multiply of two PHFEs is introduced. In Section 4, the weighted correlation coefficient is proposed and the properties of the proposed weighted correlation coefficient are discussed. The weighted correlation coefficient of two PHFEs is applied to an evaluation of the alternatives problem in Section 5. Finally, the conclusions are given in Section 6.

2. Preliminaries

In this section, several basic definitions and notations related to our research will be reviewed, mainly including HFS, correlation coefficient and the concept of PHFS, its score function and indeterminacy index function. In addition, an evaluation information integrate method is introduced.

Definition 1. *[39] Let X be a reference set, a HFS A on X is defined in terms of a function $h_A(x)$ when applied to X returns a finite subset of [0, 1].*

To be easily understand, Xia et al. [40] expressed the HFS by a mathematical symbol:

$$A = \{< x, h_A(x) > | x \in X\}. \quad (1)$$

Here, the function $h_A(x)$ is a set of some different values in [0, 1], representing the possible membership degrees of the element x in X to A. For convenience, $h_A(x)$ is called a HFE.

Example 1. *Let $X = \{x_1, x_2, x_3\}$ be a reference set, $h_A(x_1) = \{0.2, 0.4, 0.6\}$, $h_A(x_2) = \{0.3, 0.4, 0.5\}$ and $h_A(x_3) = \{0.2, 0.3, 0.5, 0.6\}$.*

Be three HFEs of $x_i (i = 1, 2, 3)$ to a set A, respectively. Then A can be considered as a HFS,

$$A = \{\langle x_1, \{0.2, 0.4, 0.6\}\rangle, \langle x_2, \{0.3, 0.4, 0.5\}\rangle, \langle x_3, \{0.2, 0.3, 0.5, 0.6\}\rangle\}.$$

Correlation coefficient is a frequently use formulas and has been applied in measure the similarity between two objects. Liao et al. [23] defined the correlation coefficient between two HFSs as follows.

Definition 2. *[23] Let* $X = \{x_1, x_2, \cdots, x_n\}$ *be a discrete universe of discourse and* F^i *be the hesitant fuzzy space containing all HFSs defined over X. For any two HFSs,* $h_A(x_i) = \left\{\gamma_{A_{i1}}, \gamma_{A_{i2}}, \cdots, \gamma_{A_{il_{A_i}}}\right\}$ *and* $h_B(x_i) = \left\{\gamma_{B_{i1}}, \gamma_{B_{i2}}, \cdots, \gamma_{B_{il_{B_i}}}\right\}$*, the correlation coefficient between them is defined as follows:*

$$\rho(A,B) = \frac{Cov(A,B)}{[D(A)D(B)]^{1/2}}, \tag{2}$$

where

$$Cov(A,B) = \frac{1}{n}\sum_{i=1}^{n}\left[\overline{h}_A(x_i) - \overline{A}\right]\cdot\left[\overline{h}_B(x_i) - \overline{B}\right],$$

$$D(A) = \frac{1}{n}\sum_{i=1}^{n}\left[\overline{h}_A(x_i) - \overline{A}\right]\cdot\left[\overline{h}_A(x_i) - \overline{A}\right],$$

$$D(B) = \frac{1}{n}\sum_{i=1}^{n}\left[\overline{h}_B(x_i) - \overline{B}\right]\cdot\left[\overline{h}_B(x_i) - \overline{B}\right].$$

and, $\overline{h}_A(x_i) = \frac{1}{l_{A_i}}\sum_{k=1}^{l_{A_i}}\gamma_{A_{ik}}$, $\overline{h}_B(x_i) = \frac{1}{l_{B_i}}\sum_{k=1}^{l_{B_i}}\gamma_{B_{ik}}$, $\overline{A} = \frac{1}{n}\sum_{i=1}^{n}\overline{h}_A(x_i)$ *and* $\overline{B} = \frac{1}{n}\sum_{i=1}^{n}\overline{h}_B(x_i)$*. Where* l_{A_i} *and* l_{B_i} *are respectively denotes the number of the elements in* $h_A(x_i)$ *and* $h_B(x_i)$*.*

Recently, Zhang et al. [31] proposed the improvement of PHFSs, which have added the partial ignorance information to PHFSs that was proposed by Zhu [30].

Definition 3. *[31] Let X be a reference set, then a PHFS P on X can be expressed by as:*

$$P = \{< x, h_x(p_x) > | x \in X\}. \tag{3}$$

Here, the function h_x is a set of several different values in [0, 1], which is described by the probability distribution p_x. Where h_x denotes the possible membership degree of element $_x$ in X to P. For convenience, $h_x(p_x)$ is called a PHFE and denoted as $h(p)$ and is indicated by

$$h(p) = \{\gamma_i(p_i) | i = 1, 2, \cdots, |h(p)|\},$$

where p_i satisfying $\sum_{i=1}^{|h(p)|} p_i \leq 1$, is the probability of the possible value γ_i and $|h(p)|$ is the number of all $\gamma_i(p_i)$ in $h(p)$. If $\sum_{i=1}^{|h(p)|} p_i < 1$, means there is some missing values in PHFE. If there is no special explanation, in this paper, we only discuss the condition $\sum_{i=1}^{|h(p)|} p_i = 1$.

Example 2. *Let* $X = \{x_1, x_2\}$ *be a reference set,* $h_1(p_1) = \{0.2(0.3), 0.4(0.2), 0.5(0.1), 0.7(0.4)\}$ *and* $h_2(p_2) = \{0.3(0.1), 0.4(0.9)\}$ *be two PHFEs of* $x_i(i = 1, 2)$ *to a set P, respectively. Then P can be considered as a PHFS,*

$$P = \{\langle x_1, \{0.2(0.3), 0.4(0.2), 0.5(0.1), 0.7(0.4)\}\rangle, \langle x_2, \{0.3(0.1), 0.4(0.9)\}\rangle\}.$$

If we ignore the probabilities of the possible values in a PHFE, then the possible values are with the same probability, in this case, PHFE turn to HFE.

In order to rank the PHFEs, Xu et al. [41] introduced the score function and indeterminacy index function of PHFEs. As a matter of fact, the score function and indeterminacy index function can be regarded as expect mean and variance of PHFEs.

Definition 4. *[41] Let $h_A(p_A) = \{\gamma_i(p_i)|i = 1, 2, \cdots, |h(p)|\}$ be a PHFE, the expect mean of it is defined as:*

$$E(A) = \sum_{i=1}^{|h(p)|} \gamma_i p_i. \tag{4}$$

It is noted that if the probabilities of PHFE are equally, that is $p_1 = p_2 = \cdots = p_{|h(p)|} = \frac{1}{|h(p)|}$, in this case, expect mean will turn to the score function of HFS that was introduced in Definition 4 in [40].

Definition 5. *[41] Let $h_A(p_A) = \{\gamma_i(p_i)|i = 1, 2, \cdots, |h(p)|\}$ be a PHFE, the variance of it is defined as:*

$$D(A) = \sum_{i=1}^{|h(p)|} (\gamma_i - E(A))^2 p_i. \tag{5}$$

Example 3. *Let $h_A(p_A) = \{0.5(0.25), 0.6(0.5), 0.7(0.25)\}$ be a PHFE, according to Definition 4 and Definition 5, we have $E(A) = 0.5 \times 0.25 + 0.6 \times 0.5 + 0.7 \times 0.25 = 0.6$, $D(A) = (0.5 - 0.6)^2 \times 0.25 + (0.6 - 0.6)^2 \times 0.5 + (0.7 - 0.6)^2 \times 0.25 = 0.005$.*

Remark 1. *If there is only one element in a PHFE, in this case, we have $D(A) = 0$.*

Remark 2. *According to statistics knowledge, there is an equivalence formula related to the variance of $h_A(p_A)$ in Definition 5, that is:*

$$D(A) = \sum_{i=1}^{|h(p)|} \gamma_i^2 p_i - E(A)^2. \tag{6}$$

In order to integrate the evaluation information obtained from decision makers in the decision-making processes, according to the total probability formula in statistics, Li et al. [42] introduced an information integrate method as follows.

Definition 6. *[42] For a reference set X, let $P = \{\langle x, h_x(p_x)|x = 1, 2, \cdots, n\rangle\}$ be a PHFS, where $h(p) = \{\gamma_i(p_i)|i = 1, 2, \cdots, |h(p)|\}$ is the PHFE indicating all possible values in P. Then, the probability of the value γ_i can be calculated as follows:*

$$P(x = \gamma_i) = \sum_{i=1}^{n} P(x = h(p))P(x = \gamma_i / x = h(p)), \tag{7}$$

where n is the number of all PHFEs in P.

Example 4. *Let $P = \{\langle x_1, \{0.4(0.6), 0.6(0.3), 0.7(0.1)\}\rangle, \langle x_2, \{0.3(0.7), 0.4(0.3)\}\rangle, \langle x_3, \{0.8(0.2), 0.9(0.8)\}\rangle\}$ be a PHFS. The probability of value 0.4 in P is calculated as follows: $P(x = 0.4) = \frac{1}{3} \times 0.6 + \frac{1}{3} \times 0.3 + \frac{1}{3} \times 0 = 0.3$.*

3. The Correlation Coefficient of PHFEs

Since every PHFE has two parts, that is: γ_i and p_i, where γ_i can be regarded as the condition of a random variable, p_i can be regarded as the corresponding probability of γ_i. Based on this consideration, every PHFE can be regarded as a discrete random variable. In the following section, covariance and the correlation coefficients of PHFEs will be introduced.

In order to obtain the correlation coefficients of PHFEs. First, the standard deviation of it is calculated. Second, the expect value related to the multiple of two PHFEs is calculated. Then, the covariance of the PHFEs is obtained. Finally, based on the standard deviation and covariance, the correlation coefficient of PHFEs can be obtained.

According to the statistics knowledge, the standard deviation of PHFEs can be obtained from the square of the deviation. And in order to obtain the expect value related to the multiply of AB, that is,

PHFE $h_A(p_A)$ multiply PHFE $h_B(p_B)$. First, the joint distribution law between $h_A(p_A)$ and $h_B(p_B)$ will be determined and then based on the joint distribution law, we can calculate the expect value related to the multiply of AB. Considering sometimes two PHFEs maybe do not mutual independent, in the following section, a method to determine the joint distribution law between $h_A(p_A)$ and $h_B(p_B)$ will be introduced.

Let $h_A(p_A) = \{\gamma_i(p_i)|i = 1,2,\cdots,|h(p)|\}$ and $h_B(p_B) = \left\{\gamma'_j(p'_j)|j = 1,2,\cdots,|l(p)|\right\}$ be two PHFEs, utilizing the northwest corner rule [38], the joint distribution law between $h_A(p_A)$ and $h_B(p_B)$ can be determined and is shown in Table 1.

Where $\sum_{j=1}^{|l(p)|} p_{ij} = p_i$, $(i = 1,2,\cdots,|h(p)|)$ and $\sum_{i=1}^{|h(p)|} p_{ij} = p'_j$, $(j = 1,2,\cdots,|l(p)|)$ and $|h(p)|$ and $|l(p)|$ are respectively denotes the number of the elements in $h_A(p_A)$ and $h_B(p_B)$.

Base on the joint distribution law, expect value related to the multiply of AB can be obtained as follows.

Table 1. Joint distribution law between $h_A(p_A)$ and $h_B(p_B)$.

	γ'_1	γ'_2	$\cdots$	γ'_j	$\cdots$	$\gamma'_{\lvert l(p)\rvert}$	$p_i\cdot$
γ_1	p_{11}	p_{12}	$\cdots$	p_{1j}	$\cdots$	$p_{1\lvert l(p)\rvert}$	p_1
γ_2	p_{21}	p_{22}	$\cdots$	p_{2j}	$\cdots$	$p_{2\lvert l(p)\rvert}$	p_2
$\vdots$	$\vdots$	$\vdots$	$\vdots$	$\vdots$	$\vdots$	$\vdots$	$\vdots$
γ_i	p_{i1}	p_{i2}	$\cdots$	p_{ij}	$\cdots$	$p_{i\lvert l(p)\rvert}$	p_i
$\vdots$	$\vdots$	$\vdots$	$\vdots$	$\vdots$	$\vdots$	$\vdots$	$\vdots$
$\gamma_{\lvert h(p)\rvert}$	$p_{\lvert h(p)\rvert 1}$	$p_{\lvert h(p)\rvert 2}$	$\cdots$	$p_{\lvert h(p)\rvert j}$	$\cdots$	$p_{\lvert h(p)\rvert\lvert l(p)\rvert}$	$p_{\lvert h(p)\rvert}$
$p\cdot_j$	p'_1	p'_2	$\cdots$	p'_j	$\cdots$	$p'_{\lvert l(p)\rvert}$	1

Definition 7. *Let $h_A(p_A) = \{\gamma_i(p_i)|i = 1,2,\cdots,|h(p)|\}$ and $h_B(p_B) = \left\{\gamma'_j(p'_j)|j = 1,2,\cdots,|l(p)|\right\}$ be two PHFEs, then expect value related to the multiply of AB is calculated as:*

$$E(AB) = \sum_{i=1}^{|h(p)|}\sum_{j=1}^{|l(p)|} \gamma_i\gamma'_j p_{ij}. \tag{8}$$

Example 5. *Let $h_A(p_A) = \{0.5(0.25), 0.6(0.5), 0.7(0.25)\}$ and $h_B(p_B) = \{0.2(0.25), 0.3(0.75)\}$ be two PHFEs, utilizing the northwest corner rule, the joint distribution law between $h_A(p_A)$ and $h_B(p_B)$ can be determined and is shown in Table 2.*

Table 2. Joint distribution law between $h_A(p_A)$ and $h_B(p_B)$ of Example 5.

	0.2	0.3	$p_i\cdot$
0.5	0.25	0	0.25
0.6	0	0.5	0.5
0.7	0	0.25	0.25
$p\cdot_j$	0.25	0.75	1

According to Definition 8, we have $E(AB) = 0.5 \times 0.2 \times 0.25 + 0.6 \times 0.3 \times 0.5 + 0.7 \times 0.3 \times 0.25 = 0.64$.

Remark 3. *If $h_A(p_A) = \{\gamma_i(p_i)|i = 1,2,\cdots,|h(p)|\}$ and $h_B(p_B) = \left\{\gamma'_j(p'_j)|j = 1,2,\cdots,|l(p)|\right\}$ be two mutual independent PHFEs, according to statistics knowledge (the properties of mutual independent discrete random variable), the joint distribution law between $h_A(p_A)$ and $h_B(p_B)$ can be determined and is shown in Table 3.*

Table 3. Joint distribution law between $h_A(p_A)$ and $h_B(p_B)$ of Remark 3.

	γ'_1	γ'_2	$\cdots$	γ'_j	$\cdots$	$\gamma'_{\lvert l(p)\rvert}$	$p_i\cdot$
γ_1	$p_1p'_1$	$p_1p'_2$	$\cdots$	$p_1p'_j$	$\cdots$	$p_1p'_{\lvert l(p)\rvert}$	p_1
γ_2	$p_2p'_1$	$p_2p'_2$	$\cdots$	$p_2p'_j$	$\cdots$	$p_2p'_{\lvert l(p)\rvert}$	p_2
$\vdots$	$\vdots$	$\vdots$	$\vdots$	$\vdots$	$\vdots$	$\vdots$	$\vdots$
γ_i	$p_ip'_1$	$p_ip'_2$	$\cdots$	$p_ip'_j$	$\cdots$	$p_ip'_{\lvert l(p)\rvert}$	p_i
$\vdots$	$\vdots$	$\vdots$	$\vdots$	$\vdots$	$\vdots$	$\vdots$	$\vdots$
$\gamma_{\lvert h(p)\rvert}$	$p_{\lvert h(p)\rvert}p'_1$	$p_{\lvert h(p)\rvert}p'_2$	$\cdots$	$p_{\lvert h(p)\rvert}p'_j$	$\cdots$	$p_{\lvert h(p)\rvert}p'_{\lvert l(p)\rvert}$	$p_{\lvert h(p)\rvert}$
$p\cdot_j$	p'_1	p'_2	$\cdots$	p'_j	$\cdots$	$p'_{\lvert l(p)\rvert}$	1

Example 6. *Let $h_A(p_A)$ and $h_B(p_B)$ be two PHFEs are shown in Example 5, if $h_A(p_A)$ and $h_B(p_B)$ are mutual independent, according to Remark 3, the joint distribution law between them can be determined and is shown in Table 4.*

Table 4. Joint distribution law between $h_A(p_A)$ and $h_B(p_B)$ of Example 6.

	0.2	0.3	$p_i\cdot$
0.5	0.0625	0.1875	0.25
0.6	0.125	0.375	0.5
0.7	0.0625	0.1875	0.25
$p\cdot_j$	0.25	0.75	1

According to Definition 8, we have $E(AB) = 0.0625 \times 0.5 \times 0.2 + 0.1875 \times 0.5 \times 0.3 + 0.125 \times 0.6 \times 0.2 + 0.375 \times 0.6 \times 0.3 + 0.0625 \times 0.7 \times 0.2 + 0.1875 \times 0.7 \times 0.3 = 0.165$.

Utilizing Definitions 4 and 7, the covariance between $h_A(p_A)$ and $h_B(p_B)$ is obtained as follows.

Definition 8. *Let $h_A(p_A) = \{\gamma_i(p_i)|i = 1, 2, \cdots, |h(p)|\}$ and $h_B(p_B) = \left\{\gamma'_j(p'_j)|j = 1, 2, \cdots, |l(p)|\right\}$ be two PHFEs, the covariance between them is obtained as:*

$$Cov(A, B) = E(A - E(A))(B - E(B)). \tag{9}$$

The covariance defined in Equation (9) has the following properties.

Property 1. *For any PHFEs $h_A(p_A)$ and $h_B(p_B)$, the covariance defined in Equation (9) satisfies:*

(1) $Cov(A, A) = D(A)$;
(2) $Cov(A, B) = Cov(B, A)$;
(3) $Cov(A, B) = E(AB) - E(A)E(B)$;
(4) If one of the PHFEs has only one element in it, in this case, we have $Cov(A, B) = 0$.

The proof of Property 1 is shown in Appendix A.

Utilizing Definitions 5 and 8, correlation coefficient between two PHFEs is obtained as follows.

Definition 9. *Let $h_A(p_A) = \{\gamma_i(p_i)|i = 1, 2, \cdots, |h(p)|\}$ and $h_B(p_B) = \left\{\gamma'_j(p'_j)|j = 1, 2, \cdots, |l(p)|\right\}$ be two PHFEs, the correlation coefficient between them is obtained as:*

$$\rho(A, B) = \frac{E(A - E(A))(B - E(B))}{\sqrt{D(A)}\sqrt{D(B)}}. \tag{10}$$

The correlation coefficient obtained in Equation (10) has the following properties.

Property 2. *For any PHFEs $h_A(p_A)$ and $h_B(p_B)$, the correlation coefficient obtained in Equation (10) satisfies:*

(1) $\rho(A, B) = \rho(B, A)$;
(2) $-1 \leq \rho(A, B) \leq 1$;
(3) if $h_A(p_A) = h_B(p_B) \Rightarrow \rho(A, B) = 1$;
(4) if $h_B(p_B) = h_A(p_A)^c \Rightarrow \rho(A, A^c) = -1$.

The proof of Property 2 is shown in Appendix B.

Remark 4. *In Property 2, if $h_A(p_A) = h_B(p_B) \Rightarrow \rho(A, B) = 1$, conversely, it is not hold. That is, if $\rho(A, B) = 1, \nRightarrow h_A(p_A) = h_B(p_B)$.*

4. Weighted Correlation Coefficient of PHFEs

Considering in some situations, the objects may be assigned with different weights. In this section, the weighted form of the expect mean, variance, covariance and correlation coefficient of PHFEs will be introduced.

Let $w = (w_1, w_2, \cdots, w_n)$ be the weight vector of $x_i \in X$, $(i = 1, 2, \cdots, n)$ with $w_i \in [0, 1]$ and $\sum_{i=1}^{n} w_i = 1$. For two PHFEs $h_A(p_A) = \{\gamma_i(p_i) | i = 1, 2, \cdots, |h(p)|\}$ and $h_B(p_B) = \left\{\gamma'_j(p'_j) | j = 1, 2, \cdots, |l(p)|\right\}$, the following definitions can be developed.

Definition 10. *Let $h_A(p_A) = \{\gamma_i(p_i) | i = 1, 2, \cdots, |h(p)|\}$ be a PHFE, the weighted probabilistic hesitant fuzzy element (WPHFE) on X is obtained as:*

$$A_w = \{(w_i\gamma_i)(p_i) | i = 1, 2, \cdots, h(p)\}. \tag{11}$$

Definition 11. *Let A_w be a WPHFE, the weighted expect mean of it is obtained as:*

$$E_w(A) = \sum_{i=1}^{|h(p)|} w_i\gamma_i p_i. \tag{12}$$

Definition 12. *Let A_w be a WPHFE, the weighted variance of it is obtained as:*

$$D_w(A) = \sum_{i=1}^{|h(p)|} (w_i\gamma_i - E_w(A))^2 p_i. \tag{13}$$

Let A_w and $B_{w'}$ be two WPHFEs, utilizing the northwest corner rule, the weight joint distribution law between them can be determined and is shown in Table 5.

Where $\sum_{j=1}^{|l(p)|} p'_{ij} = p_i$, $(i = 1, 2, \cdots, |h(p)|)$ and $\sum_{i=1}^{|h(p)|} p'_{ij} = p'_j$, $(j = 1, 2, \cdots, |l(p)|)$.

Base on the weight joint distribution law, the weighted except mean $E_w(AB)$ related to the multiply of AB can be obtained as follows.

Table 5. Weight joint distribution law between A_w and $B_{w'}$.

	$w'_1\gamma'_1$	$w'_2\gamma'_2$	$\cdots$	$w'_j\gamma'_j$	$\cdots$	$w'_{\lvert l(p)\rvert}\gamma'_{\lvert l(p)\rvert}$	$p_i\cdot$
$w_1\gamma_1$	p'_{11}	p'_{12}	$\cdots$	p'_{1j}	$\cdots$	$p'_{1\lvert l(p)\rvert}$	p_1
$w_2\gamma_2$	p'_{21}	p'_{22}	$\cdots$	p'_{2j}	$\cdots$	$p'_{2\lvert l(p)\rvert}$	p_2
$\vdots$	$\vdots$	$\vdots$	$\vdots$	$\vdots$	$\vdots$	$\vdots$	$\vdots$
$w_i\gamma_i$	p'_{i1}	p'_{i2}	$\cdots$	p'_{ij}	$\cdots$	$p'_{i\lvert l(p)\rvert}$	p_i
$\vdots$	$\vdots$	$\vdots$	$\vdots$	$\vdots$	$\vdots$	$\vdots$	$\vdots$
$w_{\lvert h(p)\rvert}\gamma_{\lvert h(p)\rvert}$	$p'_{\lvert h(p)\rvert 1}$	$p'_{\lvert h(p)\rvert 2}$	$\cdots$	$p'_{\lvert h(p)\rvert j}$	$\cdots$	$p'_{\lvert h(p)\rvert\lvert l(p)\rvert}$	$p_{\lvert h(p)\rvert}$
$p\cdot_j$	p'_1	p'_2	$\cdots$	p'_j	$\cdots$	$p'_{\lvert l(p)\rvert}$	1

Definition 13. *Let A_w and $B_{w'}$ be two WPHFEs, the weighted expect mean between them is obtained as:*

$$E_w(AB) = \sum_{i=1}^{|h(p)|} \sum_{j=1}^{|l(p)|} w_i w'_j \gamma_i \gamma'_j p'_{ij}. \tag{14}$$

Using Definitions 11 and 13, the weighted covariance between two WPHFEs can be derived as follows.

Definition 14. *Let A_w and $B_{w'}$ be two WPHFEs, the weighted covariance between them is obtained as:*

$$Cov_w(A,B) = E(A - E_w(A))(B - E_{w'}(B)). \tag{15}$$

Using Definitions 13 and 15, the weighted correlation coefficient between two WPHFEs can be calculated as follows.

Definition 15. *Let A_w and $B_{w'}$ be two WPHFEs, the weighted correlation coefficient between them is obtained as:*

$$\rho_w(A,B) = \frac{E(A - E_w(A))(B - E_{w'}(B))}{\sqrt{D_w(A)}\sqrt{D_{w'}(B)}}. \tag{16}$$

The weighted correlation coefficient obtained in Equation (16) has the following properties.

Property 3. *For any WPHFEs A_w and $B_{w'}$, the weighted correlation coefficient obtained in Equation (16) satisfies:*

(1) $\rho_w(A,B) = \rho_w(B,A)$;
(2) $-1 \leq \rho_w(A,B) \leq 1$;
(3) if $A_w = B_{w'} \Rightarrow \rho_w(A,B) = 1$.

The proof of Property 3 is similar to the proof of Property 2, so it has been omitted here.

5. Multi-Criteria Decision-Making Based on Probabilistic Hesitant Fuzzy Information

In this section, a MCDM problems within probabilistic hesitant fuzzy environment is adopted to demonstrate how to apply the proposed method.

5.1. Problems Description

For a MCDM problems, let $A = \{A_1, A_2, \cdots, A_n\}$ be a set of alternatives, $G = \{G_1, G_2, \cdots, G_m\}$ be a set of criteria, the criteria weights are completely unknown. Assume the criteria are independent to each other. $D = \{D_1, D_2, \cdots, D_t\}$ be a set of decision makers. And the evaluation of the alternative A_j with respect to the criterion G_i is represent in PHFEs.

The MCDM processes designed to find the best alternative is given by the following steps:

Step 1: Construct individual probabilistic hesitant fuzzy decision matrix

The individual probabilistic hesitant fuzzy decision matrix can be constructed and denoted as $D^k = \left(\gamma_{ij}(p_{ij})^k\right)_{m \times n}$;

Step 2: Integrate individual probabilistic hesitant fuzzy decision matrix into one

Use Equation (6) to integrate individual probabilistic hesitant fuzzy decision matrix into an overall decision matrix and denotes as $D = \left(\gamma_{ij}(p_{ij})\right)_{m \times n}$;

Step 3: Derive the criteria weights

The criteria weights can be derived utilizing the following formula:

$$w_i = \frac{\sum\limits_{j=1}^{n} E(\gamma_{ij}(p_{ij}))}{\sum\limits_{j=1}^{n}\sum\limits_{i=1}^{m} E(\gamma_{ij}(p_{ij}))}, (i = 1, 2, \cdots, m). \tag{17}$$

Step 4: Calculate the weighted correlation coefficient

By applying Equation (12), the ideal alternative A^* under the criterion G_i can be obtained as follows:

$$A^* = \max\{ A_j | w_i \gamma_{ij} p_{ij}, i = 1, 2, \cdots, m; j = 1, 2, \cdots, n\}. \tag{18}$$

Here, the weighted of w_i is obtained from Equation (17).

And then calculate the weighted correlation coefficient between any alternative A and ideal alternative A^*.

Use Equation (16), calculate the weighted correlation coefficient between A and A^* as follows:

$$\rho_w(A, A^*) = \frac{E(A - E_w(A))(A^* - E_{w'}(A^*))}{\sqrt{D_w(A)}\sqrt{D_{w'}(A^*)}}. \tag{19}$$

Step 5: Rank all alternatives

Since the higher the score of the weighted correlation coefficient obtain from Equation (19), means that the more similarity between any alternative A and the ideal alternative A^*, the better the alternative A is. Based on this consideration, the ranking result of the alternatives A_j, $(j = 1, 2, \cdots, n)$ can be obtained according to the following formula:

$$r_j = \sum_{i=1}^{m} \beta_{ij}, (j = 1, 2, \cdots, n). \tag{20}$$

Here, β_{ij} is the value of weighted correlation coefficient, obtained from Step 4.

5.2. Illustrative Example

Suppose there is an investment company, which wants to invest a sum of money in the best option, there is a panel with four possible alternatives to invest: (1) A_1 is a car company; (2) A_2 is a food company; (3) A_3 is a computer company; (4) A_4 is an arms company. The investment company must take a decision according to the following five criteria: (1) G_1 is the productivity; (2) G_2 is the technological innovation capability; (3) G_3 is the marketing capability; (4) G_4 is the management; (5) G_5 is the risk avoidance.

An expert group is formed which consists of four experts D_t $(t = 1, 2, 3, 4)$ from each strategic decision area (whose weight vector is equally). Suppose each expert consulted 10 people in the same industry through online questionnaire and the 10 people they consulted were not exactly similar. Four experts provided their preference evaluations on alternatives in the form of PHFEs, as shown in Tables 6–9, respectively. Take the evaluation values $\{0.6(0.4), 0.7(0.6)\}$ from Expert 1, for example, evaluation information is obtained from 10 people related to computer company A_3 with respect to productivity G_1. Four of them set a value of 0.6, whereas six of them set a value 0.7 and thus, the probability of the vale 0.6 is 0.4 and the probability of the vale 0.7 is 0.6. Other entries, that is, other PHFEs, in Tables 6–9 can be similarly explained. Because four experts consulted the people in the same industry may be communicate with each other, in this case, the evaluation information obtained from four experts are interact with each other.

Table 6. The evaluation information provided from D_1.

	A_1	A_2	A_3	A_4
G_1	0.5	0.3	$\{0.6(0.4), 0.7(0.6)\}$	0.4
G_2	0.7	0.4	$\{0.4(0.5), 0.5(0.5)\}$	0.3
G_3	$\{0.4(0.4), 0.3(0.6)\}$	$\{0.6(0.5), 0.5(0.5)\}$	0.3	0.5
G_4	0.6	0.7	0.3	0.5
G_5	0.5	0.6	0.4	$\{0.4(0.5), 0.5(0.5)\}$

Table 7. The evaluation information provided from D_2.

	A_1	A_2	A_3	A_4
G_1	0.5	0.4	0.6	0.3
G_2	0.6	0.5	0.6	0.4
G_3	0.4	0.4	0.5	0.5
G_4	0.7	0.6	0.3	0.5
G_5	0.3	0.6	0.4	0.7

Table 8. The evaluation information provided from D_3.

	A_1	A_2	A_3	A_4
G_1	0.5	$\{0.2(0.4), 0.3(0.6)\}$	$\{0.6(0.4), 0.7(0.6)\}$	0.6
G_2	$\{0.8(0.4), 0.7(0.6)\}$	0.4	$\{0.4(0.5), 0.5(0.5)\}$	0.3
G_3	0.4	0.5	0.6	$\{0.4(0.5), 0.3(0.5)\}$
G_4	0.7	0.4	0.5	$\{0.4(0.7), 0.5(0.3)\}$
G_5	0.4	0.6	0.7	0.2

Table 9. The evaluation information provided from D_4.

	A_1	A_2	A_3	A_4
G_1	0.7	0.3	0.4	0.2
G_2	0.7	0.5	0.5	0.6
G_3	0.6	0.8	0.2	$\{0.4(0.5), 0.3(0.5)\}$
G_4	$\{0.4(0.5), 0.3(0.5)\}$	$\{0.7(0.5), 0.6(0.5)\}$	0.1	0.6
G_5	0.8	0.4	$\{0.7(0.5), 0.6(0.5)\}$	0.5

The processes are designed to find the best alternative is given by the following steps:

Step 1: Construct individual probabilistic hesitant fuzzy decision matrix

The individual probabilistic hesitant fuzzy decision matrix has been constructed and is shown in Tables 6–9.

Step 2: Integrate individual probabilistic hesitant fuzzy decision matrix

Use Equation (7) to integrate individual probabilistic hesitant fuzzy decision matrix into an overall decision matrix and is shown in Table 10.

Step 3: Derive the criteria weights

By applying Equation (17), the weights of criteria are calculated as follows:

$$w_1 = 0.19,\ w_2 = 0.21,\ w_3 = 0.19,\ w_4 = 0.20 \text{ and } w_5 = 0.21.$$

Step 4: Calculate the weighted correlation coefficient

By applying Equation (19), the weighted correlation coefficient can be obtained and is shown in Table 11.

Table 10. Integrate individual evaluation information.

	A_1	A_2	A_3	A_4
G_1	$\{0.5(0.75), 0.7(0.25)\}$	$\{0.2(0.1), 0.3(0.65), 0.4(0.25)\}$	$\{0.4(0.25), 0.6(0.45), 0.7(0.3)\}$	$\{0.2(0.25), 0.3(0.25), 0.4(0.25), 0.6(0.25)\}$
G_2	$\{0.6(0.25), 0.7(0.65), 0.8(0.1)\}$	$\{0.4(0.5), 0.5(0.5)\}$	$\{0.4(0.25), 0.5(0.5), 0.6(0.25)\}$	$\{0.3(0.5), 0.4(0.25), 0.6(0.25)\}$
G_3	$\{0.3(0.15), 0.4(0.6), 0.6(0.25)\}$	$\{0.4(0.25), 0.5(0.375), 0.6(0.125), 0.8(0.25)\}$	$\{0.2(0.25), 0.3(0.25), 0.5(0.25), 0.6(0.25)\}$	$\{0.3(0.25), 0.4(0.25), 0.5(0.5)\}$
G_4	$\{0.3(0.125), 0.4(0.125), 0.6(0.25), 0.7(0.5)\}$	$\{0.4(0.25), 0.6(0.375), 0.7(0.375)\}$	$\{0.1(0.25), 0.3(0.5), 0.5(0.25)\}$	$\{0.4(0.175), 0.6(0.25), 0.5(0.575)\}$
G_5	$\{0.3(0.25), 0.4(0.25), 0.5(0.25), 0.8(0.25)\}$	$\{0.6(0.75), 0.4(0.25)\}$	$\{0.4(0.5), 0.6(0.125), 0.7(0.375)\}$	$\{0.2(0.25), 0.5(0.325), 0.4(0.125), 0.7(0.25)\}$

Table 11. Weight correlation coefficient.

	$\rho_w(A_1, A^*)$	$\rho_w(A_2, A^*)$	$\rho_w(A_3, A^*)$	$\rho_w(A_4, A^*)$
G_1	0.62	0.75	1	0.28
G_2	1	0.61	0.86	0.64
G_3	0.94	1	0.9	−0.03
G_4	0.95	1	0.91	0.27
G_5	−0.93	1	−0.67	−0.74

Step 5: Rank all alternatives.

The ranking result of the alternatives A_j $(j = 1, 2, \cdots, 4)$ can be obtained according to the Formula (20) as follows:

$$r_1 = 2.28, r_2 = 4.36, \ r_3 = 3, \ r_4 = 0.42.$$

Since $r_2 > r_3 > r_1 > r_4$, then $A_2 \succ A_3 \succ A_1 \succ A_4$. Hence, the most desirable alternative is A_2. That is, the food company is the best option to invest.

It is stated that in this example, the correlation coefficient is proposed in this paper lies in the interval $[-1, 1]$, which not only consider the strength of relationship between the PHFSs but also whether the PHFSs are positively or negatively related. In this illustrative example, we can also use the methods proposed in [8,36] to solve the problem illustrative in the example. However, the method proposed in [36] directly integrated the probability part into the membership degree part, this make cause a lot of information loss. For example, the positively or negatively related obtained from the proposed method. And the method proposed in [8] fail in the condition when two PHFEs are not mutual independent.

6. Conclusions

This article puts forward a framework to tackle MCDM problems within probabilistic hesitant fuzzy environments with completely unknown criteria weight information. Since every PHFE consists of two parts, that is, the membership degree of the element to the set and the corresponding probability of the membership degree, this information can be treated as a probabilistic distribution function, inspired by statistics knowledge, each PHFE can be regarded as a discrete random variable. The primary contributions of this paper are summarized as follows. (1) The correlation coefficient is proposed in this paper adopt the knowledge of statistics, the significant characteristic of the proposed formula is that it lies in the interval $[-1, 1]$. The proposed formula not only consider the strength of the PHFSs but also whether the PHFSs are positively or negatively related, it avoids the inconsistency of the decision-making result due to the loss of information; (2) The existing decision-making methods related to probabilistic hesitant fuzzy environments, very few papers discussed the condition when two random variables are not mutual independent. In this paper, the northwest corner rule to obtain the expected mean of two PHFEs multiply is introduced; (3) A novel MCDM method with the probabilistic hesitant fuzzy environment is introduced based on the proposed weighted correlation coefficient and this proposed method is applied to practical decision-making processes.

In this paper, we have applied the proposed correlation coefficient in evaluation of the alternatives. In the future, we will apply it in other aspects, such as, pattern recognition and cluster analysis. In addition, in this paper, we only discuss the correlation coefficients between two PHFSs, in future study, the proposed correlation coefficients will be extended to other extension of PHFSs, such as, interval-valued probabilistic hesitant fuzzy sets, probabilistic linguistic term sets and so on.

Acknowledgments: The authors thank the editors and anonymous reviewers for providing very thoughtful comments which have leaded to an improved version of this paper.

Author Contributions: All authors have contributed equally to this paper.

Conflicts of Interest: The authors declare no conflict of interest.

Appendix A. Proof of Property 1

Proof. The proof of (1), (2) and (3) is obvious, so it has been omitted here. And the proof of (4) will be demonstrated as follows:

Suppose $h_A(p_A)$ has only one element in it, denoted by $h_A(p_A) = \{h_1(p_1)\}$, according to Definition 3, we have $p_1 = 1$. Let $h_B(p_B) = \left\{\gamma'_j(p'_j)|j = 1, 2, \cdots, |l(p)|\right\}$, utilizing the northwest corner rule, the joint distribution law between them can be determined and is shown in Table A1.

Table A1. Joint distribution law between $h_A(p_A)$ and $h_B(p_B)$ in Property 1.

	γ'_1	γ'_2	$\cdots$	γ'_j	$\cdots$	$\gamma'_{\lvert l(p)\rvert}$	$p_i\cdot$
γ_1	p'_1	p'_2	$\cdots$	p'_j	$\cdots$	$p'_{\lvert l(p)\rvert}$	p_1
$p\cdot_j$	p'_1	p'_2	$\cdots$	p'_j	$\cdots$	$p'_{\lvert l(p)\rvert}$	1

According to Definition 8, we have

$$\begin{aligned} E(AB) &= \gamma_1\gamma'_1p'_1 + \gamma_1\gamma'_2p'_2 + \cdots + \gamma_1\gamma'_{|l(p)|}p'_{|l(p)|} \\ &= \gamma_1\left(\gamma'_1p'_1 + \gamma'_2p'_2 + \cdots + \gamma'_{|l(p)|}p'_{|l(p)|}\right) \\ &= \gamma_1p_1\left(\gamma'_1p'_1 + \gamma'_2p'_2 + \cdots + \gamma'_{|l(p)|}p'_{|l(p)|}\right) \\ &= E(A)E(B). \end{aligned}$$

Therefore, we have $Cov = E(A - E(A))(B - E(B)) = E(AB) - E(A)E(B) = 0$.
The similar proof can be obtained if $h_B(p_B)$ has only one element in it.
This completes the proof. □

Appendix B. Proof of Property 2

Proof.

(1) It is straightforward.
(2) According to Property 1, we have

$$Cov(A, B) = E(AB) - E(A)E(B) = E(A - E(A))(B - E(B)),$$

and

$$E(A-E(A))(B-E(B)) = \sum_{i=1}^{|h(p)|}\sum_{j=1}^{|l(p)|}(\gamma_i - E(A))(\gamma'_j - E(B))p_{ij}$$
$$= \Sigma_{j=1}^{|l(p)|}(\gamma_1 - E(A))(\gamma'_j - E(B))p_{1j} + \Sigma_{j=1}^{|l(p)|}(\gamma_2 - E(A))(\gamma'_j - E(B))p_{2j}$$
$$+ \cdots + \Sigma_{j=1}^{|l(p)|}(\gamma_{|h(p)|} - E(A))(\gamma'_j - E(B))p_{|h(p)|j}$$
$$= \Sigma_{j=1}^{|l(p)|}(\gamma_1 - E(A))\sqrt{p_{1j}}(\gamma'_j - E(B))\sqrt{p_{1j}} + \Sigma_{j=1}^{|l(p)|}(\gamma_2 - E(A))\sqrt{p_{2j}}(\gamma'_j - E(B))\sqrt{p_{2j}}$$
$$+ \cdots + \Sigma_{j=1}^{|l(p)|}(\gamma_{|h(p)|} - E(A))\sqrt{p_{|h(p)|j}}(\gamma'_j - E(B))\sqrt{p_{|h(p)|j}}.$$

Using the Cauchy-Schwarz inequality: $(a_1b_1 + a_2b_2 + \cdots + a_nb_n)^2 \leq (a_1^2 + a_2^2 + \cdots + a_n^2)(b_1^2 + b_2^2 + \cdots + b_n^2)$, it follows that

$$[Cov(A,B)]^2 = \left[\sum_{i=1}^{|h(p)|}\sum_{j=1}^{|l(p)|}(\gamma_i - E(A))(\gamma'_j - E(B))p_{ij}\right]^2$$
$$\leq \left[\Sigma_{j=1}^{|l(p)|}(\gamma_1 - E(A))^2 p_{1j} + \Sigma_{j=1}^{|l(p)|}(\gamma_2 - E(A))^2 p_{2j} + \cdots + \Sigma_{j=1}^{|l(p)|}(\gamma_{|h(p)|} - E(A))^2 p_{|h(p)|j}\right]$$
$$\times \left[\Sigma_{j=1}^{|l(p)|}(\gamma'_j - E(B))^2 p_{1j} + \Sigma_{j=1}^{|l(p)|}(\gamma'_j - E(B))^2 p_{2j} + \cdots + \Sigma_{j=1}^{|l(p)|}(\gamma'_j - E(B))^2 p_{|h(p)|j}\right]$$
$$= \left[(\gamma_1 - E(A))^2\Sigma_{j=1}^{|l(p)|} p_{1j} + (\gamma_2 - E(A))^2\Sigma_{j=1}^{|l(p)|} p_{2j} + \cdots + (\gamma_{|h(p)|} - E(A))^2\Sigma_{j=1}^{|l(p)|} p_{|h(p)|j}\right]$$
$$\times \left[(\gamma'_1 - E(B))^2\Sigma_{i=1}^{|h(p)|} p_{i1} + (\gamma'_2 - E(B))^2\Sigma_{i=1}^{|h(p)|} p_{i2} + \cdots + (\gamma'_{|l(p)|} - E(B))^2\Sigma_{i=1}^{|h(p)|} p_{i|l(p)|}\right]$$
$$= \left[(\gamma_1 - E(A))^2 p_1 + (\gamma_2 - E(A))^2 p_2 + \cdots + (\gamma_{|h(p)|} - E(A))^2 p_{|h(p)|}\right]$$
$$\times \left[(\gamma'_1 - E(B))^2 p'_1 + (\gamma'_2 - E(B))^2 p'_2 + \cdots + (\gamma'_{|l(p)|} - E(B))^2 p'_{|l(p)|}\right]$$
$$= \left[\Sigma_{i=1}^{|h(p)|}(\gamma_i - E(A))^2 p_i\right] \times \left[\Sigma_{j=1}^{|l(p)|}(\gamma'_j - E(B))^2 p'_j\right].$$

Taking the square root of both sides, this inequality reduces to:

$$\left|\sum_{i=1}^{|h(p)|}\sum_{j=1}^{|l(p)|}(\gamma_i - E(A))(\gamma'_j - E(B))p_{ij}\right| \leq \left[\sum_{i=1}^{|h(p)|}(\gamma_i - E(A))^2 p_i\right]^{\frac{1}{2}} \times \left[\sum_{j=1}^{|l(p)|}(\gamma'_j - E(B))^2 p'_j\right]^{\frac{1}{2}}$$

Therefore, this inequality can be rewritten as:

$$\frac{\left|\sum_{i=1}^{|h(p)|}\sum_{j=1}^{|l(p)|}(\gamma_i - E(A))(\gamma'_j - E(B))p_{ij}\right|}{\sqrt{\Sigma_{i=1}^{|h(p)|}(\gamma_i - E(A))^2 p_i} \times \sqrt{\Sigma_{j=1}^{|l(p)|}(\gamma'_j - E(B))^2 p'_j}} \leq 1.$$

Therefore, we have $-1 \leq \rho(A,B) \leq 1$.

(3) if $h_A(p_A) = h_B(p_B) \Rightarrow \gamma_i = \gamma'_j$, $p_i = p'_j$ and $|h(p)| = |l(p)|$, utilizing the northwest corner rule, the joint distribution law between them can be determined and is shown in Table A2.

Table A2. Joint distribution law between $h_A(p_A)$ and $h_A(p_A)$ in Property 2.

	γ_1	γ_2	$\cdots$	γ_i	$\cdots$	$\gamma_{\|h(p)\|}$	$p_{i\cdot}$
γ_1	p_1	0	$\cdots$	0	$\cdots$	0	p_1
γ_2	0	p_2	$\cdots$	0	$\cdots$	0	p_2
$\vdots$	$\vdots$	$\vdots$	$\vdots$	$\vdots$	$\vdots$	$\vdots$	$\vdots$
γ_i	0	0	$\cdots$	p_i	$\cdots$	0	p_i
$\vdots$	$\vdots$	$\vdots$	$\vdots$	$\vdots$	$\vdots$	$\vdots$	$\vdots$
$\gamma_{\|h(p)\|}$	0	0	$\cdots$	0	$\cdots$	$p_{\|h(p)\|}$	$p_{\|h(p)\|}$
$p_{\cdot j}$	p_1	p_2	$\cdots$	p_i	$\cdots$	$p_{\|h(p)\|}$	1

And

$$E(A^2) = \gamma_1^2 p_1 + \gamma_2^2 p_2 + \cdots + \gamma_{|h(p)|}^2 p_{|h(p)|},$$
$$E(A) = \gamma_1 p_1 + \gamma_2 p_2 + \cdots + \gamma_{|h(p)|} p_{|h(p)|}$$

Thus, we have

$$\begin{aligned}&Cov(A,A) = E(A^2) - E(A)^2\\&= \gamma_1^2 p_1 + \gamma_2^2 p_2 + \cdots + \gamma_{|h(p)|}^2 p_{|h(p)|} - (\gamma_1 p_1 + \gamma_2 p_2 + \cdots + \gamma_{|h(p)|} p_{|h(p)|})^2\\&= D(A).\end{aligned}$$

Thus, we have

$$\rho(A,A) = 1.$$

(4) If $h_B(p_B) = h_A(p_A)^c$, according to the supplement operation law was introduced in [41], we obtain: $h_A(p_A)^c = \{(1-\gamma_i)(p_i) | i = 1,2,\cdots,|h(p)|\}$, utilizing the northwest corner rule, the joint distribution law between them can be determined and is shown in Table A3.

And

Since

$$\begin{aligned}&E(A^c) = (1-\gamma_1)p_1 + (1-\gamma_2)p_2 + \cdots + \left(1-\gamma_{|h(p)|}\right)p_{|h(p)|}\\&= p_1 - \gamma_1 p_1 + p_2 - \gamma_2 p_2 + \cdots + p_{|h(p)|} - \gamma_{|h(p)|} p_{|h(p)|}\\&= p_1 + p_2 + \cdots + p_{|h(p)|} - \left(\gamma_1 p_1 + \gamma_2 p_2 + \cdots + \gamma_{|h(p)|} p_{|h(p)|}\right)\\&= 1 - E(A).\end{aligned}$$

Thus, we have

$$\begin{aligned}&Cov(A,A^c) = E(AA^c) - E(A)E(A^c)\\&= E(A) - E(A^2) - E(A)(1-E(A))\\&= -E(A^2) + E(A)^2.\end{aligned}$$

for $D(A) = \gamma_1^2 p_1 + \gamma_2^2 p_2 + \cdots + \gamma_{|h(p)|}^2 p_{|h(p)|} - E(A)^2 = E(A^2) - E(A)^2$.

$$\begin{aligned}
&D(A^c) = (1-\gamma_1)^2 p_1 + (1-\gamma_2)^2 p_2 + \cdots + \left(1-\gamma_{|h(p)|}\right)^2 p_{|h(p)|} - (1-E(A))^2 \\
&= p_1 - 2\gamma_1 p_1 + \gamma_1^2 p_1 + p_2 - 2\gamma_2 p_2 + \gamma_2^2 p_2 + \cdots + p_{|h(p)|} - 2\gamma_{|h(p)|} p_{|h(p)|} + \gamma_{|h(p)|}^2 p_{|h(p)|} - (1-E(A))^2 \\
&= p_1 + p_2 + \cdots + p_{|h(p)|} - 2\left(\gamma_1 p_1 + \gamma_2 p_2 + \cdots + \gamma_{|h(p)|} p_{|h(p)|}\right) + \gamma_1^2 p_1 + \gamma_2^2 p_2 + \cdots + \gamma_{|h(p)|}^2 p_{|h(p)|} - (1-E(A))^2 \\
&= 1 - 2E(A) + E(A^2) - (1-E(A))^2 \\
&= E(A^2) - E(A)^2.
\end{aligned}$$

Thus, we have

$$\rho(A, A^c) = \frac{Cov(A, A^c)}{\sqrt{D(A)}\sqrt{D(A^c)}} = \frac{-E^2(A) + E(A)^2}{\sqrt{E(A^2) - E(A)^2}\sqrt{E(A^2) - E(A)^2}} = -1.$$

This completes the proof. □

Table A3. Joint distribution law between $h_A(p_A)$ and $h_A(p_A)^c$ in Property 2.

	$1-\gamma_1$	$1-\gamma_2$	$\cdots$	$1-\gamma_i$	$\cdots$	$1-\gamma_{\lvert h(p)\rvert}$	$p_{i\cdot}$
γ_1	p_1	0	$\cdots$	0	$\cdots$	0	p_1
γ_2	0	p_2	$\cdots$	0	$\cdots$	0	p_2
$\vdots$	$\vdots$	$\vdots$	$\vdots$	$\vdots$	$\vdots$	$\vdots$	$\vdots$
γ_i	0	0	$\cdots$	p_i	$\cdots$	0	p_i
$\vdots$	$\vdots$	$\vdots$	$\vdots$	$\vdots$	$\vdots$	$\vdots$	$\vdots$
$\gamma_{\lvert h(p)\rvert}$	0	0	$\cdots$	0	$\cdots$	$p_{\lvert h(p)\rvert}$	$p_{\lvert h(p)\rvert}$
$p_{\cdot j}$	p_1	p_2	$\cdots$	p_i	$\cdots$	$p_{\lvert h(p)\rvert}$	1

References

1. Sałabun, W.; Piegat, A. Comparative analysis of MCDM methods for the assessment of mortality in patients with acute coronary syndrome. *Artif. Intell. Rev.* **2016**, 1–15. [CrossRef]
2. Anisseh, M.; Piri, F.; Shahraki, M.R.; Agamohamadi, F. Fuzzy extension of TOPSIS model for group decision making under multiple criteria. *Artif. Intell. Rev.* **2012**, *38*, 325–338. [CrossRef]
3. Sun, B.; Ma, W. Soft fuzzy rough sets and its application in decision making. *Artif. Intell. Rev.* **2014**, *41*, 67–80. [CrossRef]
4. Bashir, Z.; Wątróbski, J.; Rashid, T.; Sałabun, W.; Ali, J. Intuitionistic-Fuzzy goals in zero-sum multi criteria matrix games. *Symmetry* **2017**, *9*, 158. [CrossRef]
5. Liao, H.C.; Xu, Z.S. Consistency of the fused intuitionistic fuzzy preference relation in group intuitionistic fuzzy analytic hierarchy process. *Appl. Soft Comput.* **2015**, *35*, 812–826. [CrossRef]
6. Faizi, S.; Rashid, T.; Sałabun, W.; Zafar, S.; Wątróbski, J. Decision making with uncertainty using hesitant fuzzy sets. *Int. J. Fuzzy Syst.* **2017**, 1–11. [CrossRef]
7. Faizi, S.; Sałabun, W.; Rashid, T.; Atróbski, J.W.; Zafar, S. Group decision-making for hesitant fuzzy sets based on characteristic objects method. *Symmetry* **2017**, *9*, 136. [CrossRef]
8. Alcantud, J.C.R.; Torra, V. Decomposition theorems and extension principles for hesitant fuzzy sets. *Inf. Fusion* **2018**, *41*, 48–56. [CrossRef]
9. Li, J.; Wang, J.Q. An extended QUALIFLEX method under probability hesitant fuzzy environment for selecting green suppliers. *Int. J. Fuzzy Syst.* **2017**, 1–14. [CrossRef]
10. Zhu, B.; Xu, Z.S. Probability-hesitant fuzzy sets and the representation of preference relations. *Technol. Econ. Dev. Econ.* **2017**, in press.
11. Pang, Q.; Wang, H.; Xu, Z.S. Probabilistic linguistic term sets in multi-attribute group decision making. *Inf. Sci.* **2016**, *369*, 128–143. [CrossRef]
12. Lv, Z.; Zhao, J.; Liu, Y.; Wang, W. Data imputation for gas flow data in steel industry based on non-equal-length granules correlation coefficient. *Inf. Sci.* **2016**, *367–368*, 311–323. [CrossRef]

13. Bai, X.Z. Morphological center operator based infrared and visible image fusion through correlation coefficient. *Infrared Phys. Technol.* **2016**, *76*, 546–554. [CrossRef]
14. Rao, C.S.; Raju, S.V. Similarity analysis between chromosomes of Homo sapiens and monkeys with correlation coefficient, rank correlation coefficient and cosine similarity measures. *Genom. Data* **2016**, *7*, 202–209.
15. Yang, C.C. Correlation coefficient evaluation for the fuzzy interval data. *J. Bus. Res.* **2016**, *69*, 2138–2144. [CrossRef]
16. Hong, D.H. Fuzzy measures for a correlation coefficient of fuzzy numbers under TW (the weakest t-norm)-based fuzzy arithmetic operations. *Inf. Sci.* **2006**, *176*, 150–160. [CrossRef]
17. Liu, S.T.; Kao, C. Fuzzy measures for correlation coefficient of fuzzy numbers. *Inf. Sci.* **2006**, *128*, 267–275. [CrossRef]
18. Hung, W.L.; Wu, J.W. Correlation of intuitionistic fuzzy sets by centroid method. *Inf. Sci.* **2002**, *144*, 219–225. [CrossRef]
19. Ye, J. Fuzzy decision-making method based on the weighted correlation coefficient under intuitionistic fuzzy environment. *Eur. J. Oper. Res.* **2010**, *205*, 202–204. [CrossRef]
20. Ye, J. Multicriteria fuzzy decision-making method using entropy weights-based correlation coefficients of interval-valued intuitionistic fuzzy sets. *Appl. Math. Model.* **2010**, *34*, 3864–3870. [CrossRef]
21. Dong, G.P.; Kwun, Y.C.; Jin, H.P.; Park, I.Y. Correlation coefficient of interval-valued intuitionistic fuzzy sets and its application to multiple attribute group decision making problems. *Math. Comput. Model. Int. J.* **2009**, *50*, 1279–1293.
22. Chen, N.; Xu, Z.S.; Xia, M.M. Correlation coefficients of hesitant fuzzy sets and their applications to clustering analysis. *Appl. Math. Model.* **2013**, *37*, 2197–2211. [CrossRef]
23. Liao, H.C.; Xu, Z.S.; Zeng, X.J. Novel correlation coefficients between hesitant fuzzy sets and their application in decision making. *Knowl.-Based Syst.* **2015**, *82*, 115–127. [CrossRef]
24. Liao, H.C.; Xu, Z.S.; Zeng, X.J. Qualitative decision making with correlation coefficients of hesitant fuzzy linguistic term sets. *Knowl.-Based Syst.* **2015**, *76*, 127–138. [CrossRef]
25. González-Arteaga, T.; Alcantud, J.C.R.; Andrés, R.D. New correlation coefficients for hesitant fuzzy sets. In Proceedings of the 2015 Conference of the International Fuzzy Systems Association and the European Society for Fuzzy Logic and Technology (IFSA-EUSFLAT), Gijon, Spain, 30 June–3 July 2015.
26. Ye, J. Correlation coefficient of dual hesitant fuzzy sets and its application to multiple attribute decision making. *Appl. Math. Model.* **2014**, *38*, 659–666. [CrossRef]
27. Şahin, R.; Liu, P.D. Correlation coefficient of single-valued neutrosophic hesitant fuzzy sets and its applications in decision making. *Neural Comput. Appl.* **2016**, *7*, 1387–1395. [CrossRef]
28. Zhang, H.Y.; Ji, P.; Wang, J.Q.; Chen, X.H. An Improved Weighted Correlation Coefficient Based on Integrated Weight for Interval Neutrosophic Sets and Its Application in Multi-criteria Decision-making Problems. *Int. J. Comput. Intell. Syst.* **2015**, *8*, 1027–1043. [CrossRef]
29. Karaaslan, F. Correlation coefficients of single-valued neutrosophic refined soft sets and their applications in clustering analysis. *Neural Comput. Appl.* **2015**, *8*, 2781–2793. [CrossRef]
30. Zhu, B. *Decision Method for Research and Application Based on Preference Relation*; Southeast University: Nanjing, China, 2014.
31. Zhang, S.; Xu, Z.S.; He, Y. Operations and integrations of probabilistic hesitant fuzzy information in decision making. *Inf. Fusion* **2017**, *38*, 1–11. [CrossRef]
32. He, Y.; Xu, Z.S.; Jiang, W.L. Probabilistic interval reference ordering sets in multi-criteria group decision making. *Int. J. Uncertain. Fuzziness Knowl.-Based Syst.* **2017**, *25*, 189–212. [CrossRef]
33. Abdullah, L.; Najib, L. A new type-2 fuzzy set of linguistic variables for the fuzzy analytic hierarchy process. *Expert Syst. Appl.* **2014**, *41*, 3297–3305. [CrossRef]
34. Gou, X.J.; Xu, Z.S. Novel basic operational laws for linguistic terms, hesitant fuzzy linguistic term sets and probabilistic linguistic term sets. *Inf. Sci.* **2016**, *372*, 407–427. [CrossRef]
35. Bai, C.Z.; Zhang, R.; Qian, L.X.; Wu, Y.N. Comparisons of probabilistic linguistic term sets for multi-criteria decision making. *Knowl.-Based Syst.* **2017**, *119*, 284–291. [CrossRef]
36. Zhou, W.; Xu, Z.S. Probability calculation and element optimization of probabilistic hesitant fuzzy preference relations based on expected consistency. *IEEE Trans. Fuzzy Syst.* **2017**, *PP*, 1. [CrossRef]
37. Hung, W.L. Using statistical viewpoint in developing correlation of intuitionistic fuzzy sets. *Int. J. Uncertain. Fuzziness Knowl.-Based Syst.* **2001**, *9*, 509–516. [CrossRef]

38. Klinz, B.; Woeginger, G.J. The Northwest corner rule revisited. *Discret. Appl. Math.* **2013**, *159*, 1284–1289. [CrossRef]
39. Torra, V. Hesitant fuzzy sets. *Int. J. Intell. Syst.* **2010**, *25*, 529–539. [CrossRef]
40. Xia, M.M.; Xu, Z.S. Hesitant fuzzy information aggregation in decision making. *Int. J. Approx. Reason.* **2011**, *52*, 395–407. [CrossRef]
41. Xu, Z.S.; Zhou, W. Consensus building with a group of decision makers under the hesitant probabilistic fuzzy environment. *Fuzzy Optim. Decis. Mak.* **2016**, 1–23. [CrossRef]
42. Li, J.; Wang, J.Q. Multi-criteria outranking methods with hesitant probabilistic fuzzy sets. *Cogn. Comput.* **2017**, *9*, 611–625. [CrossRef]

symmetry

MDPI

Article

A New Multi-Attribute Decision-Making Method Based on m-Polar Fuzzy Soft Rough Sets

Muhammad Akram [1,*], Ghous Ali [1] and Noura Omair Alshehri [2]

1 Department of Mathematics, University of the Punjab, New Campus, Lahore 54590, Pakistan; mr.ghous1782@gmail.com

2 Department of Mathematics, Faculty of Science, AL Faisaliah, Campus, King Abdulaziz University, Jeddah 21589, Saudi Arabia; nalshehrie@kau.edu.sa

* Correspondence: m.akram@pucit.edu.pk

Received: 19 October 2017; Accepted: 1 November 2017; Published: 10 November 2017

Abstract: We introduce notions of soft rough m-polar fuzzy sets and m-polar fuzzy soft rough sets as novel hybrid models for soft computing, and investigate some of their fundamental properties. We discuss the relationship between m-polar fuzzy soft rough approximation operators and crisp soft rough approximation operators. We also present applications of m-polar fuzzy soft rough sets to decision-making.

Keywords: soft rough m-polar fuzzy sets; m-polar fuzzy soft rough sets; m-polar fuzzy soft rough approximation operators; decision-making

1. Introduction

The notion of bipolar fuzzy sets was generalized to m-polar fuzzy sets by Chen et al. [1] in 2014. Chen et al. [1] proved that bipolar fuzzy sets and 2-polar fuzzy sets are cryptomorphic mathematical tools. In many real life complicated problems, data sometimes comes from n agents ($n \geq 2$), that is, multipolar information (not just bipolar information, which corresponds to two-valued logic) exists. There are many applications of m-polar fuzzy sets to decision-making problems when it is compulsory to make assessments with a group of agreements. For example, similarity degrees of two logic formulas that are based on n logic implication operators ($n \geq 2$), ordering results of a magazine, a group of friends wants to plan to visit a country, ordering results of a university. Akram et al. [2–5] promoted the work on m-polar fuzzy graphs and introduced many new concepts. Li et al. [6] considered different algebraic operations on m-polar fuzzy graphs. In 1982, Pawlak [7] introduced the idea of rough set theory, which is an important mathematical tool to handle imprecise, vague and incomplete information. In fuzzy set theory [8], membership function plays the vital role. However, the selection of membership function is uncertain. The fuzzy set theory is an uncertain tool to solve the uncertain problems, but, in rough set theory, two precise boundary lines are established to describe the vague concepts. Consequently, the rough set theory is a mathematical tool to solve uncertain problems. Dubois and Prade [9] introduced the ideas of rough fuzzy sets and fuzzy rough sets by combining fuzzy sets and rough sets. Recently, works on granular computing are progressing rapidly. Xu and Gou [10] described an overview of interval-valued intuitionistic fuzzy information aggregation techniques, and their applications in different fields such as decision-making, entropy measure and supplier selection. Das et al. [11] introduced a robust decision-making approach using intuitionistic trapezoidal fuzzy number. Cai et al. [12] defined dynamic fuzzy sets by means of shadowed sets and proposed an analytic solution to computing the pair of thresholds by searching for a balance of uncertainty in the framework of shadowed sets. Pedrycz and Chen [13] provided various methods of fuzzy sets and granular computing, brings new concepts, architectures and practice of fuzzy decision-making with various applications.

Many real-world problems in different domains, including social sciences, physical sciences, applied sciences and life sciences contain vague and imprecise information. The classical mathematical tools and theories are unfit to handle the difficulties of the data having uncertainties, whereas a lot of theories including probability theory and fuzzy set theory [8] are very helpful mathematical tools for dealing with different types of uncertain data. Molodtsov [14] indicated the drawbacks of these theories. In order to overcome these difficulties, Molodtsov [14] introduced the concept of soft set theory. Maji et al. [15] proposed some fundamental algebraic operations for soft sets. Maji et al. [16] generalized the idea of soft sets and presented a hybrid model fuzzy soft sets. Alcantud [17–19] gave a novel approach to the problems of fuzzy soft sets based decision-making. Alcantud and Santos-Garcia [20,21] produced a completely new approach to soft set based decision-making problems when information is incomplete. They also proposed and compared an algorithmic solution with previous approaches in the literature in [20]. Feng et al. [22] gave the novel idea of rough soft sets by combining the Pawlak rough sets and soft sets. In 2011, Feng et al. [23] introduced the idea of soft rough sets. All mathematical models, including fuzzy sets, rough sets, soft sets and fuzzy soft sets have their advantages and drawbacks. One of the crucial drawbacks of all of these models is that they have a lack of a sufficient number of parameters to handle the uncertain data. In order to overcome this problem, we combine rough sets, soft sets with m-polar fuzzy sets and propose the concepts of new hybrid models called soft rough m-polar fuzzy sets and m-polar fuzzy soft rough sets. We define the lower and upper soft approximations of an m-polar fuzzy set. The idea of m-polar fuzzy soft rough sets can be utilized to solve different real-life problems. Thus, we present a new method to decision-making based on m-polar fuzzy soft rough sets.

2. Soft Rough m-Polar Fuzzy Sets

Definition 1. *An m-polar fuzzy set (mF set, for short) on a universe Y is a function* $Q = (p_1 \circ Q(z), p_2 \circ Q(z), \ldots, p_m \circ Q(z)) : Y \to [0,1]^m$, *where the i-th projection mapping is defined as* $p_i \circ Q : [0,1]^m \to [0,1]$. *Denote* $\mathbf{0} = (0,0,\cdots,0)$ *is the smallest element in* $[0,1]^m$ *and* $\mathbf{1} = (1,1,\cdots,1)$ *is the largest element in* $[0,1]^m$ *[1].*

Definition 2. *([14]) Let Y be a nonempty set called universe, T a set of parameters. A pair* (η, T) *is called a soft set over Y if* η *is a mapping given by* $\eta : T \to P(Y)$, *where* $P(Y)$ *is the collection of all subsets of Y.*

Definition 3. *([24]) Let Y be an initial universe,* (η, T) *a soft set on Y. For any* $N \subseteq Y \times T$, *the crisp soft relation N over* $Y \times T$ *is given by*

$$N = \{\langle (v,w), \varrho_N(v,w)\rangle \mid (v,w) \in Y \times T\},$$

where $\varrho_N : Y \times T \to \{0,1\}$, $\varrho_N(v,w) = \begin{cases} 1 & if\ (v,w) \in N, \\ 0 & if\ (v,w) \notin N. \end{cases}$

Definition 4. *([25]) Let Y be the universe of discourse and let T be a set of parameters. For any crisp soft relation* $\xi \subseteq Y \times T$, *a set-valued function* $\xi_s : Y \to P(T)$ *is given by*

$$\xi_s(v) = \{w \in T \mid (v,w) \in \xi\}, \quad v \in Y.$$

ξ *is referred to as serial if* $\forall\, v \in Y, \xi_s(v) \neq \emptyset$. *The pair* (Y, T, ξ) *is said to be a crisp soft approximation space. For any* $Q \subseteq T$, *the lower and upper soft approximations of Q about* (Y, T, ξ), *denoted by* $\underline{\xi}(Q)$ *and* $\overline{\xi}(Q)$, *respectively, are defined as*

$$\underline{\xi}(Q) = \{v \in Y \mid \xi_s(v) \cap Q \neq \phi\},$$

$$\overline{\xi}(Q) = \{v \in Y \mid \xi_s(v) \subseteq Q\}.$$

The pair $(\underline{\xi}(Q), \overline{\xi}(Q))$ *is said to be a crisp soft rough set and* $\underline{\xi}, \overline{\xi} : P(T) \to P(Y)$ *are, respectively, called lower and upper crisp soft rough approximation operators. Furthermore, if* $\underline{\xi}(Q) = \overline{\xi}(Q)$*, then* Q *is called a definable set.*

We now define soft rough m-polar fuzzy sets.

Definition 5. *Let* Y *be an initial universe and* T *a universe of parameters. For any crisp soft relation* ξ *over* $Y \times T$*, the pair* (Y, T, ξ) *is called a crisp soft approximation space. For an arbitrary* $Q \in m(T)$*, the lower and upper soft approximations of* Q *about* (Y, T, ξ)*, denoted by* $\underline{\xi}(Q)$ *and* $\overline{\xi}(Q)$*, respectively, are defined by*

$$\underline{\xi}(Q) = \{\langle v, Q_{\underline{\xi}}(v)\rangle \mid v \in Y\},$$
$$\overline{\xi}(Q) = \{\langle v, Q_{\overline{\xi}}(v)\rangle \mid v \in Y\},$$

where

$$Q_{\underline{\xi}}(v) = \bigwedge_{w \in \xi_s(v)} p_i \circ Q(w), \qquad Q_{\overline{\xi}}(v) = \bigvee_{w \in \xi_s(v)} p_i \circ Q(w).$$

The pair $(\underline{\xi}(Q), \overline{\xi}(Q))$ *is called the soft rough mF set of* Q *about* (Y, T, ξ)*, and* $\underline{\xi}, \overline{\xi} : m(T) \to m(Y)$ *are, respectively, said to be lower and upper soft rough mF approximation operators. Moreover, if* $\underline{\xi}(Q) = \overline{\xi}(Q)$*, then* Q *is referred to as definable.*

Example 1. *Let* $Y = \{y_1, y_2, y_3, y_4, y_5, y_6\}$ *be a universe of discourse,* $T = \{k_1, k_2, k_3, k_4\}$ *a set of parameters. Assume that a soft set on* Y *is defined by*

$$\eta(k_1) = \{y_1, y_2, y_5\}, \qquad \eta(k_2) = \{y_3, y_4, y_5\},$$
$$\eta(k_3) = \emptyset, \qquad \eta(k_4) = Y.$$

Then, a crisp soft relation ξ *over* $Y \times T$ *is given by*

$$\xi = \{(y_1, k_1), (y_2, k_1), (y_5, k_1), (y_3, k_2), (y_4, k_2), (y_5, k_2), (y_1, k_4), (y_2, k_4), (y_3, k_4), (y_4, k_4), (y_5, k_4), (y_6, k_4)\}.$$

By Definition 4, we have

$$\xi_s(y_1) = \{k_1, k_4\}, \qquad \xi_s(y_2) = \{k_1, k_4\},$$
$$\xi_s(y_3) = \{k_2, k_4\}, \qquad \xi_s(y_4) = \{k_2, k_4\},$$
$$\xi_s(y_5) = \{k_1, k_2, k_4\}, \qquad \xi_s(y_6) = \{k_4\}.$$

Consider a 3-polar fuzzy set $Q \in m(T)$ *as follows:*

$$Q = \{(k_1, 0.75, 0.25, 0.13), (k_2, 0.12, 0.7, 0.4), (k_3, 0.3, 0.85, 0.6), (k_4, 0.1, 0.3, 0.5)\}.$$

By Definition 5, we have lower and upper soft approximations:

$$Q_{\underline{\xi}}(y_1) = (0.1, 0.25, 0.13), \qquad Q_{\overline{\xi}}(y_1) = (0.75, 0.3, 0.5),$$
$$Q_{\underline{\xi}}(y_2) = (0.1, 0.25, 0.13), \qquad Q_{\overline{\xi}}(y_2) = (0.75, 0.3, 0.5),$$
$$Q_{\underline{\xi}}(y_3) = (0.1, 0.3, 0.4), \qquad Q_{\overline{\xi}}(y_3) = (0.12, 0.7, 0.5),$$
$$Q_{\underline{\xi}}(y_4) = (0.1, 0.3, 0.4), \qquad Q_{\overline{\xi}}(y_4) = (0.12, 0.7, 0.5),$$
$$Q_{\underline{\xi}}(y_5) = (0.1, 0.25, 0.13), \qquad Q_{\overline{\xi}}(y_5) = (0.75, 0.7, 0.5),$$
$$Q_{\underline{\xi}}(y_6) = (0.1, 0.3, 0.5), \qquad Q_{\overline{\xi}}(y_6) = (0.1, 0.3, 0.5).$$

Thus,

$$\underline{\xi}(Q) = \{(y_1, 0.1, 0.25, 0.13), (y_2, 0.1, 0.25, 0.13), (y_3, 0.1, 0.3, 0.4), (y_4, 0.1, 0.3, 0.4), \\ (y_5, 0.1, 0.25, 0.13), (y_6, 0.1, 0.3, 0.5)\},$$
$$\overline{\xi}(Q) = \{(y_1, 0.75, 0.3, 0.5), (y_2, 0.75, 0.3, 0.5), (y_3, 0.12, 0.7, 0.5), (y_4, 0.12, 0.7, 0.5), \\ (y_5, 0.75, 0.7, 0.5), (y_6, 0.1, 0.3, 0.5)\}.$$

Hence, the pair $(\underline{\xi}(Q), \overline{\xi}(Q))$ *is said to be a soft rough* 3*-polar fuzzy set.*

We now present properties of soft rough mF sets.

Theorem 1. *Let* (Y, T, ξ) *be a crisp soft approximation space. Then, the lower and upper soft rough mF approximation operators* $\underline{\xi}(Q)$ *and* $\overline{\xi}(Q)$*, respectively, satisfy the following properties, for any* $Q, R \in m(T)$*:*

1. $\underline{\xi}(Q) = \sim \overline{\xi}(\sim Q)$,
2. $Q \subseteq R \Rightarrow \underline{\xi}(Q) \subseteq \underline{\xi}(R)$,
3. $\underline{\xi}(Q \cap R) = \underline{\xi}(Q) \cap \underline{\xi}(R)$,
4. $\underline{\xi}(Q \cup R) \supseteq \underline{\xi}(Q) \cup \underline{\xi}(R)$,
5. $\overline{\xi}(Q) = \sim \underline{\xi}(\sim Q)$,
6. $Q \subseteq R \Rightarrow \overline{\xi}(Q) \subseteq \overline{\xi}(R)$,
7. $\overline{\xi}(Q \cup R) = \overline{\xi}(Q) \cup \overline{\xi}(R)$,
8. $\overline{\xi}(Q \cap R) \subseteq \overline{\xi}(Q) \cap \overline{\xi}(R)$,

where $\sim Q$ *denotes the compliment of* Q*.*

Proof. 1. From Definition 5, we have

$$\begin{aligned}
\sim \overline{\xi}(\sim Q) &= \Big\{\Big\langle v, (\mathbf{1} - (\sim Q)_{\overline{\xi}}(v))\Big\rangle \mid v \in Y\Big\}, \\
&= \Big\{\Big\langle v, \Big(\mathbf{1} - \bigvee_{w \in \xi_s(v)} p_i \circ (\sim Q)(w)\Big)\Big\rangle \mid v \in Y\Big\}, \\
&= \Big\{\Big\langle v, \Big(\mathbf{1} \wedge \bigwedge_{w \in \xi_s(v)} p_i \circ Q(w)\Big)\Big\rangle \mid v \in Y\Big\}, \\
&= \Big\{\Big\langle v, \bigwedge_{w \in \xi_s(v)} p_i \circ Q(w)\Big\rangle \mid v \in Y\Big\}, \\
&= \Big\{\Big\langle v, Q_{\underline{\xi}}(v)\Big\rangle \mid v \in Y\Big\}, \\
&= \underline{\xi}(Q).
\end{aligned}$$

It follows that $\underline{\xi}(Q) = \sim \overline{\xi}(\sim Q)$.

2. It can be easily proved by Definition 5.

3. By Definition 5,

$$\begin{aligned}
\underline{\xi}(Q\cap R) &= \Big\{\Big\langle v, (Q\cap R)_{\underline{\xi}}(v)\Big\rangle \mid v\in Y\Big\},\\
&= \Big\{\Big\langle v, \bigwedge_{w\in\xi_s(v)} p_i\circ(Q\cap R)(w)\Big\rangle \mid v\in Y\Big\},\\
&= \Big\{\Big\langle v, \bigwedge_{w\in\xi_s(v)} \big(p_i\circ Q(w)\wedge p_i\circ R(w)\big)\Big\rangle \mid v\in Y\Big\},\\
&= \Big\{\Big\langle v, \bigwedge_{w\in\xi_s(v)} \big(p_i\circ Q(w)\big)\wedge \bigwedge_{w\in\xi_s(v)} \big(p_i\circ R(w)\big)\Big\rangle \mid v\in Y\Big\},\\
&= \Big\{\Big\langle v, Q_{\underline{\xi}}(v)\wedge R_{\underline{\xi}}(v)\Big\rangle \mid v\in Y\Big\},\\
&= \underline{\xi}(Q)\cap\underline{\xi}(R).
\end{aligned}$$

Hence, $\underline{\xi}(Q\cap R)=\underline{\xi}(Q)\cap\underline{\xi}(R)$.

4. From Definition 5,

$$\begin{aligned}
\underline{\xi}(Q\cup R) &= \Big\{\Big\langle v, (Q\cup R)_{\underline{\xi}}(v)\Big\rangle \mid v\in Y\Big\},\\
&= \Big\{\Big\langle v, \bigwedge_{w\in\xi_s(v)} p_i\circ(Q\cup R)(w)\Big\rangle \mid v\in Y\Big\},\\
&= \Big\{\Big\langle v, \bigwedge_{w\in\xi_s(v)} \big(p_i\circ Q(w)\vee p_i\circ R(w)\big)\Big\rangle \mid v\in Y\Big\},\\
&\supseteq \Big\{\Big\langle v, \bigwedge_{w\in\xi_s(v)} \big(p_i\circ Q(w)\big)\vee \bigwedge_{w\in\xi_s(v)} \big(p_i\circ R(w)\big)\Big\rangle \mid v\in Y\Big\},\\
&= \Big\{\Big\langle v, Q_{\underline{\xi}}(v)\vee R_{\underline{\xi}}(v)\Big\rangle \mid v\in Y\Big\},\\
&= \underline{\xi}(Q)\cup\underline{\xi}(R).
\end{aligned}$$

Hence, $\underline{\xi}(Q\cup R)\supseteq\underline{\xi}(Q)\cup\underline{\xi}(R)$.

Similarly, properties (5–8) of the upper soft rough mF approximation operator $\overline{\xi}(Q)$ can be proved by using the above arguments. □

Example 2. *Let $Y=\{g_1,g_2,g_3,g_4\}$ be a universe and let $T=\{n_1,n_2,n_3\}$ be a set of parameters. Consider a soft set (η,T) over Y is defined as*

$$\eta(n_1)=\{g_1,g_2,g_4\},\qquad \eta(n_2)=\{g_3\},\qquad \eta(n_3)=Y.$$

Then, a crisp soft relation ξ on $Y\times T$ is given by

$$\xi=\big\{(g_1,n_1),(g_2,n_1),(g_4,n_1),(g_3,n_2),(g_1,n_3),(g_2,n_3),(g_3,n_3),(g_4,n_3)\big\}.$$

By Definition 4,

$$\begin{aligned}
&\xi_s(g_1)=\{n_1,n_3\}, && \xi_s(g_2)=\{n_1,n_3\},\\
&\xi_s(g_3)=\{n_2,n_3\}, && \xi_s(g_4)=\{n_1,n_3\}.
\end{aligned}$$

Consider 3-polar fuzzy sets $Q,R\in m(T)$ as follows:

$$Q=\big\{(n_1,0.5,0.2,0.3),(n_2,0.2,0.6,0.5),(n_3,0.5,0.9,0.1)\big\},$$

$$R=\big\{(n_1,0.7,0.5,0.1),(n_2,0.1,0.7,0.4),(n_3,0.3,0.8,0.6)\big\}.$$

Then,

$$
\begin{aligned}
\sim Q &= \{(n_1, 0.5, 0.8, 0.7), (n_2, 0.8, 0.4, 0.5), (n_3, 0.5, 0.1, 0.9)\},\\
Q \cup R &= \{(n_1, 0.7, 0.5, 0.3), (n_2, 0.2, 0.7, 0.5), (n_3, 0.5, 0.9, 0.6)\},\\
Q \cap R &= \{(n_1, 0.5, 0.2, 0.1), (n_2, 0.1, 0.6, 0.4), (n_3, 0.3, 0.8, 0.1)\}.
\end{aligned}
$$

By Definition 5*, we have*

$$
\begin{aligned}
\underline{\xi}(Q) &= \{(g_1, 0.5, 0.2, 0.1), (g_2, 0.5, 0.2, 0.1), (g_3, 0.2, 0.6, 0.1), (g_4, 0.5, 0.2, 0.1)\},\\
\overline{\xi}(Q) &= \{(g_1, 0.5, 0.9, 0.3), (g_2, 0.5, 0.9, 0.3), (g_3, 0.5, 0.9, 0.5), (g_4, 0.5, 0.9, 0.3)\},\\
\underline{\xi}(R) &= \{(g_1, 0.3, 0.5, 0.1), (g_2, 0.3, 0.5, 0.1), (g_3, 0.1, 0.7, 0.4), (g_4, 0.3, 0.5, 0.1)\},\\
\overline{\xi}(R) &= \{(g_1, 0.7, 0.8, 0.6), (g_2, 0.7, 0.8, 0.6), (g_3, 0.3, 0.8, 0.6), (g_4, 0.7, 0.8, 0.6)\},\\
\underline{\xi}(\sim Q) &= \{(g_1, 0.5, 0.1, 0.7), (g_2, 0.5, 0.1, 0.7), (g_3, 0.5, 0.1, 0.5), (g_4, 0.5, 0.1, 0.7)\},\\
\overline{\xi}(\sim Q) &= \{(g_1, 0.5, 0.8, 0.9), (g_2, 0.5, 0.8, 0.9), (g_3, 0.8, 0.4, 0.9), (g_4, 0.5, 0.8, 0.9)\},\\
\sim \underline{\xi}(\sim Q) &= \{(g_1, 0.5, 0.9, 0.3), (g_2, 0.5, 0.9, 0.3), (g_3, 0.5, 0.9, 0.5), (g_4, 0.5, 0.9, 0.3)\},\\
\sim \overline{\xi}(\sim Q) &= \{(g_1, 0.5, 0.2, 0.1), (g_2, 0.5, 0.2, 0.1), (g_3, 0.2, 0.6, 0.1), (g_4, 0.5, 0.2, 0.1)\},\\
\underline{\xi}(Q \cup R) &= \{(g_1, 0.5, 0.5, 0.3), (g_2, 0.5, 0.5, 0.3), (g_3, 0.2, 0.7, 0.5), (g_4, 0.5, 0.5, 0.3)\},\\
\overline{\xi}(Q \cup R) &= \{(g_1, 0.7, 0.9, 0.6), (g_2, 0.7, 0.9, 0.6), (g_3, 0.5, 0.9, 0.6), (g_4, 0.7, 0.9, 0.6)\},\\
\underline{\xi}(Q \cap R) &= \{(g_1, 0.3, 0.2, 0.1), (g_2, 0.3, 0.2, 0.1), (g_3, 0.1, 0.6, 0.1), (g_4, 0.3, 0.2, 0.1)\},\\
\overline{\xi}(Q \cap R) &= \{(g_1, 0.5, 0.8, 0.1), (g_2, 0.5, 0.8, 0.1), (g_3, 0.3, 0.8, 0.4), (g_4, 0.5, 0.8, 0.1)\}.
\end{aligned}
$$

Now,

$$
\begin{aligned}
\underline{\xi}(Q) \cup \underline{\xi}(R) &= \{(g_1, 0.5, 0.5, 0.1), (g_2, 0.5, 0.5, 0.1), (g_3, 0.2, 0.7, 0.4), (g_4, 0.5, 0.5, 0.1)\},\\
\overline{\xi}(Q) \cup \overline{\xi}(R) &= \{(g_1, 0.7, 0.9, 0.6), (g_2, 0.7, 0.9, 0.6), (g_3, 0.5, 0.9, 0.6), (g_4, 0.7, 0.9, 0.6)\},\\
\underline{\xi}(Q) \cap \underline{\xi}(R) &= \{(g_1, 0.3, 0.2, 0.1), (g_2, 0.3, 0.2, 0.1), (g_3, 0.1, 0.6, 0.1), (g_4, 0.3, 0.2, 0.1)\},\\
\overline{\xi}(Q) \cap \overline{\xi}(R) &= \{(g_1, 0.5, 0.8, 0.3), (g_2, 0.5, 0.8, 0.3), (g_3, 0.3, 0.8, 0.5), (g_4, 0.5, 0.8, 0.3)\}.
\end{aligned}
$$

From the above calculations, we observe that the following properties are satisfied:

$$
\begin{aligned}
&\sim \underline{\xi}(\sim Q) = \overline{\xi}(Q), && \sim \overline{\xi}(\sim Q) = \underline{\xi}(Q),\\
&\underline{\xi}(Q \cap R) = \underline{\xi}(Q) \cap \underline{\xi}(R), && \underline{\xi}(Q \cup R) \supseteq \underline{\xi}(Q) \cup \underline{\xi}(R),\\
&\overline{\xi}(Q \cup R) = \overline{\xi}(Q) \cup \overline{\xi}(R), && \overline{\xi}(Q \cap R) \subseteq \overline{\xi}(Q) \cap \overline{\xi}(R).
\end{aligned}
$$

Proposition 1. *Let* (Y, T, ξ) *be a crisp soft approximation space. Then, lower and upper soft rough approximations of mF sets Q and R satisfies the following laws:*

1. $\sim \left(\underline{\xi}(Q) \cup \underline{\xi}(R)\right) = \overline{\xi}(\sim Q) \cap \overline{\xi}(\sim R),$
2. $\sim \left(\underline{\xi}(Q) \cup \overline{\xi}(R)\right) = \overline{\xi}(\sim Q) \cap \underline{\xi}(\sim R),$
3. $\sim \left(\overline{\xi}(Q) \cup \underline{\xi}(R)\right) = \underline{\xi}(\sim Q) \cap \overline{\xi}(\sim R),$
4. $\sim \left(\overline{\xi}(Q) \cup \overline{\xi}(R)\right) = \underline{\xi}(\sim Q) \cap \underline{\xi}(\sim R),$
5. $\sim \left(\underline{\xi}(Q) \cap \underline{\xi}(R)\right) = \overline{\xi}(\sim Q) \cup \overline{\xi}(\sim R),$
6. $\sim \left(\underline{\xi}(Q) \cap \overline{\xi}(R)\right) = \overline{\xi}(\sim Q) \cup \underline{\xi}(\sim R),$
7. $\sim \left(\overline{\xi}(Q) \cap \underline{\xi}(R)\right) = \underline{\xi}(\sim Q) \cup \overline{\xi}(\sim R),$

8. $\sim\left(\overline{\zeta}(Q)\cap\overline{\zeta}(R)\right)=\underline{\zeta}(\sim Q)\cup\underline{\zeta}(\sim R)$.

Proof. Its proof follows immediately from the Definition 5. □

3. *m*F Soft Rough Sets

Definition 6. *Let Y be a universe of discourse, T a set of parameters and $V\subseteq T$. A pair (τ,V) is referred to as an mF soft set on Y if τ is a mapping $\tau:T\to m(Y)$.*

Definition 7. *Let (τ,V) be an mF soft set over Y. Then, an mF subset ζ of $Y\times T$ is referred to as an mF soft relation from Y to T is given by*

$$\zeta=\{\langle(x,t),p_i\circ\zeta(x,t)\rangle\mid(x,t)\in Y\times T\},$$

where $\zeta:Y\times T\to[0,1]^m$.

If $Y=\{x_1,x_2,\cdots,x_n\}$, $T=\{t_1,t_2,\cdots,t_n\}$, then an *m*F soft relation ζ over $Y\times T$ can be presented as follows:

ζ	t_1	t_2	$\cdots$	t_n
x_1	$p_i\circ(x_1,t_1)$	$p_i\circ(x_1,t_2)$	$\cdots$	$p_i\circ(x_1,t_n)$
x_2	$p_i\circ(x_2,t_1)$	$p_i\circ(x_2,t_2)$	$\cdots$	$p_i\circ(x_2,t_n)$
$\vdots$	$\vdots$	$\vdots$	$\ddots$	$\vdots$
x_n	$p_i\circ(x_n,t_1)$	$p_i\circ(x_n,t_2)$	$\cdots$	$p_i\circ(x_n,t_n)$.

Example 3. *Let $Y=\{x_1,x_2,x_3\}$ be a universe, $T=\{t_1,t_2,t_3\}$ a set of parameters. A 3-polar fuzzy soft relation $\zeta:Y\to T$ of the universe $Y\times T$ is given by*

ζ	t_1	t_2	t_3
x_1	(0.6, 0.3, 0.1)	(0.4, 0.7, 0.6)	(0.4, 0.6, 0.2)
x_2	(0.5, 0.3, 0.2)	(0.5, 0.2, 0.8)	(0.6, 0.9, 0.6)
x_3	(0.3, 0.2, 0.1)	(0.3, 0.4, 0.8)	(0.7, 0.3, 0.5).

We now define *m*-polar fuzzy soft rough sets.

Definition 8. *Let Y be a nonempty set called universe, T a universe of parameters. For any mF soft relation ζ on $Y\times T$, the pair (Y,T,ζ) is referred to as an mF soft approximation space. For an arbitrary $Q\in m(T)$, the lower and upper soft approximations of Q about (Y,T,ζ), denoted by $\underline{\zeta}(Q)$ and $\overline{\zeta}(Q)$, respectively, are defined as follows:*

$$\underline{\zeta}(Q)=\{\langle v,Q_{\underline{\zeta}}(v)\rangle\mid v\in Y\},$$
$$\overline{\zeta}(Q)=\{\langle v,Q_{\overline{\zeta}}(v)\rangle\mid v\in Y\},$$

where

$$Q_{\underline{\zeta}}(v)=\bigwedge_{w\in T}\left[(\mathbf{1}-p_i\circ Q_{\zeta}(v,w))\vee p_i\circ Q(w)\right],$$
$$Q_{\overline{\zeta}}(v)=\bigvee_{w\in T}\left(p_i\circ Q_{\zeta}(v,w)\wedge p_i\circ Q(w)\right).$$

The pair $(\underline{\zeta}(Q),\overline{\zeta}(Q))$ is called mF soft rough set of Q about (Y,T,ζ), and $\underline{\zeta}$, $\overline{\zeta}:m(T)\to m(Y)$ are, respectively, said to be lower and upper mF soft rough approximation operators. Moreover, if $\underline{\zeta}(Q)=\overline{\zeta}(Q)$, then Q is said to be definable.

Example 4. *Let $Y = \{x_1, x_2, x_3, x_4, x_5\}$ be the set of five laptops and let $T = \{t_1 = size, t_2 = beautiful, t_3 = technology, t_4 = price\}$ be the set of parameters. Consider a 3-polar fuzzy soft relation $\zeta : Y \to T$ is given by*

ζ	t_1	t_2	t_3	t_4
x_1	$(0.6, 0.3, 0.1)$	$(0.4, 0.7, 0.6)$	$(0.4, 0.6, 0.2)$	$(0.4, 0.6, 0.2)$
x_2	$(0.5, 0.3, 0.2)$	$(0.5, 0.2, 0.8)$	$(0.6, 0.9, 0.6)$	$(0.7, 0.3, 0.6)$
x_3	$(0.3, 0.2, 0.1)$	$(0.3, 0.4, 0.8)$	$(0.7, 0.3, 0.5)$	$(0.2, 0.9, 0.9)$
x_4	$(0.4, 0.3, 0.6)$	$(0.5, 0.1, 0.4)$	$(0.3, 0.1, 0.0)$	$(0.6, 0.4, 0.4)$
x_5	$(0.2, 0.7, 0.3)$	$(0.4, 0.8, 0.1)$	$(0.4, 0.0, 0.7)$	$(0.8, 0.9, 0.0)$.

Consider a 3-polar fuzzy subset Q of T as follows:

$$Q = \{(t_1, 0.3, 0.1, 0.7), (t_2, 0.3, 0.6, 0.4), (t_3, 0.5, 0.6, 0.1), (t_4, 0.9, 0.1, 0.4)\}.$$

From Definition 8, the lower and upper soft approximations are given by

$$\begin{aligned}
Q_{\underline{\zeta}}(x_1) &= (0.4, 0.4, 0.4), & Q_{\overline{\zeta}}(x_1) &= (0.4, 0.6, 0.4),\\
Q_{\underline{\zeta}}(x_2) &= (0.5, 0.6, 0.4), & Q_{\overline{\zeta}}(x_2) &= (0.7, 0.6, 0.4),\\
Q_{\underline{\zeta}}(x_3) &= (0.5, 0.1, 0.4), & Q_{\overline{\zeta}}(x_3) &= (0.5, 0.4, 0.4),\\
Q_{\underline{\zeta}}(x_4) &= (0.5, 0.6, 0.6), & Q_{\overline{\zeta}}(x_4) &= (0.6, 0.1, 0.6),\\
Q_{\underline{\zeta}}(x_5) &= (0.6, 0.1, 0.3), & Q_{\overline{\zeta}}(x_5) &= (0.8, 0.6, 0.3).
\end{aligned}$$

Now,

$$\begin{aligned}
\underline{\zeta}(Q) = \{&(x_1, 0.4, 0.4, 0.4), (x_2, 0.5, 0.6, 0.4), (x_3, 0.5, 0.1, 0.4), (x_4, 0.5, 0.6, 0.6),\\
&(x_5, 0.6, 0.1, 0.3)\},\\
\overline{\zeta}(Q) = \{&(x_1, 0.4, 0.6, 0.4), (x_2, 0.7, 0.6, 0.4), (x_3, 0.5, 0.4, 0.4), (x_4, 0.6, 0.1, 0.6),\\
&(x_5, 0.8, 0.6, 0.3)\}.
\end{aligned}$$

Hence, the pair $(\underline{\zeta}(Q), \overline{\zeta}(Q))$ is called a 3-polar fuzzy soft rough set.

We now present properties of mF soft rough sets.

Theorem 2. *Let (Y, T, ζ) be an mF soft approximation space. Then, the lower and upper soft rough mF approximation operators $\underline{\zeta}(Q)$ and $\overline{\zeta}(Q)$, respectively, satisfy the following properties, for any $Q, R \in m(T)$:*

1. $\underline{\zeta}(Q) = \sim \overline{\zeta}(\sim Q)$,
2. $Q \subseteq R \Rightarrow \underline{\zeta}(Q) \subseteq \underline{\zeta}(R)$,
3. $\underline{\zeta}(Q \cap R) = \underline{\zeta}(Q) \cap \underline{\zeta}(R)$,
4. $\underline{\zeta}(Q \cup R) \supseteq \underline{\zeta}(Q) \cup \underline{\zeta}(R)$,
5. $\overline{\zeta}(Q) = \sim \underline{\zeta}(\sim Q)$,
6. $Q \subseteq R \Rightarrow \overline{\zeta}(Q) \subseteq \overline{\zeta}(R)$,
7. $\overline{\zeta}(Q \cup R) = \overline{\zeta}(Q) \cup \overline{\zeta}(R)$,
8. $\overline{\zeta}(Q \cap R) \subseteq \overline{\zeta}(Q) \cap \overline{\zeta}(R)$,

where $\sim Q$ denotes the compliment of Q.

Proof. 1. From Definition 8,

$$\begin{aligned}
\sim \overline{\zeta}(\sim Q) &= \Big\{\Big\langle v, (\mathbf{1} - (\sim Q)_{\overline{\zeta}}(v))\Big\rangle \mid v \in Y\Big\}, \\
&= \Big\{\Big\langle v, \mathbf{1} - \bigvee_{w \in T} (p_i \circ (\sim Q)_{\zeta}(v,w) \wedge p_i \circ (\sim Q)(w))\Big\rangle \mid v \in Y\Big\}, \\
&= \Big\{\Big\langle v, \mathbf{1} \wedge \bigwedge_{w \in T} (\mathbf{1} - p_i \circ Q_{\zeta}(v,w)) \vee p_i \circ Q(w)\Big\rangle \mid v \in Y\Big\}, \\
&= \Big\{\Big\langle v, \bigwedge_{w \in T} (\mathbf{1} - p_i \circ Q_{\zeta}(v,w)) \vee p_i \circ Q(w)\Big\rangle \mid v \in Y\Big\}, \\
&= \Big\{\Big\langle v, Q_{\underline{\zeta}}(v)\Big\rangle \mid v \in Y\Big\}, \\
&= \underline{\zeta}(Q).
\end{aligned}$$

Thus, $\underline{\zeta}(Q) = \sim \overline{\zeta}(\sim Q)$.

2. It can be proved directly by Definition 8.
3. By Definition 8,

$$\begin{aligned}
\underline{\zeta}(Q \cap R) &= \Big\{\Big\langle v, (Q \cap R)_{\underline{\zeta}}(v)\Big\rangle \mid v \in Y\Big\}, \\
&= \Big\{\Big\langle v, \bigwedge_{w \in T} (\mathbf{1} - p_i \circ (Q \cap R)(v,w)) \vee p_i \circ (Q \cap R)(w)\Big\rangle \mid v \in Y\Big\}, \\
&= \Big\{\Big\langle v, \bigwedge_{w \in T} \Big(\mathbf{1} - p_i \circ (Q(v,w) \wedge R(v,w))\Big) \vee p_i \circ (Q(w) \wedge R(w))\Big\rangle \mid v \in Y\Big\}, \\
&= \Big\{\Big\langle v, Q_{\underline{\zeta}}(v) \wedge R_{\underline{\zeta}}(v)\Big\rangle \mid v \in Y\Big\}, \\
&= \underline{\zeta}(Q) \cap \underline{\zeta}(R).
\end{aligned}$$

Hence, $\underline{\zeta}(Q \cap R) = \underline{\zeta}(Q) \cap \underline{\zeta}(R)$.

4. Using Definition 8,

$$\begin{aligned}
\underline{\zeta}(Q \cup R) &= \Big\{\Big\langle v, (Q \cup R)_{\underline{\zeta}}(v)\Big\rangle \mid v \in Y\Big\}, \\
&= \Big\{\Big\langle v, \bigwedge_{w \in T} (\mathbf{1} - p_i \circ (Q \cup R)(v,w)) \vee p_i \circ (Q \cup R)(w)\Big\rangle \mid v \in Y\Big\}, \\
&\supseteq \Big\{\Big\langle v, \bigwedge_{w \in T} \Big(\mathbf{1} - p_i \circ (Q(v,w) \vee R(v,w))\Big) \vee p_i \circ (Q(w) \vee R(w))\Big\rangle \mid v \in Y\Big\}, \\
&= \Big\{\Big\langle v, Q_{\underline{\zeta}}(v) \vee R_{\underline{\zeta}}(v)\Big\rangle \mid v \in Y\Big\}, \\
&= \underline{\zeta}(Q) \cup \underline{\zeta}(R).
\end{aligned}$$

Thus, $\underline{\zeta}(Q \cup R) \supseteq \underline{\zeta}(Q) \cup \underline{\zeta}(R)$.

The properties (5–8) can be proved by using similar arguments. □

Example 5. *Let* $Y = \{w_1, w_2, w_3, w_4\}$ *be the set of four cars and let* $T = \{v_1, v_2, v_3\}$ *be the set of parameters, where*

- v_1 *denotes the Fuel efficiency,*
- v_2 *denotes the Price,*
- v_3 *denotes the Technology.*

Consider a 3-polar fuzzy soft relation $\zeta : Y \to T$ *is given by*

ζ	v_1	v_2	v_3
w_1	$(0.6, 0.3, 0.1)$	$(0.4, 0.7, 0.6)$	$(0.4, 0.6, 0.2)$
w_2	$(0.5, 0.3, 0.2)$	$(0.5, 0.2, 0.8)$	$(0.6, 0.9, 0.6)$
w_3	$(0.3, 0.2, 0.1)$	$(0.3, 0.4, 0.8)$	$(0.7, 0.3, 0.5)$
w_4	$(0.4, 0.3, 0.6)$	$(0.5, 0.1, 0.4)$	$(0.3, 0.1, 0.0)$.

Consider 3-polar fuzzy subsets Q, R *of* T *as follows:*

$$Q = \{(v_1, 0.2, 0.1, 0.9), (v_2, 0.7, 0.5, 0.3), (v_3, 0.5, 0.6, 0.1)\},$$

$$R = \{(v_1, 0.4, 0.2, 0.5), (v_2, 0.6, 0.7, 0.3), (v_3, 0.4, 0.7, 0.8)\}.$$

Then,

$$\sim Q = \{(v_1, 0.8, 0.9, 0.1), (v_2, 0.3, 0.5, 0.7), (v_3, 0.5, 0.4, 0.9)\},$$
$$Q \cup R = \{(v_1, 0.4, 0.2, 0.9), (v_2, 0.7, 0.7, 0.3), (v_3, 0.5, 0.7, 0.8)\},$$
$$Q \cap R = \{(v_1, 0.2, 0.1, 0.5), (v_2, 0.6, 0.5, 0.3), (v_3, 0.4, 0.6, 0.1)\}.$$

By Definition 8, we have

$$\begin{aligned}
\underline{\zeta}(Q) &= \{(w_1, 0.4, 0.5, 0.4), (w_2, 0.5, 0.6, 0.3), (w_3, 0.5, 0.6, 0.3), (w_4, 0.6, 0.7, 0.6)\},\\
\overline{\zeta}(Q) &= \{(w_1, 0.4, 0.6, 0.3), (w_2, 0.5, 0.6, 0.3), (w_3, 0.5, 0.4, 0.3), (w_4, 0.5, 0.1, 0.6)\},\\
\underline{\zeta}(R) &= \{(w_1, 0.4, 0.7, 0.4), (w_2, 0.4, 0.7, 0.3), (w_3, 0.4, 0.7, 0.3), (w_4, 0.6, 0.7, 0.5)\},\\
\overline{\zeta}(R) &= \{(w_1, 0.4, 0.7, 0.3), (w_2, 0.5, 0.7, 0.6), (w_3, 0.4, 0.4, 0.5), (w_4, 0.5, 0.2, 0.5)\},\\
\sim \underline{\zeta}(\sim Q) &= \{(w_1, 0.4, 0.6, 0.3), (w_2, 0.5, 0.6, 0.3), (w_3, 0.5, 0.4, 0.3), (w_4, 0.5, 0.1, 0.6)\},\\
\sim \overline{\zeta}(\sim Q) &= \{(w_1, 0.4, 0.5, 0.4), (w_2, 0.5, 0.6, 0.3), (w_3, 0.5, 0.6, 0.3), (w_4, 0.6, 0.7, 0.6)\},\\
\underline{\zeta}(Q \cup R) &= \{(w_1, 0.4, 0.7, 0.4), (w_2, 0.5, 0.7, 0.3), (w_3, 0.5, 0.7, 0.3), (w_4, 0.6, 0.7, 0.6)\},\\
\overline{\zeta}(Q \cup R) &= \{(w_1, 0.4, 0.7, 0.3), (w_2, 0.5, 0.7, 0.6), (w_3, 0.5, 0.4, 0.5), (w_4, 0.5, 0.2, 0.6)\},\\
\underline{\zeta}(Q \cap R) &= \{(w_1, 0.4, 0.5, 0.4), (w_2, 0.4, 0.6, 0.3), (w_3, 0.4, 0.6, 0.3), (w_4, 0.6, 0.7, 0.5)\},\\
\overline{\zeta}(Q \cap R) &= \{(w_1, 0.4, 0.6, 0.3), (w_2, 0.5, 0.6, 0.3), (w_3, 0.4, 0.4, 0.3), (w_4, 0.5, 0.1, 0.5)\}.
\end{aligned}$$

Now,

$$\begin{aligned}
\underline{\zeta}(Q) \cup \underline{\zeta}(R) &= \{(w_1, 0.4, 0.7, 0.4), (w_2, 0.5, 0.7, 0.3), (w_3, 0.5, 0.7, 0.3), (w_4, 0.6, 0.7, 0.6)\},\\
\overline{\zeta}(Q) \cup \overline{\zeta}(R) &= \{(w_1, 0.4, 0.7, 0.3), (w_2, 0.5, 0.7, 0.6), (w_3, 0.5, 0.4, 0.5), (w_4, 0.5, 0.2, 0.6)\},\\
\underline{\zeta}(Q) \cap \underline{\zeta}(R) &= \{(w_1, 0.4, 0.5, 0.4), (w_2, 0.4, 0.6, 0.3), (w_3, 0.4, 0.6, 0.3), (w_4, 0.6, 0.7, 0.5)\},\\
\overline{\zeta}(Q) \cap \overline{\zeta}(R) &= \{(w_1, 0.4, 0.6, 0.3), (w_2, 0.5, 0.6, 0.3), (w_3, 0.4, 0.4, 0.3), (w_4, 0.5, 0.1, 0.5)\}.
\end{aligned}$$

From the above calculations,

$$\begin{aligned}
\sim \underline{\zeta}(\sim Q) &= \overline{\zeta}(Q), & \sim \overline{\zeta}(\sim Q) &= \underline{\zeta}(Q),\\
\underline{\zeta}(Q \cap R) &= \underline{\zeta}(Q) \cap \underline{\zeta}(R), & \underline{\zeta}(Q \cup R) &\supseteq \underline{\zeta}(Q) \cup \underline{\zeta}(R),\\
\overline{\zeta}(Q \cup R) &= \overline{\zeta}(Q) \cup \overline{\zeta}(R), & \overline{\zeta}(Q \cap R) &\subseteq \overline{\zeta}(Q) \cap \overline{\zeta}(R).
\end{aligned}$$

Remark 1. *In Theorem 2, properties (1) and (5) show that the lower and upper mF soft rough approximations operators* $\underline{\zeta}$ *and* $\overline{\zeta}$*, respectively, are dual to one another.*

Proposition 2. *Let (Y, T, ζ) be an mF soft approximation space. Then, the lower and upper soft rough approximations of mF sets Q and R satisfy the following laws:*

1. $\sim\left(\underline{\zeta}(Q) \cup \underline{\zeta}(R)\right) = \overline{\zeta}(\sim Q) \cap \overline{\zeta}(\sim R)$,
2. $\sim\left(\underline{\zeta}(Q) \cup \overline{\zeta}(R)\right) = \overline{\zeta}(\sim Q) \cap \underline{\zeta}(\sim R)$,
3. $\sim\left(\overline{\zeta}(Q) \cup \underline{\zeta}(R)\right) = \underline{\zeta}(\sim Q) \cap \overline{\zeta}(\sim R)$,
4. $\sim\left(\overline{\zeta}(Q) \cup \overline{\zeta}(R)\right) = \underline{\zeta}(\sim Q) \cap \underline{\zeta}(\sim R)$,
5. $\sim\left(\underline{\zeta}(Q) \cap \underline{\zeta}(R)\right) = \overline{\zeta}(\sim Q) \cup \overline{\zeta}(\sim R)$,
6. $\sim\left(\underline{\zeta}(Q) \cap \overline{\zeta}(R)\right) = \overline{\zeta}(\sim Q) \cup \underline{\zeta}(\sim R)$,
7. $\sim\left(\overline{\zeta}(Q) \cap \underline{\zeta}(R)\right) = \underline{\zeta}(\sim Q) \cup \overline{\zeta}(\sim R)$,
8. $\sim\left(\overline{\zeta}(Q) \cap \overline{\zeta}(R)\right) = \underline{\zeta}(\sim Q) \cup \underline{\zeta}(\sim R)$.

Proof. Its proof follows immediately from Definition 8. □

Definition 9. *Let Y be a universe, $Q = \{(v, p_i \circ Q(v)) \mid v \in Y\} \in m(Y)$, and $\sigma \in [0,1]^m$. The σ-level cut set of Q and the strong σ-level cut set of Q, denoted by Q_σ and $Q_{\sigma+}$, respectively, are defined as follows:*

$$Q_\sigma = \{v \in Y \mid p_i \circ Q(v) \geq \sigma\},$$
$$Q_{\sigma+} = \{v \in Y \mid p_i \circ Q(v) > \sigma\}.$$

Definition 10. *Let ζ be an mF soft relation on $Y \times T$, we define*

$$\zeta_\sigma = \{(v, w) \in Y \times T \mid p_i \circ \zeta(v, w) \geq \sigma\},$$
$$\zeta_\sigma(v) = \{w \in T \mid p_i \circ \zeta(v, w) \geq \sigma\},$$
$$\zeta_{\sigma+} = \{(v, w) \in Y \times T \mid p_i \circ \zeta(v, w) > \sigma\},$$
$$\zeta_{\sigma+}(v) = \{w \in T \mid p_i \circ \zeta(v, w) > \sigma\}.$$

Then, ζ_σ and $\zeta_{\sigma+}$ are two crisp soft relations on $Y \times T$.

We now prove that the *m*F soft rough approximation operators can be described by crisp soft rough approximation operators.

Theorem 3. *Let (Y, T, ζ) be an mF soft approximation space and $Q \in m(T)$. Then, the upper mF soft rough approximation operator can be described as follows, $\forall\, v \in Y$:*

1.

$$Q_{\overline{\zeta}}(v) = \bigvee_{\sigma \in [0,1]^m} \left(\sigma \wedge \overline{\zeta}_\sigma(Q_\sigma)(v)\right) = \bigvee_{\sigma \in [0,1]^m} \left(\sigma \wedge \overline{\zeta}_\sigma(Q_{\sigma+})(v)\right),$$
$$= \bigvee_{\sigma \in [0,1]^m} \left(\sigma \wedge \overline{\zeta}_{\sigma+}(Q_\sigma)(v)\right) = \bigvee_{\sigma \in [0,1]^m} \left(\sigma \wedge \overline{\zeta}_{\sigma+}(Q_{\sigma+})(v)\right).$$

2. $[\overline{\zeta}(Q)]_{\sigma+} \subseteq \overline{\zeta}_{\sigma+}(Q_{\sigma+}) \subseteq \overline{\zeta}_{\sigma+}(Q_\sigma) \subseteq \overline{\zeta}_\sigma(Q_\sigma) \subseteq [\overline{\zeta}(Q)]_\sigma$.

Proof. 1. For all $v \in Y$,

$$\bigvee_{\sigma \in [0,1]^m} \left(\sigma \wedge \overline{\zeta}_\sigma(Q_\sigma)(v)\right) = \sup\{\sigma \in [0,1]^m \mid v \in \overline{\zeta}_\sigma(Q_\sigma)\},$$
$$= \sup\{\sigma \in [0,1]^m \mid \zeta_\sigma(v) \cap Q_\sigma\},$$
$$= \sup\{\sigma \in [0,1]^m \mid \exists\, w \in T[w \in \zeta_\sigma(v), w \in Q_\sigma]\},$$
$$= \sup\{\sigma \in [0,1]^m \mid \exists\, w \in T[p_i \circ Q_\zeta(v,w) \geq \sigma, p_i \circ Q(w) \geq \sigma]\},$$
$$= \bigvee_{w \in T} \left(p_i \circ Q_\zeta(v,w) \wedge p_i \circ Q(w)\right),$$
$$= Q_{\overline{\zeta}}(v).$$

By similar arguments, we can compute

$$Q_{\overline{\zeta}}(v) = \bigvee_{\sigma \in [0,1]^m} \left(\sigma \wedge \overline{\zeta}_\sigma(Q_{\sigma+})(v)\right) = \bigvee_{\sigma \in [0,1]^m} \left(\sigma \wedge \overline{\zeta}_{\sigma+}(Q_\sigma)(v)\right) = \bigvee_{\sigma \in [0,1]^m} \left(\sigma \wedge \overline{\zeta}_{\sigma+}(Q_{\sigma+})(v)\right).$$

2. By Definitions 9 and 10, we directly verified that $\overline{\zeta}_{\sigma+}(Q_{\sigma+}) \subseteq \overline{\zeta}_{\sigma+}(Q_\sigma) \subseteq \overline{\zeta}_\sigma(Q_\sigma)$. Now, it is sufficient to show that $[\overline{\zeta}(Q)]_{\sigma+} \subseteq \overline{\zeta}_{\sigma+}(Q_{\sigma+})$ and $\overline{\zeta}_\sigma(Q_\sigma) \subseteq [\overline{\zeta}(Q)]_\sigma$.
For all $v \in [\overline{\zeta}(Q)]_{\sigma+}$, we have $Q_{\overline{\zeta}}(v) > \sigma$. By Definition 8, $\bigvee_{w \in T} \left(p_i \circ Q_\zeta(v,w) \wedge p_i \circ Q(w)\right) > \sigma$.
Then, there exists $w_0 \in T$, such that $p_i \circ Q_\zeta(v, w_0) \wedge p_i \circ Q(w_0) > \sigma$, that is, $p_i \circ Q_\zeta(v, w_0) > \sigma$ and $p_i \circ Q(w_0) > \sigma$. Thus, $w_0 \in \zeta_{\sigma+}(v)$ and $w_0 \in Q_\sigma$. It follows that $\zeta_{\sigma+}(v) \cap Q_\sigma \neq \emptyset$. By Definition 4, we have $v \in \overline{\zeta}_{\sigma+}(Q_{\sigma+})$. Hence, $[\overline{\zeta}(Q)]_{\sigma+} \subseteq \overline{\zeta}_{\sigma+}(Q_{\sigma+})$.

To prove $\overline{\zeta}_\sigma(Q_\sigma) \subseteq [\overline{\zeta}(Q)]_\sigma$, let an arbitrary $v \in \overline{\zeta}_\sigma(Q_\sigma)$, we have $\overline{\zeta}_\sigma(Q_\sigma)(v) = 1$. Since $Q_{\overline{\zeta}}(v) = \bigvee_{\sigma \in [0,1]^m} [\overline{\zeta}_\sigma(Q_\sigma)(v)] \geq \sigma \wedge \overline{\zeta}_\sigma(Q_\sigma)(v) = \sigma$, we obtain $v \in [\overline{\zeta}(Q)]_\sigma$.
Hence, $\overline{\zeta}_\sigma(Q_\sigma) \subseteq [\overline{\zeta}(Q)]_\sigma$.
□

Theorem 4. *Let (Y, T, ζ) be an mF soft approximation apace. If ζ is serial, then the lower and upper mF soft rough approximation operators $\underline{\zeta}(Q)$ and $\overline{\zeta}(Q)$, respectively, satisfy the following:*

1. $\underline{\zeta}(\emptyset) = \emptyset. \overline{\zeta}(T) = Y,$
2. $\underline{\zeta}(Q) \subseteq \overline{\zeta}(Q)$, *for all* $Q \in m(T)$.

Proof. Its proof follows directly by Definition 8. □

Definition 11. *Let Q be an mF set of the universe set Y and let $\underline{\zeta}(Q)$, $\overline{\zeta}(Q)$ be the lower and upper soft rough approximation operators. Then, ring sum operation about mF sets $\underline{\zeta}(Q)$ and $\overline{\zeta}(Q)$ is defined by*

$$\underline{\zeta}(Q) \oplus \overline{\zeta}(Q) = \left\{\left(v, p_i \circ Q_{\underline{\zeta}}(v) + p_i \circ Q_{\overline{\zeta}}(v) - p_i \circ Q_{\underline{\zeta}}(v) \times p_i \circ Q_{\overline{\zeta}}(v)\right) \mid v \in Y\right\}.$$

4. Applications to Decision-Making

4.1. Selection of a Hotel

The selection of the right hotel to stay is always a difficult task. Since every person has different needs when searching for a hotel. The location of the hotel is something that is very important for an enjoyable stay. There are a number of factors to take into consideration for selecting the right hotel, whether we are looking for a great location, a great meal option or a great service. Suppose a person (Mr. Adeel) wants to stay in a hotel for a long period. There are four alternatives in his mind.

The alternatives are y_1, y_2, y_3, y_4. He wants to select the most suitable hotel. The location, meal options and services are the main parameters for the selection of a hotel.
Let $Y = \{y_1, y_2, y_3, y_4\}$ be the set of four hotels under consideration and let $T = \{z_1, z_2, z_3\}$ be the set of parameters related to the hotels in Y, where,

'z_1' represents the Location,
'z_2' represents the Meal Options,
'z_3' represents the Services.

We give more features of these parameters as follows:

- The "Location" of the hotel include close to main road, in the green surroundings, in the city center.
- The "Meal options" of the hotel include fast food, fast casual, casual dining.
- The "Services" of the hotel include Wi-Fi connectivity, fitness center, room service.

Suppose that Adeel explains the "attractiveness of the hotel" by forming a 3-polar fuzzy soft relation $\zeta : Y \to T$, which is given by

ζ	z_1	z_2	z_3
y_1	$(0.2, 0.6, 0.1)$	$(0.3, 0.4, 0.7)$	$(0.7, 0.3, 0.2)$
y_2	$(0.4, 0.5, 0.7)$	$(0.4, 0.5, 0.5)$	$(0.7, 0.4, 0.1)$
y_3	$(0.7, 0.8, 0.3)$	$(0.8, 0.9, 0.4)$	$(0.6, 0.2, 0.6)$
y_4	$(0.5, 0.6, 0.4)$	$(0.6, 0.7, 0.1)$	$(0.8, 0.5, 0.3)$.

Thus, ζ over $Y \times T$ is the 3-polar fuzzy soft relation in which location, meal option and price of the hotels are considered. For example, if we consider "Location" of the hotel, $((y_1, z_1), 0.2, 0.6, 0.1)$ means that the hotel y_1 is 20% close to the main road, 60% in the green surroundings and 10% in the city center.

We now assume that Adeel gives the optimal normal decision object Q, which is a 3-polar fuzzy subset of T as follows:

$$Q = \{(z_1, 0.5, 0.6, 0.7), (z_2, 0.7, 0.6, 0.9), (z_3, 0.9, 0.6, 0.8)\}.$$

By Definition 8,

$$\begin{aligned}
Q_{\underline{\zeta}}(y_1) &= (0.7, 0.6, 0.8), & Q_{\overline{\zeta}}(y_1) &= (0.7, 0.6, 0.7),\\
Q_{\underline{\zeta}}(y_2) &= (0.6, 0.6, 0.7), & Q_{\overline{\zeta}}(y_2) &= (0.7, 0.5, 0.7),\\
Q_{\underline{\zeta}}(y_3) &= (0.5, 0.6, 0.7), & Q_{\overline{\zeta}}(y_3) &= (0.7, 0.6, 0.6),\\
Q_{\underline{\zeta}}(y_4) &= (0.5, 0.6, 0.7), & Q_{\overline{\zeta}}(y_4) &= (0.8, 0.6, 0.4).
\end{aligned}$$

Now, 3-polar fuzzy soft rough approximation operators $\underline{\zeta}(Q)$, $\overline{\zeta}(Q)$, respectively, are given by

$$\begin{aligned}
\underline{\zeta}(Q) &= \{(y_1, 0.7, 0.6, 0.8), (y_2, 0.6, 0.6, 0.7), (y_3, 0.5, 0.6, 0.7), (y_4, 0.5, 0.6, 0.7)\},\\
\overline{\zeta}(Q) &= \{(y_1, 0.7, 0.6, 0.7), (y_2, 0.7, 0.5, 0.7), (y_3, 0.7, 0.6, 0.6), (y_4, 0.8, 0.6, 0.4)\}.
\end{aligned}$$

These operators are very close to the decision alternatives y_n, $n = 1, 2, 3, 4$.
By Definition 11, we have the choice set as follows:

$$\underline{\zeta}(Q) \oplus \overline{\zeta}(Q) = \{(y_1, 0.91, 0.84, 0.94), (y_2, 0.88, 0.8, 0.91), (y_3, 0.85, 0.84, 0.88), (y_4, 0.9, 0.84, 0.82)\}.$$

Thus, Mr. Adeel will select the hotel y_1 to stay because the optimal decision in the choice set $\underline{\zeta}(Q) \oplus \overline{\zeta}(Q)$ is y_1.

The method of selecting a suitable hotel is explained in the following Algorithm 1.

Algorithm 1: Selection of a suitable hotel

1. Input Y as universe of discourse.
2. Input T as a set of parameters.
3. Construct an mF soft relation $\zeta : Y \to T$ according to the different needs of the decision maker.
4. Give an mF subset Q over T, which is an optimal normal decision object according to the various requirements of decision maker.
5. Compute the mF soft rough approximation operators $\underline{\zeta}(Q)$ and $\overline{\zeta}(Q)$ by Definition 8.
6. Find the choice set $S = \underline{\zeta}(Q) \oplus \overline{\zeta}(Q)$ by Definition 11.
7. Select the optimal decision y_k. If $p_i \circ S(y_k) \geq M$, where $M = \bigvee_{1 \leq k \leq n} p_i \circ S(y_k)$, n is equal to the number of objects in Y, then the optimal decision will be y_k.

If there exists more than one optimal choice in step 7 of the Algorithm 1, that is, $y_{k_i} = y_{k_j}$, where $1 \leq k_i \neq k_j \leq n$, one may go back and change the optimal normal decision object Q and repeat the Algorithm 1 so that the final decision is only one.

4.2. Selection of a Place

Choosing a place to go when some people have the opportunity to travel can sometimes be very difficult task. Suppose that a group of ten peoples plan a tour to a suitable place in a country Z. There are four alternatives in their mind. The alternatives are q_1, q_2, q_3, q_4. They want to select the best place for the tour. It is a challenge to find advice in one place. The environment and cost are the main parameters for the selection of a suitable place. In the environment of the place, they want to check whether the place has availability of built environment, natural environment and social environment. The term built environment refers to the man-made surroundings. Built environment of the place includes buildings, parks and every other things that are made by human beings. Natural environment of the place includes forests, oceans, rivers, lakes, atmosphere, climate, weather, etc. The social environment includes the culture and lifestyle of the human beings. Lastly, the tour cost is an important criteria for the place selection. It includes low, medium and high.

Let $Y = \{q_1, q_2, q_3, q_4\}$ be the set of four places and $T = \{a_1, a_2\}$ be the set of parameters, where

'a_1' represents the Environment,
'a_2' represents the Tour Cost.

We give more characteristics of these parameters.

- The "Environment" of the place includes built environment, natural environment, and social environment.
- The "Tour Cost" of the place may be low, medium, or high.

Suppose that they describe the "attractiveness of the place" by constructing a 3-polar fuzzy soft relation ζ over $Y \times T$, which is given by

ζ	a_1	a_2
q_1	$(0.8, 0.8, 0.9)$	$(0.4, 0.7, 0.6)$
q_2	$(0.5, 0.7, 0.6)$	$(0.5, 0.7, 0.8)$
q_3	$(0.8, 0.6, 0.7)$	$(0.8, 0.9, 0.4)$
q_4	$(0.7, 0.9, 0.6)$	$(0.6, 0.7, 0.8)$

Thus, $\zeta : Y \to T$ is the 3-polar fuzzy soft relation in which environment and tour cost of the places are considered. For example, if we consider "Environment" of the place, $((q_1, a_1), 0.8, 0.8, 0.9)$ means that the place q_1 include 80% built environment, 80% natural environment and 90% social environment.

We now assume that they give the optimal normal decision object Q, which is a 3-polar fuzzy subset of T as follows:

$$Q = \{(a_1, 0.8, 0.7, 0.9), (a_2, 0.7, 0.6, 0.8)\}.$$

From Definition 8,

$$\begin{aligned} Q_{\underline{\zeta}}(q_1) &= (0.7, 0.6, 0.8), & Q_{\overline{\zeta}}(q_1) &= (0.8, 0.7, 0.9), \\ Q_{\underline{\zeta}}(q_2) &= (0.7, 0.6, 0.8), & Q_{\overline{\zeta}}(q_2) &= (0.5, 0.7, 0.8), \\ Q_{\underline{\zeta}}(q_3) &= (0.7, 0.6, 0.8), & Q_{\overline{\zeta}}(q_3) &= (0.8, 0.6, 0.7), \\ Q_{\underline{\zeta}}(q_4) &= (0.7, 0.6, 0.8), & Q_{\overline{\zeta}}(q_4) &= (0.7, 0.7, 0.8). \end{aligned}$$

We now have 3-polar fuzzy soft rough approximation operators $\underline{\zeta}(Q), \overline{\zeta}(Q)$, respectively, as follows:

$$\underline{\zeta}(Q) = \{(q_1, 0.7, 0.6, 0.8), (q_2, 0.7, 0.6, 0.8), (q_3, 0.7, 0.6, 0.8), (q_4, 0.7, 0.6, 0.8)\},$$
$$\overline{\zeta}(Q) = \{(q_1, 0.8, 0.7, 0.9), (q_2, 0.5, 0.7, 0.8), (q_3, 0.8, 0.6, 0.7), (q_4, 0.7, 0.7, 0.8)\}.$$

These operators are very close to the decision alternatives q_n, $n = 1, 2, 3, 4$.

By Definition 11,

$$\underline{\zeta}(Q) \oplus \overline{\zeta}(Q) = \{(q_1, 0.94, 0.88, 0.98), (q_2, 0.85, 0.88, 0.96), (q_3, 0.94, 0.84, 0.94), (q_4, 0.91, 0.88, 0.96)\}.$$

Thus, the optimal decision in the choice set $\underline{\zeta}(Q) \oplus \overline{\zeta}(Q)$ is q_1. Therefore, they will select the place q_1 for the tour.

The method of selecting a suitable place for tour is explained in the following Algorithm 2.

Algorithm 2: Selection of a suitable place

1. Input Y as universe of discourse.
2. Input T as a set of parameters.
3. Construct an mF soft relation ζ over $Y \times T$ according to the different needs of the decision makers.
4. Give an mF subset Q of T, which is an optimal normal decision object according to the various requirements of decision makers.
5. Compute the mF soft rough approximation operators $\underline{\zeta}(Q)$ and $\overline{\zeta}(Q)$ by Definition 8.
6. Find the choice set $S = \underline{\zeta}(Q) \oplus \overline{\zeta}(Q)$ by Definition 11.
7. Select the optimal decision q_k. If $p_i \circ S(q_k) \geq M$, where $M = \bigvee_{1 \leq k \leq n} p_i \circ S(q_k)$, n is equal to the number of objects in Y, and then the optimal decision will be q_k.

If there exists more than one optimal choice in step 7 of the Algorithm 2, that is, $q_{k_i} = q_{k_j}$ where $1 \leq k_i \neq k_j \leq n$, one may go back and change the optimal normal decision object Q and repeat the Algorithm 2 so that the final decision is only one.

4.3. Selection of a House

Buying a house is an exhilarating time in many people lives, but it is also a very difficult task to those who are not particularly real estate savvy. There are a number of factors to take into consideration for buying the house such as location of the house, size of the house and price of the house. These factors among many others influence house buyers before they even get to start thinking about buying a new house. Suppose a person (Mr. Ali) wants to buy a house. The alternatives in his mind are u_1, u_2, u_3. The size, location and price are the main parameters for the selection of a suitable house.

Let $Y = \{u_1, u_2, u_3\}$ be the set of three houses and let $T = \{t_1, t_2, t_3\}$ be the set of parameters related to the houses in Y, where

't_1' represents the Size,
't_2' represents the Location,
't_3' represents the Price.

We give further characteristics of these parameters.

- The "Size" of the house include small , large, and very large.
- The "Location" of the house include close to the main road, in the green surroundings, and in the city center.
- The "Price" of the house includes low, medium, and high.

Suppose that Ali describes the "attractiveness of the house" by forming a 3-polar fuzzy soft relation $\zeta : Y \to T$, which is given by

ζ	t_1	t_2	t_3
u_1	$(0.5, 0.7, 0.9)$	$(0.7, 0.6, 0.8)$	$(0.5, 0.6, 0.9)$
u_2	$(0.8, 0.9, 0.1)$	$(0.6, 0.8, 0.9)$	$(0.8, 0.4, 0.2)$
u_3	$(0.9, 0.7, 0.6)$	$(0.9, 0.8, 0.9)$	$(0.4, 0.6, 0.3)$.

Thus, ζ over $Y \times T$ is the 3-polar fuzzy soft relation in which size, location and price of the houses are considered. For example, if we consider "Location" of the house, $((u_2, t_1), 0.8, 0.9, 0.1)$ means that the house u_1 is, 80% close to the main road, 90% in the green surroundings and 10% in the city center.

We now assume that Ali gives the optimal normal decision object Q, which is a 3-polar fuzzy subset of T as follows:

$$Q = \{(t_1, 0.6, 0.8, 0.7), (t_2, 0.5, 0.8, 0.8), (t_3, 0.9, 0.8, 0.7)\}.$$

By Definition 8,

$$\begin{aligned} Q_{\underline{\zeta}}(u_1) &= (0.5, 0.8, 0.7), & Q_{\overline{\zeta}}(u_1) &= (0.5, 0.7, 0.8), \\ Q_{\underline{\zeta}}(u_2) &= (0.5, 0.8, 0.8), & Q_{\overline{\zeta}}(u_2) &= (0.8, 0.8, 0.8), \\ Q_{\underline{\zeta}}(u_3) &= (0.5, 0.8, 0.7), & Q_{\overline{\zeta}}(u_3) &= (0.6, 0.8, 0.8). \end{aligned}$$

Now, 3-polar fuzzy soft rough approximation operators $\underline{\zeta}(Q)$, $\overline{\zeta}(Q)$, respectively, are given by

$$\underline{\zeta}(Q) = \{(u_1, 0.5, 0.8, 0.7), (u_2, 0.5, 0.8, 0.8), (u_3, 0.5, 0.8, 0.7)\},$$
$$\overline{\zeta}(Q) = \{(u_1, 0.5, 0.7, 0.8), (u_2, 0.8, 0.8, 0.8), (u_3, 0.6, 0.8, 0.8)\}.$$

These operators are very close to the decision alternatives u_n, $n = 1, 2, 3$.

Using Definition 11,

$$\underline{\zeta}(Q) \oplus \overline{\zeta}(Q) = \{(u_1, 0.75, 0.94, 0.94), (u_2, 0.9, 0.96, 0.96), (u_3, 0.8, 0.96, 0.94)\}.$$

Hence, Ali will buy the house u_2 because the optimal decision in the choice set $\underline{\zeta}(Q) \oplus \overline{\zeta}(Q)$ is u_2. The method of selecting a suitable house is explained in the following Algorithm 3.

Algorithm 3: Selection of a suitable house

1. Input Y as universe of discourse.
2. Input T as a set of parameters.
3. Construct an mF soft relation $\zeta : Y \rightarrow T$ according to the different needs of the decision maker.
4. Give an mF subset Q over T, which is an optimal normal decision object according to the various requirements of the decision maker.
5. Compute the mF soft rough approximation operators $\underline{\zeta}(Q)$ and $\overline{\zeta}(Q)$ by Definition 8.
6. Find the choice set $S = \underline{\zeta}(Q) \oplus \overline{\zeta}(Q)$ by Definition 11.
7. Select the optimal decision u_k. If $p_i \circ S(u_k) \geq M$, where $M = \bigvee_{1 \leq k \leq n} p_i \circ S(u_k)$, n is equal to the number of objects in Y, and then the optimal decision will be u_k.

If there exist too many optimal choices in step 7 of Algorithm 3, that is, $u_{k_i} = u_{k_j}$, where $1 \leq k_i \neq k_j \leq n$, change the optimal normal decision object Q and repeat the Algorithm 3 so that the final decision is only one.

5. Conclusions

The theory of mF sets plays a vital role in decision-making problems, when multiple information is given. An mF soft rough set is a combination of an mF set, soft set and rough set. In this paper, we have presented the concepts of two new hybrid models called soft rough mF sets and mF soft rough sets, which provide more exactness and compatibility with a system when compared with other hybrid mathematical models. We have discussed the properties of both hybrid models. We have examined the relationship between mF soft rough approximation operators and crisp soft rough approximation operators. We have discussed some applications of mF soft rough sets in real-life decision-making problems. We are expanding our research work to (1) soft rough mF graphs; (2) soft rough mF hypergraphs; (3) mF soft rough graphs; (4) mF soft rough hypergraphs; and a (5) decision support system based on mF soft rough hypergraphs.

Acknowledgments: The authors are thankful to the anonymous referees for their valuable comments and suggestions.

Author Contributions: Muhammad Akram, Ghous Ali and Noura Omair Alshehri conceived and designed the experiments; Ghous Ali performed the experiments; Noura Omair Alshehri analyzed the data; Muhammad Akram contributed reagents/materials/analysis tools; Noura Omair Alshehri and Ghous Ali wrote the paper.

Conflicts of Interest: The authors declare no conflict of interest.

References

1. Chen, J.; Li, S.; Ma, S.; Wang, X. m-polar fuzzy sets: An extension of bipolar fuzzy sets. *Sci. World J.* **2014**, doi:10.1155/2014/416530.
2. Akram, M.; Younas, H.R. Certain types of irregular m-polar fuzzy graphs. *J. Appl. Math. Comput.* **2017**, *53*, 365–382.
3. Akram, M.; Adeel, A. m-polar fuzzy labeling graphs with application. *Math. Comput. Sci.* **2016**, *10*, 387–402.
4. Akram, M.; Waseem, N. Certain metrics in m-polar fuzzy graphs. *New Math. Natl. Comput.* **2016**, *12*, 135–155.
5. Akram, M.; Sarwar, M. Novel applications of m-polar fuzzy competition graphs in decision support system. *Neural Comput. Appl.* **2017**, 1–21, doi:10.1007/s00521-017-2894-y.
6. Li, S.; Yang, X.; Li, H.; Miao, M.A. Operations and decompositions of m-polar fuzzy graphs. *Basic Sci. J. Text. Univ. Fangzhi Gaoxiao Jichu Kexue Xuebao* **2017**, *30*, 149–162.
7. Pawlak, Z. Rough sets. *Int. J. Comput. Inf. Sci.* **1982**, *11*, 145–172.
8. Zadeh, L.A. Fuzzy sets. *Inf. Control* **1965**, *8*, 338–353.

9. Dubois, D.; Prade, H. Rough fuzzy sets and fuzzy rough sets. *Int. J. Gen. Syst.* **1990**, *17*, 191–209.
10. Xu, Z.; Gou, X. An overview of interval-valued intuitionistic fuzzy information aggregations and applications. *Granul. Comput.* **2017**, *2*, 13–39.
11. Das, S.; Kar, S.; Pal, T. Robust decision making using intuitionistic fuzzy numbers. *Granul. Comput.* **2017**, *2*, 41–54.
12. Cai, M.; Li, Q.; Lang, G. Shadowed sets of dynamic fuzzy sets. *Granul. Comput.* **2017**, *2*, 85–94.
13. Pedrycz, W.; Chen, S.M. *Granular Computing and Decision-Making: Interactive and Iterative Approaches*; Springer: Heidelberg, Germany, 2015.
14. Molodtsov, D.A. Soft set theory—First results. *Comput. Math. Appl.* **1999**, *37*, 19–31.
15. Maji, P.K.; Biswas, R.; Roy, A.R. Soft set theory. *Comput. Math. Appl.* **2003**, *45*, 555–562.
16. Maji, P.K.; Biswas, R.; Roy, A.R. Fuzzy soft sets. *J. Fuzzy Math.* **2001**, *9*, 589–602.
17. Alcantud, J.C.R. A novel algorithm for fuzzy soft set based decision making from multiobserver input parameter data set. *Inf. Fusion* **2016**, *29*, 142–148.
18. Alcantud, J.C.R. Fuzzy soft set based decision making: A novel alternative approach. In Proceedings of the 16th World Congress of the International Fuzzy Systems Association (IFSA) and 9th Conference of the European Society for Fuzzy Logic and Technology (EUSFLAT), Gijón, Spain, 30 June–3 July 2015.
19. Alcantud, J.C.R. Fuzzy soft set decision making algorithms: Some clarifications and reinterpretations. In Proceedings of the Spanish Association for Artificial Intelligence, Salamanca, Spain, 14–16 September 2016; Springer: Basel, Switzerland, 2016; pp. 479–488.
20. Alcantud, J.C.R.; Santos-Garcia, G. A new criterion for soft set based decision making problems under incomplete information. *Int. J. Comput. Intell. Syst.* **2017**, *10*, 394–404.
21. Alcantud, J.C.R.; Santos-Garcia, G. Incomplete soft sets: New solutions for decision making problems. In *Decision Economics: In Commemoration of the Birth Centennial of Herbert A. Simon 1916–2016 (Nobel Prize in Economics 1978)*; Springer: Cham, Switzerland, 2016; pp. 9–17.
22. Feng, F.; Li, C.X.; Davvaz, B.; Ali, M.I. Soft sets combined with fuzzy sets and rough sets: A tentative approach. *Soft Comput.* **2010**, *14*, 899–911.
23. Feng, F.; Liu, X.; Leoreanu-Fotea, V.; Jun, Y.B. Soft sets and soft rough sets. *Inf. Sci.* **2011**, *181*, 1125–1137.
24. Cagman, N.; Enginoglu, S. Soft matrix theory and decision making. *Comput. Math. Appl.* **2010**, *59*, 3308–3314.
25. Zhang, H.; Shu, L.; Liao, S. Intuitionistic fuzzy soft rough set and its application in decision-making. *Abstr. Appl. Anal.* **2014**, doi:10.1155/2014/287314.

symmetry

MDPI

Article

Group Decision-Making for Hesitant Fuzzy Sets Based on Characteristic Objects Method

Shahzad Faizi [1], Wojciech Sałabun [2,*], Tabasam Rashid [1], Jarosław Wątróbski [3] and Sohail Zafar [1]

[1] Department of Mathematics, University of Management and Technology, Lahore 54770, Pakistan; shahzadfaizi@gmail.com (S.F.); tabasam.rashid@gmail.com (T.R.); sohailahmad04@gmail.com (S.Z.)

[2] Department of Artificial Intelligence method and Applied Mathematics in the Faculty of Computer Science and Information Technology, West Pomeranian University of Technology, Szczecin, 71-210, Poland; wsalabun@wi.zut.edu.pl

[3] Department of Web Systems Analysis and Data Processing in the Faculty of Computer Science and Information Technology, West Pomeranian University of Technology, Szczecin, 71-210, Poland; jwatrobski@wi.zut.edu.pl

* Correspondence: wsalabun@wi.zut.edu.pl; Tel.: +48-503-417-373

Received: 28 June 2017; Accepted: 25 July 2017; Published: 29 July 2017

Abstract: There are many real-life problems that, because of the need to involve a wide domain of knowledge, are beyond a single expert. This is especially true for complex problems. Therefore, it is usually necessary to allocate more than one expert to a decision process. In such situations, we can observe an increasing importance of uncertainty. In this paper, the Multi-Criteria Decision-Making (MCDM) method called the Characteristic Objects Method (COMET) is extended to solve problems for Multi-Criteria Group Decision-Making (MCGDM) in a hesitant fuzzy environment. It is a completely new idea for solving problems of group decision-making under uncertainty. In this approach, we use L-R-type Generalized Fuzzy Numbers (GFNs) to get the degree of hesitancy for an alternative under a certain criterion. Therefore, the classical COMET method was adapted to work with GFNs in group decision-making problems. The proposed extension is presented in detail, along with the necessary background information. Finally, an illustrative numerical example is provided to elaborate the proposed method with respect to the support of a decision process. The presented extension of the COMET method, as opposed to others' group decision-making methods, is completely free of the rank reversal phenomenon, which is identified as one of the most important MCDM challenges.

Keywords: hesitant fuzzy sets; L-R-type generalized fuzzy numbers; Multi-Criteria Group Decision-Making (MCGDM); Characteristic Objects Method (COMET)

1. Introduction

For human activities and their problems, the Multi-Criteria Group Decision-Making (MCGDM) is an important tool [1,2]. In complex real-world conditions, it is not possible for a single Decision-Maker (DM) to recognize all of the relevant aspects of a decision-making problem [3]. Thus, the decision-making procedure requires considering many DMs or experts from different fields. In many group decision-making problems, a group is established by various DMs from different fields, including work experience, education backgrounds and knowledge structure [4]. It could be implemented to select the most suitable alternative from a given set of decision variants or their subset [5,6]. The essential prerequisite of the MCGDM is the combination of experts' preferences and judgments about the candidate alternatives versus the conflicting criteria [7], which is a popular trend of present research to develop new group MCDM methods [8–11].

In the decision-making, the problems of uncertainty and hesitancy usually turn out to be unavoidable. To express the DMs' evaluation information more objectively, several tools have been developed, such as fuzzy set [1,12], interval-valued fuzzy set [13,14], linguistic fuzzy set [15–17], which allow one to present an element's membership function as a set denoted by a fuzzy number, an interval fuzzy number, a linguistic variable and a fuzzy set, respectively. Intuitionistic fuzzy set [18] and fuzzy multiset [19,20] are another two generalizations of the fuzzy set. Whilst the former contains three types of information (the membership, the non-membership and the hesitancy), the latter permits the elements to repeat more than once.

In many practical problems, sometimes, it is difficult to define the membership grade of an element, because of a set of possible membership values [21]. This issue is very important in MCGDM problems, when the DMs do not support the same membership grade for an element [22,23]. In this case, the difficulty of establishing a common membership grade is caused not by the margin of error (as happens in Intuitionistic Fuzzy Set (IFS)) or some possible distribution values (as happens in Type-2 Fuzzy Sets), but by the fact that several membership values are possible [10]. To deal with these cases, the Hesitant Fuzzy Set (HFS) was introduced [24] as a new generalization of fuzzy sets. Many MCDM methods have been extended by using the HFS theory, e.g., the ELECTRE family methods [25], Viekriterijumsko Kompromisno Rangiranje (VIKOR) [26] or prospect theory [27]. There was also established a number of new methods [13,28–30] or aggregation operators [31,32], which are based on the HFS concept. Presently, group decision-making problems are solved for hesitant fuzzy sets and with aggregation operators in [33–36]. Interval-valued hesitant fuzzy sets have been used in the applications of group decision-making in [28,37–40]. MCGDM with hesitant two-tuple linguistic information and by using trapezoidal valued HFSs is discussed in [41,42]. Yu [43] gave the concept of triangular hesitant fuzzy sets and used it for the solution of decision-making problems. Unfortunately, all of the mentioned group decision-making methods are susceptible to the occurrence of the rank reversal phenomenon paradox, which lies at the heart of the main MCDM challenges.

The Characteristic Objects Method (COMET) is a useful technique in dealing with Multi-Criteria Decision-Making (MCDM) problems [44–48]. It helps a DM to organize the structure of the problems to be solved and carry out the analysis, comparisons and ranking of the alternatives, where the complexity of the algorithm is completely independent of the alternatives' number [49,50]. Additionally, comparisons between the Characteristic Objects (COs) are easier than comparisons between alternatives. However, the most important merit of the COMET method is the fact that this method is completely free of the rank reversals phenomenon [51] because the final ranking is constructed based on COs and fuzzy rules.

In this study, we extend the COMET concept to develop a methodology for solving multi-criteria group decision-making problems under uncertainty. The proposed method allows a group of DMs to make their opinion independent of linguistic terms by using HFS. The proposed method is designed for modeling uncertainty from different sources, which are related to expert knowledge. The main motivation of this research is the fact that the presented extension is also completely free of the rank reversals paradox as the classical version.

The group version of the HFS COMET method can be used in various research fields and disciplines such as economics [29,30,32], resource management [51], production [52], transport [53], game theory (Nash equilibrium) [54–63], medical problems [48,64], sustainability manufacturing [65] or web systems [66]; especially in decision situations requiring the involvement of many experts [67].

The rest of this paper is organized as follows. In Section 2, we introduced some basic concepts related to the hesitant fuzzy sets, L-R-type Generalized Fuzzy Numbers (GFNs), the fuzzy rule, the rule base and the t-norm. In Section 3, we established a group decision-making method based on COMET to deal with the uncertainty environment. In Section 4, an illustrative example is given to demonstrate the practicality and effectiveness of the proposed approach. Finally, we conclude the paper and give some remarks in Section 5.

2. Preliminaries

The HFS [24], as a generalization of the fuzzy set, maps the membership degree of an element to a set presented as several possible values between zero and one, which can better describe the situations where people have hesitancy in providing their preferences over objects in the process of decision-making.

In this section, we recall some important concepts that are necessary to understand our proposed decision-making method.

Definition 1. *A hesitant fuzzy set A on X is a function h^A that when applied to X returns a finite subset of $[0,1]$, which can be represented as the following mathematical symbol [24]:*

$$A = \{(x, h^A(x)) | x \in X\},$$

where $h^A(x)$ is a set of some values in $[0,1]$, denoting the possible membership degrees of the element $x \in X$ to the set A. For convenience, Xia and Xu [68] named $h^A(x)$ a Hesitant Fuzzy Element (HFE).

Definition 2. *For an HFS represented by its membership function h, we define its complement as follows [24]:*

$$h^c(x) = \bigcup_{\gamma \in h(x)} \{1 - \gamma\}.$$

Definition 3. *In reference [68], for an HFE h, $Sc(h) = \frac{1}{l_h} \sum_{\gamma \in h} \gamma$, is called the score function of h, where l_h is the number of elements in h and $Sc(h) \in [0,1]$. For two HFEs h_1 and h_2, if $Sc(h_1) > Sc(h_2)$, then $h_2 \prec h_1$, if $TODOSc(h_1) = Sc(h_2)$, then $h_1 \approx h_2$.*

Xia and Xu [68] define some operations on the HFEs (h, h_1 and h_2) and the scalar number k :

1. $kh = \bigcup_{\gamma \in h} \{1 - (1-\gamma)^k\}$;
2. $h_1 \oplus h_2 = \bigcup_{\gamma_1 \in h_1, \gamma_2 \in h_2} \{\gamma_1 + \gamma_2 - \gamma_1\gamma_2\}$;
3. $h_1 \otimes h_2 = \bigcup_{\gamma_1 \in h_1, \gamma_2 \in h_2} \{\gamma_1\gamma_2\}$.

Definition 4. *Let L and R both be decreasing, shape functions from $\Re^+ = [0,\infty)$ to $[0,1]$ with $L(0) = \omega$; $L(x) < \omega$ for all $x < 1$; $L(1) = 0$ or ($L(x) > 0$ for all x and $L(+\infty) = 0$) (and the same for R). A GFN is called the L-Rtype if there are real numbers m, $\alpha > 0, \beta > 0$ and ω $(0 \le \omega \le 1)$ with [69]:*

$$\mu_{\tilde{A}}(x) = \begin{cases} \omega L(\frac{m-x}{\alpha}), & x \le m \\ \omega R(\frac{x-m}{\beta}), & x \ge m \end{cases}$$

where m is called the mean value of $\tilde{A}$ and α and β are called the left and right spreads, respectively. The L-R-type GFN $\tilde{A}$ is symbolically denoted by $\tilde{A} = (m, \alpha, \beta; \omega)_{LR}$. If $\omega = 1$, then $\tilde{A}$ is called the L-R-type fuzzy number and simply denoted by $\tilde{A} = (m, \alpha, \beta)_{LR}$.

For an L-R-type GFN $\tilde{A} = (m, \alpha, \beta; \omega)_{LR}$, if L and R are of the form:

$$T(x) = \begin{cases} 1 - x, & 0 \le x \le 1 \\ 0, & \text{otherwise} \end{cases}$$

Then, $\tilde{A}$ is called a generalized triangular fuzzy number denoted by $\tilde{A} = (m, \alpha, \beta; \omega)_T$. Similarly, for $\omega = 1$, $\tilde{A}$ is simply called a triangular fuzzy number denoted by $\tilde{A} = (m, \alpha, \beta)_T$.

A fuzzy number $\tilde{A}$ is called an L-R-type generalized trapezoidal fuzzy number if there are real numbers $m_1, m_2, \alpha > 0$ and $\beta > 0$ with the following membership function:

$$\mu_{\tilde{A}}(x) = \begin{cases} \omega L(\frac{m_1 - x}{\alpha}), & x \leq m_1 \\ \omega, & m_1 \leq x \leq m_2 \\ \omega R(\frac{x - m_2}{\beta}), & x \geq m_2 \end{cases}$$

where m_1 and m_2 are called the mean values of $\tilde{A}$ and α, β are called the left and right spreads, respectively. Symbolically, $\tilde{A}$ is denoted by $(m_1, m_2, \alpha, \beta; \omega)_{LR}$. The L-R-type generalized trapezoidal fuzzy number $\tilde{A} = (m_1, m_2, \alpha, \beta; \omega)_{LR}$ divides into three parts: left part, middle part and right part. The left, middle and right parts include the intervals $[m_1 - \alpha, m_1], [m_1, m_2]$ and $[m_2, m_2 + \beta]$, respectively.

If we take L and R to be of the form as mentioned in Equation (4), then $\tilde{A}$ is called the generalized trapezoidal fuzzy number denoted by $(m_1, m_2, \alpha, \beta; \omega)_T$. A generalized trapezoidal fuzzy number $\tilde{A}(m_1, m_2, \alpha, \beta; \omega)_T$ is simply called a trapezoidal fuzzy number denoted by $\tilde{A}(m_1, m_2, \alpha, \beta)_T$ when $\omega = 1$.

We know that the L-R-type fuzzy numbers are used to present real numbers in a fuzzy environment, and trapezoidal fuzzy numbers are used to present fuzzy intervals that are widely applied in linguistics, knowledge representation, control systems, database, and so forth [21,70–72]. Similarly, the L-R-type GFNs are very general and allow one to represent the different types of information. For example, the L-R-type GFN $\tilde{B} = (m, m, 0, 0; \omega)_{LR}$ with $m \in \Re = (-\infty, \infty)$ is used to denote a real number $\tilde{B}$, and the L-R-type GFN $\tilde{C} = (m_1, m_2, 0, 0; \omega)_{LR}$ with $m_1, m_2 \in \Re$ and $m_1 < m_2$ is used to denote an interval $\tilde{C}$.

Definition 5. *For a triangular fuzzy number $\tilde{A}$, we define:*

1. *The support of $\tilde{A}$ is $S(\tilde{A}) = \{x : \mu_{\tilde{A}}(x) > 0\}$.*
2. *The core of $\tilde{A}$ is $C(\tilde{A}) = \{x : \mu_{\tilde{A}}(x) = 1\}$.*

Definition 6. *The fuzzy rule [73,74]:*

The single fuzzy rule can be based on the modus ponens tautology [73,74]. The reasoning process uses logical connectives IF-THEN, OR and AND.

Definition 7. *The rule base [75]:*

The rule base consists of logical rules determining causal relationships existing in the system between the fuzzy sets of its inputs and outputs [75].

Definition 8. *In reference [76], a triangular norm (t-norm) is a binary operation $T : [0,1] \times [0,1] \to [0,1]$ satisfying $\forall x, y, z \in [0,1]$:*

1. *$T(x, y) = T(y, x)$ (commutativity),*
2. *$T(x, y) \leq T(x, z)$, if $y \leq z$ (monotonicity),*
3. *$T(x, T(y, z)) = T(T(x, y), z)$ (associativity),*
4. *$T(x, 1) = x$ (neutrality of one).*

Throughout this paper, only the product is used as a t-norm operator, i.e., $P(\mu_{\alpha_1}(x), \mu_{\alpha_2}(y)) = \mu_{\alpha_1}(x).\mu_{\alpha_2}(y)$.

3. COMET for MCGDM Using HFS

Consider an MCGDM problem in which the ratings of the alternative evaluations are expressed as HFSs. The solution procedure for the proposed MCGDM approach is described below.

Let A_j $(j = 1, 2, ..., m)$ be the set of alternatives and suppose a group of DMs $D = \{d_1, d_2, ..., d_k\}$ is asked to evaluate the given alternatives with respect to several criteria C_i $(i = 1, 2, ..., n)$. The ranking algorithm of the COMET has the following five steps:

Step 1: Define the space of the problem as follows:

Let $\mathcal{F}$ be the collection of all L-R-type GFNs and $F_i^{1\delta}, F_i^{2\delta}, ..., F_i^{q\delta}$ be different families of subsets of $\mathcal{F}$ selected by a DM d_δ $(\delta = 1, 2, ..., k)$ for each criterion C_i $(i = 1, 2, ..., n)$ where

$F_i^{1\delta} = \{F_{i1}^{1\delta}, F_{i2}^{1\delta}, ..., F_{ic_i}^{1\delta}\};$

$F_i^{2\delta} = \{F_{i1}^{2\delta}, F_{i2}^{2\delta}, ..., F_{ic_i}^{2\delta}\};$

$\vdots$

$F_i^{q\delta} = \{F_{i1}^{q\delta}, F_{i2}^{q\delta}, ..., F_{ic_i}^{q\delta}\}.$

In this way, the following result is obtained:

$C_1 = \left\{F_{11}^{b\delta}, F_{12}^{b\delta}, ..., F_{1c_1}^{b\delta}\right\};$

$C_2 = \left\{F_{21}^{b\delta}, F_{22}^{b\delta}, ..., F_{2c_2}^{b\delta}\right\};$

$\vdots$

$C_n = \left\{F_{n1}^{b\delta}, F_{n2}^{b\delta}, ..., F_{nc_n}^{b\delta}\right\};$

where $1 \leq b \leq q$ and $c_1, c_2, ..., c_n$ are the numbers of fuzzy numbers in each family $F_i^{b\delta}$ $(1 \leq b \leq q, 1 \leq i \leq n)$ for all criteria.

Initially, suppose each alternative is assessed by all DMs by means of n criteria in the form of a single family of TFNs F_i^t $(1 \leq i \leq n)$ with their fuzzy semantics as shown in Figures 1–6. Suppose each DM further provides the hesitant information of an alternative for each criterion in the form of L-R-type GFNs. Note that, in this method, the observations already provided by all of the DMs for each criterion in the form of the single family of TFNs set F_i^t $(1 \leq i \leq n)$ is a necessary part of all of the family of the remaining L-R-type GFNs set during the computation. The core of each criterion is defined as the core of each F_i^t $(1 \leq i \leq n)$, i.e.,

$C(C_1) = \left\{C(F_{11}^t), C(F_{12}^t), ..., C(F_{1c_1}^t)\right\};$

$C(C_2) = \left\{C(F_{21}^t), C(F_{22}^t), ..., C(F_{2c_2}^t)\right\};$

$\vdots$

$C(C_n) = \{C(F_{n1}^t), C(F_{n2}^t), ..., C(F_{nc_n}^t)\}.$

Step 2: Generate the characteristic objects:

By using the Cartesian product of all TFNs cores, the COs can be obtained as follows:

$CO = C(C_1) \times C(C_2) \times ... \times C(C_n)$

As the result of this, the ordered set of all COs is obtained:

$CO_1 = \{C(F_{11}^t), C(F_{21}^t), ..., C(F_{n1}^t)\};$

$CO_2 = \{C(F_{11}^t), C(F_{21}^t), ..., C(F_{n2}^t)\};$

$\vdots$

$CO_s = \left\{C(F_{1c_1}^t), C(F_{2c_2}^t), ..., C(F_{nc_n}^t)\right\};$

where $s = \prod_{i=1}^{n} c_i$ is a number of COs.

Step 3: Rank and evaluate the characteristic objects:

A comparison of COs is obtained by adding the opinion of DMs. After this, determine the Matrix of Expert Judgment (MEJ) as follows:

$$\mathbf{ME}J = \begin{bmatrix} \tilde{h}_{11} & \tilde{h}_{12} & \cdots & \tilde{h}_{1s} \\ \tilde{h}_{21} & \tilde{h}_{22} & \cdots & \tilde{h}_{2s} \\ \vdots & \vdots & \ddots & \vdots \\ \tilde{h}_{s1} & \tilde{h}_{s2} & \cdots & \tilde{h}_{ss} \end{bmatrix}$$

where $\tilde{h}_{\alpha\beta} = \{\tilde{h}_{\alpha\beta}^{\omega}, \omega = 1, 2, ..., l_{\tilde{h}_{\alpha\beta}}\}$ is the HFE containing preferences of all DMs and is obtained as a result of comparing CO_α and CO_β. The more preferred CO obtains a stronger preference degree, and the second object obtains a weaker one. If the preferences are balanced, then both objects obtain a preference degree denoted by HFE $\tilde{h}_f = \{0.5\}$. The selection of $\tilde{h}_{\alpha\beta}$ depends solely on the knowledge and opinion of the experts. Mathematically, $\tilde{h}_{\alpha\beta}$ should satisfy the following conditions:

1. $\tilde{h}_{\alpha\beta}^{\sigma(\omega)} + \tilde{h}_{\beta\alpha}^{\sigma(l_{\tilde{h}_{\alpha\beta}}-\omega+1)} = 1, \alpha, \beta = l, 2, ..., s;$
2. $\tilde{h}_{\alpha\alpha} = \{0.5\}, \alpha = l, 2, ..., s;$
3. $l_{\tilde{h}_{\alpha\beta}} = l_{\tilde{h}_{\beta\alpha}}, \alpha, \beta = l, 2, ..., s.$

where the values in $\tilde{h}_{\alpha\beta}$ are assumed to be arranged in increasing order for convenience, and let $\tilde{h}_{\alpha\beta}^{\sigma(\omega)} (\omega = 1, 2, ..., l_{\tilde{h}_{\alpha\beta}})$ denote the ωth smallest value in $\tilde{h}_{\alpha\beta}$ and $l_{\tilde{h}_{\alpha\beta}}$ the number of the values in $\tilde{h}_{\alpha\beta}$.

The last equation indicates that the sum of the ωth smallest value in $\tilde{h}_{\alpha\beta}$ and the ωth largest value in $\tilde{h}_{\beta\alpha}$ should be equivalent to one, which is the complement condition as introduced by Torra in [24] (see Definition 2). In other words, if $\tilde{h}_{\alpha\beta} = \{\tilde{h}_{\alpha\beta}^{\omega}, \omega = 1, 2, ..., l_{\tilde{h}_{\alpha\beta}}\}$ is known, then we can obtain $\tilde{h}_{\beta\alpha}$, which is given by $\tilde{h}_{\beta\alpha} = \{1 - \tilde{h}_{\alpha\beta}^{\omega}, \omega = 1, 2, ..., l_{\tilde{h}_{\alpha\beta}}\}$. The second equation indicates that the diagonal elements in **MEJ** should be equivalent to $\{0.5\}$, which implies the balanced preference degrees of CO_α and CO_β. The third equation indicates that the number of elements in $\tilde{h}_{\alpha\beta}$ and $\tilde{h}_{\beta\alpha}$ should be the same.

Suppose $\tilde{H}_\alpha = \oplus_{\beta=1}^{s} \tilde{h}_{\alpha\beta}$, where each $\tilde{H}_\alpha$ is an HFE. Afterward, we get a vertical vector SJ of the summed judgments where $SJ_\alpha = Sc(\tilde{H}_\alpha) = \frac{1}{l_{\tilde{H}_\alpha}} \sum_{\gamma \in \tilde{H}_\alpha} \gamma$ (see Definition 3). To assign the approximate value of preference to each CO, we use the same MATLAB code as used by Salabun in [45]. As a result, we get a vertical vector P, where the α-th component of P represents the approximate value of preference for CO_α.

Step 4: The rule base:

Each CO and value of preference is converted to a fuzzy rule as follows:

$IF \quad CO_\alpha \quad THEN \quad P_\alpha$

$IF \quad C(F_{1\alpha}^t) \quad AND \quad C(F_{2\alpha}^t) \ AND ... THEN \quad P_\alpha$

In this way, the complete fuzzy rule base is obtained, which can be presented as follows:

$IF \quad CO_1 \quad THEN \quad P_1$

$IF \quad CO_2 \quad THEN \quad P_2$

$\vdots$

$IF \quad CO_s \quad THEN \quad P_s$

Step 5: Inference in a fuzzy model and final ranking:

Each alternative activates the specified number of fuzzy rules, where for each one, the fulfillment degree of the conjunctive complex premise is determined. The fulfillment degrees of each activated rule corresponding to each element of F_i^b $(1 \leq b \leq q, 1 \leq i \leq n)$ always sum to one. Each alternative is a set of crisp numbers, corresponding to criteria $C_1, C_2, ..., C_n$. It can be presented as follows:

$A_j = \{a_{1j}, a_{2j}, ..., a_{nj}\}$, where the following conditions must be satisfied:

$a_{1j} \in [C(F_{11}^t), C(F_{1c_1}^t)];$

$a_{2j} \in [C(F_{21}^t), C(F_{2c_2}^t)];$

$\vdots$

$a_{nj} \in [C(F_{n1}^t), C(F_{nc_n}^t)].$

To infer the final ranking of the alternatives corresponding to each criterion, we proceed as follows:

For each $j = 1, 2, ..., m$,

$a_{1j} \in [C(F_{1k_1}^t), C(F_{1(k_1+1)}^t)];$

$a_{2j} \in [C(F_{2k_2}^t), C(F_{2(k_2+1)}^t)];$

$\vdots$

$a_{nj} \in [C(F^t_{nk_n}), C(F^t_{n(k_n+1)})]$;
where $k_i = 1, 2, ..., (c_i - 1)$, $(1 \le i \le n)$. The activated rules (COs), i.e., the group of those COs where the membership function of each alternative A_j $(1 \le j \le m)$ is non-zero, are:

$\left(C(F^t_{1k_1}), C(F^t_{2k_2}), ..., C(F^t_{nk_n})\right)$;

$\left(C(F^t_{1k_1}), C(F^t_{2k_2}), ..., C(F^t_{n(k_n+1)})\right)$;

$\vdots$

$\left(C(F^t_{1(k_1+1)}), C(F^t_{2(k_2+1)}), ..., C(F^t_{n(k_n+1)})\right)$.

The number of COs are obviously 2^n and $1 \le 2^n \le s$.

Let $p_1, p_2, ..., p_{2^n}$ be the approximate values of the preference of the activated rules (COs), which were already calculated in Step 3, where p_η's $(1 \le \eta \le 2^n)$ are some values in P_α's $(1 \le \alpha \le s)$. We denote the HFE at the point $x \in A_j$ $(1 \le j \le m)$ provided by a DM d_δ $(\delta = 1, 2, ..., k)$ as:

$$h^\delta_{ij}(x) = \{F^{1\delta}_{ij}(x), F^{2\delta}_{ij}(x), ..., F^{q\delta}_{ij}(x)\}$$

for each criterion C_i $(i = 1, 2, .., n)$.

To aggregate the information in the form of HFEs from every DM, in order to achieve a single HFE, which summarizes all of the information provided by the different DMs, there are several aggregation operators that are available in the literature. However, in this paper, we simply use the average operator to get the average of the membership values obtained from LR-type GFNs provided by the DMs in the form of HFE corresponding to each $a_{ij} \in A_j$ $(1 \le i \le n, 1 \le j \le m)$. Suppose $h_{ij}(x)$ is an HFE obtained as a result of aggregating the HFEs $h^\delta_{ij}(x)$, $(\delta = 1, 2, ..., k)$ where:

$$h_{ij}(x) = \{F^1_{ij}(x), F^2_{ij}(x), ..., F^q_{ij}(x)\}$$

Let $\mathbf{A}_j$ be HFE, which is computed as the sum of the products of all activated rules, as their fulfillment degrees and their values of the preference, i.e.,

$$\mathbf{A}_j = p_1\left(h_{1k_1}(a_{1j}) \otimes h_{2k_2}(a_{2j}) \otimes \ldots h_{nk_n}(a_{nj})\right) \oplus p_2\left(h_{1k_1}(a_{1j}) \otimes h_{2k_2}(a_{2j}) \otimes \ldots h_{n(k_n+1)}(a_{nj})\right)$$
$$\oplus \ldots p_{2^n}\left(h_{1(k_1+1)}(a_{1j}) \otimes h_{2(k_2+1)}(a_{2j}) \otimes \ldots h_{n(k_n+1)}(a_{nj})\right)$$

The preference of each alternative A_j $(1 \le j \le m)$ can be found by finding the score of the corresponding HFE $\mathbf{A}_j$ $(1 \le j \le m)$ as follows:

$$Sc(\mathbf{A}_j) = \frac{1}{l_{\mathbf{A}_j}} \sum_{y \in \mathbf{A}_j} y$$

The final ranking of alternatives is obtained by sorting the preference of alternatives. The greater the preference value, the better the alternative A_j $(1 \le j \le m)$.

As the summary of this section, Figure 1 presents the stepwise procedure of the proposed extension of the COMET method. After initiating the decision process, the procedure starts by modeling the structure of a considered decision problem. At this point, each expert determine generalized fuzzy numbers for each criterion. This is followed by generating characteristic objects in Step 2, evaluating the preferences of the characteristic objects in Step 3 and generating the fuzzy rule base in Step 4. The procedure ends by computing the assessment for each alternative from the considered set. The set of alternatives can be ranked according to the descending order of the computed assessments.

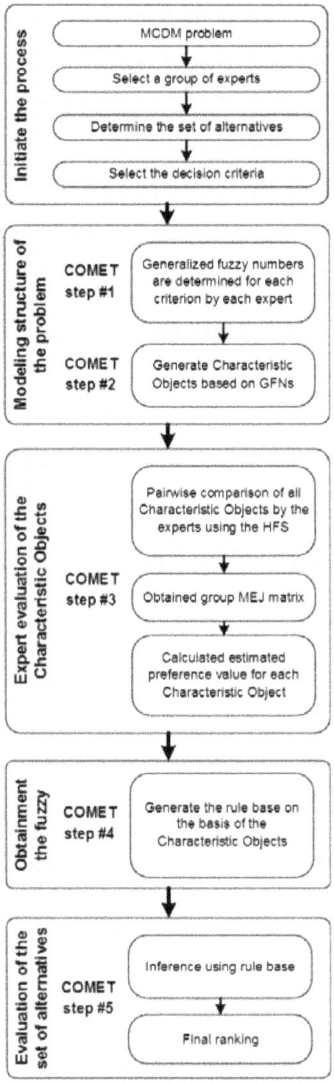

Figure 1. The procedure of the proposed extension of COMET to group decision-making.

4. An Illustrative Example

In this section, an example is given to understand our approach. We used the method proposed in Section 3 to get the most desirable alternative, as well as to rank the alternatives from the best to the worst or vice versa.

Let us consider a factory, whose maximum capacity of using mobile units is a total of 1000 per month, which intends to select a new mobile company. Four companies A_1, A_2, A_3 and A_4 are available, and three DMs are asked to consider two criteria C_1 (fixed line rent) and C_2 (rates per unit) to decide which mobile company to choose. The fixed line rent, rates per unit and the original ranking order of the feasible mobile companies are shown in Table 1.

Table 1. Original ranking of the alternatives, where LR - fixed line rent and R/U - rates per unit.

Alternatives	C_1 (LR)	C_2 (R/U)	Bill Amount	Original Rank
A_1	150	1500	1650	2
A_2	50	2000	2050	3
A_3	250	1250	1500	1
A_4	30	2150	2180	4

A set of TFNs and trapezoidal fuzzy numbers for both criteria C_1 and C_2 set by three DMs are shown in Tables 2 and 3. The average of the membership values obtained from LR-type GFNs for both the criteria are shown in Table 4.

Table 2. LR-type Group Fuzzy Numbers (GFNs) selected by the Decision-Makers (DMs) for criteria C_1.

DM1	$\{(30,30,200),(30,200,300),(200,300,300)\}$ $\{(30,30,30,170),(30,170,220,300),(220,300,300,300)\}$
DM2	$\{(30,30,200),(30,200,300),(200,300,300)\}$ $\{(30,30,30,180),(30,180,230,300),(230,300,300,300)\}$
DM3	$\{(30,30,200),(30,200,300),(200,300,300)\}$ $\{(30,30,30,160),(30,160,215,300),(215,300,300,300)\}$

Table 3. LR-type GFNs selected by the DMs for criteria C_2.

DM1	$\{(1200,1200,1800),(1200,1800,2500),(1800,2500,2500)\}$ $\{(1200,1200,1200,1600),(1200,1600,1900,2500),(1900,2500,2500,2500)\}$
DM2	$\{(1200,1200,1800),(1200,1800,2500),(1800,2500,2500)\}$ $\{(1200,1200,1200,1700),(1200,1700,1900,2500),(1900,2500,2500,2500)\}$
DM3	$\{(1200,1200,1800),(1200,1800,2500),(1800,2500,2500)\}$ $\{(1200,1200,1200,1650),(1200,1650,1950,2500),(1950,2500,2500,2500)\}$

Table 4. Average of the membership values obtained from LR-type GFNs for criteria C_1.

Average of the Membership Values Obtained from LR-Type GFNs for Criterion C_2			
30	50	150	250
$(1,0,0)$	$(0.8824,0.1176,0)$	$(0.2941,0.7059,0)$	$(0,0.5000,0.5000)$
$(1,0,0)$	$(0.8567,0.1433,0)$	$(0.8567,0.1433,0)$	$(0,0.6425,0.3575)$
1250	1500	2000	2150
$(0.9167,0.0833,0)$	$(0.5000,0.5000,0)$	$(0,0.7143,0.2857)$	$(0,0.5000,0.5000)$
$(0.8880,0.1120,0)$	$(0.3278,0.6722,0)$	$(0,0.8586,0.1414)$	$(0,0.6010,0.3990)$

The graphical representation of L-R-type GFNs selected by the DMs for both the criteria C_1 and C_2 are shown in Figures 2–7, respectively.

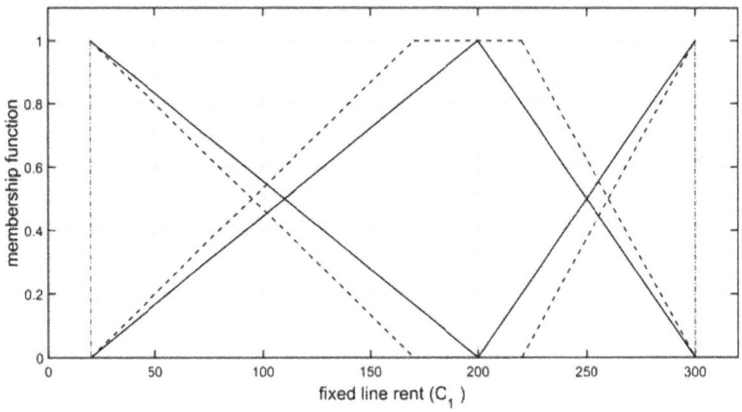

Figure 2. Graphical representation of *LR*-type GFNs selected by DM1 for the criterion C_1.

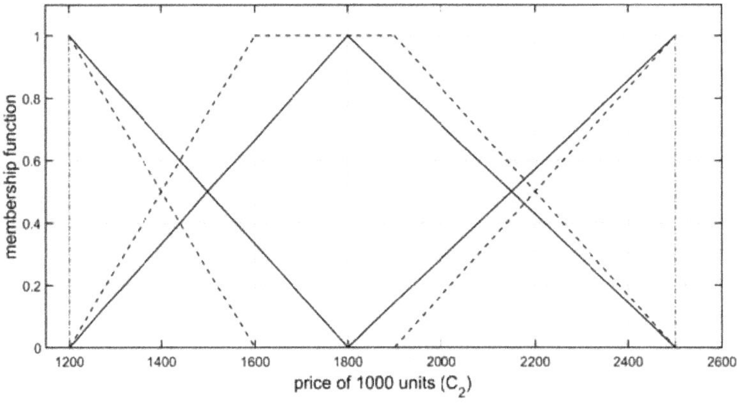

Figure 3. Graphical representation of *LR*-type GFNs selected by DM1 for the criterion C_2.

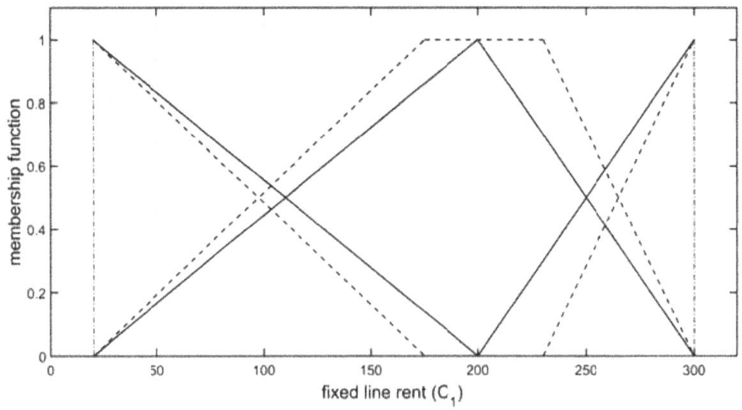

Figure 4. Graphical representation of *LR*-type GFNs selected by DM2 for the criterion C_1.

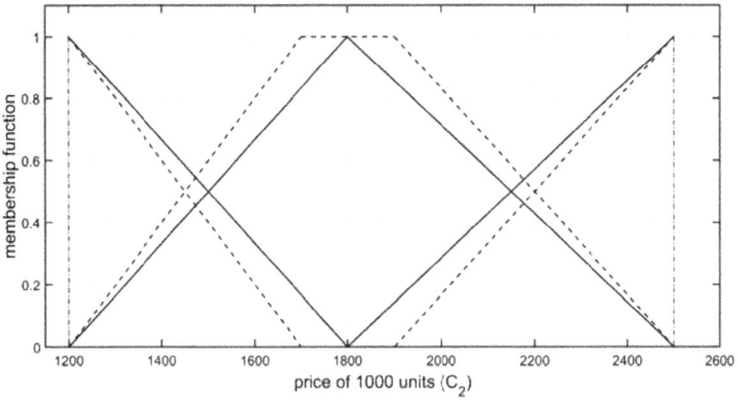

Figure 5. Graphical representation of *LR*-type GFNs selected by DM2 for the criterion C_2.

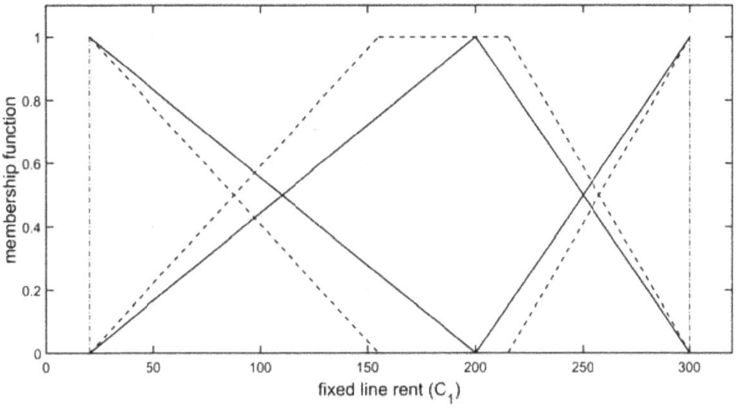

Figure 6. Graphical representation of *LR*-type GFNs selected by DM3 for the criterion C_1.

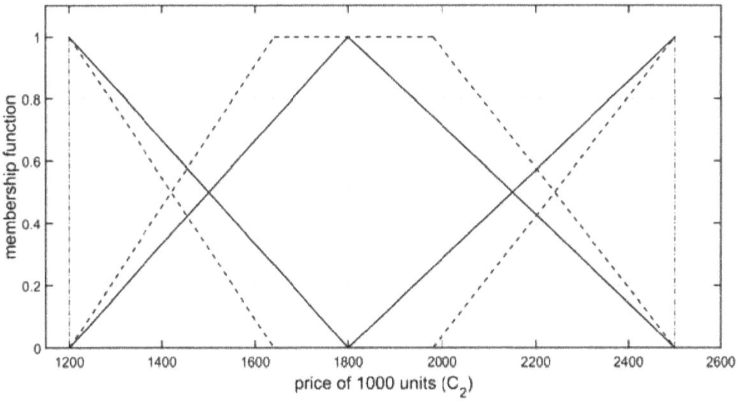

Figure 7. Graphical representation of *LR*-type GFNs selected by DM3 for the criterion C_2.

The cores of the family of TFNs for both the criteria C_1 and C_2 are respectively $\{30, 200, 300\}$ and $\{1200, 1800, 2500\}$. The solution of the COMET is obtained for different number of COs. The simplest solution involves the use of nine COs, which are presented as follows:

$$CO_1 = \{30, 1200\},\ CO_2 = \{30, 1800\},\ CO_3 = \{30, 2500\}$$
$$CO_4 = \{200, 1200\},\ CO_5 = \{200, 1800\},\ CO_6 = \{200, 2500\}$$
$$CO_7 = \{300, 1200\},\ CO_8 = \{300, 1800\},\ CO_9 = \{300, 2500\}$$

To rank and evaluate the COs, suppose the three DMs give their assessments by providing the HFEs as shown in Tables 5 and 6, and therefore, the Matrix of Expert Judgment (MEJ) is as follows:

Table 5. Matrix of Expert Judgment (MEJ).

	CO_1	CO_2	CO_3	CO_4	CO_5
CO_1	$\{0.5\}$	$\{0.8, 1\}$	$\{0.8, 0.9\}$	$\{0.7, 0.8\}$	$\{0.8, 0.9, 1\}$
CO_2	$\{0, 0.2\}$	$\{0.5\}$	$\{0.8, 1\}$	$\{0, 0.1, 0.2\}$	$\{0.9, 1\}$
CO_3	$\{0.1, 0.2\}$	$\{0, 0.2\}$	$\{0.5\}$	$\{0, 0.2, 0.3\}$	$\{0.1, 0.2\}$
CO_4	$\{0.2, 0.3\}$	$\{0.8, 0.9, 1\}$	$\{0.7, 0.8, 1\}$	$\{0.5\}$	$\{0.8, 0.9, 1\}$
CO_5	$\{0, 0.1, 0.2\}$	$\{0, 0.1\}$	$\{0.8, 0.9\}$	$\{0, 0.1, 0.2\}$	$\{0.5\}$
CO_6	$\{0, 0.2\}$	$\{0, 0.2\}$	$\{0, 0.1\}$	$\{0, 0.2\}$	$\{0, 0.1, 0.2\}$
CO_7	$\{0, 0.2\}$	$\{0.8, 1\}$	$\{0, 0.8\}$	$\{0, 0.2\}$	$\{0.8, 0.9, 1\}$
CO_8	$\{0, 0.1, 0.2\}$	$\{0.1, 0.2\}$	$\{0.7, 0.8, 1\}$	$\{0.1, 0.2\}$	$\{0, 0.1, 0.2\}$
CO_9	$\{0, 0.2\}$	$\{0, 0.1\}$	$\{0.2, 0.3\}$	$\{0, 0.1, 0.2\}$	$\{0, 0.2\}$

Table 6. Matrix of Expert Judgment (MEJ).

	CO_6	CO_7	CO_8	CO_9	SJ
CO_1	$\{0.8, 1\}$	$\{0.8, 1\}$	$\{0.8, 0.9, 1\}$	$\{0.8, 1\}$	0.999999
CO_2	$\{0.8, 1\}$	$\{0, 0.2\}$	$\{0.8, 0.9\}$	$\{0.9, 1\}$	0.999980
CO_3	$\{0.9, 1\}$	$\{0, 0.2\}$	$\{0, 0.2, 0.3\}$	$\{0.7, 0.8\}$	0.995033
CO_4	$\{0.8, 1\}$	$\{0.8, 1\}$	$\{0.8, 0.9\}$	$\{0.8, 0.9, 1\}$	0.999998
CO_5	$\{0.8, 0.9, 1\}$	$\{0, 0.1, 0.2\}$	$\{0.8, 0.9, 1\}$	$\{0.8, 1\}$	0.999751
CO_6	$\{0.5\}$	$\{0, 0.1, 0.2\}$	$\{0.2, 0.3\}$	$\{0.8, 0.9, 1\}$	0.968745
CO_7	$\{0.8, 0.9, 1\}$	$\{0.5\}$	$\{0.8, 1\}$	$\{0.8, 0.9\}$	0.999970
CO_8	$\{0.7, 0.8\}$	$\{0, 0.2\}$	$\{0.5\}$	$\{0.7, 0.8, 0.9\}$	0.996636
CO_9	$\{0, 0.1, 0.2\}$	$\{0.1, 0.2\}$	$\{0.1, 0.2, 0.3\}$	$\{0.5\}$	0.841614

The vector SJ on the basis of MEJ is obtained as follows:

$$SJ = [0.999999, 0.999980, 0.995033, 0.999998, 0.999751, 0.968745, 0.999970, 0.996636, 0.841614]^T$$

A vertical vector P is obtained by using a MATLAB code (see [45]) as follows:

$$P = [1.0000, 0.7500, 0.2500, 0.8750, 0.5000, 0.1250, 0.6250, 0.3750, 0]^T$$

Each component of the vector P represents the approximate values of the preference for the generated COs. Each CO and the value of preference p_i is converted to a fuzzy rule, as follows:

$$
\begin{array}{llllllll}
IF & LR \sim 30 & AND & R/U \sim 1200 & THEN & P_1 \sim 1.0000 \\
IF & LR \sim 30 & AND & R/U \sim 1800 & THEN & P_2 \sim 0.7500 \\
IF & LR \sim 30 & AND & R/U \sim 2500 & THEN & P_3 \sim 0.2500 \\
IF & LR \sim 200 & AND & R/U \sim 1200 & THEN & P_4 \sim 0.8750 \\
IF & LR \sim 200 & AND & R/U \sim 1800 & THEN & P_5 \sim 0.5000 \\
IF & LR \sim 200 & AND & R/U \sim 2500 & THEN & P_6 \sim 0.1250 \\
IF & LR \sim 300 & AND & R/U \sim 1200 & THEN & P_7 \sim 0.6250 \\
IF & LR \sim 300 & AND & R/U \sim 1800 & THEN & P_8 \sim 0.3750 \\
IF & LR \sim 300 & AND & R/U \sim 2500 & THEN & P_9 \sim 0.0000
\end{array}
$$

With respect to Model 4, for the alternative $A_1 = \{150, 1500\}$, we have nine rules (COs), but the activated rules are CO_1, CO_2, CO_4, CO_5. The approximate values of the preference of corresponding COs are $p_1 \sim 1$, $p_2 \sim 0.7500$, $p_3 \sim 0.8750$, $p_4 \sim 0.5000$. The HFE $\mathbf{A}_1$ and the preference value of the corresponding alternative A_1 are computed respectively as follows:

$$
\mathbf{A}_1 = p_1 h_{11}(150) \otimes h_{21}(1500) \oplus p_2 h_{11}(150) \otimes h_{22}(1500) \oplus p_3 h_{12}(150) \otimes h_{21}(1500) \oplus \\
p_4 h_{12}(150) \otimes h_{22}(1500)
$$

$$
Sc(\mathbf{A}_1) = \frac{1}{l_{\mathbf{A}_1}} \sum_{y \in \mathbf{A}_1} y = 0.5725
$$

Similarly, we can find the preference values for the rest of the alternatives and their ranking, which are shown in Table 7.

Table 7. Comparison of the original ranking with the ranking obtained using the proposed method.

Alternatives	C_1 (LR)	C_2 (R/U)	Original Ranking	Preference Values	New Ranking
A_1	150	1500	2	0.5725	3
A_2	50	2000	3	0.6236	2
A_3	250	1250	1	0.6272	1
A_4	30	2150	4	0.5281	4

The best choice is the alternative A_3 followed by A_2, A_1 and A_4. The worst choice is the alternative A_4. The extrema elements are consistent with the original ranking. However, the ranking obtained by the COMET method is not perfect. The main reason is that this problem was solved under an uncertain environment by a group of decision-makers. In other words, it is extremely difficult to make a reliable decision using uncertain data, but we believe that it is possible. This example also illustrates how hard it is to make a group decision under uncertainty. Notwithstanding, the COMET method shows the best and the worst decision.

The main contribution of the proposed approach can be expressed by the most important properties of this extension, i.e., the proposed approach is completely free of the rank reversal phenomenon and obtains not only a discrete value of priority, but the mathematical function, which can be used to calculate the priority for all alternatives from the space of the problem. Quantitative expression of efficiency is a very difficult task because a large number of assumptions is needed. Additionally, the reference ranking of the alternatives set is needed in this task, but the reference rank is almost always unknown. However, the problem of quantitative effectiveness assessment is a very important and interesting direction for further research.

5. Conclusions

The hesitant fuzzy sets theory is a useful tool to deal with uncertainty in multi-criteria group decision-making problems. Various sources of uncertainty can be a challenge to make a reliable decision. The paper presented the extension of the COMET method, which was proposed for solving real-life problems under the opinions of experts in a hesitant fuzzy environment. Therefore, the proposed approach successfully helps to deal with group decision-making under uncertainty. The basic concept of the proposed method is based on the distance of alternatives from the nearest characteristic objects and their values of preference. The characteristic objects are obtained from the crisp values of all of the considered fuzzy numbers for each criterion. The proposed method is different from all of the previous techniques for MCGDM due to the fact that it uses hesitant fuzzy sets theory and the modification of the COMET method. The prominent feature of the proposed method is that it could provide a useful and flexible way to efficiently facilitate DMs under a hesitant fuzzy environment. The related calculations are simple and have a low computational complexity. Hence, it enriched and developed the theories and methods of MCGDM problems and also provided a new idea for solving MCGDM problems. Finally, a practical example was given to verify the developed approach and to demonstrate its practicality and effectiveness.

During the research, some possible areas of improvement of the proposed approach were identified. From a formal way, the COMET method can be extended over intuitionistic fuzzy sets, hesitant intuitionistic fuzzy sets, hesitant intuitionistic fuzzy linguistic term sets or other uncertain forms. Additionally, analysis and improvement of the accuracy of the presented extension of the COMET method should be performed. The future works may cover the practical usage of the proposed approach in the different decision-making domains.

Acknowledgments: The work was supported by the National Science Centre, Decision No. DEC-2016/23/N/HS4/01931, and by the Faculty of Computer Science and Information Technology, West Pomeranian University of Technology, Szczecin statutory funds.

Author Contributions: This paper is a result of common work of the authors in all aspects

Conflicts of Interest: The authors declare no conflict of interest

References

1. Pedrycz, W.; Ekel, P.; Parreiras, R. *Fuzzy Multicriteria Decision-Making: Models, Methods and Applications*; John Wiley & Sons: Hoboken, NJ, USA, 2010.
2. Yeh, T.M.; Pai, F.Y.; Liao, C.W. Using a hybrid MCDM methodology to identify critical factors in new product development. *Neural Comput. Appl.* **2014**, *24*, 957–971.
3. Xu, Z. On consistency of the weighted geometric mean complex judgement matrix in AHP. *Eur. J. Oper. Res.* **2000**, *126*, 683–687.
4. Lu, N.; Liang, L. Correlation Coefficients of Extended Hesitant Fuzzy Sets and Their Applications to Decision Making. *Symmetry* **2017**, *9*, 47.
5. Zhou, L.; Chen, H. A generalization of the power aggregation operators for linguistic environment and its application in group decision-making. *Knowl. Based Syst.* **2012**, *26*, 216–224.
6. Zhou, L.; Chen, H.; Liu, J. Generalized power aggregation operators and their applications in group decision-making. *Comput. Ind. Eng.* **2012**, *62*, 989–999.
7. Sengupta, A.; Pal, T.K. *Fuzzy Preference Ordering of Interval Numbers in Decision Problems*; Springer: Berlin, Germany, 2009.
8. Fang, Z.; Ye, J. Multiple Attribute Group Decision-Making Method Based on Linguistic Neutrosophic Numbers. *Symmetry* **2017**, *9*, 111.
9. Gao, J.; Liu, H. Reference-Dependent Aggregation in Multi-AttributeGroup Decision-Making. *Symmetry* **2017**, *9*, 43.
10. Wang, Z.-L.; You, J.-X.; Liu, H.-C. Uncertain Quality Function Deployment Using a Hybrid Group Decision Making Model. *Symmetry* **2016**, *8*, 119.
11. You, X.; Chen, T.; Yang, Q. Approach to Multi-Criteria Group Decision-Making Problems Based on the Best-Worst-Method and ELECTRE Method. *Symmetry* **2016**, *8*, 95.
12. Zadeh, L.A. Fuzzy sets. *Inf. Control* **1965**, *8*, 338–353.

13. Gitinavard, H.; Mousavi, S.M.; Vahdani, B. Soft computing-based new interval-valued hesitant fuzzy multi-criteria group assessment method with last aggregation to industrial decision problems. *Soft Comput.* **2017**, *21*, 3247–3265.
14. Zadeh, L.A. The concept of a linguistic variable and its applications to approximate reasoning-Part I. *Inf. Sci.* **1975**, *8*, 199–249.
15. Herrera, F.; Herrera-Viedma, E.; Verdegay, J.L. A model of consensus in group decision-making under linguistic assessments. *Fuzzy Sets Syst.* **1996**, *78*, 73–87.
16. Xu, Z.S. A method based on linguistic aggregation operators for group decision-making with linguistic preference relations. *Inf. Sci.* **2004**, *166*, 19–30.
17. Xu, Z.S. Deviation measures of linguistic preference relations in group decision-making. *Omega* **2005**, *33*, 249–254.
18. Atanassov, K.T. Intuitionistic fuzzy sets. *Fuzzy Sets Syst.* **1986**, *20*, 87–96.
19. Liu, Z.Q.; Miyamoto, S. *Soft Computing and Human-Centered Machines*; Springer: Berlin, Germany, 2000; pp. 9–33.
20. Yager, R.R. On the theory of bags. *Int. J. Gen. Syst.* **1986**, *13*, 23–37.
21. Ding, Z.; Wu, Y. An Improved Interval-Valued Hesitant Fuzzy Multi-Criteria Group Decision-Making Method and Applications. *Math. Comput. Appl.* **2016**, *21*, 22.
22. Wibowo, S.; Deng, H.; Xu, W. Evaluation of cloud services: A fuzzy multi-criteria group decision making method. *Algorithms* **2016**, *9*, 84.
23. Zhang, J.; Hegde, G.G.; Shang, J.; Qi, X. Evaluating Emergency Response Solutions for Sustainable Community Development by Using Fuzzy Multi-Criteria Group Decision Making Approaches: IVDHF-TOPSIS and IVDHF-VIKOR. *Sustainability* **2016**, *8*, 291.
24. Torra, V. Hesitant fuzzy sets. *Int. J. Intell. Syst.* **2010**, *25*, 529–539.
25. Jin, B. ELECTRE method for multiple attributes decision making problem with hesitant fuzzy information. *J. Intell. Fuzzy Syst.* **2015**, *29*, 463–468.
26. Zhang, N.; Wei, G. Extension of VIKOR method for decision making problem based on hesitant fuzzy set. *Appl. Math. Model.* **2013**, *37*, 4938–4947.
27. Peng, J.J.; Wang, J.Q.; Wu, X.H. Novel multi-criteria decision-making approaches based on hesitant fuzzy sets and prospect theory. *Int. J. Inf. Technol. Decis. Mak.* **2016**, *15*, 621–643.
28. Peng, X.; Yang, Y. Interval-valued hesitant fuzzy soft sets and their application in decision making. *Fundam. Inform.* **2015**, *141*, 71–93.
29. Yu, D. Hesitant fuzzy multi-criteria decision making methods based on Heronian mean. *Technol. Econ. Dev. Econ.* **2017**, *23*, 296–315.
30. Zhu, B.; Xu, Z. Extended hesitant fuzzy sets. *Technol. Econ. Dev. Econ.* **2016**, *22*, 100–121.
31. Wu, Y.; Xu, C.; Zhang, H.; Gao, J. Interval Generalized Ordered Weighted Utility Multiple Averaging Operators and Their Applications to Group Decision-Making. *Symmetry* **2017**, *9*, 103.
32. Yu, D.; Zhang, W.; Huang, G. Dual hesitant fuzzy aggregation operators. *Technol. Econ. Dev. Econ.* **2016**, *22*, 194–209.
33. Liu, J.; Sun, M. Generalized power average operator of hesitant fuzzy numbers and its application in multiple attribute decision-making. *J. Comput. Inf. Syst.* **2013**, *9*, 3051–3058.
34. Xia, M.; Xu, Z.; Chen, N. Some hesitant fuzzy aggregation operators with their application in group decision-making. *Group Decis. Negot.* **2013**, *22*, 259–279.
35. Yu, D.; Wu, Y.; Zhou, W. Multi-criteria decision-making based on Choquet integral under hesitant fuzzy environment. *J. Comput. Inf. Syst.* **2011**, *7*, 4506–4513.
36. Zhang, Z. Hesitant fuzzy power aggregation operators and their application to multiple attribute group decision-making. *Inf. Sci.* **2013**, *234*, 150–181.
37. Chen, N.; Xu, Z.; Xia, M. Interval-valued hesitant preference relations and their applications to group decision-making. *Knowl. Based Syst.* **2013**, *37*, 528–540.
38. Peng, D.H.; Wang, T.D.; Gao, C.Y.; Wang, H. Continuous hesitant fuzzy aggregation operators and their application to decision-making under interval-valued hesitant fuzzy setting. *Sci. World J.* **2014**, *2014*, 897304.
39. Peng, J.J.; Wang, J.Q.; Wang, J.; Chen, X.H. Multi-criteria decision-making approach with hesitant interval-valued intuitionistic fuzzy sets. *Sci. World J.* **2014**, *2014*, 868515.

40. Wei, G.; Zhao, X.; Lin, R. Some hesitant interval-valued fuzzy aggregation operators and their applications to multiple attribute decision-making. *Knowl. Based Syst.* **2013**, *46*, 43–53.
41. Beg, I.; Rashid, T. Hesitant 2-tuple linguistic information in multiple attributes group decision-making. *J. Intell. Fuzzy Syst.* **2016**, *30*, 143–150.
42. Rashid, T.; Husnine, S.M. Multi-criteria Group decision-making by Using Trapezoidal Valued Hesitant Fuzzy Sets. *Sci. World J.* **2014**, *2014*, 304834.
43. Yu, D. Triangular hesitant fuzzy set and its application to teaching quality evaluation. *J. Inf. Comput. Sci.* **2013**, *10*, 1925–1934.
44. Piegat, A.; Sałabun, W. Nonlinearity of human multi-criteria in decision-making. *J. Theor. Appl. Comput. Sci.* **2012**, *6*, 36–49.
45. Sałabun, W. The Characteristic Objects Method, A new distance based approach to multi-criteria decision-making problems. *J. Multi Criteria Decis. Anal.* **2015**, *22*, 37–50.
46. Sałabun, W. The Characteristic Objects Method: A new approach to Identify a multi-criteria group decision-making problems. *Int. J. Comput. Technol. Appl.* **2014**, *5*, 1597–1602.
47. Sałabun, W. Application of the fuzzy multi-criteria decision-making method to identify nonlinear decision models. *Int. J. Comput. Appl.* **2014**, *89*, 1–6.
48. Sałabun, W.; Piegat, A. Comparative analysis of MCDM methods for the assessment of mortality in patients with acute coronary syndrome. *Artif. Intell. Rev.* **2016**, 1–15, doi:10.1007/s10462-016-9511-9.
49. Faizi, S.; Rashid, T.; Sałabun, W.; Zafar, S.; Wątróbski, J. Decision Making with Uncertainty Using Hesitant Fuzzy Sets. *Int. J. Fuzzy Syst.* **2017**, 1–11, doi:10.1007/s40815-017-0313-2.
50. Piegat, A.; Sałabun, W. Identification of a multicriteria decision-making model using the characteristic objects method. *Appl. Comput. Intell. Soft Comput.* **2014**, *2014*, 1–14.
51. Sałabun, W.; Ziemba, P.; Wątróbski, J. The Rank Reversals Paradox in Management Decisions: The Comparison of the AHP and COMET Methods. In *Intelligent Decision Technologies*; Springer: Berlin, Germany, 2016; pp. 181–191.
52. Wątróbski, J.; Jankowski, J. Guideline for MCDA method selection in production management area. In *New Frontiers in Information and Production Systems Modelling and Analysis*; Springer: Berlin, Germany, 2016, pp. 119–138.
53. Sałabun, W.; Ziemba, P. Application of the Characteristic Objects Method in Supply Chain Management and Logistics. In *Recent Developments in Intelligent Information and Database Systems*; Springer: Berlin, Germany, 2016; pp. 445–453.
54. Bector, C.R.; Chandra, S. *Fuzzy mathematical programming and fuzzy matrix games*; Springer: Berlin, Germany, 2005; Volume 169.
55. Chakeri, A.; Dariani, A.N.; Lucas, C. How can fuzzy logic determine game equilibriums better? In Proceedings of the 4th International IEEE Conference Intelligent Systems, 2008 IS'08, Varna, Bulgaria, 6–8 September 2008; Volume 1, pp. 2–51.
56. Chakeri, A.; Habibi, J.; Heshmat, Y. Fuzzy type-2 Nash equilibrium. In Proceedings of the 2008 International Conference on Computational Intelligence for Modelling Control & Automation, Vienna, Austria, 10–12 December 2008; pp. 398–402.
57. Chakeri, A.; Sadati, N.; Dumont, G.A. Nash Equilibrium Strategies in Fuzzy Games. In *Game Theory Relaunched*; InTech: Exton, PA, USA, 2013.
58. Chakeri, A.; Sadati, N.; Sharifian, S. Fuzzy Nash equilibrium in fuzzy games using ranking fuzzy numbers. In Proceedings of the 2010 IEEE International Conference on Fuzzy Systems (FUZZ), Barcelona, Spain, 18–23 July 2010; pp. 1–5.
59. Chakeri, A.; Sheikholeslam, F. Fuzzy Nash equilibriums in crisp and fuzzy games. *IEEE Trans. Fuzzy Syst.* **2013**, *21*, 171–176.
60. Garagic, D.; Cruz, J.B. An approach to fuzzy noncooperative nash games. *J. Optim. Theory Appl.* **2003**, *118*, 475–491.
61. Sharifian, S.; Chakeri, A.; Sheikholeslam, F. Linguisitc representation of Nash equilibriums in fuzzy games. In Proceedings of the 2010 Annual Meeting of the North American Fuzzy Information Processing Society (NAFIPS), Toronto, ON, Canada, 12–14 July 2010; pp. 1–6.
62. Sharma, R.; Gopal, M. Hybrid game strategy in fuzzy Markov-game-based control. *IEEE Trans. Fuzzy Syst.* **2008**, *16*, 1315–1327.

63. Tan, C.; Jiang, Z.Z.; Chen, X.; Ip, W.H. A Banzhaf function for a fuzzy game. *IEEE Trans. Fuzzy Syst.* **2014**, *22*, 1489–1502.
64. Piegat, A.; Sałabun, W. Comparative analysis of MCDM methods for assessing the severity of chronic liver disease. In Proceedings of the 14th International Conference on Artificial Intelligence and Soft Computing, Zakopane, Poland, 14–18 June 2015; pp. 228–238.
65. Wątróbski, J.; Sałabun, W. The characteristic objects method: a new intelligent decision support tool for sustainable manufacturing. In *Sustainable Design and Manufacturing*; Springer: Berlin, Germany, 2016; pp. 349–359.
66. Jankowski, J.; Sałabun, W.; Wątróbski, J. Identification of a multi-criteria assessment model of relation between editorial and commercial content in web systems. In *Multimedia and Network Information Systems*; Springer: Berlin, Germany, 2017; pp. 295–305.
67. Velasquez, M.; Hester, P.T. An Analysis of Multi-Criteria Decision Making Methods. *Int. J. Oper. Res.* **2013**, *10*, 56–66.
68. Xia, M.; Xu, Z. Hesitant fuzzy information aggregation in decision-making. *Int. J. Approx. Reason.* **2011**, *52*, 395–407.
69. Zimmermann, H.J. *Fuzzy Set Theory—And Its Applications*; Springer Science and Business Media: Berlin, Germany, 2001.
70. Chen, J.; Huang, X. Dual Hesitant Fuzzy Probability. *Symmetry* **2017**, *9*, 52.
71. Liu, X.; Wang, Z.; Zhang, S. A Modification on the Hesitant Fuzzy Set Lexicographical Ranking Method. *Symmetry* **2016**, *8*, 153.
72. Zhang, X.; Xu, Z.; Liu, M. Hesitant trapezoidal fuzzy QUALIFLEX method and its application in the evaluation of green supply chain initiatives. *Sustainability* **2016**, *8*, 952.
73. Piegat, A. *Fuzzy Modeling and Control*; Springer: New York, NY, USA, 2001.
74. Wang, G.; Wang, H. Non-fuzzy versions of fuzzy reasoning in classical logics. *Inf. Sci.* **2001**, *138*, 211–236.
75. Ross, T.J. *Fuzzy Logic with Engineering Applications*; John Wiley & Sons: Hoboken, NJ, USA, 2010.
76. Klement, E.P.; Mesiar, R.; Pap, E. *Triangular Norms*; Kluwer Academic Publishers: Dordrecht, The Netherlands, 2000.

symmetry

MDPI

Article

Discrete Optimization with Fuzzy Constraints

Primož Jelušič * and Bojan Žlender

Faculty of Civil Engineering, Transportation Engineering and Architecture, University of Maribor, Smetanova ulica 17, 2000 Maribor, Slovenia; bojan.zlender@um.si

* Correspondence: primoz.jelusic@um.si; Tel.: +386-31267229

Academic Editor: José Carlos R. Alcantud

Received: 8 May 2017; Accepted: 13 June 2017; Published: 16 June 2017

Abstract: The primary benefit of fuzzy systems theory is to approximate system behavior where analytic functions or numerical relations do not exist. In this paper, heuristic fuzzy rules were used with the intention of improving the performance of optimization models, introducing experiential rules acquired from experts and utilizing recommendations. The aim of this paper was to define soft constraints using an adaptive network-based fuzzy inference system (ANFIS). This newly-developed soft constraint was applied to discrete optimization for obtaining optimal solutions. Even though the computational model is based on advanced computational technologies including fuzzy logic, neural networks and discrete optimization, it can be used to solve real-world problems of great interest for design engineers. The proposed computational model was used to find the minimum weight solutions for simply-supported laterally-restrained beams.

Keywords: uncertainty; discrete optimization; neuro-fuzzy technique; structural optimization

1. Introduction

The theory of fuzzy sets can be used to model imprecision, ambiguity or fuzziness in the formulation of structural optimization problems. In the formulation of such problems, a major source of imprecision, or fuzziness, occurs in the evaluation of constraints. In traditional optimization algorithms, constraints are satisfied with a tolerance defined by a crisp or non-fuzzy number. In reality, in common engineering practice, this evaluation involves many sources of approximations [1]. A design of structure is considered satisfactory when the several constraints are satisfied within a given predetermined tolerance. However, when an optimization algorithm satisfies the constraints precisely (for a defined small tolerance degree of numerical computations), it can miss the true optimum design within the confines of practical and realistic approximations.

Adeli [2] demonstrated that by taking into account the fuzziness and imprecision in the constraints and employing fuzzy set theory, it is possible to reduce the objective function further and substantially increase the probability of finding the actual global optimum solution. The goal of the present research, carried out by several authors, was to model the effects of fuzziness in the formulation of a genetic algorithm (GA)-based structural design optimization problem [3–6]. Another objective of the authors' research was to improve the convergence and efficiency of GAs through the use of fuzzy set theory [7–9]. Several articles have been published on the fuzzy optimization of structures [10,11], with the objective of reducing the number of iterations and the total computer processing time needed.

Uncertainty exists in almost every real-world problem. In general, uncertainty is inseparable from measurement. It emerges from a combination of the limits of measurement with instruments and unavoidable errors in measurement. In this paper, the fuzziness was considered as part of the constraints. The constraints were developed using the neuro-fuzzy technique, which was based on past experience, recommendations and measurements. Fuzzy constraints were then used in discrete optimization along with other crisp constraints. This allowed us to include experience, recommendations and experimental measurements in the optimization problem. Optimization using non-linear programming (NLP) and fuzzy constraints has been done by Jelusic [12], however without the discrete optimization approach. While NLP deals with the continuous optimization of structures, mixed-integer linear programming (MILP) performs continuous-discrete optimization, where the structural topology, discrete materials (steel) and standard dimensions (steel sections) are known.

In order to find the minimum weight solutions for simply-supported laterally-restrained beams, the appropriate deflection limit should be specified. The comparison of different design codes showed that the deflection limits are too liberal. This paper defines the soft constraint for the deflection limit based on experiential rules acquired from experts and utilizing recommendations. This newly-developed soft constraint obtained with an adaptive network-based fuzzy inference system (ANFIS) is then used in the optimization model. ANFIS can learn from examples and is fault tolerant in the sense that it is able to handle noisy or incomplete data. The expert evaluations for the deflection limit are very subjective; therefore, the data for approximation function are expected to be vague, imprecise, incomplete or even contradictory. Additionally, the proposed ANFIS techniques include fuzzy clustering (FCM), which searches for patterns in data points. The study suggests that deflection limits could be reconsidered in the future by the experts who have a prolonged or intense experience through practice.

2. Structural Optimization and Fuzzy Set Theory

A crisp non-fuzzy structural optimization is formulated as follows: find the vector of the design variables x such that the objective function $F(x)$ is minimized subject to the equality and inequality constraints:

$$\min z = F(x) \tag{1}$$

s.t.

$$\begin{gathered} h_i(x) = 0,\ i = 1,2,\ldots,N_{ec} \\ g_i^l(x) \leq g_i(x) \leq g_i^u(x),\ i = 1,2,\ldots,N_{iec} \end{gathered} \tag{2}$$

where N_{ec} is the number of equality constraints and N_{iec} is the number of inequality constraints; $g_i^u(x)$ is the upper bound on the constraint $g_i(x)$, and $g_i^l(x)$ is the lower bound on the constraint $g_i(x)$. If vagueness is considered in the objective function and constraints, then the variables ($\mathbf{x}$) can be obtained from a fuzzy decision D, such that the membership function μ_D for the fuzzy decision D can be obtained from the intersection of the fuzzy membership functions for the objective function and constraints; see Equation (3):

$$\mu_D = \mu_F(x) \cap \left[\bigcap_{i=1,2,\ldots,N_{iec}} \mu_{g_i}(x) \right] \tag{3}$$

where $\mu_F(x)$ is the membership functions for the objective function and $\mu_{g_i}(x)$ is the membership functions for the i-th inequality design constraint. From this fuzzy decision, the optimum solution (x^*) for the variable $\mathbf{x}$ can be obtained by using the max-min procedure [13]; see Equation (4):

$$\mu_D(x^*) = \text{maximize } \mu_D(x) \tag{4}$$

where:

$$\mu_D(x) = \min\left[\mu_F(x), \min_{i=1,2,\ldots,N_{iec}} \mu_{g_i}(x) \right] \tag{5}$$

The max-min procedure can be solved by maximizing a scalar parameter λ (overall satisfaction parameter) [14]; see Equations (6)–(9):

$$\text{Max}\lambda$$
$$\text{s.t:}$$
$$\lambda \leq \mu_F(\boldsymbol{x}) \tag{6}$$
$$\lambda \leq \mu^u_{g_i}(\boldsymbol{x}),\ i = 1,\ 2, \ldots, N_{iec} \tag{7}$$
$$\lambda \leq \mu^l_{g_i}(\boldsymbol{x}),\ i = 1,\ 2, \ldots, N_{iec} \tag{8}$$
$$0 \leq \lambda \leq 1 \tag{9}$$

where $\mu^u_{g_i}(\boldsymbol{x})$ and $\mu^l_{g_i}(\boldsymbol{x})$ are the membership functions for the upper and lower bounds of the inequality constraints $\mu_{g_i}(\boldsymbol{x})$ (Equation (2)), respectively. The equality constraints, in Equation (1), are not included in the fuzzy formulations because they have to be satisfied strictly.

2.1. ANFIS Architecture for the Development of Soft Constraint Functions

For a Sugeno fuzzy model, a rule set with n fuzzy "if-then" is as follows:
Rule 1: If x is A_1 and y is B_1, then:

$$f_1 = a^1_0 + a^1_1 x + a^1_2 y \tag{10}$$

Rule i: If x is A_i and y is B_i, then:

$$f_i = a^i_0 + a^i_1 x + a^i_2 y \tag{11}$$

Rule n: If x is A_n and y is B_n, then:

$$f_n = a^n_0 + a^n_1 x + a^n_2 y \tag{12}$$

where $a^1_0,\ a^1_1, a^1_2, a^i_0, a^i_1, a^i_2, a^n_0, a^n_1, a^n_2$ are consequent parameters and x and y are input variables. The output of each rule is equal to the constant, and the final output is the weighted average of each rule's output.

$$f = \sum_{i=1}^{n} \overline{w}_i \cdot f_i = \sum_{i=1}^{n} \overline{w}_i \cdot \left(a^i_0 + a^i_1 x + a^i_2 y\right) \tag{13}$$

The weights are obtained from a Gaussian membership function.

$$\mu(x) = \exp\left[-\left(\frac{x - c}{\sigma \cdot \sqrt{2}}\right)^2\right] \tag{14}$$

where c is the position of the center of the curve's peak and σ is the width of the curve. Parameters c and σ are premise parameters. The first membership grade of the fuzzy set (A_i, B_i, C_i, D_i) is calculated with the following equations:

$$\mu_{A_i}(x) = \exp\left[-\left(\frac{x - c_{A_i}}{\sigma_{A_i} \cdot \sqrt{2}}\right)^2\right] \tag{15}$$

$$\mu_{B_i}(y) = \exp\left[-\left(\frac{y - c_{B_i}}{\sigma_{B_i} \cdot \sqrt{2}}\right)^2\right] \tag{16}$$

where x and y are the input variables in the Gaussian membership function. After this, the product of the membership function for every rule is calculated:

$$w_i = \mu_{A_i}(x) \cdot \mu_{B_i}(y) \tag{17}$$

where w_i represents the fire strength of the rule i. The ratio of the i-th rule's firing strength to the sum of all of the rule's firing strengths is defined with:

$$\overline{w_i} = \frac{w_i}{w_1 + \ldots + w_i + \ldots + w_n} \text{for } i = 1, 2, \ldots, n. \tag{18}$$

In order to achieve the desired input-output mapping, the consequent and premise parameters need to be updated according to the given training data and the hybrid learning procedure. This hybrid learning procedure [15] is composed of a forward pass and backward pass. In the forward pass, the algorithm uses the least-squares method to identify the consequent parameters. In the backward pass, the errors are propagated backward, and the premise parameters are updated by gradient descent.

3. Discrete Optimization

Exhaustive enumeration (EE) is the simplest of the discrete optimization techniques. It evaluates an optimum solution for all combinations of the discrete variables. The best solution is obtained by scanning the list of all feasible solutions for the minimum value. The total number of evaluations is:

$$n_e = \prod_{i=1}^{n_d} p_i \tag{19}$$

where n_d is the number of discrete variables and p_i is the pre-established set of discrete values. If either n_d or p_i (or both) are large, it shows that much work will be required. It also shows an exponential growth in the calculations with the number of discrete variables. In a mixed optimization problem, this would involve the continuous optimum solution of a reduced model. It is not a serious problem if the mathematical model and its computer calculations are easy to implement. If the mathematical model requires extensive calculations, then some concerns may arise. Programming exhaustive enumeration is straightforward. The processing speed, large available desktop-memory and easy programming, through software like MATLAB [16], make exhaustive enumeration a very good idea today. This program is also ideal because the solution is a global optimum. The most important step is translating the mathematical model into a program code.

The number of design variables in the model is reduced by the number of discrete variables. Model reduction is involved in enumeration techniques.

The algorithm with feasibility requirements is as follows:

Step 1. $s^* = inf$, $\mathbf{K} = [0, 0, \ldots, 0]$
For every allowable combination of $(y_1, y_2, \ldots, y_{nd}) \Rightarrow (Y_b)$
Solve optimization problem (solution $\mathbf{K}^*$)
  If $h(\mathbf{K}^*, \mathbf{Y_b}) = [0]$ *and*
    If $g(\mathbf{K}^*, \mathbf{Y_b}) \leq [0]$ *and*
    If $f(\mathbf{K}^*, \mathbf{Y_b}) < s^*$
      Then $s^* \leftarrow f(\mathbf{K}^*, \mathbf{Y_b})$
      $\mathbf{K} \leftarrow \mathbf{K}^*$
      $\mathbf{Y} \leftarrow \mathbf{Y_b}$
      End If
    End If
End If
End For

This application example presents how soft constraints are included in discrete optimization. ANFIS is used to integrate recommendations of deflection limits into an optimization model.

4. Example Design of a Simply-Supported Laterally-Restrained Beam Application

The basic design process is formed by determining the design loads acting on the structure, determining the design loads on individual elements and calculating the bending moments, shear forces and deflections of the beams. Generally, laterally-restrained beams should be checked for their ultimate limit state (ULS) and serviceability limit state (SLS). As this article is about the design of steel structural elements, the following were examined:

(1) resistance of the cross-section to bending (ULS),
(2) resistance to shear buckling (ULS),
(3) resistance to flange-induced buckling (ULS),
(4) resistance of the web to transverse forces (ULS) and
(5) deflection (SLS).

The beam is loaded by a uniformly-distributed dead load g_k and a uniformly-distributed imposed load q_k, as shown in Figure 1.

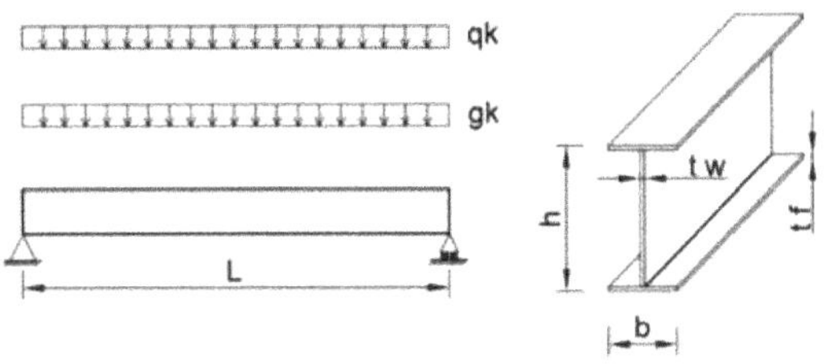

Figure 1. Simply-supported beam and steel section.

The civil engineer must evaluate the possible future levels of loading (self-weight, snow, wind), which the structure may be subjected to during its design life. Then, using hand calculations or computer methods, the loads acting on individual construction elements can be evaluated. The loads are used to calculate the shear forces, bending moments and deflections at critical sections along the construction elements. Finally, sufficient dimensions for the construction element can be defined.

4.1. Design Loads

The loads acting on a beam are divided into imposed and dead loads. For each type of loading there will be characteristic values and design values that must be estimated. In addition to this, the designer will have to determine the particular combination of loading that is likely to produce the most adverse effect on the structure in terms of shear forces, bending moments and deflections. The design loads are obtained by multiplying the characteristic loads by the partial safety factor for loads. However, before flexural members, such as beams, can be sized, the design bending moments and shear forces must be evaluated. Design shear forces and moments in beams are calculated using standard equations (Equations (21) and (22)). Having calculated the shear force and design bending moment, all that now remains to be done is to estimate the dimensions and strength of the beam required. Yield strength and section classification are used for the initial choice of the section. If the section is thick, i.e., has thick flanges and web, it can sustain the formation of a plastic hinge. On the other hand, a slender section, for example with thin flanges and web, will fail by local buckling before the yield stress can be reached.

Design action:

$$F_{Ed} = (\gamma_G \cdot g_k + \gamma_Q \cdot q_k) \cdot l \tag{20}$$

Design bending moment:

$$M_{Ed} = F_{Ed} \cdot l/8 \tag{21}$$

Design shear force:

$$V_{Ed} = F_{Ed}/2 \tag{22}$$

Strength classification:

$$\varepsilon = \left(235/f_y\right)^{0.5} \tag{23}$$

Section classification:

$$c/t_f \leq 10 \cdot \varepsilon \tag{24}$$

$$c^*/t_w \leq 83 \cdot \varepsilon \tag{25}$$

where:

$$c = (b - t_w - 2 \cdot r)/2 \tag{26}$$

and:

$$c^* = d \tag{27}$$

4.2. Resistance of Steel Cross-Sections

The structural design of steel beams primarily involves predicting the strength of their members. This requires the designer to imagine all of the ways in which each member may fail during its design life. Common modes of failure associated with beams are local buckling, shear, shear buckling, web bearing and buckling, lateral torsional buckling, bending and deflection.

4.2.1. Bending Moment

When shear force is absent or of a low value, the design value of the bending moment, M_{Ed}, at each section should satisfy the following:

$$M_{Ed}/M_{c,Rd} \leq 1.0 \tag{28}$$

where $M_{c,Rd}$ is the design resistance for bending around one principal axis, taken as follows:
(a) the plastic design resistance moment of the gross section:

$$M_{pl,Rd} = W_{pl} \cdot f_y/\gamma_{M0} \tag{29}$$

where W_{pl} is the plastic section modulus, for Class 1 and 2 sections only;
(b) the elastic design resistance moment of the gross section:

$$M_{el,Rd} = W_{el,min} \cdot f_y/\gamma_{M0} \tag{30}$$

where $W_{el,min}$ is the minimum elastic section modulus for Class 3 sections;
(c) the local buckling design resistance moment of the gross section:

$$M_{c,Rd} = W_{eff,min} \cdot f_y/\gamma_{M0} \tag{31}$$

where $W_{eff,min}$ is the minimum effective section modulus for Class 4 cross-sections only;
(d) the ultimate design resistance moment of the net section at bolt holes $M_{u,Rd}$, if this is less than the appropriate values above. In calculating this value, fastener holes in the compression zone do not need to be considered; they would need to be if they were oversized, slotted or filled by fasteners. In the tension zone, holes do not need to be considered, provided that:

$$A_{f,net} \cdot 0.9 \cdot \mathrm{f_u}/\gamma_{M2} \geq A_f \cdot f_y/\gamma_{M0} \tag{32}$$

4.2.2. Shear

The design value of the shear force V_{Ed} at each cross-section should satisfy the following:

$$V_{Ed}/V_{c,Rd} \leq 1.0 \quad (33)$$

where $V_{c,Rd}$ is the design shear resistance. For the plastic design, $V_{c,Rd}$ is taken as the design plastic shear resistance, $V_{pl,Rd}$, given by:

$$V_{Pl,Rd} = A_v \cdot \left(f_y/\sqrt{3}\right)/\gamma_{M0} \quad (34)$$

where A_v is the shear area, which, for the rolled I and H sections, loaded parallel to the web, is:

$$A_v = A - 2bt_f + (t_w + 2r)t_f \geq \eta h_w t_f \quad (35)$$

where:

b overall breadth
r root radius
t_f flange thickness
t_w web thickness
h_w depth of the web
η conservatively taken as 1.0
A cross-sectional area

Fastener holes in the web do not have to be considered in shear verification. Shear buckling resistance for unstiffened webs must additionally be considered when:

$$h_w/t_w > 72\varepsilon/\eta \quad (36)$$

For a stiffened web, shear buckling resistance will need to be considered when:

$$h_w/t_w > 31\varepsilon\sqrt{k_\tau}/\eta \quad (37)$$

where k_τ is the buckling factor for shear and is given by:

$$\text{for } a/h_w < 1;\ k_\tau = 4 + 5.34(h_w/a)^2 \quad (38)$$

$$\text{for } a/h_w \geq 1;\ k_\tau = 5.34 + 4(h_w/a)^2 \quad (39)$$

4.2.3. Resistance of Cross-Section-Bending and Shear

The plastic resistance moment of the section is reduced by the presence of shear force. When the design value of the shear force, V_{Ed}, exceeds 50 percent of the plastic shear design resistance, $V_{pl,Rd}$, the design resistance moment of the section, $M_{v,Rd}$, should be calculated using a reduced yield strength taken as:

$$\rho = \left(2V_{Ed}/V_{pl,Rd} - 1\right)^2 \quad (40)$$

Thus, for the rolled I and H sections, the reduced design resistance moment for the section around the major axis, $M_{y,v,Rd}$, will be given by:

$$M_{y,v,Rd} = f_y\left(W_{pl,y} - \rho A_v^2/4t_w\right)/\gamma_{M0} \leq M_{y,c,Rd} \quad (41)$$

4.2.4. Shear Buckling Resistance

As noted above, the shear buckling resistance of unstiffened beam webs has to be checked when:

$$h_w/t_w > 72\varepsilon/\eta \quad (42)$$

The value of 1.0 for η for all steel grades up to and including S460 is recommended. For standard rolled beams and columns, this check is rarely necessary. However, as h_w/t_w is usually less than 72ε, it was not discussed in this section.

4.2.5. Flange-Induced Buckling

To prevent the possibility of the compression flange buckling in the plane of the web, Eurocode 3–5 [17] requires that the ratio h_w/t_w of the web should satisfy the following criterion:

$$h_w/t_w \leq \mathrm{k}{\cdot}\mathrm{E}/\mathrm{f_{yf}}\sqrt{\mathrm{A_w}/\mathrm{A_{fc}}} \tag{43}$$

where:

A_w is the area of the web = $(h-2{\cdot}t_f){\cdot}t_w$
A_{fc} is the area of the compression flange = $b{\cdot}t_f$
f_{yf} is the yield strength of the compression flange

The factor k assumes the following values: plastic rotation utilized, i.e., Class 1 flanges: 0.3; plastic moment resistance utilized, i.e., Class 2 flanges: 0.4; elastic moment resistance utilized, i.e., Class 3 or Class 4 flanges: 0.55.

4.2.6. Resistance of the Web to Transverse Forces

Eurocode 3–5 [17] categorize between two types of forces applied through a flange to the web:

(a) forces resisted by shear in the web (loading Types (a) and (c)).
(b) forces transferred through the web directly to the other flange (loading Type (b)).

For loading Types (a) and (c), the web is likely to fail as a result of:

(i) crushing of the web close to the flange accompanied by yielding of the flange; the combined effect is sometimes referred to as web crushing
(ii) localized buckling and crushing of the web beneath the flange; the combined effect is sometimes referred to as web crippling.

For loading Type (b) the web is likely to fail as a result of:

(i) web crushing
(ii) buckling of the web over most of the depth of the member.

Provided that the compression flange is sufficiently restrained in the lateral direction, the design resistance of webs of beams under transverse forces can be determined in accordance with the recommendations in Eurocode 3 [17].

In Eurocode 3 [17], it is stated that the design resistance of webs to local buckling is given by:

$$F_{Rd} = f_y{\cdot}L_{eff}{\cdot}t_w/\gamma_{M1} \tag{44}$$

where:

f_{yw} is the yield strength of the web
t_w is the thickness of the web
γ_{M1} is the partial safety factor = 1.0
L_{eff} is the effective length of the web that resists transverse forces = $\chi_F l_y$, in which χ_F is the reduction factor due to local buckling.
l_y is the effective loaded length, appropriate to the length of the stiff bearing s_s. As stated in Clause 6.3 of Eurocode 3–5 [17], s_s should be taken as the distance over which the applied load is effectively distributed at a slope of 1:1, but $s_s \leq h_w$.

Reduction factor, χ_F: The reduction factor χ_F is given by:

$$\chi_F = 0.5/\overline{\lambda}_F \leq 1 \tag{45}$$

where:

$$\overline{\lambda}_F = \sqrt{l_y \cdot t_w \cdot f_{yw}/F_{cr}} \tag{46}$$

in which:

$$F_{cr} = 0.9 \cdot k_F \cdot E \cdot t_w^3/h_w \tag{47}$$

Effective load length, l_y: As stated in Clause 6.5 [17] for loading Types (a) and (b), the effective load length, l_y, is given by:

$$l_y = s_s + 2 \cdot t_f \cdot \left(1 + \sqrt{m_1 + m_2}\right) \leq a \tag{48}$$

where:

$$m_1 = f_{yt} \cdot b_f / \left(f_{yw} \cdot t_w\right) \tag{49}$$

and if:

$$\overline{\lambda}_F > 0.5; m_2 = 0.02 \cdot \left(h_w/t_f\right)^2 \tag{50}$$

or if:

$$\overline{\lambda}_F \leq 0.5; m_2 = 0 \tag{51}$$

For loading Type (c), l_y is taken as the smallest value obtained from Equations (52) and (53), as follows:

$$l_y = l_e + t_f \cdot \sqrt{m_2/2 + \left(l_e/t_f\right)^2 + m_2} \tag{52}$$

$$l_y = l_e + t_f \cdot \sqrt{m_1 + m_2} \tag{53}$$

where:

$$l_e = k_F \cdot E \cdot t_w^2 / \left(2 \cdot f_{yw} \cdot h_w\right) \leq s_s + c \tag{54}$$

4.3. *Deflections*

Several vertical deflections are defined in the Eurocode [18]. However, the National Annex to Eurocode 3 [17] recommends that verification of vertical deflections, δ, under unfactored imposed loads should be carried out. The designer is responsible for specifying appropriate limits of vertical deflections, which should be agreed upon with the client. However, like British Standard (BS) 5950 [19], the National Annex to Eurocode 3 [17] also recommends that verifications be made on vertical deflections, δ, under unfactored imposed loads. It suggests that in the absence of other limits, the recommendations in the Eurocode may be used.

The recommendations were examined and used for the building of the model and to predict the limits of vertical deflection. Two parameters with the biggest influence on the vertical deflection limits were considered. The influence of each parameter was determined on the basis of recommendations and engineering judgment. The comparison of different design codes (Eurocode, American Institute of Timber Construction (AITC) [20]) showed, that the deflection limits are very different. The study also suggests that deflection limits could be reconsidered in the future by the designers who have a prolonged or intense experience through practice.

4.3.1. ANFIS for the Development of the Constraint Function

A model to limit the vertical deflection based on recommendations was developed. The ANFIS model has two inputs: applied live load *LL* (kN/m^2) and classification *CLASS* (-); and it has one output: *LIMIT*. The ANFIS-LIMIT model was proposed in order to calculate the deflection limits.

MATLAB [16] and a Fuzzy Logic Toolbox were used as an interface for mathematical modeling and data handling.

One of the most important stages in the ANFIS technique is the collection of data. The training data were chosen based on the recommendations in AITC [20], Eurocode (Table 1), professional experience and past projects (Table 2). The classification used is separated into three groups. The first group is reserved for railway bridge stringers and beams used for commercial and institutional buildings with plaster ceilings. The second group is reserved for highway bridge stringers and beams used for commercial and institutional buildings without plaster ceilings. The third group is reserved for industrial roof beams.

Table 1. Deflection limit according to the recommendations.

Use Classification	Deflection Limit
Roof beams (industrial)	L/180
Roof beams (commercial and institutional without plaster ceiling)	L/240
Roof beams (commercial and institutional with plaster ceiling)	L/360
Floor beams (ordinary usage)	L/360
Highway bridge stringers	L/200 to L/300
Railway bridge stringers	L/300 to L/400
$LL < 2.5$ kN/m^2	L/480
2.5 kN/m$^2 < LL < 4.0$ kN/m^2	L/420
$LL > 4.0$ kN/m^2	L/360

Table 2. Training data for the ANFIS-LIMIT model.

Inputs		Output
Applied Live Load *LL* (kN/m^2)	Classification *CLASS* *	Deflection Limit *LIMIT*
20	1	360
10	1	360
4	1	360
3.5	1	420
3	1	420
2.5	1	480
2	1	480
1.5	1	480
1	1	480
0.5	1	480
0	1	480
20	2	240
10	2	240
4	2	240
3.5	2	280
3	2	280
2.5	2	320
2	2	320
1.5	2	320
1	2	320
0.5	2	320
0	2	320
20	3	180
10	3	180
4	3	180
3.5	3	210
3	3	210
2.5	3	240
2	3	240
1.5	3	240
1	3	240
0.5	3	240
0	3	240

* 1, Railway bridge stringers, beams used for commercial and institutional buildings with plaster ceiling; 2, highway bridge stringers and beams used for commercial and institutional buildings without plaster ceiling; 3, industrial roof beams.

The applied load (*LL*) and the classification system (*CLASS*) were taken as input parameters; whereas, the deflection limit (*LIMIT*) was considered as an output parameter. The training dataset (see Table 2) can be improved by adding additional recommendations and more past experience. These values can be assigned to the parameters and deflection limit. In this model, 33 evaluations were defined for a different applied live load and classification.

For the Sugeno fuzzy model [21], a rule set with *i*, *i* I, I = {1, 2} and fuzzy "if-then" rules were defined by Equations (55) and (56):

Rule 1: If LL is A_1 and CLASS is B_1, then:

$$LIMIT_1 = a_0^1 + a_1^1 \cdot LL + a_2^1 \cdot CLASS \tag{55}$$

Rule 2: If LL is A_2 and CLASS is B_2, then:

$$LIMIT_2 = a_0^2 + a_1^2 \cdot LL + a_2^2 \cdot CLASS \tag{56}$$

where a_0^1, $a_1^1, a_2^1, a_0^2, a_1^2, a_2^2$ are consequent parameters and *LL* and *CLASS* are input variables. The calculation procedure of the ANFIS models is as follows:

1. the membership grade of the fuzzy set (A_i, B_i, C_i, D_i) is calculated;
2. the product of membership function for each rule is calculated;
3. the ratio between the *i*-th rule's firing strength and the sum of all rules' firing strengths is calculated;
4. the output of each rule is calculated; and
5. the weighted average of each rule's output is calculated.

The first membership grade of the fuzzy set (A_i, B_i, C_i, D_i) is calculated with Equations (57) and (58):

$$\mu_{A_i}(LL) = \exp\left[-\left(\frac{LL - c_{A_i}}{\sigma_{A_i} \cdot \sqrt{2}}\right)^2\right] \tag{57}$$

$$\mu_{B_i}(CLASS) = \exp\left[-\left(\frac{CLASS - c_{B_i}}{\sigma_{B_i} \cdot \sqrt{2}}\right)^2\right] \tag{58}$$

where *LL* and *CLASS* are inputs to Gaussian membership functions, and the parameters c_{Ai}, c_{Bi}, σ_{Ai}, σ_{Bi} are premise parameters. In addition to this, the products between the membership functions for every rule are calculated; see Equations (59) and (60):

$$w_1 = \mu_{A_1}(LL) \cdot \mu_{B_1}(CLASS) \tag{59}$$

$$w_2 = \mu_{A_2}(LL) \cdot \mu_{B_2}(CLASS) \tag{60}$$

where w_1 and w_2 represent the firing strength of the each rule. The weighted average of each rules' output is defined as the ratio between the *i*-th rule's firing strength and the sum of all of the rule's firing strengths; see Equation (61):

$$\overline{w}_i = \frac{w_i}{w_1 + w_2}, \text{ for } i = 1, 2. \tag{61}$$

The output of each rule is finally determined as the sum of products between the weighted average of each rule's output and the linear combination between input variables and consequent parameters:

$$LIMIT = \sum_{i=1}^{2} \overline{w}_i \cdot LIMIT_i = \sum_{i=1}^{2} \overline{w}_i \cdot \left(a_0^i + a_1^i \cdot LL + a_2^i \cdot CLASS\right) \tag{62}$$

For the model, the following values were evaluated: premise parameters, consequent parameters, firing strengths and weighted averages of rules outputs.

The structure of the model is shown in Figure 2. While the nodes on the left side represent the input data, the right node stands for the output. The model includes two inputs, the applied live load *LL* (kN/m^2) and the classification *CLASS* (-), as well as a single output deflection limit *LIMIT* (-).

In a conventional fuzzy inference system, the number of rules is decided by the researcher/engineer who is familiar with the system to be modeled. There are no simple ways of determining in advance the minimum number of membership functions to achieve a desired performance level. In the present attempt, the number of membership functions assigned to each input variable was chosen empirically by examining the desired input-output data and by trial and error. For the deflection limit model, two membership functions were chosen for each input. Figure 2 shows the membership functions for *LL* and *CLASS* for the deflection model *LIMIT*. Note that all of the membership functions used were Gaussian membership functions, defined by Equations (57) and (58).

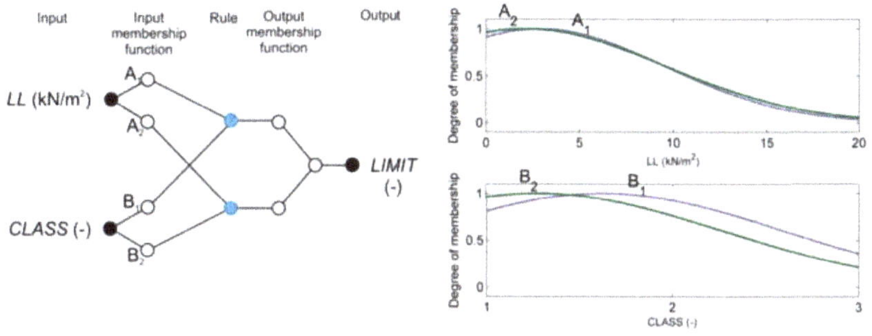

Figure 2. Fuzzy inference system implemented within the framework of adaptive networks ANFIS-LIMIT.

After the numbers of membership functions associated with each input were fixed, the initial values of premise parameters were set in such a way that membership functions were equally spaced along the operating range of each input variable. The *LIMIT* model contained two rules with two membership functions being assigned to each input variable. The total number of fitting parameters was 14, composed of eight premise parameters and six consequent parameters. These parameters were obtained by using a hybrid algorithm. MATLAB [16] was used as an interface for mathematical modeling. Premise and consequent parameters are presented in Table 3.

Table 3. Premise and consequent parameters of the ANFIS-LIMIT model.

Membership Function	Premise Parameters		Consequent Parameters	
i	σ_i	c_i		-
A_1	6.61768494044089	2.90117735987428	$a_0{}^1$	−771.211045670957
A_2	7.47150990794045	2.08316049335067	$a_1{}^1$	8.40911494744558
B_1	0.962309787332703	1.62121248482762	$a_2{}^1$	172.024353054372
B_2	0.979723951565027	1.27590144219536	$a_0{}^2$	1535.72815330126
-	-	-	$a_1{}^2$	−29.7774593495827
-	-	-	$a_2{}^2$	−158.884574516486

5. Fuzzy Optimization Model: Beam Implementation

In accordance with the exhaustive enumeration (EE) problem formulation, an EE optimization model for the optimization of a simply-supported beam (BEAM) was developed. Since the model BEAM is proposed to be used in the design of steel elements, all decisive design constraints were involved in the model. The model enables optimization of the system for various spans, loads and different material properties. For mathematical modeling and data input/output, a high level language, MATLAB [16], was used. The proposed optimization model includes input data, variables and the BEAM system's weight objective function, which is subjected to the structure's crisp and soft constraints; see Appendix A.

Choosing an I-beam from the list also identifies all of the design variables. This then becomes a single variable problem, the variable being the particular item from the list of beams.

The input data (constants) represent various design data for the optimization, i.e., constants/coefficients, which are involved in the objective function and (in) equality constraints. The design data are comprised of the span length L (m), the characteristic dead g_k (kN/m) and imposed q_k (kN/m) loads, the yield strength of the steel f_y (MPa), the modulus of elasticity E (MPa), the density of the steel *gam* (kg/m^3), bearings width *ss* (mm) and the allowable deflection of the beam *lim* (-). In addition to this, the data also include the coefficients involved in the design inequality constraints: safety factor for dead loads *SFg* (-), safety factor for imposed loads *SFq* (-), partial factor for resistance of cross-sections *SFm0*, partial factor for resistance of members to instability *SFm1*, modification factor k (-), non-dimensional slenderness *lamflim* (-) and the reduction factor for the relevant buckling curve *ksiflim* (-).

The objective variable f defines the weight of the steel beam. The aim of optimization is to find a steel beam with the minimum weight that satisfies all of the design constraints.

Design constraints that enable the ULS and SLS are satisfied by the following conditions:

- Condition 1, resistance of the cross-section to bending (ULS): verified by Equation (28), by which the design bending moment M_{Ed} (kNm) must not exceed the bending moment resistance M_{Rd} (kNm).
- Condition 2, resistance of the cross-section to shear (ULS): verified by Equation (33), by which the design shear force V_{Ed} (kN) must not exceed the shear resistance V_{Rd} (kNm).
- Condition 3, deflection (SLS) is considered: the calculated vertical deflection of the steel beam must be less than specified by the ANFIS-LIMIT model.
- Condition 4, resistance to flange-induced buckling (ULS): to prevent the possibility of the compression flange buckling in the plane of the web.
- Condition 5, Condition 6, Condition 7 and Condition 8, resistance of the web to transverse forces (ULS): to prevent the possibility of the local buckling of webs.

This is a single-variable problem, the variable being the particular item from the list of beams. In this case, there are no necessary geometrical constraints or side constraints. A complete enumeration can be performed on the selected beams from the stock list (see Appendix B), and the best beam can easily be identified.

In this example, the beam was used as a horizontal member. The objective was to design a minimum mass beam that would not fail, according to recommendations, under bending, shear and specified deflection. The length L of the beam was specified as 25 m. The steel beam was loaded by uniformly-distributed loading g_k = 5 kN/m and q_k = 15 kN/m, as shown below. Steel was chosen from the material of the beam. The modulus of elasticity of the steel was E = 210 GPa. The weight of the steel was ρ = 7850 kg/m^3. The yield strength in tension was f_y = 335 MPa. The specified applied load LL was 3 kN/m^2 and used a classification of 1. A safety factor of 1.35 on the permanent load and 1.50 on the variable load were assumed. The results of an exhaustive enumeration computer code showed that the optimal section is HE 1000 × 393, with a weight of 9816.4 kg.

6. Conclusions

The article presents how recommendations can be implemented in a discrete optimization model. Good engineering judgment should be integrated into construction design; therefore, a mathematical model was developed based on engineering judgment and past experience. For this purpose, the theory of fuzzy sets was used. The advantages of the proposed fuzzy algorithm ANFIS are acknowledged and incorporated into the model based on the imprecision and fuzziness in the code-based design constraints. The comparison of different design codes showed that the deflection limits are very different and too liberal. The expert evaluations for the deflection limit are subjective; therefore, the data obtained from experts are expected to be vague, imprecise, incomplete or even contradictory. Additionally, the proposed ANFIS techniques include fuzzy clustering (FCM), which searches for patterns in data points [22]. The study also suggests that deflection limits could be reconsidered in the future by the experts who have a prolonged or intense experience through practice [23]. The advantages of using the exhaustive enumeration are reduced optimum weight values and obtaining a solution that is at a global optimum. The proposed computational model was used to find minimum weight solutions for simply-supported laterally-restrained beams. For selected design variables, an optimum steel section was found based on steel sections found on the market, according to the European beams tables. The computational model is based on advanced computational technologies, including fuzzy logic, neural networks and discrete optimization. It was developed to solve real-world problems that are of great interest to design engineers.

Author Contributions: The manuscript was written by both authors. The Assistant Professor Dr. Primož Jelušič developed the computer code. The Associate Professor Dr. Bojan Žlender developed the constraint function for limit deflection.

Conflicts of Interest: The authors declare no conflicts of interest.

Appendix A

Computer code for discrete optimization of fully laterally restrained beams with soft constraint ANFIS-LIMIT.

```
% Optimization of Fully Laterally Restrained Beams with soft constrain
% Simply supported steel beam
%------------------------------------------------
%%%%%%%%%%%%%%%%%%%%%%%%%%%%%%%%%%%%%%%%%%%%%%%%%%%%
% Discrete Optimization with soft constrain
% Dr. P. Jelusic
% See Text for Problem description
% The Beam Properties are loaded from the file
% BeamPropertiesEU.m
%****************************************


%%%
clear
clear global
clc
close
format compact
warning off

%%% Run File %%%%%%%%%
BeamPropertiesEU
%%%%%%%%%%%%%%%%%%%%%%%%
%%% monitor cpu time
starttime = cputime;

fprintf('\n*******************************')
fprintf('\nSimply supported steel beam (Enumeration)')
fprintf
```

```
% %******************************
% % Computer Code
% %*******************************

%Span, safety factors and loads
L = 25;                         % span (m)
SFg = 1.35;                     % safety factor for dead load (-)
SFq = 1.50;                     % safety factor for imposed load (-)
gk = 5;                         % dead load(kN/m)
qk = 15;                        % imposed load(kN/m)
ss = 100;                       % bearings width (mm)
eta = 1;                        % shear factor eta (-)
k = 0.3;                        % factor k (-)
lamflim = 0.5;                  % factor lamflim (-)
ksiflim = 1;                    % factor ksiflim (-)
CLASS = 1;                      % use classification (-)
LL = 3;                         % Applied live load (kN/m^2)

%ANFIS model coefficients
sigA1 = 6.61768494044089;
sigA2 = 7.47150990794045;
sigB1 = 0.962309787332703;
sigB2 = 0.979723951565027;
cA1 = 2.90117735987428;
cA2 = 2.08316049335067;
cB1 = 1.62121248482762;
cB2 = 1.27590144219536;
a01 = -771.211045670957;
a11 = 8.40911494744558;
a21 = 172.024353054372;
a02 = 1535.72815330126;
a12 = -29.7774593495827;
a22 = -158.884574516486;

%Material properties
fy = 355;                       % yield strength (MPa)
E = 210000;                     % modulus of elasticity (MPa)
SFmo = 1.00;                    % safety factor for material bending (-)
SFm1 = 1.05;                    % safety factor for elastic resistance deflection (-)
gam = 7850;                     % density (kg/m^3)

%%%Start Exhaustive Enumeration :
fstar = inf;
xstar = [inf inf inf inf];
gstar = [inf inf inf];
istar = 1;
% %
%
fprintf('\n---------------------------')
fprintf('\nFeasible Beams')
fprintf('\n----------------------------\n\n')
for i = 1:length(RolledSteelBeamSI)
```

```
    x1 = RolledSteelBeamSI(i).D;
    x2 = RolledSteelBeamSI(i).B;
    x3 = RolledSteelBeamSI(i).tw;
    x4 = RolledSteelBeamSI(i).tf;

    %*****************************

    A = RolledSteelBeamSI(i).Area;
    D = RolledSteelBeamSI(i).D;
    B = RolledSteelBeamSI(i).B;
    tw = RolledSteelBeamSI(i).tw;
    tf = RolledSteelBeamSI(i).tf;
    Rr = RolledSteelBeamSI(i).Rr;
    dd = RolledSteelBeamSI(i).dd;
    Ix = RolledSteelBeamSI(i).Ix;
    Welx = RolledSteelBeamSI(i).Welx;
    Wplx = RolledSteelBeamSI(i).Wplx;

%Design action. The reason for discrete optimization is to choose off-the-shelf I-beam which will keep the cost
and production time down. Several mills provide information on standard rolling stock they manufacture.
    Fed = (SFg*(gk+A*gam*9.81/10000000) +
SFq*qk)*L;                                             %design action (kN)
    Med = Fed*L/8;                                     %design bending moment (kNm)
    Ved = Fed/2;                                       %design shear force (kN)

    %Section resistance
    Mrd = Wplx*fy/(SFmo*1000);                         %moment resistance (kNm)
    Av = A*100-2*B*tf + (tw + 2*Rr)*tf;                %shear area(mm^2)
    Vrd = Av*(fy/(3)^0.5)/(SFmo*1000);                 %design shear resistance(kN)

    %Deflection
    Mmax = (gk+qk)*L^2/8;                              %maximum bending moment due to working load
                                                       (kNm)
    Mcrd = Welx*fy/(SFm1*1000);                        %elastic resistance (kNm)
    u = 5*qk*(L*1000)^4/(384*E*Ix*10000);              %deflection (mm)

    %ANFIS calculation procedure
    A1ev = exp(-0.5*(((LL-cA1)/(sigA1))^2));
    A2ev = exp(-0.5*(((LL-cA2)/(sigA2))^2));
    B1ev = exp(-0.5*(((CLASS-cB1)/(sigB1))^2));
    B2ev = exp(-0.5*(((CLASS-cB2)/(sigB2))^2));
    w1 = A1ev*B1ev;
    w2 = A2ev*B2ev;
    w1n = w1/(w1+w2);
    w2n = w2/(w1+w2);
    fun1 = a01+a11*LL+a21*CLASS;
    fun2 = a02+a12*LL+a22*CLASS;
    nfun1 = w1n*fun1;
    nfun2 = w2n*fun2;
    lim = nfun1+nfun2;
    uult = L*1000/lim;                                 %permissible deflection (mm)
```

```
%Section classification
eps = (235/fy)^0.5;                                  %factor eps (-)
c = (B-tw-2*Rr)/2;                                   %depth between fillets (mm)
hw = D-2*tf;                                         %depth between flanges (mm)

%Flange-induced buckling
Aw = (D-2*tf)*tw;                                    %area of the web (mm2)
Afc = B*tf;                                          %area of the compression flange (mm2)
Fib = hw/tw;                                         %criteria ratio of flange-induced buckling(-)
Fibalw = k*(E/fy)*(Aw/Afc)^0.5;                      %criteria ratio (-)

%Web buckling
kf = 2+6*(ss/hw);                                    %buckling coefficient (-)
kfalw = 6;                                           %limit of buckling coefficient(-)
Fcr = (0.9*kf*E*tw^3)/hw;                            %elastic critical buckling load(N)
m1 = fy*B/(fy*tw);                                   %coefficient m1(-)
m2 = 0.02*(hw/tf)^2;                                 %coefficient m2(-)
le = min(kf*E*tw^2/(2*fy*hw),ss);                    %effective loaded length(mm)
ly = min(le+tf*(m1/2+(le/tf)^2+m2)^0.5,le +
tf*(m1+m2)^0.5);                                     %(mm)
lamf = (ly*tw*fy/Fcr)^0.5;                           %reduction factor lamf (-)
lamflim = 0.5;                                       %permissible reduction factor lamflim(-)
ksif = 0.5/lamf;                                     %reduction factor ksif(-)
leff = ksif*ly;                                      %effective length of web(mm)
Frdweb = fy*leff*tw/1000;                            %design resistance of web(kN)

%Objective function
f = gam*L*A/10000;                                   %weight of steel beam (kg)

%Constraints
g1 = Med - Mrd;                                      %bending (kNm)
g2 = Ved - Vrd;                                      %shear (kN)
g3 = u - uult;                                       %deflection(mm)
g4 = Fib - Fibalw;                                   %flange-induced buckling (-)
g5 = kf - kfalw;                                     %web buckling constraint 1 (-)
g6 = lamflim - lamf;                                 %web buckling constraint 2 (-)
g7 = ksif - 1;                                       %web buckling constraint 3(-)
g8 = Ved - Frdweb;                                   %resistance of web constraint (kN)


%%% total constraint vector
G = [g1 g2 g3 g4 g5 g6 g7 g8];

if (g1 <= 0) & (g2 <= 0) & (g3 <= 0)
    if (g4 <= 0) & (g5 <= 0) & (g6 <= 0)
        if (g7 <= 0) & (g8 <= 0)
            if (f <= fstar)
            xstar = [x1 x2 x3 x4];
            fstar = f
            Gstar = G;
            istar = i
```

```
                    end
                end
            end
        end
end

fprintf('\n****************************************')
fprintf('\nOptimum Fully Laterally Restrained Beam')
fprintf('\n****************************************\n\n')
fprintf('Rolled Beam Designation : '),disp(RolledSteelBeamSI(istar).Name)
fprintf('Depth(mm) Width(mm) Web Thickness(mm) Flange Thickness (mm)\n')
fprintf('%8.5f %8.5f %8.5f %8.3f\n',xstar)
fprintf('\nObjective Function(kg): '),disp(fstar)
fprintf('\nConstraints\n')
fprintf('---------------\n')
fprintf('Bending Stress Constraint - g1 (kNm): '),disp(Gstar(1))
fprintf('Shear Stress Constraint - g2 (kN): '),disp(Gstar(2))
fprintf('Deflection Constraint - g3 (mm): '),disp(Gstar(3))
fprintf('flange-induced buckling - g4 (-): '),disp(Gstar(4))
fprintf('web buckling constraint 1 - g5 (-): '),disp(Gstar(5))
fprintf('web buckling constraint 2 - g6 (-): '),disp(Gstar(6))
fprintf('web buckling constraint 3 - g7 (-): '),disp(Gstar(7))
fprintf('resistance of web constraint - g8 (kN): '),disp(Gstar(8))

%%% print time
totaltime = cputime - starttime;
fprintf('\n\nTotal cpu time (s)= %7.4f \n\n',totaltime)
```

The companion file for the problem of fully-laterally-restrained beams is a file that contains beam properties for standard steel beams.

```
% EE - Exhaustive Enumeration
% For constrained optimization of fully laterally restrained beams
% Dr. P. Jelusic
% University of Maribor, Faculty of Civil Engineering
%%%%%%%%%%%%%%%%%%%%%%%%%%%%%%%%%%%%%%%%%%%%%%%%%%%%%%%%%%%%%%%%%%%%
%%% File UniversalbeamsEU.m
%%%%%%%%%%%%%%%%%%%%%%%%%%%%%%%%%%%%%%%%%%%%%%%%%%%%%%%%%%%%%%%%%%%%
% Discrete Variables
%--------------------------------------------------
% See Text for Problem description
%******************************************
%%% COMPANION FILE FOR PROBLEM Fully laterally restrained beams
%%% This file contains Beam Properties for universal beams
%%% beams in SI Units
%******************************************

%%%%%%%%%%%%%%%%%%%%%%%%%%%%%%%%%%%%%%%%%%%%%%
%%% Define the section properties
%%%%%%%%%%%%%%%%%%%%%%%%%%%%%%%%%%%%%%%%%%%%%%

RolledSteelBeamSI(1).Name ='IPE AA 80';                    %beam identifier (-)
```

```
RolledSteelBeamSI(1).Area = 6.31;                    %area (cm2)
RolledSteelBeamSI(1).D = 78;                         %Depth of section (mm)
RolledSteelBeamSI(1).B = 46;                         %width of section (mm)
RolledSteelBeamSI(1).tw = 3.2;                       %web thickness (mm)
RolledSteelBeamSI(1).tf = 4.2;                       %flange thickness (mm)
RolledSteelBeamSI(1).Rr = 5;                         %root radius (mm)
RolledSteelBeamSI(1).dd = 59.6;                      %depth between fillets(mm)
RolledSteelBeamSI(1).Ix = 64.1;                      %second moment of area Ixx (cm4)
RolledSteelBeamSI(1).Welx = 16.4;                    %elastic modulus Welx (cm3)
RolledSteelBeamSI(1).Wplx = 18.9;                    %plastic modulus Wplx (cm3)

RolledSteelBeamSI(2) = struct('Name','IPE A 80','Area',6.38, ...
    'D',78,'B',46,'tw',3.3,'tf',4.2, ...
    'Rr',5,'dd',59.6,'Ix',64.4, ...
    'Welx',16.5,'Wplx',19);

RolledSteelBeamSI(3) = struct('Name','IPE 80','Area',7.64, ...
    'D',80,'B',46,'tw',3.8,'tf',5.2, ...
    'Rr',5,'dd',59.6,'Ix',80.1, ...
    'Welx',20,'Wplx',23.2);



RolledSteelBeamSI(75) = struct('Name','HE 1000 X 584','Area',743.7, ...
    'D',1056,'B',314,'tw',36,'tf',64, ...
    'Rr',30,'dd',868,'Ix',1246100, ...
    'Welx',23600,'Wplx',28039);

return;
```

The results are given in the following form:

```
****************************************
Optimum Fully Laterally Restrained Beam
****************************************

Rolled Beam Designation : HE 1000 X 393
Depth(mm)          Width(mm)          Web Thickness(mm)     Flange Thickness (mm)
     1016.00000         303.00000          24.40000              43.900

Objective Function(kg):                                     9.8164e+003

Constraints
---------------
Bending Stress Constraint                 - g1 (kNm):           -3.8903e+003
Shear Stress Constraint                   - g2 (kN):            -5.1282e+003
Deflection Constraint                     - g3 (mm):            -12.9339
flange-induced buckling                   - g4 (-):             -193.5248
web buckling constraint 1                 - g5 (-):             -3.3536
web buckling constraint 2                 - g6 (-):             -0.0742
web buckling constraint 3                 - g7 (-):             -0.1293
resistance of web constraint              - g8 (kN):            -1.8169e+003

Total cpu time (s)=                                         0.2184
```

Appendix B

Table A1. Dimensions and properties of steel beams (European beams).

Designation Serial Size	A cm^2	h mm	b mm	t_w mm	t_f mm	r mm	d mm	I_y cm^4	$W_{el.y}$ cm^3	$W_{pl.y}$ cm^3	Designation Serial Size	A cm^2	h mm	b mm	t_w mm	t_f mm	r mm	d mm	I_y cm^4	$W_{el.y}$ cm^3	$W_{pl.y}$ cm^3
IPE AA 80	6.31	78	46	3.2	4.2	5	59.6	64.1	16.4	18.9	IPE O 360	84.1	364	172	9.2	14.7	18	298.6	19,050	1047	1186
IPE A 80	6.38	78	46	3.3	4.2	5	59.6	64.4	16.5	19	IPE A 400	73.1	397	180	7	12	21	331	20,290	1022	1144
IPE 80	7.64	80	46	3.8	5.2	5	59.6	80.1	20	23.2	IPE 400	84.5	400	180	8.6	13.5	21	331	23,130	1160	1307
IPE AA 100	8.56	97.6	55	3.6	4.5	7	74.6	136	27.9	31.9	IPE O 400	96.4	404	182	9.7	15.5	21	331	26,750	1324	1502
IPE A 100	8.8	98	55	3.6	4.7	7	74.6	141	28.8	33	IPE A 450	85.6	447	190	7.6	13.1	21	378.8	29,760	1331	1494
IPE 100	10.3	100	55	4.1	5.7	7	74.6	171	34.2	39.4	IPE 450	98.8	450	190	9.4	14.6	21	378.8	33,740	1500	1702
IPE AA 120	10.7	117	64	3.8	4.8	7	93.4	244	41.7	47.6	IPE O 450	118	456	192	11	17.6	21	378.8	40,920	1795	2046
IPE A 120	11	117.6	64	3.8	5.1	7	93.4	257	43.8	49.9	IPE A 500	101	497	200	8.4	14.5	21	426	42,930	1728	1946
IPE 120	13.2	120	64	4.4	6.3	7	93.4	318	53	60.7	IPE 500	116	500	200	10.2	16	21	426	48,200	1930	2194
IPE AA 140	12.8	136.6	73	3.8	5.2	7	112.2	407	59.7	67.6	IPE O 500	137	506	202	12	19	21	426	57,780	2284	2613
IPE A 140	13.4	137.4	73	3.8	5.6	7	112.2	435	63.3	71.6	IPE A 550	117	547	210	9	15.7	24	467.6	59,980	2193	2475
IPE 140	16.4	140	73	4.7	6.9	7	112.2	541	77.3	88.3	IPE 550	134	550	210	11.1	17.2	24	467.6	67,120	2440	2787
IPE AA 160	15.4	156.4	82	4	5.6	7	131.2	646	82.6	93.3	IPE O 550	156	556	212	12.7	20.2	24	467.6	79,160	2847	3263
IPE A 160	16.2	157	82	4	5.9	9	127.2	689	87.8	99.1	IPE A 600	137	597	220	9.8	17.5	24	514	82,920	2778	3141
IPE 160	20.1	160	82	5	7.4	9	127.2	869	109	124	IPE 600	156	600	220	12	19	24	514	92,080	3070	3512
IPE AA 180	19	176.4	91	4.3	6.2	9	146	1020	116	131	IPE O 600	197	610	224	15	24	24	514	118,300	3879	4471
IPE A 180	19.6	177	91	4.3	6.5	9	146	1063	120	135	IPE 750 × 134	171	750	264	12	15.5	17	685	150,700	4018	4644
IPE 180	23.9	180	91	5.3	8	9	146	1317	146	166	IPE 750 × 147	188	753	265	13.2	17	17	685	166,100	4411	5110
IPE O 180	27.1	182	92	6	9	9	146	1505	165	189	IPE 750 × 173	221	762	267	14.4	21.6	17	685	205,800	5402	6218
IPE AA 200	22.9	196.4	100	4.5	6.7	12	159	1533	156	176	IPE 750 × 196	251	770	268	15.6	25.4	17	685	240,300	6241	7174
IPE A 200	23.5	197	100	4.5	7	12	159	1591	162	182	HE 100 A	21.2	96	100	5	8	12	56	349.2	72.76	83.01
IPE 200	28.5	200	100	5.6	8.5	12	159	1943	194	221	HE 100 B	26	100	100	6	10	12	56	449.5	89.91	104.2
IPE O 200	32	202	102	6.2	9.5	12	159	2211	219	249	HE 120 A	25.3	114	120	5	8	12	74	606.2	106.3	119.5
IPE AA 220	27	216.4	110	4.7	7.4	12	177.6	2219	205	230	HE 120 B	34	120	120	6.5	11	12	74	864.4	144.1	165.2
IPE A 220	28.3	217	110	5	7.7	12	177.6	2317	214	240	HE 140 A	31.4	133	140	5.5	8.5	12	92	1033	155.4	173.5
IPE 220	33.4	220	110	5.9	9.2	12	177.6	2772	252	285	HE 140 B	43	140	140	7	12	12	92	1509	215.6	245.4
IPE O 220	37.4	222	112	6.6	10.2	12	177.6	3134	282	321	HE 300 A	112.5	290	300	8.5	14	27	208	18,260	1260	1383
IPE AA 240	31.7	236.4	120	4.8	8	15	190.4	3154	267	298	HE 300 B	149.1	300	300	11	19	27	208	25,170	1678	1869
IPE A 240	33.3	237	120	5.2	8.3	15	190.4	3290	278	312	HE 300 M	303.1	340	310	21	39	27	208	59,200	3482	4078
IPE 240	39.1	240	120	6.2	9.8	15	190.4	3892	324	367	HE 700 A	260.5	690	300	14.5	27	27	582	215,300	6241	7032
IPE O 240	43.7	242	122	7	10.8	15	190.4	4369	361	410	HE 700 B	306.4	700	300	17	32	27	582	256,900	7340	8327
IPE A 270	39.2	267	135	5.5	8.7	15	219.6	4917	368	413	HE 800 AA	218.5	770	300	14	18	30	674	208,900	5426	6225
IPE 270	45.9	270	135	6.6	10.2	15	219.6	5790	429	484	HE 800 A	285.8	790	300	15	28	30	674	303,400	7682	8699
IPE O 270	53.8	274	136	7.5	12.2	15	219.6	5947	507	575	HE 900 AA	252.2	870	300	15	20	30	770	301,100	6923	7999
IPE A 300	46.5	297	150	6.1	9.2	15	248.6	7173	483	542	HE 900 A	320.5	890	300	16	30	30	770	422,100	9485	10,810

Table A1. *Cont.*

Designation Serial Size	A cm^2	h mm	b mm	t_w mm	t_f mm	r mm	d mm	I_y cm^4	$W_{el.y}$ cm^3	$W_{pl.y}$ cm^3	Designation Serial Size	A cm^2	h mm	b mm	t_w mm	t_f mm	r mm	d mm	I_y cm^4	$W_{el.y}$ cm^3	$W_{pl.y}$ cm^3
IPE 300	53.8	300	150	7.1	10.7	15	248.6	8356	557	628	HE 900 × 466	593.7	938	312	30	54	30	770	814,900	17,380	20,380
IPE O 300	62.8	304	152	8	12.7	15	248.6	9994	658	744	HE 1000 AA	282.2	970	300	16	21	30	868	406,500	8380	9777
IPE A 330	54.7	327	160	6.5	10	18	271	10,230	626	702	HE 1000 A	346.8	990	300	16.5	31	30	868	553,800	11,190	12,820
IPE 330	62.6	330	160	7.5	11.5	18	271	11,770	713	804	HE 1000 × 393	500.2	1016	303	24.4	43.9	30	868	807,700	15,900	18,540
IPE O 330	72.6	334	162	8.5	13.5	18	271	13,910	833	943	HE 1000 × 415	528.7	1020	304	26	46	30	868	853,100	16,728	19,571
IPE A 360	64	357.6	170	6.6	11.5	18	298.6	14,520	812	907	HE 1000 × 438	556	1026	305	26.9	49	30	868	909,200	17,720	20,750
IPE 360	72.7	360	170	8	12.7	18	298.6	16,270	904	1019	HE 1000 × 494	629.1	1036	309	31	54	30	868	1,028,000	19,845	23,413

References

1. Adeli, H.; Sarma, K.C. *Cost Optimization of Structures: Fuzzy Logic, Genetic Algorithms, and Parallel Computing*; John Wiley & Sons: Chichester, UK, 2006.
2. Adeli, H. *Knowledge Engineering: Fundamentals*; McGraw-Hill: New York, NY, USA, 1990.
3. Soh, C.K.; Yang, J. Fuzzy controlled genetic algorithm search for shape optimization. *J. Comput. Civ. Eng.* **1996**, *10*, 143–150. [CrossRef]
4. Smith, A.E.; Tate, D.M. Genetic optimization using a penalty function. In Proceedings of the Fifth International Conference on Genetic Algorithms, Champaign, IL, USA, 17–21 July 1993; pp. 499–505.
5. Kim, J.H.; Myung, H. Evolutionary programming techniques for constrained optimization problems. *IEEE Trans. Evolut. Comput.* **1997**, *1*, 129–140.
6. Adeli, H.; Cheng, N.T. Augmented Lagrangian genetic algorithm for structural optimization. *J. Aerosp. Eng.* **1994**, *7*, 104–118. [CrossRef]
7. Adeli, H.; Park, H.S. Optimization of space structures by neural dynamics model. *Neural Netw.* **1995**, *8*, 769–781. [CrossRef]
8. Adeli, H.; Park, H.S. *Neurocomputing for Design Automation*; CRC Press: Boca Raton, FL, USA, 1998.
9. Wang, G.; Wang, W. Fuzzy optimum design of aseismic structures. *Earthq. Eng. Struct. Dyn.* **1985**, *13*, 827–837.
10. Rao, S.S. Description and optimum design of fuzzy mechanical systems. *J. Mech. Transm. Autom. Des.* **1987**, *109*, 126–132. [CrossRef]
11. Yeh, Y.; Hsu, D. Structural optimization with fuzzy parameters. *Comput. Struct.* **1990**, *37*, 917–924.
12. Jelusic, P. Soil compaction optimization with soft constrain. *J. Intell. Fuzzy Syst.* **2015**, *29*, 955–962. [CrossRef]
13. Bellman, R.E.; Zadeh, L.A. Decision-making in a fuzzy environment. *Manag. Sci.* **1970**, *17*, B141–B164. [CrossRef]
14. Zimmermann, H.J. Fuzzy programming and linear programming with several objective functions. *Fuzzy Sets Syst.* **1978**, *1*, 45–55. [CrossRef]
15. Jang, J.S.R. ANFIS: Adaptive-network-based fuzzy inference system. *IEEE Trans. Syst. Man Cybern.* **1993**, *23*, 665–685.
16. MathWorks. *MATLAB Function Reference*; The MathWorks, Inc.: Natick, MA, USA, 2010.
17. European Committee for Standardization. *Eurocode 3: Design of Steel Structures*; EN 1993; CEN: Brussels, Belgium, 2002.
18. European Committee for Standardization. *Eurocode: Basis of Structural Design*; EN 1990; CEN: Brussels, Belgium, 2002.
19. British Standard. *Structural Use of Steelwork in Building—Part 1: Code of Practice for Design—Rolled and Welded Sections*; BS 5950-1; BSI: London, UK, 2000.
20. American Institute of Timber Construction. *Timber Construction Manual*, 5th ed.; John Wiley and Sons: Hoboken, NJ, USA, 2005.
21. Sugeno, M. *Industrial Applications of Fuzzy Control*; Elsevier Science Pub. Co: New York, NY, USA, 1985.
22. Jelušič, P.; Žlender, B. Soil-nail wall stability analysis using ANFIS. *Acta Geotech. Slov.* **2013**, *10*, 61–73.
23. Jelušič, P.; Žlender, B. An adaptive network fuzzy inference system approach for site investigation. *Geotech. Test. J.* **2014**, *37*, 400–411. [CrossRef]

symmetry

MDPI

Article

Possibility/Necessity-Based Probabilistic Expectation Models for Linear Programming Problems with Discrete Fuzzy Random Variables

Hideki Katagiri [1,*], Kosuke Kato [2] and Takeshi Uno [3]

1 Department of Industrial Engineering, Faculty of Engineering, Kanagawa University, 3-27-1 Rokkakubashi, Yokohama-shi, Kanagawa 221-8686, Japan
2 Department of Computer Science, Hiroshima Institute of Technology, 2-1-1 Miyake, Saeki-ku, Hiroshima 731-5193, Japan; k.katoh.me@it-hiroshima.ac.jp
3 Department of Mathematical Science, Graduate School of Technology, Industrial and Social Science, Tokushima University, 2-1, Minamijosanjima-cho, Tokushima-shi, Tokushima 770-8506, Japan; uno.takeshi@tokushima-u.ac.jp
* Correspondence: katagiri@kanagawa.ac.jp; Tel.: +81-45-481-5661

Received: 7 September 2017; Accepted: 6 October 2017; Published: 30 October 2017

Abstract: This paper considers linear programming problems (LPPs) where the objective functions involve discrete fuzzy random variables (fuzzy set-valued discrete random variables). New decision making models, which are useful in fuzzy stochastic environments, are proposed based on both possibility theory and probability theory. In multi-objective cases, Pareto optimal solutions of the proposed models are newly defined. Computational algorithms for obtaining the Pareto optimal solutions of the proposed models are provided. It is shown that problems involving discrete fuzzy random variables can be transformed into deterministic nonlinear mathematical programming problems which can be solved through a conventional mathematical programming solver under practically reasonable assumptions. A numerical example of agriculture production problems is given to demonstrate the applicability of the proposed models to real-world problems in fuzzy stochastic environments.

Keywords: discrete fuzzy random variable; linear programming; possibility measure; necessity measure; expectation model; Pareto optimal solution

---

1. Introduction

One of the traditional tools for taking into consideration uncertainty of parameters involved in mathematical programming problems is stochastic programming [1,2]. Stochastic programming approaches implicitly assume that uncertain parameters involved in problems can be expressed as random variables. For example, demanding amounts of products are often mathematically modeled as random variables. In this case, realized values of random parameters under event occurrence are assumed to be represented with deterministic values such as real values.

On the other hand, random variables are not always suitable to estimate parameters of problems, when human judgments and/or knowledge have to be mathematically handled. It is worth utilizing not only historical or past data but also experts' knowledge or judgments involving ambiguity or vagueness which are often represented as fuzzy sets.

Simultaneous consideration of fuzziness and randomness is highly important in modeling decision making problems, because decision making by humans in stochastic environments is intrinsically based not only on randomness but also on fuzziness. In the last decade, mathematical models which take into consideration both fuzziness and randomness have considerably drawn attentions in the research field of decision making such as linear programming [3–12], integer programming [13], inventory [14,15], transportation [16], facility layout [17], flood management [18] and network optimization [19,20].

In this paper, we focus on mathematical optimization models in fuzzy stochastic decision making situations where possible realized values of random parameters in linear programming problems (LPPs) are ambiguously estimated by experts as fuzzy sets or fuzzy numbers. Such fuzzy set-valued random variables, namely, random parameters whose realized values are represented with fuzzy sets, can be expressed as fuzzy random variables [21–26].

Previous studies on fuzzy random LPPs have mainly focused on the case where the coefficients of the objective function and the constraints are expressed by continuous fuzzy random variables, which is an extended concept of continuous random variables. Fuzzy random optimization models were firstly developed by Luhandjula and his colleagues [27,28] as LPPs with fuzzy random variable coefficients, and further studied by Liu [29,30], Katagiri et al. [4,6] and Yano [11] and so on. A brief survey of major fuzzy stochastic programming models including mathematical programming models using fuzzy random variables was found in the paper by Luhandjula [8].

On the other hand, there are a few studies [13,31,32] on LPPs with discrete fuzzy random variables. As will be discussed later in more details, it is quite important to propose more general fuzzy random LPP models in order to widen the range of application of fuzzy stochastic programming, which motivates this article to provide new generalized mathematical programming models with discrete fuzzy random variables.

This paper is organized as follows: In Section 2, the definitions of fuzzy random variables are introduced. Some types of fuzzy random variables are newly defined. Section 3 focuses on discrete fuzzy random variables and defines some types of discrete fuzzy random variables. Section 4 formulates a single/multiple objective LPP where the coefficients of the objective function(s) are discrete fuzzy random variables. In Section 5, we construct new optimization criteria for optimization problems with discrete fuzzy random variables, which are based both on possibility theory and on probability theory. Section 6 proposes decision making models using optimization criteria introduced in Section 5, and defines (weak) Pareto optimal solutions of the proposed models in the multi-objective case. Section 7 discusses how the proposed model can be solved and construct an algorithm for obtaining a Pareto optimal solution of the proposed models. In Section 8, we execute a numerical experiment with an example of agriculture production problems in order to demonstrate the applicability of the proposed model to real-world decision making problems. It is shown that the R [33] language can be used to solve the problems with hundreds of decision variables in a practical computational time. Finally, Section 9 summarizes this paper and discusses future research works.

2. Preliminaries

In this section, we review some mathematical concepts related to discrete fuzzy random variables, such as convex fuzzy sets and fuzzy numbers. Definitions of fuzzy random variables are also provided.

2.1. Fuzzy Set and Fuzzy Number

As a preparation for the introduction of fuzzy random variables, we firstly introduce the definition of fuzzy sets.

Definition 1. *(Normal convex fuzzy set)*
A normal convex fuzzy set is characterized by a membership function $\mu_{\tilde{A}} : \mathbb{R} \to [0,1]$, *that is,* $\mu_{\tilde{A}}(x) \in [0,1]$, *for all* $x \in \mathbb{R}$, *such that* A_α *is a nonempty compact interval*

$$A_\alpha = \begin{cases} \{x \in \mathbb{R} \mid \mu_{\tilde{A}}(x) \geq \alpha\} & \text{if } \alpha \in (0,1] \\ cl(supp\ \mu_{\tilde{A}}) & \text{if } \alpha = 0, \end{cases}$$

where $cl(supp\ \mu_{\tilde{A}})$ *denotes the closure of set* $supp\ \mu_{\tilde{A}}$*, and* $supp\ \mu_{\tilde{A}}$ *denotes a support of membership function* $\mu_{\tilde{A}}$.

An *L-R* fuzzy number was introduced by Dubois and Prade [34] and is defined based on a normal convex fuzzy set.

Definition 2. *(L-R fuzzy number)*
A normal convex fuzzy set $\tilde{F}$ *is said to be an L-R fuzzy number, denoted by* $(d, \beta, \gamma)_{LR}$*, if its membership function* $\mu_{\tilde{F}}$ *is defined as follows:*

$$\mu_{\tilde{F}}(\tau) = \begin{cases} L\left(\dfrac{d-\tau}{\beta}\right) & \text{if } \tau \leq d \\ R\left(\dfrac{\tau-d}{\gamma}\right) & \text{if } \tau > d, \end{cases} \quad (1)$$

where L and R are reference functions satisfying the following conditions:

1. $L(t)$ *and* $R(t)$ *are nonincreasing for any* $t > 0$.
2. $L(0) = R(0) = 1$.
3. $L(t) = L(-t)$ *and* $R(t) = R(-t)$ *for any* $t \in \mathbb{R}$.
4. *There exists a* $t_0^L > 0$ *such that* $L(t) = 0$ *holds for any t larger than* t_0^L*. Similarly, there exists a* $t_0^R > 0$ *such that* $R(t) = 0$ *holds for any t larger than* t_0^R.

Fuzzy numbers are regarded as extended concepts of real numbers because $\tilde{F}$ is reduced to a real number d if $\beta = \gamma = 0$ in Definition 2. Fuzzy numbers are a useful tool for representing human knowledge and/or estimation. Figure 1 shows a typical membership function of an *L-R* fuzzy number.

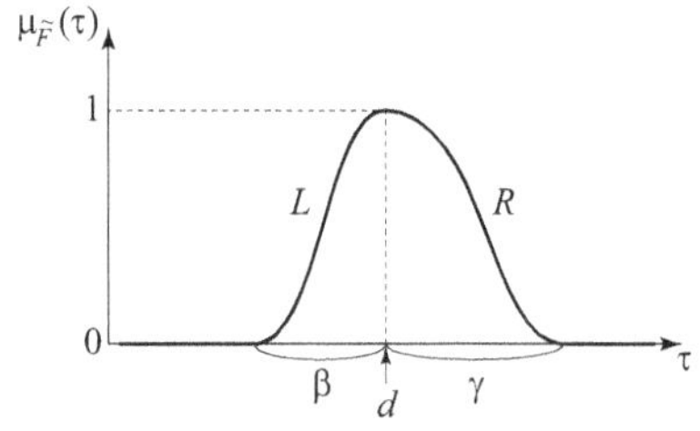

Figure 1. *L-R* fuzzy number.

In Definition 2, if $L = R$ and $\beta = \gamma$, we call such an *L-R* fuzzy number an *L* fuzzy number. In other words, *L* fuzzy numbers are symmetric fuzzy numbers defined as follows:

Definition 3. *(L fuzzy number)*
A normal convex fuzzy set $\tilde{F}$ *is said to be an L fuzzy number if its membership function* $\mu_{\tilde{F}}$ *is defined as follows:*

$$\mu_{\tilde{F}}(\tau) = L\left(\frac{d-\tau}{\beta}\right), \quad (2)$$

where L is a reference function satisfying the following conditions:

1. $L(t)$ *is nonincreasing for any* $t > 0$.
2. $L(0) = 1$.
3. $L(t) = L(-t)$ *for any* $t \in \mathbb{R}$.
4. *There exists a* $t_0^L > 0$ *such that* $L(t) = 0$ *holds for any t larger than* t_0^L.

Figure 2 shows a typical membership function of an L fuzzy number.

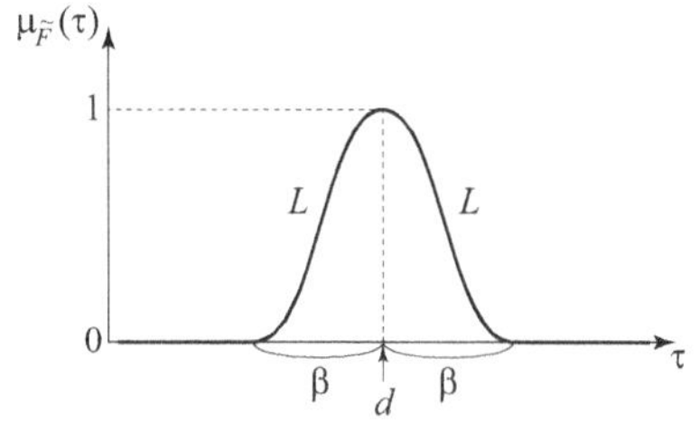

Figure 2. L fuzzy number.

In Definition 2, if $L(t) = R(t) = \max\{0, 1 - |t|\}$, we call such an L-R fuzzy number a triangular fuzzy number.

Definition 4. *(Triangular fuzzy number)*
An L-R fuzzy number $\tilde{F}$ *is said to be a triangular fuzzy number, denoted by* $(d, \beta, \gamma)_{tri}$, *if the reference functions L and R of an L-R fuzzy number are given as* $L(t) = R(t) = \max(1 - |t|, 0)$. *In other words, a triangular fuzzy number* $\tilde{F}$ *is characterized by the following piece-wise linear membership function:*

$$\mu_{\tilde{F}}(\tau) = \begin{cases} \max\left\{1 - \dfrac{|d-\tau|}{\beta}, 0\right\} & \text{if } \tau \leq d \\ \max\left\{1 - \dfrac{|\tau-d|}{\gamma}, 0\right\} & \text{if } \tau > d. \end{cases} \quad (3)$$

Figure 3 shows a typical membership function of a triangular fuzzy number.

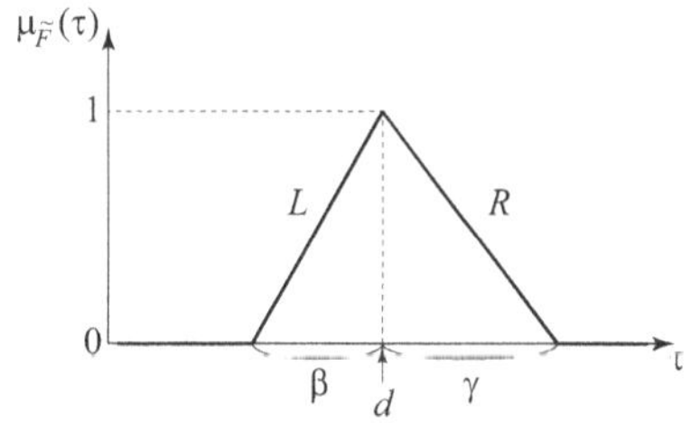

Figure 3. Triangular fuzzy number.

2.2. Fuzzy Random Variable

In this section, we review and define some important concepts underlying fuzzy random programming problems. There are mainly two definitions of fuzzy random variables. A fuzzy random variable was firstly defined by Kwakernaak [24] as an extended concept of random variables in the sense that the realized values for given events or scenarios are not real but fuzzy numbers. Kruse and Meyer [35] provided some concepts similar to the model by Kwakernaak. Puri and Ralescu [25] defined fuzzy random variables as random fuzzy sets and developed a mathematical basis of fuzzy random variables with Klemment [23]. Overviews of fuzzy random variables have been provided by Gil et al. [22] and Shapiro [26].

We introduce a general definition of fuzzy random variables, which is based on the works of Kwakernaak [24], Kruse and Meyer [35] and Gil et al. [21]:

Definition 5. *(Fuzzy random variable)*
Let $(\Omega, \mathcal{F}, P)$ be a probability space and $F(\mathbb{R})$ denote the set of all fuzzy numbers in $\mathbb{R}$, where $F(\mathbb{R})$ denotes a class of normal convex fuzzy subsets of $\mathbb{R}$ having compact α level set for $\alpha \in [0,1]$. A fuzzy random variable is a mapping $\tilde{\bar{A}} : \Omega \to F(\mathbb{R})$ such that for any $\alpha \in [0,1]$ and all $\omega \in \Omega$, the real-valued mapping

$$\inf \tilde{\bar{A}}_\alpha : \Omega \to \mathbb{R}, \text{ satisfying } \inf \tilde{A}_\alpha(\omega) = \inf(\tilde{A}(\omega))_\alpha$$

and

$$\sup \tilde{\bar{A}}_\alpha : \Omega \to \mathbb{R}, \text{ satisfying } \sup \tilde{A}_\alpha(\omega) = \sup(\tilde{A}(\omega))_\alpha$$

are real-valued random variables, that is, Borel measurable real-valued functions. $(\tilde{A}(\omega))_\alpha$ is a nonempty compact interval defined by

$$(\tilde{A}(\omega))_\alpha = \begin{cases} \{x \in \mathbb{R} | \ \mu_{\tilde{A}(\omega)}(x) \geq \alpha\} & \text{if } \alpha \in (0,1] \\ cl(supp\ \mu_{\tilde{A}(\omega)}) & \text{if } \alpha = 0, \end{cases}$$

where $\mu_{\tilde{A}(\omega)}$ is the membership function of a fuzzy set $\tilde{A}(\omega)$, $cl(supp\ \mu_{\tilde{A}(\omega)})$ denotes the closure of set $supp\ \mu_{\tilde{A}(\omega)}$, and $supp\ \mu_{\tilde{A}(\omega)}$ denotes a support of function $\mu_{\tilde{A}(\omega)}$.

2.3. Special Types of Fuzzy Random Variables Used in Decision Making

For the purpose of applying fuzzy random variables to decision making problems, Katagiri et al. [4,6,20,36,37] introduced some special types of fuzzy random variables where the realized values of random variables for given events or scenarios are *L-R* fuzzy numbers or triangular fuzzy numbers. Since these fuzzy random variables are useful for modeling various decision making problems, we categorize these fuzzy random variables into several types such as *L-R* fuzzy random variable, *L* fuzzy random variable and triangular fuzzy random variable, together with their examples which were originally introduced in the previous papers.

First, we define *L-R* fuzzy random variable as follows:

Definition 6. *(L-R fuzzy random variable)*
Let $\bar{d}$, $\bar{\beta}$ and $\bar{\gamma}$ be random variables whose realization for a given event $\omega \in \Omega$ are $d(\omega)$, $\beta(\omega)$ and $\gamma(\omega)$, respectively, where Ω is a sample space, and $\beta(\omega)$ and $\gamma(\omega)$ are positive constants for any $\omega \in \Omega$. Then, a fuzzy random variable $\tilde{\bar{F}}$ is said to be an L-R fuzzy random variable, denoted by $(\bar{d}, \bar{\beta}, \bar{\gamma})_{LR}$, if its realized values $\tilde{F}(\omega) = (d(\omega), \beta(\omega), \gamma(\omega))_{LR}$ for any event $\omega \in \Omega$ are L-R fuzzy numbers defined as

$$\mu_{\tilde{F}(\omega)}(\tau) = \begin{cases} L\left(\dfrac{d(\omega) - \tau}{\beta(\omega)}\right) & \text{if } \tau \leq d(\omega) \\ R\left(\dfrac{\tau - d(\omega)}{\gamma(\omega)}\right) & \text{if } \tau > d(\omega). \end{cases} \tag{4}$$

L-R fuzzy random variables were introduced to decision making problems such as a portfolio selection problem [38], an LPP [36] and a multi-objective programming problem [4]. In these studies, the coefficients of objective functions are represented as a special type of *L-R* fuzzy random variables in which the spread parameters β and γ are constants, not random variables, as shown in the following example:

Example 1. *In Definition 6, let $\bar{g}$ be a Gaussian (normal) random variable $N(m, \sigma^2)$ where m is the mean and σ is the standard deviation. Also, let $\bar{\beta}$ and $\bar{\gamma}$ be positive constants, not random variables. Then, $\tilde{\bar{F}}$ is a kind of L-R fuzzy random variables if the membership function of the realization of $\tilde{\bar{F}}$ is defined as*

$$\mu_{\tilde{F}(\omega)}(\tau) = \begin{cases} L\left(\dfrac{g(\omega)-\tau}{\beta}\right) & \text{if } \tau \le g(\omega) \\ R\left(\dfrac{\tau - g(\omega)}{\gamma}\right) & \text{if } \tau > g(\omega), \end{cases} \tag{5}$$

where $g(\omega)$ is a realized value of $\bar{g}$ for a given event $\omega \in \Omega$, and Ω is a sample space.

Another example of *L-R* fuzzy random variables is shown in the study on a multi-objective LPP [6] as follows:

Example 2. *Let $\bar{d}$, $\bar{\beta}$ and $\bar{\gamma}$ be random variables expressed as*

$$\bar{d} = d_1 \cdot \bar{t} + d_2,\ \bar{\beta} = \beta_1 \cdot \bar{t} + \beta_2,\ \bar{\gamma} = \gamma_1 \bar{t} + \gamma_2,$$

where $\bar{t}$ is a random variable whose mean and variance are m and σ^2, respectively, and $d_1, d_2, \beta_1, \beta_2, \gamma_1$ and γ_2 are constant values. Then, $\tilde{\bar{A}}$ is a kind of L-R fuzzy random variables if the membership function of the realization of $\tilde{\bar{A}}$ is defined as

$$\mu_{\tilde{A}(\omega)}(\tau) = \begin{cases} L\left(\dfrac{d_1 \cdot t(\omega) + d_2 - \tau}{\beta_1 \cdot t(\omega) + \beta_2}\right) & \text{if } \tau \le d_1 \cdot t(\omega) + d_2 \\ R\left(\dfrac{\tau - d_1 \cdot t(\omega) + d_2}{\gamma_1 \cdot t(\omega) + \gamma_2}\right) & \text{if } \tau > d_1 \cdot t(\omega) + d_2, \end{cases} \tag{6}$$

where $t(\omega)$ is a realized value of $\bar{t}$ for a given event $\omega \in \Omega$, and Ω is a sample space.

When the reference functions of left-hand and right-hand sides are the same in Definition 6, namely, if it holds $L = R$, we call such an *L-R* fuzzy random variable an *L* fuzzy random variable defined as follows:

Definition 7. *(L fuzzy random variable)*
Let $\bar{d}$ and $\bar{\beta}$ be random variables whose realization for a given event $\omega \in \Omega$ are $d(\omega)$ and $\beta(\omega)$, respectively, where Ω is a sample space, and $\beta(\omega)$ is a positive constant for any $\omega \in \Omega$. Then, a fuzzy random variable $\tilde{\bar{F}}$ is said to be an L fuzzy random variable if its realized values for any event $\omega \in \Omega$ are L-R fuzzy numbers defined as

$$\mu_{\tilde{F}(\omega)}(\tau) = L\left(\frac{d(\omega)-\tau}{\beta(\omega)}\right). \tag{7}$$

L fuzzy random variables were introduced in network optimization problems such as bottleneck minimum spanning tree problems [20,39]. In these studies, the cost for constructing each edge in an optimal network construction problem was expressed as an *L* fuzzy random variable shown in the following example:

Example 3. *In Definition 7, let $\bar{g}$ be a Gaussian (normal) random variable $N(m, \sigma^2)$ where m is the mean and σ is the standard deviation. Also, let β be a positive constant, not a random variable. Then, $\tilde{\bar{F}}$ is a kind of L fuzzy random variables if the membership function of the realization of $\tilde{\bar{F}}$ is defined as*

$$\mu_{\tilde{F}(\omega)}(\tau) = L\left(\frac{g(\omega)-\tau}{\beta}\right), \tag{8}$$

where $g(\omega)$ *is a realized value of* $\bar{g}$ *for a given event* $\omega \in \Omega$, *and* Ω *is a sample space.*

The L fuzzy random variable shown in Example 3 can be interpreted as a "hybrid number." The hybrid number, which was originally introduced by Kaufman and Gupta [40], is composed of a series of fuzzy numbers, and is obtained by shifting fuzzy numbers in a random way along the abscissa.

Especially if $L(t) = R(t) = \max\{0, 1 - |t|\}$ in Definition 6, we call such an L-R fuzzy random variable a triangular fuzzy random variable.

Definition 8. *(Triangular fuzzy random variable)*
An L-R fuzzy random variable $\tilde{\bar{F}}$ *is said to be a triangular fuzzy random variable, denoted by* $(\bar{d}, \bar{\beta}, \bar{\gamma})_{tri}$, *if the realization* $\tilde{F}(\omega)$ *for each* $\omega_k \in \Omega$ *is represented by a triangular fuzzy number* $(d(\omega), \beta(\omega), \gamma(\omega))_{tri}$, *where* Ω *is a sample space. In other words, a discrete triangular fuzzy random variable* $\tilde{\bar{F}}$ *is a discrete fuzzy random variable whose realization for each event* ω *is a triangular fuzzy number characterized by the following membership function:*

$$\mu_{\tilde{F}(\omega)}(\tau) = \begin{cases} \max\left\{1 - \dfrac{|d(\omega) - \tau|}{\beta(\omega)}, 0\right\} & \text{if } \tau \le d(\omega) \\ \max\left\{1 - \dfrac{|\tau - d(\omega)|}{\gamma(\omega)}, 0\right\} & \text{if } \tau > d(\omega). \end{cases} \tag{9}$$

Triangular fuzzy random variables were introduced in the study on a multi-objective LPP [37]. In this study, spread parameters $\bar{\beta}$ and $\bar{\gamma}$ are not random variables but constant values as shown in the following example:

Example 4. *In Definition 8, let* $\bar{\beta}$ *and* $\bar{\gamma}$ *be positive constants, not random variables. Then,* $\tilde{\bar{F}}$ *is a kind of triangular fuzzy random variables if the membership function of the realization of* $\tilde{\bar{F}}$ *is defined as*

$$\mu_{\tilde{F}(\omega)}(\tau) = \begin{cases} \max\left\{1 - \dfrac{|d(\omega) - \tau|}{\beta}, 0\right\} & \text{if } \tau \le d(\omega) \\ \max\left\{1 - \dfrac{|\tau - d(\omega)|}{\gamma}, 0\right\} & \text{if } \tau > d(\omega), \end{cases} \tag{10}$$

where $d(\omega)$ *is a realized value of* $\bar{d}$ *for a given event* $\omega \in \Omega$, *and* Ω *is a sample space.*

3. Discrete Fuzzy Random Variable

In this section, we discuss discrete fuzzy random variable as for a preparation for proposing a new framework of LPPs with discrete fuzzy random variables.

Firstly, we review the definition of discrete fuzzy random variable given by Kawakernaak [41]. Secondly, we provide the definition of discrete L-R fuzzy random variable and that of discrete triangular fuzzy random variable which was applied to a network optimization problem [31], an LPP [32] and a multi-objective 0-1 programming problem [13].

Definitions of Discrete Fuzzy Random Variables

In the 1970s, Kwakernaak [41] originally proposed a concept of discrete fuzzy random variable. In this paper, we provide the definition of discrete fuzzy random variable as follows:

Definition 9. *(Discrete fuzzy random variable)*
Let Ω *be a set of events such that the occurrence probability of each event* $\omega_k \in \Omega$ *is* p_k *and that* $\sum_k p_k = 1$. *Let* $\tilde{F}_k$ *be a fuzzy set characterized by a membership function* $\mu_{\tilde{F}_k}$, *and let* $\mathcal{F}$ *be a set of* F_k, $\forall k \in K$, *where* K *is*

an index set of k. Let $\tilde{\tilde{F}}$ *be a mapping from* Ω *to* $\mathcal{F}$ *such that* $\tilde{\tilde{F}}(\omega_k) \triangleq \tilde{F}_k$*. Then, a mapping* $\tilde{\tilde{F}}$ *is said to be a discrete fuzzy random variable.*

Considering the applicability of the discrete fuzzy random variables in real-world decision making, we define discrete *L-R* fuzzy random variables as a special type of discrete fuzzy random variables.

Definition 10. *(Discrete L-R fuzzy random variable)*
A discrete fuzzy random variable $\tilde{\tilde{F}}$ *is said to be a discrete L-R fuzzy random variable, denoted by* $(\bar{d}, \bar{\beta}, \bar{\gamma})_{LR}$*, if the realization of* $\tilde{\tilde{F}} = (\bar{d}, \bar{\beta}, \bar{\gamma})_{LR}$ *for any event* $\omega_k \in \Omega$ *is an L-R fuzzy number* $\tilde{F}_k = (d_k, \beta_k, \gamma_k)_{LR}$*, where* d_k*,* β_k *and* γ_k *are the realized values of* $\bar{d}$*,* $\bar{\beta}$*, and* $\bar{\gamma}$ *for a given event* $\omega_k \in \Omega$*, respectively, and* Ω *is a sample space. Then,* $\tilde{\tilde{F}}$ *is an L-R fuzzy random variable in which the membership function of the realization* $\tilde{F}_k$ *for each event* $\omega_k \in \Omega$ *is defined as*

$$\mu_{\tilde{F}_k}(\tau) = \begin{cases} L\left(\dfrac{d_k - \tau}{\beta_k}\right) & \text{if } \tau \leq d_k \\ R\left(\dfrac{\tau - d_k}{\gamma_k}\right) & \text{if } \tau > d_k. \end{cases} \tag{11}$$

The following is an example of discrete *L-R* fuzzy random variables:

Example 5. *Consider* β_k *and* γ_k *vary dependent on events or scenarios. Then,* $\tilde{\tilde{F}}$ *is a discrete L-R fuzzy random variable in which the membership functions of the realized fuzzy numbers of* $\tilde{F}_k$*,* $k = 1, 2, 3$ *are defined as follows:*

$$\mu_{\tilde{F}_1}(\tau) = \begin{cases} L\left(\dfrac{300 - \tau}{35}\right) & \text{if } \tau \leq 300 \\ R\left(\dfrac{\tau - 300}{20}\right) & \text{if } \tau > 300, \end{cases}$$

$$\mu_{\tilde{F}_2}(\tau) = \begin{cases} L\left(\dfrac{200 - \tau}{25}\right) & \text{if } \tau \leq 200 \\ R\left(\dfrac{\tau - 200}{10}\right) & \text{if } \tau > 200, \end{cases} \tag{12}$$

$$\mu_{\tilde{F}_3}(\tau) = \begin{cases} L\left(\dfrac{100 - \tau}{30}\right) & \text{if } \tau \leq 100 \\ R\left(\dfrac{\tau - 100}{15}\right) & \text{if } \tau > 100. \end{cases}$$

Figure 4 shows a typical membership function of a discrete *L-R* fuzzy random variable.

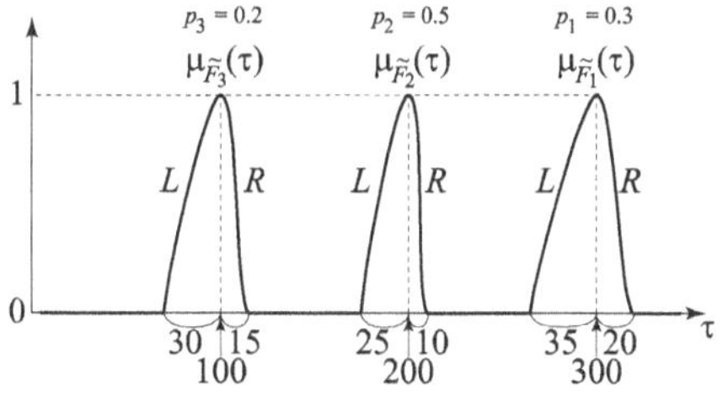

Figure 4. Discrete *L-R* fuzzy random variable.

In particular, if $L(t) = R(t) = \max\{0, 1 - |t|\}$ in Definition 10, we call such a discrete L-R fuzzy random variable a discrete triangular fuzzy random variable.

Definition 11. *(Discrete triangular fuzzy random variable)*
A discrete L-R fuzzy random variable $\tilde{\bar{F}}$ *is said to be a discrete triangular fuzzy random variable, denoted by* $(\bar{d}, \bar{\beta}, \bar{\gamma})_{tri}$*, if the realization* $\tilde{F}_k$ *for each* $\omega_k \in \Omega$ *is represented by a triangular fuzzy number* $(d_k, \beta_k, \gamma_k)_{tri}$*, where* Ω *is a sample space. In other words, a discrete triangular fuzzy random variable* $\tilde{\bar{F}}$ *is a discrete fuzzy random variable whose realization for each event* ω_k *is a triangular fuzzy number characterized by the following membership function:*

$$\mu_{\tilde{F}_k}(\tau) = \begin{cases} \max\left\{1 - \dfrac{|d_k - \tau|}{\beta_k}, 0\right\} & \text{if } \tau \leq d_k \\ \max\left\{1 - \dfrac{|\tau - d_k|}{\gamma_k}, 0\right\} & \text{if } \tau > d_k. \end{cases} \tag{13}$$

Example 6. *When it holds that* $L(t) = R(t) = \max\{0, 1 - |t|\}$ *in Example 5,* $\tilde{\bar{F}}$ *is a discrete triangular fuzzy random variable whose realized values* $\tilde{F}_k$*,* $k = 1, 2, 3$ *are characterized by the following membership function:*

$$\mu_{\tilde{F}_1}(\tau) = \begin{cases} \max\left\{1 - \dfrac{|300 - \tau|}{35}, 0\right\} & \text{if } \tau \leq 300 \\ \max\left\{1 - \dfrac{|\tau - 300|}{20}, 0\right\} & \text{if } \tau > 300, \end{cases}$$

$$\mu_{\tilde{F}_2}(\tau) = \begin{cases} \max\left\{1 - \dfrac{|200 - \tau|}{25}, 0\right\} & \text{if } \tau \leq 200 \\ \max\left\{1 - \dfrac{|\tau - 200|}{10}, 0\right\} & \text{if } \tau > 200, \end{cases} \tag{14}$$

$$\mu_{\tilde{F}_3}(\tau) = \begin{cases} \max\left\{1 - \dfrac{|100 - \tau|}{30}, 0\right\} & \text{if } \tau \leq 100 \\ \max\left\{1 - \dfrac{|\tau - 100|}{15}, 0\right\} & \text{if } \tau > 100. \end{cases}$$

Figure 5 shows a typical membership function of a discrete triangular fuzzy random variable.

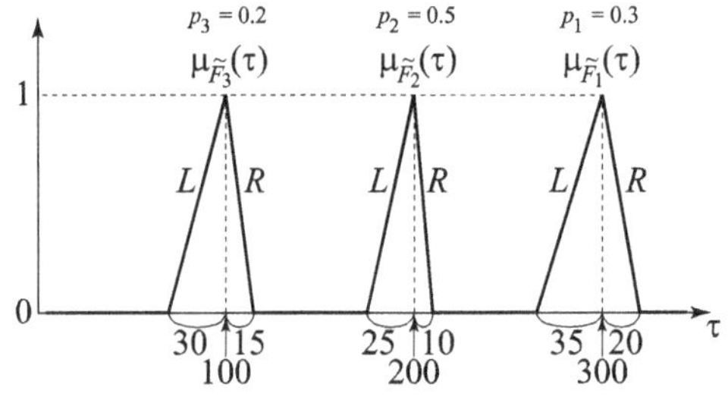

Figure 5. Discrete triangular fuzzy random variable.

Discrete triangular fuzzy random variables were firstly introduced in some of previous studies on a network optimization problem [31], an LPP [32] and a multi-objective 0-1 programming problem [13]. In these studies, the spread parameters β_k and γ_k do not vary with events ω_k but they are fixed as constants for any events. To the author's best knowledge, there has been no study on linear programming model where the spread parameters β_k and γ_k of discrete fuzzy random variables vary with different events $\omega_k \in \Omega$.

In the next section we shall propose new linear programming models with discrete fuzzy random variables in which spread parameters vary with stochastic events.

4. Problem Formulation

Assuming that the coefficients of the objective functions are given as discrete fuzzy random variables, we consider the following fuzzy random programming problem:

$$\left.\begin{array}{ll}\text{minimize} & \tilde{\bar{C}}_l \boldsymbol{x},\ l = 1, 2, \ldots, q \\ \text{subject to} & A\boldsymbol{x} \leq \boldsymbol{b},\ \boldsymbol{x} \geq \boldsymbol{0},\end{array}\right\} \tag{15}$$

where $\tilde{\bar{\boldsymbol{C}}}_l = (\tilde{\bar{C}}_{l1}, \ldots, \tilde{\bar{C}}_{ln})$, $l = 1, 2, \ldots, q$ are n dimensional coefficient row vectors whose elements are discrete fuzzy random variables, $\boldsymbol{x}$ is an n dimensional decision variable column vector, A is an $m \times n$ coefficient matrix, and $\boldsymbol{b}$ is an m dimensional column vector. When the number of objective functions is equal to 1 ($q = 1$), then problem (15) becomes a single-objective fuzzy random programming problem; otherwise, when $q \geq 2$, (15) is a multi-objective fuzzy random programming problem. In problem (15), all the objective functions are to be minimized. Without loss of generality, this paper considers minimization problems, because any maximization problems can be transformed into minimization problems by multiplying the original objective function in the maximization problem by -1.

4.1. Model Using Discrete L-R Fuzzy Random Variables

In problem (15), we firstly consider the case where each element $\tilde{\bar{C}}_{lj}$ of the coefficient vectors $\tilde{\bar{\boldsymbol{C}}}_l = (\tilde{\bar{C}}_{l1}, \ldots, \tilde{\bar{C}}_{ln})$, $l = 1, 2, \ldots, q$ in (15) is a discrete L-R fuzzy random variable $(\bar{d}_{lj}, \bar{\beta}_{lj}, \bar{\gamma}_{lj})_{LR}$ whose realization for a given event $\omega_{lk} \in \Omega_l$ is an L-R fuzzy number $\tilde{C}_{ljk} \triangleq (d_{ljk}, \beta_{ljk}, \gamma_{ljk})_{LR}$, $l = 1, 2, \ldots, q$, $j = 1, 2, \ldots, n$, $k = 1, 2, \ldots, r_l$ with the membership function defined as

$$\mu_{\tilde{C}_{ljk}}(\tau) = \begin{cases} L\left(\dfrac{d_{ljk} - \tau}{\beta_{ljk}}\right) & \text{if } \tau \leq d_{ljk} \\ R\left(\dfrac{\tau - d_{ljk}}{\gamma_{ljk}}\right) & \text{if } \tau > d_{ljk}, \end{cases} \tag{16}$$

where $\Omega_l \triangleq \{\omega_{l1}, \omega_{l2}, \ldots, \omega_{lr_l}\}$ denotes a set of events related to the lth objective function. In (16), the values of d_{ljk}, β_{ljk} and γ_{ljk} are constant, and β_{ljk} and γ_{ljk} are positive. The probability that each event ω_{lk} occurs is given as p_{lk}, where $\sum_{k=1}^{r_l} p_{lk} = 1, \forall l \in \{1, 2, \ldots, q\}$. Figure 6 shows that a typical membership function of an L-R fuzzy number defined by (16).

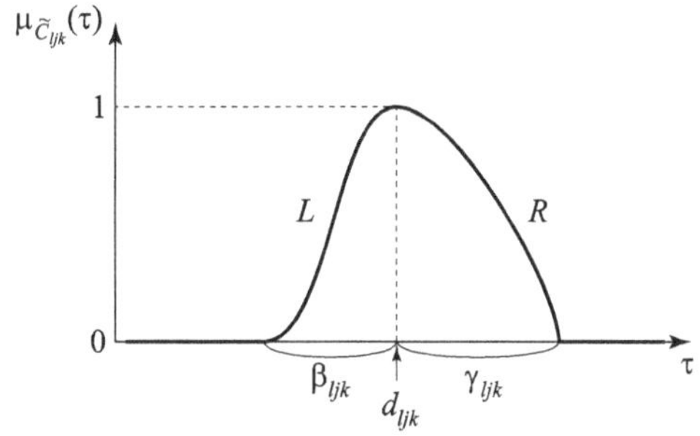

Figure 6. Realized values $\tilde{C}_{ljk}$ for the kth event of a discrete L-R fuzzy random variable $\tilde{\bar{C}}_{lj}$.

Through the extended sum of fuzzy numbers [42] based on the Zadeh's extension principle [43], the objective function $\tilde{\bar{C}}_l x$ is represented by a single fuzzy random variable whose realized value for an event or scenario ω_{lk} is an *L*-*R* fuzzy number $\tilde{C}_{lk}x = (\boldsymbol{d}_{lk}\boldsymbol{x}, \boldsymbol{\beta}_{lk}\boldsymbol{x}, \boldsymbol{\gamma}_{lk}\boldsymbol{x})_{LR}$ characterized by the membership function

$$\mu_{\tilde{C}_{lk}\boldsymbol{x}}(v) = \begin{cases} L\left(\dfrac{\boldsymbol{d}_{lk}\boldsymbol{x} - v}{\boldsymbol{\beta}_{lk}\boldsymbol{x}}\right) & \text{if } v \leq \boldsymbol{d}_{lk}\boldsymbol{x} \\ R\left(\dfrac{v - \boldsymbol{d}_{lk}\boldsymbol{x}}{\boldsymbol{\gamma}_{lk}\boldsymbol{x}}\right) & \text{if } v > \boldsymbol{d}_{lk}\boldsymbol{x}, \end{cases} \tag{17}$$

where $\boldsymbol{d}_{lk}$, $\boldsymbol{\beta}_{lk}$ and $\boldsymbol{\gamma}_{lk}$ are n dimensional column vectors whose values vary dependent on events $\omega_{lk} \in \Omega_l$, $l \in \{1, 2, \ldots, q\}$. Figure 7 shows that the membership function of an *L*-*R* fuzzy number defined by (17).

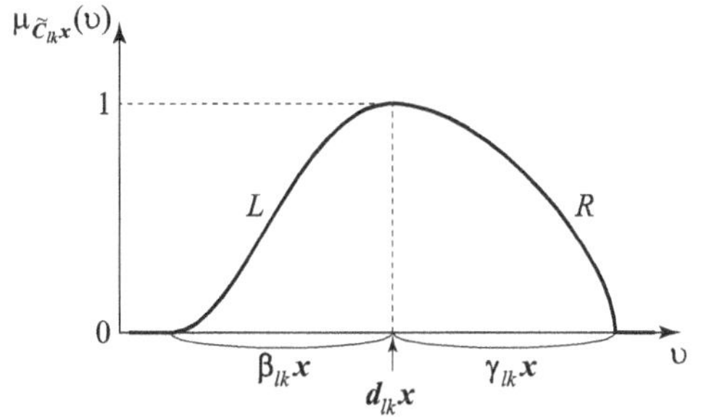

Figure 7. Realized values $\tilde{C}_{ljk}$ for the *k*th event of a discrete *L*-*R* fuzzy random variable $\tilde{\bar{C}}_{lj}$.

4.2. Model Using Discrete Triangular Fuzzy Random Variables

As a special type of discrete *L*-*R* fuzzy random variable defined in (17), we also consider the case where $\tilde{\bar{C}}_{lj}$ is a discrete triangular fuzzy random variable in which its realized values for events or scenarios are triangular fuzzy numbers $\tilde{C}_{ljk} = (d_{ljk}, \beta_{ljk}, \gamma_{ljk})_{tri}$ for $\omega_{kl} \in \Omega_l$, $l = 1, 2, \ldots, q$, $j = 1, 2, \ldots, n$, $k = 1, 2, \ldots, r_l$ with the following membership function:

$$\mu_{\tilde{C}_{ljk}}(\tau) = \begin{cases} \max\left\{1 - \dfrac{|d_{ljk} - \tau|}{\beta_{ljk}}, 0\right\} & \text{if } \tau \leq d_{ljk} \\ \max\left\{1 - \dfrac{|\tau - d_{ljk}|}{\gamma_{ljk}}, 0\right\} & \text{if } \tau > d_{ljk}. \end{cases} \tag{18}$$

Then, through the Zadeh's extension principle, the realized value of each objective function $\tilde{\bar{C}}_l x$ for a given event ω_{lk} is represented by a single triangular fuzzy number $(\boldsymbol{d}_{lk}\boldsymbol{x}, \boldsymbol{\beta}_{lk}\boldsymbol{x}, \boldsymbol{\gamma}_{lk}\boldsymbol{x})_{tri}$ which is characterized by

$$\mu_{\tilde{C}_{lk}\boldsymbol{x}}(v) = \begin{cases} \max\left\{1 - \dfrac{|\boldsymbol{d}_{lk}\boldsymbol{x} - v|}{\boldsymbol{\beta}_{lk}\boldsymbol{x}}, 0\right\} & \text{if } v \leq \boldsymbol{d}_{lk}\boldsymbol{x} \\ \max\left\{1 - \dfrac{|v - \boldsymbol{d}_{lk}\boldsymbol{x}|}{\boldsymbol{\gamma}_{lk}\boldsymbol{x}}, 0\right\} & \text{if } v > \boldsymbol{d}_{lk}\boldsymbol{x}. \end{cases} \tag{19}$$

Figures 8 and 9 show the membership functions of $\tilde{C}_{ljk}$ and $\tilde{C}_{lk}\boldsymbol{x}$.

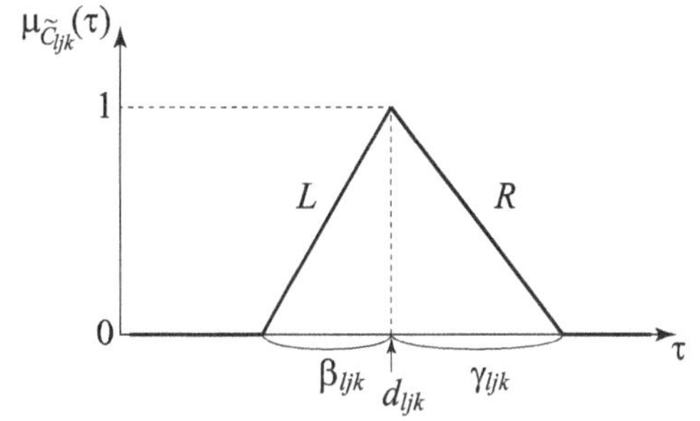

Figure 8. Realized value $\tilde{C}_{ljk}$ for the kth event of a discrete triangular fuzzy random variable $\tilde{\bar{C}}_{lj}$.

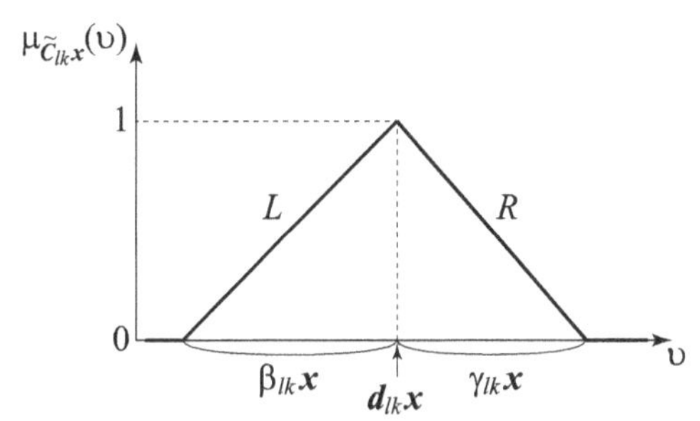

Figure 9. Realized value $\tilde{C}_{lk}x$ for the kth event of a discrete triangular fuzzy random variable $\tilde{\bar{C}}_l x$.

5. Possibility/Necessity-Based Probabilistic Expectation

This section is devoted to discussing optimization criteria to solve problem (15) with discrete fuzzy random variables whose realized values are given as L-R fuzzy numbers defined by (17) or triangular fuzzy numbers defined by (19).

It should be noted here that problem (15) is not a well-defined mathematical programming problem because, even when a decision vector x is determined, the objective function value $\tilde{C}_{lk}x$ is not determined as a constant due to both randomness and fuzziness of $\tilde{C}_{lk}$. In other words, a certain optimization criterion is needed to compare the value of fuzzy random objective function.

In this section, we propose some useful optimization criteria based on both possibility and probability measures, called a possibility/necessity-based probabilistic expectation.

5.1. Preliminary: Possibility and Necessity Measures

As a preparation for optimization criteria in fuzzy stochastic decision making environments, we review the definition of possibility and necessity measures, and discuss how the measures are applied to our problems with discrete fuzzy random variables.

5.1.1. Possibility Measure

Considering that membership functions of fuzzy sets can be regarded as possibilistic distributions of possibilistic variables [44], a definition of possibility measure is given [34,44] as follows:

Definition 12. *(Possibility measure)*
Let $\tilde{A}$ and $\tilde{B}$ be fuzzy sets characterized by membership functions $\mu_{\tilde{A}}$ and $\mu_{\tilde{B}}$, respectively. Then, under a possibilistic distribution of $\mu_{\tilde{A}}$ of a possibilistic variable α, possibility measure of the event that α is in a fuzzy set $\tilde{B}$ is defined as follows:

$$\Pi_{\tilde{A}}(\tilde{B}) \stackrel{\triangle}{=} \sup_{v} \min\left(\mu_{\tilde{A}}(v),\ \mu_{\tilde{B}}(v)\right). \quad (20)$$

In decision making situations where the objective function is to be minimized, decision makers (DMs) often have a fuzzy goal such as "the objective function value $\tilde{C}_{lk}\boldsymbol{x}$ is substantially less than or equal to a certain value f_l," which is expressed by $\tilde{\bar{C}}_{lk}\boldsymbol{x} \lesssim f_l$, where $\lesssim$ denotes "substantially less than or equal to" defined in (12). Let $\mu_{\tilde{G}_l}$ be a membership function of fuzzy set $\tilde{G}_l$ such that the degree of y being substantially less than or equal to a certain value f_l is represented with $\mu_{\tilde{G}_l}(y)$.

Assume that a certain event ω_{lk} has occurred, on the basis of possibility theory and notations (20). Then, the degree of possibility that $\tilde{C}_{lk}\boldsymbol{x}$ satisfies fuzzy goal $\tilde{G}$ (namely, the degree of possibility that the objective function value $\tilde{\bar{C}}_{lk}\boldsymbol{x}$ for any event $\omega_{lk} \in \Omega_l$ is substantially less than or equal to a certain aspiration level f_l) is defined as

$$\begin{aligned} \Pi\left(\tilde{C}_{lk}\boldsymbol{x} \lesssim f_l\right) &\triangleq \Pi_{\tilde{C}_{lk}\boldsymbol{x}}(\tilde{G}_l) \\ &= \sup_y \min\left\{\mu_{\tilde{C}_{lk}\boldsymbol{x}}(y),\ \mu_{\tilde{G}_l}(y)\right\},\ l = 1,2,\ldots,q,\ k = 1,2,\ldots,r_l. \end{aligned} \tag{21}$$

Figure 10 illustrates the degree of possibility defined by (21) for a fixed event ω_{lk}, which is the ordinate of the crossing point between the membership functions of fuzzy goal $\tilde{G}_l$ and the objective function $\tilde{C}_{lk}\boldsymbol{x}$.

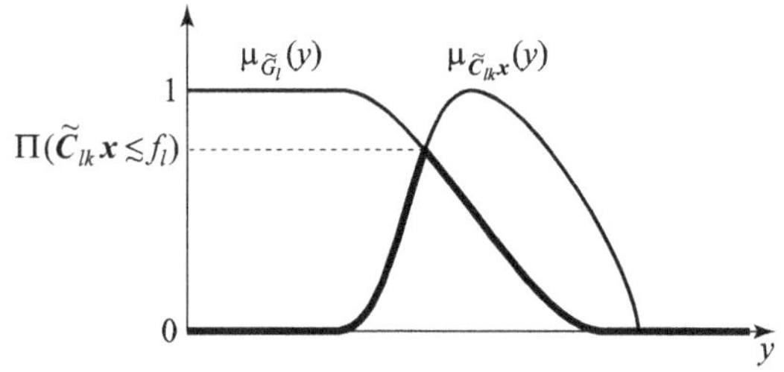

Figure 10. Degree of possibility $\Pi\left(\tilde{C}_{lk}\boldsymbol{x} \lesssim f_l\right)$.

5.1.2. Necessity Measure

For DMs who make decisions from pessimistic view points, a necessity measure is recommended. The necessity measure defined by Zadeh [44] and Dubois and Prade [34] is as follows:

Definition 13. *(Necessity measure)*
Let $\tilde{A}$ and $\tilde{B}$ be fuzzy sets characterized by membership functions $\mu_{\tilde{A}}$ and $\mu_{\tilde{B}}$, respectively. Then, under a possibilistic distribution of $\mu_{\tilde{A}}$ of a possibilistic variable α, the necessity measure of the event that α is in a fuzzy set $\tilde{B}$ is defined as follows:

$$N_{\tilde{A}}(\tilde{B}) \triangleq \inf_v \max\left(1 - \mu_{\tilde{A}}(v),\ \mu_{\tilde{B}}(v)\right). \tag{22}$$

Then, in view of (22), the degree of necessity that the objective function value $\tilde{\bar{C}}_{lk}\boldsymbol{x}$ for any event $\omega_{lk} \in \Omega_l$ satisfies the fuzzy goal $\tilde{G}_l$ is defined as

$$\begin{aligned} N\left(\tilde{C}_{lk}\boldsymbol{x} \lesssim f_l\right) &\triangleq N_{\tilde{C}_{lk}\boldsymbol{x}}(\tilde{G}_l) \\ &= \inf_y \max\left\{1 - \mu_{\tilde{C}_{lk}\boldsymbol{x}}(y),\ \mu_{\tilde{G}_l}(y)\right\},\ l = 1,2,\ldots,q. \end{aligned} \tag{23}$$

Figure 11 illustrates the degree of necessity defined by (23), which is the ordinate of the crossing point between the membership functions of fuzzy goal $\tilde{G}_l$ and the upside-down of the membership function of the objective function $\tilde{C}_{lk}\boldsymbol{x}$.

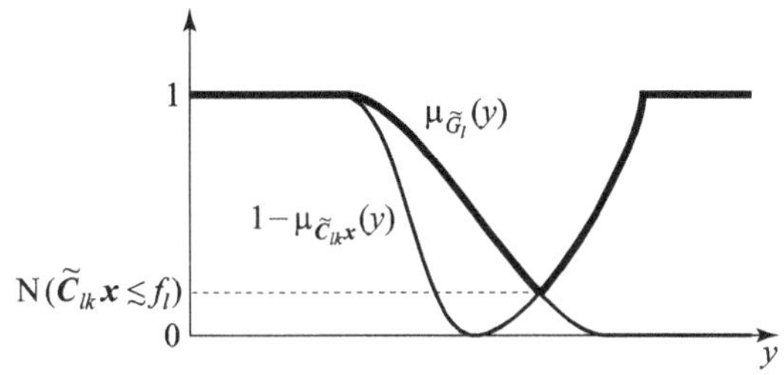

Figure 11. Degree of necessity $N\left(\tilde{C}_{lk}x \lesssim f_l\right)$.

Possibilistic programming [45,46] is one of the most promising tools for handling mathematical optimization problems with ambiguous parameters.

5.2. Optimization Criteria in Fuzzy Random Environments

Possibilistic programming approaches cannot directly be applied to solving problems with discrete fuzzy random variables. This is because the degrees of possibility or necessity defined in (21) or (23) are not constants but vary dependent on events ω_{lk}.

In this section, taking into consideration both fuzziness and randomness involved in the coefficients of the problems, we newly propose some useful optimization criteria for problems with discrete fuzzy random variables. As novel optimization criteria, we provide possibility-based probabilistic expectation (PPE) and necessity-based probabilistic expectation (NPE) as follows:

Definition 14. *(Possibility-based probabilistic expectation (PPE))*
Let $\tilde{\bar{C}}_l = (\tilde{\bar{C}}_{l1}, \ldots, \tilde{\bar{C}}_{ln})$, $l = 1, 2, \ldots, q$ *be n dimensional coefficient row vectors of fuzzy random variables in (multi-objective) LPP (15). Suppose that the realized value of* $\tilde{\bar{C}}_{lj}$ *is a fuzzy set (or fuzzy number as a special case)* $\tilde{C}_{ljk}$. *Let* $\Pi\left(\tilde{C}_{lk}x \lesssim f_l\right)$ *be the degree of possibility for a fixed event* ω_{lk} *defined in (21). By using* p_{lk} *which is the probability that an event or scenario* w_{lk} *occurs, the optimization criterion called a possibility-based probabilistic expectation (PPE) is defined and calculated as follows:*

$$
\begin{aligned}
E\left[\Pi\left(\tilde{\bar{C}}_l x \lesssim f_l\right)\right] &\triangleq \sum_{k=1}^{r_l} p_{lk} \cdot \Pi\left(\tilde{C}_{lk}x \lesssim f_l\right) \\
&= \sum_{k=1}^{r_l} p_{lk} \cdot \Pi_{\tilde{C}_{lk}x}(\tilde{G}_l) \\
&= \sum_{k=1}^{r_l} p_{lk} \cdot \sup_y \min\left\{\mu_{\tilde{C}_{lk}x}(y),\ \mu_{\tilde{G}_l}(y)\right\},\ l = 1, 2, \ldots, q,
\end{aligned} \tag{24}
$$

where $E[\cdot]$ *denotes a probabilistic expectation.*

Possibility measures are recommended to optimistic DMs. On the other hand, since DMs are not always optimistic in general, we introduce the following new optimization criterion based on necessity measures in order to construct an optimization criterion for pessimistic DMs:

Definition 15. *(Necessity-based probabilistic expectation (NPE))*
Let $\tilde{\bar{C}}_l = (\tilde{\bar{C}}_{l1}, \ldots, \tilde{\bar{C}}_{ln})$, $l = 1, 2, \ldots, q$ *be n dimensional coefficient row vectors of fuzzy random variables in (multi-objective) LPP (15). Suppose that the realized value of* $\tilde{\bar{C}}_{lj}$ *is a fuzzy set (or fuzzy number as a special*

case) $\tilde{\bar{C}}_{ljk}$. *Let* $N\left(\tilde{C}_{lk}\boldsymbol{x} \lesssim f_l\right)$ *be the degree of necessity for a fixed event* ω_{lk} *defined in (21). Then, the following optimization criterion is said to be necessity-based probabilistic expectation (NPE):*

$$\begin{aligned} E\left[N\left(\tilde{\bar{C}}_l\boldsymbol{x} \lesssim f_l\right)\right] &\triangleq \sum_{k=1}^{r_l} p_{lk}\cdot N\left(\tilde{C}_{lk}\boldsymbol{x} \lesssim f_l\right) \\ &= \sum_{k=1}^{r_l} p_{lk}\cdot N_{\tilde{C}_{lk}\boldsymbol{x}}(\tilde{G}_l) \\ &= \sum_{k=1}^{r_l} p_{lk}\cdot \inf_y \max\left\{1-\mu_{\tilde{C}_{lk}\boldsymbol{x}}(y),\, \mu_{\tilde{G}_l}(y)\right\},\ l=1,2,\ldots,q. \end{aligned} \tag{25}$$

6. Discrete Fuzzy Random Linear Programming Models Using Possibility/Necessity-Based Probabilistic Expectation

On the basis of the new optimization criteria defined as (24) or (25) in the previous section, we propose new linear programming-based decision making models in fuzzy stochastic environments.

6.1. Possibility-Based Probabilistic Expectation (PPE) Model

When the DM is optimistic, it is reasonable to use the model based on PPE. Then, we consider the following problem to maximize the probabilistic expectation of the degree of possibility:

[Possibility-based probabilistic expectation model (PPE model)]

$$\left.\begin{array}{ll} \text{maximize} & E\left[\Pi\left(\tilde{\bar{C}}_l\boldsymbol{x} \lesssim f_l\right)\right],\ l=1,2,\ldots,q \\ \text{subject to} & \boldsymbol{x}\in X, \end{array}\right\} \tag{26}$$

where the objective functions of problem (26) are given as (24).

In general, problem (26) is a multi-objective programming problem. Especially in the case of $q=1$, (26) becomes a single-objective programming problem, and the optimal solution is a feasible solution which maximizes the objective function. On the other hand, when $q \neq 1$, the problem to be solved has multiple objective functions, which means there does not generally exist a complete solution that simultaneously maximizes all the objective functions. In such multi-objective cases, one of reasonable solution approaches to (26) is to seek a solution satisfying Pareto optimality, called a Pareto optimal solution. We define Pareto optimal solutions of (26). Firstly, we introduce the concepts of weak Pareto optimal solution as follows:

Definition 16. *(Weak Pareto optimal solution of PPE model)*
$\boldsymbol{x}^* \in X$ *is said to be a weak Pareto optimal solution of the possibility-based probabilistic expectation model if and only if there is no* $\boldsymbol{x}\in X$ *such that*
$E\left[\Pi\left(\tilde{\bar{C}}_l\boldsymbol{x} \lesssim f_l\right)\right] > E\left[\Pi\left(\tilde{\bar{C}}_l\boldsymbol{x}^* \lesssim f_l\right)\right]$ *for all* $l \in \{1,2,\ldots,q\}$.

As a stronger concept than a weak Pareto optimal solution, a (strong) Pareto optimal solution of (26) is defined as follows:

Definition 17. *((Strong) Pareto optimal solution of PPE model)*
$\boldsymbol{x}^* \in X$ *is said to be a (strong) Pareto optimal solution of the possibility-based probabilistic expectation model if and only if there is no* $\boldsymbol{x}\in X$ *such that* $E\left[\Pi\left(\tilde{\bar{C}}_l\boldsymbol{x} \lesssim f_l\right)\right] \geq E\left[\Pi\left(\tilde{\bar{C}}_l\boldsymbol{x}^* \lesssim f_l\right)\right]$ *for all* $l \in \{1,2,\ldots,q\}$, *and that* $E\left[\Pi\left(\tilde{\bar{C}}_l\boldsymbol{x} \lesssim f_l\right)\right] > E\left[\Pi\left(\tilde{\bar{C}}_l\boldsymbol{x}^* \lesssim f_l\right)\right]$ *for at least one* $l \in \{1,2,\ldots,q\}$.

In order to obtain a (weak/strong) Pareto optimal solution of PPE model, we consider the following maximin problem, which is one of scalarization methods for obtaining a (weak/strong) Pareto optimal solution of multi-objective programming problems [47]:

[Maximin problem for PPE model]

$$\left.\begin{array}{ll}\text{maximize} & \min\limits_{l\in\{1,2,\ldots,q\}} E\left[\Pi\left(\tilde{\bar{C}}_l \boldsymbol{x} \lesssim f_l\right)\right] \\ \text{subject to} & \boldsymbol{x} \in X.\end{array}\right\} \tag{27}$$

In the theory of multi-objective optimization, it is known that an optimal solution of the maximin problem assures at least weak Pareto optimality. Then, we show the following proposition:

Proposition 1. *(Weak Pareto optimality of the maximin problem for PPE model)*
Let $\boldsymbol{x}^*$ be an optimal solution of problem (27). Then, $\boldsymbol{x}^*$ is a weak Pareto optimal solution of problem (26), namely, a weak Pareto optimal solution for PPE model.

Proof. Assume that an optimal solution $\boldsymbol{x}^*$ of (27) is not a weak Pareto optimal solution of PPE model defined in Definition 16. Then, there exists a feasible solution $\hat{\boldsymbol{x}} \in X$ of (27) such that $E\left[\Pi\left(\tilde{\bar{C}}_l \hat{\boldsymbol{x}} \lesssim f_l\right)\right] > E\left[\Pi\left(\tilde{\bar{C}}_l \boldsymbol{x}^* \lesssim f_l\right)\right]$ for all $l \in \{1,2,\ldots,q\}$. Then, it follows

$$\min_l E\left[\Pi\left(\tilde{\bar{C}}_l \hat{\boldsymbol{x}} \lesssim f_l\right)\right] > \min_l E\left[\Pi\left(\tilde{\bar{C}}_l \boldsymbol{x}^* \lesssim f_l\right)\right].$$

This contradicts the fact that $\boldsymbol{x}^*$ is an optimal solution of (27). □

Since an optimal solution of (27) is not always a (strong) Pareto optimal solution but only a weak Pareto optimal solution in general, we consider the following augmented maximin problems in order to find a solution satisfying strong Pareto optimality instead of weak Pareto optimality.

[Augmented maximin problem for PPE model]

$$\left.\begin{array}{ll}\text{maximize} & z^{\Pi}(\boldsymbol{x}) \triangleq \min\limits_{l\in\{1,2,\ldots,q\}} E\left[\Pi\left(\tilde{\bar{C}}_l \boldsymbol{x} \lesssim f_l\right)\right] + \rho \sum\limits_{l=1}^{q} E\left[\Pi\left(\tilde{\bar{C}}_l \boldsymbol{x} \lesssim f_l\right)\right] \\ \text{subject to} & \boldsymbol{x} \in X,\end{array}\right\} \tag{28}$$

where ρ is a sufficiently small positive constant, say 10^{-6}.

In the theory of multi-objective optimization [47], it is known that an optimal solution of the augmented maximin problem assures (strong) Pareto optimality. Then, we obtain the following proposition:

Proposition 2. *((Strong) Pareto optimality of augmented maximin problem for PPE model)*
Let $\boldsymbol{x}^*$ be an optimal solution of problem (28). Then, $\boldsymbol{x}^*$ is a (strong) Pareto optimal solution of (26), namely, a (strong) Pareto optimal solution for PPE model.

Proof. Assume that an optimal solution of (28), denoted by $\boldsymbol{x}^*$, is not (strong) Pareto optimal solution of PPE model. Then, there exists $\hat{\boldsymbol{x}}$ such that

$E\left[\Pi\left(\tilde{\tilde{C}}_l x \lesssim f_l\right)\right] \geq E\left[\Pi\left(\tilde{\tilde{C}}_l x^* \lesssim f_l\right)\right]$ for all $l \in \{1,2,\ldots,q\}$, and that $E\left[\Pi\left(\tilde{\tilde{C}}_l x \lesssim f_l\right)\right] > E\left[\Pi\left(\tilde{\tilde{C}}_l x^* \lesssim f_l\right)\right]$ for at least one $l \in \{1,2,\ldots,q\}$. Then, it follows

$$\min_{l\in\{1,2,\ldots,q\}} E\left[\Pi\left(\tilde{\tilde{C}}_l \hat{x} \lesssim f_l\right)\right] \geq \min_{l\in\{1,2,\ldots,q\}} E\left[\Pi\left(\tilde{\tilde{C}}_l x^* \lesssim f_l\right)\right]$$
$$\rho\sum_{l=1}^{q} E\left[\Pi\left(\tilde{\tilde{C}}_l \hat{x} \lesssim f_l\right)\right] > \rho\sum_{l=1}^{q} E\left[\Pi\left(\tilde{\tilde{C}}_l x^* \lesssim f_l\right)\right].$$

Therefore, it holds that

$$\begin{aligned} &\min_{l\in\{1,2,\ldots,q\}} E\left[\Pi\left(\tilde{\tilde{C}}_l \hat{x} \lesssim f_l\right)\right] + \rho\sum_{l=1}^{q} E\left[\Pi\left(\tilde{\tilde{C}}_l \hat{x} \lesssim f_l\right)\right] \\ > &\min_{l\in\{1,2,\ldots,q\}} E\left[\Pi\left(\tilde{\tilde{C}}_l x^* \lesssim f_l\right)\right] + \rho\sum_{l=1}^{q} E\left[\Pi\left(\tilde{\tilde{C}}_l x^* \lesssim f_l\right)\right]. \end{aligned}$$

This contradicts the fact that x^* is an optimal solution of the augmented minimax problem. □

6.2. Necessity-Based Probabilistic Expectation Model (NPE Model)

Unlike the case discussed in the previous section, when the DM is pessimistic, the NPE model is recommended, instead of the PPE model. This section is devoted to addressing how the necessity-based probabilistic expectation (NPE) model based on (25) can be solved in the case of linear membership functions.

Using the necessity-based probabilistic mean defined in (25), we consider another new decision making model called NPE model and formulate the mathematical programming problem as follows:

[Necessity-based probabilistic expectation model (NPE model)]

$$\left.\begin{array}{ll} \text{maximize} & E\left[N\left(\tilde{\tilde{C}}_l x \lesssim f_l\right)\right],\ l = 1,2,\ldots,q \\ \text{subject to} & x \in X. \end{array}\right\} \tag{29}$$

When $q = 1$, (29) is a single-objective problem. Otherwise, namely, when $q \geq 2$, (29) is a multi-objective problem in which a solution satisfying (strong) Pareto optimality, called a (strong) Pareto optimal solution, is considered to be a reasonable optimal solution. We define (strong) Pareto optimal solutions of (29). The concept of weak Pareto optimal solution for NPE model is defined as follows:

Definition 18. *(Weak Pareto optimal solution of NPE model)*
$x^* \in X$ is said to be a weak Pareto optimal solution of the necessity-based probabilistic expectation model if and only if there is no $x \in X$ such that
$E\left[N\left(\tilde{\tilde{C}}_l x \lesssim f_l\right)\right] > E\left[N\left(\tilde{\tilde{C}}_l x^* \lesssim f_l\right)\right]$ *for all $l \in \{1,2,\ldots,q\}$.*

As a stronger concept than weak Pareto optimal solutions, (strong) Pareto optimal solutions of (29) is defined as follows:

Definition 19. *((Strong) Pareto optimal solution of NPE model)*
$x^* \in X$ is said to be a (strong) Pareto optimal solution of the necessity-based probabilistic expectation model if and only if there is no $x \in X$ such that $E\left[N\left(\tilde{\tilde{C}}_l x \lesssim f_l\right)\right] \geq E\left[N\left(\tilde{\tilde{C}}_l x^* \lesssim f_l\right)\right]$ for all $l \in \{1,2,\ldots,q\}$, and that $E\left[N\left(\tilde{\tilde{C}}_l x \lesssim f_l\right)\right] > E\left[N\left(\tilde{\tilde{C}}_l x^* \lesssim f_l\right)\right]$ for at least one $l \in \{1,2,\ldots,q\}$.

Scalarization-Based Problems for Obtaining a Pareto Optimal Solution

In order to obtain a (weak) Pareto optimal solution of NPE model, we consider the following maximin problem, which is one of well-known scalarization methods for solving multi-objective optimization problems:

[Maximin problem for NPE model]

$$\left.\begin{array}{ll}\text{maximize} & \min\limits_{l\in\{1,2,\ldots,q\}} E\left[N\left(\tilde{\bar{C}}_l x \lesssim f_l\right)\right] \\ \text{subject to} & x \in X.\end{array}\right\} \tag{30}$$

Similar to the case of PPE model discussed in the previous section, we obtain the following proposition:

Proposition 3. *(Weak Pareto optimality of the maximin problem for NPE model)*
Let x^* be an optimal solution of problem (30). Then, x^* is a weak Pareto optimal solution of (29), namely, a weak Pareto optimal solution for NPE model.

Since the proof of Proposition 3 is very similar to that of Proposition 1, we omit its proof. Similar to the property of the optimal solution of problem (29), an optimal solution of (30) is not always a (strong) Pareto optimal solution but only a weak Pareto optimal solution in general.

To find a solution satisfying (strong) Pareto optimality instead of weak Pareto optimality, we consider the following augmented maximin problem.

[Augmented maximin problem for NPE model]

$$\left.\begin{array}{ll}\text{maximize} & \min\limits_{l\in\{1,2,\ldots,q\}} E\left[N\left(\tilde{\bar{C}}_l x \lesssim f_l\right)\right] + \rho \sum\limits_{l=1}^{q} E\left[N\left(\tilde{\bar{C}}_l x \lesssim f_l\right)\right] \\ \text{subject to} & x \in X,\end{array}\right\} \tag{31}$$

where ρ is a sufficiently small positive constant, say 10^{-6}.

Then, we obtain the following proposition:

Proposition 4. *((Strong) Pareto optimality of augmented maximin problem for NPE model)*
Let x^* be an optimal solution of problem (31). Then, x^* is a (strong) Pareto optimal solution of (29), namely, a (strong) Pareto optimal solution for NPE model.

We omit the proof of Proposition 4 because it is similar to that of Proposition 2.

7. Solution Algorithms

7.1. Solution Algorithm for the PPE Model

Now we discuss how to solve problem (28) in order to obtain a (strong) Pareto optimal solution for the PPE model. Here, we focus on the case where all the membership functions of fuzzy numbers and fuzzy goals are represented by linear membership functions. To be more specific, we restrict ourselves to considering the case that the coefficients of objective function in (15) are triangular fuzzy random variables defined in (11), and that the membership function of the fuzzy goal $\tilde{G}_l$ for the lth objective function is the following piecewise linear membership function, called a linear membership function:

$$\mu_{\tilde{G}_l}(y) = \begin{cases} 1 & \text{if } y < f_l^1,\ l = 1,2,\ldots,q \\ \dfrac{y - f_l^0}{f_l^1 - f_l^0} & \text{if } f_l^1 \leq y \leq f_l^0 \\ 0 & \text{if } y > f_l^0, \end{cases} \tag{32}$$

where f_l^0 and f_l^1 are parameters whose values are determined by a DM. Figure 12 shows the membership of fuzzy goal $\tilde{G}_l$ which is expressed by a linear membership function.

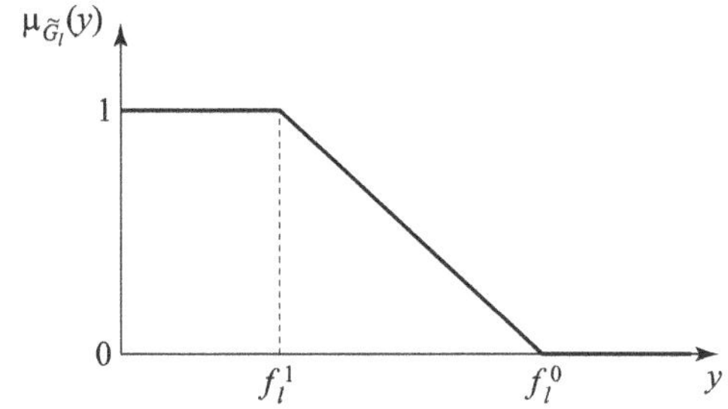

Figure 12. Linear membership function $\mu_{\tilde{G}_l}$ of a fuzzy goal $\tilde{G}_l$.

From a practical aspect, it is important to show how to determine the values of parameters f_l^0 and f_l^1 in the linear membership functions (32) of fuzzy goals $\tilde{G}_l, l = 1,2,\ldots,q$. When a DM can easily set the values of f_l^0 and $f_l^1, l = 1,2,\ldots,q$, these values should be determined by the DM's own idea or choice. On the other hand, when it is difficult for a DM to determine the parameter values of fuzzy goals, we recommend that the values of f_l^1 and f_l^0 are determined as follows:

$$\left.\begin{aligned} f_l^1 &= \sum_{j=1}^{n}\sum_{k=1}^{r_l} p_{lk} d_{ljk} \hat{x}_j^{l*} \\ f_l^0 &= \max_{r \in \{1,2,\ldots,q\}} \sum_{j=1}^{n}\sum_{k=1}^{r_l} p_{lk} d_{ljk} \hat{x}_j^{r*} \end{aligned}\right\} \text{for } l = 1,2,\ldots,q, \tag{33}$$

where $\hat{x}^{l*}$ denotes an optimal solution of the following lth optimization problem which has a single objective function:

$$\left.\begin{aligned} &\text{minimize} \quad \sum_{j=1}^{n}\sum_{k=1}^{r_l} p_{lk} d_{ljk} x_j \\ &\text{subject to} \quad \boldsymbol{x} \in X \end{aligned}\right\} \text{for } l = 1,2,\ldots,q. \tag{34}$$

The above calculation method is similar to the Zimmermann's method [48] which was originally introduced for fuzzy (non-stochastic) linear programming.

We consider the case where the coefficients of the objective function are given as discrete triangular fuzzy random variables. Assuming that $\mu_{\tilde{C}_{ljk}}$ and $\mu_{\tilde{G}_l}$ are given by (18) and (32), respectively, we can show that the following theorem holds:

Theorem 1. *Assume that $\tilde{\bar{C}}_{lj}$ is a discrete triangular fuzzy random variable whose realized values for events are triangular fuzzy numbers characterized by (18), and that the membership function of each fuzzy goal $\tilde{G}_l$ is characterized by (32) and (33). Then, the possibility-based probabilistic expectation (PPE) is calculated as*

$$E\left[\Pi\left(\tilde{\bar{C}}_l x \lesssim f_l\right)\right] = \sum_{k=1}^{r_l} p_{lk} \cdot \min\left[1,\ \max\left\{0,\ g_{lk}^{\Pi}(x)\right\}\right],\ l = 1,2,\ldots,q, \tag{35}$$

where

$$g_{lk}^{\Pi}(x) \triangleq \frac{\sum_{j=1}^{n}(\beta_{ljk} - d_{ljk})x_j + f_l^0}{\sum_{j=1}^{n}\beta_{ljk}x_j - f_l^1 + f_l^0},\ l = 1,2,\ldots,q,\ k = 1,2,\ldots,r_l. \tag{36}$$

Proof. The calculation of $\Pi\left(\tilde{C}_{lk}x \lesssim f_l\right)$ is done by dividing into three cases, namely, 1) Case 1: If $\boldsymbol{d}_{lk}\boldsymbol{x} < f_l^1$, 2) Case 2: If $f_l^1 \le \boldsymbol{d}_{lk}\boldsymbol{x} \le f_l^0 + \gamma_{lk}\boldsymbol{x}$, 3) Case 3: If $\boldsymbol{d}_{lk}\boldsymbol{x} > f_l^0 + \gamma_{lk}\boldsymbol{x}$.

1. Case 1: If $\boldsymbol{d}_{lk}\boldsymbol{x} < f_l^1$ the value of $\Pi\left(\tilde{C}_{lk}x \lesssim f_l\right)$ is equal to 1, as shown in Figure 13.
2. Case 2: If $f_l^1 \le \boldsymbol{d}_{lk}\boldsymbol{x} \le f_l^0 + \gamma_{lk}\boldsymbol{x}$, the value of $\Pi\left(\tilde{C}_{lk}x \lesssim f_l\right)$ is calculated as the ordinate of the crossing point between the membership function of fuzzy goal $\tilde{G}_l$ and the objective function $C_{lk}x$, as shown in Figure 14. The abscissa of the crossing point of two functions ($\mu_{\tilde{C}_{lk}x}$ and $\mu_{\tilde{G}_l}$) is obtained by solving the equation

$$1 - \frac{\boldsymbol{d}_{lk}\boldsymbol{x} - y}{\boldsymbol{\beta}_{lk}\boldsymbol{x}} = \frac{y - f_l^0}{f_l^1 - f_l^0},\ l = 1,2,\ldots,q,\ k = 1,2,\ldots,r_l. \tag{37}$$

   Then, the solution $y_{lk}^{\Pi*}$ of (37) is

$$y_{lk}^{\Pi*} = \frac{f_l^1\sum_{j=1}^{n}\beta_{ljk}x_j + (f_l^0 - f_l^1)\sum_{j=1}^{n}d_{ljk}x_j}{\sum_{j=1}^{n}\beta_{ljk}x_j - f_l^1 + f_l^0},\ l = 1,2,\ldots,q,\ k = 1,2,\ldots,r_l.$$

   Consequently, the ordinate of the crossing point is calculated as

$$\mu_{\tilde{G}_l}\left(y_{lk}^{\Pi*}\right) = \mu_{\tilde{C}_{lk}x}(y_{lk}^{\Pi*}) = \frac{\sum_{j=1}^{n}(\beta_{ljk} - d_{ljk})x_j + f_l^0}{\sum_{j=1}^{n}\beta_{ljk}x_j - f_l^1 + f_l^0}\ \left(\triangleq g_{lk}^{\Pi}(x)\right)$$

   for $l = 1,2,\ldots,q$, $k = 1,2,\ldots,r_l$.
3. Case 3: If $\boldsymbol{d}_{lk}\boldsymbol{x} > f_l^0 + \gamma_{lk}$, the value of $\Pi\left(\tilde{C}_{lk}x \lesssim f_l\right)$ is equal to 0, as shown in Figure 15.

Therefore, the computational results of the above three cases can be integrated and represented as a single form

$$\begin{aligned}\Pi\left(\tilde{C}_{lk}\boldsymbol{x} \lesssim f_l\right) &\triangleq \Pi_{\tilde{C}_{lk}\boldsymbol{x}}(\tilde{G}_l)\\ &= \begin{cases} 1 & \text{if } \boldsymbol{d}_{lk}\boldsymbol{x} < f_l^1 \\ g_{lk}^{\Pi}(\boldsymbol{x}) & \text{if } f_l^1 \leq \boldsymbol{d}_{lk}\boldsymbol{x} \leq f_l^0 + \gamma_{lk}\boldsymbol{x} \\ 0 & \text{if } \boldsymbol{d}_{lk}\boldsymbol{x} > f_l^0 + \gamma_{lk}\boldsymbol{x} \end{cases}\\ &= \min\left[1,\ \max\left\{0,\ g_{lk}^{\Pi}(\boldsymbol{x})\right\}\right].\end{aligned}$$

Consequently, $E\left[\Pi\left(\tilde{\bar{C}}_l\boldsymbol{x} \lesssim f_l\right)\right]$ is calculated based on Definition 14 as follows:

$$\begin{aligned}E\left[\Pi\left(\tilde{\bar{C}}_l\boldsymbol{x} \lesssim f_l\right)\right] &\triangleq \sum_{k=1}^{r_l} p_{lk}\cdot\Pi\left(\tilde{C}_{lk}\boldsymbol{x} \lesssim f_l\right)\\ &= \sum_{k=1}^{r_l} p_{lk}\cdot\Pi_{\tilde{C}_{lk}\boldsymbol{x}}(\tilde{G}_l)\\ &= \sum_{k=1}^{r_l} p_{lk}\cdot\min\left[1,\ \max\left\{0,\ g_{lk}^{\Pi}(\boldsymbol{x})\right\}\right],\ l = 1,2,\ldots,q.\end{aligned}$$

□

Figures 13–15 illustrate the degrees of possibility that the fuzzy goal $\tilde{G}_l$ is fulfilled under the possibility distribution $\mu_{\tilde{C}_{lk}\boldsymbol{x}}$, each of which is corresponding to Case 1, Case 2 and Case 3, respectively. In each figure, the bold line expresses the value of $\min\left\{\mu_{\tilde{C}_{lk}\boldsymbol{x}}(y),\ \mu_{\tilde{G}_l}(y)\right\}$. In Figure 13, the maximum of the bold line is 1. In Figure 14, the maximum of the bold line is between 0 and 1. In Figure 15, the maximum of the bold line is 0.

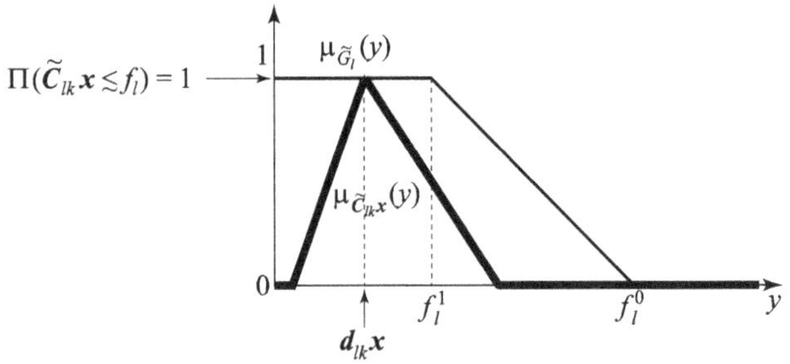

Figure 13. Case 1 in the proof of Theorem 1.

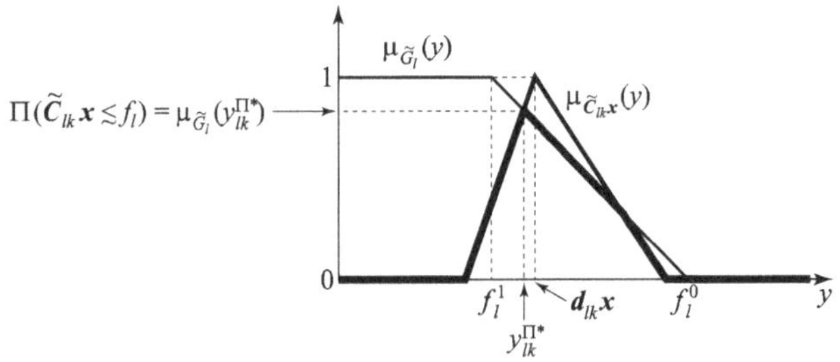

Figure 14. Case 2 in the proof of Theorem 1.

$\Pi(\widetilde{C}_{lk}x \lesssim f_l) = 0$

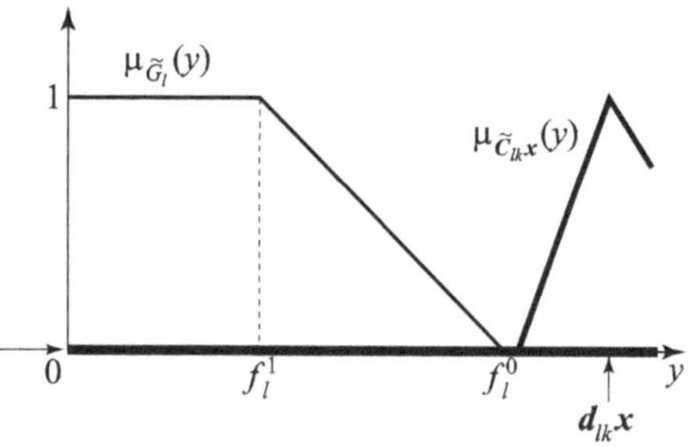

Figure 15. Case 3 in the proof of Theorem 1.

From Proposition 2, the optimal solution of augmented maximin problem (28) is a (strong) Pareto optimal solution. In the case of linear membership functions, the augmented maximin problem (28) for the PPE model is formulated using (35) and (36) as follows:

[Augmented maximin problem for PPE model (linear membership function case)]

$$\left.\begin{array}{l} \text{maximize } z^{\Pi}(x) \overset{\triangle}{=} \min\limits_{l\in\{1,2,\ldots,q\}} \sum\limits_{k=1}^{r_l} p_{lk}\cdot\min\left[1,\ \max\left\{0,\ g_{lk}^{\Pi}(x)\right\}\right] \\ \qquad +\rho\sum\limits_{l=1}^{q}\sum\limits_{k=1}^{r_l} p_{lk}\cdot\min\left[1,\ \max\left\{0,\ g_{lk}^{\Pi}(x)\right\}\right] \\ \text{subject to } x\in X, \end{array}\right\} \tag{38}$$

where g_{lk}^{Π} is given by (36), namely,

$$g_{lk}^{\Pi}(x) \overset{\triangle}{=} \frac{\sum\limits_{j=1}^{n}(\beta_{ljk}-d_{ljk})x_j + f_l^0}{\sum\limits_{j=1}^{n}\beta_{ljk}x_j - f_l^1 + f_l^0},\ l=1,2,\ldots,q,\ k=1,2,\ldots,r_l,$$

and ρ is a sufficiently small positive constant.

Now we summarize an algorithm for obtaining a (strong) Pareto optimal solution of possibility-based probabilistic expectation model (PPE model) in the multi-objective case.

[An algorithm for obtaining a (strong) Pareto optimal solution of PPE model (linear membership function case)]

Step 1: (Calculation of possible objective function values)

Using a linear programming technique, solve individual minimization problems (34) for $l=1,2,\ldots,q$, namely

$$\left.\begin{array}{ll} \text{minimize} & \sum\limits_{j=1}^{n}\sum\limits_{k=1}^{r_l} p_{lk}d_{ljk}x_j \\ \text{subject to} & x\in X \end{array}\right\}\ \text{for } l=1,2,\ldots,q,$$

and obtain optimal solutions x^{l*} of the lth minimization problems for $l=1,2,\ldots,q$.

Step 2: (Setting of membership functions of fuzzy goals)

Ask the DM to specify the values of f_l^0 and f_l^1, $l=1,2,\ldots,q$. If the DM has no idea of how f_l^0

and $f_l^1, l = 1, 2, \ldots, q$ are determined, then the DM can set the following values calculated by (33) as

$$\left.\begin{array}{rcl} f_l^1 &=& \sum_{j=1}^{n}\sum_{k=1}^{r_l} p_{lk} d_{ljk} \hat{x}_j^{l*} \\ f_l^0 &=& \max_{r\in\{1,2,\ldots,q\}} \sum_{j=1}^{n}\sum_{k=1}^{r_l} p_{lk} d_{ljk} \hat{x}_j^{r*} \end{array}\right\} \text{for } l = 1, 2, \ldots, q,$$

using the optimal solutions x^{l*} obtained in Step 1.

Step 3: (Derivation of a strong Pareto optimal solution of PPE model)
Using a nonlinear programming technique, solve the following augmented maximin problem (38):

$$\begin{array}{ll} \text{maximize} & \min_{l\in\{1,2,\ldots,q\}} \sum_{k=1}^{r_l} p_{lk} \cdot \min\left[1, \max\left\{0, g_{lk}^{\Pi}(x)\right\}\right] \\ & +\rho \sum_{l=1}^{q}\sum_{k=1}^{r_l} p_{lk} \cdot \min\left[1, \max\left\{0, g_{lk}^{\Pi}(x)\right\}\right] \\ \text{subject to} & x \in X, \end{array}$$

where

$$g_{lk}^{\Pi}(x) \triangleq \frac{\sum_{j=1}^{n}(\beta_{ljk} - d_{ljk})x_j + f_l^0}{\sum_{j=1}^{n} \beta_{ljk} x_j - f_l^1 + f_l^0}, \; l = 1, 2, \ldots, q, \; k = 1, 2, \ldots, r_l,$$

and ρ is a sufficiently small positive constant.

It should be noted here that (38) is a nonlinear programming problem (NLPP) with linear constraints in which the objective function has points at which the gradient is not calculated. In such a case, a certain heuristic or metaheuristic algorithm can be used to solve the problem. Another applicable solution method is the Nelder-Mead method [49] which can solve a linear-constrained NLPPs without any information on the derivative of the objective function and constraints.

7.2. Solution Algorithm for the NPE Model

When a DM is pessimistic for the attained objective function values, a necessity-based probabilistic expectation (NPE) model is recommended. In a manner similar to Theorem 1 which holds for the PPE model, we obtain the following theorem with respect to the NPE model:

Theorem 2. *Assume that $\tilde{\bar{C}}_{lj}$ is a discrete triangular fuzzy random variable whose realized values for events are triangular fuzzy numbers characterized by (18), and that the membership function of each fuzzy goal $\tilde{G}_l$ is characterized by (32) and (33). Then, the necessity-based probabilistic expectation defined in (25) is calculated as*

$$E\left[N\left(\tilde{\bar{C}}_l x \lesssim f_l\right)\right] = \sum_{k=1}^{r_l} p_{lk} \cdot \min\left[1, \max\left\{0, g_{lk}^{N}(x)\right\}\right], \; l = 1, 2, \ldots, q, \tag{39}$$

where

$$g_{lk}^{N}(x) \triangleq \frac{-\sum_{j=1}^{n} d_{ljk} x_j + f_l^0}{\sum_{j=1}^{n} \gamma_{ljk} x_j - f_l^1 + f_l^0}, \; l = 1, 2, \ldots, q, \; k = 1, 2, \ldots, r_l. \tag{40}$$

Proof. From the definition of necessity measure, the calculation of $N\left(\tilde{C}_{lk}\boldsymbol{x} \lesssim f_l\right)$ is done by dividing by three cases, namely, 1) Case 1: If $\boldsymbol{d}_{lk}\boldsymbol{x} < f_l^1$, 2) Case 2: If $f_l^1 \leq \boldsymbol{d}_{lk}\boldsymbol{x} \leq f_l^0 + \gamma_{lk}\boldsymbol{x}$, 3) Case 3: If $\boldsymbol{d}_{lk}\boldsymbol{x} > f_l^0 + \gamma_{lk}\boldsymbol{x}$.

1. Case 1: If $(\boldsymbol{d}_{lk} + \gamma_{lk})\boldsymbol{x} < f_l^1$, the value of $N\left(\tilde{C}_{lk}\boldsymbol{x} \lesssim f_l\right)$ is equal to 1, as shown in Figure 16.
2. Case 2: If $f_l^1 - \gamma_{lk}\boldsymbol{x} \leq \boldsymbol{d}_{lk}\boldsymbol{x} \leq f_l^0 + \gamma_{lk}\boldsymbol{x}$, the value of $N\left(\tilde{C}_{lk}\boldsymbol{x} \lesssim f_l\right)$ is calculated as the ordinate of the crossing point between the membership functions of fuzzy goal $\tilde{G}_l$ and the objective function $C_{lk}\boldsymbol{x}$, as shown in Figure 17. The abscissa of the crossing point of two functions ($\mu_{\tilde{C}_{lk}\boldsymbol{x}}$ and $\mu_{\tilde{G}_l}$) is obtained by solving the equation

$$\frac{y - \boldsymbol{d}_{lk}\boldsymbol{x}}{\gamma_{lk}\boldsymbol{x}} = \frac{y - f_l^0}{f_l^1 - f_l^0},\ l = 1,2,\ldots,q,\ k = 1,2,\ldots,r_l. \tag{41}$$

Then, the solution of (41) is

$$y_{lk}^{N*} = \frac{f_l^0 \sum_{j=1}^{n} \gamma_{ljk}x_j + (f_l^0 - f_l^1)\sum_{j=1}^{n} d_{ljk}x_j}{\sum_{j=1}^{n} \gamma_{ljk}x_j - f_l^1 + f_l^0},\ l = 1,2,\ldots,q,\ k = 1,2,\ldots,r_l.$$

Consequently, the ordinate of the crossing point is calculated as

$$\mu_{\tilde{G}_l}\left(y_{lk}^{N*}\right) = 1 - \mu_{\tilde{C}_{lk}\boldsymbol{x}}(y_{lk}^{N*}) = \frac{-\sum_{j=1}^{n} d_{ljk}x_j + f_l^0}{\sum_{j=1}^{n} \gamma_{ljk}x_j - f_l^1 + f_l^0} \left(\stackrel{\triangle}{=} g_{lk}^N(\boldsymbol{x})\right)$$

for $l = 1,2,\ldots,q, k = 1,2,\ldots,r_l$.
3. Case 3: If $\boldsymbol{d}_{lk}\boldsymbol{x} > f_l^0$, the value of $N\left(\tilde{C}_{lk}\boldsymbol{x} \lesssim f_l\right)$ is equal to 0, as shown in Figure 18.

The computational results of the three cases above can be integrated and expressed as a single form

$$\begin{aligned} N\left(\tilde{C}_{lk}\boldsymbol{x} \lesssim f_l\right) &\stackrel{\triangle}{=} N_{\tilde{C}_{lk}\boldsymbol{x}}(\tilde{G}_l) \\ &= \begin{cases} 1 & \text{if } \boldsymbol{d}_{lk}\boldsymbol{x} < f_l^1 - \gamma_{lk}\boldsymbol{x} \\ g_{lk}^N(\boldsymbol{x}) & \text{if } f_l^1 - \gamma_{lk}\boldsymbol{x} \leq \boldsymbol{d}_{lk}\boldsymbol{x} \leq f_l^0 + \gamma_{lk}\boldsymbol{x} \\ 0 & \text{if } \boldsymbol{d}_{lk}\boldsymbol{x} > f_l^0 \end{cases} \\ &= \min\left[1,\ \max\left\{0,\ g_{lk}^N(\boldsymbol{x})\right\}\right]. \end{aligned}$$

Consequently, the necessity-based probabilistic expectation defined in (25) is calculated as

$$\begin{aligned} E\left[N\left(\bar{\tilde{C}}_l\boldsymbol{x} \lesssim f_l\right)\right] &\stackrel{\triangle}{=} \sum_{k=1}^{r_l} p_{lk} \cdot N\left(\tilde{C}_{lk}\boldsymbol{x} \lesssim f_l\right) \\ &= \sum_{k=1}^{r_l} p_{lk} \cdot N_{\tilde{C}_{lk}\boldsymbol{x}}(\tilde{G}_l) \\ &= \sum_{k=1}^{r_l} p_{lk} \cdot \min\left[1,\ \max\left\{0,\ g_{lk}^N(\boldsymbol{x})\right\}\right],\ l = 1,2,\ldots,q. \end{aligned}$$

□

Figures 16–18 illustrate the degrees of necessity that the fuzzy goal $\tilde{G}_l$ is fulfilled under the possibility distribution $\mu_{\tilde{\boldsymbol{C}}_{lk}\boldsymbol{x}}$, each of which is corresponding to Case 1, Case 2 and Case 3, respectively. In each figure, the bold line expresses the values of $\max\{1-\mu_{\tilde{\boldsymbol{C}}_{lk}\boldsymbol{x}}(y_l),\ \mu_{\tilde{G}_l}(y_l)\}$. In Figure 16, the minimum of the bold line is 1. In Figure 17, the minimum of the bold line is between 0 and 1. In Figure 18, the minimum of the bold line is 0.

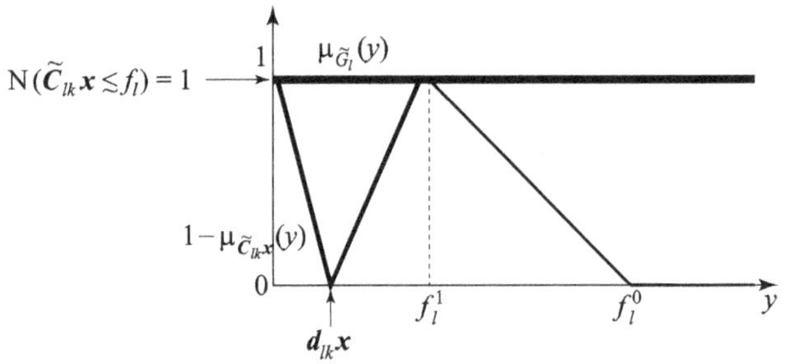

Figure 16. Case 1 in the proof of Theorem 2.

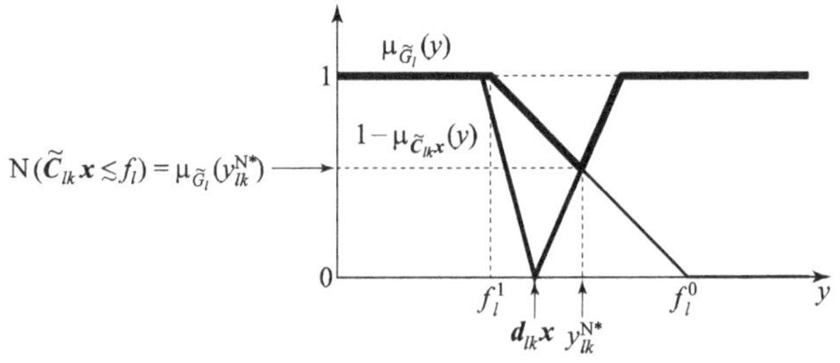

Figure 17. Case 2 in the proof of Theorem 2.

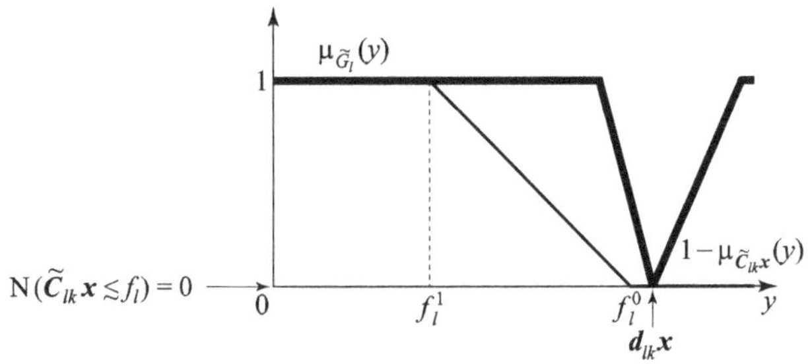

Figure 18. Case 3 in the proof of Theorem 2.

From Proposition 4, the optimal solution of augmented maximin problem (31) is a (strong) Pareto optimal solution of NPE model. In the case of linear membership functions, the augmented maximin problem (42) for NPE model is formulated using (39) and (40) as follows:

[Augmented maximin problem for the NPE model (linear membership function case)]

$$\left.\begin{aligned} &\text{maximize } z^N(\boldsymbol{x}) \triangleq \min_{l\in\{1,2,\ldots,q\}} \sum_{k=1}^{r_l} p_{lk}\cdot\min\left[1,\ \max\left\{0,\ g_{lk}^N(\boldsymbol{x})\right\}\right] \\ &\qquad +\rho\sum_{l=1}^{q}\sum_{k=1}^{r_l} p_{lk}\cdot\min\left[1,\ \max\left\{0,\ g_{lk}^N(\boldsymbol{x})\right\}\right] \\ &\text{subject to } \boldsymbol{x}\in X, \end{aligned}\right\} \tag{42}$$

where $g_{lk}^N(\boldsymbol{x})$ is given by (40), namely,

$$g_{lk}^N(\boldsymbol{x}) \triangleq \frac{-\sum_{j=1}^{n} d_{ljk}x_j + f_l^0}{\sum_{j=1}^{n}\gamma_{ljk}x_j - f_l^1 + f_l^0},\ l=1,2,\ldots,q,\ k=1,2,\ldots,r_l,$$

and ρ is a sufficiently small positive constant.

Now we summarize an algorithm for obtaining a (strong) Pareto optimal solution of the necessity-based probabilistic expectation model (NPE model) in the multi-objective case.

[An algorithm for obtaining a (strong) Pareto optimal solution of NPE model (linear membership function case)]

Step 1: (Calculation of possible objective function values)
By using a linear programming technique, solve individual minimization problems (34) for $l=1,2,\ldots,q$, namely

$$\left.\begin{aligned} &\text{minimize} && \sum_{j=1}^{n}\sum_{k=1}^{r_l} p_{lk}d_{ljk}x_j \\ &\text{subject to} && \boldsymbol{x}\in X \end{aligned}\right\}\ \text{for } l=1,2,\ldots,q,$$

and obtain optimal solutions $\boldsymbol{x}^{l*}$ of the lth minimization problems for $l=1,2,\ldots,q$.

Step 2: (Setting of membership functions of fuzzy goals)
Ask the DM to specify the values of f_l^0 and $f_l^1, l=1,2,\ldots,q$. If the decision has no idea of how f_l^0 and $f_l^1, l=1,2,\ldots,q$ are determined, the DM could set the values calculated by (33) as

$$\left.\begin{aligned} f_l^1 &= \sum_{j=1}^{n}\sum_{k=1}^{r_l} p_{lk}d_{ljk}\hat{x}_j^{l*} \\ f_l^0 &= \max_{r\in\{1,2,\ldots,q\}} \sum_{j=1}^{n}\sum_{k=1}^{r_l} p_{lk}d_{ljk}\hat{x}_j^{r*} \end{aligned}\right\}\ \text{for } l=1,2,\ldots,q,$$

using the optimal solutions $\boldsymbol{x}^{l*}$ obtained in Step 1.

Step 3: (Derivation of a (strong) Pareto optimal solution of the NPE model)
Solve the following augmented maximin problem (42) using a nonlinear programming technique:

$$\begin{aligned} &\text{maximize} \min_{l\in\{1,2,\ldots,q\}} \sum_{k=1}^{r_l} p_{lk}\cdot\min\left[1,\ \max\left\{0,\ g_{lk}^N(\boldsymbol{x})\right\}\right] \\ &\qquad +\rho\sum_{l=1}^{q}\sum_{k=1}^{r_l} p_{lk}\cdot\min\left[1,\ \max\left\{0,\ g_{lk}^N(\boldsymbol{x})\right\}\right] \\ &\text{subject to } \boldsymbol{x}\in X, \end{aligned}$$

where $g_{lk}^{N}(\boldsymbol{x})$ is given by (40), namely,

$$g_{lk}^{N}(\boldsymbol{x}) \triangleq \frac{-\sum_{j=1}^{n} d_{ljk}x_j + f_l^0}{\sum_{j=1}^{n} \gamma_{ljk}x_j - f_l^1 + f_l^0}, \ l = 1,2,\ldots,q, \ k = 1,2,\ldots,r_l,$$

and ρ is a sufficiently small positive constant.

Similar to the case of PPE model in the previous section, (42) is a nonlinear programming problem with linear constraints, which can be solved by a certain nonlinear programming technique.

8. Numerical Experiments

In order to demonstrate feasibility and efficiency of the proposed model, we consider an example of agriculture production problems. One of classical approaches to crop planning problems is stochastic programming [1] using several stochastic events or scenarios related to climate and/or economic conditions. However, it is sometimes difficult to definitely estimate the exact values of the profit and the working time in crop planning problems because of lack of data and/or some factors such as human skills. Zeng et al. [50] considered a fuzzy multi-objective programming approach to a crop planning problem. In this section, we apply the proposed model to solve a crop planning problem in a fuzzy stochastic environment where the profit and the working times are given as discrete fuzzy random variables.

In order to solve the problem, we employ 'constrOptim' function which is prepared as a standard function in the R language [33] and is often used as a solver for NLPPs with linear constraints. It should be stressed here that some state-of-the-art algorithms based on heuristics or metaheuristics [51] may solve problems more efficiently. Nonetheless, we do not propose a specific solution algorithm for the proposed model in this article, because the proposal of a specific solution algorithm is not a purpose of this paper. The R language is easy to use for many researchers and practical persons, even if they are not good at writing their own programming codes.

8.1. Crop Area Planning Problem Under a Fuzzy Random Environment

Assume that an agricultural company (DM) produces 5 kinds of summer vegetables (bell pepper, cucumber, eggplant, tomato and watermelon). We consider the following fuzzy random LPP with bi-objective functions ($q = 2$), 5 decision variables ($n = 5$) and 5 constraints ($m = 5$):

$$\left.\begin{array}{ll}
\text{maximize} & \tilde{\bar{C}}_{11}x_1 + \tilde{\bar{C}}_{12}x_2 + \tilde{\bar{C}}_{13}x_3 + \tilde{\bar{C}}_{14}x_4 + \tilde{\bar{C}}_{15}x_5 \\
\text{minimize} & \tilde{\bar{C}}_{21}x_1 + \tilde{\bar{C}}_{22}x_2 + \tilde{\bar{C}}_{23}x_3 + \tilde{\bar{C}}_{24}x_4 + \tilde{\bar{C}}_{25}x_5 \\
\text{subject to} & a_{11}x_1 + a_{12}x_2 + a_{13}x_3 + a_{14}x_4 + a_{15}x_5 \leq b_1 \\
& a_{21}x_1 + a_{22}x_2 + a_{23}x_3 + a_{24}x_4 + a_{25}x_5 \leq b_2 \\
& a_{31}x_1 + a_{32}x_2 + a_{33}x_3 + a_{34}x_4 + a_{35}x_5 \leq b_3 \\
& a_{41}x_1 + a_{42}x_2 + a_{43}x_3 + a_{44}x_4 + a_{45}x_5 \leq b_4 \\
& a_{51}x_1 + a_{52}x_2 + a_{53}x_3 + a_{54}x_4 + a_{55}x_5 \leq b_5 \\
& x_j \geq 0, \ j = 1,2,\ldots,5,
\end{array}\right\} \quad (43)$$

where the first objective function represents the profit ($\times$10 thousand yen) earned by producing and selling vegetables, and the second one expresses the total working time ($\times$8 h). Let x_j, $j = 1,2,\ldots,5$ denote the growing area ($\times 10^3$ m^2) of vegetables $j = 1$ (bell pepper), $j = 2$ (cucumber), $j = 3$ (eggplant), $j = 4$ (tomato) and $j = 5$ (watermelon), respectively.

In the objective function, let $\tilde{\bar{C}}_{1j}$ and $\tilde{\bar{C}}_{2j}$ be the profit and the working time per unit of vegetables $j = 1,2,\ldots,5$, respectively. Assume that $\tilde{\bar{C}}_{1j}$ and $\tilde{\bar{C}}_{2j}$, $j = 1,2,\ldots,5$ are estimated as discrete triangular fuzzy random variables. On the basis of the research results on relationships between vegetable diseases and humidity [52], we assume that the number of events (scenarios) related to the 1st objective function and the second one are 5 ($r_1 = r_2 = 5$). To be more specific, the set of events are given as $\Omega_1 = \{\omega_{11}, \omega_{12}, \ldots, \omega_{15}\}$ and $\Omega_2 = \{\omega_{21}, \omega_{22}, \ldots, \omega_{25}\}$ as shown in Table 1.

Table 1. Events related to the 1st and 2nd objective functions.

Event	Probability	Situation
ω_{11}	$p_{11} = 0.50$	average annual temperature is normal.
ω_{12}	$p_{12} = 0.25$	average annual temperature is high.
ω_{13}	$p_{13} = 0.15$	average annual temperature is low.
ω_{14}	$p_{14} = 0.06$	it happens an epidemic disease for cucurbitaceous vegetables such as cucumber and watermelon, due to a very high-temperature.
ω_{15}	$p_{15} = 0.04$	it happens an epidemic disease for solanaceae vegetables such as bell pepper, eggplant and tomato, due to a very low-temperature.
ω_{21}	$p_{21} = 0.50$	average annual humidity is normal.
ω_{22}	$p_{22} = 0.20$	average annual humidity is high.
ω_{23}	$p_{23} = 0.16$	average annual humidity is low.
ω_{24}	$p_{24} = 0.08$	it happens an epidemic disease for cucurbitaceous vegetables such as cucumber and watermelon, caused by very low-humidity.
ω_{25}	$p_{25} = 0.06$	it happens an epidemic disease for solanaceae vegetables such as bell pepper, eggplant and tomato, due to a very low-temperature.

Tables 2 and 3 show the parameter values of the realized fuzzy numbers $\tilde{C}_{1jk} = (d_{1jk}, \beta_{1jk}, \gamma_{1jk})_{tri}$ and $\tilde{C}_{2jk} = (d_{2jk}, \beta_{2jk}, \gamma_{2jk})_{tri}$, $j = 1, 2, \ldots, 5$, $k = 1, 2, \ldots, 5$, which characterize fuzzy random variables $\tilde{\bar{C}}_{1j}$ and $\tilde{\bar{C}}_{2j}$, $j = 1, 2, \ldots, 5$, respectively. The values of d_{1jk} are given in Table 2, each of which is based on the statistical data in 2007 by the Japanese Ministry of Agriculture, Forestry and Fisheries (JMAFF) [53]. The values of d_{2jk} are given in Table 3, each of which is based on the report of Mekonnen et al. [54]. By taking into consideration the degree of risk of producing different vegetables, it is assumed that the parameter values of β_{ljk} and γ_{ljk} for $j = 1, 2, 5$ are larger than those for $j = 3, 4$.

Table 2. Parameters of $\tilde{C}_{1jk}$ in the 1st objective function.

Parameter	$k = 1$	$k = 2$	$k = 3$	$k = 4$	$k = 5$
d_{11k}	89.50	95.10	83.80	97.50	61.80
d_{12k}	118.50	118.80	117.90	79.60	117.00
d_{13k}	122.60	123.60	125.60	113.30	83.40
d_{14k}	90.30	82.60	93.10	85.10	66.10
d_{15k}	25.80	28.90	24.40	21.50	23.70
γ_{11k}	8.20	8.60	7.40	10.20	5.70
γ_{12k}	10.70	10.90	10.60	8.50	11.20
γ_{13k}	9.00	8.70	8.80	8.70	5.90
γ_{14k}	8.10	7.60	8.40	7.30	5.80
γ_{15k}	2.60	3.20	2.50	2.10	2.20
β_{11k}	11.40	11.80	11.30	12.20	8.60
β_{12k}	10.70	10.30	10.20	7.10	9.70
β_{13k}	9.70	9.10	9.80	8.60	5.10
β_{14k}	6.40	6.20	6.40	5.90	5.00
β_{15k}	3.90	4.20	3.60	3.30	3.60

Table 3. Parameters of $\tilde{C}_{2jk}$ in the 2nd objective function.

Parameter	$k=1$	$k=2$	$k=3$	$k=4$	$k=5$
d_{21k}	97.00	100.20	90.30	102.50	124.50
d_{22k}	116.50	114.60	119.50	172.50	121.90
d_{23k}	131.10	133.80	128.10	146.40	172.50
d_{24k}	88.60	86.10	93.50	89.90	139.70
d_{25k}	27.60	23.10	28.10	31.40	29.50
β_{21k}	18.40	18.80	16.90	20.20	23.60
β_{22k}	11.70	11.40	12.20	17.90	13.20
β_{23k}	14.70	15.60	12.90	15.90	21.80
β_{24k}	5.40	5.20	5.80	5.30	7.20
β_{25k}	5.10	4.80	5.40	6.30	5.70
γ_{21k}	6.80	7.10	6.80	8.10	11.70
γ_{22k}	19.10	19.20	19.90	27.80	20.50
γ_{23k}	6.60	7.20	6.60	7.00	8.80
γ_{24k}	12.30	12.70	12.10	12.60	26.70
γ_{25k}	3.30	2.90	3.70	3.80	3.50

As shown in problem (43), there are five constraints in the crop planning problem. Tables 4 and 5 shows the coefficients of these constraints. The 1st constraint reflects that there is the upper limit of the total cost of cropping, sales, etc. The unit of a_{1j} and b_1 in the 1st constraint is converted from a unit area to 10 thousand yen, based on the statistical data in 2007 by JMAFF [53]. The 2nd constraint and the 3rd one represents the upper limit and the lower limit of the total growing area of vegetables, respectively. The 4th and 5th constraints represent that the agricultural company signs contracts with two major customers for selling certain amounts of specific vegetables. In these two constraints, the unit for these constraints is converted from area to kilo gram, based on the statistical data in 2007 by JMAFF [53].

Table 4. Left-hand side coefficients in constraints.

LHS Value	$j=1$	$j=2$	$j=3$	$j=4$	$j=5$
a_{1j}	53.20	58.80	57.70	63.70	33.00
a_{2j}	1.00	1.00	1.00	1.00	1.00
a_{3j}	−1.00	−1.00	−1.00	−1.00	−1.00
a_{4j}	−53.90	−80.50	−75.30	0.00	0.00
a_{5j}	0.00	0.00	0.00	−75.00	−48.40

Table 5. Right-hand side values in constraints.

RHS Value	$i=1$	$i=2$	$i=3$	$i=4$	$i=5$
b_i	30,000.00	500.00	−300.00	−10,500.00	−7000.00

In problem (43), the 1st objective function is the profit to be maximized. Since the algorithm proposed in Section 5 is valid for minimization problems, we transform the maximization problem into a minimization problem by multiplying the original 1st objective function by -1 as follows:

$$\left.\begin{array}{ll}\text{minimize} & -\tilde{\tilde{C}}_{11}x_1 - \tilde{\tilde{C}}_{12}x_2 - \tilde{\tilde{C}}_{13}x_3 - \tilde{\tilde{C}}_{14}x_4 - \tilde{\tilde{C}}_{15}x_5 \\ \text{minimize} & \tilde{\tilde{C}}_{21}x_1 + \tilde{\tilde{C}}_{22}x_2 + \tilde{\tilde{C}}_{23}x_3 + \tilde{\tilde{C}}_{24}x_4 + \tilde{\tilde{C}}_{25}x_5 \\ \text{subject to} & a_{11}x_1 + a_{12}x_2 + a_{13}x_3 + a_{14}x_4 + a_{15}x_5 \leq b_1 \\ & a_{21}x_1 + a_{22}x_2 + a_{23}x_3 + a_{24}x_4 + a_{25}x_5 \leq b_2 \\ & a_{31}x_1 + a_{32}x_2 + a_{33}x_3 + a_{34}x_4 + a_{35}x_5 \leq b_3 \\ & a_{41}x_1 + a_{42}x_2 + a_{43}x_3 + a_{44}x_4 + a_{45}x_5 \leq b_4 \\ & a_{51}x_1 + a_{52}x_2 + a_{53}x_3 + a_{54}x_4 + a_{55}x_5 \leq b_5 \\ & x_j \geq 0,\ j = 1, 2, \ldots, 5. \end{array}\right\} \quad (44)$$

In order to utilize the results obtained in the previous sections, we transform maximization into minimization of the 1st objective function by setting $\tilde{\bar{C}}'_{ij} \triangleq -\tilde{\bar{C}}_{ij}$ as follows:

$$\left.\begin{array}{ll}\text{minimize} & \tilde{\bar{C}}'_{11}x_1 + \tilde{\bar{C}}'_{12}x_2 + \tilde{\bar{C}}'_{13}x_3 + \tilde{\bar{C}}'_{14}x_4 + \tilde{\bar{C}}'_{15}x_5 \\ \text{minimize} & \tilde{\bar{C}}_{21}x_1 + \tilde{\bar{C}}_{22}x_2 + \tilde{\bar{C}}_{23}x_3 + \tilde{\bar{C}}_{24}x_4 + \tilde{\bar{C}}_{25}x_5 \\ \text{subject to} & a_{11}x_1 + a_{12}x_2 + a_{13}x_3 + a_{14}x_4 + a_{15}x_5 \leq b_1 \\ & a_{21}x_1 + a_{22}x_2 + a_{23}x_3 + a_{24}x_4 + a_{25}x_5 \leq b_2 \\ & a_{31}x_1 + a_{32}x_2 + a_{33}x_3 + a_{34}x_4 + a_{35}x_5 \leq b_3 \\ & a_{41}x_1 + a_{42}x_2 + a_{43}x_3 + a_{44}x_4 + a_{45}x_5 \leq b_4 \\ & a_{51}x_1 + a_{52}x_2 + a_{53}x_3 + a_{54}x_4 + a_{55}x_5 \leq b_5 \\ & x_j \geq 0,\ j = 1, 2, \ldots, 5, \end{array}\right\} \quad (45)$$

where $\tilde{\bar{C}}'_{ij}$ are discrete triangular fuzzy random variables expressed as $\tilde{C}'_{ijk} = (d'_{ijk}, \beta'_{ijk}, \gamma'_{ijk})_{tri}$. Then, the following remark should be noted:

Remark 1. *Let $\tilde{\bar{C}}$ and $\tilde{\bar{C}}'_{ij}$ be L-R fuzzy random variables expressed as $\tilde{C}_{ijk} = (d_{ijk}, \beta_{ijk}, \gamma_{ijk})_{tri}$ and $\tilde{C}'_{ijk} = (d'_{ijk}, \beta'_{ijk}, \gamma'_{ijk})_{tri}$. If $\tilde{\bar{C}}'_{ij} = -\tilde{\bar{C}}_{ij}$, it holds that*

$$d'_{ijk} = -d_{ijk},\ \beta'_{ijk} = \gamma_{ijk},\ \gamma'_{ijk} = \beta_{ijk}.$$

Then, the values of parameters in the 1st objective function as shown in Table 2 can be replaced by Table 6, where we use the property of triangular fuzzy random variables described in Remark 1.

Based on the algorithm proposed in the previous section, a Pareto optimal solution in the crop planning problem is obtained. Firstly, the fuzzy goals for each objective function are given, by solving LPPs in Step 1 and computing f_l^1 and f_l^0 for $l = 1, 2$ in Step 2, as $(f_1^1, f_1^0) = (-57026.56, -19396.41)$ and $(f_2^1, f_2^0) = (20447.14, 63438.03)$. In Step 3, based on the DM's preference, the augmented maximin problems (38) and/or (42) are solved, which corresponds to the possibility-based probabilistic expectation model (PPE model) and the necessity-based probabilistic expectation model (NPE model), respectively. Since the obtained solutions through the function in the R language do not always satisfy global optimality but local optimality, we apply this function to 100 initial solutions that are randomly generated, and select the best solution among 100 local optimal solutions. Thus, we obtain the following optimal solutions for PPE model and NPE model:

$$\boldsymbol{x}^{\Pi} = (65.74, 240.25, 0.00, 4.87, 189.10),\ z^{\Pi}(\boldsymbol{x}^{\Pi}) = 0.5693,$$

$$\boldsymbol{x}^{N} = (0.13, 163.08, 50.35, 114.48, 134.89),\ z^{N}(\boldsymbol{x}^{N}) = 0.4668,$$

where $\boldsymbol{x}^{\Pi}$ and $\boldsymbol{x}^{N}$ are optimal solutions of (38) and (42), respectively, and $z^{\Pi}(\boldsymbol{x}^{\Pi})$ and $z^{N}(\boldsymbol{x}^{N})$ are their objective function values, respectively. From the computational results, the possibility-based probabilistic expectation model (PPE model) tends to crop high-risk high-return vegetables such as bell pepper, cucumber and watermelon, and few areas are assigned to other vegetables. On the other hand, the necessity-based probabilistic expectation model (NPE model) has a tendency to increase the cropping areas of low-risk low-return vegetables such as tomato and to decrease those of high-risk low-return vegetables such as bell pepper.

Table 6. Parameters of $\tilde{C}'_{1jk} (\triangleq -\tilde{C}_{1jk})$ in the 1st objective function.

Parameter	$k = 1$	$k = 2$	$k = 3$	$k = 4$	$k = 5$
d'_{11k}	−89.50	−95.10	−83.80	−97.50	−61.80
d'_{12k}	−118.50	−118.80	−117.90	−79.60	−117.00
d'_{13k}	−122.60	−123.60	−125.60	−113.30	−83.40
d'_{14k}	−90.30	−82.60	−93.10	−85.10	−66.10
d'_{15k}	−25.80	−28.90	−24.40	−21.50	−23.70
β'_{11k}	11.40	11.80	11.30	12.20	8.60
β'_{12k}	10.70	10.30	10.20	7.10	9.70
β'_{13k}	9.70	9.10	9.80	8.60	5.10
β'_{14k}	6.40	6.20	6.40	5.90	5.00
β'_{15k}	3.90	4.20	3.60	3.30	3.60
γ'_{11k}	8.20	8.60	7.40	10.20	5.70
γ'_{12k}	10.70	10.90	10.60	8.50	11.20
γ'_{13k}	9.00	8.70	8.80	8.70	5.90
γ'_{14k}	8.10	7.60	8.40	7.30	5.80
γ'_{15k}	2.60	3.20	2.50	2.10	2.20

8.2. Computational Times for Different Size Problems

As we mentioned before, we do not propose a specific solution algorithm for the proposed model, because it may take much time and efforts for researchers or practical persons to write programming codes even if we propose new solution algorithms.

Instead of proposing a specific solution algorithm, we use the R language and show the R language can solve problems with hundreds of decision variables in a practical computational time. We expect that the use of the R language can promote the use of our model for solving real-world problems in fuzzy stochastic environments.

In order to show the applicability of the R language to our model in terms of computational time, we conduct additional experiments using 5 numerical examples in which the number of decision variables and that of constraint are different. To be more precise, the numbers of decision variables in 7 examples are $10, 30, 60, 100, 150, 200, 250$, respectively. The number of constraints in each example is set to be the half number of decision variables.

To focus on the effect of the number of decision variables and that of constraints, the number of the objective functions and that of events (scenarios) are fixed. To be more specific, we fix the number of the objective functions and that of events (scenarios) as 5 ($q = 5$) and 10 ($r_l = 10,\ l = 1, 2, \ldots, 5$), respectively.

The values of parameters d_{ljk} are randomly chosen in $\{-5, -4, \ldots, 5\}$, β_{ljk} and γ_{ljk} are the absolute values of the products of d_{ljk} and values randomly chosen in $[0.1, 0.2]$ for $l = 1, 2, \ldots, 5$, $j = 1, 2, \ldots, n$, $k = 1, 2, \ldots, 10$. As for the constraints, the values of a_{ij} in matrix A are randomly chosen in $\{1, 2, \ldots, 10\}$, and the values of b_i are given as the sum of elements in $\boldsymbol{a}_i$ for any $i = 1, 2, \ldots, m$, $j = 1, 2, \ldots, n$. Similar to the experiment in Section 8.1, we used constrOptim function in the R language and conducted 30 runs in which initial solutions are randomly generated. We conducted this numerical experiment using R version 3.2.0 on iMac (OS X Yosemite version 10.10.3, CPU: 3.4 GHz Intel Core i7, RAM: 32 GB 1600 MHz DDR3). Table 7 shows computational times for solving 7 problem instances.

Table 7. Computational times for different size problems.

No. of Decision Variable	10	30	60	100	150	200	250
CPU Times (s)	34.08	106.26	116.95	309.49	515.94	741.12	769.94

Figure 19 shows the relationship between the number of decision variables and computational times obtained in Table 7. It is shown from this graph that the computational time linearly increases as the numbers of decision variables and constraints increase.

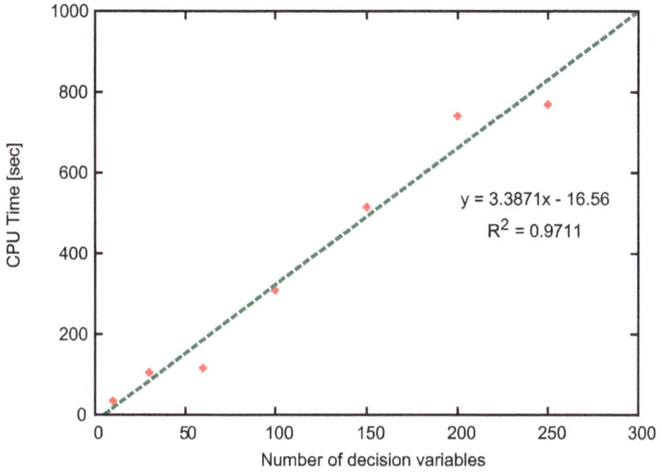

Figure 19. Relationship between the number of decision variables and computational times.

9. Conclusions

In this paper, we have considered LPPs in which the coefficients of the objective functions are discrete fuzzy random variables. Incorporating possibility and necessity measures into a probability measure, we have proposed new decision making models in fuzzy stochastic environments, called possibility/necessity-based probabilistic expectation model (PPE/NPE model), which is to maximize the expectation of the degree of possibility or necessity that the objective function values satisfy the given fuzzy goals. It has been shown that the formulated problems based on the proposed models can be transformed into deterministic nonlinear (multi-objective) programming problems, especially, into more simple problems when linear membership functions are used. In addition, we have defined (strong) Pareto optimal solutions of the proposed models in the multi-objective case, and proposed an algorithm for obtaining a solution satisfying (strong) Pareto optimality. In order to show how the proposed model can be applied to real-world problems, we have conducted a numerical experiment with an agriculture production problem. We also have demonstrated that a standard function in the R language is applicable to solve the problems with hundreds of decision variables in a practical computational time.

In the near future, we will show a generalized variance minimization model which is an extended version discussed in the previous study [32]. Furthermore, some applications of the proposed models to real-world problems will be discussed elsewhere.

Acknowledgments: This work was supported by JSPS KAKENHI Grant Number 17K01276.

Author Contributions: Hideki Katagiri proposed the method and wrote the paper. Kosuke Kato and Takeshi Uno analyzed the validity of the method and performed the numerical experiments.

Conflicts of Interest: The authors declare no conflict of interest.

References

1. Birge, J.R.; Louveaux, F. *Introduction to Stochastic Programming*, 2nd ed.; Springer: Berlin/Heidelberg, Germany, 2011.
2. Dantzig, G.B. Linear programming under uncertainty. *Manag. Sci.* **1955**, *1*, 197–206.
3. Aiche, F.; Abbas, M.; Dubois, D. Chance-constrained programming with fuzzy stochastic coefficients. *Fuzzy Optim. Decis. Mak.* **2013**, *12*, 125–152.
4. Katagiri, H.; Sakawa, M. Interactive multiobjective fuzzy random programming through the level set-based probability model. *Inf. Sci.* **2011**, *181*, 1641–1650.

5. Inuiguchi, M.; Kato, K.; Katagiri, H. Fuzzy Multi-Criteria Optimization: Possibilistic and Fuzzy/Stochastic Approaches. In *Multiple Criteria Decision Analysis*; Springer: Berlin/Heidelberg, Germany, 2016; pp. 851–902.
6. Katagiri, H.; Sakawa, M.; Kato, K.; Nishizaki, I. Interactive multiobjective fuzzy random linear programming: Maximization of possibility and probability. *Eur. J. Oper. Res.* **2008**, *188*, 530–539.
7. Katagiri, H.; Uno, T.; Kato, K.; Tsuda, H.; Tsubaki, H. Random fuzzy multiobjective programming: Optimization of possibilistic Value at Risk (pVaR). *Expert Syst. Appl.* **2013**, *40*, 563–574.
8. Luhandjula, M.K. Fuzzy stochastic linear programming: Survey and future research directions. *Eur. J. Oper. Res.* **2006**, *174*, 1353–1367.
9. Luhandjula, M.K.; Joubert, J.W. On some optimisation models in a fuzzy-stochastic environment. *Eur. J. Oper. Res.* **2010**, *207*, 1433–1441.
10. Nematian, J. A random fuzzy multi-objective linear programming approach through possibility theory. *Eur. J. Ind. Eng.* **2015**, *9*, 512–537.
11. Yano, H. Fuzzy decision making for fuzzy random multiobjective linear programming problems with variance covariance matrices. *Inf. Sci.* **2014**, *272*, 111–125.
12. Yano, H. Multiobjective programming problems involving fuzzy coefficients, random variable coefficients and fuzzy random variable coefficients. *Int. J. Uncertain. Fuzziness Knowl. Based Syst.* **2015**, *23*, 483–504.
13. Katagiri, H.; Sakawa, M.; Kato, K.; Nishizaki, I. A fuzzy random multiobjective 0-1 programming based on the expectation optimization model using possibility and necessity measures. *Math. Comput. Model.* **2004**, *40*, 411–421.
14. Dutta, P.; Chakraborty, D.; Roy, A.R. A single-period inventory model with fuzzy random variable demand. *Math. Comput. Modell.* **2005**, *41*, 915–922.
15. Wang, X. Continuous review inventory model with variable lead time in a fuzzy random environment. *Expert Syst. Appl.* **2011**, *38*, 11715–11721.
16. Giri, P.K.; Maiti, M.K.; Maiti, M. Entropy based solid transportation problems with discounted unit costs under fuzzy random environment. *OPSEARCH* **2014**, *51*, 479–532.
17. Nematian, J. A robust single row facility layout problem with fuzzy random variables. *Int. J. Adv. Manuf. Technol.* **2014**, *72*, 255–267.
18. Wang, S.; Huang, G.; Baetz, B.W. An inexact probabilistic-possibilistic optimization framework for flood management in a hybrid uncertain environment. *IEEE Trans. Fuzzy Syst.* **2015**, *23*, 897–908.
19. Katagiri, H.; Kato, K.; Hasuike, T. A random fuzzy minimum spanning tree problem through a possibility-based value at risk model. *Expert Syst. Appl.* **2012**, *39*, 205–220.
20. Katagiri, H.; Sakawa, S.; Ishii, H. Fuzzy random bottleneck spanning tree problems using possibility and necessity measures. *Eur. J. Oper. Res.* **2004**, *152*, 88–95.
21. Gil, M.A. Fuzzy random variables: Development and state of the art. In Proccedings of the 25th Linz Seminar on Fuzzy Set Theory, Linz, Austria, 3–7 February 2004; pp. 11–15.
22. Gil, M.A.; Lopez-Diaz, M.; Ralescu, D.A. Overview on the development of fuzzy random variables. *Fuzzy Sets Syst.* **2006**, *157*, 2546–2557.
23. Klement, E.P.; Puri, M.L.; Ralescu, D.A. Limit theorems for fuzzy random variables. *Proc. R. Soc. Lond.* **1986**, *407*, 171–182.
24. Kwakernaak, H. Fuzzy random variables-1. Definitions and theorems. *Inf. Sci.* **1978**, *15*, 1–29.
25. Puri, M.L.; Ralescu, D.A. Fuzzy random variables. *J. Math. Anal. Appl.* **1986**, *14*, 409–422.
26. Shapiro, A.F. Fuzzy random variables. *Insur. Math. Econ.* **2009**, *44*, 307–314.
27. Luhandjula, M.K. Fuzziness and randomness in an optimization framework. *Fuzzy Sets Syst.* **1996**, *77*, 291–297.
28. Luhandjula, M.K.; Gupta, M.M. On fuzzy stochastic optimization. *Fuzzy Sets Syst.* **1996**, *81*, 47–55.
29. Liu, B. Fuzzy random chance-constrained programming. *IEEE Trans. Fuzzy Syst.* **2001**, *9*, 713–720.
30. Liu, Y.-K.; Liu, B. A class of fuzzy random optimization: Expected value models. *Inf. Sci.* **2003**, *155*, 89–102.
31. Katagiri, H.; Mermri, E.B.; Sakawa, M.; Kato, K.; Nishizaki, I. A possibilistic and stochastic programming approach to fuzzy random MST problems. *IEICE Trans. Inf. Syst.* **2005**, *88*, 1912–1919.
32. Katagiri, H.; Sakawa, M. A study on fuzzy random linear programming problems based on possibility and necessity measures. In *Fuzzy Sets and Systems-IFSA 2003*; Springer: Berlin/Heidelberg, Germany, 2003; pp. 725–732.

33. R Core Team. *R: A Language and Environment for Statistical Computing*; R Foundation for Statistical Computing: Vienna, Austria, 2014. Available online: http://www.R-project.org/ (accessed on 6 September 2017).
34. Dubois, D.; Prade, H. *Possibility Theory*; John Wiley & Sons: Hoboken, NJ, USA, 1988.
35. Kruse, R.; Meyer, K.D. *Statistics with Vague Data*; D. Reidel Publishing Company: Dordrecht, The Netherlands, 1987.
36. Katagiri, H.; Sakawa, M.; Ishii, H. Studies of stochastic programming models using possibility and necessity measures for linear programming problems with fuzzy random variable coefficients. *Electron. Commun. Jpn. Part III* **2005**, *88*, 68–75.
37. Katagiri, H.; Sakawa, M.; Kato, K.; Ohsaki, S. An interactive fuzzy satisficing method based on the fractile optimization model using possibility and necessity measures for a fuzzy random multiobjective linear programming problem. *Electron. Commun. Jpn. Part III* **2005**, *88*, 20–28.
38. Katagiri, H.; Sakawa, M.; Ishii, H. A study on fuzzy random portfolio selection problems based on possibility and necessity measures. *Sci. Math. Jpn.* **2005**, *61*, 361–370.
39. Katagiri, H.; Ishii, H. Chance constrained bottleneck spanning tree problem with fuzzy random edge costs. *J. Oper. Res. Soc. Jpn.* **2000**, *43*, 128–137.
40. Kaufmann, A.; Gupta, M.M. *Introduction to Fuzzy Arithmetic. Theory and Applications*; Van Nostrand Reinhold: New York, NY, USA, 1985.
41. Kwakernaak, H. Fuzzy random variables-II. Algorithms and examples for the discrete case. *Inf. Sci.* **1979**, *17*, 253–278.
42. Dubois, D.; Prade, H. *Fuzzy Sets and Systems: Theory and Application*; Academic Press: New York, NY, USA, 1980.
43. Zadeh, L.A. Fuzzy sets. *Inf. Control* **1965**, *8*, 338–353.
44. Zadeh, L.A. Fuzzy sets as a basis for a theory of possibility. *Fuzzy Sets Syst.* **1978**, *1*, 3–28.
45. Inuiguchi, M.; Ichihasi, H. Relative modalities and their use in possibilistic linear programming. *Fuzzy Sets Syst.* **1990**, *35*, 303–323.
46. Inuiguchi, M.; Ramik, J. Possibilistic linear programming: A brief review of fuzzy mathematical programming and a comparison with stochastic programming in portfolio selection problem. *Fuzzy Sets Syst.* **2000**, *111*, 3–28.
47. Branke, J.; Deb, K.; Miettinen, K.; Slowinski, R. *Multiobjective Optimization: Interactive and Evolutionary Approaches*; Springer: Berlin/Heidelberg, Germany, 2008.
48. Zimmermann, H.J. Fuzzy programming and linear programming with several objective functions. *Fuzzy Sets Syst.* **1978**, *1*, 5–55.
49. Lange, K. *Numerical Analysis for Statisticians*, 2nd ed.; Springer: Berlin/Heidelberg, Germany, 2010.
50. Zeng, X.; Kang, S.; Li, F.; Zhang, L.; Guo, P. Fuzzy multi-objective linear programming applying to crop area planning. *Agric. Water Manag.* **2010**, *98*, 134–142.
51. Katagiri, H.; Hayashida, T.; Nishizaki, I.; Guo, Q. A hybrid algorithm based on tabu search and ant colony optimization for k-minimum spanning tree problems. *Expert Syst. Appl.* **2012**, *39*, 5681–5686.
52. Sherf, A.F. *Vegetable Diseases and Their Control*; John Wiley & Sons: Hoboken, NJ, USA, 1986.
53. Japanese Ministry of Agriculture, Forestry and Fisheries. Data for business statistics by agricultural products in 2007 (Japanese version only). Available online: http://www.e-stat.go.jp/SG1/estat/List.do?lid=000001061833 (accessed on 6 September 2017).
54. Mekonnen, M.M.; Hoekstra, A.Y. The green, blue and grey water footprint of crops and derived crop products. *Hydrol. Earth Syst. Sci.* **2011**, *15*, 1577–1600.

symmetry

MDPI

Article

A Method for Fuzzy Soft Sets in Decision-Making Based on an Ideal Solution

Zhicai Liu [1,2], Keyun Qin [1,*] and Zheng Pei [2]

1 College of Mathematics, Southwest Jiaotong University, Chengdu 610031, China; liuzc.xhu@foxmail.com
2 Center for Radio Administration & Technology Development, Xihua University, Chengdu 610039, China; pqyz@263.net
* Correspondence: keyunqin@263.net

Received: 25 September 2017; Accepted: 18 October 2017; Published: 23 October 2017

Abstract: In this paper, a decision model based on a fuzzy soft set and ideal solution approaches is proposed. This new decision-making method uses the divide-and-conquer algorithm, and it is different from the existing algorithm (the choice value based approach and the comparison score based approach). The ideal solution is generated according to each attribute (pros or cons of the attributes, with or without constraints) of the fuzzy soft sets. Finally, the weighted Hamming distance is used to compute all possible alternatives and get the final result. The core of the decision process is the design phase, the existing decision models based on soft sets mostly neglect the analysis of attributes and decision objectives. This algorithm emphasizes the correct expression of the purpose of the decision maker and the analysis of attributes, as well as the explicit decision function. Additionally, this paper shows the fact that the rank reversal phenomenon occurs in the comparison score algorithm, and an example is provided to illustrate the rank reversal phenomenon. Experiments indicate that the decision model proposed in this paper is efficient and will be useful for practical problems. In addition, as a general model, it can be extended to a wider range of fields, such as classifications, optimization problems, etc.

Keywords: fuzzy sets; soft sets; fuzzy soft sets; decision making; rank reversal; ideal solution

1. Introduction

The complicated problems in economics, engineering, environmental science and social science are full of imprecision and vagueness. For the various types of uncertainties presented in these problems, the methods in classical mathematics are not always successful. There are some mathematical tools for dealing with uncertainties. Some of them are probability theory, fuzzy set theory, rough set theory, and interval mathematics, but all these theories have their own difficulties. In 1999, Molodtsov [1] introduced the concept of soft sets, which can be considered as a new mathematical tool for dealing with uncertainties. It has proven useful in many fields such as decision-making, data analysis, forecasting and texture classification [2].

Research works on soft set theory and its applications in various fields are progressing rapidly, and many significant results have been achieved. Maji et al. [3] initiated the study of hybrid structures involving fuzzy sets and soft sets and introduced the notion of fuzzy soft set. Qin et al. [4] combine interval sets and soft sets. Zhang [5] studies interval soft sets. Shao and Qin [6] define fuzzy soft lattices and discuss their structure. Basu [7] introduce the structure and form of soft set theory. Li et al. [8] investigates roughness of fuzzy soft sets, introduced the concept of fuzzy soft rough sets. Bustince [9] proved that fuzzy sets are intuitionistic fuzzy sets. Torra et al. [10] extended this theory by introducing hesitant fuzzy sets. An extension of traditional fuzzy sets that permit the membership degree of an element to be a set of several possible values in $[0,1]$ and whose main purpose is to model the uncertainty produced by human doubt when eliciting information [11].

There is vast literature on fuzzy soft sets and their applications, including many successful generalizations. The comparison score and choice value are two different approaches applying soft set theory to decision-making problems. Maji et al. [3] pioneered soft set based decision-making and firstly proposed the choice value based approach. They established the criterion that an object could be selected if it maximizes the choice value of the problem. The comparison score based approach is proposed by Roy et al. [12] to dealing with the fuzzy soft set based decision-making problems. In this approach, they compare the membership values of two objects with respect to a common attribute to determine which one relatively possesses that attribute. Rodríguez et al. [13] overviewed on fuzzy modeling of complex linguistic preferences in decision-making and pointed out the different points of view used in each proposal to model these complex preferences. Kong et al. [14] revised this method, and their revision (the fuzzy choice value based method) has been proved as another method based on the maximum fuzzy choice value. Feng et al. [15] presented a novel approach to fuzzy soft set based decision-making problems by using level soft sets. They investigated the fuzzy soft set based decision-making problems more deeply, and their new method can be successfully applied to some decision-making problems.

The core of the decision process is the design phase. Firstly, the purpose of the decision maker should be expressed very clearly. Secondly, the data set should be analyzed accurately. Thirdly, choose the correct and efficient decision-making function. In general, the traditional decision-making algorithm (the score based method and the choice value based method) based on soft sets have some shortcomings. Firstly, the purpose of the decision maker is ignored, and it is generally assumed that the greater the value of each attribute, the better. Secondly, the analysis of the data set is ignored. In the fuzzy soft sets $(\widetilde{F}, A)$, all the attributes are treated uniformly, that is, all attributes are treated as good attributes. Sometimes, the value of the attribute is not the bigger the better, for example, expensive. Thirdly, there is ambiguity of the decision function such as comparison score algorithm because of reversal phenomenon occurred in this algorithm, which can lead to unacceptable choices in practice. It is unrealistic to use a fixed method to deal with the ever-changing problems. Therefore, based on the above factors, a decision model based on the ideal solution is proposed for the fuzzy soft set decision problem.

We will shortly describe the algorithm. In this study, we use the divide-and-conquer algorithm to design a decision-making model, and the model can dynamically adjust the ideal solution according to each attribute (positive attributes, negative attributes, and constraint attributes) of the fuzzy soft set. In addition, the weighted Hamming distance is used to compute all possible choices and get the final result. In other words, the $(\widetilde{F}, A)$ is a fuzzy soft set, according to the membership function of each attribute, and the ideal solution u_{goal} can be generated. By measuring the similarity between object u_x and u_{goal}, the object that is the most similar to u_{goal} is the optimal choice. The algorithm emphasizes the correct expression of the purpose of the decision maker at the design stage and emphasizes the analysis of attributes, as well as the explicit decision function. This clear decision-making structure makes fuzzy soft sets more practical in decision-making.

The rest of this paper is organized as follows. Section 2 describes the basic concept of soft set theory. Section 3 gives an analysis of previous soft set-based decision-making algorithms and their limitations. Section 4 presents an alternative approach to the decision model by 'ideal solution' algorithm, and Section 5 shows the real-life applications of the proposed algorithm. Section 6 presents conclusions and future work.

2. Fuzzy Sets, Soft Sets and Fuzzy Soft Sets

In this section, we recall some fundamental notions of fuzzy sets, soft sets, and fuzzy soft sets, their relation to decision-making, and existing research.

2.1. Fuzzy Sets

In 1965, Zadeh [16] created a mathematical method of describing the fuzzy phenomenon in mathematics-fuzzy set theory.

Definition 1. *([16]) Let U be a set, called a universe. A fuzzy set μ on U is defined by a membership function $\mu : U \to [0,1]$. For any $x \in U$, the $\mu(x)$ represents the extent to which the x belongs to the fuzzy set μ.*

The fuzzy sets $\mu(x)$ is denoted as follows:

$$\mu(x) = \{(x, \mu(x)), x \in U\}. \tag{1}$$

A fuzzy set can be discrete or continuous. For discrete fuzzy sets, $\mu(x)$ can be expressed as follows:

$$\mu(x) = \sum_{i=1}^{n} (\mu(x_i)/x_i). \tag{2}$$

n is the number of elements in U.

There are several forms of operations on fuzzy sets. According to maximum-minimal operator Zadeh proposed by [16], the intersection, union, and complement on fuzzy sets are defined as follows:

$$(\mu \cap v)(x) = \mu(x) \wedge v(x),$$
$$(\mu \cup v)(x) = \mu(x) \vee v(x),$$
$$\mu^c(x) = 1 - \mu(x).$$

The decision-making theory plays a fundamental role in many scientific branches, such as AI (Artificial Intelligence), robots and big data. It is mainly developed in the setting of fuzzy decision theory. In 1965, fuzzy sets were proposed to confront the problems of linguistic or uncertain information. With the successful applications in the field of automatic control, fuzzy sets have been incorporated into fuzzy decision-making for dealing with decision-making problems. The idea of applying fuzzy sets in decision sciences comes from the seminal paper of Bellman and Zadeh. The application of the Bellman-Zadeh approach to decision-making in the fuzzy environment proposed in [17].

2.2. Soft Sets and Fuzzy Soft Sets

We review some fundamental notions of soft sets and fuzzy soft sets. Let U be the universe set and E be the set of all possible parameters under consideration with respect to U. Usually, parameters are attributes, characteristics, or properties of objects in U. (U, E) will be called a soft space. Molodtsov defined the notion of a soft set in the following way:

Definition 2. *([1]) A pair (F, A) is called a soft set over U, where $A \subseteq E$ and F is a mapping given by $F : A \to P(U)$.*

In other words, a soft set over U is a parameterized family of subsets of U. A is called the parameter set of the soft set (F, A). For $e \in A$, $F(e)$ may be considered as the set of e-approximate elements of (F, A). For illustration, we consider the following example of soft set.

Example 1. *Suppose that there are six houses in the universe U given by $U = \{h_1, h_2, h_3, h_4, h_5, h_6\}$ and $E = \{e_1, e_2, e_3, e_4, e_5\}$ is the set of parameters. e_1, e_2, e_3, e_4 and e_5 stand for the parameters 'expensive', 'beautiful', 'wooden', 'cheap' and 'in the green surroundings', respectively.*

In this case, to define a soft set means to point out expensive houses, beautiful houses, and so on. The soft set (F, E) may describe the 'attractiveness of the houses' that Mr.X is going to buy. Suppose that $F(e_1) = \{h_2, h_4\}$, $F(e_2) = \{h_1, h_3\}$, $F(e_3) = \{h_3, h_4, h_5\}$, $F(e_4) = \{h_1, h_3, h_5\}$, $F(e_5) = \{h_1\}$. Then, the soft set (F, E) is a parameterized family $\{F(e_i); 1 \le i \le 5\}$ of subsets of U and give us a collection of approximate descriptions of an object. $F(e_1) = \{h_2, h_4\}$ means 'houses h_2 and h_4' are 'expensive'.

Maji et al. [18] introduced the concept of fuzzy soft sets by combining soft set and fuzzy set.

Definition 3. *([18]) Let (U, E) be a soft space. A pair $(\widetilde{F}, A)$ is called a fuzzy soft set over U, where $A \subseteq E$ and $\widetilde{F}$ is a mapping given by $\widetilde{F} : A \rightarrow \widetilde{F}(U)$, $\widetilde{F}(U)$ is the set of all fuzzy subsets on U.*

Let us denote $\mu_{\widetilde{F}(e)}(x)$ the membership degree that object x holds attribute e where $x \in U$ and $e \in A$. Then, $\widetilde{F}(e)$ can be written as $\widetilde{F}(e) = \{< x, \mu_{\widetilde{F}(e)}(x) > | x \in U\}$.

Definition 4. *([18]) Let $(\widetilde{F}, A)$ and $(\widetilde{G}, B)$ be a fuzzy soft set over a common universe U.*

(1) *$(\widetilde{F}, A)$ is said to be a fuzzy soft subset of $(\widetilde{G}, B)$, denoted by $(\widetilde{F}, A) \subseteq (\widetilde{G}, B)$, if $A \subseteq B$ and $\forall e \in A$, $\widetilde{F}(e) \subseteq \widetilde{G}(e)$.*
(2) *$(\widetilde{F}, A)$ is said to be a null fuzzy soft set, denoted by $\emptyset_A$, if $\widetilde{F}(e) = \emptyset$ for any $e \in A$.*
(3) *$(\widetilde{F}, A)$ is said to be a absolute fuzzy soft set, denoted by U_A, if $\widetilde{F}(e) = U$ for any $e \in A$.*

Definition 5. *([19]) For any fuzzy soft set $(\widetilde{F}, E)$ over U, a pair $(\widetilde{F}^{-1}, E)$ is called an induced fuzzy soft set over E of $(\widetilde{F}, E)$, where $\widetilde{F}^{-1}(x) = \{e \in E, x \in \widetilde{F}(e)\}$ for each $x \in U$.*

Definition 6. *The quadruple (U, A, F, V) is called an information system, where $U = \{x_1, ..., x_n\}$ is a universe containing all interested objects, $A = \{a_1, ..., a_n\}$ is a set of attributes, $V = \bigcup_{i=1}^{m} V_i$ where V_j is the value set of the attribute a_j, and $F = \{f_1, ..., f_m\}$ where $f_j : U \rightarrow V_j$.*

Information systems can represent fuzzy sets, soft sets, and fuzzy soft sets. If (F, A) is a soft set over the universe U, then (F, A) is a Boolean-valued information system $S = (U, A, V_{\{0,1\}}, f)$. As shown in Table 1.

A soft set is a simple information system in which the attributes only take two values 0 and 1, and partition-type soft sets and information systems are the same formal structures.

Table 1. Soft set (F, A) represented as a boolean-valued information system.

U	e_1	e_2	e_3	e_4
x_1	0	0	0	0
x_2	0	1	0	1
x_3	0	1	1	1
x_4	1	0	0	0
x_5	1	0	1	0

If $(\widetilde{F}, A)$ is a fuzzy soft set over the universe U, then $(\widetilde{F}, A)$ is a real-valued information system $S = (U, A, V_{[0,1]}, f)$, as shown in Table 2.

Table 2. Fuzzy soft set $(\widetilde{F}, A)$ represented as a real-valued information system.

U	e_1	e_2	e_3	e_4	e_5	e_6	e_7
x_1	0.2	0.4	0.1	0.5	0.8	0.1	0.1
x_2	0.3	0.2	0.3	0.6	0.3	0.9	0.6
x_3	0.3	0.1	0.6	0.7	0.8	0.8	0.3
x_4	0.3	0.7	0.9	0.9	0.1	0.4	0.5
x_5	0.3	0.9	0.1	0.3	0.2	0.2	0.3
x_6	0.3	0.9	0.1	0.3	0.9	0.7	0.8
x_7	0.3	0.9	0.1	0.3	0.2	0.8	0.9
x_8	0.3	0.9	0.1	0.3	0.1	0.4	0.2

3. Fuzzy Soft Set Based Decision-Making and Their Limitations

The decision-making is a process of choosing among alternative courses of action for the purpose of attaining a goal or goals. The decision-making problems based on fuzzy soft sets actually is multi attributes decision-making problems. Two different approaches applying soft set theory to decision-making problems: the choice value based approach and the comparison score based approach. Maji et al. [3] proposed the choice value algorithm for the application of soft set theory in decision-making problems. Roy and Maji [12] proposed the comparison score based approach to solving fuzzy soft set based decision-making problems.

3.1. The Choice Value Algorithm (Algorithm 1)

Let (F, A) be a soft set, (F, A) can be expressed as a binary table. Let h_{ij} be the entries in the table, and if $h_i \in F(e_i)$, then $h_{ij} = 1$. Otherwise, $h_{ij} = 0$. The choice value c_i of an object h_i is computed by $c_i = \sum_j h_{ij}$, the object with the maximum choice value is selected as the optimal decision. The algorithm is as follows:

Algorithm 1 The choice value algorithm

1: Input the soft set (F, A).
2: Compute the choice values c_i for each object h_i, where $c_i = \sum_j h_{ij}$.
3: The decision is $h_i \quad if \quad c_i = max_j c_j$.
4: If i has more than one value then any one of h_i may be chosen.

For decision-making problems using soft sets, the choice value of an object precisely represents the number of 'good' attributes possessed by the object. Hence, it is reasonable to select the object with the maximum choice value as the optimal alternative.

Example 2. *From Table 3, it can be seen that Mr. X will select the house h_1 or h_6.*

Table 3. A soft set (F, A) with choice values.

U	e_1	e_2	e_3	e_4	Choice Value
h_1	1	1	1	1	4
h_2	1	1	1	0	3
h_3	1	0	1	1	3
h_4	1	0	1	0	2
h_5	1	0	0	0	1
h_6	1	1	1	1	4

In real decision-making problems, the choice parameters are not entirely of the equal importance. To cope with such problems, we can impose different weights to different decision parameters. Additionally, it has been generalized to deal with the fuzzy soft set. In this case, the choice value will be computed by: $c_i = \sum_j F(e_j)(h_i)$, where $F(e_j)(h_i)$ is the membership value of h_i with respect to fuzzy set $F(e_j)$. Tables 3 and 4 are examples of the soft sets and weighted soft sets with choice values, respectively.

Table 4. A weighted soft set (F, A) with choice values.

U	$e_1, w_1 = \frac{1}{2}$	$e_2, w_2 = \frac{1}{4}$	$e_3, w_3 = \frac{1}{8}$	$e_4, w_4 = \frac{1}{16}$	**Choice Value**
h_1	1	1	1	1	0.9375
h_2	1	1	1	0	0.8750
h_3	1	0	1	1	0.6875
h_4	1	0	1	0	0.6250
h_5	1	0	0	0	0.5000
h_6	1	1	1	1	0.9375

Example 3. *From Table 4, it can be seen that Mr. X will select the house* h_1 *or* h_6.

Remark 1. *The choice value algorithm is essentially a weighted sorting algorithm, the logic is rational and understandable and the computation processes are straightforward. Algorithm 1, which returns the maximum value in an array with size of n and it takes O(n) times. The time complexity of the algorithm is O(n).*

However, there is a prerequisite for using this method, that is, all attributes are 'good' descriptions, and the greater the value, the better. However, in practice, attributes may be 'good', 'bad', and 'constrained', so this algorithm needs to be further improved according to the actual problem.

3.2. The Comparison Score Algorithm (Algorithm 2)

In the comparison score algorithm, rather than utilizing the concept of choice values designed for crisp soft sets, it compares the membership values of two objects concerning a common attribute to determine which one relatively possesses that attribute. The algorithm is as follows:

Algorithm 2 The comparison score algorithm

1: Input the fuzzy soft sets $(\widetilde{F}, A)$.
2: Construct the comparison-table of the fuzzy soft sets $(\widetilde{F}, A)$ and compute r_i and t_i for o_i, $\forall i$.
3: Compute the score of o_i, $\forall i$.
4: The decision is S_k if, $S_k = max_i S_i$.
5: If k has more than one value then any one of o_k may be chosen.

The comparison table of a fuzzy soft set $(\widetilde{F}, A)$ is a square table in which rows and columns are both labeled by the objects $o_1, o_2, ..., o_n$ of the universe. The entries c_{ij} indicate the number of parameters for which the membership value of o_i exceeds or equal to the membership value of o_j. The c_{ij} is computed by

$$c_{ij} = |\{e \in A; F(e)(o_i) \geq F(e)(o_j)\}|. \tag{3}$$

The row-sum r_i of object o_i is computed by

$$r_i = \sum_{j=1}^{n} c_{ij}. \tag{4}$$

The column-sum t_j of object o_j is computed by

$$t_i = \sum_{i=1}^{n} c_{ij}. \tag{5}$$

The score s_i of object o_i is defined as

$$s_i = r_i - t_i. \tag{6}$$

The objects with the maximum score computed from the comparison table will be regarded as the optimal decision.

Example 4. *We consider fuzzy soft set* $(\widetilde{F}, A)$ *given in Table 2. The comparison table and the comparison score table of* $(\widetilde{F}, A)$ *are given in Tables 5 and 6, respectively. From Table 6, it is seen that Mr. X will select the house* h_2.

Table 5. The comparison table of fuzzy soft set $(\widetilde{F}, A)$.

U	x_1	x_2	x_3	x_4	x_5	x_6	x_7	x_8
x_1	7	2	2	1	3	2	3	3
x_2	5	7	4	4	6	4	5	6
x_3	6	4	7	3	6	4	5	6
x_4	6	4	5	7	5	3	3	6
x_5	5	2	3	3	7	4	5	6
x_6	6	4	4	5	7	7	5	7
x_7	5	3	4	5	7	6	7	7
x_8	5	2	2	4	5	4	4	7

Table 6. The comparison score table of fuzzy soft set $(\widetilde{F}, A)$.

	Row-Sum (r_i)	**Column- Sum** (t_i)	**Comparison Score** (s_i)
h_1	23	45	−22
h_2	41	28	13
h_3	41	31	10
h_4	39	32	7
h_5	35	46	−11
h_6	45	34	11
h_7	44	37	7
h_8	33	48	−15

Remark 2. *The number of objects in the fuzzy soft set* $(\widetilde{F}, A)$ *is assumed to be* n. *For calculating each entry of the comparison table, the objects need to compare with each other, and the complexity of computing the comparison table is* $O(n^2)$. *The complexity of computing each score of each object is* $O(2n)$, *and the complexity of selecting the max value is* $O(n)$. *Thus, the complexity of Algorithm 2 is* $O(n^2)$.

However, the comparison score algorithm presents certain limitations. Alcantud [20] shows that Algorithm 2 may result in a loss of information along the construction of a resultant fuzzy soft set from the multi-observer data. The main novelty in his proposal regarding Algorithm 2 is in the definition of the comparison matrix. Our concerns are as follows:

1. *Rank reversal occurs in the comparison score algorithm. In this phenomenon, the objects' order of preference changes when an object is added to or removed from the decision problem. We will illustrate this phenomenon in Section 3.3.*
2. *Add/delete an object, and the comparison matrix needs to be recalculated. This means that a new comparison table has to be recalculated when the attributes/objects need to be added/deleted, which indicates that plenty of recalculations should be involved to get a new solution set.*
3. *Attribute importance is considered to be the equal importance, and then the option cannot be distinguished according to the importance of the attribute.*

3.3. Rank Reversal in the Comparison Score Algorithm

In a decision-making problem, the rank reversal means a change in the rank ordering of the preferability of possible alternative decisions when the method of choosing changes or the set of other available alternatives changes. Such a phenomenon was first pointed out by Belton and Gear [21].

Some decision-making algorithms have been criticized for the possible rank reversal phenomenon caused by the addition or deletion of an alternative [21–24].

There are strong arguments on which a fuzzy soft sets based decision-making method is more reasonable than others. The purpose of this paper is not to contribute further to that debate, but to point out problems and analyze the causes, and prepare for further improvements. Here, an example is provided to illustrate the rank reversal phenomenon in the comparison score algorithm.

Example 5. *Let $(\widetilde{F}, A)$ be a fuzzy soft set; it can be expressed in Equation (7). By the comparison score algorithm, we can get comparison table (8) and comparison score (9):*

$$(\widetilde{F}, A) = \begin{bmatrix} 0.2 & 0.3 & 0.6 & 0.3 & 0.9 & 0.6 \\ 0.9 & 0.1 & 0.3 & 0.9 & 0.7 & 0.8 \\ 0.4 & 0.1 & 0.5 & 0.8 & 0.1 & 0.1 \end{bmatrix}, \tag{7}$$

$$Comparison\ Table = \begin{bmatrix} 6 & 3 & 4 \\ 3 & 6 & 5 \\ 2 & 2 & 6 \end{bmatrix}, \tag{8}$$

$$Comparison\ Score = \begin{bmatrix} 13 & 11 & 2 \\ 14 & 11 & 3 \\ 10 & 15 & -5 \end{bmatrix}. \tag{9}$$

From the comparison score (9), it is seen that h_2 will be chosen, and we have a sorted sequence $h_2 \succ h_1 \succ h_3$.

Example 6. *We add an object $h_4 = (0.9, 0.3, 0.3, 0.2, 0.8, 0.9)$ to $(\widetilde{F}, A)$, as Equation (10), and then we can get comparison table (11) and comparison score (12):*

$$(\widetilde{F}, A) = \begin{bmatrix} 0.2 & 0.3 & 0.6 & 0.3 & 0.9 & 0.6 \\ 0.9 & 0.1 & 0.3 & 0.9 & 0.7 & 0.8 \\ 0.4 & 0.1 & 0.5 & 0.8 & 0.1 & 0.1 \\ 0.9 & 0.3 & 0.3 & 0.2 & 0.8 & 0.9 \end{bmatrix}, \tag{10}$$

$$Comparison\ Table = \begin{bmatrix} 6 & 3 & 4 & 4 \\ 3 & 6 & 5 & 3 \\ 2 & 2 & 6 & 2 \\ 3 & 5 & 4 & 6 \end{bmatrix}, \tag{11}$$

$$Comparison\ Score = \begin{bmatrix} 17 & 14 & 3 \\ 17 & 16 & 1 \\ 12 & 19 & -7 \\ 18 & 15 & 3 \end{bmatrix}. \tag{12}$$

From the comparison score (12), it is seen that h_1 and h_4 will be chosen, and we have a sorted sequence $(h_1 = h_4) \succ h_2 \succ h_3$.

Remark 3. *Examples 5 and 6 show that rank reversal phenomenon occurs in the comparison score algorithm; it caused by the addition or deletion of an object. As can be seen from Examples 5 and 6, the ranking between h_1 and h_2 is $h_2 \succ h_1$ before h_4 is introduced, but becomes $h_1 \succ h_2$ after h_4 is added, where the symbol '$\succ$' means 'is superior to.' The ranking is reversed after the addition of alternative h_4. Such a phenomenon is referred to as rank reversal, which may occur not only when a copy of an alternative is added, but also when a new alternative is added as well as when an existing alternative is removed. In some cases, this may lead to total rank reversal, where the order of preferences is entirely inverted. That is, that the alternative considered the best, with the inclusion or removal of an alternative, then becomes the worst. Such a phenomenon in many cases may not be acceptable, for example, the ranking of candidates in recruitment, choosing the best students according to their* *grades, and so on. In practice, we can construct special test problems to test the validity of the decision-making algorithm. If the solution shows some logical contradictions, then one might argue that there is a problem with the method that derives them.*

In classical mathematics, the decision-making problems description is $(A, \Theta, \Xi, \kappa, D)$ (Grabisch et al.) [25]. The A is the set of alternatives or possible actions. The Θ is a set of states of the environment in which decisions are taken. The Ξ is a set of consequences resulting from the choice of a particular alternative. The κ is a mapping $A \times \Theta \to \Xi$ specifying a consequence for each element of the environment. The space $A \times \Theta$ defines the solution space. The D is the decision function $D : \Theta \to R$ reflects the preference structure of the decision maker.

Definition 7. *([26]) The decision function D incorporates the goals of the decision maker. It induces a preference ordering on the set of consequences Ξ such that*

$$\xi_i \succ \xi_j \quad iff \quad D(\xi_i) \succ D(\xi_j), \tag{13}$$

where $\xi_i, \xi_j \in \Xi$, and $\succ$ is the preference relation, i.e., consequence is preferred to consequence.

Let $(\widetilde{F}, A)$ be two fuzzy soft sets on the universe U. Suppose that o_i and o_j are objects in the universe U. In the fuzzy soft set $(\widetilde{F}, A)$, let D be the decision function, and $D(o_i) \succ D(o_j)$. Add an object o_k to $(\widetilde{F}, A)$, let D' be the decision function, but $D'(o_j) \succ D'(o_i)$. This means $D \neq D'$, the decision function changed caused by the addition of an object.

From Examples 5 and 6, we can see $D(o_2) \succ D(o_1)$ in Example 5, but $D'(o_1) \succ D'(o_2)$ in Example 6. The instability of the selection result indicates that the 'decision rule' is ambiguous, that is, the decision function will be changed according to the addition/deletion of objects.

In practice, suppose k students $U = \{n_1, n_2..., n_k\}$ participated in a competition, by the comparison score algorithm, you can choose the desired candidate n_x and $n_x \in U$. When a new student n_{k+1} participation in the competition, $U' = U \cup \{n_{k+1}\}$, by the comparison score algorithm you can choose the desired candidate n_y, and $(n_y \in U \quad and \quad n_y \neq n_x)$. This means the ranking of n_x and n_y is reversed when candidate n_{k+1} participation in the competition. In real life, a decision maker's preference ordering between two alternatives should remain unchanged if an additional alternative added or removed. Usually, if $n_x < n_{k+1}$, then n_{k+1} will be chosen, else n_x still be selected.

4. Improved Decision-Making Algorithm Based on Fuzzy Soft Set and Ideal Solution

Most of our real-life problems are imprecise in nature. The classical crisp mathematical tools are not capable of dealing with such problems. The fuzzy set theory has been used quite extensively to deal with such imprecisions. In general, the traditional decision-making algorithm based on soft sets have some shortcomings. All the attributes are treated uniformly, that is, all attributes are treated as 'good' attributes. Sometimes, the value of the attribute is not the bigger the better, for example, 'expensive'. The purpose of the decision maker ignored, and it usually assumed that the bigger (the value of each attribute) the better. It is unrealistic to use a fixed method to deal with the ever-changing problems.

In order to overcome these shortcomings, this paper proposes improvement from the following aspects, as shown in Figure 1:

Firstly, the ideal solution is introduced in the design phase, that is, a very clear decision objective is formulated. We will illustrate this in Section 4.1.

Secondly, make clear the meaning of attributes. The attribute will be divided into 'pros' and 'cons' attributes. The 'pros attribute' is a 'good' description of the object, and the 'cons attribute' is a 'bad' description of the object. At the same time, whether attributes contain constraints is also taken into account. We will illustrate this in Section 4.2.

Finally, when an ideal solution is generated, the decision is made by comparing the similarity between the object and the ideal solution. We will illustrate this in Section 4.3.

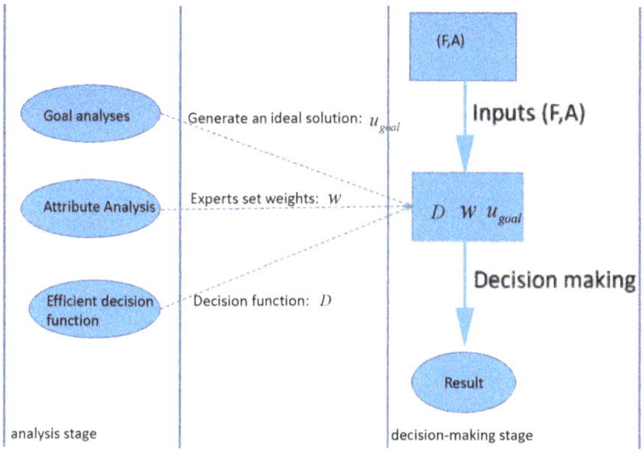

Figure 1. The decision-making model.

4.1. The Ideal Solution Method

Yoon and Hwang [27] developed the algorithm for order preference by similarity to the ideal solution in 1981. The ideal solution method aims to obtain the best compromise solution, which is the one that is the closest to the ideal solution, that is, it has the shortest distance from the ideal solution. Let $S = (S_1, S_2, ..., S_n)$, $S_i = (s_{i1}, s_{il}, ..., s_{im})$, $i = 1, ..., n$ be a solution of a decision-making problem from the ith group member, m be the number of objectives ($m > 1$). Let $S_0 = (s_{01}, s_{0l}, ..., s_{0m})$ be the ideal solution. The ideal solution method is formulated as follows:

$$\begin{aligned} & find \quad p \\ & s.t. \quad d^* = d_p = min\{d_i; i = 1, 2, ..., n\} \\ & = min\{\sum_{j=1}^{m} |s'_{ij} - s_{0j}|; i = 1, 2, ..., n\}, \end{aligned} \tag{14}$$

where

$$\begin{aligned} & s'_{ij} = \begin{cases} \frac{s_{ij}}{\tilde{s}_j}, & if \quad \tilde{s}_j \neq 0, \\ 0, & if \quad \tilde{s}_j = 0, \end{cases} \quad i = 1, 2, ..., n, \quad j = 1, 2, ..., m, \\ & \tilde{s}_j = max\{s_{ij}; i = 1, 2, ..., n\}, j = 1, 2, ..., m. \end{aligned} \tag{15}$$

When an ideal solution $S_0 = (s_{01}, s_{0l}, ..., s_{0m})$ is generated, the algorithm starts to measure the distance of the ideal solution to the other candidates. A distance matrix D for each objective of solutions to the ideal solution is thus established:

$$D = \begin{bmatrix} (d_{11}) & (d_{12}) & \dots & (d_{1m}) \\ (d_{21}) & (d_{22}) & \dots & (d_{2m}) \\ \vdots & \vdots & & \vdots \\ (d_{n1}) & (d_{n2}) & \dots & (d_{nm}) \end{bmatrix}, \tag{16}$$

where $d_{ij} = |s'_{ij} - s_{0j}|, i = 1, ..., n, j = 1, ..., m$.

The distances from different objective values of each solution are obtained:

$$d_i = \sum_{j=1}^{m} d_{ij}, i = 1, ..., n. \tag{17}$$

The final solution that has the shortest distance is then found from

$$\begin{aligned} & find \quad p \\ & s.t. \quad d^* = d_p = min\{d_i, i = 1, 2, ..., n\}, 1 \le p \le n, \end{aligned} \tag{18}$$

where d^* is the shortest total-distance between the solutions and the ideal solution, and the pth solution is the closest solution as the final compromise solution of this decision-making problem.

4.2. The Ideal Solution of Each Attribute

In practice, when we use a soft set to solve a problem, the attribute can be a 'good' description of an object or a 'bad' one. Likewise, attributes sometimes contain constraints, and sometimes do not contain constraints. In this situation, a choice value based approach and comparison score based approach are not useable. In other words, there are two prerequisites for the choice value based approach and the comparison score based approach, that is, on the universal U, all attributes are positive descriptions and are unconstrained. For each attribute, a bigger value indicates a better candidate. In reality, this is not always reasonable.

Bellman and Zadeh proposed [17] a fuzzy decision model in 1970, and discuss how to apply these concepts to the decision-making process under a fuzzy environment.

Definition 8. *([17]) The X represents all possible strategies, and the fuzzy objective $\widetilde{G}$ is a fuzzy set on the X, the membership function $\mu_{\widetilde{G}} : X \to [0, 1]$. The objective function is the reaction of the decision-maker to a certain ambiguity of the target. The $\mu_{\widetilde{G}}$ response strategy x achieves satisfaction with target $\widetilde{G}$.*

Definition 9. *([17]) Let $\widetilde{G}$ and $\widetilde{C}$ be fuzzy targets and fuzzy constraints in universal X, the fuzzy decision $\widetilde{D}$ is also a fuzzy set of X, and it is defined as the intersection of $\widetilde{G}$ and $\widetilde{C}$, that is, $\widetilde{D} = \widetilde{G} \cap \widetilde{C}$, the membership function is*

$$\mu_{\widetilde{D}}(x) = min\{\mu_{\widetilde{G}}(x), \mu_{\widetilde{C}}(x)\}.$$

In fuzzy decision-making, the membership function of $\mu_{\widetilde{G}}(x)$ that achieves maximum value strategy is called maximizing strategy, and the membership function is

$$\mu_{\widetilde{D}}(x^*) = max_{x \in X} min\{\mu_{\widetilde{G}}(x), \mu_{\widetilde{C}}(x)\}.$$

For the attribute with constraints, we can use Bellman and Zadeh's model to find the best solution. For the attribute without constraints, it is a maximum/minimum problem.

Let $U = \{u_1, u_2, ..., u_n\}$ and $(\widetilde{F}, A)$ be a fuzzy soft set of dimension k over U, $e_j \in A$. For attribute e_j, let μ_j be the membership function and $j \in \{1, 2, ..., k\}$. $\widetilde{F}(e_j) = \{\mu_j(u_1)/u_1, \mu_j(u_2)/u_2, ..., \mu_j(u_n)/u_n\}$. Let $\Re_{e_j}$ be the maximum target of attribute e_j, $\Re_{e_j} = \mu_{j\widetilde{D}}(x^*)$.

Definition 10. *Let $(\widetilde{F}, A)$ be a fuzzy soft set, and $\mu_{j\widetilde{G}}(x)$ is the membership function of attribute e_j. $\mu_{j\widetilde{C}}(x)$ is the constraint function. The ideal solution of e_j is formulated as follows:*

$$\begin{cases} \Re_{e_j} = max_{x \in X} min\{\mu_{j\widetilde{G}}(x), \mu_{j\widetilde{C}}(x)\}, \\ s.t. \quad x \in U, \end{cases} \tag{19}$$

as shown in Figure 2.

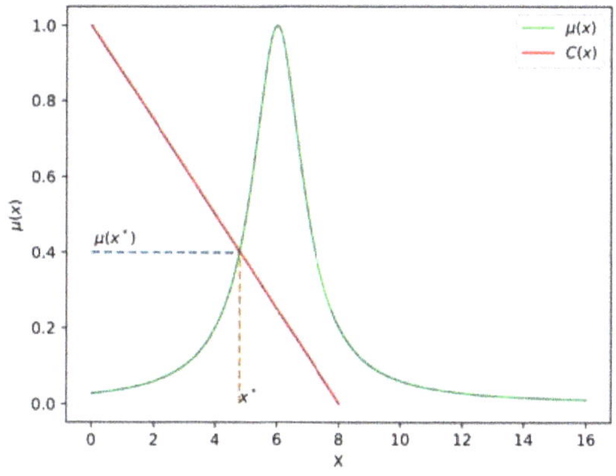

Figure 2. Attribute with constraint $\widetilde{D} = \widetilde{G} \cap \widetilde{C}$.

Let $(\widetilde{F}, A)$ be a fuzzy soft set,$\mu_{j\widetilde{G}}(x)$ is the membership function of attribute e_j and without constraints function.

Definition 11. *The attribute e_j without constraints is a 'good' description, and the ideal solution is the maximum value of $\mu_j(x)$. The ideal solution of e_j is formulated as follows:*

$$\begin{cases} \Re_{e_j} = max\mu_j(x), \\ s.t. \quad x \in U, \end{cases} \tag{20}$$

as shown in Figure 3a.

Definition 12. *The attribute e_j without constraints is a 'bad' description, and the ideal solution is the minimum value of $\mu_j(x)$. The ideal solution of e_j is formulated as follows:*

$$\begin{cases} \Re_{e_j} = min\mu_j(x), \\ s.t. \quad x \in U, \end{cases} \tag{21}$$

as shown in Figure 3b.

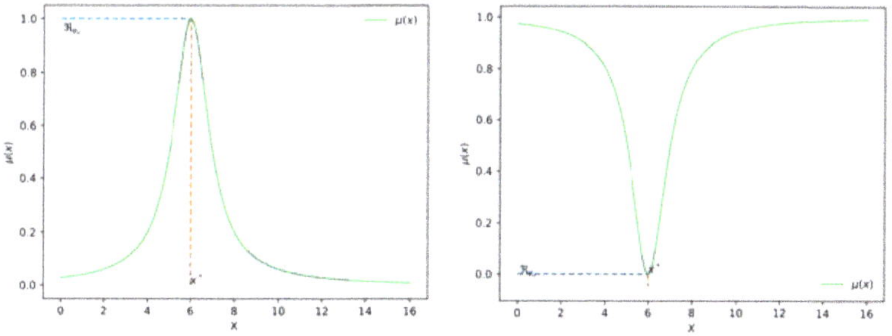

(**a**) (**b**)

Figure 3. Attribute without constraints. (**a**) 'Good' attribute without constraints; (**b**) 'Bad' attribute without constraints.

The ideal solution u_{goal} of $(\widetilde{F}, A)$ is the combination of each attribute $u_{goal} = \{\Re_{e_1}, \Re_{e_2}, ..., \Re_{e_k}\}$.

4.3. The Decision Function—Hamming Distance

The decision function is used to determine the similarity between u_x and u_{goal} in the fuzzy soft set. Many algorithms can be used as efficient decision functions, especially when the fuzzy soft set $(\widetilde{F}, A)$ has many objects, such as fuzzy S-trees, signature trees, t-concept lattice, Artificial Bee Colony (ABC) algorithm [28–31], and so on. Here, we use the widely used Hamming distance.

The normalized Hamming distance is a useful technique for calculating the differences between two elements, two sets, etc [32]. For two sets A and B, it can be defined as follows.

Definition 13. *A normalized Hamming distance of dimension n is a mapping $d_H : R_n \times R_n \rightarrow R$ such that:*

$$d_H(A, B) = \frac{1}{n}(\sum_{i=1}^{n} |a_i - b_i|),$$

where a_i and b_i are the ith arguments of the sets $A = \{a_1, a_2, ..., a_n\}$ and $B = \{b_1, b_2, ..., b_n\}$, respectively.

Let $(\widetilde{F}, A)$ be a fuzzy soft set over U. All attributes have the same degree of importance, u_{goal} is the ideal solution, and u_i is the object. The decision-making problem becomes the optimization problem:

$$\begin{cases} min \quad d_H(u_{goal}, u_i), \\ s.t. \quad u_i \in U, \end{cases} \quad i = 1, 2, ..., n. \tag{22}$$

Definition 14. *A weighted Hamming distance of dimension n is a mapping $d_{WH} : R_n \times R_n \rightarrow R$ that has an associated weighting vector W of dimension n such that the sum of the weights is 1 and $w_j \in [0, 1]$. Then:*

$$d_{WH}(A, B) = (\sum_{i=1}^{n} \omega_i |a_i - b_i|),$$

where a_i and b_i are the ith arguments of the sets $A = \{a_1, a_2, ..., a_n\}$ and $B = \{b_1, b_2, ..., b_n\}$, respectively.

Almost all methods of decision-making problems require information regarding the relative importance of each attribute. The relative importance is usually given by a set of weights that are normalized to sum to one. In the case of n attributes, a weight set is $\omega = (\omega_1, \omega_2, ..., \omega_n)$ and $\sum_{j=1}^{n} \omega_j = 1$. The weights can be assigned by the decision maker directly, or calculated using the eigenvector method or the weighted least square method. The IOWA operator was introduced by Yager and Filev [33].

Definition 15. *([33]) An IOWA operator of dimension n is a mapping $f : R^n \rightarrow R$ that has an associated weighting vector ω of dimension n such that the sum of the weights is 1 and $w_j \in [0, 1]$. Then,*

$$f_{IOWA}(< u_1, a_1 >, < u_2, a_2 >, ..., < u_n, a_n >) = \sum_{j=1}^{n} w_j b_j,$$

where b_j is the a_i value of the IOWA pair u_i ,a_i having the jth largest u_i, u_i is the order inducing variable and a_i is the argument variable.

Let $(\widetilde{F}, A)$ be a fuzzy soft set over U, all attributes have the same degree of importance, the attribute has a weight of ω, u_{goal} is the optimal target, and u_i is the object. The decision-making problem becomes the optimization problem:

$$\begin{cases} min \quad d_{WH}(u_{goal}, u_i), \\ s.t. \quad u_i \in U, \end{cases} \quad i = 1, 2, ..., n. \tag{23}$$

4.4. The Decision-Making Algorithm Based on Fuzzy Soft Sets and Ideal Solution

Let $U = \{u_1, u_2, ..., u_n\}$ and $(\widetilde{F}, A)$ be a fuzzy soft set with k attributes $A = \{e_1, e_2, ..., e_k\}$, as Equation (24):

$$(\widetilde{F}, A) = \begin{bmatrix} (u_{11}) & (u_{12}) & \dots & (u_{1k}) \\ (u_{21}) & (u_{22}) & \dots & (u_{2k}) \\ \vdots & \vdots & & \vdots \\ (u_{n1}) & (u_{n2}) & \dots & (u_{nk}) \end{bmatrix}. \tag{24}$$

We can analyze each attribute e_x independently, and the ideal solution $\Re_{e_x}$ of each attribute can be obtained. The analysis and processing are described in Section 4.2.

By combining the ideal solution of each attribute, we can get the ideal solution u_{goal} of the fuzzy soft sets:

$$u_{goal} = \{\Re_{e_1}, \Re_{e_2}, ..., \Re_{e_k}\}.$$

The decision-making fuzzy soft set $(\widetilde{FD}, A)$ can be expressed in matrix form as Equation (25):

$$(\widetilde{FD}, A) = \begin{bmatrix} (\Re_{e_1}) & (\Re_{e_2}) & \dots & (\Re_{e_k}) \\ (u_{11}) & (u_{12}) & \dots & (u_{1k}) \\ (u_{21}) & (u_{22}) & \dots & (u_{2k}) \\ \vdots & \vdots & & \vdots \\ (u_{n1}) & (u_{n2}) & \dots & (u_{nk}) \end{bmatrix}. \tag{25}$$

In the fuzzy soft set $(\widetilde{F}, A)$, we first establish the ideal solution and then find the object closest to the ideal solution through the choice algorithm, which is the result of selection. The decision-making algorithm based on fuzzy soft sets and ideal solution (Algorithm 3) is formulated as follows:

$$\begin{aligned} & find \quad p, \\ & s.t. \quad d^* = d_p = min\{d_i; i = 1, 2, ..., n\}, \end{aligned} \tag{26}$$

where $d_i = d_{H_i}(u_{goal}, u_i); i = 1, 2, ..., n$ when using the normalized Hamming distance and $d_i = d_{WH_i}(u_{goal}, u_i); i = 1, 2, ..., n$ when using the weighted Hamming distance, respectively.

The algorithm is as follows:

Algorithm 3 The decision-making algorithm based on fuzzy soft sets and ideal solution

1: Input the fuzzy soft set $(\widetilde{F}, A)$.
2: Sort the attributes $(e_1, e_2, ..., e_n)$ in descending order according to its weight, and set the IOWA operator ω according to the purpose of the decision maker.
3: Compute the optimization target $u_{goal} = \{\Re_1, \Re_2, ..., \Re_k\}$ according to membership function of each attribute.
4: Compute the Hamming distance $d_i(u_{goal}, u_i), i = 1, 2, ..., n$.
5: The decision is d_k if, $d_k = min_k d_i$.
6: If k has more than one value, then any one of o_k may be chosen.

Remark 4. *If $u_{goal} = \{1,1,...,1\}$ and $d_H(u_{goal}, u_i)$ is used, it is a choice value decision-making model. If $u_{goal} = \{1,1,...,1\}$ and $d_{WH}(u_{goal}, u_i)$ is used, it is a weighted choice value decision-making model. It should be noted that $u_{goal} = \{1,1,...,1\}$ is not always reasonable in practical problems.*

As can be seen from Table 7, the fuzzy soft sets and ideal solution based algorithm focuses on the modular structure, and it emphasizes the analysis of decision objectives, attributes analysis and the flexible decision function. In Algorithm 3, each object needs to compare with the ideal solution, and, to deal with n items, its algorithm complexity is $O(n)$.

Table 7. Features of the fuzzy soft set based decision-making algorithms.

Algorithm	Time Complexity	Subjective Weights	Attribute Analysis	Decision Function
[3]	$O(N)$	Yes	No	Choice value
[12]	$O(N^2)$	No	No	Comparison matrix
[14]	$O(N)$	Yes	No	Fuzzy choice value
[15]	$O(N)$	Yes	No	Choice value of level soft set
[20]	$O(N^2)$	No	No	New relative comparison matrix
Algorithm 3	$O(N)$	Yes	Yes	Similarity measure & Substitutable

5. Numerical Experiments

We provide numerical experiments in this section. We will use an example to illustrate Algorithm 3, see Section 5.1. In Section 5.2, we will use Hwang and Yoon [27]'s example to illustrate the algorithm proposed in this paper. As a comparison, the traditional method and the algorithm proposed in this paper are applied to this example. We've added Python programs and validation data to validate examples in this article easily [34].

5.1. Example of Fuzzy Soft Sets and Ideal Solution Based Decision-Making Algorithm

Let $U = \{u_1, u_2, ..., u_8\}$ and $(\widetilde{F}, A)$ be a fuzzy soft set with seven attributes. Then, we add the u_{goal} to $(\widetilde{F}, A)$ and the decision fuzzy soft set as Equation (27).

Assuming that all attributes are 'good' description and without constraints. By Equation (20) , $\mu_{goal} = \{1,1,...,1\}$:

$$(\widetilde{FD}, A) = \begin{bmatrix} & e_1 & e_2 & e_3 & e_4 & e_5 & e_6 & e_7 \\ u_{goal} & 1 & 1 & 1 & 1 & 1 & 1 & 1 \\ u_1 & 0.2 & 0.4 & 0.1 & 0.5 & 0.8 & 0.1 & 0.1 \\ u_2 & 0.3 & 0.2 & 0.3 & 0.6 & 0.3 & 0.9 & 0.6 \\ u_3 & 0.3 & 0.1 & 0.6 & 0.7 & 0.8 & 0.8 & 0.3 \\ u_4 & 0.3 & 0.7 & 0.9 & 0.9 & 0.1 & 0.4 & 0.5 \\ u_5 & 0.3 & 0.9 & 0.1 & 0.3 & 0.2 & 0.2 & 0.3 \\ u_6 & 0.3 & 0.9 & 0.1 & 0.3 & 0.9 & 0.7 & 0.8 \\ u_7 & 0.3 & 0.9 & 0.1 & 0.3 & 0.2 & 0.8 & 0.9 \\ u_8 & 0.3 & 0.9 & 0.1 & 0.3 & 0.1 & 0.4 & 0.2 \end{bmatrix}, \tag{27}$$

$$d_H(u_{goal}, u_1) = \frac{4.8}{7}, \quad d_H(u_{goal}, u_2) = \frac{3.8}{7},$$
$$d_H(u_{goal}, u_3) = \frac{3.4}{7}, \quad d_H(u_{goal}, u_4) = \frac{3.2}{7},$$
$$d_H(u_{goal}, u_5) = \frac{4.7}{7}, \quad d_H(u_{goal}, u_6) = \frac{3.0}{7},$$
$$d_H(u_{goal}, u_7) = \frac{3.5}{7}, \quad d_H(u_{goal}, u_8) = \frac{4.7}{7}.$$

From the normalized Hamming distance, it is seen that u_6 will be chosen.

Letting $\omega = \{\frac{1}{2}, \frac{1}{4}, \frac{1}{8}, \frac{1}{16}, \frac{1}{32}, \frac{1}{64}, 0\}$,

$$\begin{aligned}
d_{WH}(u_{goal}, u_1) &= 0.714, & d_{WH}(u_{goal}, u_2) &= 0.686, \\
d_{WH}(u_{goal}, u_3) &= 0.653, & d_{WH}(u_{goal}, u_4) &= 0.481, \\
d_{WH}(u_{goal}, u_5) &= 0.569, & d_{WH}(u_{goal}, u_6) &= 0.539, \\
d_{WH}(u_{goal}, u_7) &= 0.559, & d_{WH}(u_{goal}, u_8) &= 0.569.
\end{aligned}$$

From the weighted Hamming distance, it is seen that u_4 will be chosen.

5.2. Algorithm Comparison

A country decided to purchase a fleet of jet fighters from the U.S. The Pentagon officials offered the characteristic information of four models that may be sold to that country. The Air Force analyst team of that country agreed that six characteristics (attributes) should be considered. They are maximum speed (X_1) ferry range (X_2), maximum payload (X_3), purchasing cost (X_4), reliability (X_5), and maneuverability (X_6). The measurement units for the attributes are mach, miles, pounds, dollars (in millions), high-low scale, and high-low scale, respectively. The decision matrix for the fighter aircraft selection problem, then, is:

$$D = \begin{bmatrix} & X_1 & X_2 & X_3 & X_4 & X_5 & X_6 \\ A_1 & 2.0 & 1500 & 20000 & 5.5 & average & veryhigh \\ A_2 & 2.5 & 2700 & 18000 & 6.5 & low & average \\ A_3 & 1.8 & 2000 & 21000 & 4.5 & high & high \\ A_4 & 2.2 & 1800 & 20000 & 5.0 & average & average \end{bmatrix}. \quad (28)$$

5.2.1. The Traditional Decision-Making Method

Attribute ratings are usually normalized to eliminate computational problems caused by different measurement units in a decision matrix. Linear normalization is a simple procedure that divides the ratings of a certain attribute by its maximum value. The normalized value of x_{ij} is given as

$$r_{ij} = \frac{x_{ij}}{x_j^*} \quad i = 1, ..., m; j = 1, ..., n, \quad (29)$$

where x_j^* is the maximum value of the jth attribute. Clearly, the attribute is more satisfactory as r_{ij} approaches 1, $(0 \leq r_{ij} \leq 1)$:

$$D = \begin{bmatrix} & X_1 & X_2 & X_3 & X_4 & X_5 & X_6 \\ A_1 & 0.8 & 0.56 & 0.95 & 0.82 & 0.71 & 1.0 \\ A_2 & 1.0 & 1.0 & 0.86 & 0.69 & 0.43 & 0.56 \\ A_3 & 0.72 & 0.74 & 1.0 & 1.0 & 1.0 & 0.78 \\ A_4 & 0.88 & 0.64 & 0.95 & 0.9 & 0.71 & 0.56 \end{bmatrix}. \quad (30)$$

The key idea of the weighting method is to transform the multiple objectives in the decision-making problem into weighted single objective functions, which are described as follows (Zadeh, 1963) [17]:

$$\begin{cases} max \quad wf(x) = \sum_{i=1}^{k} \omega_i f_i(x), \\ s.t. \quad x \in X, \end{cases}$$

where $\omega = \{\omega_1, \omega_2, ..., \omega_k\}$ is a vector of weighting coefficients assigned to the objective functions.

Let $(\widetilde{F}, A) = D$, and e_j the jth attribute. Let $\omega_1 = \omega_2 = \omega_3 = \omega_4 = \omega_5 = \omega_6 = \frac{1}{6}$.

The decision table is as shows in the following:

$$\widetilde{D} = \begin{bmatrix} & X_1 & X_2 & X_3 & X_4 & X_5 & X_6 \\ A_1 & 0.8 & 0.56 & 0.95 & 0.82 & 0.71 & 1.0 \\ A_2 & 1.0 & 1.0 & 0.86 & 0.69 & 0.43 & 0.56 \\ A_3 & 0.72 & 0.74 & 1.0 & 1.0 & 1.0 & 0.78 \\ A_4 & 0.88 & 0.64 & 0.95 & 0.9 & 0.71 & 0.56 \end{bmatrix} \quad Score = \begin{bmatrix} \omega f(x) \\ 0.8067 \\ 0.7567 \\ 0.8733 \\ 0.7733 \end{bmatrix}. \tag{31}$$

From Equation (31), A_3 will be chosen.

Remark 5. *Let* $(\widetilde{F}, A) = D$, *and* e_j *the jth attribute. By using linear normalization,*

$$\mu_{e_j}(x) = \frac{x}{x_j^*}, j = 1, ..., n,$$

where x_j^* *is the maximum value of the jth attribute.*

In other words, in the traditional decision-making mode, the membership function of the attribute is always established in the form of a linear function. This is not always accurate and feasible.

5.2.2. The Decision-Making Based on Fuzzy Soft Sets and Ideal Solution

In this subsection, we illustrate the decision process with the following examples. Suppose three groups of air force analyst team make the following goals:

Team 1: "Spare no expense to buy a jet fighter, and the jet fighter that is the fastest, most stable and has the best maneuverability".

Team 2: "Buy a jet fighter with a budget of 5 million, and a jet fighter that is stable and has the best maneuverability".

Team 3: "Spend the least money to buy the indicators of a relatively good jet fighter".

Let $A = \{e_1 =$ 'maximum speed', $e_2 =$ 'ferry range', $e_3 =$ 'maximum payload', $e_4 =$ 'purchasing cost', $e_5 =$ 'reliability', $e_6 =$ 'maneuverability'$\}$. The attributes $\{e_1 =$ 'maximum speed', $e_2 =$ 'ferry range', $e_3 =$ 'maximum payload', $e_5 =$ 'reliability', $e_6 =$ 'maneuverability'$\}$ are 'pros' attributes, and they are all positive descriptions of jet fighters. For the attribute $\{e_4 =$ 'purchasing cost'$\}$, of course, the cheaper, the better.

For Team 1, the attribute $\{e_4 =$ 'purchasing cost'$\}$ is a factor that doesn't need to be considered, no matter how expensive it is. The attribute $\{e_1 =$ 'maximum speed'$\}$ is the primary consideration, $\{e_5 =$ 'reliability'$\}$ second, and finally consider $\{e_6 =$ 'maneuverability'$\}$. Other factors are relatively unimportant. Therefore, the degree of importance is: $e_1 > e_5 > e_6 > (e_2 = e_3) > e_4$.

For Team 2, the attribute $\{e_4 =$ 'purchasing cost'$\}$ is the primary consideration. This is a user constraint, and, by Equation (19), we can get the ideal solution of e_4. The attributes $\{e_5 =$ 'reliability'$\}$ and $\{e_6 =$ 'maneuverability'$\}$ are relatively important attributes, and $e_5 > e_6$. Therefore, the degree of importance is: $e_4 > e_5 > e_6 > (e_1 = e_2 = e_3)$.

For Team 3, the attribute $\{e_4 =$ 'purchasing cost'$\}$ is the primary consideration, and the cheaper the better. All other attributes are secondary attributes that are equally important, which is: $e_4 > (e_1 = e_2 = e_3 = e_5 = e_6)$.

The attributes are normalized by a small number of samples, and a rigorous decision maker needs to analyze each indicator carefully. To determine its membership function through investigation and research (the definition of membership function is subjective, and the optimal membership function is not the problem discussed in this paper), we can obtain the optimal goal of our decision more accurately.

Suppose the membership function of each attribute is formulated as follows.

Let $\mu_1(x)$ be the membership function of fast jet fighters:

$$\mu_1(x) = \begin{cases} 1, & x \geq 3.5, \\ \frac{1}{1+e^{-2.1\times(x-1.5)}}, & 0.5 < x < 3.5, \\ 0, & x \leq 0.5. \end{cases}$$

Let $\mu_2(x)$ be the membership function of ferry range:

$$\mu_2(x) = \begin{cases} 1, & x \geq 3200, \\ \frac{1}{1+e^{-0.003\times(x-2000)}}, & 1000 < x < 3200, \\ 0, & x \leq 1000. \end{cases}$$

Let $\mu_3(x)$ be the membership function of maximum payload:

$$\mu_3(x) = \begin{cases} 1, & x \geq 32{,}000, \\ \frac{1}{e^{-0.0005\times(x-20{,}000)}}, & 15{,}000 < x < 32{,}000, \\ 0, & x \leq 15{,}000. \end{cases}$$

Let $\mu_4(x)$ be the membership function of the expensive jet fighter:

$$\mu_4(x) = \begin{cases} \frac{1}{1+e^{-(x-3)}}, & 0 < x, \\ 0, & x \leq 0. \end{cases}$$

Let $\mu_5(x)$ be the membership function of reliability:

$$\mu_5(x) = \begin{cases} \frac{1}{4}, & 0 \leq x < 0.25, \\ \frac{2}{4}, & 0.25 \leq x < 0.5, \\ \frac{3}{4}, & 0.5 \leq x < 0.75, \\ 1, & 0.75 \leq x \leq 1. \end{cases}$$

Let $\mu_6(x)$ be the membership function of maneuverability:

$$\mu_6(x) = \begin{cases} \frac{1}{4}, & 0 \leq x < 0.25, \\ \frac{2}{4}, & 0.25 \leq x < 0.5, \\ \frac{3}{4}, & 0.5 \leq x < 0.75, \\ 1, & 0.75 \leq x \leq 1. \end{cases}$$

The decision table for the fighter aircraft selection can be changed to a fuzzy soft set, as Equation (32):

$$(\widetilde{F}, A) = \begin{bmatrix} & e_1 & e_2 & e_3 & e_4 & e_5 & e_6 \\ A_1 & 0.741 & 0.182 & 0.5 & 0.924 & 0.5 & 1 \\ A_2 & 0.891 & 0.891 & 0.269 & 0.971 & 0 & 0.5 \\ A_3 & 0.652 & 0.5 & 0.622 & 0.818 & 0.75 & 0.75 \\ A_4 & 0.813 & 0.354 & 0.5 & 0.881 & 0.5 & 0.5 \end{bmatrix}. \quad (32)$$

Example 7. *Team 1: Spare no expense to buy a jet fighter, and the jet fighter that is the fastest, most stable and has the best maneuverability.*

Without user constraints, the weight of attributes is: $e_1 > e_5 > e_6 > (e_2 = e_3) > e_4$.

Let $\omega = \{\omega_1 = \frac{1}{2}, \omega_5 = \frac{1}{4}, \omega_6 = \frac{1}{8}, \omega_2 = \frac{1}{16}, \omega_3 = \frac{1}{16}, \omega_4 = 0\}$, $\sum_{i=1}^{6} \omega_i = 1$.

$\Re_1 = 1, \Re_2 = 1, \Re_3 = 1, \Re_4 = 0, \Re_5 = 1, \Re_6 = 1, u_{goal} = \{1,1,1,0,1,1\},$

$$(\widetilde{FD}, A) = \begin{bmatrix} & e_1 & e_2 & e_3 & e_4 & e_5 & e_6 \\ A_{goal} & 1 & 1 & 1 & 0 & 1 & 1 \\ A_1 & 0.741 & 0.182 & 0.5 & 0.924 & 0.5 & 1 \\ A_2 & 0.891 & 0.891 & 0.269 & 0.971 & 0 & 0.5 \\ A_3 & 0.652 & 0.5 & 0.622 & 0.818 & 0.75 & 0.75 \\ A_4 & 0.813 & 0.354 & 0.5 & 0.881 & 0.5 & 0.5 \end{bmatrix}, \tag{33}$$

$$d_{WH}(A_{goal}, A_1) = 0.337, \quad d_{WH}(A_{goal}, A_2) = 0.419,$$
$$d_{WH}(A_{goal}, A_3) = 0.323, \quad d_{WH}(A_{goal}, A_4) = 0.353.$$

From the weighted Hamming distance, it is seen that A_3 will be the choice because it is the closest object to A_{goal}. In addition, $d_{A_2} \succ d_{A_4} \succ d_{A_1} \succ d_{A_3}$.

Example 8. *Team 2: Buy a jet fighter with a budget of 5 million, and a jet fighter that is stable and has the best maneuverability.*

The prices include user constraints: $x^* = 5$, then, $\Re_4 = \mu_4(5) = 0.881$.

The weight of attributes is: $e_4 > (e_5 = e_6) > (e_1 = e_2 = e_3)$.

Let $\omega = \{\omega_4 = \frac{14}{60}, \omega_5 = \frac{11}{60}, \omega_6 = \frac{11}{60}, \omega_1 = \frac{8}{60}, \omega_2 = \frac{8}{60}, \omega_3 = \frac{8}{60}\}, \sum_{i=1}^{6} \omega_i = 1$.

$\Re_1 = 1, \Re_2 = 1, \Re_3 = 1, \Re_4 = 0.881, \Re_5 = 1, \Re_6 = 1, u_{goal} = \{1,1,1,0.881,1,1\}.$

$$(\widetilde{FD}, A) = \begin{bmatrix} & e_1 & e_2 & e_3 & e_4 & e_5 & e_6 \\ A_{goal} & 1 & 1 & 1 & 0.881 & 1 & 1 \\ A_1 & 0.741 & 0.182 & 0.5 & 0.924 & 0.5 & 1 \\ A_2 & 0.891 & 0.891 & 0.269 & 0.971 & 0 & 0.5 \\ A_3 & 0.652 & 0.5 & 0.622 & 0.818 & 0.75 & 0.75 \\ A_4 & 0.813 & 0.354 & 0.5 & 0.881 & 0.5 & 0.5 \end{bmatrix}, \tag{34}$$

$$d_{WH}(A_{goal}, A_1) = 0.312, \quad d_{WH}(A_{goal}, A_2) = 0.423,$$
$$d_{WH}(A_{goal}, A_3) = 0.270, \quad d_{WH}(A_{goal}, A_4) = 0.361.$$

From the weighted Hamming distance, it is seen that A_3 will be the choice. In addition, $d_{A_2} \succ d_{A_4} \succ d_{A_1} \succ d_{A_3}$.

Example 9. *Team 3: Spend the least money to buy the indicators of a relatively good jet fighter".*

Without user constraints, the weight of attributes is:$e_4 > (e_1 = e_2 = e_3 = e_5 = e_6)$.

Let $\omega = \{\omega_4 = \frac{5}{15}, \omega_5 = \frac{2}{15}, \omega_6 = \frac{2}{15}, \omega_2 = \frac{2}{15}, \omega_3 = \frac{2}{15}, \omega_1 = \frac{2}{15}\}, \sum_{i=1}^{6} \omega_i = 1$.

$\Re_1 = 1, \Re_2 = 1, \Re_3 = 1, \Re_4 = 0, \Re_5 = 1, \Re_6 = 1, u_{goal} = \{1,1,1,0,1,1\}.$

$$(\widetilde{FD}, A) = \begin{bmatrix} & e_1 & e_2 & e_3 & e_4 & e_5 & e_6 \\ A_{goal} & 1 & 1 & 1 & 0 & 1 & 1 \\ A_1 & 0.741 & 0.182 & 0.5 & 0.924 & 0.5 & 1 \\ A_2 & 0.891 & 0.891 & 0.269 & 0.971 & 0 & 0.5 \\ A_3 & 0.652 & 0.5 & 0.622 & 0.818 & 0.75 & 0.75 \\ A_4 & 0.813 & 0.354 & 0.5 & 0.881 & 0.5 & 0.5 \end{bmatrix} \tag{35}$$

$$d_{WH}(A_{goal}, A_1) = 0.585 \quad d_{WH}(A_{goal}, A_2) = 0.650$$
$$d_{WH}(A_{goal}, A_3) = 0.503 \quad d_{WH}(A_{goal}, A_4) = 0.605$$

From the weighted Hamming distance, it is seen that A_3 will be the choice. In addition, $d_{A_2} \succ d_{A_4} \succ d_{A_1} \succ d_{A_3}$.

Remark 6. *The decision-making based on fuzzy soft sets and the ideal solution has some advantages. Firstly, the soft set model can be combined with other mathematical models. When it is combined with fuzzy decision-making, the soft set is a natural multi-attribute decision making model, which holds a wide range of application prospects in decision-making and analysis. Secondly, in the traditional ideal solution algorithm, it can be seen that the normalization process is the process of establishing the membership function $\mu(x)$. There are many commonly used normalization methods, i.e., Equation (29), but few of them can reflect the nature of the problem. The fuzzy soft set already contains the membership function $\mu(x)$, which can be used well. Thirdly, from the analysis of the attributes of the fuzzy soft sets, we can see that the attributes themselves are associated with each other. For example, 'price' and other attributes are related to each other, usually because of their high performance and therefore high pricing. Therefore, some attributes have less impact on the decision results because other attributes already contain information about that attribute.*

6. Conclusions

The decision-making is a significant problem, and a good decision system will undoubtedly play a huge role in promoting economy, management, and society. However, it is hard for us to expect a single mathematical model to accomplish such a difficult task. There are mainly two different approaches applying soft set theory to decision-making problems. One is based on choice value, and the other is based on comparison score. This paper analyzes the existing problems of these two methods. The choice value algorithm is not always reasonable in practice because it lacks the analysis of attributes. At the same time, we point out that the comparison score algorithm has the phenomenon of rank reversal, which can be further analyzed and improved. This paper is dedicated to the analysis of these approaches and proposes a new decision-making algorithm. We focus on the application of fuzzy soft set and ideal solutions in decision-making problems. From the decision-making process, we have found that the core of the decision process is the design phase, which is to formulate a model for an identified decision problem. Therefore, this paper emphasizes the analysis of decision objectives, attributes analysis and explicit decision function. Based on these results, we can further probe the practical applications of soft set theory in decision-making problems. Thanks to this modular structure, we can design more efficient decision functions for this model. Moreover, how to avoid rank reversal of the comparison score algorithm is another promising research topic.

Acknowledgments: The authors are very grateful to two anonymous referees and the editor for their valuable comments and constructive suggestions that significantly improved the quality of this paper. This work has been partially supported by the National Natural Science Foundation of China (Grant Nos. 61175044, 61372187, 61473239) and the Fundamental Research Funds for the Central Universities of China (Grant No. SWJTU11ZT29), and the open research fund of key laboratory of intelligent network information processing, Xihua University (SZJJ2012-032, SZJJ2014-052, 16224221).

Author Contributions: All authors have contributed equally to this paper.

Conflicts of Interest: The authors declare no conflict of interest.

References

1. Molodtsov, D. Soft set theory-first results. *Comput. Math. Appl.* **1999**, *37*, 19–31.
2. Molodtsov, D. The Theory of Soft Sets. *URSS Publ. Moscow.* **2004**. (In Russian)
3. Maji, P.; Roy, A.R.; Biswas, R. An application of soft sets in a decision making problem. *Comput. Math. Appl.* **2002**, *44*, 1077–1083.
4. Qin, K.; Meng, D.; Pei, Z.; Xu, Y. Combination of interval set and soft set. *Int. J. Comput. Intell. Syst.* **2013**, *6*, 370–380.
5. Zhang, X. On interval soft sets with applications. *Int. J. Comput. Intell. Syst.* **2014**, *7*, 186–196.
6. Shao, Y.; Qin, K. Fuzzy soft sets and fuzzy soft lattices. *Int. J. Comput. Intell. Syst.* **2012**, *5*, 1135–1147.
7. Basu, K.; Deb, R.; Pattanaik, P.K. Soft sets: An ordinal formulation of vagueness with some applications to the theory of choice. *Fuzzy Sets Syst.* **1992**, *45*, 45–58.
8. Li, Z.; Xie, T. Roughness of fuzzy soft sets and related results. *Int. J. Comput. Intell. Syst.* **2015**, *8*, 278–296.
9. Bustince, H.; Burillo, P. Vague sets are intuitionistic fuzzy sets. *Fuzzy Sets Syst.* **1996**, *79*, 403–405.
10. Torra, V. Hesitant fuzzy sets. *Int. J. Comput. Intell. Syst.* **2010**, *25*, 529–539.
11. Rodríguez, R.M.; Bedregal, B.; Bustince, H.; Dong, Y.; Farhadinia, B.; Kahraman, C.; Martínez, L.; Torra, V.; Xu, Y.; Xu, Z.; et al. A position and perspective analysis of hesitant fuzzy sets on information fusion in decision making. Towards high quality progress. *Inf. Fusion* **2016**, *29*, 89–97.
12. Roy, A.R.; Maji, P. A fuzzy soft set theoretic approach to decision making problems. *J. Comput. Appl. Math.* **2007**, *203*, 412–418.
13. Rodríguez, R.M.; Labella, A.; Martínez, L. An overview on fuzzy modelling of complex linguistic preferences in decision making. *Int. J. Comput. Intell. Syst.* **2016**, *9*, 81–94.
14. Kong, Z.; Gao, L.; Wang, L. Comment on "A fuzzy soft set theoretic approach to decision making problems". *J. Comput. Appl. Math.* **2009**, *223*, 540–542.
15. Feng, F.; Jun, Y.B.; Liu, X.; Li, L. An adjustable approach to fuzzy soft set based decision making. *J. Comput. Appl. Math.* **2010**, *234*, 10–20.
16. Zadeh, L.A. Fuzzy sets. *Inf. Control* **1965**, *8*, 338–353.
17. Bellman, R.E.; Zadeh, L.A. Decision-making in a fuzzy environment. *Manag. Sci.* **1970**, *17*, B-141–B-164.
18. Maji, P.K.; Biswas, R.; Roy, A.R. Intuitionistic fuzzy soft sets. *J. Fuzzy Math.* **2001**, *9*, 677–692.
19. Liu, Z.; Qin, K.; Pei, Z.; Liu, J. On Induced Soft Sets and Topology for the Parameter Set of a Soft Set. Ubiquitous Computing and Communications; Dependable, Autonomic and Secure Computing; Pervasive Intelligence and Computing (CIT/IUCC/DASC/PICOM). In Proceedings of the 2015 IEEE International Conference on Computer and Information Technology, Liverpool, UK, 26–28 October 2015; pp. 1349–1353.
20. Alcantud, J.C.R. A novel algorithm for fuzzy soft set based decision making from multiobserver input parameter data set. *Inf. Fusion* **2016**, *29*, 142–148.
21. Belton, V.; Gear, T. On a short-coming of Saaty's method of analytic hierarchies. *Omega* **1983**, *11*, 228–230.
22. Barzilai, J.; Golany, B. AHP rank reversal, normalization and aggregation rules. *Inf. Syst. Oper. Res.* **1994**, *32*, 57–64.
23. Belton, V.; Gear, T. The legitimacy of rank reversal-a comment. *Omega* **1985**, *13*, 143–144.
24. Belton, V.; Gear, T. On the meaning of relative importance. *J. Multi-Criteria Decis. Anal.* **1997**, *6*, 335–338.
25. Grabisch, M. Fuzzy integral in multicriteria decision making. *Fuzzy Sets Syst.* **1995**, *69*, 279–298.
26. Sousa, J.M. *Fuzzy Decision Making in Modeling and Control*; World Scientific: Singapore, 2002; Volume 27.
27. Hwang, C.L.; Lai, Y.J.; Liu, T.Y. A new approach for multiple objective decision making. *Comput. Oper. Res.* **1993**, *20*, 889–899.
28. Medina, J.; Ojeda-Aciego, M. Multi-adjoint t-concept lattices. *Inf. Sci.* **2010**, *180*, 712–725.
29. Pozna, C.; Minculete, N.; Precup, R.E.; Kóczy, L.T.; Ballagi, Á. Signatures: Definitions, operators and applications to fuzzy modelling. *Fuzzy Sets Syst.* **2012**, *201*, 86–104.
30. Nowaková, J.; Prílepok, M.; Snášel, V. Medical image retrieval using vector quantization and fuzzy S-tree. *J. Med. Syst.* **2017**, *41*, 18.
31. Kumar, A.; Kumar, D.; Jarial, S.K. A hybrid clustering method based on improved artificial bee colony and fuzzy C-means algorithm. *Int. J. Artif. Intell.* **2017**, *15*, 40–60.
32. Hamming, R.W. Error detecting and error correcting codes. *Bell Labs Tech. J.* **1950**, *29*, 147–160.

33. Yager, R.R.; Filev, D.P Induced ordered weighted averaging operators. *IEEE Trans. Syst. Man Cybern. Part B* **1999**, *29*, 141–150.
34. Liu, Z. Fuzzy-Soft-Sets-Ideal-Solution. Available online: https://github.com/idle010/Fuzzy-Soft-Sets-Ideal-Solution (accessed on 9 October 2017).

symmetry

MDPI

Article

New Applications of m-Polar Fuzzy Matroids

Musavarah Sarwar and Muhammad Akram *

Department of Mathematics, University of the Punjab, New Campus, Lahore 54590, Pakistan; musavarah656@gmail.com

* Correspondence: makrammath@yahoo.com or m.akram@pucit.edu.pk

Received: 1 November 2017; Accepted: 11 December 2017; Published: 18 December 2017

Abstract: Mathematical modelling is an important aspect in apprehending discrete and continuous physical systems. Multipolar uncertainty in data and information incorporates a significant role in various abstract and applied mathematical modelling and decision analysis. Graphical and algebraic models can be studied more precisely when multiple linguistic properties are dealt with, emphasizing the need for a multi-index, multi-object, multi-agent, multi-attribute and multi-polar mathematical approach. An m-polar fuzzy set is introduced to overcome the limitations entailed in single-valued and two-valued uncertainty. Our aim in this research study is to apply the powerful methodology of m-polar fuzzy sets to generalize the theory of matroids. We introduce the notion of m-polar fuzzy matroids and investigate certain properties of various types of m-polar fuzzy matroids. Moreover, we apply the notion of the m-polar fuzzy matroid to graph theory and linear algebra. We present m-polar fuzzy circuits, closures of m-polar fuzzy matroids and put special emphasis on m-polar fuzzy rank functions. Finally, we also describe certain applications of m-polar fuzzy matroids in decision support systems, ordering of machines and network analysis.

Keywords: m-polar fuzzy matroid; m-polar fuzzy uniform matroid; m-polar fuzzy linear matroid; m-polar fuzzy partition matroid; m-polar fuzzy cycle matroid; m-polar fuzzy rank function; closure; m-polar fuzzy circuit

MSC: 05C65; 05C85; 05C90; 03E72

---

1. Introduction

Matroid theory had its foundations laid in 1935 after the work of Whitney [1]. This theory constitutes a useful approach for linking major ideas of linear algebra, graph theory, combinatorics and many other areas of Mathematics. Matroid theory has been a focus of active research during the last few decades.

Zadeh's fuzzy set theory [2,3] handles real life data having non-statistical uncertainty and vagueness. Petković et al. [4] investigated the accuracy of an adaptive neuro-fuzzy computing technique in precipitation estimation. Various applications of fuzzy sets in the field of automotive and railway level crossings for safety improvements are studied in [5,6]. The fuzzy set plays a vital role to solve various multi-criteria decision making problems. Some applications of fuzzy theory in multi-criteria models are discussed in [7,8]. Zhang [9] extended fuzzy set theory to bipolar fuzzy sets and discusses the bipolar behaviour of objects. The idea which lies behind such a description is connected with the existence of "bipolar information". For illustration, profit and loss, hostility and friendship, competition and cooperation, conflicted interests and common interests , unlikelihood and likelihood, feedback and feedforward, and so on, are generally two sides in coordination and decision making. Just like that, bipolar fuzzy set theory indeed has considerable impacts on many fields, including computer science, artificial intelligence, information science, decision science, cognitive science, economics, management science, neural science, medical science and social science. Recently, bipolar fuzzy set theory has been applied and studied speedily and increasingly. Thus, bipolar fuzzy sets not only have applications in mathematical theories but also in real-world problems [10–12].

In a number of real world problems, data come from m sources or agents $(m \geq 2)$, that is, multi-indexed information arises which cannot be mathematically expressed by means of the existing approaches of classical set theory, the crisp theory of graphs, fuzzy systems and bipolar fuzzy systems. The research presented in this paper is mainly developed to handle the lack of a mathematical approach towards multi-index, multipolar and multi-attribute data. Nowadays, analysts believe that the natural world is approaching the ideas of multipolarity. Multipolarity in data and information plays an important role in various domains of science and technology. In information technology, multipolar technology can be oppressed to operate large scale systems. In neurobiology, multipolar neurons in brain assemble a lot of information from other neurons. For instance, over a noisy channel, a communication channel may have a different network range, radio frequency, bandwidth and latency. In a food web, species may be of different types including strong, weak, vegetarian and non-vegetarian, and preys may be energetic, harmful and digestive. In a social network, the influence rate of different people may be different with respect to socialism, proactiveness, and trading relationship. A company may have different market power from others according to its product quality, annum profit, price control of its product, etc. These are multipolar information which are fuzzy in nature. To discuss such network models, we need mathematical and theoretical approaches which deal with multipolar information.

In view of this motivation, Chen et al. [13] extended bipolar fuzzy set theory and introduced the powerful idea of m-polar fuzzy sets. The membership value of an object, in an m-polar fuzzy set, belongs to $[0,1]^m$, which represents m different attributes of the object. Considering the idea of graphic structures, m-polar fuzzy sets can be used to describe the relationship among several individuals. In particular, m-polar fuzzy sets have found applications in the adaptation of accurate problems if it is necessary to make decisions and judgements with a number of agreements. For instance, the exact value of telecommunication safety of human beings is a point which lies in $[0,1]^m (m \approx 7 \times 10^9)$, since different people are monitored in different times. Some other applications include ordering and evaluation of alternatives and m-valued logic. m-polar fuzzy sets are shown to be useful to explore weighted games, cooperative games and multi-valued relations. In decision making issues, m-polar fuzzy sets are helpful for multi-criteria selection of objects in view of multipolar data. For example, m-polar fuzzy sets can be implemented when a country elects its political leaders, a company decides to manufacture an item or product, a group of friends wants to visit a country with multiple alternatives. In wireless communication, it can be used to discuss the conflicts and confusions of communication signals. Thus, m-polar fuzzy sets not only have applications in mathematical theories but also in real-world problems.

Akram and Younas [14] implemented the concept of m-polar fuzzy set into graph theory and discussed irregularity in m-polar fuzzy graphs. Several researchers have been applying this technique to explore various applications of m-polar fuzzy theory including grouping of objects [15], detecting human trafficking suspects [16], finding minimum number of locations [17] and decision support systems [18]. In 1988, Goetschel [19] studied the approach to the fuzzification of matroids and discussed the uncertain behaviour of matroids. The same authors [20] introduced the concept of bases of fuzzy matroids, fuzzy matroid structures and greedy algorithm in fuzzy matroids. Akram and Sarwar [15,21] have also discussed m-polar fuzzy hypergraphs, product formulae of distance for various types of m-polar fuzzy graphs and applications of m-polar fuzzy competition graphs in different domains. Akram and Waseem [22] constructed antipodal and self-median m-polar fuzzy graphs. Li et al. [23] considered different algebraic operations on m-polar fuzzy graphs. Hsueh [24] discussed independent axioms of matroids which preserve basic operational properties. Fuzzy matroids can be used to study the uncertain behaviour of objects but if the data have multipolar information to be dealt with, fuzzy matroids cannot give appropriate results. For this reason, we need the theory of m-polar fuzzy matroids to handle data and information with multiple uncertainties. In this research paper, we present the notion of m-polar fuzzy matroids and study various types of m-polar fuzzy matroids. We apply the concept of m-polar fuzzy matroids to graph theory, linear algebra and discuss their fundamental properties. We present the notion of closure of an m-polar fuzzy matroid and give special focus to the m-polar fuzzy rank function. We also describe certain applications of m-polar fuzzy matroids. We have used basic concepts and terminologies in this paper. For other notations, terminologies and applications not mentioned in the paper, the readers are referred to [22,25–34].

Throughout this research paper, we will use the notation "mF set" for an m-polar fuzzy set, denote the elements of an m-polar fuzzy set A as $(y, A(y))$ and use A^* as a crisp set and A as an m-polar fuzzy set.

2. Preliminaries

The term crisp matroid has various equivalent definitions. We use here the simplest definition of matroid.

Definition 1. *If Y is a non-empty universe and I is a subset of $P(Y)$, power set of Y, satisfying the following conditions,*

1. *If $D_1 \in I$ and $D_2 \subset D_1$ then , $D_2 \in I$,*
2. *If $D_1, D_2 \in I$ and $|D_1| < |D_2|$ then there exists $D_3 \in I$ such that $D_1 \subset D_3 \subseteq D_1 \cup D_2$.*

The pair $M = (Y, I)$ is a matroid and I is known as the family of independent sets of M.

Definition 2. *([19]) If $M = (Y, I)$ is a matroid then the mapping $R : P(Y) \to \{0, 1, 2, \ldots, |Y|\}$ defined by*

$$R(D) = \max\{|F| : F \subseteq D, F \in I\}$$

is a rank function for M. If $D \in P(Y)$, R is known as rank of D.

Definition 3. *([19]) For any non-empty universe Y, a mapping $\mu : P(Y) \to [0, \infty)$ is called submodular if for each, $D, F \in P(Y)$,*

$$\mu(D) + \mu(F) \geq \mu(D \cup F) + \mu(D \cap F).$$

Definition 4. *([2,3]) A fuzzy set τ in a non-empty universe Y is a mapping $\tau : Y \to [0, 1]$. A fuzzy relation on Y is a fuzzy subset δ in $Y \times Y$. If τ is a fuzzy set in Y and δ is a fuzzy relation on Y then we can say that δ is a fuzzy relation on τ if $\delta(y, z) \leq \min\{\tau(y), \tau(z)\}$ for all $y, z \in Y$.*

Definition 5. *([19]) If $\mathcal{F}(Y)$ is a power set of fuzzy subsets on Y and $\mathcal{I} \subseteq \mathcal{F}(X)$ which satisfy the following conditions,*

1. *If $\tau_1 \in \mathcal{I}$ and $\tau_2 \subset \tau_1$ then, $\tau_2 \in \mathcal{I}$,*
   where, $\tau_2 \subset \tau_1 \Rightarrow \tau_2(y) < \tau_1(y)$, for every $y \in X$.
2. *If $\tau_1, \tau_2 \in \mathcal{I}$ and $|supp(\tau_1)| < |supp(\tau_2)|$ then there exists $\tau_3 \in \mathcal{I}$ such that*
   a. *$\tau_1 \subset \tau_3 \subseteq \tau_1 \cup \tau_2$, for any $y \in X$, $\tau_1 \cup \tau_2(y) = \max\{\tau_1(y), \tau_2(y)\}$,*
   b. *$m(\tau_3) \geq \min\{m(\tau_1), m(\tau_2)\}$ where, $m(\nu) = \min\{\nu(y) : y \in supp(\nu)\}$.*

The pair $\mathrm{M} = (X, \mathcal{I})$ is called a fuzzy matroid. $\mathcal{I}$ is known as the collection of independent fuzzy sets of M.

Definition 6. *([13]) An mF set C on a non-empty set Y is a mapping $C = (P_1 \circ C(z), P_2 \circ C(z), \ldots, P_m \circ C(z)) : Y \to [0, 1]^m$ where, the jth projection mapping is defined as $P_j \circ C : [0, 1]^m \to [0, 1]$.*

Definition 7. *([22]) An mF relation $D = (P_1 \circ D, P_2 \circ D, \ldots, P_m \circ D)$ on C is a function $D : C \to C$ such that, $D(yz) \leq \inf\{C(y), C(z)\}$, for all $y, z \in Y$. That is, for all $y, z \in Y$, $P_j \circ D(yz) \leq \inf\{P_j \circ C(y), P_j \circ$*

$C(z)\}$, *for each* $1 \le j \le m$, *where* $P_j \circ C(z)$ *and* $P_j \circ D(yz)$ *represent the jth membership values of the element* z *and the relation* yz.

Definition 8. *([13,22]) An mF graph* $G = (C, D)$ *in a universe* Y *consists of two mappings* $C : Y \to [0,1]^m$ *and* $D : Y \times Y \to [0,1]^m$ *such that,* $D(yz) \le \inf\{C(y), C(z)\}$, *for all* $y, z \in Y$. *That is,* $P_j \circ D(yz) \le \inf\{P_j \circ C(y), P_j \circ C(z)\}$, *for each* $1 \le j \le m$. *Note that* $P_j \circ D(yz) = 0$ *for all* $yz \in Y \times Y - E$, $1 \le j \le m$ *where, E is the set of edges having non-zero degree of membership. mF relation, D, is called symmetric if* $P_j \circ D(yz) = P_j \circ D(zy)$ *for all* $y, z \in Y$.

3. Matroids Based on mF Sets

In this section, we define mF vector spaces, mF matroids and study their properties.

Definition 9. *An mF vector space over a field K is defined as a pair* $\widetilde{Y} = (Y, C_v)$ *where,* $C_v : Y \to [0,1]^m$ *is a mapping and Y is a vector space over K such that for all* $c, d \in F$ *and* $y, z \in Y$ $C_v(cy + dz) \ge \inf\{C_v(y), C_v(z)\}$, *i.e., for all* $1 \le i \le m$,

$$P_i \circ C_v(cy + dz) \ge \inf\{P_i \circ C_v(y), P_i \circ C_v(z)\}.$$

Example 1. *Let Y be a vector space of* 2×1 *column vectors over* $\mathbb{R}$. *Define a mapping* $C_v : Y \to [0,1]^3$ *such that for each* $\mathbf{z} = \begin{bmatrix} x & y \end{bmatrix}^t$,

$$C_v(z) = \begin{cases} (1,1,1), & z = \begin{bmatrix} 0 & 0 \end{bmatrix}^t, \\ (1, \frac{1}{3}, \frac{2}{3}), & z = \begin{bmatrix} x & 0 \end{bmatrix}^t \text{ or } z = \begin{bmatrix} 0 & y \end{bmatrix}^t, \\ (1,1,1), & x \neq 0 \text{ and } y \neq 0. \end{cases}$$

It remains only to show that $\widetilde{Y} = (Y, C_v)$ *is a 3-polar fuzzy vector space. For* $\mathbf{z} = \begin{bmatrix} 0 & 0 \end{bmatrix}^t$, *the case is trivial. So the following cases are to be discussed.*
Case 1: *Consider two column vectors* $\mathbf{z} = \begin{bmatrix} x & y \end{bmatrix}^t$ *and* $\mathbf{u} = \begin{bmatrix} u & v \end{bmatrix}^t$ *then, for any scalars c and d,*

$$C_v(c\mathbf{z} + d\mathbf{u}) = C_v\left(\begin{bmatrix} cx + du \\ cy + dv \end{bmatrix}\right).$$

If either exactly one of c or d is zero or both are non-zero then, $cx + du \neq 0$ *and* $cy + dv \neq 0$ *and so* $C_v(c\mathbf{z} + d\mathbf{u}) = (1,1,1) = \inf\{C_v(\mathbf{z}), C_v(\mathbf{u})\}$. *Also if* $c = 0$ *and* $d = 0$ *then ,* $C_v(c\mathbf{z} + d\mathbf{u}) = (1,1,1)$.
Case 2: *If* $\mathbf{z} = \begin{bmatrix} x & 0 \end{bmatrix}^t$ *and* $\mathbf{u} = \begin{bmatrix} 0 & v \end{bmatrix}^t$ *then,* $c\mathbf{z} + d\mathbf{u} = \begin{bmatrix} cx & dv \end{bmatrix}^t$. *If either both c and d are zero or both are non-zero then,* $C_v(c\mathbf{z} + d\mathbf{u}) = (1,1,1) > \inf\{C_v(\mathbf{z}), C_v(\mathbf{u})\}$. *If exactly one of c or d is zero then,* $C_v(c\mathbf{z} + d\mathbf{u}) = (1, \frac{1}{3}, \frac{2}{3}) = \inf\{C_v(\mathbf{z}), C_v(\mathbf{u})\}$. *Hence* $\widetilde{Y}$ *is a 3-polar fuzzy vector space.*

Definition 10. *Let* $\widetilde{Y} = (Y, C_v)$ *be an mF vector space over K. A set of vectors* $\{\mathbf{x}_k\}_{k=1}^n$ *is known as mF linearly independent in* $\widetilde{Y}$ *if*

1. $\{\mathbf{x}_k\}_{k=1}^n$ *is linearly independent,*
2. $C_v(\sum_{k=1}^{n} c_k\mathbf{x}_k) = \bigwedge_{k=1}^{n} C_v(c_k\mathbf{x}_k)$ *for all* $\{c_k\}_{k=1}^n \subset K$.

Definition 11. *A set of vectors* $\mathcal{B} = \{\mathbf{x}_k\}_{k=1}^n$ *is known to be an mF basis in* $\widetilde{Y}$ *if* $\mathcal{B}$ *is a basis in Y and condition 2 of Definition 10 is satisfied.*

Proposition 1. *If $\widetilde{Y} = (Y, C_v)$ is an mF vector space then any set of vectors with distinct jth, for each $1 \leq j \leq m$, degree of membership is linearly independent and mF linearly independent.*

Proposition 2. *Let $\widetilde{Y} = (Y, C_v)$ be an mF vector space then,*

1. $C_v(\mathbf{0}) = \sup_{y \in Y} C_v(\mathbf{y})$,
2. $C_v(a\mathbf{y}) = C_v(\mathbf{y})$ for all $a \in K \setminus \{0\}$ and $\mathbf{y} \in Y$,
3. If $C_v(\mathbf{y}) \neq C_v(\mathbf{z})$ for some $\mathbf{y}, \mathbf{z} \in Y$ then $C_v(\mathbf{y} + \mathbf{z}) = C_v(\mathbf{y}) \wedge C_v(\mathbf{z})$.

Remark 1. *If $\mathcal{B}$ is an mF basis of $\widetilde{Y}$ then the membership value of every element of Y can be calculated from the membership values of basis elements, i.e., if $\mathbf{u} = \sum_{k=1}^{n} c_k \mathbf{u}_k$ then,*

$$C_v(\mathbf{u}) = C_v(\sum_{k=1}^{n} c_k \mathbf{u}_k) = \bigwedge_{k=1}^{n} C_v(c_k \mathbf{u}_k) = \bigwedge_{k=1}^{n} C_v(\mathbf{u}_k).$$

We now come to the main idea of this research paper called mF matroids.

Definition 12. *Let Y be a non-empty finite set of elements and $\mathcal{C} \subseteq \mathcal{P}(Y)$ be a family of mF subsets, $\mathcal{P}(Y)$ is an mF power set of Y, satisfying the following the conditions,*

1. *If $\eta_1 \in \mathcal{C}$, $\eta_2 \in \mathcal{P}(Y)$ and $\eta_2 \subset \eta_1$ then, $\eta_2 \in \mathcal{C}$,*
   where, $\eta_2 \subset \eta_1 \Rightarrow \eta_2(y) < \eta_1(y)$ for every $y \in Y$.
2. *If $\eta_1, \eta_2 \in \mathcal{C}$ and $|supp(\eta_1)| < |supp(\eta_2)|$ then there exists $\eta_3 \in \mathcal{C}$ such that*
   a. $\eta_1 \subset \eta_3 \subseteq \eta_1 \cup \eta_2$,
   where for any $y \in Y$, $(\eta_1 \cup \eta_2)(y) = \sup\{\eta_1(y), \eta_2(y)\}$,
   b. $m(\eta_3) \geq \inf\{m(\eta_1), m(\eta_2)\}$,
   $m(\eta_i) = \inf\{\eta_i(x) | x \in supp(\eta_i)\}$, $i = 1, 2, 3$.

Then the pair $\mathcal{M}(Y) = (Y, \mathcal{C})$ is called an mF matroid on Y, and $\mathcal{C}$ is a family of independent mF subsets of $\mathcal{M}(Y)$.

$\{\delta : \delta \in \mathcal{P}(Y), \delta \notin \mathcal{C}\}$ is the family of dependent mF subsets in $\mathcal{M}(Y)$. A minimal mF dependent set is called an *m-polar fuzzy circuit*. The family of all mF circuits is denoted by $C_r(\mathcal{M})$. An mF circuit having n number of elements is called an mF n-circuit. An mF matroid can be uniquely determined from $C_r(\mathcal{M})$ because the elements of $\mathcal{C}$ are those members of $\mathcal{P}(Y)$ that contain no member of $C_r(\mathcal{M})$. Therefore, the members of $C_r(\mathcal{M})$ can be characterized with the following properties:

1. $\emptyset \notin C_r(\mathcal{M})$,
2. If δ_1 and δ_2 are distinct and $\delta_1 \subseteq \delta_2$ then, $supp(\delta_1) = supp(\delta_2)$,
3. If $\delta_1, \delta_2 \in C_r(G)$ and for $A \in \mathcal{P}(Y)$, $A(e) = \inf\{\delta_1(e), \delta_2(e)\}$, $e \in supp(\delta_1 \cap \delta_2)$ then there exists δ_3 such that $\delta_3 \subseteq \delta_1 \cup \delta_2 - \{(e, A(e)\}$.

Proposition 3. *If $\widetilde{Y} = (Y, C_v)$ is an mF vector space of $p \times q$ column vectors over $\mathbb{R}$, and $\mathcal{C}$ is the family of linearly independent mF subsets η_i in $\widetilde{Y}$ then $(Y, \mathcal{C})$ is an mF matroid on Y.*

Proposition 4. *If $\mathcal{M} = (Y, \mathcal{C})$ is an mF matroid and y is an element of Y such that $\mathcal{C} \cup \{(y, A(y))\}$, $A \in \mathcal{P}(Y)$ is dependent. Then $\mathcal{M}(Y)$ has a unique mF circuit contained in $\mathcal{C} \cup \{(y, A(y))\}$ and this mF circuit contains $\{(y, A(y))\}$.*

Definition 13. *Let Y be a non-empty universe. For any mF matroid, the mF rank function $\mu_r : \mathcal{P}(Y) \to [0, \infty)^m$ is defined as,*

$$\mu_r(\xi) = \sup\{|\eta| : \eta \subseteq \xi \text{ and } \eta \in \mathcal{C}\}$$

where, $|\eta| = \sum_{y \in Y} \eta(y)$. *Clearly the mF rank function of an mF matroid possesses the following properties:*

1. *If* $\eta_1, \eta_2 \in \mathcal{P}(Y)$ *and* $\eta_1 \subseteq \eta_2$ *then* $\mu_r(\eta_1) \leq \mu_r(\eta_2)$,
2. *If* $\eta \in \mathcal{P}(Y)$ *then,* $\mu_r(\eta) \leq |\eta|$,
3. *If* $\eta \in \mathcal{C}$ *then,* $\mu_r(\eta) = |\eta|$.

We now describe the concept of mF matroids by various examples.

1. A trivial example of an mF matroid is known as an *mF uniform matroid* which is defined as,

$$\mathcal{C} = \{\eta \in \mathcal{P}(Y) : |supp(\eta)| \leq l\}.$$

It is denoted by $\mathcal{U}_{l,n} = (Y, \mathcal{C})$ where, l is any positive integer and $|Y| = n$. The mF circuit of $\mathcal{U}_{l,n}$ contains those mF subsets δ such that $|supp(\delta)| = l + 1$.
Consider the example of a 2-polar fuzzy uniform matroid $\mathcal{M} = (Y, \mathcal{C})$ where, $Y = \{e_1, e_2, e_3\}$ and $\mathcal{C} = \{\eta \in \mathcal{P}(Y) : |supp(\eta)| \leq 2\}$ such that for any $\eta \in \mathcal{P}(Y)$, $\eta(y) = \tau(y)$, for all $y \in Y$ where,

$$\tau(y) = \begin{cases} (0.2, 0.3), & y = e_1 \\ (0.4, 0.5), & y = e_2 \\ (0.1, 0.3), & y = e_3 \end{cases}.$$

$$\begin{aligned} \mathcal{C} = & \{\emptyset, \{(e_1, 0.2, 0.3)\}, \{(e_2, 0.4, 0.5)\}, \{(e_3, 0.1, 0.3)\}, \{(e_1, 0.2, 0.3), (e_2, 0.4, 0.5)\}, \\ & \{(e_2, 0.4, 0.5), (e_3, 0.1, 0.3)\}, \{(e_1, 0.2, 0.3), (e_3, 0.1, 0.3)\}\}. \end{aligned}$$

The 2-polar fuzzy circuit of $\mathcal{M}$ is $C_r(\mathcal{M}) = \{(e_1, 0.2, 0.3), (e_2, 0.4, 0.5), (e_3, 0.1, 0.3)\}$. For $\eta = \{(e_2, 0.4, 0.5), (e_1, 0.2, 0.3)\}$, $\mu_r(\eta) = (0.6, 0.8)$.

2. *mF linear matroid* is derived from an mF matrix. Assume that Y represents the column labels of an mF matrix and η_x denotes an mF submatrix having those columns labelled by Y. It is defined as,

$$\mathcal{C} = \{\eta_x \in \mathcal{P}(Y) : \text{ columns of } \eta_x \text{ are } m- \text{ polar fuzzy linearly independent}\}.$$

For any $\eta_x \in \mathcal{P}(Y)$, $|\eta_x| = \sum_{k=1}^{r} \sup\{\eta_x(a_{k1}), \eta_x(a_{k2}), \ldots, \eta_x(a_{kc})\}$, $\eta_x^* = [a_{ij}]_{r \times c}$.

Let $A = \{\mathbf{1}, \mathbf{2}, \mathbf{3}, \mathbf{4}\}$ be a set of 3-polar fuzzy 2×1 vectors over $\mathbb{R}$ such that for any $\eta_x \in \mathcal{P}(Y)$, $\eta_x(y) = A(y)$ where,

$$A = \begin{bmatrix} \mathbf{1} & \mathbf{2} & \mathbf{3} & \mathbf{4} \\ (0.1, 0.2, 0.3) & (0.3, 0.4, 0.5) & (0.5, 0.6, 0.7) & (0.7, 0.8, 0.9) \\ (0.2, 0.3, 0.4) & (0.4, 0.5, 0.6) & (0.6, 0.7, 0.8) & (0.8, 0.9, 1.0) \end{bmatrix}.$$

Take $\mathcal{C} = \{\emptyset, \{\mathbf{1}\}, \{\mathbf{2}\}, \{\mathbf{4}\}, \{\mathbf{1}, \mathbf{2}\}, \{\mathbf{2}, \mathbf{4}\}\}$ then, $\mathcal{M}(A) = (A, \mathcal{C})$ is a 3-polar fuzzy matroid on A. The family of dependent 3-polar fuzzy subsets of matroid $\mathcal{M}(A)$ is $\{\{\mathbf{3}\}, \{\mathbf{1}, \mathbf{3}\}, \{\mathbf{1}, \mathbf{4}\}, \{\mathbf{2}, \mathbf{3}\}, \{\mathbf{3}, \mathbf{4}\}\} \cup \{\boldsymbol{\eta} : \boldsymbol{\eta} \subseteq A, |supp(\boldsymbol{\eta})| \geq 3\}$. For $\boldsymbol{\eta} = \{\mathbf{2}, \mathbf{4}\}$, $\mu_r(\boldsymbol{\eta}) = (1.5, 1.7, 1.9)$.

3. An *mF partition matroid* in which the universe Y is partitioned into mF sets $\alpha_1, \alpha_2, \ldots, \alpha_r$ such that

$$\mathcal{C} = \{\eta \in \mathcal{P}(Y) : |supp(\eta) \cap supp(\alpha_i)| \leq l_i, \text{ for all } 1 \leq i \leq r\}$$

for given positive integers $l_1, l_2, \ldots, l_r$. The circuit of an *mF partition matroid* is the family of those mF subsets δ such that $|supp(\delta) \cap supp(\alpha_i)| = l_i + 1$.

4. The very important class of mF matroids are derived from mF graphs. The detail is discussed in Proposition 5. The mF matroid derived using this method is known as *m-polar fuzzy cycle matroid*,

denoted by $\mathcal{M}(G)$. Clearly $\mathcal{C}$ is an independent set in G if and only if for each $\eta \in \mathcal{C}$, $supp(\equiv)$ is not edge set of any cycle. Equivalently, the members of $\mathcal{M}(G)$ are mF graphs η such that $supp(\eta)$ is a forest.

Consider the example of an mF fuzzy cycle matroid $(Y, \mathcal{C})$ where, $Y = \{y_1, y_2, y_3, y_4, y_5\}$ and for any, $\eta \in \mathcal{C}$, $\beta(y) = D(y)$, (C, D) is an mF multigraph on Y as shown in Figure 1.

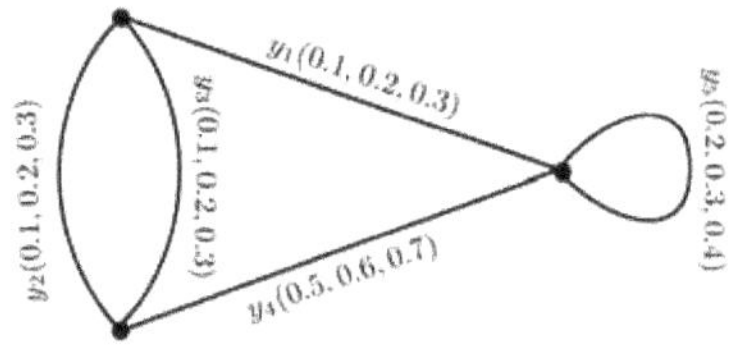

Figure 1. 3-polar fuzzy multigraph.

By Proposition 5, $C_r(G) = \{\{(y_5, 0.2, 0.3, 0, 4)\}, \{(y_2, 0.1, 0.2, 0.3), (y_3, 0.1, 0.2, 0.3)\}, \{(y_1, 0.1, 0.2, 0.3), (y_2, 0.1, 0.2, 0.3), (y_4, 0.5, 0.6, 0.7)\}, \{(y_1, 0.1, 0.2, 0.3), (y_3, 0.1, 0.2, 0.3), (y_4, 0.5, 0.6, 0.7)\}\}$.

$$\begin{aligned}\mathcal{C} =& \{\emptyset, \{(y_1, 0.1, 0.2, 0.3)\}, \{(y_2, 0.1, 0.2, 0.3)\}, \{(y_3, 0.1, 0.2, 0.3)\}, \{(y_1, 0.1, 0.2, 0.3), \\ &(y_2, 0.1, 0.2, 0.3)\}, \{(y_1, 0.1, 0.2, 0.3), (y_4, 0.5, 0.6, 0.7)\}, \{(y_4, 0.5, 0.6, 0.7)\}, \{(y_2, 0.1, 0.2, 0.3), \\ &(y_4, 0.5, 0.6, 0.7)\}, \{(y_1, 0.1, 0.2, 0.3), (y_3, 0.1, 0.2, 0.3)\}, \{(y_3, 0.1, 0.2, 0.3), (y_4, 0.5, 0.6, 0.7)\}\}\end{aligned}$$

For $\eta = \{(y_2, 0.1, 0.2, 0.3), (y_4, 0.5, 0.6, 0.7)\}$, $\mu_r(\eta) = (0.6, 0.8, 1.0)$.

Proposition 5. *For any mF graph $G = (C, D)$ on Y, if C_r is the family of mF edge sets δ such that $supp(\delta)$ is the edge set of a cycle in G^*. Then C_r is the family of mF circuits of an mF matroid on Y.*

Proof. Clearly conditions 1 and 2 of Definition 12 hold. To prove condition 3, let δ_1 and δ_2 be mF edge sets of distinct cycles that have yz as a common edge. Clearly, $\delta_3 = \delta_1 \cup \delta_2 - \{(yz, D(yz))\}$ is an mF edge set of a cycle and so condition 3 is satisfied. □

Example 2. *For any mF graph $G = (C, D)$ and $\mathbf{0} \le \mathbf{t} \le \mathbf{1}$ define,*
$E_t = \{yz \in supp(D) | D(yz) \ge \mathbf{t}\}$,
$F_t = \{H | H$ *is a forest in the crisp graph* $(Y, E_t)\}$,
$C_t = \{E(F) | F \in F_t\}$, $E(F)$ *is the edge set of* F.
Clearly (E_t, C_t) is a matroid for each $\mathbf{0} \le \mathbf{t} \le \mathbf{1}$. Define $\mathcal{D} = \{\eta \in \mathcal{P}(Y) | \eta_t \in C_t, \mathbf{0} \le \mathbf{t} \le \mathbf{1}\}$ then, $(Y, \mathcal{D})$ is an mF cycle matroid.

Theorem 1. *Let $\mathcal{M} = (Y, \mathcal{C})$ be an mF matroid and, for each $\mathbf{0} \le \mathbf{t} \le \mathbf{1}$, define $\mathcal{C}_t = \{\eta_t | \eta \in \mathcal{C}\}$. Then $(Y, \mathcal{C}_t)$ is a matroid on Y.*

Proof. We prove conditions 1 and 2 of Definition 12. Assume that $\eta_{1t} \in \mathcal{C}_t$ and $\alpha \subseteq \eta_{1t}$. Define an mF set $\eta_2 \in \mathcal{P}(Y)$ by

$$\eta_2(y) = \begin{cases} t & y \in \alpha, \\ 0 & \text{otherwise}. \end{cases}$$

Clearly $\eta_2 \subseteq \eta_1$,$\eta_2 \in \mathcal{C}$ and $\eta_{2t} = \alpha$ therefore, $\alpha \in \mathcal{C}_t$. To prove condition 2, let $\alpha_1, \alpha_2 \in \mathcal{C}_t$ and $|\alpha_1| < |\alpha_2|$. Then there exist η_1 and η_2 such that $\eta_{1t} = \alpha_1$ and $\eta_{2t} = \alpha_2$. Define $\hat{\eta}_1$ and $\hat{\eta}_2$ by

$$\hat{\eta}_1(y) = \begin{cases} t & y \in \eta_1, \\ 0 & \text{otherwise}. \end{cases} \qquad \hat{\eta}_2(y) = \begin{cases} t & y \in \eta_2, \\ 0 & \text{otherwise}. \end{cases}$$

It is clear that $supp(\hat{\eta}_1) < supp(\hat{\eta}_2)$. Since $\mathcal{M}$ is an mF matroid, there exists η_3 such that $\hat{\eta}_1 \subseteq \eta_3 \subseteq \hat{\eta}_1 \cup \hat{\eta}_2$. Since

$$\hat{\eta}_1 \cup \hat{\eta}_2(y) = \begin{cases} t & y \in \alpha_1 \cup \alpha_2, \\ 0 & \text{otherwise}. \end{cases}$$

Therefore, there exists a set α_3 such that

$$\eta_3(y) = \begin{cases} t & y \in \alpha_3, \\ 0 & \text{otherwise}. \end{cases}$$

Also, $\alpha_1 \subseteq \alpha_3 \subseteq \alpha_1 \cup \alpha_2$, $\alpha_3 \in \mathcal{C}_t$. Hence $\mathcal{M}_t$ is a matroid on Y. □

Remark 2. *Let $\mathcal{M} = (Y, \mathcal{C})$ be an mF matroid and, for each $\mathbf{0} \leq \mathbf{t} \leq \mathbf{1}$, $\mathcal{M}_t = (Y, \mathcal{C}_t)$ be the matroid on a finite set Y as given in Theorem 1. As Y is finite therefore, there is a finite sequence $\mathbf{0} < \mathbf{t}_1 < \mathbf{t}_2 < \ldots < \mathbf{t}_n$ such that $\mathcal{M}_{t_i} = (Y, \mathcal{C}_{t_i})$ is a crisp matroid, for each $1 \leq i \leq n$, and*

1. $t_0 = \mathbf{0}$, $\quad t_n \leq \mathbf{1}$,
2. *$\mathcal{C}_w \neq \emptyset$ if $\mathbf{0} < w \leq t_n$ and $\mathcal{C}_w = \emptyset$ if $w > t_n$,*
3. *If $t_i < w, s < t_{i+1}$ then, $\mathcal{C}_w = \mathcal{C}_s$, $0 \leq i \leq n-1$,*
4. *If $t_i < w < t_{i+1} < s < t_{i+2}$ then, $\mathcal{C}_w \supset \mathcal{C}_s$, $0 \leq i \leq n-2$.*

The sequence $\mathbf{0}, \mathbf{t}_1, \mathbf{t}_2, \ldots, \mathbf{t}_n$ is known as fundamental sequence of $\mathcal{M}$. Let $\bar{\mathbf{t}}_i = \frac{1}{2}(\mathbf{t}_{i-1} + \mathbf{t}_i)$ for $1 \leq i \leq n$. The decreasing sequence of crisp matroids $\mathcal{M}_{t_1} \supset \mathcal{M}_{t_2} \supset \ldots \supset \mathcal{M}_{t_n}$ is known as $\mathcal{M}$-indeced matroid sequence.

Theorem 2. *If Y is a finite set and $\mathbf{0} = t_0 < t_1 < t_2 < \ldots < t_n \leq \mathbf{1}$ is a finite sequence such that $(Y, \mathcal{C}_{t_1})$, $(Y, \mathcal{C}_{t_2}), \ldots, (Y, \mathcal{C}_{t_n})$ is a sequence of crisp matroids. For each m-tuple $\mathbf{t}$, where, $t_{i-1} < \mathbf{t} \leq t_i$ $(1 \leq i \leq n)$, assume that $\mathcal{C}_t = \mathcal{C}_{t_i}$ and $\mathcal{C}_t = \emptyset$ if $t_n < \mathbf{t} \leq \mathbf{1}$.*

Define $\mathcal{C}^* = \{\eta \in \mathcal{P}(Y) | \eta_t \in \mathcal{C}_t, \mathbf{0} < \mathbf{t} \leq \mathbf{1}\}$ then $\mathcal{M} = (Y, \mathcal{C}^*)$ is an mF matroid.

Proof. Let $\eta_1 \in \mathcal{C}^*$, $\eta_2 \in \mathcal{P}(Y)$, and $\eta_2 \subseteq \eta_1$. Clearly $\eta_{1t} \in \mathcal{C}_t$, $\eta_{2t} \subseteq \eta_{1t}$, and since $(Y, \mathcal{C}_t)$ is a crisp matroid therefore, $\eta_{2t} \in \mathcal{C}_t$, so $\eta_2 \in \mathcal{C}^*$.

Assume that $\eta_1, \eta_2 \in \mathcal{C}^*$ and $|supp(\eta_2)| < |supp(\eta_1)|$. Define

$$\beta = \inf\{\inf_{y \in supp(\eta_1)} \mathcal{C}^*(y), \inf_{y \in supp(\eta_2)} \mathcal{C}^*(y)\}.$$

It is easy to see that $supp(\eta_1), supp(\eta_2) \in \mathcal{C}_\beta$. Since $\mathcal{C}_\beta$ is the family of independent sets of a crisp matroid therefore, there exists an independent set $A \in \mathcal{C}_\beta$ such that

$$supp(\eta_2) \subset A \subseteq supp(\eta_1) \cup supp(\eta_2).$$

Let

$$\eta_3(y) = \begin{cases} \eta_2(y) & y \in supp(\eta_2), \\ \beta & y \in A \setminus supp(\eta_2), \\ 0 & \text{otherwise}. \end{cases}$$

The mF set η_3 satisfies condition 2 of Definition 12 and hence $(Y, \mathcal{C}^*)$ is an mF matroid. □

Theorem 3. *Let $\mathcal{M} = (Y, \mathcal{C})$ be an mF matroid and for each $\mathbf{0} < \mathbf{t} \leq \mathbf{1}$, $\mathcal{M}_{\mathbf{t}} = (Y, \mathcal{C}_{\mathbf{t}})$ is a crisp matroid by Theorem 2. Let $\mathcal{C}^* = \{\eta \in \mathcal{P}(Y) : \eta_{\mathbf{t}} \in \mathcal{C}_{\mathbf{t}}, \mathbf{0} < \mathbf{t} \leq \mathbf{1}\}$. Then $\mathcal{C} = \mathcal{C}^*$.*

Proof. It is clear from the definition of $\mathcal{C}^*$ that $\mathcal{C} \subseteq \mathcal{C}^*$. To prove the converse part, we proceed on the following steps.

Suppose that $\{\boldsymbol{\alpha}_1, \boldsymbol{\alpha}_2, \ldots, \boldsymbol{\alpha}_p\}$ is the non-zero range of $\eta \in \mathcal{C}$ such that $\boldsymbol{\alpha}_1 > \boldsymbol{\alpha}_2 > \ldots > \boldsymbol{\alpha}_p > \mathbf{0}$. For each $1 \leq i \leq p$, $\eta_{\boldsymbol{\alpha}_i} \in \mathcal{C}_{\boldsymbol{\alpha}_i}$ and $\eta_{\boldsymbol{\alpha}_{i-1}} \subset \eta_{\boldsymbol{\alpha}_i}$. Define $\boldsymbol{f}_i \in \mathcal{P}(Y)$ by

$$f_i(y) = \begin{cases} \boldsymbol{\alpha}_i & \text{if } y \in \eta_{\boldsymbol{\alpha}_i}, \\ \mathbf{0} & \text{otherwise}. \end{cases}$$

Since $\eta_{\boldsymbol{\alpha}_i} \in \mathcal{C}_{\boldsymbol{\alpha}_i}$ therefore, $f_i \in \mathcal{C}$ and $\bigcup_{i=1}^{q} f_i = \eta$. Assume that $supp(\eta) = \{y_1, y_2, \ldots, y_{n_p}\}$. We use the induction method to show that $\eta \in \mathcal{C}$. Since $f_1 \in \mathcal{C}$ therefore, it remains to show that if $\bigcup_{i=1}^{l-1} f_i \in \mathcal{C}$ then, $\bigcup_{i=1}^{l} f_i \in \mathcal{C}$, for each $l < p$. Define

$$g_1(y) = \begin{cases} \boldsymbol{\alpha}_l & \text{if } y \in \{y_1, y_2, \ldots, y_{n_{l-1}}, y_{n_{l-1}+1}\}, \\ \mathbf{0} & \text{otherwise}. \end{cases}$$

Since for each $1 \leq i \leq l-1$, $\boldsymbol{\alpha}_i > \boldsymbol{\alpha}_l$ therefore, $g_1 \subseteq f_l$ which implies that $g_1 \in \mathcal{C}$. Define $h_1 \in \mathcal{P}(Y)$ by

$$h_1(y) = \begin{cases} \eta(y_{n_{l-1}+1}) = \boldsymbol{\alpha}_l & \text{if } y = y_{n_{l-1}+1}, \\ \mathbf{0} & \text{otherwise}. \end{cases}$$

Since by induction method $\bigcup_{i=1}^{l-1} f_i \in \mathcal{C}$ and $supp(\bigcup_{i=1}^{l-1} f_i) = \{y_1, y_2, \ldots, y_{n_{l-1}}\}$, $m(\bigcup_{i=1}^{l-1} f_i) > \boldsymbol{\alpha}_l$ therefore, condition 2(b) of Definition 12 implies that $\bigcup_{i=1}^{l-1} f_i \cup h_1 \in \mathcal{C}$. If $n_{l-1} + 1 = n_l$ then, $\bigcup_{i=1}^{l} f_i \in \mathcal{C}$ and we are done. But if on the other hand, $n_{l-1} + 1 < n_l$ then define,

$$g_2(y) = \begin{cases} \boldsymbol{\alpha}_l & \text{if } y \in \{y_1, y_2, \ldots, y_{n_{l-1}}, y_{n_{l-1}+1}, y_{n_{l-1}+2}\}, \\ \mathbf{0} & \text{otherwise}. \end{cases}$$

Since for each $1 \leq i \leq l-1$, $\boldsymbol{\alpha}_i > \boldsymbol{\alpha}_l$ therefore, $g_2 \subseteq f_l$ which implies that $g_2 \in \mathcal{C}$. Define $h_2 \in \mathcal{P}(Y)$ by

$$h_2(y) = \begin{cases} \eta(y_{n_{l-1}+2}) = \boldsymbol{\alpha}_l & \text{if } y = y_{n_{l-1}+2}, \\ \mathbf{0} & \text{otherwise}. \end{cases}$$

Since $supp(\bigcup_{i=1}^{l-1} f_i \cup h_1) = \{y_1, y_2, \ldots, y_{n_{l-1}}, y_{n_{l-1}+1}\}$, $m(\bigcup_{i=1}^{l-1} f_i \cup h_1) > \boldsymbol{\alpha}_l$ therefore, condition 2(b) of Definition 12 implies that $\bigcup_{i=1}^{l-1} f_i \cup h_1 \cup h_2 \in \mathcal{C}$. If $n_{l-1} + 1 = n_l$ then, $\bigcup_{i=1}^{l} f_i \in \mathcal{C}$ and we are done. If o $n_{l-1} + 2 < n_l$ then we continue the process and obtain an mF set $\beta_n = \bigcup_{i=1}^{l-1} f_i \cup h_1 \cup h_2 \cup \ldots \cup h_n$ such that $\beta_n = \bigcup_{i=1}^{l} f_i$ which completes the induction procedure and the proof. □

The submodularity of an mF rank function μ_r is quiet difficult and it depends on Theorem 3 and the following definition.

Definition 14. *Let $\mathbf{t}_0, \mathbf{t}_1, \ldots, \mathbf{t}_n$ be the fundamental sequence of an mF matroid. For any m-tuple $\mathbf{t}, \mathbf{0} < \mathbf{t} \leq \mathbf{1}$, define $\overline{\mathcal{C}}_{\mathbf{t}} = \mathcal{C}_{\overline{\mathbf{t}}_i}$ where, $\mathbf{t}_{i-1} < \mathbf{t} \leq \mathbf{t}_i$ and $\overline{\mathbf{t}}_i = \frac{1}{2}(\mathbf{t}_{i-1} + \mathbf{t}_i)$. If $\mathbf{t} > \mathbf{t}_n$ take $\overline{\mathcal{C}}_{\mathbf{t}} = \mathcal{C}_{\mathbf{t}}$. Define*

$$\overline{\mathcal{C}} = \{\eta \in \mathcal{P}(Y) : \eta_{\mathbf{t}} \in \overline{\mathcal{C}}_{\mathbf{t}}, \text{ for each } \mathbf{t},\ \mathbf{0} < \mathbf{t} \leq \mathbf{1}\}.$$

Then $\overline{\mathcal{M}} = (Y, \overline{\mathcal{C}})$ is known as closure of $\mathcal{M} = (Y, \mathcal{C})$.

Example 3. *We now explain the concept of closure by an example of a 3-polar fuzzy uniform matroid $\mathcal{M} = (Y, \mathcal{C})$ where, $Y = \{y_1, y_2, y_3\}$ and $\mathcal{C} = \{\eta \in \mathcal{P}(Y) : |supp(\eta)| \leq 1\}$ such that for any $\eta \in \mathcal{P}(Y)$, $\eta(y) = \tau(y)$, for all $y \in Y$ where,*

$$\tau(y) = \begin{cases} (0.1, 0.2, 0.3), & y = y_1 \\ (0.2, 0.3, 0.4), & y = y_2 \\ (0.3, 0.4, 0.5), & y = y_3 \end{cases}.$$

$$\mathcal{C} = \{\emptyset, \{(y_1, 0.1, 0.2, 0.3)\}, \{(y_2, 0.2, 0.3, 0.4)\}, \{(y_3, 0.3, 0.4, 0.5)\}\}.$$

The fundamental sequence of $\mathcal{M}$ is $\{\mathbf{t}_0 = \mathbf{0}, \mathbf{t}_1 = (0.1, 0.2, 0.3), \mathbf{t}_2 = (0.2, 0.3, 0.4), \mathbf{t}_3 = (0.3, 0.4, 0.5)\}$. From routine calculations, $\overline{\mathbf{t}}_1 = (0.05, 0.1, 0.15)$, $\overline{\mathbf{t}}_2 = (0.15, 0.25, 0.35)$, $\overline{\mathbf{t}}_3 = (0.25, 0.35, 0.45)$. Since for any $\mathbf{0} < \mathbf{t} \leq \mathbf{1}$, $\overline{\mathcal{C}}_{\mathbf{t}} = \mathcal{C}_{\overline{\mathbf{t}}_i}$, $1 \leq i \leq 3$, therefore, $\mathcal{C}_{\overline{\mathbf{t}}_1} = \mathcal{C}_{\mathbf{t}_1}$, $\mathcal{C}_{\overline{\mathbf{t}}_2} = \{\{y_2\}, \{y_3\}\}$, $\mathcal{C}_{\overline{\mathbf{t}}_2} = \{\{y_3\}\}$. Hence the closure of $\mathcal{C}$ can be defined as,

$$\begin{aligned}\overline{\mathcal{C}} =& \{\emptyset, \{(y_1, 0.1, 0.2, 0.3)\}, \{(y_2, 0.2, 0.3, 0.4)\}, \{(y_3, 0.3, 0.4, 0.5)\}, \{(y_1, 0.1, 0.2, 0.3), (y_2, 0.2, 0.3, 0.4)\}, \\ & \{(y_1, 0.1, 0.2, 0.3), (y_3, 0.3, 0.4, 0.5)\}, \{(y_2, 0.2, 0.3, 0.4), (y_3, 0.3, 0.4, 0.5)\}\}.\end{aligned}$$

Theorem 4. *The closure $\overline{\mathcal{M}} = (Y, \overline{\mathcal{C}})$ of an mF matroid $\mathcal{M} = (Y, \mathcal{C})$ is also an mF matroid.*

The proof of this theorem is a clear consequence of Theorem 1.

Definition 15. *An mF matroid with fundamental sequence $\mathbf{t}_0, \mathbf{t}_1, \ldots, \mathbf{t}_n$ is known as a closed mF matroid if for each $\mathbf{t}_{i-1} < \mathbf{t} \leq \mathbf{t}_i$, $\mathcal{C}_{\mathbf{t}} = \mathcal{C}_{\mathbf{t}_i}$.*

Remark 3. *Note that the closure of an mF matroid is closed and that it is the smallest closed mF matroid containing $\mathcal{M}$. Also the fundamental sequence of $\mathcal{M}$ and $\overline{\mathcal{M}}$ is same.*

Lemma 1. *If $\overline{\mu}_r$ and μ_r are mF rank functions of $\overline{\mathcal{M}} = (Y, \overline{\mathcal{C}})$ and $\mathcal{M} = (Y, \mathcal{C})$, respectively then $\overline{\mu}_r = \mu_r$.*

Assume that $\mathcal{M} = (Y, \mathcal{C})$ is an mF matroid with fundamental sequence $\mathbf{t}_0, \mathbf{t}_1, \ldots, \mathbf{t}_n$ and rank function μ_r. To prove that μ_r is submodular, we now define a function $\hat{\mu}_r : \mathcal{P}(Y) \to [0, \infty)^m$ which is also submodular.

For any $\eta \in \mathcal{P}(Y)$, let $\mathbf{0} < \boldsymbol{\alpha}_1 < \boldsymbol{\alpha}_2 < \ldots < \boldsymbol{\alpha}_p$ be the non-zero range of η and $\boldsymbol{\beta}_1 < \boldsymbol{\beta}_2 < \ldots < \boldsymbol{\beta}_q$ be the common refinement of $\mathbf{t}_i'$s and $\boldsymbol{\alpha}_j'$s defined as,

$$\{\boldsymbol{\beta}_1, \boldsymbol{\beta}_2, \ldots, \boldsymbol{\beta}_q\} = \{\boldsymbol{\alpha}_1, \boldsymbol{\alpha}_2, \ldots, \boldsymbol{\alpha}_p\} \cup \{\mathbf{t}_1, \mathbf{t}_2, \ldots, \mathbf{t}_n\}.$$

R_i is the rank function of crisp matroid $\mathcal{M}_{\mathbf{t}_i} = (Y, \mathcal{C}_{\mathbf{t}_i})$, for all $1 \leq i \leq n$. For each integer j, there is an integer i, $1 \leq i \leq n$, such that $\mathbf{t}_{i-1} \leq \boldsymbol{\beta}_{j-1} < \boldsymbol{\beta}_j \leq \mathbf{t}_i$. Then (i, j) is known as a *correspondence pair*. For each correspondence pair (i, j), define

$$\gamma_j(\eta) = \begin{cases} (\boldsymbol{\beta}_j - \boldsymbol{\beta}_{j-1}) R_i(\eta_{\boldsymbol{\beta}_j}) & \text{if } \boldsymbol{\beta}_j \leq \mathbf{t}_n, \\ \mathbf{0} & \text{if } \boldsymbol{\beta}_j > \mathbf{t}_n. \end{cases}$$

Since for each $\beta_{j-1} < \beta < \beta_j$, $\eta_\beta = \eta_{\beta_j}$. Define a new function $\hat{\mu}_r : \mathcal{P}(Y) \to [0, \infty)^m$ by

$$\hat{\mu}_r = \sum_{j=1}^{q} \gamma_j(\eta). \tag{1}$$

Lemma 2. *Assume that* $\mathbf{0} < \rho_1 < \rho_2 < \ldots < \rho_p$ *and* $\{\beta_1, \beta_2, \ldots, \beta_q\} \subseteq \{\rho_1, \rho_2, \ldots, \rho_p\}$. *For each* i, $1 \leq i \leq n$, *let* (i, j) *be the correspondence pair if* $\mathbf{t}_{i-1} \leq \rho_{j-1} < \rho_j \leq \mathbf{t}_i$. *For each correspondence pair* (i, j), *define* $\gamma_j^* : \mathcal{P}(Y) \to \mathbb{R}^m$ *by*

$$\gamma_j^*(\eta) = \begin{cases} (\rho_j - \rho_{j-1}) R_i(\eta_{\rho_j}) & \text{if } \rho_j \leq \mathbf{t}_n, \\ \mathbf{0} & \text{if } \rho_j > \mathbf{t}_n. \end{cases}$$

Then $\sum_{j=1}^{q} \gamma_j(\eta) = \sum_{j=1}^{q} \gamma_j^*(\eta)$.

Theorem 5. *If* $\mathbf{t}_0, \mathbf{t}_1, \ldots, \mathbf{t}_n$ *is the fundamental sequence of an mF matroid* $\mathcal{M} = (Y, \mathcal{C})$ *and* $\hat{\mu}_r$ *is defined by (1) then,* $\hat{\mu}_r$ *is submodular.*

Proof. Let $\eta_1, \eta_2 \in \mathcal{P}(Y)$ and $\{\alpha_1, \alpha_2, \ldots, \alpha_s\}$, $\{\beta_1, \beta_2, \ldots, \beta_r\}$ be the non-zero ranges of η_1 and η_2, respectively. Define

$$\{\rho_1, \rho_2, \ldots, \rho_p\} = \{\alpha_1, \alpha_2, \ldots, \alpha_s\} \cup \{\beta_1, \beta_2, \ldots, \beta_r\} \cup \{\mathbf{t}_0, \mathbf{t}_1, \ldots, \mathbf{t}_n\}.$$

Lemma 2 implies that $\hat{\mu}_r = \sum_{j=1}^{q} \gamma_j^*(\eta)$. Since $\rho_j - \rho_{j-1} > \mathbf{0}$, for each j therefore, by the submodularity of the crisp rank function R_i,

$$\sum_{j=1}^{p} (\rho_j - \rho_{j-1}) R_i(\eta_{1t_j}) - \sum_{j=1}^{p} (\rho_j - \rho_{j-1}) R_i(\eta_{2t_j}) \geq \sum_{j=1}^{p} (\rho_j - \rho_{j-1}) R_i(\eta_{1t_j} \cup \eta_{2t_j})$$
$$+ \sum_{j=1}^{p} (\rho_j - \rho_{j-1}) R_i(\eta_{1t_j} \cap \eta_{2t_j}).$$
$$\Rightarrow \hat{\mu}_r(\eta_1) + \hat{\mu}_r(\eta_1) \geq \hat{\mu}_r(\eta_1 \cup \eta_2) + \hat{\mu}_r(\eta_1 \cap \eta_2).$$

□

Example 4. *Consider a 3-polar fuzzy matroid given in Example 3. For* $\eta = \{(y_2, 0.2, 0.3, 0.4)\}$, *the non-zero range of* η *is* $\{\alpha_1 = (0.2, 0.3, 0.4)\}$. *Define*

$$\{\beta_1, \beta_2, \beta_3\} = \{\mathbf{t}_0, \mathbf{t}_1, \mathbf{t}_2, \mathbf{t}_3\} \cup \{\alpha_1\} = \{\beta_1 = (0.1, 0.2, 0.3), \beta_2 = (0.2, 0.3, 0.4), \beta_3 = (0.3, 0.4, 0.5)\}.$$

Since $\mathbf{t}_1 = \beta_1 < \beta_2 = \mathbf{t}_2$ *therefore,* $(2, 2)$ *is correspondence pair. Similarly* $(3, 3)$ *is also a correspondence pair. Now* $\gamma_1(\eta) = \mathbf{0}$,

$$\gamma_2(\eta) = (\beta_2 - \beta_1) R_2(\eta_{\beta_2}) = (0.1, 0.1, 0.1), \qquad \gamma_3(\eta) = (\beta_3 - \beta_2) R_3(\eta_{\beta_3}) = (0, 0, 0).$$

Thus $\hat{\mu}_r(\eta) = (0.1, 0.1, 0.1)$.

Theorem 6. *For any mF matroid,* $\mu_r \geq \hat{\mu}_r$.

Proof. Since $\mu_r = \overline{\mu}_r$ therefore, assume that $\mathcal{M}$ is a closed mF matroid and $\mu_r(\eta_1) \neq \mathbf{0}$ for some $\eta_1 \in \mathcal{P}(Y)$. Suppose that there exists $\eta_2 \in \mathcal{C}$ $\eta_2 \subseteq \eta_1$ such that $\mu_r(\eta_1) = |\eta_2|$. We will show that $\hat{\mu}_r(\eta_1) \leq |\eta_2|$.

Take $\mathbf{t}_0 < \mathbf{t}_1 < \ldots < \mathbf{t}_n$ as the fundamental sequence of $\mathcal{M}$ and $\boldsymbol{\alpha}_1 < \boldsymbol{\alpha}_2 < \ldots < \boldsymbol{\alpha}_p$ as the non-zero range of η_1. Let $\boldsymbol{\beta}_1 < \boldsymbol{\beta}_2 < \ldots < \boldsymbol{\beta}_q$ be defined by

$$\{\boldsymbol{\beta}_1, \boldsymbol{\beta}_2, \ldots, \boldsymbol{\beta}_q\} = \{\boldsymbol{\alpha}_1, \boldsymbol{\alpha}_2, \ldots, \boldsymbol{\alpha}_p\} \cup \{\mathbf{t}_0, \mathbf{t}_1, \ldots, \mathbf{t}_n\}.$$

For each $\mathbf{0} < \boldsymbol{\beta} \leq \mathbf{1}$, define

$$\mathcal{C}^{\eta_1}_{\boldsymbol{\beta}} = \{C \in \mathcal{C}_{\boldsymbol{\beta}} : C \subseteq \eta_{1\boldsymbol{\beta}}\}, \qquad \boldsymbol{\beta}^* = \sup\{\boldsymbol{\beta} : \mathcal{C}^{\eta_1}_{\boldsymbol{\beta}} \neq \emptyset\}.$$

Remark 2 implies that $\boldsymbol{\beta}^* = \boldsymbol{\beta}_{i^*}$, for some $\boldsymbol{\beta}_{i^*} \in \{\boldsymbol{\beta}_j\}_{j=1}^{q}$. The following properties of $\boldsymbol{\beta}_{i^*}$ always hold:

(i) $\boldsymbol{\beta}_{i^*} \leq \mathbf{t}_n$, $\hat{\mu}_r(\eta_1) = \sum_{i=1}^{i^*} \gamma_i(\eta_1)$.
(ii) For $\eta_2 \in \mathcal{C}, \eta_2 \subseteq \eta_1$ we have, $\mathbf{0} < \eta_2(y) \leq \boldsymbol{\beta}_{i^*}$ for each $y \in supp(\eta_2)$.

For each integer $i \leq i^*$, let $|C_{\boldsymbol{\beta}_i}| = R_j(\eta_{\boldsymbol{\beta}_i})$ where, $A_{\boldsymbol{\beta}_i} \in \mathcal{C}^{\eta_1}_{\boldsymbol{\beta}_i}$, $\mathbf{t}_{i-1} \leq \boldsymbol{\beta}_{j-1} < \boldsymbol{\beta}_j \leq \mathbf{t}_i$ and R_i is rank function of $M_{\mathbf{t}_i}$. Clearly, $|C_{\boldsymbol{\beta}_{i^*}}| < |C_{\boldsymbol{\beta}_{i^*-1}}| < \ldots < |C_{\boldsymbol{\beta}_1}|$ and define a new sequence $D_{\boldsymbol{\beta}_{i^*}} \subseteq D_{\boldsymbol{\beta}_{i^*-1}} \subseteq \ldots \subseteq D_{\boldsymbol{\beta}_1}$ such that $D_{\boldsymbol{\beta}_{i^*}} = C_{\boldsymbol{\beta}_{i^*}}$ and

$$D_{\boldsymbol{\beta}_{i^*-1}} = \begin{cases} D_{\boldsymbol{\beta}_{i^*}} & \text{if } |D_{\boldsymbol{\beta}_{i^*}}| = |C_{\boldsymbol{\beta}_{i^*-1}}|, \\ C'_{\boldsymbol{\beta}_{i^*-1}} & \text{if } |D_{\boldsymbol{\beta}_{i^*}}| < |C_{\boldsymbol{\beta}_{i^*-1}}|, \end{cases}$$

where, $|C'_{\boldsymbol{\beta}_{i^*-1}}| = |C_{\boldsymbol{\beta}_{i^*-1}}|$ and $D_{\boldsymbol{\beta}_{i^*}} \subseteq C'_{\boldsymbol{\beta}_{i^*-1}}$ which is by condition 2 of Definition 12. Proceeding in this way, we can find a sequence $\{D_{\boldsymbol{\beta}_{i^*}}\}_{i=1}^{i^*}$ such that

(i) $D_{\boldsymbol{\beta}_i}$ is maximal in $(Y, \mathcal{C}^{\eta_1}_{\boldsymbol{\beta}_i})$
(ii) $|D_{\boldsymbol{\beta}_i}| = R_j(\eta_{\boldsymbol{\beta}_i})$ where, i and j are such that $\mathbf{t}_{i-1} \leq \boldsymbol{\beta}_{j-1} < \boldsymbol{\beta}_j \leq \mathbf{t}_i$.

For each positive integer i, $1 \leq i \leq i^*$, define η_{2i} as mF set such that $supp(\eta_{2i}) = D_{\boldsymbol{\beta}_i}$ with non-zero range $\{\boldsymbol{\beta}_i\}$. Let $\eta_2 = \bigcup_{i=1}^{i^*} \eta_{2i}$. Since $\eta_2 \subseteq \eta_1$ and $\eta_2 \in \mathcal{C}^*$ therefore, by Theorem 3,

$$\mu_r(\eta_1) = |\eta_2| \geq \sum_{i=1}^{i^*} (\boldsymbol{\beta}_i - \boldsymbol{\beta}_{i-1}) |D_{\boldsymbol{\beta}_i}| = \hat{\mu}_r(\eta_1).$$

□

4. Applications

mF matroids have interesting applications in graph theory, combinatorics and algebra. mF matroids are used to discuss the uncertain behaviour of objects if the data have multipolar information and have many applications in addition to Mathematics.

4.1. Decision Support Systems

mF matroids can be used in decision support systems to find the ordering of n tasks if each task constitutes m linguistic values. All tasks are available at 0 time and each task has a profit p associated with its m properties and a deadline d. The profit p_j can be gained if each mF task j is completed at the deadline d_j. The problem is to find the mF ordering of tasks to maximize the total profit. mF matroids can also be used in the secret sharing problem to share parts of secret information among different participants such that we have multipolar information about each participant.

It doesn't look like an mF matroid problem because the mF matroid problem asks to find an optimal mF subset, but this problem requires one to find an optimal schedule. However, this is an mF matroid problem. The profit, penalty and expense of any ordering can be determined by an mF subset of tasks that are on or before time. For an mF subset S of deadlines $\{d_1, d_2, \ldots, d_n\}$ corresponding to tasks $T = \{t_1, t_2, \ldots, t_n\}$, if there is a ordering such that every task in S is on or before time, and all tasks out of S are late. The procedure for the selection of tasks has net time complexity is $O(n2^n)$.

4.2. *Ordering of Machines/Workers for Certain Tasks*

An important application is to divide a set of workers into different groups to perform a specific task for which they are eligible. Consider the example of allocating a collection of tasks to a set of workers $W_1, W_2, \ldots, W_7$ who are eligible to perform that task. The problem is to assign a task to a group of workers to be fulfilled in required time, accuracy and cost. The 3-polar fuzzy set of workers is,

$$W' = \{(W_1, 0.8, 0.9, 0.9), (W_2, 0.7, 0.9, 0.7), (W_3, 0.7, 0.7, 0.6), (W_4, 0.7, 0.9, 0.8), (W_5, 0.6, 0.9, 0.8), (W_6, 0.6, 0.8, 0.75), (W_7, 0.7, 0.7, 0.6)\}.$$

The degree of membership of each worker shows the time taken by them, the accuracy of the output if they work on the task and cost of the worker for service. The problem is to determine a collection of workers for tasks T_1 and T_2 such that,

$$T_1 = \{(W_i, W'(W_i)) \mid P_1 \circ W_i \leq 0.7, P_2 \circ W_i \geq 0.7, P_3 \circ W_i \leq 0.7\},$$
$$T_2 = \{(W_i, W'(W_i)) \mid P_1 \circ W_i \leq 0.8, P_2 \circ W_i \geq 0.9, P_3 \circ W_i \leq 0.9\}.$$

The 3-polar fuzzy set of workers for both the tasks are,

$$T_1 = \{(W_2, 0.7, 0.9, 0.7), (W_3, 0.7, 0.7, 0.6), (W_6, 0.6, 0.8, 0.75), (W_7, 0.7, 0.7, 0.6)\},$$
$$T_2 = \{(W_1, 0.8, 0.9, 0.9), (W_3, 0.7, 0.7, 0.6), (W_4, 0.7, 0.9, 0.8)\}.$$

The workers W_2, W_3, W_6, W_7 are preferable for task T_1 and W_1, W_3, W_4 are preferable for task T_2.

4.3. *Network Analysis*

mF models can be used in network analysis problems to determine the minimum number of connections for wireless communication. The procedure for the selection of minimum number of locations from a wireless connection is explained in the following steps.

1. Input the n number of locations $L_1, L_2, \ldots, L_n$ of wireless communication network.
2. Input the adjacency matrix $\xi = [L_{ij}]_{n^2}$ of membership values of edges among locations.
3. From this adjacency matrix, arrange the membership values in increasing order.
4. Select an edge having minimum membership value.
5. Repeat Step 4 so that the selected edge does not create any circuit with previous selected edges.
6. Stop the procedure if the connection between every pair of locations is set up.

Here we explain the use of mF matroids in network analysis. The 2-polar fuzzy graph in Figure 2 represents the wireless communication between five locations L_1, L_2, L_3, L_4, L_5.

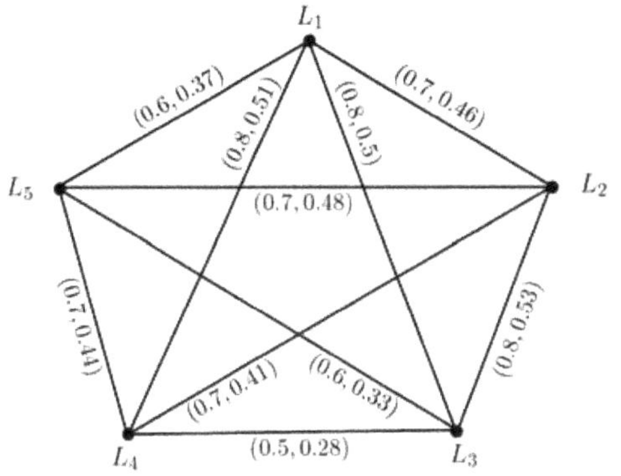

Figure 2. Wireless communication.

The degree of membership of each edge shows the time taken and cost for sending a message from one location to the other. Each pair of vertices is connected by an edge. However, in general we do not need connections among all the vertices because the vertices linked indirectly will also have a message service between them, i.e., if there is a connection from L_2 to L_3 and L_3 to L_4, then we can send a message from L_2 to L_4, even if there is no edge between L_2 and L_4. The problem is to find a set of edges such that we are able to send message between every two vertices under the condition that time and cost is minimum. The procedure is as follows. Arrange the membership values of edges in increasing order as, $\{(0.5, 0.28), (0.6, 0.33), (0.6, 0.37), (0.7, 0.41), (0.7, 0.44), (0.7, 0.46), (0.7, 0.48), (0.8, 0.5), (0.8, 0.51), (0.8, 0.53)\}$. At each step, select an edge having minimum membership value so that it does not create any circuit with previous selected edges. The 2-polar fuzzy set of selected edges is,

$$\{(L_3L_4, 0.5, 0.28), (L_3L_5, 0.6, 0.33), (L_1L_5, 0.6, 0.37), (L_2L_4, 0.7, 0.41), (L_1L_4, 0.7, 0.46)\}.$$

The communication network with minimum number of locations and cost is shown in Figure 3.

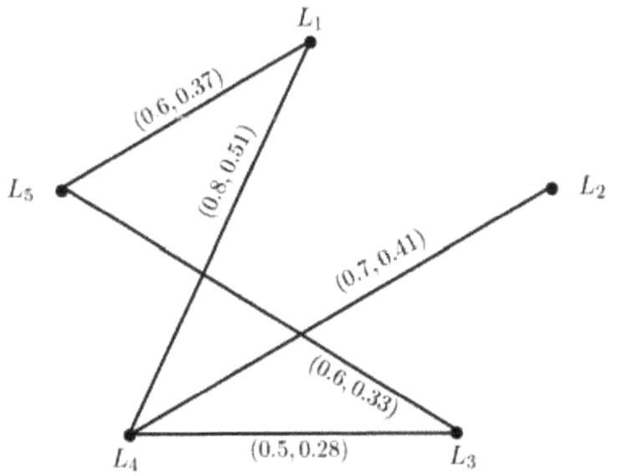

Figure 3. Communication network with minimum connections.

Figure 3 shows that only five connections are needed to communicate among given locations in order to minimize the cost and improve the network communication.

5. Conclusions

In this research paper, we have applied the powerful technique of mF sets to extend the theory of vector spaces and matroids. The mF models give more accuracy, precision and compatibility to the system when more than one agreements are to be dealt with. We have mainly introduced the idea of the mF matroid, implemented this concept to graph theory, linear algebra and have studied various examples including the mF uniform matroid, mF linear matroid, mF partition matroid and mF cycle matroid. We have also presented the idea of mF circuit, closure of mF matroid and put special emphasis on mF rank function. The paper is concluded with some real life applications of mF matroids in decision support system, ordering of machines to perform specific tasks and detection of minimum number of locations in wireless network in order to motivate the idea presented in this research paper. We are extending our work to (1) decision support systems based on intuitionistic fuzzy soft circuits, (2) fuzzy rough soft circuits, (3) and neutrosophic soft circuits.

Author Contributions: Musavarah Sarwar and Muhammad Akram conceived of the presented idea. Musavarah Sarwar developed the theory and performed the computations. Muhammad Akram verified the analytical methods.

Conflicts of Interest: The authors declare no conflict of interest.

References

1. Whitney, H. On the abstract properties of linear dependence. *Am. J. Math.* **1935**, *57*, 509–533.
2. Zadeh, L.A. Fuzzy sets. *Inf. Control* **1965**, *8*, 338–353.
3. Zadeh, L.A. Similarity relations and fuzzy orderings. *Inf. Sci.* **1971**, *3*, 177–200.
4. Petković, D.; Gocić, M.; Shamshirband, S. Adaptive neuro-fuzzy computing technique for precipitation estimation. *Facta Univ. Ser. Mech. Eng.* **2016**, *14*, 209–218.
5. Ćirović, G.; Pamučar, D. Decision support model for prioritizing railway level crossings for safety improvements. *Expert Syst. Appl.* **2013**, *91*, 89–106.
6. Radu-Emil, P.; Stefan, P.; Claudia-Adina, B.-D. Automotive applications of evolving Takagi-Sugeno-Kang fuzzy models. *Facta Univ. Ser. Mech. Eng.* **2017**, *15*, 231–244.
7. Lukovac, V.; Pamučar, D.; Popović, M.; Đorović, B. Portfolio model for analyzing human resources: An approach based on neuro-fuzzy modeling and the simulated annealing algorithm. *Expert Syst. Appl.* **2017**, *90*, 318–331.
8. Pamučar, D.; Petrović, I.; Ćirović, G. Modification of the BestWorst and MABAC methods: A novel approach based on interval-valued fuzzy-rough numbers. *Expert Syst. Appl.* **2018**, *91*, 89–106.
9. Zhang, W.-R. Bipolar fuzzy sets and relations: A computational framework for cognitive modeling and multiagent decision analysis. In Proceedings of the IEEE Conference Fuzzy Information Processing Society Biannual Conference, San Antonio, TX, USA, 18–21 December 1994; pp. 305–309.
10. Akram, M.; Feng, F.; Borumand Saeid, A.; Fotea, V. A new multiple criteria decision-making method based on bipolar fuzzy soft graphs. *Iran. J. Fuzzy Syst.* **2017**, doi:10.22111/IJFS.2017.3394.
11. Sarwar, M.; Akram, M. Novel concepts bipolar fuzzy competition graphs. *J. Appl. Math. Comput.* **2016**, *54*, 511–547.
12. Sarwar, M.; Akram, M. Certain algorithms for computing strength of competition in bipolar fuzzy graphs. *Int. J. Uncertain. Fuzziness Knowl. Based Syst.* **2017**, *25*, 877–896.
13. Chen, J.; Li, S.; Ma, S.; Wang, X. m-polar fuzzy sets: An extension of bipolar fuzzy sets. *Sci. World J.* **2014**, 416530, doi:10.1155/2014/416530.
14. Akram, M.; Younas, H.R. Certain types of irregular m-polar fuzzy graphs. *J. Appl. Math. Comput.* **2017**, *53*, 365–382.
15. Akram, M.; Sarwar, M. Novel applications of m-polar fuzzy hypergraphs. *J. Intell. Fuzzy Syst.* **2017**, *32*, 2747–2762.
16. Sarwar, M.; Akram, M. Novel applications of m-polar fuzzy concept lattice. *New Math. Nat. Comput.* **2017**, *13*, 261–287.
17. Akram, M.; Sarwar, M. Transversals of m-polar fuzzy hypergraphs with applications. *J. Intell. Fuzzy Syst.* **2016**, *32*, 351–364.

18. Akram, M.; Sarwar, M. Novel applications of *m*-polar fuzzy competition graphs in decision support system. *Neural Comput. Appl.* **2017**, doi:10.1007/s00521-017-2894-y.
19. Goetschel, R.; Voxman, W. Fuzzy matroids. *Fuzzy Sets Syst.* **1988**, *27*, 291–302.
20. Goetschel, R.; Voxman, W. Bases of fuzzy matroids. *Fuzzy Sets Syst.* **1989**, *31*, 253–261.
21. Sarwar, M.; Akram, M. Representation of graphs using *m*-polar fuzzy environment. *Ital. J. Pure Appl. Math.* **2017**, *38*, 291–312.
22. Akram, M.; Waseem, N. Certain metric in *m*-polar fuzzy graphs. *New Math. Nat. Comput.* **2016**, *12*, 135–155.
23. Li, S.; Yang, X.; Li, H.; Ma, M. Operations and decompositions of m-polar fuzzy graphs. *Basic Sci. J. Text. Univ./Fangzhi Gaoxiao Jichu Kexue Xuebao* **2017**, *30*, 149–162.
24. Hsueh, Y.-C. On fuzzification of matroids. *Fuzzy Sets Syst.* **1993**, *53*, 319–327.
25. Akram, M.; Adeel, A. *m*-polar fuzzy labeling graphs with application. *Math. Comput. Sci.* **2016**, *10*, 387–402.
26. Akram, M.; Akmal, R. Certain concepts in *m*-polar fuzzy graph structures. *Discret. Dyn. Nat. Soc.* **2016**, *2016*, 6301693.
27. Akram, M.; Waseem, N.; Dudek, W.A. Certain types of edge *m*-polar fuzzy graphs. *Iran. J. Fuzzy Syst.* **2017**, *14*, 27–50.
28. Koczy, K.T. Fuzzy graphs in the evaluation and optimization of networks. *Fuzzy Sets Syst.* **1992**, *46*, 307–319.
29. Mathew, S.; Sunitha, M.S. Types of arcs in a fuzzy graph. *Inf. Sci.* **2009**, *179*, 1760–1768.
30. Novak, L.A. A comment om "Bases of fuzzy matroids". *Fuzzy Sets Syst.* **1997**, *87*, 251–252.
31. Sarwar, M.; Akram, M. An algorithm for computing certain metrics in intuitionistic fuzzy graphs. *J. Intell. Fuzzy Syst.* **2016**, *30*, 2405–2416.
32. Wilson, R.J. An Introduction to matroid theory. *Am. Math. Mon.* **1973**, *80*, 500–525.
33. Zafar, F.; Akram, M. A novel decision making method based on rough fuzzy information. *Int. J. Fuzzy Syst.* **2017**, doi:10.1007/s40815-017-0368-0.
34. Mordeson, J.N.; Nair, P.S. *Fuzzy Graphs and Fuzzy Hypergraphs*, 2nd ed.; Physica Verlag: Heidelberg, Germany, 2001.

symmetry

MDPI

Article

Multi-objective Fuzzy Bi-matrix Game Model: A Multicriteria Non-Linear Programming Approach

Wei Zhang, Yumei Xing and Dong Qiu *

College of Science, Chongqing University of Post and Telecommunication, Chongqing 400065, China; zhangwei@cqupt.edu.cn (W.Z.); x_ym2015@163.com (Y.X.)

* Correspondence: dongqiumath@163.com; Tel.: +86-15123126186; Fax: +86-23-62461179

Received: 13 July 2017; Accepted: 11 August 2017; Published: 15 August 2017

Abstract: A multi-objective bi-matrix game model based on fuzzy goals is established in this paper. It is shown that the equilibrium solution of such a game model problem can be translated into the optimal solution of a multi-objective, non-linear programming problem. Finally, the results of this paper are demonstrated through a numerical example.

Keywords: multi-objective fuzzy bi-matrix game; equilibrium solution; multi-objective nonlinear programming

1. Introduction

When the bet is a small amount of money, the multi-objective bi-matrix game model is accurate. In real life, however, the interests of the relationship are more complex, particularly in some areas of the economy where the interests of the two players is precisely opposite. It is well known that these games are two person non-zero-sum games, called multi-objective bi-matrix games. Therefore, the research of multi-objective bi-matrix game problems has become more and more widespread in recent years.

The fuzzy set theory was introduced initially in 1965 by Zadeh [1]. The fuzziness occurring in game problems is categorized as fuzzy game problems. Single objective fuzzy game problems and related problems attached a wide range of research [2–9]. Tan et al. [5] presented a concept of the potential function for solving fuzzy games problems. They also reached a conclusion that the solution of fuzzy games and the marginal value of potential functions are equivalent. Chakeri et al. [10] used fuzzy logic to determine the priority of the pay-off based on the linguistic preference relation and proposed the notion of linguistic Nash equilibriums. Fuzzy preference relation has been widely used in fuzzy game theory [7–9,11]. At the same time, they [11] utilize the same method [10] to determine the priority of the pay-off based on fuzzy preference relation. In order to deal with this game model, a new approach was put forward. Moreover, Sharifian et al. [6] also applied fuzzy linguistic preference relation to fuzzy game theory.

The notions of max–min and min–max values were the earliest applied to solve the multi-objective game model in [12]. Roy et al. [13] presented solution procedures in view of the multi-objective bi-matrix game model. Besides, they [14] applied fuzzy optimization means to solve the fuzzy multicriteria bi-matrix game model. Nishizaki et al. [15–17] solved the multi-objective bi-matrix game via the resolution approach. Chen et al. [18,19] proposed an alternative technique for solving fuzzy multi-objective bi-matrix game problems through genetic algorithms in [20]. Angelov [21] proposed a new concept of the optimization problem based on degrees of satisfaction. Precup [22] introduced a new optimisation criteria in the development of fuzzy controllers with dynamics based on an attractive development method. In order to solve numerical optimization problems, a new algorithm was introduced in [23]. Ghosn et al. [24] investigated the use of parallel genetic algorithms in order to discuss the open-shop scheduling problem. Roy et al. [25] provided a mathematical optimization model for solving the multiple objective bi-matrix goal game problem on account of the entropy circumstance. Additionally, to solve the formulated mathematical model, they proposed a solution procedure of the fuzzy optimization method.

Since Wierzbicki [26] proposed equilibrium solutions for game problems, he analysed multi-objective game models based on pay-offs related to scalarising functions. There is a debate about the existence of equilibrium solutions of multicriteria bi-matrix games put forward by Borm et al. [27]. Nishizaki et al. [17] studied an equilibrium solution of multi-objective bi-matrix games. Qiu et al. [28] discussed the relationship of two fuzzy numbers via the lower limit$-\frac{1}{2}$ of the possibility degree. They also concluded that the equilibrium solution of multiple objective fuzzy games and the optimal solution of multi-objective linear optimization problems are of equal value. Bector et al. [2] only considered a single objective bi-matrix game based on fuzzy goals. Having gained enlightenment from [2,29–31], we will consider a multiple objective bi-matrix game based on fuzzy goals, so as to obtain better results.

The outline of this paper is as follows. Section 2 is about basic definitions and recalls results with regard to a crisp multi-objective bi-matrix game. In Section 3, a multi-objective bi-matrix game model based on fuzzy goals is established. Section 4 presents a kind of multicriteria, non-linear programming problem in some special cases. The results of this paper are demonstrated through a numerical example in Section 5.

2. Preliminaries

In this section, we recall some basic definitions and preliminaries. Further, we shall describe a crisp multi-objective bi-matrix game model in [29].

Definition 1. *[32] The set of mixed strategies for Player* I *is denoted by:*

$$S^m = \{x = (x_1, x_2, \cdots, x_m)^T \in \mathbb{R}^m | \sum_{i=1}^{m} x_i = 1, x_i \geq 0, i = 1, 2, \cdots, m.\} \tag{1}$$

Similarly, the set of mixed strategies for Player II *is denoted by:*

$$S^n = \{y = (y_1, y_2, \cdots, y_n)^T \in \mathbb{R}^n | \sum_{j=1}^{n} y_j = 1, y_j \geq 0, j = 1, 2, \cdots, n.\} \tag{2}$$

where x^T is the transposition of x, $\mathbb{R}^m$ and $\mathbb{R}^n$ are m- and n-dimensional Euclidean spaces.

The multiple pay-off matrices of Player I and Player II in multi-objective bi-matrix games are denoted by [29]:

$$A^1 = \begin{pmatrix} a^1_{11} & \cdots & a^1_{1n} \\ \vdots & \ddots & \vdots \\ a^1_{m1} & \cdots & a^1_{mn} \end{pmatrix}, \cdots, A^r = \begin{pmatrix} a^r_{11} & \cdots & a^r_{1n} \\ \vdots & \ddots & \vdots \\ a^r_{m1} & \cdots & a^r_{mn} \end{pmatrix} \tag{3}$$

and

$$B^1 = \begin{pmatrix} b^1_{11} & \cdots & b^1_{1n} \\ \vdots & \ddots & \vdots \\ b^1_{m1} & \cdots & b^1_{mn} \end{pmatrix}, \cdots, B^s = \begin{pmatrix} b^s_{11} & \cdots & b^s_{1n} \\ \vdots & \ddots & \vdots \\ b^s_{m1} & \cdots & b^s_{mn} \end{pmatrix} \tag{4}$$

respectively. Here, Player I and Player II have r and s objectives, respectively. Without any loss of generality, we assume that the Player I and Player II are both maximized players.

A multi-objective bi-matrix game $(MOBG)$ model is taken as:

$$MOBG = (S^m, S^n, A^k(1,2,\cdots,r), B^l(l=1,2,\cdots,s)).$$

Definition 2. *[25] Let $A = (A^1, A^2, \cdots, A^r)$. When Player I chooses a mixed strategy $x \in S^m$, the expected pay-off of Player I is denoted by:*

$$\begin{aligned} E(x,y) &= x^T A y = [E_1(x,y), E_2(x,y), \cdots, E_r(x,y)] \\ &= [x^T A^1 y, x^T A^2 y, \cdots, x^T A^r y] \\ &= \left[\sum_{i=1}^{m}\sum_{j=1}^{n} a_{ij}^1 x_i y_j, \sum_{i=1}^{m}\sum_{j=1}^{n} a_{ij}^2 x_i y_j, \cdots, \sum_{i=1}^{m}\sum_{j=1}^{n} a_{ij}^r x_i y_j\right] \end{aligned} \tag{5}$$

Similarly, let $B = (B^1, B^2, \cdots, B^s)$, when Player II chooses a mixed strategy $y \in S^n$, the expected pay-off of Player II is denoted by:

$$\begin{aligned} E(x,y) &= x^T B y = [E_1(x,y), E_2(x,y), \cdots, E_s(x,y)] \\ &= [x^T B^1 y, x^T B^2 y, \cdots, x^T B^s y] \\ &= \left[\sum_{i=1}^{m}\sum_{j=1}^{n} b_{ij}^1 x_i y_j, \sum_{i=1}^{m}\sum_{j=1}^{n} b_{ij}^2 x_i y_j, \cdots, \sum_{i=1}^{m}\sum_{j=1}^{n} b_{ij}^s x_i y_j\right] \end{aligned} \tag{6}$$

Definition 3. *[29] Suppose $D^k = \{x^T A^k y : (x,y) \in S^m \times S^n\} \subseteq \mathbb{R}$ be the domain of kth pay-offs of Player I. Then a fuzzy goal $\tilde{g}_I^k$ of Player I corresponding to the kth pay-offs is a fuzzy set on D^k, whose the membership function is defined by:*

$$u_{\tilde{g}_I^k} : D^k \to [0,1], \tag{7}$$

Similarly, suppose $D^l = \{x^T B^l y : (x,y) \in S^m \times S^n\} \subseteq \mathbb{R}$ be the domain of lth payoff of Player II. Then a fuzzy goal $\tilde{g}_{II}^l$ of Player II corresponding to the lth pay-offs is a fuzzy set on D^l, whose the membership function is defined by:

$$v_{\tilde{g}_{II}^l} : D^l \to [0,1]. \tag{8}$$

3. A Multi-objective Bi-matrix Game with Fuzzy Goals

In this section, we first introduce the concepts of fuzzy sets and fuzzy numbers.

A fuzzy set $\tilde{F}$ of $\mathbb{R}$ is characterized by a membership function $u_{\tilde{F}} : \mathbb{R} \to [0,1]$ [1]. An α-level set of $\tilde{F}$ is given as $[\tilde{F}]_\alpha = \{x \in \mathbb{R} : u_{\tilde{F}}(x) \geq \alpha\}$ for each $\alpha \in (0,1]$. A strict α-level set of $\tilde{F}$ is given by $(\tilde{F})_\alpha = \{x \in \mathbb{R} : u_{\tilde{F}}(x) > \alpha\}$ for each $\alpha \in (0,1]$. We define the set $[\tilde{F}]_0$ by $[\tilde{F}]_0 = \overline{\{x \in \mathbb{R} : u_{\tilde{F}}(x) > 0\}}$, where $\overline{F}$ denotes the closure of a crisp set F. A fuzzy set $\tilde{F}$ is said to be a fuzzy number if it satisfies the following conditions [33]:

(1) $\tilde{F}$ is normal, i.e., there exists an $x_0 \in \mathbb{R}$ such that $u_{\tilde{F}}(x_0) = 1$;
(2) $\tilde{F}$ is convex, i.e., $u_{\tilde{F}}(\lambda x_1 + (1-\lambda)x_2) \geq \min\{u_{\tilde{F}}(x_1), u_{\tilde{F}}(x_2)\}$, for all $x_1, x_2 \in \mathbb{R}$ and $\lambda \in [0,1]$;
(3) $\tilde{F}$ is upper semi-continuous;
(4) $[\tilde{F}]_0$ is compact.

In the following, we establish a multi-objective bi-matrix game model in the fuzzy environment. Suppose S^m, S^n, A^k $(k = 1,2,\cdots,r)$, and B^l $(l = 1,2,\cdots,s)$ be as introduced in Section 2.

Definition 4. *Let $A = (A^1, A^2, \cdots, A^r)$. When Player I chooses a mixed strategy $x \in S^m$, an aspiration level of Player I with respect to the kth pay-offs is denoted by:*

$$V_0^k = \max_{y \in S^n} E_k(x,y) = \max_{y \in S^n} x^T A^k y, \quad (k = 1,2,\cdots,r).$$

Similarly, let $B = (B^1, B^2, \cdots, B^s)$*, when Player* II *chooses a mixed strategy* $y \in S^n$*, an aspiration level of Player* II *with respect to the* l^{th} *pay-offs is denoted by:*

$$W_0^l = \max_{x \in S^m} E_l(x, y) = \max_{x \in S^m} x^T B^l y, \quad (l = 1, 2, \cdots, s).$$

Therefore, we obtain that the multi-objective bi-matrix game based on fuzzy goals, denoted by $MOBGFG$, can be presented as:

$$MOBGFG = (S^m, S^n, A^k, B^l, V_0^k, \gtrsim, W_0^l, \lesssim, (k = 1, 2, \cdots, r; l = 1, 2, \cdots, s)) \tag{9}$$

where $\gtrsim$ and $\lesssim$ are the fuzzified versions of symbols $\geq$ and $\leq$, respectively in [34].

Let t, a, and $p \in \mathbb{R}$ $(p > 0)$, then the membership function of the fuzzy set $\tilde{F}$ defining the fuzzy inequality $t \gtrsim_p a$, where this fuzzy inequality $t \gtrsim_p a$ can be interpreted as "t essentially greater than or equal to a with tolerance error p", can be defined by [2]:

$$u_{\tilde{F}}(t) = \begin{cases} 1, & t \geq a, \\ 1 - (\frac{a-t}{p}), & a - p \leq t \leq a, \\ 0, & t < a - p. \end{cases} \tag{10}$$

Based on the above discussion, let p_0^k and $p_0^{'k}$ $(k = 1, 2, \cdots, r)$ (respectively, q_0^l and $q_0^{'l}$ $(l = 1, 2 \cdots, s)$) be the positive tolerance errors of Player I (respectively, Player II) about the fuzzy inequalities, with respect to k^{th} pay-offs (respectively, l^{th} pay-offs). Thus the game $MOBGFG$ model becomes:

$$MOBGFG = (S^m, S^n, A^k, B^l, V_0^k, p_0^k, p_0^{'k}, W_0^l, q_0^l, q_0^{'l}, \gtrsim, \lesssim, (k = 1, 2, \cdots, r; l = 1, 2, \cdots, s)) \tag{11}$$

Definition 5. $(\bar{x}, \bar{y}) \in S^m \times S^n$ *is called a pair of equilibrium solution of the game* $(MOBGFG)$ *model if:*

$$\begin{array}{l} x^T A^k \bar{y} \lesssim_{p_0^k} V_0^k, k = 1, 2, \cdots, r; \forall x \in S^m, \\ \bar{x}^T B^l y \lesssim_{q_0^l} W_0^l, l = 1, 2, \cdots, s; \forall y \in S^n, \\ \bar{x}^T A^k \bar{y} \gtrsim_{p_0^{'k}} V_0^k, k = 1, 2, \cdots, r, \\ \bar{x}^T B^l \bar{y} \gtrsim_{q_0^{'l}} W_0^l, l = 1, 2, \cdots, s. \end{array} \tag{12}$$

In order to deal with the above game $(MOBGFG)$ model, we can get the following theorem.

Theorem 1. *Suppose* $(\bar{x}, \bar{y}, \bar{\lambda})$ *is an optimal solution of the problem* $(MONLP2)$ *if and only if we have that* $(\bar{x}, \bar{y})$ *is a pair of equilibrium solution of the game* $(MOBGFG)$ *model. Additionally,* $\bar{\lambda}$ *is the security level of satisfaction of Player I and Player II.* V_0^k $(k = 1, 2, \cdots, r)$ *and* W_0^l $(l = 1, 2, \cdots, s)$ *are the aspiration levels of Player I and II, respectively.*

$$(MONLP2) \qquad \max \quad \lambda \tag{13}$$

$$\begin{array}{ll} \text{subject to} & A_i^k y + (\lambda - 1) p_0^k \leq V_0^k, (k = 1, 2, \cdots, r; i = 1, 2, \cdots m), \\ & B_j^{l^T} x + (\lambda - 1) q_0^l \leq W_0^l, (l = 1, 2, \cdots, s; j = 1, 2, \cdots n), \\ & x^T A^k y + (1 - \lambda) p_0^{'k} \geq V_0^k, (k = 1, 2, \cdots, r), \\ & x^T B^l y + (1 - \lambda) q_0^{'l} \geq W_0^l, (l = 1, 2, \cdots, s), \\ & 0 \leq \lambda \leq 1, \\ & x \in S^m, y \in S^n. \end{array}$$

Proof. Since $(\bar{x}, \bar{y})$ is a pair of equilibrium solutions of the game $(MOBGFG)$ model. By using Definition 5, we can get that the equilibrium solution of the game $(MOBGFG)$ model and the following multiple objective fuzzy optimization problem $(MOFOP)$ are of equal value.

$$(MOFOP) \qquad \text{Find} \quad (x,y) \in S^m \times \in S^n \quad \text{subject to:}$$

$$\begin{array}{l} A_i^k y \lesssim_{p_0^k} V_0^k, (k = 1,2,\cdots,r; i = 1,2,\cdots m), \\ B_j^{T^l} x \lesssim_{q_0^l} W_0^l, (l = 1,2,\cdots,s; j = 1,2,\cdots n), \\ x^T A^k y \gtrsim_{p_0^{'k}} V_0^k, (k = 1,2,\cdots,r), \\ x^T B^l y \gtrsim_{q_0^{'l}} W_0^l, (l = 1,2,\cdots,s), \end{array} \tag{14}$$

where A_i^k $(i = 1,2,\cdots,m)$ is the ith row of the matrix A^k and B_j^l $(j = 1,2,\cdots,n)$ is the jth column of the matrix B^l.

By using (9), we obtain that membership functions $u_i^k(A_i^k y)$, $(i = 1,2,\cdots,m)$ (respectively, $v_j^l(B_j^{l^T} x)$, $(j = 1,2,\cdots,n)$) of fuzzy inequalities $A_i^k y \lesssim_{p_0^k} V_0^k$ $(\forall y \in S^n)$ (respectively, $B_j^{l^T} x \lesssim_{q_0^l} W_0^l$ $(\forall x \in S^m)$) can be presented as:

$$u_i^k(A_i^k y) = \begin{cases} 1, & A_i^k y \leq V_0^k, \\ 1 - \frac{A_i^k y - V_0^k}{p_0^k}, & V_0^k \leq A_i^k y \leq V_0^k + p_0^k, \\ 0, & A_i^k y \geq V_0^k + p_0^k, \end{cases} \tag{15}$$

and

$$v_j^l(B_j^{l^T} x) = \begin{cases} 1, & B_j^{l^T} x \leq W_0^l, \\ 1 - \frac{B_j^{l^T} x - W_0^l}{q_0^l}, & W_0^l \leq B_j^{l^T} x \leq W_0^l + q_0^l, \\ 0, & B_j^{l^T} x \geq W_0^l + q_0^l. \end{cases} \tag{16}$$

respectively.

Similarly, we have that the non-linear membership functions of the fuzzy inequalities $x^T A^k y \gtrsim_{p_0^{'k}} V_0^k$ (respectively, $x^T B^l y \gtrsim_{q_0^{'l}} W_0^l$) can be expressed as:

$$u_{\tilde{g}_I^k}(x^T A^k y) = \begin{cases} 1, & x^T A^k y \geq V_0^k, \\ 1 - \frac{V_0^k - x^T A^k y}{p_0^{'k}}, & V_0^k \geq x^T A^k y \geq V_0^k - p_0^{'k}, \\ 0, & x^T A^k y \leq V_0^k - p_0^{'k}, \end{cases} \tag{17}$$

and

$$v_{\tilde{g}_{II}^l}(x^T B^l y) = \begin{cases} 1, & x^T B^l y \geq W_0^l, \\ 1 - \frac{W_0^l - x^T B^l y}{q_0^{'l}}, & W_0^l \geq x^T B^l y \geq W_0^l - q_0^{'l}, \\ 0, & x^T B^l y \leq W_0^l - q_0^{'l}. \end{cases} \tag{18}$$

respectively.

Inspired by [35], by combining (10)–(13) we obtain that the problem $(MOFOP)$ model is equivalent to the multicriteria non-linear programming $(MONLP1)$ problem.

$$
(MONLP1) \qquad \max \quad \lambda
$$

$$
\begin{aligned}
\text{Subject to} \quad & \lambda \le 1 - \frac{A_i^k y - V_0^k}{p_0^k}, (k = 1, 2, \cdots, r; i = 1, 2, \cdots m), \\
& \lambda \le 1 - \frac{B_j^{l^T} x - W_0^l}{q_0^l}, (l = 1, 2, \cdots, s; j = 1, 2, \cdots n), \\
& \lambda \le 1 + \frac{x^T A^k y - V_0^k}{p_0^{'k}}, (k = 1, 2, \cdots, r), \\
& \lambda \le 1 + \frac{x^T B^l y - W_0^l}{q_0^{'l}}, (l = 1, 2, \cdots, s), \\
& 0 \le \lambda \le 1, \\
& x \in S^m, y \in S^n.
\end{aligned} \tag{19}
$$

That is, by simplifying the above problem, that is equal to:

$$
(MONLP2) \qquad \max \quad \lambda
$$

$$
\begin{aligned}
\text{subject to} \quad & A_i^k y + (\lambda - 1) p_0^k \le V_0^k, (k = 1, 2, \cdots, r; i = 1, 2, \cdots m), \\
& B_j^{l^T} x + (\lambda - 1) q_0^l \le W_0^l, (l = 1, 2, \cdots, s; j = 1, 2, \cdots n), \\
& x^T A^k y + (1 - \lambda) p_0^{'k} \ge V_0^k, (k = 1, 2, \cdots, r), \\
& x^T B^l y + (1 - \lambda) q_0^{'l} \ge W_0^l, (l = 1, 2, \cdots, s), \\
& 0 \le \lambda \le 1, \\
& x \in S^m, y \in S^n.
\end{aligned} \tag{20}
$$

Then, we have that $(\bar{x}, \bar{y})$ is a pair of equilibrium solutions of the game $(MOBGFG)$ model if and only if $(\bar{x}, \bar{y}, \bar{\lambda})$ is an optimal solution of the problem $(MONLP2)$.

$$
(MONLP2) \qquad \max \quad \lambda
$$

$$
\begin{aligned}
\text{subject to} \quad & A_i^k y + (\lambda - 1) p_0^k \le V_0^k, (k = 1, 2, \cdots, r; i = 1, 2, \cdots m), \\
& B_j^{l^T} x + (\lambda - 1) q_0^l \le W_0^l, (l = 1, 2, \cdots, s; j = 1, 2, \cdots n), \\
& x^T A^k y + (1 - \lambda) p_0^{'k} \ge V_0^k, (k = 1, 2, \cdots, r), \\
& x^T B^l y + (1 - \lambda) q_0^{'l} \ge W_0^l, (l = 1, 2, \cdots, s), \\
& 0 \le \lambda \le 1, \\
& x \in S^m, y \in S^n.
\end{aligned} \tag{21}
$$

□

Remark 1. *Let $\bar{\lambda} = 1$ and suppose $(\bar{x}, \bar{y}, \bar{\lambda})$ is an optimal solution of the problem $(MONLP2)$. Then, we obtain that the game $(MOBG)$ model is a special case of the game $(MOBGFG)$ model.*

Remark 2. *Let $\bar{\lambda} = 1$ and suppose $(\bar{x}, \bar{y}, \bar{\lambda})$ is an optimal solution of the problem $(MONLP2)$. Then the problem $(MONLP2)$ model changes into:*

$$
(MONLP3) \qquad \max \quad \lambda
$$

$$
\begin{aligned}
\text{subject to} \quad & A_i^k y \le V_0^k, (k = 1, 2, \cdots, r; i = 1, 2, \cdots m), \\
& B_j^{l^T} x \le W_0^l, (l = 1, 2, \cdots, s; j = 1, 2, \cdots n), \\
& x^T A^k y \ge V_0^k, (k = 1, 2, \cdots, r), \\
& x^T B^l y \ge W_0^l, (l = 1, 2, \cdots, s), \\
& 0 \le \lambda \le 1, \\
& x \in S^m, y \in S^n.
\end{aligned} \tag{22}
$$

4. Special Case:

In this section, we present a multicriteria non-linear programming problem in some special cases.

Theorem 2. *Let* $V_0^k = \overline{a}^k$, $p_0^k = p_0^{'k} = \overline{a}^k - \underline{a}^k$, $W_0^l = \overline{b}^l$, *and* $q_0^l = q_0^{'l} = \overline{b}^l - \underline{b}^l$. *Suppose* $(\bar{x}, \bar{y})$ *are a pair of equilibrium solutions of the game* $(MOBGFG)$ *model if and only if* $(\bar{x}, \bar{y}, \bar{\lambda})$ *is an optimal solution of the problem* $(MONLP4)$.

$$(MONLP4) \qquad \max \quad \lambda$$

$$\text{subject to} \quad \begin{array}{l} A_i^k y + (\lambda - 1)(\overline{a}^k - \underline{a}^k) \leq \overline{a}^k, (k = 1, 2, \cdots, r; i = 1, 2, \cdots m), \\ B_j^{l^T} x + (\lambda - 1)(\overline{b}^l - \underline{b}^l) \leq \overline{b}^l, (l = 1, 2, \cdots, s; j = 1, 2, \cdots n), \\ x^T A^k y + (1 - \lambda)(\overline{a}^k - \underline{a}^k) \geq \overline{a}^k, (k = 1, 2, \cdots, r), \\ x^T B^l y + (1 - \lambda)(\overline{b}^l - \underline{b}^l) \geq \overline{b}^l, (l = 1, 2, \cdots, s), \\ x \in S^m, y \in S^n, \end{array} \tag{23}$$

where

$$\underline{a}^k = \min_{x \in S^m} \min_{y \in S^n} x^T A^k y = \min_{x \in S^m} \min_{y \in S^n} a_{ij}^k, \quad \overline{a}^k = \max_{x \in S^m} \max_{y \in S^n} x^T A^k y = \max_{x \in S^m} \max_{y \in S^n} a_{ij}^k,$$

$$\underline{b}^l = \min_{x \in S^m} \min_{y \in S^n} x^T B^l y = \min_{x \in S^m} \min_{y \in S^n} b_{ij}^l, \quad \overline{b}^l = \max_{x \in S^m} \max_{y \in S^n} x^T B^l y = \max_{x \in S^m} \max_{y \in S^n} b_{ij}^l.$$

Proof. Since $(\bar{x}, \bar{y})$ is a pair of equilibrium solutions of the game $(MOBGFG)$ model and $V_0^k = \overline{a}^k$, and $p_0^k = p_0^{'k} = \overline{a}^k - \underline{a}^k$, $W_0^l = \overline{b}^l$, $q_0^l = q_0^{'l} = \overline{b}^l - \underline{b}^l$. By using Definition 5 and Theorem 1, we can get that the equilibrium solutions of the game $(MOBGFG)$ model and the following multiple objective fuzzy optimization problem $(MOFOP1)$ are of equal value.

$$(MOFOP1) \qquad \text{Find} \quad (x, y) \in S^m \times \in S^n \quad \text{subject to}$$

$$\begin{array}{l} A_i^k y \lesssim_{\overline{a}^k - \underline{a}^k} \overline{a}^k, (k = 1, 2, \cdots, r; i = 1, 2, \cdots m), \\ B_j^{T^l} x \lesssim_{\overline{b}^l - \underline{b}^l} \overline{b}^l, (l = 1, 2, \cdots, s; j = 1, 2, \cdots n), \\ x^T A^k y \gtrsim_{\overline{a}^k - \underline{a}^k} \overline{a}^k, (k = 1, 2, \cdots, r), \\ x^T B^l y \gtrsim_{\overline{b}^l - \underline{b}^l} \overline{b}^l, (l = 1, 2, \cdots, s), \end{array} \tag{24}$$

Inspired by [2,29], now combining (10), (11), (12) and (13), we take membership functions $u_i^k(A_i^k y)$ $(i = 1, 2, \cdots, m)$, $v_j^l(B_j^{l^T} x)$ $(j = 1, 2, \cdots, n)$, $u_{\tilde{g}_I^k}(x^T A^k y)$ and $v_{\tilde{g}_{II}^l}(x^T B^l y)$ $(k = 1, 2, \cdots, r; l = 1, 2, \cdots, s)$ as:

$$u_i^k(A_i^k y) = \begin{cases} 1, & A_i^k y \leq \overline{a}^k, \\ 1 - \frac{A_i^k y - \overline{a}^k}{\overline{a}^k - \underline{a}^k}, & \overline{a}^k \leq A_i^k y \leq 2\overline{a}^k - \underline{a}^k, \\ 0, & A_i^k y \geq 2\overline{a}^k - \underline{a}^k, \end{cases} \tag{25}$$

$$v_j^l(B_j^{l^T} x) = \begin{cases} 1, & B_j^{l^T} x \leq \overline{b}^l, \\ 1 - \frac{B_j^{l^T} x - \overline{b}^l}{\overline{b}^l - \underline{b}^l}, & \overline{b}^l \leq B_j^{l^T} x \leq 2\overline{b}^l - \underline{b}^l, \\ 0, & B_j^{l^T} x \geq 2\overline{b}^l - \underline{b}^l. \end{cases} \tag{26}$$

$$u_{\tilde{g}_I^k}(x^T A^k y) = \begin{cases} 1, & x^T A^k y \geq \overline{a}^k, \\ 1 - \frac{\overline{a}^k - x^T A^k y}{\overline{a}^k - \underline{a}^k}, & \overline{a}^k \geq x^T A^k y \geq \underline{a}^k, \\ 0, & x^T A^k y \leq \underline{a}^k, \end{cases} \tag{27}$$

and

$$v_{\tilde{g}_{1l}^{l}}(x^T B^l y) = \begin{cases} 1, & x^T B^l y \geq \overline{b}^l, \\ 1 - \frac{\overline{b}^l - x^T B^l y}{q_0^{'l}}, & \overline{b}^l \geq x^T B^l y \geq \underline{b}^l, \\ 0, & x^T B^l y \leq \underline{b}^l. \end{cases} \tag{28}$$

Similarly, we obtain that the problem $(MOFOP1)$ model changes into:

$$\begin{aligned} (MONLP5) \qquad & max \qquad \lambda \\ subject\ to \quad & \lambda \leq 1 - \frac{A_i^k y - \overline{a}^k}{\overline{a}^k - \underline{a}^k}, (k = 1, 2, \cdots, r; i = 1, 2, \cdots m), \\ & \lambda \leq 1 - \frac{B_j^{l^T} x - \overline{b}^l}{\overline{b}^l - \underline{b}^l}, (l = 1, 2, \cdots, s; j = 1, 2, \cdots n), \\ & \lambda \leq 1 + \frac{x^T A^k y - \overline{a}^k}{\overline{a}^k - \underline{a}^k}, (k = 1, 2, \cdots, r), \\ & \lambda \leq 1 + \frac{x^T B^l y - \overline{b}^l}{\overline{b}^l - \underline{b}^l}, (l = 1, 2, \cdots, s), \\ & 0 \leq \lambda \leq 1, \\ & x \in S^m, y \in S^n. \end{aligned} \tag{29}$$

That is, the problem $(MONLP5)$ model is equal to:

$$\begin{aligned} (MONLP4) \qquad & max \qquad \lambda \\ subject\ to \quad & A_i^k y + (\lambda - 1)(\overline{a}^k - \underline{a}^k) \leq \overline{a}^k, (k = 1, 2, \cdots, r; i = 1, 2, \cdots m), \\ & B_j^{l^T} x + (\lambda - 1)(\overline{b}^l - \underline{b}^l) \leq \overline{b}^l, (l = 1, 2, \cdots, s; j = 1, 2, \cdots n), \\ & x^T A^k y + (1 - \lambda)(\overline{a}^k - \underline{a}^k) \geq \overline{a}^k, (k = 1, 2, \cdots, r), \\ & x^T B^l y + (1 - \lambda)(\overline{b}^l - \underline{b}^l) \geq \overline{b}^l, (l = 1, 2, \cdots, s), \\ & x \in S^m, y \in S^n. \end{aligned} \tag{30}$$

Then, we have that $(\bar{x}, \bar{y})$ is a pair of equilibrium solutions of the game $(MOBGFG)$ model if and only if $(\bar{x}, \bar{y}, \bar{\lambda})$ is an optimal solution of the problem $(MONLP4)$.

$$\begin{aligned} (MONLP4) \qquad & max \qquad \lambda \\ subject\ to \quad & A_i^k y + (\lambda - 1)(\overline{a}^k - \underline{a}^k) \leq \overline{a}^k, (k = 1, 2, \cdots, r; i = 1, 2, \cdots m), \\ & B_j^{l^T} x + (\lambda - 1)(\overline{b}^l - \underline{b}^l) \leq \overline{b}^l, (l = 1, 2, \cdots, s; j = 1, 2, \cdots n), \\ & x^T A^k y + (1 - \lambda)(\overline{a}^k - \underline{a}^k) \geq \overline{a}^k, (k = 1, 2, \cdots, r), \\ & x^T B^l y + (1 - \lambda)(\overline{b}^l - \underline{b}^l) \geq \overline{b}^l, (l = 1, 2, \cdots, s), \\ & x \in S^m, y \in S^n. \end{aligned} \tag{31}$$

□

Theorem 3. *Suppose $(\bar{x}, \bar{y}, \bar{\lambda})$ are an optimal solution of the problem $(MONLP2)$. Let $V_0^k = \overline{a}^k$, and $p_0^k = p_0^{'k} = \overline{a}^k - \underline{a}^k$, $W_0^l = \overline{b}^l$, $q_0^l = q_0^{'l} = \overline{b}^l - \underline{b}^l$. Then the problem $(MONLP2)$ model changes into the following problem $(MONLP4)$.*

Proof. Since $(\bar{x}, \bar{y}, \bar{\lambda})$ is an optimal solution of the problem $(MONLP2)$, then we can get:

$$
\begin{array}{llll}
(MONLP2) & & max & \lambda \\
& subject\ to & A_i^k y + (\lambda - 1)p_0^k \le V_0^k, (k = 1, 2, \cdots, r; i = 1, 2, \cdots m), & \\
& & B_j^{l^T} x + (\lambda - 1)q_0^l \le W_0^l, (l = 1, 2, \cdots, s; j = 1, 2, \cdots n), & \\
& & x^T A^k y + (1 - \lambda)p_0^{'k} \ge V_0^k, (k = 1, 2, \cdots, r), & \\
& & x^T B^l y + (1 - \lambda)q_0^{'l} \ge W_0^l, (l = 1, 2, \cdots, s), & \\
& & 0 \le \lambda \le 1, & \\
& & x \in S^m, y \in S^n. &
\end{array}
\tag{32}
$$

Now, let $V_0^k = \bar{a}^k$, $p_0^k = p_0^{'k} = \bar{a}^k - \underline{a}^k$, $W_0^l = \bar{b}^l$, and $q_0^l = q_0^{'l} = \bar{b}^l - \underline{b}^l$. Hence, we obtain that the problem $(MONLP2)$ model changes into:

$$
\begin{array}{llll}
(MONLP4) & & max & \lambda \\
& subject\ to & A_i^k y + (\lambda - 1)(\bar{a}^k - \underline{a}^k) \le \bar{a}^k, (k = 1, 2, \cdots, r; i = 1, 2, \cdots m), & \\
& & B_j^{l^T} x + (\lambda - 1)(\bar{b}^l - \underline{b}^l) \le \bar{b}^l, (l = 1, 2, \cdots, s; j = 1, 2, \cdots n), & \\
& & x^T A^k y + (1 - \lambda)(\bar{a}^k - \underline{a}^k) \ge \bar{a}^k, (k = 1, 2, \cdots, r), & \\
& & x^T B^l y + (1 - \lambda)(\bar{b}^l - \underline{b}^l) \ge \bar{b}^l, (l = 1, 2, \cdots, s), & \\
& & x \in S^m, y \in S^n. &
\end{array}
\tag{33}
$$

Then, we have that $(\bar{x}, \bar{y}, \bar{\lambda})$ is an optimal solution of the problem $(MONLP4)$. □

5. Example

Now, we consider the following multi-objective fuzzy bi-matrix game $(MOBGFG)$ model.

Example 1. *A the multi-objective bi-matrix game is considered. The multiple pay-off matrices of the Player I and Player II are taken as:*

$$
A^1 = \begin{pmatrix} 6 & 3 & 4 \\ 3 & 6 & 8 \\ 7 & 3 & 4 \end{pmatrix}, \qquad A^2 = \begin{pmatrix} 9 & 2 & 7 \\ 4 & 5 & 8 \\ 2 & 7 & 3 \end{pmatrix}, \qquad A^3 = \begin{pmatrix} 5 & 1 & 2 \\ 3 & 4 & 8 \\ 1 & 8 & 1 \end{pmatrix},
$$

and

$$
B^1 = \begin{pmatrix} 9 & 1 & 4 \\ 0 & 6 & 3 \\ 5 & 2 & 8 \end{pmatrix}, \qquad B^2 = \begin{pmatrix} 1 & 6 & 7 \\ 8 & 2 & 3 \\ 4 & 9 & 3 \end{pmatrix}, \qquad B^3 = \begin{pmatrix} 8 & 2 & 3 \\ -5 & 6 & 0 \\ -3 & 1 & 6 \end{pmatrix}
$$

respectively.

We now solve this problem with the above model. Thus, by Theorem 2, we have:

$$
\underline{a}^1 = \min_{x \in S^3} \min_{y \in S^3} a_{ij}^1 = 3,\ V_0^1 = \bar{a}^1 = \max_{x \in S^3} \max_{y \in S^3} a_{ij}^1 = 8,\ p_0^1 = p_0^{'1} = \bar{a}^1 - \underline{a}^1 = 5; \tag{34}
$$

$$
\underline{a}^2 = \min_{x \in S^3} \min_{y \in S^3} a_{ij}^2 = 2,\ V_0^2 = \bar{a}^2 = \max_{x \in S^3} \max_{y \in S^3} a_{ij}^2 = 9, p_0^2 = p_0^{'2} = \bar{a}^2 - \underline{a}^2 = 7; \tag{35}
$$

$$
\underline{a}^3 = \min_{x \in S^3} \min_{y \in S^3} a_{ij}^3 = 1,\ V_0^3 = \bar{a}^3 = \max_{x \in S^3} \max_{y \in S^3} a_{ij}^3 = 8, p_0^3 = p_0^{'3} = \bar{a}^3 - \underline{a}^3 = 7; \tag{36}
$$

$$
\underline{b}^1 = \min_{x \in S^3} \min_{y \in S^3} b_{ij}^1 = 0,\ W_0^1 = \bar{b}^1 = \max_{x \in S^3} \max_{y \in S^3} b_{ij}^1 = 9,\ q_0^1 = q_0^{'1} = \bar{b}^1 - \underline{b}^1 = 9; \tag{37}
$$

$$\underline{b}^2 = \min_{x \in S^3} \min_{y \in S^3} b_{ij}^2 = 1,\ W_0^2 = \overline{b}^2 = \max_{x \in S^3} \max_{y \in S^3} b_{ij}^2 = 9,\ q_0^2 = q_0^{\prime 2} = \overline{b}^2 - \underline{b}^2 = 8; \tag{38}$$

$$\underline{b}^3 = \min_{x \in S^3} \min_{y \in S^3} b_{ij}^3 = -5,\ W_0^3 = \overline{b}^3 = \max_{x \in S^3} \max_{y \in S^3} b_{ij}^3 = 8,\ q_0^3 = q_0^{\prime 3} = \overline{b}^3 - \underline{b}^3 = 13. \tag{39}$$

By the above numerical values, we can get that the equilibrium solutions of the above model and the following multiple objective fuzzy optimization problem $(MOFOP2)$ are of equal value.

$$(MOFOP2) \qquad \text{Find } (x, y) \in S^3 \times \in S^3$$

$$\begin{array}{lll}
\textit{subject to} & 6y_1 + 3y_2 + 4y_3 \lesssim_5 8, & 9x_1 + 0x_2 + 5x_3 \lesssim_9 9, \\
& 3y_1 + 6y_2 + 8y_3 \lesssim_5 8, & 1x_1 + 6x_2 + 2x_3 \lesssim_9 9, \\
& 7y_1 + 3y_2 + 4y_3 \lesssim_5 8, & 4x_1 + 3x_2 + 8x_3 \lesssim_9 9, \\
& 9y_1 + 2y_2 + 7y_3 \lesssim_7 9, & 1x_1 + 8x_2 + 4x_3 \lesssim_8 9, \\
& 4y_1 + 5y_2 + 8y_3 \lesssim_7 9, & 6x_1 + 2x_2 + 9x_3 \lesssim_8 9, \\
& 2y_1 + 7y_2 + 3y_3 \lesssim_7 9, & 7x_1 + 3x_2 + 3x_3 \lesssim_8 9, \\
& 5y_1 + 1y_2 + 2y_3 \lesssim_7 8, & 8x_1 - 5x_2 - 3x_3 \lesssim_{13} 8, \\
& 3y_1 + 4y_2 + 8y_3 \lesssim_7 8, & 2x_1 + 6x_2 + 1x_3 \lesssim_{13} 8, \\
& 1y_1 + 8y_2 + 1y_3 \lesssim_7 8, & 3x_1 + 0x_2 + 6x_3 \lesssim_{13} 8,
\end{array}$$
$$\begin{array}{l}
6x_1y_1 + 3x_2y_1 + 7x_3y_1 + 3x_1y_2 + 6x_2y_2 + 3x_3y_2 + 4x_1y_3 + 8x_2y_3 + 4x_3y_3 \gtrsim_5 8, \\
9x_1y_1 + 4x_2y_1 + 2x_3y_1 + 2x_1y_2 + 5x_2y_2 + 7x_3y_2 + 7x_1y_3 + 8x_2y_3 + 3x_3y_3 \gtrsim_7 9, \\
5x_1y_1 + 3x_2y_1 + 1x_3y_1 + 1x_1y_2 + 4x_2y_2 + 8x_3y_2 + 2x_1y_3 + 8x_2y_3 + 1x_3y_3 \gtrsim_7 8, \\
9x_1y_1 + 0x_2y_1 + 5x_3y_1 + 1x_1y_2 + 6x_2y_2 + 2x_3y_2 + 4x_1y_3 + 3x_2y_3 + 8x_3y_3 \gtrsim_9 9, \\
1x_1y_1 + 8x_2y_1 + 4x_3y_1 + 6x_1y_2 + 2x_2y_2 + 9x_3y_2 + 7x_1y_3 + 3x_2y_3 + 3x_3y_3 \gtrsim_8 9, \\
8x_1y_1 - 5x_2y_1 - 3x_3y_1 + 2x_1y_2 + 6x_2y_2 + 1x_3y_2 + 3x_1y_3 + 0x_2y_3 + 6x_3y_3 \gtrsim_{13} 8.
\end{array} \tag{40}$$

Now we get the following membership functions based on the above fuzzy inequalities.

$$u_1^1(6y_1 + 3y_2 + 4y_3) = \begin{cases} 1, & 6y_1 + 3y_2 + 4y_3 \leq 8, \\ 1 - \frac{6y_1 + 3y_2 + 4y_3 - 8}{5}, & 8 \leq 6y_1 + 3y_2 + 4y_3 \leq 13, \\ 0, & 6y_1 + 3y_2 + 4y_3 \geq 13, \end{cases} \tag{41}$$

$$u_2^1(3y_1 + 6y_2 + 8y_3) = \begin{cases} 1, & 3y_1 + 6y_2 + 8y_3 \leq 8, \\ 1 - \frac{3y_1 + 6y_2 + 8y_3 - 8}{5}, & 8 \leq 3y_1 + 6y_2 + 8y_3 \leq 13, \\ 0, & 3y_1 + 6y_2 + 8y_3 \geq 13, \end{cases} \tag{42}$$

$$u_3^1(7y_1 + 3y_2 + 4y_3) = \begin{cases} 1, & 7y_1 + 3y_2 + 4y_3 \leq 8, \\ 1 - \frac{7y_1 + 3y_2 + 4y_3 - 8}{5}, & 8 \leq 7y_1 + 3y_2 + 4y_3 \leq 13, \\ 0, & 7y_1 + 3y_2 + 4y_3 \geq 13, \end{cases} \tag{43}$$

$$u_1^2(9y_1 + 2y_2 + 7y_3) = \begin{cases} 1, & 9y_1 + 2y_2 + 7y_3 \leq 9, \\ 1 - \frac{9y_1 + 2y_2 + 7y_3 - 9}{7}, & 9 \leq 9y_1 + 2y_2 + 7y_3 \leq 16, \\ 0, & 9y_1 + 2y_2 + 7y_3 \geq 16, \end{cases} \tag{44}$$

$$u_2^2(4y_1 + 5y_2 + 8y_3) = \begin{cases} 1, & 4y_1 + 5y_2 + 8y_3 \leq 9, \\ 1 - \frac{4y_1 + 5y_2 + 8y_3 - 9}{7}, & 9 \leq 4y_1 + 5y_2 + 8y_3 \leq 16, \\ 0, & 4y_1 + 5y_2 + 8y_3 \geq 16, \end{cases} \tag{45}$$

$$u_3^2(2y_1 + 7y_2 + 3y_3) = \begin{cases} 1, & 2y_1 + 7y_2 + 3y_3 \leq 9, \\ 1 - \frac{2y_1 + 7y_2 + 3y_3 - 9}{7}, & 9 \leq 2y_1 + 7y_2 + 3y_3 \leq 16, \\ 0, & 2y_1 + 7y_2 + 3y_3 \geq 16, \end{cases} \tag{46}$$

$$u_1^3(5y_1+1y_2+2y_3)=\begin{cases}1, & 5y_1+1y_2+2y_3\le 8,\\ 1-\frac{5y_1+1y_2+2y_3-8}{7}, & 8\le 5y_1+1y_2+2y_3\le 15,\\ 0, & 5y_1+1y_2+2y_3\ge 15,\end{cases} \tag{47}$$

$$u_2^3(3y_1+4y_2+8y_3)=\begin{cases}1, & 3y_1+4y_2+8y_3\le 8,\\ 1-\frac{3y_1+4y_2+8y_3-8}{7}, & 8\le 3y_1+4y_2+8y_3\le 15,\\ 0, & 3y_1+4y_2+8y_3\ge 15,\end{cases} \tag{48}$$

$$u_3^3(1y_1+8y_2+1y_3)=\begin{cases}1, & 1y_1+8y_2+1y_3\le 8,\\ 1-\frac{1y_1+8y_2+1y_3-8}{7}, & 8\le 1y_1+8y_2+1y_3\le 15,\\ 0, & 1y_1+8y_2+1y_3\ge 15,\end{cases} \tag{49}$$

$$v_1^1(9x_1+0x_2+5x_3)=\begin{cases}1, & 9x_1+0x_2+5x_3\le 9,\\ 1-\frac{9x_1+0x_2+5x_3-9}{9}, & 9\le 9x_1+0x_2+5x_3\le 18,\\ 0, & 9x_1+0x_2+5x_3\ge 18.\end{cases} \tag{50}$$

$$v_2^1(1x_1+6x_2+2x_3)=\begin{cases}1, & 1x_1+6x_2+2x_3\le 9,\\ 1-\frac{1x_1+6x_2+2x_3-9}{9}, & 9\le 1x_1+6x_2+2x_3\le 18,\\ 0, & 1x_1+6x_2+2x_3\ge 18.\end{cases} \tag{51}$$

$$v_3^1(4x_1+3x_2+8x_3)=\begin{cases}1, & 4x_1+3x_2+8x_3\le 9,\\ 1-\frac{4x_1+3x_2+8x_3-9}{9}, & 9\le 4x_1+3x_2+8x_3\le 18,\\ 0, & 4x_1+3x_2+8x_3\ge 18.\end{cases} \tag{52}$$

$$v_1^2(1x_1+8x_2+4x_3)=\begin{cases}1, & 1x_1+8x_2+4x_3\le 9,\\ 1-\frac{1x_1+8x_2+4x_3-9}{8}, & 9\le 1x_1+8x_2+4x_3\le 17,\\ 0, & 1x_1+8x_2+4x_3\ge 17.\end{cases} \tag{53}$$

$$v_2^2(6x_1+2x_2+9x_3)=\begin{cases}1, & 6x_1+2x_2+9x_3\le 9,\\ 1-\frac{6x_1+2x_2+9x_3-9}{8}, & 9\le 6x_1+2x_2+9x_3\le 17,\\ 0, & 6x_1+2x_2+9x_3\ge 17.\end{cases} \tag{54}$$

$$v_3^2(7x_1+3x_2+3x_3)=\begin{cases}1, & 7x_1+3x_2+3x_3\le 9,\\ 1-\frac{7x_1+3x_2+3x_3-9}{8}, & 9\le 7x_1+3x_2+3x_3\le 17,\\ 0, & 7x_1+3x_2+3x_3\ge 17.\end{cases} \tag{55}$$

$$v_1^3(8x_1-5x_2-3x_3)=\begin{cases}1, & 8x_1-5x_2-3x_3\le 8,\\ 1-\frac{8x_1-5x_2-3x_3-8}{13}, & 8\le 8x_1-5x_2-3x_3\le 21,\\ 0, & 8x_1-5x_2-3x_3\ge 21.\end{cases} \tag{56}$$

$$v_2^3(2x_1+6x_2+1x_3)=\begin{cases}1, & 2x_1+6x_2+1x_3\le 8,\\ 1-\frac{2x_1+6x_2+1x_3-8}{13}, & 8\le 2x_1+6x_2+1x_3\le 21,\\ 0, & 2x_1+6x_2+1x_3\ge 21.\end{cases} \tag{57}$$

$$v_3^3(3x_1+0x_2+6x_3)=\begin{cases}1, & 3x_1+0x_2+6x_3\le 8,\\ 1-\frac{3x_1+0x_2+6x_3-8}{13}, & 8\le 3x_1+0x_2+6x_3\le 21,\\ 0, & 3x_1+0x_2+6x_3\ge 21.\end{cases} \tag{58}$$

$$u_{\tilde{g}_l^1}(x^TA^1y)=\begin{cases}1, & x^TA^1y\ge 8,\\ 1-\frac{8-x^TA^1y}{5}, & 8\ge x^TA^1y\ge 3,\\ 0, & x^TA^1y\le 3,\end{cases} \tag{59}$$

$$u_{\tilde{g}_I^2}(x^TA^2y) = \begin{cases} 1, & x^TA^2y \geq 9, \\ 1 - \frac{9-x^TA^2y}{7}, & 9 \geq x^TA^2y \geq 2, \\ 0, & x^TA^2y \leq 2, \end{cases} \tag{60}$$

$$u_{\tilde{g}_I^3}(x^TA^3y) = \begin{cases} 1, & x^TA^3y \geq 8, \\ 1 - \frac{8-x^TA^3y}{7}, & 8 \geq x^TA^3y \geq 1, \\ 0, & x^TA^3y \leq 1, \end{cases} \tag{61}$$

$$v_{\tilde{g}_{II}^1}(x^TB^1y) = \begin{cases} 1, & x^TB^1y \geq 9, \\ 1 - \frac{9-x^TB^1y}{9}, & 9 \geq x^TB^1y \geq 0, \\ 0, & x^TB^1y \leq 0, \end{cases} \tag{62}$$

$$v_{\tilde{g}_{II}^2}(x^TB^2y) = \begin{cases} 1, & x^TB^2y \geq 9, \\ 1 - \frac{9-x^TB^2y}{8}, & 9 \geq x^TB^2y \geq 1, \\ 0, & x^TB^2y \leq 1, \end{cases} \tag{63}$$

and

$$v_{\tilde{g}_{II}^3}(x^TB^3y) = \begin{cases} 1, & x^TB^3y \geq 8, \\ 1 - \frac{8-x^TB^3y}{13}, & 8 \geq x^TB^3y \geq -5, \\ 0, & x^TB^3y \leq -5, \end{cases} \tag{64}$$

where

$$\begin{aligned} x^TA^1y &= 6x_1y_1 + 3x_2y_1 + 7x_3y_1 + 3x_1y_2 + 6x_2y_2 + 3x_3y_2 + 4x_1y_3 + 8x_2y_3 + 4x_3y_3, \\ x^TA^2y &= 9x_1y_1 + 4x_2y_1 + 2x_3y_1 + 2x_1y_2 + 5x_2y_2 + 7x_3y_2 + 7x_1y_3 + 8x_2y_3 + 3x_3y_3, \\ x^TA^3y &= 5x_1y_1 + 3x_2y_1 + 1x_3y_1 + 1x_1y_2 + 4x_2y_2 + 8x_3y_2 + 2x_1y_3 + 8x_2y_3 + 1x_3y_3, \\ x^TB^1y &= 9x_1y_1 + 0x_2y_1 + 5x_3y_1 + 1x_1y_2 + 6x_2y_2 + 2x_3y_2 + 4x_1y_3 + 3x_2y_3 + 8x_3y_3, \\ x^TB^2y &= 1x_1y_1 + 8x_2y_1 + 4x_3y_1 + 6x_1y_2 + 2x_2y_2 + 9x_3y_2 + 7x_1y_3 + 3x_2y_3 + 3x_3y_3, \\ x^TB^3y &= 8x_1y_1 - 5x_2y_1 - 3x_3y_1 + 2x_1y_2 + 6x_2y_2 + 1x_3y_2 + 3x_1y_3 + 0x_2y_3 + 6x_3y_3. \end{aligned} \tag{65}$$

Now using (21) and (22), we have the following multiple objective non-liner programming problem $(MONLP6)$.

$(MONLP6)$ $\quad max \quad \lambda$

$$\begin{aligned} \textit{subject to} \quad & 6y_1 + 3y_2 + 4y_3 + (\lambda-1)5 \leq 8,\ 9x_1 + 0x_2 + 5x_3 + (\lambda-1)9 \leq 9, \\ & 3y_1 + 6y_2 + 8y_3 + (\lambda-1)5 \leq 8,\ 1x_1 + 6x_2 + 2x_3 + (\lambda-1)9 \leq 9, \\ & 7y_1 + 3y_2 + 4y_3 + (\lambda-1)5 \leq 8,\ 4x_1 + 3x_2 + 8x_3 + (\lambda-1)9 \leq 9, \\ & 9y_1 + 2y_2 + 7y_3 + (\lambda-1)7 \leq 9,\ 1x_1 + 8x_2 + 4x_3 + (\lambda-1)8 \leq 9, \\ & 4y_1 + 5y_2 + 8y_3 + (\lambda-1)7 \leq 9,\ 6x_1 + 2x_2 + 9x_3 + (\lambda-1)8 \leq 9, \\ & 2y_1 + 7y_2 + 3y_3 + (\lambda-1)7 \leq 9,\ 7x_1 + 3x_2 + 3x_3 + (\lambda-1)8 \leq 9, \\ & 5y_1 + 1y_2 + 2y_3 + (\lambda-1)7 \leq 8,\ 8x_1 - 5x_2 - 3x_3 + (\lambda-1)13 \leq 8, \\ & 3y_1 + 4y_2 + 8y_3 + (\lambda-1)7 \leq 8,\ 2x_1 + 6x_2 + 1x_3 + (\lambda-1)13 \leq 8, \\ & 1y_1 + 8y_2 + 1y_3 + (\lambda-1)7 \leq 8,\ 3x_1 + 0x_2 + 6x_3 + (\lambda-1)13 \leq 8, \\ & 6x_1y_1 + 3x_2y_1 + 7x_3y_1 + 3x_1y_2 + 6x_2y_2 + 3x_3y_2 + 4x_1y_3 + 8x_2y_3 + 4x_3y_3 + (1-\lambda)5 \geq 8, \\ & 9x_1y_1 + 4x_2y_1 + 2x_3y_1 + 2x_1y_2 + 5x_2y_2 + 7x_3y_2 + 7x_1y_3 + 8x_2y_3 + 3x_3y_3 + (1-\lambda)7 \geq 9, \\ & 5x_1y_1 + 3x_2y_1 + 1x_3y_1 + 1x_1y_2 + 4x_2y_2 + 8x_3y_2 + 2x_1y_3 + 8x_2y_3 + 1x_3y_3 + (1-\lambda)7 \geq 8, \\ & 9x_1y_1 + 0x_2y_1 + 5x_3y_1 + 1x_1y_2 + 6x_2y_2 + 2x_3y_2 + 4x_1y_3 + 3x_2y_3 + 8x_3y_3 + (1-\lambda)9 \geq 9, \\ & 1x_1y_1 + 8x_2y_1 + 4x_3y_1 + 6x_1y_2 + 2x_2y_2 + 9x_3y_2 + 7x_1y_3 + 3x_2y_3 + 3x_3y_3 + (1-\lambda)8 \geq 9, \\ & 8x_1y_1 - 5x_2y_1 - 3x_3y_1 + 2x_1y_2 + 6x_2y_2 + 1x_3y_2 + 3x_1y_3 + 0x_2y_3 + 6x_3y_3 + (1-\lambda)13 \geq 8, \\ & 0 \leq \lambda \leq 1,\ x \in S^3, y \in S^3. \end{aligned} \tag{66}$$

For some sample values of λ, we obtain the optimal solutions of the problem $(MONLP6)$ for Player I and Player II in Table 1. Similarly, for other values of $\lambda \in [0,1]$, we can obtain the optimal solutions of the problem $(MONLP6)$ model through the same approach.

In particular, let $\bar{\lambda} = 0.2885$, then we can have that $(\bar{x}_1 = 0.1, \bar{x}_2 = 0.1, \bar{x}_3 = 0.8)$ and $(\bar{y}_1 = 0.6, \bar{y}_2 = 0.3, \bar{y}_3 = 0.1)$ are the mixed strategies of Player I and Player II, respectively.

Table 1. Strategies of Example 1.

Strategies	$\bar{\lambda}$	$\bar{x}_1$	$\bar{x}_2$	$\bar{x}_3$	$\bar{y}_1$	$\bar{y}_2$	$\bar{y}_3$
1	0.2285	0.1	0.2	0.7	0.4	0.2	0.4
2	0.2685	0.2	0.3	0.5	0.7	0.1	0.2
3	0.2720	0.6	0.2	0.2	0.2	0.5	0.3
4	0.2885	0.1	0.1	0.8	0.6	0.3	0.1
5	0.2971	0.5	0.1	0.4	0.5	0.2	0.3
6	0.3000	0.4	0.1	0.5	0.3	0.4	0.3

6. Conclusions

In this paper, we have presented a multi-objective bi-matrix game with a fuzzy goals $(MOBGFG)$ model. The inspiration of the model is from [2,29,30,36] and we have solved the game $(MOBGFG)$ model via a multi-objective non-linear programming method. We will discuss a situation where the elements of matrices $A^k(l = 1, 2, \cdots, r)$ and $B^l(l = 1, 2, \cdots, s)$ of the game $(MOBGFG)$ model become fuzzy numbers in our future research. We have also concluded that the game model with entropy is becoming more and more significant and it is related to practical problems of our real life [13,14,37]. Inspired by [37], we will extend the some results of this paper to the game $(MOBGFG)$ model in an entropy or fuzzy entropy environment.

Acknowledgments: This work was supported by The National Natural Science Foundations of China (Grant No. 11671001 and 61472056).

Author Contributions: All authors contributed equally to the writing of this paper. All authors read and approved the final manuscript.

Conflicts of Interest: The authors declare that they have no competing interests.

References

1. Zadeh, L.A. Information and control. *Fuzzy sets* **1965**, *8*, 338–353.
2. Bector, C.R.; Chandra, S. *Fuzzy Mathematical Programming and Fuzzy Matrix Games*; Springer: Berlin, Germany, 2005.
3. Chen, B.S.; Tseng, C.S.; Uang, H.J. Fuzzy differential games for nonlinear stochastic systems: Suboptimal approach. *IEEE Trans. Fuzzy Syst.* **2002**, *10*, 222–233.
4. Garagic, D.; Cruz, J.B., Jr. An approach to fuzzy noncooperative nash games. *J. Optim. Theory Appl.* **2003**, *118*, 475–491.
5. Tan, C.; Jiang, Z.Z.; Chen, X.; Ip, W.H. A Banzhaf function for a fuzzy game. *IEEE Trans. Fuzzy Syst.* **2014**, *22*, 1489–1502.
6. Sharifian, S.; Chakeri, A.; Sheikholeslam, F. Linguisitc representation of Nash equilibriums in fuzzy games. In Proceedings of the IEEE 2010 Annual Meeting of the North American Fuzzy Information Processing Society (NAFIPS), Toronto, ON, Canada, 12–14 July 2010; pp. 1–6.
7. Chakeri, A.; Sadati, N.; Sharifian, S. Fuzzy Nash equilibrium in fuzzy games using ranking fuzzy numbers. In Proceedings of the 2010 IEEE International Conference on Fuzzy Systems (FUZZ), Barcelona, Spain, 18–23 July 2010; pp. 1–5.
8. Chakeri, A.; Sheikholeslam, F. Fuzzy Nash equilibriums in crisp and fuzzy games. *IEEE Trans. Fuzzy Syst.* **2013**, *21*, 171–176.
9. Chakeri, A.; Sadati, N.; Dumont, G.A. Nash equilibrium strategies in fuzzy games. In *Game Theory Relaunched*; InTech: Rijeka, Croatia, 2013.

10. Chakeri, A.; Habibi, J.; Heshmat, Y. Fuzzy type-2 Nash equilibrium. In Proceedings of the 2008 International Conference on Computational Intelligence for Modelling Control and Automation, Vienna, Austria, 10–12 December 2008; pp. 398–402.
11. Chakeri, A.; Dariani, A.N.; Lucas, C. How can fuzzy logic determine game equilibriums better? In Proceedings of the IS'08, 4th International IEEE Conference on Intelligent Systems, Varna, Bulgaria, 6–8 September 2008; pp. 251–256.
12. Blackwell, D. An analog of the minimax theorem for vector payoffs. *Pac. J. Math.* **1956**, *6*, 1–8.
13. Roy, S.K.; Biswal, M.P.; Tiwari, R.N. An approach to multi-objective bimatrix games for Nash equilibrium solutions. *Ric. Oper.* **2001**, *30*, 56–63.
14. Roy, S. K. Fuzzy programming approach to two-person multicriteria bimatrix games. *J. Fuzzy Math.* **2007**, *15*, 141–153.
15. Nishizaki, I.; Sakawa, M. Two-person zero-sum games with multiple fuzzy goals. *J. Japan Soc. Fuzzy Theory Syst.* **1992**, *4*, 504–511.
16. Nishizaki, I.; Sakawa, M. Max-min solution for fuzzy multi-objective matrix games. *J. Japan Soc. Fuzzy Theory Syst.* **1993**, *5*, 505–515.
17. Nishizaki, I.; Sakawa, M. Equilibrium solution for multiobjective bimatrix games incorporating fuzzy goals. *J. Optim. Theory Appl.* **1995**, *86*, 433–458.
18. Chen, Y.W. An alternative approach to the bimatrix non-cooperative game with fuzzy multiple objectives. *J. Chin. Inst. Eng.* **2002**, *19*, 9–16.
19. Chen, Y.W.; Shieh, H.E. Fuzzy multi-stage de-novo programming problem. *Appl. Math. Comput.* **2006**, *181*, 1139–1147.
20. Michalewicz, Z. *Genetic Algorithm + Data Structure = Evolution Programs*; Springer: Berlin, Germany; New York, NY, USA, 1999.
21. Angelov, P.P. Optimization in an intuitionistic fuzzy environment. *Fuzzy Sets Syst.* **1997**, *86*, 299–306.
22. Precup, R.E.; Preitl, S. Optimisation criteria in development of fuzzy controllers with dynamics. *Eng. Appl. Artif. Intel.* **2004**, *17*, 661–674.
23. Kiran, M.S.; Findik, O. A directed artificial bee colony algorithm. *Appl. Soft Comput.* **2015**, *26*, 454–462.
24. Ghosn, S.B.; Drouby, F.; Harmanani, H.M. A parallel genetic algorithm for the open-shop scheduling problem using deterministic and random moves. *Eng. Appl. Artif. Intel.* **2016**, *14*, 130–144.
25. Roy, S.K.; Das, C.B. Multicriteria entropy bimatrix goal game: A Fuzzy programming approach. *J. Uncertain Syst.* **2013**, *7*, 108–117.
26. Wierzbicki, A.P. *Multiple Criteria Solutions in Noncooperative Game-Theory Part III: Theoretical Foundations*; Discussion Paper 288; Kyoto Institute of Economic Research: Kyoto, Japan, 1990.
27. Borm, P.E.M.; Tijs, S.H.J.; van den Aarssen, C.M. Pareto equilibria in multiobjective games. In *Methods of Operations Research*, Fuchsstein, B., Lengauer, T., Skaka, H.J., Eds.; Verlag Anton Hain Meisenheim GmbH: Frankfurt, Germany, 1988; pp. 303–312.
28. Qiu, D.; Xing, Y.; Chen, S. Solving multi-objective matrix games with fuzzy payoffs through the lower limit of the possibility degree. *Symmetry* **2017**, *9*, 130.
29. Nishizaki, I.; Sakawa, M. *Fuzzy and Multiobjective Games for Conflict Resolution*; Kluwer Academic Publishers: Amsterdam, The Netherlands, 2003.
30. Mangasarian, O.L.; Stone, H. Two-person non-zero-sum games and quadratic programming. *J. Math. Anal. Appl.* **1964**, *9*, 348–355.
31. Nishizaki, I.; Sakawa, M. Equilibrium solutions in multiobjective bimatrix games with fuzzy payoffs and fuzzy goals. *Fuzzy Sets Syst.* **2000**, *111*, 99–116.
32. Fernandez, F.R.; Puerto, J. Vector linear programming in zero-sum multicriteria matrix games. *J. Optim. Theory Appl.* **1996**, *89*, 115–127.
33. Dubois, D.; Prade, H. *Fuzzy Sets and Systems-Theory and Application*; Academic Press: New York, NY, USA, 1980.
34. Zimmermann, H.J. *Fuzzy Set Theory and Its Applications*, 3rd ed.; Kluwer Academic Publishers: Nowell, MA, USA, 1996.
35. Zimmermann, H.J. Fuzzy programming and linear programming with several objective functions. *Fuzzy Sets Syst.* **1978**, *1*, 45–55.

36. Pal, B.B.; Moitra, B.N. A goal programming procedure for solving problems with multiple fuzzy goals using dynamic programming. *Eur. J. Oper. Res.* **2003**, *144*, 480–491.
37. Das, C.B.; Roy, S.K. Fuzzy based GA to multi-objective entropy bimatrix game. *Opsearch* **2013**, *50*, 125–140.

symmetry

MDPI

Article

Valuation Fuzzy Soft Sets: A Flexible Fuzzy Soft Set Based Decision Making Procedure for the Valuation of Assets

José Carlos R. Alcantud [1,*], Salvador Cruz Rambaud [2] and María J. Muñoz Torrecillas [2]

1 BORDA Research Unit and Multidisciplinary Institute of Enterprise (IME), University of Salamanca, E37007 Salamanca, Spain
2 Department of Economics and Business, University of Almería, E04120 Almería, Spain; scruz@ual.es (S.C.R.); mjmtorre@ual.es (M.J.M.T.)
* Correspondence: jcr@usal.es; Tel.: +34-923-294666

Received: 22 September 2017; Accepted: 23 October 2017; Published: 27 October 2017

Abstract: Zadeh's fuzzy set theory for imprecise or vague data has been followed by other successful models, inclusive of Molodtsov's soft set theory and hybrid models like fuzzy soft sets. Their success has been backed up by applications to many branches like engineering, medicine, or finance. In continuation of this effort, the purpose of this paper is to put forward a versatile methodology for the valuation of goods, particularly the assessment of real state properties. In order to reach this target, we develop the concept of (partial) valuation fuzzy soft set and introduce the novel problem of data filling in partial valuation fuzzy soft sets. The use of fuzzy soft sets allows us to quantify the qualitative attributes involved in an assessment context. As a result, we illustrate the effectiveness and validity of our valuation methodology with a real case study that uses data from the Spanish real estate market. The main contribution of this paper is the implementation of a novel methodology, which allows us to assess a large variety of assets where data are heterogeneous. Our technique permits to avoid the appraiser's subjectivity (exhibited by practitioners in housing valuation) and the well-known disadvantages of some alternative methods (such as linear multiple regression).

Keywords: fuzzy soft set; linear regression; valuation of goods; data filling; decision making

1. Introduction

Zadeh's [1] fuzzy set theory deals with impreciseness or vagueness of evaluations by associating degrees to which objects belong to a set. Its appearance boosted the rise of many related theories that attempt to model specific decision problems. In particular, the hybridization of fuzzy sets with soft sets as proposed by Molodtsov [2] (see also Maji et al. [3,4]) yields the notion of fuzzy soft set (Alcantud [5], Ali [6], Ali and Shabir [7], Maji et al. [8]). Decision-making methodologies and applications have proliferated and are the subject of relevant analyses on a regular basis. Among the most recent papers that exemplify noteworthy fuzzy decision-making trends, we can cite Alcantud et al. [9,10], Faizi et al. [11,12] and Zhang and Xu [13] in hesitant fuzzy sets, Zhan and Zhu [14] in (fuzzy) soft sets and rough soft sets, Alcantud [15] in fuzzy soft sets, Ma et al. [16] in hybrid soft set models, Chen and Ye [17] and Ye [18] in neutrosophic sets, Peng et al. [19] and Peng and Yang [20] in interval-valued fuzzy soft sets, and Fatimah et al. [21] in (dual) probabilistic soft sets. With respect to applications, in a clinical environment, Chang [22] uses the fuzzy sets theory and the so-called VIKOR (VIsekriterijumska optimizacija i KOmpromisno Resenje) method to evaluate the service quality of two public and three private medical centres in Taiwan, in the same context of uncertainty, subjectivity and linguistic variables as our study; Espinilla et al. [23] apply a decision analysis tool for the early detection of preeclampsia in women at risk by using the data of a sample of pregnant women with high risk of this disease; and Alcantud et al. [24] give a methodology for glaucoma diagnosis. On the other hand, in the field of management, Zhang and Xu [13] deal with the problem of choosing material suppliers by a manufacturer to purchase key components in order to reach a competitive advantage in the market of watches; Xu and Xia [25] provide a management case study by using the hesitant fuzzy elements to estimate the degree to which an alternative satisfies a criterion in a decision-making process; Taş et al. [26] present new applications of a soft set theory and a fuzzy soft set theory to the effective management of stock-out situations. In the field of finance, Xu and Xiao [27] apply the soft set theory to select financial ratios for business failure prediction by using real data sets from Chinese listed firms; Kalaichelvi [28] and Özgür and Taş [29] apply fuzzy soft sets to solve the investment decision making problem.

In this work, we introduce the notion of partial valuation fuzzy soft set as a tool to perform valuations of assets. Then, we apply a suitable valuation methodology, based on fuzzy soft sets, to a real case study. Fuzzy soft sets, with their ability to codify partial membership with respect to a predefined list of attributes, seem to be a useful tool to make decisions in this context. Unlike the standard approach, which selects one from a set of possible alternatives, our decision is the valuation that should be rightfully attached with some of the assets. In passing, we introduce rating procedures as well as the problem of data filling in partial valuation fuzzy soft sets.

Our application concerns the real estate market. There is ample variety of real estate valuation methods. Following [30], we classify them into traditional and advanced. Traditional methods are: comparison method, investment/income capitalization method, profits method, development/residual method, contractors/cost method, multiple regression method, and stepwise regression method. As advanced valuation methods, we can cite: artificial neural networks (ANNs), hedonic pricing method, spatial analysis methods, fuzzy logic, and autoregressive integrated moving average (ARIMA).

In Spain, real estate valuation is regulated by *Orden ECO/805/2003* (Ministerio de Economía, 2003) [31], which recommends the use of four of the previously mentioned methods: comparison, investment/income capitalization, residual, and cost methods. There are some interesting works that compare certain methods used in real estate appraisal, e.g., [32] compare fuzzy logic to multiple regression analysis, or [33] compare artificial-intelligence methods with non-traditional regression methods. We also find new hybrid methodologies, e.g., [34], which relies on the introduction of fuzzy mathematics in a spatial error model.

We contribute to this growing literature by proposing a flexible mechanism that can be specialized in several ways. The input data is a partial valuation fuzzy soft set that characterizes the problem. The practitioner can select one from a sample of rating procedures in order to start the algorithm. Then, a suitable regression analysis permits filling the missing data in the original partial valuation fuzzy soft set. The structure of the available data often allows the researcher to perform sophisticated regression analysis beyond the standard, linear case.

This paper is organized as follows. Section 2 recalls some notation and definitions related to soft sets and fuzzy soft sets. Section 3 presents the main new notions in this paper, namely, valuation and partial valuation fuzzy soft sets. We also define rating procedures for fuzzy soft sets and prove some useful fundamental properties of these concepts. Section 4 briefly introduces data filling for partial valuation fuzzy soft sets, and a flexible methodology is proposed in order to implement that concept. Then, in Section 5, we take advantage of such design in order to valuate goods through a fictitious streamlined example. In Section 6, we present an application to a real case study on the Spanish real estate market. We also examine its traits in comparison with other standard methodologies. We conclude in Section 7.

2. Notation and Definitions

Let X denote a set. Then, $\mathcal{P}(X)$ is the set of all subsets of X. A fuzzy subset (also, FS) A of X is a function $\mu_A : X \to [0,1]$. For each $x \in X$, $\mu_A(x) \in [0,1]$ is the degree of membership of x in that subset. The set of all fuzzy subsets on X will be denoted by **FS**(X).

Now, we are going to recall some basic concepts such as soft sets and fuzzy soft sets.

2.1. Soft Sets and Fuzzy Soft Sets

In soft set theory, we refer to a universe of objects U, and to a universal set of parameters E.

Definition 1 ([2])**.** *Let A be a subset of E. The pair (F, A) is a soft set over U if $F : A \longrightarrow \mathcal{P}(U)$.*

The pair (F, A) in Definition 1 is a parameterized family of subsets of U, and A represents the parameters. Then, for every parameter $e \in A$, we interpret that $F(e)$ is the subset of U approximated by e, also called the set of e-approximate elements of the soft set.

Other interesting investigations expanded the knowledge about soft sets. The notions of soft equalities, intersections and unions of soft sets and soft subsets and supersets are defined in [4]. Various types of soft subsets and soft equal relations are studied in [35]. Soft set based decision-making was initiated by [3]. Further applications of soft sets in decision-making contexts were given, for example, in [24,36,37].

The concept of soft set can be expanded so as to include fuzzy subsets approximated by parameters:

Definition 2 ([8])**.** *Let A be a subset of E. The pair (F, A) is a fuzzy soft set over U if $F : A \longrightarrow$ **FS**(U), where **FS**(U) denotes the set of all fuzzy sets on U.*

The set of all fuzzy soft sets over U will be denoted as $\mathcal{FS}(\mathrm{U})$. Due to the natural identification of subsets of U with FSs of U, any soft set can be considered a fuzzy soft set (cf., [5]). If, for example, our universe of options are films that are parameterized by attributes, then fuzzy soft sets permit to deal with properties like "funny" or "scary" for which partial memberships are almost compulsory. However, soft sets are suitable only when properties are categorical, e.g., "Oscar awarded", "3D version available", or "silent movie".

In real practice, both U and A use to be finite. Then, k and n will denote the respective number of elements of $U = \{o_1, \ldots, o_k\}$ and $A = \{e_1, \ldots, e_n\}$. In such case, soft sets can be represented either by $k \times n$ matrices or in their tabular form (cf., [38]). The k rows are associated with the objects, and the n columns are associated with the parameters. Both practical representations are binary, that is to say, all cells are either 0 or 1. One can proceed in a similar way in fuzzy soft sets, but now the possible values in the cells lie in the interval $[0,1]$.

A matrix representation of a soft set is shown in the following Example 1:

Example 1. *Let $U = \{h_1, h_2, h_3\}$ be a universe of houses. Let $A = \{e_1, e_2, e_3, e_4\}$ be the set of parameters, attributes or house characteristics (e.g., "centrally located" or "includes a garage"). Define a soft set (F, A) as follows:*

1. $h_1 \in F(e_1) \cap F(e_4)$, $h_3 \notin F(e_2) \cup F(e_3)$.
2. $h_2 \in F(e_1) \cap F(e_3)$, $h_1 \notin F(e_2) \cup F(e_4)$.
3. $h_3 \in F(e_2)$, $h_2 \notin F(e_1) \cup F(e_3) \cup F(e_4)$.

Table 1 captures the information defining (F, A).

Table 1. Tabular representation of the soft set (F, A) in Example 1.

	e_1	e_2	e_3	e_4
h_1	1	0	0	1
h_2	1	0	1	0
h_3	0	1	0	0

Suppose that a soft set (F, A) can be expressed by the $k \times n$ matrix $(t_{ij})_{i,j}$. Then, the *choice value* of object $o_i \in U$ is defined as $c_i = \sum_{j=1}^{n} t_{ij}$. According to Maji, Biswas and Roy [3], an optimal choice can be made by selecting any object o_i such that $c_i = \max_{j=1,\ldots,k} c_j$. Put differently, any choice-value maximizer is an acceptable solution to the problem.

However, fuzzy soft set decision making is far more complex thus controversial. Approaches to that problem are included in [15,39–41].

Anyhow, the example above shows that fuzzy soft sets are a suitable tool to capture the characteristics of complex representations of assets. Section 6 below clarifies this argument with a real example.

2.2. Basic Operations

Basic operations among soft sets were established in Ali et al. [42]:

Definition 3 ([42]). *Let (F, A) and (G, B) be soft sets over U, such that $A \cap B \neq \emptyset$. The restricted intersection of (F, A) and (G, B) is denoted by $(F, A) \cap_{\mathscr{R}} (G, B)$ and it is defined as $(F, A) \cap_{\mathscr{R}} (G, B) = (H, A \cap B)$, where $H(e) = F(e) \cap G(e)$ for all $e \in A \cap B$.*

Definition 4 ([42]). *The extended intersection of the soft sets (F, A) and (G, B) over U is the soft set (H, C), where $C = A \cup B$, and $\forall e \in C$,*

$$H(e) = \begin{cases} F(e), \text{if } e \in A \setminus B, \\ G(e), \text{if } e \in B \setminus A, \\ F(e) \cap G(e), \text{if } e \in A \cap B. \end{cases}$$

It is denoted by $(F, A) \cap_{\mathscr{E}} (G, B) = (H, C)$.

Definition 5 ([42]). *Let (F, A) and (G, B) be soft sets over U, such that $A \cap B \neq \emptyset$. The restricted union of (F, A) and (G, B) is denoted by $(F, A) \cup_{\mathscr{R}} (G, B)$ and it is defined as $(F, A) \cup_{\mathscr{R}} (G, B) = (H, C)$, where $C = A \cap B$ and for all $e \in C, H(e) = F(e) \cup G(e)$.*

Definition 6 ([42]). *The extended union of two soft sets (F, A) and (G, B) over U is the soft set (H, C), where $C = A \cup B$, and $\forall e \in C$,*

$$H(e) = \begin{cases} F(e), \text{if } e \in A \setminus B, \\ G(e), \text{if } e \in B \setminus A, \\ F(e) \cup G(e), \text{if } e \in A \cap B. \end{cases}$$

It is denoted by $(F, A) \cup_{\mathscr{E}} (G, B) = (H, C)$.

Maji et al. [8] defined some relations and similar operations for fuzzy soft sets as follows.

Definition 7 ([8]). *Let (F, A) and (G, B) be fuzzy soft sets over U. We say that (F, A) is a fuzzy soft subset of (G, B) if $A \subset B$ and $F(e)$ is a fuzzy subset of $G(e)$ for all $e \in A$.*

When (F, A) is a fuzzy soft subset of (G, B) and (G, B) is a fuzzy soft subset of (F, A) we say that (F, A) and (G, B) are fuzzy soft equal.

Definition 8 ([8]). *Let (F, A) and (G, B) be fuzzy soft sets over U.*

Their intersection is (H,C) *where* $C = A \cap B$ *and* $H(e) = F(e) \cap G(e)$ *for all* $e \in C = A \cap B$.
Their union is (H', C'), *where* $C' = A \cup B$, *and* $\forall e \in C'$,

$$H'(e) = \begin{cases} F(e), \text{if } e \in A \setminus B, \\ G(e), \text{if } e \in B \setminus A, \\ F(e) \cup G(e), \text{if } e \in A \cap B. \end{cases}$$

3. Some Novel Concepts Related to Valuation Fuzzy Soft Sets

In this section, we are going to introduce the main new notions in this paper, namely, valuation and partial valuation fuzzy soft sets. We also prove some fundamental properties of them.

Valuation and Partial Valuation Fuzzy Soft Sets

In order to define our novel notions, we refer to a universe U of k objects, and to a universal set of parameters E.

Definition 9. *Let* A *be a subset of* E. *The triple* (F, A, V) *is a valuation fuzzy soft set over* U *when* (F, A) *is a fuzzy soft set over* U *and* $V = (V_1, \ldots, V_k) \in \mathbb{R}^k$. *Henceforth, we abbreviate a valuation fuzzy soft set by VFSS. We denote by* **V**(*U*) *the set of all valuation fuzzy soft sets over* U.

If we restrict Definition 9 to soft sets over U, a particular concept of *valuation soft set* is naturally produced.

The motivation for valuation (fuzzy) soft sets is that, in many natural situations, option o_i from U is associated with a valuation, appraisal or assessment V_i, in addition to the standard parameterization of U as a function of the attributes in A. For example, in the usual example where the options are houses, this valuation may be the market price.

Such valuation can also be defined through elements from fuzzy soft set theory, or otherwise. We proceed to formalize these ideas.

Definition 10. *A rating procedure for fuzzy soft sets with attributes* A *on a universe* U *is a mapping*

$$\Pi : \mathcal{FS}(\mathrm{U}) \longrightarrow \mathbf{V}(\mathrm{U}).$$

Every rating procedure associates each FSS over U with a VFSS over U. For example, one can use scores associated with decision making mechanisms from the literature (e.g., fuzzy choice values, the scores computed in [39], or the refined scores computed in [15]), in order to produce particularly noteworthy rating procedures. We formalize them in the following definitions:

Definition 11. *The fuzzy choice value rating procedure is defined by the expression* $\Pi_c(F, A) = (F, A, V^{\Pi_c} = (\Pi_c^1, \ldots, \Pi_c^k))$, *where* $\Pi_c^i = c_i = \sum_{j=1}^{n} t_{ij}$ *for each* $i = 1, \ldots, c$. *Recall that* c_i *is the fuzzy choice value of option* i.

Definition 12. *Roy and Maji's rating procedure is defined by the expression* $\Pi_r(F, A) = (F, A, V^{\Pi_r} = (\Pi_r^1, \ldots, \Pi_r^k))$, *where* Π_r^i *is the score* S_i *associated with option* i *in the Algorithm in Section 3.1 of [39] (or alternatively,* s_i *in Algorithm 1 in [15]).*

Definition 13. *Alcantud's rating procedure is defined by the expression* $\Pi_a(F, A) = (F, A, V^{\Pi_a} = (\Pi_a^1, \ldots, \Pi_a^k))$, *where* $\Pi_a^i = S_i$ *is the score associated with option* i *in Algorithm 2 of [15].*

In this paper, we are especially concerned with Definition 13. In order to make this paper self-contained, we proceed to recall its construction.

We describe our fuzzy soft set (F, A) on k alternatives $o_1, \ldots, o_k$ in tabular form. Let t_{ij} denote its cell (i, j) for each possible i, j. Now, for each parameter $j = 1, \ldots, q$, let M_j be the maximum membership value of any object ($M_j = \max_{i=1,\ldots,k} t_{ij}$). Then, we construct a $k \times k$ comparison matrix $A = (a_{ij})_{k\times k}$, where for each i, j, a_{ij} is the sum of the non-negative values in the finite sequence

$$\frac{t_{i1} - t_{j1}}{M_1}, \frac{t_{i2} - t_{j2}}{M_2}, \ldots, \frac{t_{iq} - t_{jq}}{M_q}.$$

Of course, such matrix can also be expressed as a comparison table.

For each $i = 1, \ldots, k$, let R_i be the sum of the elements in row i of A, and T_i be the sum of the elements in column i of A. Finally, for each $i = 1, \ldots, k$, the score of object i is $S_i = R_i - T_i$.

The following toy example illustrates the notions above.

Example 2. *Consider the fuzzy soft set (G, B) in Section 3.3 of [40]. Its tabular form is in Table 2.*

Table 2. Tabular representation of the fuzzy soft set (G, B) in Example 2.

	e_1	e_2	e_3	e_4	e_5
o_1	0.9	0.1	0.2	0.1	0.3
o_2	0.19	0.3	0.4	0.3	0.4

The application of Definitions 11, 12 and 13 to (G, B) produces three respective VFSSs, namely,

$$(G, B, (\Pi_c^1, \Pi_c^2)) = (G, B, (1.60, 1.59)),$$

$$(G, B, (\Pi_r^1, \Pi_r^2)) = (G, B, (-3, 3)),$$

and

$$(G, B, (\Pi_a^1, \Pi_a^2)) = (G, B, (-1.29, 1.29)).$$

In order to obtain these results, we note that the fuzzy choice values c_i in Definition 11, and the scores s_i in Definition 12, are calculated in Table 6 of [40]. In addition, the scores S_i associated with the options by Definition 13 are computed in Figure 2 of [15].

These c_i, s_i and S_i scores produce the respective VFSSs above. Such VFSSs are summarized in Table 3.

Table 3. Summary of the tabular representations of the three VFSS in Example 2.

	e_1	e_2	e_3	e_4	e_5	$\Pi_c^i = c_i$	$\Pi_r^i = s_i$	$\Pi_a^i = S_i$
o_1	0.9	0.1	0.2	0.1	0.3	1.60	−3	−1.29
o_2	0.19	0.3	0.4	0.3	0.4	1.59	3	1.29

In order to select a suitable rating procedure, Definition 13 is the natural choice that we recommend for the valuation of fuzzy soft sets. Our advice is based on the following arguments. Firstly, most authors agree that it seems untenable to use Definition 11, which is a simple adapted version of choice values. Although choice values are widely acceptable in soft set theory, they cannot capture the subtleties of the more general model by FSSs. Therefore, we discard Definition 11. Secondly, Definition 12 does not capture whether an alternative beats another one by a narrow or a large margin, while Definition 13 explicitly rewards more ample differences in the degree of satisfaction of the characteristics. In applications like real estate valuation, the wide range of the feasible assessments demands a method that incorporates these differences. Otherwise, the results will be affected by the odd fact that alternatives with striking differences in their characteristics should be equally valuated, which is clearly a blunt mistake. For these reasons, we must discard Definition 12 and recommend Definition 13.

For practical purposes, the following definition will be very useful. It concerns the cases where some of the valuations are unknown.

Definition 14. *Let A be a subset of E. The triple (F, A, V^*) is a partial valuation fuzzy soft set over U when (F, A) is a fuzzy soft set over U and $V^* \in (\mathbb{R} \cup \{*\})^k$. We abbreviate partial valuation fuzzy soft set by PVFSS.*

The set of all partial valuation fuzzy soft sets over U will be denoted as $\mathbf{V}^*(\mathrm{U})$. If we restrict Definition 14 to soft sets over U, then we define the particular concept of *partial valuation soft set*. As in the case of VFSSs, each option from U is associated with a valuation in addition to the standard parameterization of U as a function of the attributes in A. However, in PVFSSs, it may happen that some of the valuations is unknown or missing. In such case we represent the unknown information or missing valuation data by the $*$ symbol.

Now the motivation for PVFSSs goes as follows. Quite often some options o_i from U have an intrinsic valuation $V_i \in \mathbb{R}$ (for example, market price), whereas the valuation of other options o_j are unknown (for example, because it is our own house that we want to put up for sale). The model by PVFSSs permits collecting all that information in a concise format.

4. Data Filling in Partial Valuation Fuzzy Soft Sets

Valuation is an abstract concept that can be specialized in many ways. We can take advantage of this issue in order to fill missing data in PVFSSs through an adjustable approach. The motivation for this novel problem is the following. If there are missing valuation data in a PVFSS, then the assessments of the corresponding alternatives are unknown. Should we need them (for example, because the valuation of our house means the market price that we might expect when we put it up for sale), then we must fill the missing data.

Therefore the problem of data filling in PVFSSs is associated with solving an original decision making problem for alternatives that are characterized by FSSs.

Remark 1. *The idea of partial valuations should not be mistaken with the well-known notion of incomplete (fuzzy) soft sets [36,43–46]. In the latter case, the parameterization has missing values. In our model, the parameterization of the universe is complete, whereas the valuation is not necessarily complete. Thus, the statement of our data filling problem is original with this paper.*

We proceed to define our class of procedures for the valuation of goods when (*a*) these goods are characterized by parameters, and (*b*) there are comparable goods that are characterized by the same parameters. In other words, we have information about the goods in the form of PVFSSs.

Our methodology is very direct. It works as follows. Let us select a rating procedure (cf., Definition 10).

1. Let us input our PVFSS, namely, (F, A, V^*).
2. We use the rating procedure in order to associate a unique number with each alternative. In this way, we obtain a VFSS (F, A, W) associated with the same FSS (F, A) as the original PVFSS.
3. Now, as long as there are two values in V^* that belong to $\mathbb{R}$ (i.e., two valuations that are not missing in the input data), we calculate a regression equation to fill the missing valuation data.

   In order to run the regression, the independent variables (or abscissas) are the values W_i given by the rating procedure that has been singled out, and the dependent variables (their respective ordinates) are the corresponding $V_i \in \mathbb{R}$ valuations.
4. Once the regression function has been calculated, we can estimate the real values of the missing valuations $V_i = *$ by its evaluations in the corresponding W_i values.

   This procedure solves our original data filling problem.

This methodology is flexible because we can use any rating procedure to produce the abscissas of the data plots and also because we can use regression models other than linear regression in order to fill the missing data (step 3). Observe that in such cases we need a larger set of non-missing data.

The flowchart in Figure 1 summarizes the steps in our solution to the problem of data filling for PVFSSs.

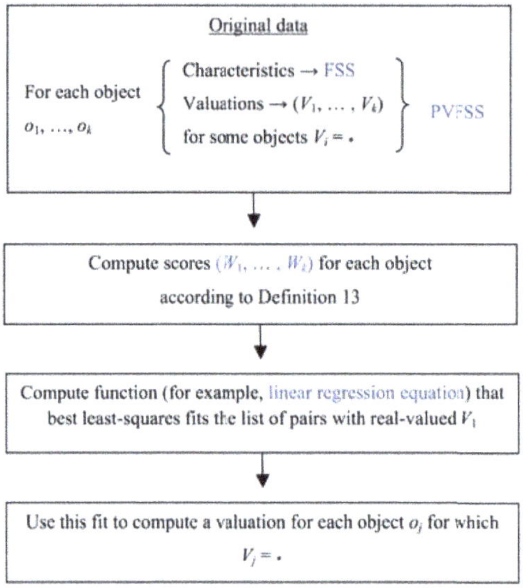

Figure 1. Flow diagram with the solution to the data filling problem in Section 4.

Section 5 below presents an illustrative example in the context of valuation of goods. Later on, in Section 6, we apply the current proposal to a real case study on the Spanish real estate market.

5. Valuation of Goods: An Example

For a given a list of options characterized by a PVFSS, the information on the known values can be used to fill the missing data. Observe that at the end of the process we are valuating the assets with missing values. Therefore, we can use the data filling procedures described in Section 4 in order to make decisions e.g., as to which prizes should be attached to properties that are put into the market.

In this section, we explain this possibility with the following fully developed example. Table 4 represents a PVFSS denoted (F, A, V^*). It uses the input data of Table 6 of [15], which we complement with valuations of some of the six alternatives.

Table 4. Tabular representation of the partial valuation fuzzy soft set (F, A, V^*) in Section 5. All V_is are expressed in thousands of euros.

	p_1	p_2	p_3	p_4	p_5	p_6	p_7	V_i
o_1	0.036	0.015	0.064	0.216	0.048	0.054	0.405	137
o_2	0.144	0.084	0.360	0.045	0.036	0.020	0.175	109
o_3	0.120	0.084	0.180	0.030	0.096	0.021	0.294	97
o_4	0.504	0.192	0.108	0.006	0.048	0.048	0.096	*
o_5	0.084	0.245	0.036	0.096	0.270	0.200	0.140	192
o_6	0.216	0.315	0.042	0.108	0.224	0.126	0.135	198

We can interpret Table 4 as follows. We are interested in selling property o_4, whose market value we want to assess ourselves. The options o_i include our property and other real state properties for sale, and they are all characterized by the p_j attributes. An inspection of the market shows that recent purchases in the same area or street amounted to the respective V_i's. With this practical information, we are ready to valuate our property.

Let us select the rating procedure Π_a. We are ready to apply the remaining steps in Section 4. As explained above, Π_a valuates the options by using Alcantud's scores. Therefore,

$$W = V^{\Pi_a} = (-1.3, -3.2, -3.78, -2.24, 5.24, 5.26)$$

because these figures are computed in Table 8 of [15]. Hence, the VFSS that we obtain at step 2 of our data filling solution in Section 4 is (F, A, W).

We now compute the linear regression equation from the bivariate data that combine the known valuations $(137, 109, 97, 192, 198)$ at (F, A, V^*) with the corresponding components of our rating procedure in the abscissas, which yields

$$((-1.3, 137), (-3.2, 109), (-3.78, 97), (5.24, 192), (5.26, 198)).$$

Observe that the $4th$ values have been discarded because the valuation of the 4th alternative is missing.

Some easy computations (see for example [47]) show that the regression line equation in step 3 is

$$y = 142.02204039129 + 10.310719839438x,$$

with a coefficient of determination $R^2 = 0.9854$. Figure 2 displays these computations.

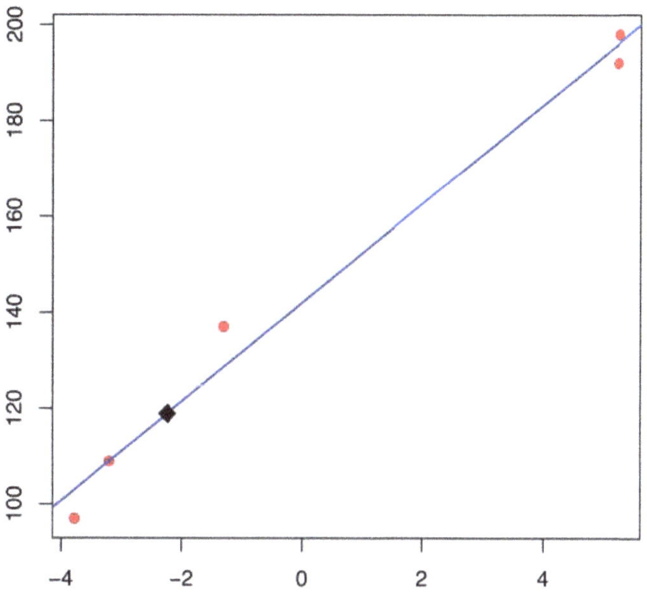

Figure 2. The regression line in Section 5. The black square shows the valuation of the missing option o_4, with score -2.24 at the horizontal axis.

Finally, in step 4, we evaluate such function at the score value $x = -2.24$ associated with o_4, which produces the evaluation value 118.92602795094888.

In conclusion, option o_4 should be valuated by 118,926.03 euros.

6. A Real Case Study

In this section, we propose a method to appraise real estate based on fuzzy logic as we concur with [48] on "the applicability of fuzzy logic for expressing the inherent imprecision in the way that people think and make decisions about the pricing of real state". As far as we know, no methodology based on fuzzy soft sets has ever been applied to real estate valuation using real data from the market. We here use the novel procedure that relies on the new notion of data filling in PVFSSs (as applied in Section 5) in order to provide an assessment of a property based on real data from Almería, in Southeast Spain. The data were obtained by one of the coauthors who acted as appraiser in 2016. Lastly, we compare our existing methodologies.

To be precise, for such real application, we intend to assess an apartment (the subject property) using the data of six apartments (the comparable properties), with known sale prices. We also know the values of four selected attributes (cf., Table 5): surface, number of bathrooms, quality, and number of bedrooms. The values for the attribute "surface" are expressed in square meters. Apartment 6 has 1.5 bathrooms, which means that it has a complete bathroom with a bathtub and another bathroom without a bathtub. There are other attributes, like location and age that we did not include in the table because the six comparable properties and the subject property had similar values (they were located in the same area and were built approximately in the same year).

The property we have to assess has a surface of 114.44 square meters, one bathroom, two bedrooms and a "good" quality.

To apply the method explained in Sections 4 and 5, we first adapt the data to fuzzy soft set format, and, for this purpose, we perform the following adjustments:

1. The maximum surface in our sample of seven apartments is 114.44 square meters. We have divided the surface of each apartment by this maximum figure.
2. We have divided the number of bathrooms of each apartment by two, the maximum number of bathrooms per apartment in our sample.
3. In order to rank the attribute "quality", we have considered four levels of quality: bad, normal, good, and luxury. We assign the values $0, 1/3, 2/3$ and 1 to each level, respectively.
4. For the attribute "number of bedrooms", we have divided the actual number of bedrooms by the maximum number of bedrooms, which, in our sample, is four.

Table 6 shows the PVFSS that captures the statement of our real valuation problem.

Table 5. Attributes of the comparable six apartments. Source: Real data from the Spanish real estate market (Almería, Spain, 2016).

Item	Surface (sq. m.)	No. of Bathrooms	Quality	No. of Bedrooms
h_1	75	1	Normal	3
h_2	105	2	Normal	4
h_3	75	1	Normal	2
h_4	90	2	Normal	3
h_5	90	1.5	Normal	3
h_6	105	2	Normal	3

Table 6. The PVFSS in the real case study in Section 6. Sale prices are given in thousands of euros.

Item	Surface (sq. m.)	No. of Bathrooms	Quality	No. of Bedrooms	Price
h_1	0.52	0.5	1/3	0.75	95
h_2	0.73	1	1/3	1	157
h_3	0.52	0.5	1/3	0.5	115
h_4	0.62	1	1/3	0.75	132
h_5	0.62	0.75	1/3	0.75	132
h_6	0.73	1	1/3	0.75	157
h_7	1	0.5	2/3	0.5	*

6.1. Evaluation of the Apartment

If we select the rating procedure Π_a, then we first compute the comparison table associated with Table 6. Such item is given in Table 7 (the values have been rounded off for the purpose of presentation).

Table 7. Comparison table and scores associated with Table 6. The values have been rounded off.

	h_1	h_2	h_3	h_4	h_5	h_6	h_7	S_i
h_1	0	0	0.2500	0	0	0	0.2500	−3.1038
h_2	0.9577	0	1.2077	0.3538	0.6038	0.2500	1	3.6000
h_3	0	0	0	0	0	0	0	−4.8538
h_4	0.6038	0	0.8538	0	0.2500	0	0.7500	1.1231
h_5	0.3538	0	0.6038	0	0	0	0.5000	−0.6269
h_6	0.7077	0	0.9577	0.1038	0.3538	0	0.7500	1.8500
h_7	0.9808	0.7731	0.9808	0.8769	0.8769	0.7731	0	2.0114

By Definition 13, we obtain:

$$W = V^{\Pi_a} = (-3.1038, 3.6000, -4.8538, 1.1231, -0.6269, 1.8500, 2.0114).$$

These values are given by Alcantud's scores S_i in Table 7.

We now need to calculate the linear regression equation from the bivariate data that combine the known valuations with the corresponding components of our rating procedure, which are

$$((-3.1038, 95), (3.6, 157), (-4.8538, 115), (1.1231, 132), (-0.6269, 132), (1.85, 157)).$$

Some easy computations show that the regression line equation is

$$y = 133.54 + 6.5722x,$$

where y are prices, x represent the scores, and the coefficient of determination takes an acceptable value ($R^2 = 0.7516$). Hence, for $x = 2.0114$, the value of the y variable is 146.75948, and we conclude that the property should be priced at 146, 759.48 euros because prices were given in thousands of euros. Figure 3 displays these computations.

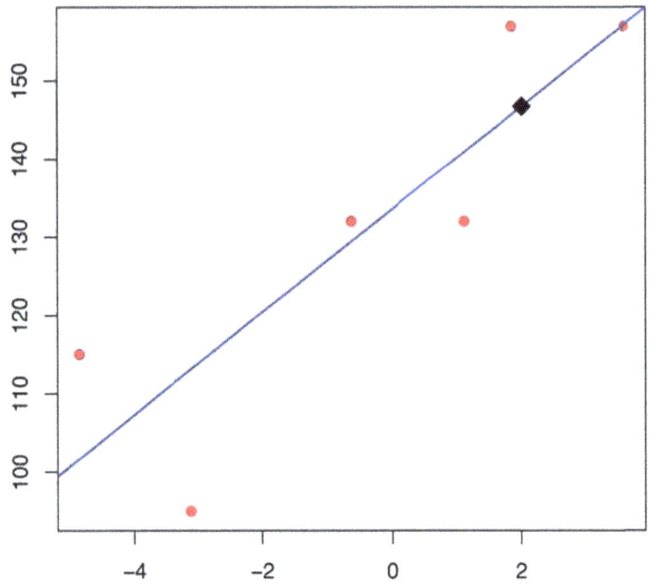

Figure 3. The regression line in the case study. The black square shows the valuation of the missing option h_7, with score 2.0114 at the horizontal axis.

6.2. Sensitivity Analysis

Prices in the real estate market are subject to volatility. In addition, the appraiser can select the small sample in accordance to the existing regulations. Therefore, we have to account for some degree of uncertainty in the valuation of the subject property.

Sensitivity analysis studies how the uncertainty in the output of a mathematical model can be associated with different sources of uncertainty in its inputs [49]. The techniques of sensitivity analysis are sundry and the choice of a suitable methodology is often dictated by the structure of the model. Since we work with given data, we can screen for submodels and check how much the selection of a subsample affects the output. We proceed to check that under such variations the differences in the outputs are small, which allows us to conclude that our valuation model is fairly robust.

1. When the first apartment is suppressed from the analysis, the remaining data produce a new comparison table and scores. With such data, we obtain $\mathcal{V} = (5.8693, -2.5846, 3.3923, 1.6423, 4.1193, 4.2807)$. The regression line equation for the observations

$$((5.8693, 157), (-2.5846, 115), (3.3923, 132), (1.6423, 132), (4.1193, 157))$$

is $y = 125.87 + 5.1152x$ and substituting $x = 4.2807$ produces a figure of 147,766.48 euros. Thus, the difference with respect to the original valuation, in absolute value, is only 0.68%.
2. When the second apartment is suppressed from the analysis, the remaining data produce $\mathcal{V} = (-2.1462, -3.6462, 1.4769, -0.0231, 2.100, 3.0114)$. The regression line equation for the observations

$$((-2.1462, 95), (-3.6462, 115), (1.4769, 132), (-0.0231, 132), (2.100, 157))$$

is $y = 129.65 + 7.7015x$ and substituting $x = 3.0114$ produces a figure of 152,842.48 euros. Therefore, the difference with respect to the original valuation, in absolute value, is only 3.98%.

3. When the sixth apartment is suppressed from the analysis, the remaining data produce $\mathcal{V} = (-2.3962, 3.6000, -3.8962, 1.2269, -0.2731, 2.7614)$. The regression line equation for the observations

$$((-2.3962, 95), (3.6000, 157), (-3.8962, 115), (1.2269, 132), (-0.2731, 132))$$

is $y = 128.54 + 6.7363x$ and substituting $x = 2.7614$ produces a figure of 147,141.78 euros. Therefore, the difference with respect to the original valuation, in absolute value, is only 0.26%.

6.3. A Description of Existing Methodologies

In this section, we briefly sketch the standard methodologies applied by practitioners. Then, in Section 6.4 below, we compare them with the new methodology proposed in Section 6.1.

In order to estimate the price of a house, the *comparison method* and the *multiple linear regression method* are most used in real estate practice in Spain. Both techniques are based on a list of variables (either qualitative or quantitative) that describe the characteristics of the houses in the sample. These attributes of the properties may be of different types:

1. Quantitative and continuous variables, such as the surface of a house.
2. Quantitative and discrete variables, such as:
   - Number of complete bathrooms.
   - Number of incomplete bathrooms.
   - Age of the building.
   - Number of rooms.
   - Level in which is the apartment situated in the building (floor).
   - Number of outward-facing rooms, etc.
3. Qualitative variables, such as quality of the construction.
4. Dummy variables, such as:
   - The building has a garage.
   - The house has a balcony.
   - Repairing and renovation works were made in the house, etc.

The *comparison method* assigns a positive or a negative weight to each modality presented by every attribute, depending on whether the variable must positively or negatively influence the housing price. The values of these weights are, of course, in agreement with the level of the corresponding attribute. Indeed, a given attribute k can be a discrete or a continuous quantitative variable, a qualitative variable, or a "dummy" variable.

To each value, interval modality or "dummy" value of Table 8, we can assign a weight $\omega_{k1}, \omega_{k2}, \omega_{k3}, \ldots, \omega_{km_k}$.

Table 8. Possible modalities depending on the type of attribute.

Quantitative		Qualitative	"Dummy"
Discrete	**Continuous**		
x_{k1}	$[0, x_{k1}]$	q_{k1}	0
x_{k2}	$[x_{k1}, x_{k2}]$	q_{k2}	1
x_{k3}	$[x_{k2}, x_{k3}]$	q_{k3}	2
$\vdots$	$\vdots$	$\vdots$	$\vdots$
x_{km_k}	$[x_{km_{k-1}}, x_{km_k}]$	q_{km_k}	m_k

If $p_i, i = 1, 2, \ldots, n$ denotes the price of the i-th apartment, S_i its surface, and $\omega_{ik^*}, k = 1, 2, \ldots, m$, represents the weight assigned to the k-th attribute of the i-th element in the sample, the normalized average price of a square meter is:

$$\overline{m} = \frac{1}{n} \sum_{i=1}^{n} \frac{p_i}{S_i \left(1 + \sum_{k=1}^{m} \omega_{ik^*}\right)}.$$

Therefore,

$$p_0 = \overline{m} S_0 \left(1 + \sum_{k=1}^{m} \omega_{0k^*}\right), \tag{1}$$

where the subscript 0 corresponds to the house to be assessed.

The *multiple linear regression method* consists of regressing the variable P (price of the house) on the rest of variables involved in the valuation process, e.g.:

- X_1: "Surface".
- X_2: "Number of bathrooms".
- X_3: "Quality".
- X_4: "Number of bedrooms".
- Etcetera.

Thus, we are able to obtain a regression hyperplane in the following form:

$$P = \beta_0 + \beta_1 X_1 + \beta_2 X_2 + \cdots + \beta_n X_n + \epsilon.$$

The concrete values of variables $X_1, X_2, \ldots, X_n$ corresponding to the house to be appraised will allow us to estimate P. Indeed, the main advantage of this method is the objectivity of the result. Unfortunately, this procedure may present two noteworthy drawbacks. First, the goodness of fit in the regression analysis can be very low, which means that the result that produces is not significant at a certain level. Second, a concrete coefficient β_k may exhibit a "wrong" sign in the way that the estimated sign on a variable is the opposite of what we anticipated it should be. For example, it is expected that the price of a house is ceteris paribus inversely related to its age, but the practical implementation of this technique can lead to a positive coefficient for this variable. A possible solution could be to restrict the coefficients to the set of positive or negative real numbers, but, unfortunately, the coefficient of determination does not measure, in this case, the goodness of the fit.

When the linear relationship between the dependent variable (the price) and the regressors (the rest of variables involved in the analysis) has been computed, the values of the attributes of the house to be assessed are included in the equation, which produces an estimated market price. Indeed, a drawback of this method is that the goodness of fit (given by the coefficient of determination) may be very low. This is the reason why it seems convenient to find an appropriate sample (composed by, at least, six houses) for which the fitting should be acceptable.

We return to these issues in the next section.

6.4. Evaluation with Alternative Procedures and Discussion

The routine application of the multiple regression technique to the data included in Table 5 leads to the following three-dimensional hyperplane:

$$p = -22.4 + 2.013x_1 + 1.6x_2 - 10x_4,$$

which leads to a price $p = 189{,}605.86$ euros. Although the coefficient of determination is very high ($R^2 = 0.9656$), this outcome is rather disappointing for two main reasons. The coefficient of x_4 has the wrong sign because ceteris paribus the price of an apartment should increase both with the number of bedrooms. Moreover, observe that the coefficient of x_3 vanishes. The reason is that all the apartments with a valuation in the sample have a "normal" quality, even though there are other values for this attribute in the sample. This is another important drawback of this method since, in our real case study, such zero value implies the fact that the apartment being assessed has a "good" quality cannot be used to increase its valuation.

Let us now approximate the price of the apartment according to the weights shown in Table 9. According to the *Orden ECO/805/2003* (Ministerio de Economía, 2003) [31], the coefficients to standardize the value of the square meter of each element in the sample will be chosen by applying the criteria suitable for the house to be assessed. Nevertheless, this procedure, called the *homogenization method*, "has got some problems which should be solved and that are related, for example, with valuer's subjectivity" [50]. Therefore, in this paper, we have implemented the standard weights used by practitioners, which are based on a proposal by González-Nebreda et al. [51].

Table 9. Attributes, modalities and assigned weights.

Surface		Bathrooms		Quality		Bedrooms	
Interval	Weight	Number	Weight	Level	Weight	Number	Weight
[0, 10]	0.00	0	0.00	Bad	0.04	1	0.03
[10, 20]	0.06	1	0.04	Low	0.08	2	0.06
[20, 30]	0.08	2	0.08	Normal	0.12	3	0.09
[30, 40]	0.10	3	0.12	Good	0.16	4	0.12
[40, 50]	0.12	4	0.16	Luxury	0.20	5	0.14
[50, 60]	0.14	5	0.20	–	–	6	0.16
[60, 70]	0.16	–	–	–	–	7	0.17
[70, 80]	0.18	–	–	–	–	8	0.18
[80, 90]	0.20	–	–	–	–	9	0.19
[90, 100]	0.22	–	–	–	–	10	0.20
[100, 110]	0.24	–	–	–	–	–	–
[110, 120]	0.26	–	–	–	–	–	–
[120, 130]	0.28	–	–	–	–	–	–
[130, 140]	0.30	–	–	–	–	–	–

The application of this information to the characteristics of the apartments in the sample and the apartment to be assessed produces Table 10, where, for completeness, the price of each apartment is shown too.

Table 10. Weights assigned to all apartments.

Item	Price	Surface		Bathrooms		Quality		Bedrooms	
		Value	Weight	Number	Weight	Level	Weight	Number	Weight
h_1	95,000	75	0.18	1	0.04	Normal	0.12	3	0.09
h_2	157,000	105	0.24	2	0.04	Normal	0.12	4	0.12
h_3	115,000	75	0.18	1	0.04	Normal	0.12	2	0.06
h_4	132,000	90	0.20	2	0.08	Normal	0.12	3	0.09
h_5	132,000	90	0.20	1.5	0.06	Normal	0.12	3	0.09
h_6	157,000	105	0.24	2	0.08	Normal	0.12	3	0.09
h_7	$*$	114.44	0.26	1	0.04	Good	0.16	2	0.06

By applying Formula (1), we obtain a price of 171,747.78 euros.

All in all, along this paper, we have compared three methodologies to approximate the value of an apartment from the information supplied by the housing market about the characteristics (price, surface, number of bathrooms, number of bedrooms and quality) of a sample composed by six other apartments.

The first method uses the linear multiple regression where the dependent variable is the price and the independent variables are the rest of the characteristics of the apartments. This technique is subject to at least three noteworthy inconveniences, which dramatically reduce the validity of its results:

1. The possible existence of coefficients with the wrong sign (in our case, the coefficient of variable x_4).
2. The possibility that a coefficient vanishes (in this section, the coefficient of variable x_3 is zero). In such case, the characteristic associated with the corresponding variable is of no use for evaluation purposes.
3. The coefficient of determination may be small (although, in the example in this section, R^2 is pretty high).

The second procedure uses the weights assigned by practitioners to highlight the "good" characteristics of all apartments and to penalize their "bad" figures. This methodology presents an important disadvantage, viz. the enormous subjectivity in choosing these weights.

The third technique is new with this paper. It produces a much more reasonable result, which is partially due to the reduction of subjectivity in the weights.

Despite the disparities between prices, let us stress that the price finally agreed in the transaction of this apartment was much closer to the value given by the proposed methodology.

7. Conclusions

In this work, we propose the new notion of partial valuation fuzzy soft sets and we briefly introduce the problem of data filling in that setting (cf., Sections 3 and 4). The use of fuzzy soft sets permits quantifying qualitative attributes, such as the finish of housing construction or the quality of materials used in the construction of a house. Therefore, we can apply these ideas in real estate valuations. By doing so, we depart from fuzzy soft sets and extend their scope with the target of real applications. In our approach, we first use a rating procedure in order to associate a unique number (score) with each alternative and then we apply regression for the purpose of data filling in partial valuation fuzzy soft sets (cf., Section 4). We have explained our model both algorithmically and with a flow diagram.

Then, we have shown how this new methodology works in a fictitious (cf., Section 5) and a real case study (cf., Section 6). With these examples, we have proved the implementability and feasibility of our methodology. We have also performed a sensitivity analysis in order to avail its robustness. The real case study concerns apartments. Obviously, it can be also applied in the valuation of other kind of assets, such as rural properties, cars, etc. We have obtained a very reasonable price for the house under valuation, which proves the feasibility and implementability of our suggestion. On the other hand, the two alternative methods (that were based on the linear multiple regression and used by practitioners) exhibit serious troubles that restrict their ability to fit real situations.

To conclude, let us point out that our technique can be useful for practitioners using other models of uncertain behavior. For example, the idea that scores can be used to perform a regression can easily be exported to models based on hesitant fuzzy sets [11,12,52–55] for which scores are already available [56–58]. It seems also feasible to export it to other hybrid soft set models (cf., Ali et al. [59], Fatimah et al. [60], Ma et al. [16], Zhan and Zhu [14], and Zhan et al. [61]).

Acknowledgments: We are grateful for the constructive comments by three anonymous referees. The second and third authors acknowledge the financial support from the Spanish Ministry of Economy, Industry and Competitiveness, and the European Regional Development Fund-ERDF/FEDER-UE (National R & D Project ECO2015-66504 and National R & D Project DER2016-76053-R).

Author Contributions: The authors contributed equally to this work.

Conflicts of Interest: The authors declare no conflict of interest.

Abbreviations

The following abbreviations are used in this manuscript:

MDPI	Multidisciplinary Digital Publishing Institute
DOAJ	Directory of Open Access Journals
FS	Fuzzy set
PVFSS	Partial valuation fuzzy soft set
VFSS	Valuation fuzzy soft set
VIKOR	VIsekriterijumska optimizacija i KOmpromisno Resenje

References

1. Zadeh, L. Fuzzy sets. *Inf. Control* **1965**, *8*, 338–353.
2. Molodtsov, D. Soft set theory–first results. *Comput. Math. Appl.* **1999**, *37*, 19–31, doi:10.1016/S0898-1221(99)00056-5.
3. Maji, P.K.; Biswas, R.; Roy, A.R. An application of soft sets in a decision making problem. *Comput. Math. Appl.* **2002**, *44*, 1077–1083, doi:10.1016/S0898-1221(02)00216-X.
4. Maji, P.K.; Biswas, R.; Roy, A.R. Soft set theory. *Comput. Math. Appl.* **2003**, *45*, 555–562, doi:10.1016/S0898-1221(03)00016-6.
5. Alcantud, J.C.R. Some formal relationships among soft sets, fuzzy sets, and their extensions. *Int. J. Approx. Reason.* **2016**, *68*, 45–53, doi:10.1016/j.ijar.2015.10.004.
6. Ali, M.I. A note on soft sets, rough soft sets and fuzzy soft sets. *Appl. Soft Comput.* **2011**, *11*, 3329–3332, doi:10.1016/j.asoc.2011.01.003.
7. Ali, M.I.; Shabir, M. Logic connectives for soft sets and fuzzy soft sets. *IEEE Trans. Fuzzy Syst.* **2014**, *22*, 1431–1442, doi:10.1109/TFUZZ.2013.2294182.
8. Maji, P.K.; Biswas, R.; Roy, A.R. Fuzzy soft sets. *J. Fuzzy Math.* **2001**, *9*, 589–602.
9. Alcantud, J.C.R.; de Andrés Calle, R.; Torrecillas, M.J.M. Hesitant fuzzy worth: An innovative ranking methodology for hesitant fuzzy subsets, *Appl. Soft Comput.* **2016**, *38*, 232–243, doi:10.1016/j.asoc.2015.09.035.
10. Alcantud, J.C.R.; de Andrés Calle, R. A segment-based approach to the analysis of project evaluation problems by hesitant fuzzy sets. *Int. J. Comput. Intell. Syst.* **2016**, *29*, 325–339.
11. Faizi, S.; Rashid, T.; Sałabun, W.; Zafar, S.; Wątróbski, J. Decision making with uncertainty using hesitant fuzzy sets. *Int. J. Fuzzy Syst.* **2017**, doi:10.1007/s40815-017-0313-2.
12. Faizi, S.; Sałabun, W.; Rashid, T.; Wątróbski, J.; Zafar, S. Group decision-making for hesitant fuzzy sets based on characteristic objects method. *Symmetry* **2017**, *9*, 136.
13. Zhang, X.; Xu, Z. Consensus model-based hesitant fuzzy multiple criteria group decision analysis. In *Hesitant Fuzzy Methods for Multiple Criteria Decision Analysis, Studies in Fuzziness and Soft Computing*; Zhang, X., Xu, Z., Eds.; Springer: Berlin, Germany, 2017; Volume 345, pp. 143–157.
14. Zhan, J.; Zhu, K. Reviews on decision making methods based on (fuzzy) soft sets and rough soft sets. *J. Intell. Fuzzy Syst.* **2015**, *29*, 1169–1176, doi:10.3233/IFS-151732.
15. Alcantud, J.C.R. A novel algorithm for fuzzy soft set based decision making from multiobserver input parameter data set. *Inf. Fusion* **2016**, *29*, 142–148, doi:10.1016/j.inffus.2015.08.007.
16. Ma, X.; Liu, Q.; Zhan, J. A survey of decision making methods based on certain hybrid soft set models. *Artif. Intell. Rev.* **2017**, *47*, 507–530, doi:10.1007/s10462-016-9490-x.
17. Chen, J; Ye, J. Some single-valued neutrosophic Dombi weighted aggregation operators for multiple attribute decision-making. *Symmetry* **2017**, *9*, 82, doi:10.3390/sym9060082.
18. Ye, J. Multiple Attribute Decision-Making Method Using Correlation Coefficients of Normal Neutrosophic Sets. *Symmetry* **2017**, *9*, 80, doi:10.3390/sym9060080.
19. Peng, X.; Dai, J.; Yuan, H. Interval-valued fuzzy soft decision making methods based on MABAC, similarity measure and EDAS. *Fundam. Inform.* **2017**, *152*, 373–396, doi:10.3233/FI-2017-1525.
20. Peng, X.; Yang, Y. Algorithms for interval-valued fuzzy soft sets in stochastic multi-criteria decision making based on regret theory and prospect theory with combined weight. *Appl. Soft Comput.* **2017**, *54*, 415–430, doi:10.1016/j.asoc.2016.06.036.

21. Fatimah, F.; Rosadi, D.; Hakim, R.B.F.; Alcantud, J.C.R. Probabilistic soft sets and dual probabilistic soft sets in decision-making. *Neural Comput. Appl.* **2017**, doi:10.1007/s00521-017-3011-y.
22. Chang, T.-H. Fuzzy VIKOR method: A case study of the hospital service evaluation in Taiwan. *Inf. Sci.* **2014**, *271*, 196–212, doi:10.1016/j.ins.2014.02.118.
23. Espinilla, M.; Medina, J.; García-Fernández, Á.L.: Campaña, S.; Londoño, J. Fuzzy intelligent system for patients with preeclampsia in wearable devices. *Mob. Inf. Syst.* **2017**, *2017*, 7838464, doi:10.1155/2017/7838464.
24. Alcantud, J.C.R.; Santos-García, G.; Galilea, E.H. Glaucoma diagnosis: A soft set based decision making procedure. In *Advances in Artificial Intelligence*; Puerta, J.M., Ed.; Springer: Cham, Switzerland, 2015; Volume 9422, pp. 49–60.
25. Xu, Z.S.; Xia, M.M. Distance and similarity measures for hesitant fuzzy sets. *Inf. Sci.* **2011**, *181*, 2128–2138, doi:10.1016/j.ins.2011.01.028.
26. Taş, N.; Yilmaz Özgür, N.; Demir, P. An application of soft set and fuzzy soft set theories to stock management. *J. Nat. Appl. Sci.* **2017**, doi:10.19113/sdufbed.82887.
27. Xu, W.; Xiao, Z.; Dang, X.; Yang, D.; Yang, X. Financial ratio selection for business failure prediction using soft set theory. *Knowl.-Based Syst.* **2014**, *63*, 59–67. doi:10.1016/j.knosys.2014.03.007.
28. Kalaichelvi, A.; Haritha Malini, P. Application of fuzzy soft sets to investment decision making problem. *Int. J. Math. Sci. Appl.* **2011**, *1*, 1583–1586.
29. Özgür, Y.; Taş, N. A note on "Application of fuzzy soft sets to investment decision making problem". *J. New Theory* **2015**, *1*, 1–10.
30. Pagourtzi, E.; Assimakopoulos, V.; Hatzichristos, T.; French, N. Real estate appraisal: A review of valuations methods. *J. Prop. Invest. Finance* **2003**, *21*, 383–401, doi:10.1108/14635780310483656.
31. Ministerio de Economía (Spain), Orden ECO/805/2003, de 27 De Marzo, Sobre Normas De Valoración De Bienes Inmuebles Y De Determinados Derechos Para Ciertas Finalidades Financieras. Available online: https://www.boe.es/buscar/doc.php?id=BOE-A-2003-7253 (accessed on 4 October 2016).
32. González, M.A.S.; Formoso, C.T. Mass appraisal with genetic fuzzy rule-based systems. *Prop. Manag.* **2006**, *24*, 20–30, doi:10.1108/02637470610643092.
33. Zurada, J.; Levitan, A.; Guan, J. A comparison of regression and artificial intelligence methods in a mass appraisal context. *J. Real Estate Res.* **2011**, *33*, 349–387, doi:10.5555/rees.33.3.q6890722u7375871.
34. Zhang, R.; Du, Q.; Geng, J.; Liu, B.; Huang, Y. An improved spatial error model for the mass appraisal of commercial real estate based on spatial analysis: Shenzhen as a case study. *Habitat Int.* **2015**, *46*, 196–205, doi:10.1016/j.habitatint.2014.12.001.
35. Feng, F.; Li, Y. Soft subsets and soft product operations. *Inf. Sci.* **2013**, *232*, 44–57, doi:10.1016/j.ins.2013.01.001.
36. Qin, H.; Ma, X.; Herawan, T.; Zain, J. Data filling approach of soft sets under incomplete information. In *Intelligent Information and Database Systems*; Nguyen, N., Kim, C.-G., Janiak, A., Eds.; Springer: Berlin/Heidelberg, Germany, 2011; pp. 302–311.
37. Qin, H.; Ma, X.; Zain, J.M.; Herawan, T. A novel soft set approach in selecting clustering attribute. *Knowl.-Based Syst.* **2012**, *36*, 139–145, doi:10.1016/j.knosys.2012.06.001.
38. Yao, Y.Y. Relational interpretations of neighbourhood operators and rough set approximation operators. *Inf. Sci.* **1998**, *111*, 239–259, doi:10.1016/S0020-0255(98)10006-3.
39. Roy, A.R.; Maji, P.K. A fuzzy soft set theoretic approach to decision making problems. *J. Comput. Appl. Math.* **2007**, *203*, 412–418, doi:10.1016/j.cam.2006.04.008.
40. Feng, F.; Jun, Y.; Liu, X.; Li, L. An adjustable approach to fuzzy soft set based decision making. *J. Comput. Appl. Math.* **2010**, *234*, 10–20, doi:10.1016/j.cam.2009.11.055.
41. Alcantud, J.C.R. Fuzzy soft set based decision making: A novel alternative approach. In Proceedings of the 2015 Conference of the International Fuzzy Systems Association and the European Society for Fuzzy Logic and Technology, Gijón, Spain, 30 June–3 July 2015; Atlantics Press, 2015; pp. 106–111, doi:10.2991/ifsa-eusflat-15.2015.194.
42. Ali, M.I.; Feng, F.; Liu, X.; Min, W.K.; Shabir, M. On some new operations in soft set theory. *Comput. Math. Appl.* **2009** *57*, 1547–1553.
43. Han, B.-H.; Li, Y.; Liu, J.; Geng, S.; Li, H. Elicitation criterions for restricted intersection of two incomplete soft sets. *Knowl.-Based Syst.* **2014**, *59*, 121–131, doi:10.1016/j.knosys.2014.01.015.

44. Alcantud, J.C.R.; Santos-García, G. Incomplete soft sets: New solutions for decision making problems. In *Advances in Intelligent Systems and Computing*; Bucciarelli, E., Ed.; Springer: Cham, Switzerland, 2016; Volume 475, pp. 9–17.
45. Alcantud, J.C.R.; Santos-García, G. A new criterion for soft set based decision making problems under incomplete information. *Int. J. Comput. Intell. Syst.* **2017**, *10*, 394–404, doi:10.2991/ijcis.2017.10.1.27.
46. Zou, Y.; Xiao, Z. Data analysis approaches of soft sets under incomplete information. *Knowl.-Based Syst.* **2008**, *21*, 941–945, doi:10.1016/j.knosys.2008.04.004.
47. Aiken, L.S.; West, S.G. *Testing and Interpreting Interactions*; SAGE Publications: Thousand Oaks, CA, USA, 1991.
48. Bagnoli, C.; Smith, H.C. The theory of fuzzy logic and its application to real estate valuation. *J. Real Estate Res.* **1998**, *16*, 169–200.
49. Saltelli, A. Sensitivity analysis for importance assessment. *Risk Anal.* **2002**, *22*, 1–12.
50. Aznar, J.; Guijarro, F. Housing valuation in Spain. Homogenization method and alternative methodologies. *Finance Markets Valuat.* **2016**, *2*, 91–125.
51. González-Nebreda, P.; Turmo-de-Padura, J.; Villaronga-Sánchez, E. *La Valoración Inmobiliaria. Teoría y Práctica*; Wolters Kluwer España, S.A.: Madrid, Spain, 2006.
52. Alcantud, J.C.R.; Torra, V. Decomposition theorems and extension principles for hesitant fuzzy sets. *Inf. Fusion* **2018**, *41*, 48–56, doi:10.1016/j.inffus.2017.08.005.
53. Torra, V. Hesitant fuzzy sets. *Int. J. Intell. Syst.* **2010**, *25*, 529–539.
54. Kobza, V.; Janiš, V.; Montes, S. Divergence measures on hesitant fuzzy sets. *J. Intell. Fuzzy Syst.* **2017**, *33*, 1589–1601, doi:10.3233/JIFS-161430.
55. Torra, V.; Narukawa, Y. On hesitant fuzzy sets and decision. In Proceedings of the 2009 IEEE International Conference on Fuzzy Systems, Jeju Island, Korea, 20–24 August 2009; pp. 1378–1382.
56. Xia, M.; Xu, Z. Hesitant fuzzy information aggregation in decision making. *Int. J. Approx. Reason.* **2011**, *52*, 395–407, doi:10.1016/j.ijar.2010.09.002.
57. Farhadinia, B. A series of score functions for hesitant fuzzy sets. *Inf. Sci.* **2014**, *277*, 102–110, doi:10.1016/j.ins.2014.02.009.
58. Farhadinia, B. A novel method of ranking hesitant fuzzy values for multiple attribute decision-making problems. *Int. J. Intell. Syst.* **2013**, *28*, 752–767, doi:10.1002/int.21600.
59. Ali, M.I.; Mahmood, T.; Rehman, M.M.U.; Aslam, M.F. On lattice ordered soft sets. *Appl. Soft Comput.* **2015**, *36*, 499–505, doi:10.1016/j.asoc.2015.05.052.
60. Fatimah, F.; Rosadi, D.; Hakim, R.B.F.; Alcantud, J.C.R. N-soft sets and their decision making algorithms. *Soft Comput.* **2017**, doi:10.1007/s00500-017-2838-6.
61. Zhan, J.; Ali, M.I.; Mehmood, N. On a novel uncertain soft set model: Z-soft fuzzy rough set model and corresponding decision making methods. *Appl. Soft Comput.* **2017**, *56*, 446–457, doi:10.1016/j.asoc.2017.03.038.

symmetry

MDPI

Article

Asymmetries in the Maintenance Performance of Spanish Industries before and after the Recession

María del Carmen Carnero [1,2]

[1] Business Administration Department, University of Castilla-la Mancha, Ciudad Real 13071, Spain; carmen.carnero@uclm.es; Tel.: +34-926-295-300

[2] Engineering and Management Department, University of Lisbon, Lisbon 1049-001, Portugal

Received: 26 June 2017; Accepted: 16 August 2017; Published: 20 August 2017

Abstract: Until the last few decades, maintenance has not been considered of special importance by organisations. Thus, the number of studies that assess maintenance performance in a country is still very small, despite the relevance this area has to the level of national competitiveness. This article describes a multicriteria model integrating the fuzzy analytic hierarchy process (FAHP) with Multi-Attribute Utility Theory (MAUT) to assess the maintenance performance of large, medium and small enterprises in Spain, before and after the recession, as well as the asymmetries in the state of maintenance between different activity sectors. The weightings are converted to utility functions which allow the final utility of an alternative to be calculated via a Multi-Attribute Utility Function. From the Spanish maintenance data for different industrial sectors in 2005 and 2010, 2400 discrete probability distributions have been produced. Finally, a Monte Carlo simulation is applied for the estimation of the uncertainty. The results show that the economic crisis experienced by Spain since 2008 has negatively affected the level of maintenance applied, rather than it being considered an area that could deliver cost reductions and improvements in productivity and quality to organisations.

Keywords: maintenance performance; recession; fuzzy analytic hierarchy process (FAHP); Multi-Attribute Utility Theory (MAUT)

1. Introduction

Maintenance is attaining a more important role in organisations because it can affect productivity and profitability [1–3], the useful lifespan of the facilities, the quality of the processes [4] and the fulfilment of safety and environmental standards. This has brought about increasing concern over the performance maintenance measurement [5], as shown by the abundant literature that analyses the matter (see [6–17]).

Different countries carry out surveys through their national maintenance associations. In the case of Spain, the Spanish Maintenance Association (SMA) carries out surveys every five years; these suggest how maintenance can contribute to improvement in the most immediate weaknesses of the Spanish productive sector, such as the lack of competitiveness and innovation [18]. Other results from Spanish companies can be consulted in Conde [19] and Álvarez [20] in the chemical industry, and in Paredes [21] in manufacturing. Although these national surveys intended to promote continuous improvement via benchmarking are applied extensively in the United States, Canada and New Zealand, in Spain its application has hardly begun [22].

The literature analyses maintenance in a country mainly based on a set of KPIs, which are held to be of equal importance; however, some KPIs influence the competitiveness of a company while other only have slight implications for cost. A multicriteria model, then, allows for a more accurate assessment of the real situation in applied maintenance. Via a multicriteria model it is possible to obtain a grade for the overall state of maintenance, and for each criterion analysed. This shows the development over time of applied maintenance, the criteria with the highest valuation and those where there are deficiencies. Also, as described in Komonen [23], the benchmarking procedure generally applied is the comparison of mean values of different indicators for a specific company with those of its industrial sector; however, in the area of maintenance this type of benchmarking is of little use [23,24]. Although maintenance benchmarking is recognised as a key element in achieving world-class maintenance performance levels [25] and for the continuous improvement process [26], only 11% of the literature reviewed by Simões et al. [27] relates benchmarking to maintenance performance measurement.

There are very few precedents that build a model or framework to analyse maintenance performance or practices by means of indicators. Among these, Macchi and Fumagalli [28] develop a scoring method for maturity assessment with five levels for evaluating maintenance practices in organisations and to improve the maintenance management system. On the same lines, Nachtmann et al. [29] describe using a balanced scorecard for flight line maintenance activities in the U.S. Air Force. Van Horenbeek and Pintelon [30] set out a maintenance performance measurement framework using the analytic network process (ANP) to assist maintenance managers in their choice of the relevant maintenance performance indicators. The model has been applied to five companies of different types. Muchiri et al. [31] propose a framework for assessing and ranking maintenance practices. The framework comprises five-level evaluation criteria qua maintenance practices in any company with a maintenance department to be ranked, and the results compared with others.

In the fuzzy environment, the number of contributions is even more restricted. Carnero [32] describes a fuzzy multicriteria model by which maintenance benchmarking can be applied among small businesses. Stefanovic et al. [33] use fuzzy sets and genetic algorithms to design a model for ranking and optimisation of maintenance performance indicators in small and medium enterprises. There is, however, no model that analyses, via a multicriteria model in a fuzzy environment, the evolution of the state of maintenance before and after recession in a country.

Kubler et al. [34], in their literature review of FAHP applications, concluded that it was predominantly applied in the areas of selection and evaluation and in the categories of manufacturing, industry and government. It was also seen that a large number of studies combine FAHP with other tools, mostly with Technique for Order of Preference by Similarity to Ideal Solution (TOPSIS), Quality Function Deployment (QFD) and Analytic Network Process (ANP). This can be justified by the natural flexibility of FAHP that enables it to be combined with a wide range of techniques and for very different purposes. However, this survey does not cover studies combining fuzzy analytic hierarchy process (FAHP) and Multi-Attribute Utility Theory (MAUT). Similar results can be seen in other multicriteria literature reviews carried out in the field of applications for solving energy management problems [35], or aging-dam management [36]. There are, therefore, very few precedents in the literature combining FAHP and MAUT. Among those that do exist are the following: Ashour and Okudan [37] developed a triage algorithm that integrates FAHP and MAUT to rank the waiting emergency department patients according to their characteristics: chief complaint, gender, age, pain level and vital signs (temperature, breathing rate, pulse and blood pressure). The intuitive judgement and preferences of triage nurses have to be considered in this decision and therefore there are uncertainties involved; this is the reason that utility theory has been selected [38]. A single utility function has been constructed for each criterion taking into account the risk attitude of the triage nurse for each attribute. The exponential distribution has been used, as the best for this approach, and multiplicative forms are applied to aggregate the single utility functions. In a latter study, Ashour and Okudan [39] compared two triage systems using Discrete Event Simulation (DES): the typical Emergency Severity Index (ESI) and the proposed algorithm integrating FAHP and MAUT. As a result, it is seen that the FAHP-MAUT algorithm performs better in terms of minimizing the number of patients with longer than the allotted upper limits of waiting times, but it also reduces potential bias and errors in decision making in clinical settings. Johal and Sandhu [40] constructed utility functions associated with the attributes: bandwidth, security, monetary cost and power consumption levels of the candidate network available for handover. The proposed algorithm uses FAHP to assign weights to the attributes and applies the utility functions to rank the alternatives using a simple weighted sum of the parameters with the objective of the level of satisfaction served by each network. With the same objective, Goyal et al. [41] designed parameterised utility functions to model the different quality of service attributes, but in this case the network selection process considers three different applications: voice, video, and best-effort applications. Different attributes are considered depending on the application. To avoid the problems in obtaining weights caused by Chang's extent analysis method, a min-max optimisation problem is presented to derive consistent weights. Final ranking is calculated with Simple Additive Weighting (SAW), TOPSIS and Multiplicative Exponential Weighting (MEW) methods. The results show that the utility-based MEW method gives more suitable final scores for each network than the utility based SAW and TOPSIS methods.

In [42] a model is designed to apply benchmarking in large buildings integrating FAHP and MAUT. The maintenance department of a hospital and a department store were compared, with the results obtained from the building sector in the case of more than 500 workers for the years 2000, 2005 and 2010. This research used 50 subcriteria; however, some of them do not provide relevant information about the state of a maintenance department because they are related to the maintenance manager's opinion about future trends in maintenance costs, outsourcing, collaboration of production workers, etc. These opinions can be applied or not in the future, but they are not considered to be relevant attributes in this model. Chang's extent analysis method has been applied and the fuzzy scale used to make judgements has six values of preference. This led to problems in the process of obtaining judgements from the decision makers, since the preferences were very close. In the case of [38,39] only five fuzzy number for linguistic variables have been used.

Kubler et al. [34] note in their survey that Chang's extent analysis method is the most popular methodology in spite of a number of criticisms in recent years. Criticisms relate to the appearance of irrational zero weights and the fact that important criteria could not be considered in the decision-making process [43]. Therefore, the relative importance of criteria or alternatives is not calculated appropriately, which may lead to poor robustness, unreasonable priorities and information loss in the models. The research presented in this paper applies the geometric-mean method suggested by Buckley [44], because of its ease of application and comprehension in comparison with other methods [45] and it provides a unique solution to the reciprocal comparison matrix [46], avoiding the criticisms applicable to Chang's extent analysis. It also uses a different means of obtaining the utility functions from the previous literature. That is because the utility functions are associated to attributes with constructed descriptors that have from two to ten qualitative scale levels, depending on the attribute. The data available via surveys of maintenance questions allow a probability to be associated with each scale level of a descriptor. MAUT [47] allows scores to be turned into utility functions if the sum of the weighting is unity. Therefore, this research does not use a decision maker to find the probability value such that there is no difference between two choices, as for example in [38] since the probabilities are calculated from the surveys. In these cases, the utility function is constructed form the mean value and in the current study the full data are used, without mean values.

Future trends in applying FAHP are related to [34]:

(a) Comparing existing fuzzy pairwise comparison matrix weighting derivation methods with regard to efficiency and ease of use.
(b) Verifying mathematically that FAHP improves the results provided by AHP.
(c) FAHP could be combined with other pre-structure planning methods, such as Delphi, to identify all the relevant decision criteria to solve complex problems. All in an easy-to-use framework.

There is also a clear trend towards a hybridisation process, combining two or more Multi-Criteria Decision-Making methods, and, a fuzzification of these same models [36]. This study is framed within both trends.

This article describes a multicriteria model that applies the analytic hierarchy process (AHP), fuzzy analytic hierarchy process (FAHP) and Multi-Attribute Utility Theory (MAUT) to assess the state of maintenance in Spain. FAHP allows weightings for criteria and subcriteria, and a hierarchy, to be obtained. The weightings are turned into utility functions that permit the final utility of an alternative to be calculated by a Multi-Measure Utility Function. From the data on the state of maintenance in Spain for different industrial sectors in the years 2005 and 2010, 2400 discrete probability distributions were derived. These distributions determine the behaviour of a sector with respect to a subcriterion. Finally, a Monte Carlo simulation is applied to estimate the uncertainty of a complex function resulting from several probability distributions. In this way, the level of excellence in applied maintenance in Spain has been determined, before and after the recession. Asymmetries in performance between different activity sectors have been identified, all analysed by different company sizes: up to 200 workers, from 201 to 500 workers, and over 500 workers. This, then, is the first multicriteria model used to assess the state of maintenance in a country.

The original contributions of this research are:

(1) An assessment of maintenance performance before and after the recession. There are no previous research studies that analyse this question.
(2) The prior assessment is carried out in 10 activity sectors and in large, medium and small enterprises in Spain. Therefore, the large number of scenarios used allows results and conclusions to be obtained with great precision and in great detail.
(3) The model constructed integrates AHP, FAHP and MAUT multicriteria techniques, guaranteeing that the criteria and subcriteria are relevant for maintenance assessment of companies. The Buckley method has been used instead of Chang's extent analysis method to guarantee that priorities obtained are accurate and to avoid loss of information in the results.
(4) The model proposed uses the complete data from surveys about maintenance in Spain, rather than using mean values by activity sector or size of enterprise, as in the remaining studies about maintenance in other countries.

This article is structured as follows. Section 2 reviews the state of maintenance in different countries. Section 3 presents the FAHP methodology applied in this study. Section 4 describes the fuzzy multicriteria model to evaluate the maintenance performance in Spanish industries. Section 5 shows the results of the application of the model by company sizes in the years 2005 and 2010. Section 6 presents the discussion.

2. Related Work

There are studies that analyse the state of maintenance in different countries. These show the deficiencies that exist in, for example, Saudi Arabia, where there is a lack of application of scientific principles in using real management support for maintenance departments, optimizing spare part provision, introducing computers into the maintenance systems and improving control of maintenance tasks by applying working orders and report production [48]. Later, Assaf et al. [49] analyse the efficiency of maintenance units in petrochemical companies in Saudi Arabia. They do this with three indicators and data envelopment analysis (DEA) approach that has enabled low- and high-performance maintenance units to be characterised. Jonsson [50] and Alsyouf [51], in their analysis of the state of maintenance management in companies in Sweden show the limited recognition that the function of maintenance receives. In Swedish companies, maintenance is seen as a source of expense, a third of maintenance time is spent on unplanned tasks, and there are deficiencies in the application of Total Productive Maintenance (TPM) or Reliability Centred Maintenance (RCM) and inefficiencies in the planning and programming of maintenance, which makes it harder to reach the set goals and so obtain a competitive advantage. Companies in Denmark, Norway, Sweden and Finland show deficiencies in maintenance resources, with a tendency to operate in the short term as opposed to considering the long-term planning of activity, and a lack of integration between corporate strategy and maintenance systems, and between production and maintenance departments [52]; while studies exclusively in Norwegian organisations show that the percentage of applied maintenance time is still around 40%, with functional maintenance at 62.3% and functional development (which includes development of new systems and functional perfective maintenance) at 37.7% [53]. The maintenance culture in a Nordic nuclear power plant has also been analysed, measuring perceived values, psychological characteristics of the job, individual conceptions of work and the organisation and perceptions of maintenance tasks [54]. In Belgian and Dutch companies, it is seen that companies with different competitive priorities apply different maintenance strategies and that the most competitive apply more preventive and predictive maintenance policies, better planning and control systems and decentralised maintenance organisation structures when compared to the others [55]. The use of Key Performance Indicators (KPI) in Belgian industries is analysed in Muchiri et al. [56], which shows that equipment, maintenance cost and safety performance are the most commonly used indicators, while those related to maintenance work are less widely used; also, there is no correlation between the established maintenance objectives and the KPIs used, and in very few cases are the results of the KPIs used in decision making. It is precisely the effective use of KPIs, the use of predictive and proactive maintenance, TPM and RCM, with little corrective maintenance and applying high operational involvement in autonomous maintenance for root cause analysis, which most significantly determines an effective maintenance management programme among manufacturing companies in the United Kingdom [57]. Forty percent of these companies apply good maintenance and are aware of the benefits it brings; also, on average U.K. companies apply some good maintenance practices and obtain advantages from them, but they still need to carry out improvements. In U.S. companies, it was found that there was no correlation between the structure of an organisation and the application of advanced maintenance methods, the wider application of technical methods such as vibration analysis, lubricants, etc., as opposed to methods that require the use of personnel, such as RCM, and deficiencies were also found in the planning and programming of maintenance [58]. Related to RCM, Reliabilityweb [59] shows the results obtained for the application of RCM, among which is that only 18.30% of the 601 companies surveyed completed a project to introduce this policy and obtained the expected results. Wireman [60,61] shows in detail how maintenance in the USA has evolved, giving performances of low range, high range and best practice in maintenance cost, maintenance labour cost, work order coverage, preventive maintenance compliance, stores investment, productivity rates, etc. The survey carried out by Tse [62] in 21 companies in Hong Kong shows that corrective and preventive maintenance are most common, while there is a lack of advanced maintenance practices; also, in general, investment in maintenance is much lower than the assets of the company and the profits obtained. This situation is also found in medium-sized and large enterprises in Brazil, among which there is a need to reduce corrective maintenance, increase well-planned maintenance and introduce more predictive maintenance and RCM to gain competitive advantage [63]. The situation in the maquiladora industry in Mexico is even worse, with a tendency to reactive maintenance, while preventive and predictive maintenance are not found in most of the companies surveyed [64]. Modgil and Sharma [65] analyse the impact of TPM and total quality management (TQM) on operational performance in Indian pharmaceutical plants. It is seen that TPM practices have a significant impact on plant-level operational performance, R&D, product innovation and technology management. TQM, on the other hand, gives significant support to a TPM programme. They show TPM assists in reducing the cost of quality through reduced scrap and fewer defective products. In Muchiri et al. [66] a global evaluation index is obtained for the level of maintenance practices carried out by Kenyan companies. The results are that processes are partially planned, and performance depends on the operators' competence and experience.

3. Fuzzy Analytic Hierarchy Process

Fuzzy numbers are usually used to capture the ambiguity, fuzziness or imprecision of the parameters related to the topic [67] and, decision makers usually feel more confident in giving interval judgements rather than fixed value judgements [68].

The model proposed in this research uses FAHP. $\widetilde{A}$ represents a fuzzified reciprocal n-by-n judgment matrix con $\widetilde{a}_{ij}$ the pairwise comparisons between the element i and j $\widetilde{a}_{ij} \forall\, i, j \in \{1, 2, \ldots, n\}$.

$$\widetilde{A} = \begin{bmatrix} (1,1,1) & \widetilde{a}_{12} & \ldots & \widetilde{a}_{1n} \\ \widetilde{a}_{21} & (1,1,1) & \ldots & \widetilde{a}_{2n} \\ & & \ldots & \\ \cdot & \cdot & & \cdot \\ \cdot & \cdot & & \cdot \\ \cdot & \cdot & & \cdot \\ \widetilde{a}_{n1} & \widetilde{a}_{n2} & \ldots & (1,1,1) \end{bmatrix} \quad (1)$$

A triangular fuzzy number $\widetilde{a} = (l, m, u)$ is defined on $\Re$ by the membership function $\mu_{\widetilde{a}}(x) : \Re \rightarrow [0, 1]$ [69]

$$\mu_{\widetilde{a}}(x) = \begin{cases} \frac{x}{m-l} - \frac{l}{m-l}, & x \in [l, m] \\ \frac{x}{m-u} - \frac{u}{m-u}, & x \in [m, u] \\ 0, & \text{otherwise} \end{cases} \quad (2)$$

With $l \leq m \leq u$. l and u are the lower and upper bounds of the fuzzy number and m the modal value. If $l = m = u$ then it is considered a crisp number by convention.

The operational laws for two triangular fuzzy numbers $\widetilde{a}_1 = (l_1, m_1, u_1)$ and $\widetilde{a}_2 = (l_2, m_2, u_2)$ are the following [69–71]:

$$\widetilde{a}_1 \oplus \widetilde{a}_2 = (l_1 + l_2, m_1 + m_2, u_1 + u_2) \quad (3)$$

$$\widetilde{a}_1 \ominus \widetilde{a}_2 = (l_1 - u_2, m_1 - m_2, u_1 - l_2) \quad (4)$$

$$\widetilde{a}_1 \otimes \widetilde{a}_2 \approx (l_1 l_2, m_1 m_2, u_1 u_2) \quad (5)$$

$$\widetilde{a}_1 \oslash \widetilde{a}_2 \approx (l_1/u_2, m_1/m_2, u_1/l_2) \quad (6)$$

$$\widetilde{a}_1^{-1} \approx (1/u_1, 1/m_1, 1/l_1) \text{ for } l, m, u > 0 \quad (7)$$

$$k \otimes \widetilde{a}_1 \approx (kl_1, km_1, ku_1), \text{ k} > 0, \text{ k} \in \text{R} \quad (8)$$

There are different fuzzy AHP methods, a description of which can be found in [46,72]. However, this paper will apply the geometric mean method suggested by Buckley [44], because of its ease of application and comprehension in comparison with other methods [44] and it provides a unique solution to the reciprocal comparison matrix [46].

To calculate the fuzzy weights of each criterion/subcriterion is applied [43,73]:

$$\widetilde{r}_i = [\widetilde{a}_{i1} \otimes \widetilde{a}_{i2} \otimes \ldots \otimes \widetilde{a}_{in}]^{\frac{1}{n}} \; \forall\, i = 1, 2, \ldots, n \quad (9)$$

$$\widetilde{w}_i = \frac{\widetilde{r}_i}{[\widetilde{r}_1 \otimes \widetilde{r}_2 \otimes \ldots \otimes \widetilde{r}_n]} \quad (10)$$

Then the $\widetilde{w}_i$ must be defuzzified. Defuzzification is an inverse transformation that maps the output from the fuzzy domain back onto the crisp domain. This is done through a centroid method [74,75]:

$$w_i = l_i + \frac{(m_i - l_i) + (u_i - l_i)}{3} \; i = 1, 2, \ldots, n. \quad (11)$$

The Consistency Index (*CI*) is used as a measurement of the consistency of the judgements expressed [76]:

$$CI = (\lambda_{\max} - n)/(n-1). \tag{12}$$

The consistency ratio (RC) is defined as the quotient of the consistency index and the random consistency index (ICR) for a matrix of similar size [77]. The judgements given are considered consistent if the RC is lower than 5% for a 3 × 3 matrix, 9% for a 4 × 4 matrix and 10% for larger matrices. To calculate the *CI* with fuzzy numbers the central value of $\lambda_{\max}$ will be used because in the symmetry of a fuzzy number, the central value corresponds to the centroid of the triangular area [78].

The FAHP-derived model is now described.

4. The Proposed Model

4.1. Description

From surveys carried out by the Spanish Maintenance Association a model structuring process has been developed in eight criteria: quality, environment and safety standards, maintenance organisation, maintenance cost, outsourcing maintenance, control, maintenance computerisation, training and maintenance management. Within each criterion there are a number of subcriteria. There is an associated descriptor for each subcriterion.

The weighting process uses FAHP to get the weightings for the criteria and subcriteria, from the judgements of two experts in maintenance. Fuzzy numbers were used to assign weightings to the criteria and subcriteria. To get the fuzzy numbers, two experts in maintenance were used as decision makers; these experts have approximately twenty years of experience in maintenance and knowledge of different sectors. The decision makers were asked to assess the importance of the criteria and subcriteria applying the triangular number scale set out in Table 1.

Table 1. Fuzzy scale.

Definition of Every Fuzzy Number	Fuzzy Numbers	Fuzzy Reciprocal Numbers
Equally important	$\tilde{1} = (1,1,1)$	(1, 1, 1)
Judgment values between equally and moderately	$\tilde{2} = (1,2,3)$	(1/3, 1/2, 1)
Moderately more important	$\tilde{3} = (2,3,4)$	(1/4, 1/3, 1/2)
Judgment values between moderately and strongly	$\tilde{4} = (3,4,5)$	(1/5, 1/4, 1/3)
Strongly more important	$\tilde{5} = (4, 5,6)$	(1/6, 1/5, 1/4)
Judgment values between strongly and very strongly	$\tilde{6} = (5,6,7)$	(1/7, 1/6, 1/5)
Very strongly more important	$\tilde{7} = (6,7,8)$	(1/8, 1/7, 1/6)
Judgment values between very strongly and extremely	$\tilde{8} = (7,8,9)$	(1/9, 1/8, 1/7)
Extremely more important	$\tilde{9} = (8,9,9)$	(1/9, 1/9, 1/8)

AHP has been used to obtain the weightings between the levels of each descriptor. These weightings have been transformed to convert measure levels to utilities because measure levels are based on scales with different units, so before they can be combined they are converted to common scales with a range from 0 to 1. The utility of the preferred alternative of a criterion is 1 and the level for the least preferred alternative is 0.

To define the alternatives the probabilities of appearance of each of the scale measure levels is calculated, for each descriptor and each industrial sector; this used answers given in the 2005 and 2010 SMA surveys for different activity sectors. Because of the economic recession the survey was not carried out in 2015. The data used, therefore, are from 2005 to 2010. These probabilities are used to build discrete probability distributions for each descriptor. A discrete distribution has probabilities defined for several different levels so that the probabilities add up to 1.

Figure 1 shows a flow diagram with the detailed procedure followed in this research.

There follows an explanation of each stage of the building of the model.

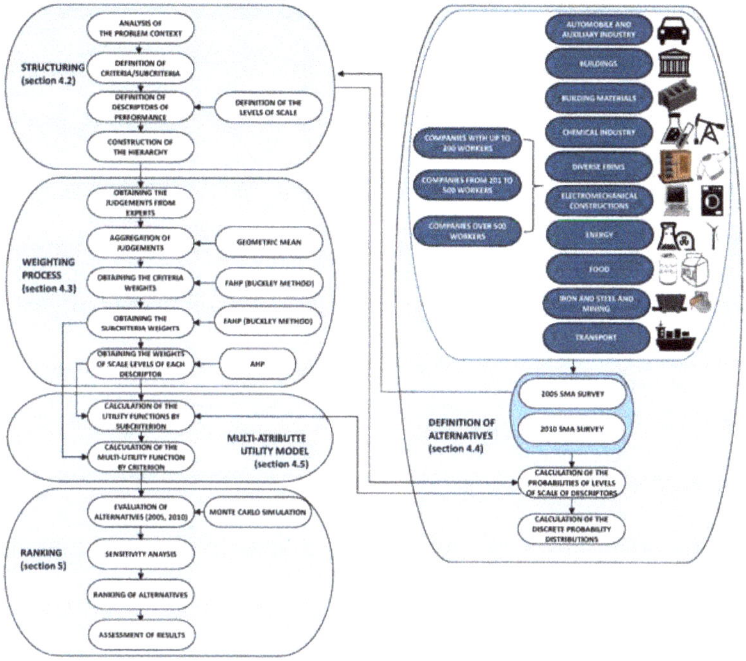

Figure 1. Flow diagram.

4.2. Structuring

There follows a description of the procedure used in the model to carry out the structuring stage:

Step 1. Analysis of the problem context.

Step 2. Analysis of the 2005 and 2010 SMA surveys.

Step 3. Selection, modification or linking of questions to get relevant and independent criteria and subcriteria.

Step 4. Definition of a descriptor for each subcriterion.

Step 5. Definition of levels of scale by subcriterion.

Step 6. Construction of a hierarchy.

The first step for structuring the multicriteria model was to choose the decision criteria and subcriteria for assessing the state of maintenance in Spain. To this end, the starting point was 64 questions from the survey carried out by the SMA [79]. These 64 questions have been modified and turned into a form that may be considered decision criteria. These criteria are exhaustive, concise, non-redundant and independent [80].

Within each criterion, a set of relevant subcriteria were grouped.

The final structure of the benchmarking model for maintenance evaluation has one goal, eight criteria and 40 subcriteria. The criteria and subcriteria used are:

C_1. Budget preparation and measurement of annual maintenance cost and distribution. This comprises the subcriteria:

- C_{11}. Total annual maintenance costs of the company.
- C_{12}. Percentage of annual cost related to in-house staff.
- C_{13}. Percentage of annual cost related to outsourcing jobs.
- C_{14}. Percentage of annual cost related to spare parts and consumables.

C_2. Certification of the company to international standards and compliance with Spanish regulations on health and safety at work. It comprises these subcriteria:

- C_{21}. Existence of a department in the company for compliance with Spanish regulations on health and safety at work. The values are whether there exists a department within the company, it is being set up, or it does not exist.
- C_{22}. Company certified to standard ISO 9000. The values here are whether the company has been audited by an external body to certify compliance with standard ISO 9000, it is in the process of gaining the certificate, or it is not certified.
- C_{23}. Company certified to standard ISO 14000. This assess whether the company has been audited by an external body to certify compliance with standard ISO 14000 so as to have introduced an effective environmental management system with the aim of reducing its impact on the environment and complying with the relevant legislation.

C_3. Control of maintenance activity and efficiency:

- C_{31}. Organisation of work in work orders. Valuing of the use of work orders that include assignment of priority to the activities, material and labour required for each fault or breakdown and the time spent on each activity.
- C_{32}. Control indices used in systematic monitoring of maintenance management.
- C_{33}. Delay in receipt of information on costs. Time lag between spending on maintenance till account data is obtained about these costs and the rest of the departmental budget.
- C_{34}. Regularity of receipt of information on maintenance costs.
- C_{35}. Percentage of maintenance work carried out internally compared to outsourced.
- C_{36}. Percentage of urgent work received.
- C_{37}. Pending work. Time that would be needed to finish the maintenance jobs in progress and carry out the jobs pending.

C_4. Characteristics of the head of maintenance and maintenance tasks carried out outside working hours. This includes the following subcriteria:

- C_{41}. Length of time as maintenance manager.
- C_{42}. Academic qualifications of head of maintenance.
- C_{43}. Length of time as member of a maintenance department.
- C_{44}. Frequency with which the head of maintenance is required to attend outside working hours to resolve maintenance incidents that cannot be sorted out by others, because they are not there or because of the difficulty of the incident.
- C_{45}. Remuneration for work in overtime or outside working hours. Existence and type of remuneration giving for working overtime, holidays, or being on call outside normal working hours.
- C_{46}. Attendance at conferences, talks, seminars, etc. on maintenance.
- C_{47}. Consulting Spanish technical journals on maintenance.
- C_{48}. Use of the internet to search for information to solve maintenance problems.
- C_{49}. Consulting international technical journals on maintenance.

C_5. Organisational characteristics of the maintenance department. This comprises the following subcriteria:

- C_{51}. Existence of a maintenance department. This considers whether the company has a specific department or section whose main purpose is to take care of maintenance.
- C_{52}. Dependence on maintenance department manager. This considers who the maintenance manager is directly accountable to, for example the General Manager, the Production Manager, etc.

- C_{53}. Responsibilities of the maintenance department.
- C_{54}. Number of employees in the maintenance department.
- C_{55}. Incidents outside working hours. This identifies how maintenance problems are dealt with when they happen outside working hours.
- C_{56}. Collaboration of the production staff in maintenance activities.

C_6. Characteristics of computerisation of maintenance in the organisation and the efficiency level. This includes the following subcriteria:

- C_{61}. Assessment of satisfaction in the application of computerisation.
- C_{62}. Number of activities in which computerisation is used.
- C_{63}. Type of computerised maintenance management system used.
- C_{64}. Hardware on which maintenance software runs.

C_7. Importance and level of acceptance of outsourced maintenance carried out in the organisation. This comprises the following subcriteria:

- C_{71}. Percentage of corrective maintenance outsourced.
- C_{72}. Percentage of preventive maintenance outsourced.
- C_{73}. Percentage of work from programmed stoppages outsourced.
- C_{74}. Quality of work outsourced. The quality of outsourced maintenance services is evaluated.
- C_{75}. Percentage of outsourced personnel.

C_8. Maintenance training given by the organisation and its results:

- C_{81}. Training courses. There are specified, current training programs for the staff of the maintenance department.
- C_{82}. Versatility of personnel. Ability of maintenance operatives to regularly carry out activities from two or more specialities.

The hierarchy is shown in Figure 2.

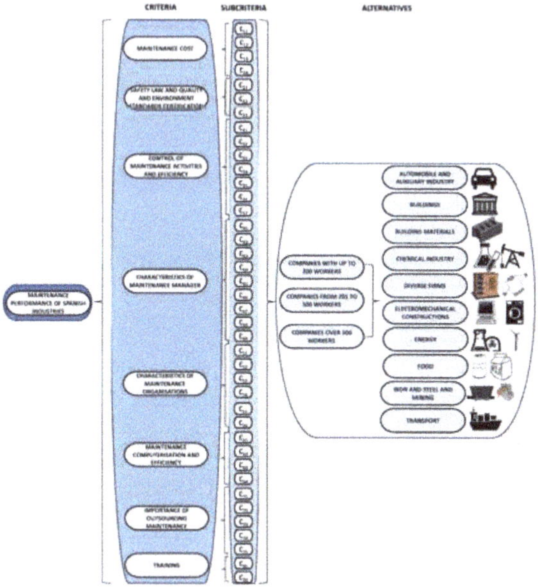

Figure 2. Hierarchy.

To each subcriterion there is associated a descriptor. A descriptor is an ordered set of impact levels that can measure quantitatively or qualitatively the level of fulfilment of a criterion [81]. Descriptor levels are used to describe plausible impacts of alternatives with respect to each criterion. Descriptors may be direct (the levels of the descriptor directly measure the effects) or indirect (the descriptor levels show causes rather than effects). When the criterion is intrinsically subjective or is made up of a set of interrelated, interdependent, elementary areas, the previously described descriptors are not suitable, and constructed descriptors are used. The levels of a constructed descriptor may be qualitative, quantitative or mixed, and they may be created using verbal descriptions of expected consequences, visual representations, indices, etc. [82].

The descriptors applied in this model are constructed and generally qualitative although in some cases they are quantitative. Table 2 shows the descriptors and measurement levels used in the criterion C_4 Characteristics of the head of maintenance and maintenance tasks carried out outside working hours.

Table 2. Descriptors and scale levels of criterion C_4 (characteristics of head of maintenance and maintenance tasks performed outside working hours).

Code/Level of Performance	Descriptor/Scale Levels
C_{41}	Length of time as maintenance manager.
L1 (highest level of performance)	More than 20 years
L2	From 13 to 20 years
L3	From 8 to 12 years
L4	From 4 to 7 years
L5	From 1 to 3 years
L6 (lowest level of performance)	Less than 1 year
C_{42}	Academic qualification of maintenance manager.
L1 (highest level of performance)	Industrial engineer
L2	Aeronautical, mining, naval, telecommunications etc. engineer
L3	Architect
L4	Technical industrial engineer
L5	Technical architect
L6	Naval technician
L7	Industrial technician
L8 (lowest level of performance)	No further education
C_{43}	Experience in maintenance positions.
L1 (highest level of performance)	More than 25 years
L2	From 21 to 25 years
L3	From 16 to 20 years
L4	From 11 to 15 years
L5	From 6 to 10 years
L6 (lowest level of performance)	Up to 5 years
C_{44}	Frequency of attendance outside working hours.
L1 (highest level of performance)	Never
L2	Rarely (1 to 3 times a year)
L3	Irregularly (4 to 10 times a year)
L4	Occasionally (1 time a month)
L5	Frequently (2 to 3 times a month)
L6 (lowest level of performance)	Continually (1 or more times a week)
C_{45}	Remuneration for work in overtime or outside working hours.
L1 (highest level of performance)	Yes
L2	Made up for with rest or holiday time
L3 (lowest level of performance)	No
C_{46}	Attendance at conferences, talks, seminars, etc. on maintenance.
L1 (highest level of performance)	Regularly (1 or more a year)
L2	Occasionally (less than 1 a year)
L3	Attendance rare due to lack of time
L4 (lowest level of performance)	Attendance rare due to not considering them important
C_{47}	Consults Spanish technical journals on maintenance.
L1 (highest level of performance)	Yes
L2 (lowest level of performance)	No
C_{48}	Use of internet to search for information to solve maintenance problems
L1 (highest level of performance)	Yes
L2 (lowest level of performance)	No
C_{49}	Consults international technical journals on maintenance.
L1 (highest level of performance)	Yes
L2 (lowest level of performance)	No

4.3. Weighting Process

The steps used for the weighting phase are the following:

Step 1. Select the fuzzy scale.

Step 2. Select maintenance experts to provide the judgements in the decision-making process.

Step 3. Explain to the maintenance experts the process required to obtain crisp judgements (between the levels of scale of each subcriterion) and fuzzy judgements (between criteria and subcriteria) and provide support during the process.

Step 4. Check the consistency ratios.

Step 5. Aggregate the crisp and fuzzy judgements by geometric mean.

Step 6. Construct a program in Excel to calculate the fuzzy weights by criteria and subcriteria following the Buckley technique.

Step 7. Apply AHP to get weights associated with the levels of scale of each descriptor.

Step 8. Transform the crisp weights in utility functions.

Step 9. Obtain fuzzy weights by criteria and subcriteria. Defuzzify the weights and apply normalisation.

The scale of Table 1 was chosen because it fits better with the original preference scale of the crisp AHP [83].

The pairwise comparison matrices of the criteria provided by the decision makers are as follows:

To aggregate the judgements the geometric mean was applied to l_{ijk}, m_{ijk} and u_{ijk} (see Equation (13)) [84]; where (l_{ijk}, m_{ijk}, u_{ijk}) is a fuzzy number associated with each decision maker k ($k = 1, 2, \ldots, K$).

$$l_{ij} = \left(\prod_{k=1}^{K} l_{ijk}\right)^{1/K}, \; m_{ij} = \left(\prod_{k=1}^{K} m_{ijk}\right)^{1/K}, \; u_{ij} = \left(\prod_{k=1}^{K} u_{ijk}\right)^{1/K} \tag{13}$$

The resulting matrix can be seen in Table 3.

Table 3. Pairwise comparison matrix of criteria.

	C_1	C_2	C_3	C_4	C_5	C_6	C_7	C_8
C_1	(1, 1, 1)	(1.414, 2.449, 3.464)	(1.732, 2.828, 3.873)	(1.414, 2.449, 3.464)	(3, 4, 5)	(4, 5, 6)	(4.899, 5.916, 6.928)	(5.477, 6.481, 7.483)
C_2	(0.289, 0.408, 0.707)	(1, 1, 1)	(1, 1, 1)	(0.577, 1, 1.732)	(1.414, 2.449, 3.464)	(2.449, 3.464, 4.472)	(3.873, 4.899, 5.916)	(2.828, 3.873, 4.899)
C_3	(0.258, 0.354, 0.577)	(1, 1, 1)	(1, 1, 1)	(0.500, 0.816, 1.225)	(1.414, 1.732, 2)	(2.449, 3.464, 4.472)	(3.464, 4.472, 5.477)	(2.828, 3.873, 4.899)
C_4	(0.289, 0.408, 0.707)	(0.577, 1, 1.732)	(0.816, 1.225, 2)	(1, 1, 1)	(1.414, 2.449, 3.464)	(2.828, 3.873, 4.899)	(3.162, 4.243, 5.292)	(3.464, 4.472, 5.477)
C_5	(0.200, 0.250, 0.333)	(0.289, 0.408, 0.707)	(0.500, 0.577, 0.707)	(0.289, 0.408, 0.707)	(1, 1, 1)	(1, 2, 3)	(2.000, 3.162, 4.243)	(2, 3, 4)
C_6	(0.167, 0.200, 0.250)	(0.224, 0.289, 0.408)	(0.224, 0.289, 0.408)	(0.204, 0.258, 0.354)	(0.333, 0.500, 1)	(1, 1, 1)	(1.414, 1.732, 2)	(1.414, 2.449, 3.464)
C_7	(0.144, 0.169, 0.204)	(0.169, 0.204, 0.258)	(0.183, 0.224, 0.289)	(0.189, 0.236, 0.316)	(0.236, 0.316, 0.500)	(0.500, 0.577, 0.707)	(1, 1, 1)	(0.500, 0.816, 1.225)
C_8	(0.134, 0.154, 0.183)	(0.204, 0.258, 0.354)	(0.204, 0.258, 0.354)	(0.183, 0.224, 0.289)	(0.250, 0.333, 0.500)	(0.289, 0.408, 0.707)	(0.816, 1.225, 2)	(1, 1, 1)

As shown in Buckley [44], if the pairwise comparison matrices given for each decision maker are consistent, then the matrix resulting from the aggregation of the judgements is consistent. The matrices given for each decision maker have consistency ratios of 0.017 and 0.031. Therefore, the aggregated judgement matrix is consistent.

Applying Equations (9) and (10) to the pairwise comparison matrix of the experts' aggregated judgements gives the fuzzy weights of the criteria: $\widetilde{w}_1 = (0.177, 0.313, 0.520)$, $\widetilde{w}_2 = (0.093, 0.162, 0.283)$, $\widetilde{w}_3 = (0.089, 0.147, 0.244)$, $\widetilde{w}_4 = (0.092, 0.168, 0.312)$, $\widetilde{w}_5 = (0.048, 0.087, 0.161)$, $\widetilde{w}_6 = (0.032, 0.053, 0.097)$, $\widetilde{w}_7 = (0.021, 0.034, 0.059)$ and $\widetilde{w}_8 = (0.022, 0.036, 0.066)$.

To get the weightings as a crisp number Equation (11) is applied, giving, after normalisation, the results: $w_1 = 0.305$, $w_2 = 0.162$, $w_3 = 0.145$, $w_4 = 0.173$, $w_5 = 0.089$, $w_6 = 0.055$, $w_7 = 0.035$, $w_8 = 0.037$.

A similar process is followed for the subcriteria associated with each criterion, giving the results shown in Table 4.

Table 4. Final non-fuzzy weights of subcriteria.

Subcriterion	$\widetilde{r}_i$	Weights after Defuzzification and Normalisation	CR
C_{11}	$\widetilde{r}_{11} = (2.213, 2.800, 3.281)$	$w_{11} = 0.530$	
C_{12}	$\widetilde{r}_{12} = (0.972, 1.286, 1.622)$	$w_{12} = 0.250$	0.044
C_{13}	$\widetilde{r}_{13} = (0.518, 0.678, 0.885)$	$w_{13} = 0.134$	0.074
C_{14}	$\widetilde{r}_{14} = (0.337, 0.410, 0.565)$	$w_{14} = 0.085$	
C_{21}	$\widetilde{r}_{21} = (3.166, 3.538, 3.888)$	$w_{21} = 0.741$	
C_{22}	$\widetilde{r}_{22} = (0.794, 0.911, 1.038)$	$w_{22} = 0.192$	0.014 0.034
C_{23}	$\widetilde{r}_{23} = (0.278, 0.310, 0.354)$	$w_{23} = 0.066$	
C_{31}	$\widetilde{r}_{31} = (3.830, 4.563, 5.255)$	$w_{31} = 0.422$	
C_{32}	$\widetilde{r}_{32} = (2.266, 2.805, 3.390)$	$w_{32} = 0.263$	
C_{33}	$\widetilde{r}_{33} = (0.999, 1.272, 1.592)$	$w_{33} = 0.120$	
C_{34}	$\widetilde{r}_{34} = (0.679, 0.766, 0.866)$	$w_{34} = 0.071$	0.012 0.082
C_{35}	$\widetilde{r}_{35} = (0.453, 0.539, 0.645)$	$w_{35} = 0.051$	
C_{36}	$\widetilde{r}_{36} = (0.359, 0.423, 0.510)$	$w_{36} = 0.040$	
C_{37}	$\widetilde{r}_{37} = (0.310, 0.352, 0.416)$	$w_{37} = 0.033$	
C_{41}	$\widetilde{r}_{41} = (4.239, 5.038, 5.764)$	$w_{41} = 0.359$	
C_{42}	$\widetilde{r}_{42} = (2.353, 2.984, 3.648)$	$w_{42} = 0.216$	
C_{43}	$\widetilde{r}_{43} = (1.758, 2.207, 2.748)$	$w_{43} = 0.162$	
C_{44}	$\widetilde{r}_{44} = (0.831, 1.074, 1.352)$	$w_{44} = 0.079$	
C_{45}	$r_{45} = (0.506, 0.637, 0.784)$	$w_{45} = 0.046$	0.021 0.048
C_{46}	$\widetilde{r}_{46} = (0.476, 0.593, 0.715)$	$w_{46} = 0.043$	
C_{47}	$\widetilde{r}_{47} = (0.381, 0.448, 0.550)$	$w_{47} = 0.033$	
C_{48}	$\widetilde{r}_{48} = (0.381, 0.448, 0.550)$	$w_{48} = 0.033$	
C_{49}	$\widetilde{r}_{49} = (0.286, 0.370, 0.519)$	$w_{49} = 0.029$	
C_{51}	$\widetilde{r}_{51} = (2.932, 3.674, 4.349)$	$w_{51} = 0.429$	
C_{52}	$\widetilde{r}_{52} = (1.335, 1.884, 2.495)$	$w_{52} = 0.228$	
C_{53}	$\widetilde{r}_{53} = (0.874, 1.140, 1.495)$	$w_{53} = 0.139$	0.024
C_{54}	$\widetilde{r}_{54} = (0.677, 0.888, 1.183)$	$w_{54} = 0.109$	0.029
C_{55}	$\widetilde{r}_{55} = (0.341, 0.448, 0.598)$	$w_{55} = 0.055$	
C_{56}	$\widetilde{r}_{56} = (0.261, 0.318, 0.423)$	$w_{56} = 0.040$	
C_{61}	$\widetilde{r}_{61} = (2.276, 2.847, 3.374)$	$w_{61} = 0.560$	
C_{62}	$\widetilde{r}_{62} = (0.539, 0.654, 0.799)$	$w_{62} = 0.131$	0.089
C_{63}	$\widetilde{r}_{63} = (0.785, 0.965, 1.178)$	$w_{63} = 0.193$	0.026
C_{64}	$\widetilde{r}_{64} = (0.446, 0.557, 0.733)$	$w_{64} = 0.115$	
C_{71}	$\widetilde{r}_{71} = (1.335, 1.578, 1.769)$	$w_{71} = 0.293$	
C_{72}	$\widetilde{r}_{72} = (1.196, 1.374, 1.585)$	$w_{72} = 0.260$	
C_{73}	$\widetilde{r}_{73} = (0.758, 0.922, 1.084)$	$w_{73} = 0.174$	0 0.053
C_{74}	$\widetilde{r}_{74} = (0.696, 0.789, 0.922)$	$w_{74} = 0.151$	
C_{75}	$\widetilde{r}_{75} = (0.565, 0.634, 0.749)$	$w_{75} = 0.122$	
C_{81}	$\widetilde{r}_{81} = (1.682, 1.968, 2.213)$	$w_{81} = 0.790$	0
C_{82}	$\widetilde{r}_{82} = (0.452, 0.508, 0.595)$	$w_{82} = 0.210$	0

Finally, a pairwise comparison matrix was produced between the scale levels of each descriptor. In this case crisp numbers have been used, obtaining the weightings of each scale level for each descriptor; these were turned into utility vectors, of which some examples can be seen in Table 5; each component of the utility vector is associated with a level of the descriptor as shown in Table 1.

Table 5. Utility vectors for the scale levels of the descriptors of the criterion C_4 (characteristics of the head of maintenance and maintenance tasks performed outside working hours).

Descriptor	Utility Vector
C_{41}: Length of time as maintenance manager	(1, 0.583, 0.379, 0.173, 0.059, 0)
C_{42}: Academic qualifications of maintenance manager	(1, 0.682, 0.390, 0.219, 0.219, 0.095, 0.036, 0)
C_{43}: Experience in maintenance positions	(1, 0.583, 0.358, 0.176, 0.065, 0)
C_{44}: Frequency of attendance outside working hours	(1, 0.652, 0.418, 0.174, 0.056, 0)
C_{45}: Remuneration for work in overtime or outside working hours	(1, 0.219, 0)
C_{46}: Attendance at conferences, talks, seminars, etc. on maintenance	(1, 0.387, 0.121, 0)
C_{47}: Consults Spanish technical journals on maintenance	(1, 0)
C_{48}: Use of internet to search for information to help solve maintenance problems	(1, 0)
C_{49}: Consults international technical journals on maintenance	(1, 0)

All the pairwise comparison matrices used in the multicriteria model have consistency ratios below 10%.

4.4. Definition of Alternatives

The alternatives are the industrial sectors assessed in the questionnaire [79]: Automobiles and auxiliary industry, Buildings, Building materials, Chemical industry, Diverse firms, Electromechanical constructions, Energy, Food, Iron and steel and Mining and Transport. In addition, in each sector, the companies were classified by size into: companies with up to 200 workers, from 201 to 500 workers, and over 500 workers.

The model uses the data collected by the SMA in the 2005 and 2010 surveys to construct the discrete probability distributions. In the 2005 survey, 2343 questionnaires were sent out, and 254 were completed and returned. If these, 113 were from companies with up to 200 workers, 85 from companies with from 201 to 500 workers, and 56 from companies with over 500 workers. In the 2010 survey 1648 questionnaires were sent out, of which 152 were completed and returned. Of these, 74 were from companies with up to 200 workers, 36 were from companies with from 201 to 500 workers, and 42 were from large companies with over 500 workers. The questionnaires were filled out by the heads of the maintenance department in each company.

The information obtained from the survey was turned into probabilities, and so each measurement level of a subcriterion is associated with the probability of appearance of each answer in the sector analysed. Since there are 40 subcriteria, 10 activity sectors and three company sizes, 1200 discrete probability distributions were calculated for each year evaluated (2400 discrete probability distributions in all). Figure 3 shows the discrete probability distribution for the subcriterion length of time as maintenance manager (C_{41}) in companies with from 201 to 500 workers for the year 2010.

Next the steps for developing the definition of alternatives are set out:

Step 1. Review the 2005 and 2010 SMA surveys.

Step 2. Check the data sample.

Step 3. Define the alternatives (10 activity sectors and three company sizes by sector, in total 30 alternatives).

Step 4. Calculate the probabilities by level of scale of each descriptor by alternative.

Step 5. Construct discrete probabilities distributions by alternative.

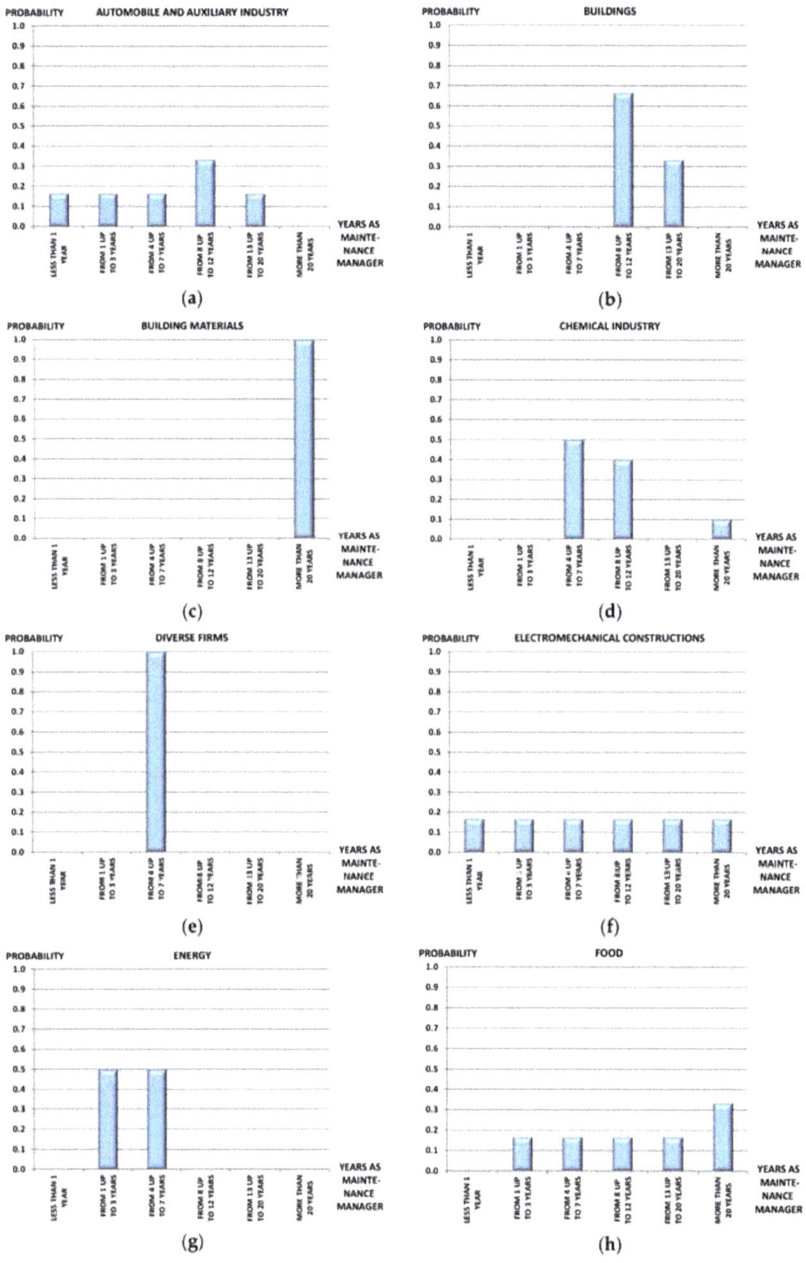

Figure 3. *Cont.*

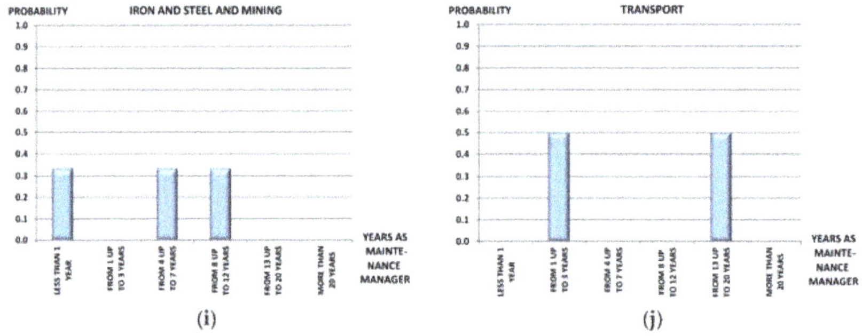

Figure 3. Discrete probability distributions for the subcriterion Length of time as maintenance manager (C_{41}) in companies with from 201 to 500 workers in 2010. (**a**) Automobile and auxiliary industry; (**b**) Buildings; (**c**) Building materials; (**d**) Chemical industry; (**e**) Diverse forms; (**f**) Electromechanical constructions; (**g**) Energy; (**h**) Food; (**i**) Iron and steel and mining; (**j**) Transport.

4.5. Multi-Attribute Utility Model

Multi-Attribute Utility Theory (MAUT) [47] allows scores to be turned into utility functions. If w_i is the weighting associated with the criterion i and $\sum_i^n w_i = 1$ is satisfied, an alternative has an additive utility function U:

$$U = \sum_i^n w_i u_i(x_i) \tag{14}$$

where x_i is typically normalised to a range from the worst to best possible values of an descriptor, and u_i ranging from 0 to 1 reflects the decision maker's attitude to risk within criterion i.

If $p(x)$ is the probability associated with each scale level of an descriptor in an alternative and, $U(x)$ is the utility associated with that scale level, the value of equivalent certainty for each alternative is obtained from the expected utility summing $p(x) \times U(x)$ for all levels x with non-zero probability for the probability distribution. The equivalent certainty is the level estimated in which the utility function of the result $U(y)$, is equal to the expected utility of the random utility. The final utility of a criterion in an alternative is calculated by a Multi-Measure Utility Function. The Multi-Measure Utility Function is obtained by multiplying the weights of each subcriterion by the $U(y)$ previously obtained for each subcriterion.

The procedure followed in this stage, to obtain the final results is:

Step 1. Calculate the utility function by subcriterion.
Step 2. Calculate the Multi-Measure Utility Function by criterion.
Step 3. Construct a model with the intermediate results obtained in Logical Decisions.
Step 4. Apply a Monte Carlos simulation to get the final by alternative.
Step 5. Perform sensitivity analysis.
Step 6. Analyse the results.

The utility function of the subcriterion length of time as maintenance manager (C_{41}), $U_{C_{41}}(y)$, is calculated:

$U_{C_{41}}(y)$ = utility (best scale level) $\times$ probability (appearance of best scale level in sector) + ... + utility (worst scale level) $\times$ probability (appearance of worst scale level in sector).

The best scale level of the subcriterion C_{41} (see Table 1) consists in the years of experience of the head of maintenance being more than 20; the worst level is less than one year of experience. Figure 3 shows that a company with between 201 and 500 employees belonging to the Automobile and auxiliary industry in 2010 has a probability of 0.167 that the experience of the head of maintenance in that post is less than one year; a similar probability is found for experience between one and three years, between four and seven years, and between 13 and 20 years. The probability of having a head of maintenance with between eight and 12 years of experience is 0.333. The utility associated with each measurement level for each descriptor is shown in Table 5. Thus, the resulting $U_{C_{41}}(y)$ is:

$$U_{C_{41}}(y) = 1 * 0 + 0.583 * 0.167 + 0.379 * 0.333 + 0.173 * 0.167$$
$$+0.059 * 0.167 + 0 * 0.167 = 0.262$$

A similar process is followed with the subcriteria C_{42}, C_{43}, C_{44}, C_{45}, C_{46}, C_{47}, C_{48} and C_{49} giving (see utilities in Table 5):

$$U_{C_{42}}(y) = 1 * 0.333 + 0.682 * 0 + 0.390 * 0 + 0.219 * 0.5 + 0.219 * 0 + 0.095 * 0 + 0.036 * 0.167$$
$$+0 * 0 = 0.449$$

$$U_{C_{43}}(y) = 1 * 0.333 + 0.583 * 0 + 0.358 * 0.5 + 0.176 * 0 + 0.065 * 0 + 0 * 0.167 = 0.512$$

$$U_{C_{44}}(y) = 1 * 0 + 0.652 * 0.167 + 0.418 * 0.667 + 0.174 * 0 + 0.056 * 0.167 + 0 * 0 = 0.397$$

$$U_{C_{45}}(y) = 1 * 1 + 0.219 * 0 + 0 * 0 = 1$$

$$U_{C_{46}}(y) = 1 * 0.5 + 0.387 * 0.333 + 0.121 * 0 + 0 * 0.167 = 0.629$$

$$U_{C_{47}}(y) = 1 * 0.5 + 0 * 0.5 = 0.5$$

$$U_{C_{48}}(y) = 1 * 1 + 0 * 0 = 1$$

$$U_{C_{49}}(y) = 1 * 0.167 + 0 * 0.833 = 0.167$$

The weightings of the subcriteria included in criterion C_4 (Characteristics of the head of maintenance and maintenance tasks performed outside working hours) are (see Table 4): $w_{41} = 0.359$, $w_{42} = 0.216$, $w_{43} = 0.162$, $w_{44} = 0.079$, $w_{45} = 0.046$, $w_{46} = 0.043$, $w_{47} = 0.033$, $w_{48} = 0.033$ and $w_{49} = 0.029$, respectively. The utility in the criterion characteristics of the head of maintenance and maintenance tasks performed outside working hours is calculated as follows:

$$\text{Utility (Characteristics of the head of maintenance}\ldots) = w_{41} * U_{C_{41}}(y) + w_{42} * U_{C_{42}}(y)$$
$$+w_{43} * U_{C_{43}}(y) + w_{44} * U_{C_{44}}(y) + w_{45} * U_{C_{45}}(y) + w_{46} * U_{C_{46}}(y) + w_{47} * U_{C_{47}}(y) + w_{48} * U_{C_{48}}(y)$$
$$+w_{49} * U_{C_{49}}(y) = 0.359 * 0.262 + 0.216 * 0.449 + 0.162 * 0.512 + 0.079 * 0.397$$
$$+0.046 * 1 + 0.043 * 0.629 + 0.033 * 0.5 + 0.033 * 1 + 0.029 * 0.167 = 0.433$$

A similar process is used for the other criteria.

The calculations related to the application of FAHP and the obtaining of the discrete probability distributions have been carried out using a program in Excel. The results obtained in both cases have been included in a model constructed by means of Logical Decision software. In this way it was possible to apply the Monte Carlo simulation to calculate the global results and the uncertainty associated.

The Monte Carlo simulation allows the estimation of the uncertainty of a number that is a complex function of one or more probability distributions. The Monte Carlo simulation uses random numbers to provide an estimation of the distribution. A generator of random numbers is used to produce random samples of the probability levels. Each set of samples is used to calculate the utility of a possible result of the uncertainties of each scale of the descriptor. The execution of a certain number of trials is used as an estimation of the accumulated probability distribution of the desired utility.

To apply the Monte Carlo simulation a different number of trials have been considered. After 5000 trials the results no longer change.

A sensitivity analysis was performed, increasing and decreasing the weightings of the decision criteria used in the model by 5%. The sensitivity analysis shows that there is only a change in the classification of alternatives in companies of up to 200 workers. The change appears when the weighting associated with the criteria Maintenance costs and Certification to international standards and compliance with the law on health and safety at work is reduced by 5%; in this case the alternatives Automobile and auxiliary industry and Building materials change positions, taking the fourth and fifth positions, respectively. In the case of increasing the weighting of the criterion Maintenance management by 5% a similar exchange is observed between the alternatives Automobile and auxiliary industry and Building materials. Companies with from 201 to 500 workers and companies with more than 500 workers see no change in the full classification of alternatives.

There are only eight modifications in the classification of alternatives (in all sizes of company) when the variation in the weightings of the criteria is 10%. In all cases the variation in the classification is simply a permutation of alternatives occupying adjacent positions. It can therefore be stated that the model is robust.

There follows an analysis of the development of maintenance in each size of company, the sectors with the best and worst performance before and after the economic crisis, and the results will be compared with those of the same sector in companies of different sizes.

5. Results

5.1. Companies with up to 200 Employees

Figure 4 shows the results for all sectors for companies with up to 200 workers in 2005 and 2010. In the results for 2005 it can be seen that the Automobile and auxiliary industry sector is in first place with a utility of 0.6896, followed by Electromechanical constructions (0.6683) and the Food sector (0.6630). The Iron and Steel and mining, Chemical industry and Food sectors are those that best apply maintenance in 2010, with utilities of 0.6701, 0.6350 and 0.6214, respectively.

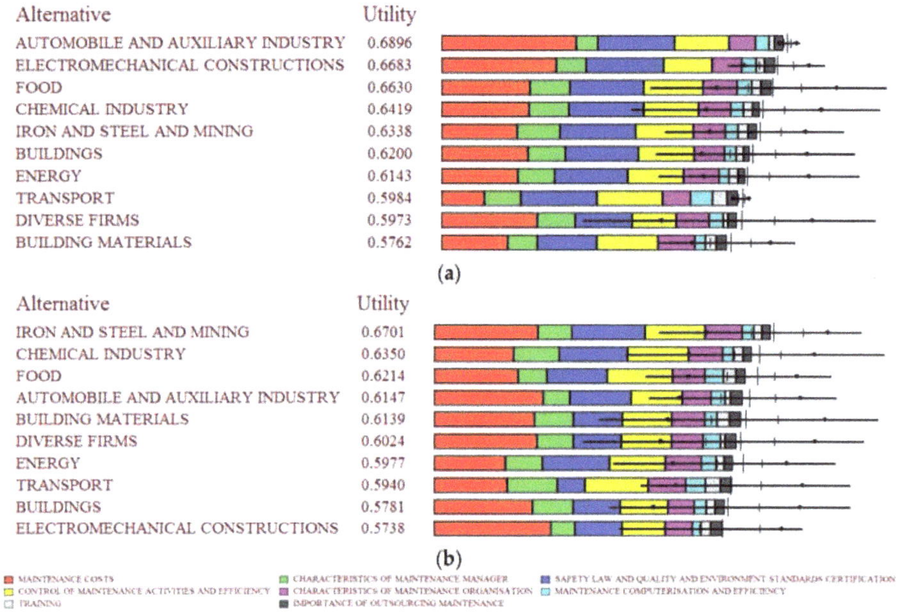

Figure 4. Classification of the state of maintenance by industrial sectors: (**a**) Description of companies with up to 200 workers in 2005; (**b**) Description of companies with up to 200 workers in 2010.

It can be seen that there is a decrease in most sectors in the level of maintenance applied in Spain in 2010 with respect to 2005; Building Materials has an increase in utility of 6.54%, Iron and steel and mining have 5.72% and Diverse firms have 0.85%. The sectors Electromechanical constructions, Automobile and auxiliary industry and Buildings have had the greatest decrease in maintenance (−14.14%, −10.86% and −6.75%, respectively).

The sector Building Materials (cement, building steel, facilities, furniture, carpentry and medical apparatus), has an average number of workers of 25, that is, small businesses [85]. This sector has been seriously affected by the economic recession, as it is directly related to the building industry. For example, consumption of cement has dropped by 50% in Spain since the second quarter of 2007. The improvement in maintenance undergone by the sector could be due to the essential optimisation carried out by companies which are still operating, which includes the area of maintenance.

Total maintenance costs is a criterion that clearly differentiates sectors. Building materials shows a rise in utility (reduction in total maintenance costs) in 2010 of 239.96%. In 2005, 75% of the companies surveyed in this sector had maintenance costs between €2,001,000 and €4,500,000; in 2010, however, 60% of companies in the sector were not above €900,000. The Iron and steel and Mining, Buildings, Transport, Diverse firms and Electromechanical constructions sectors had an increase in utility in this criterion of 87.57%, 38.93%, 31.58%, 24.73% and 13.29%. The most significant decrease in utility in total maintenance costs was in the Chemical industry with −22.72%, followed by the Automobile and auxiliary industry with −21%. In the Chemical industry, in 2005, 82.61% of companies had less than two million euros a year in maintenance costs; in 2010, however, only 58.82% of companies were below this figure.

The Building materials, Automobile and auxiliary industry, Electromechanical construction, Iron and Steel and mining sectors had a utility of 1.000 in 2005 and 2010 with respect to certification to standard ISO 9000. The Energy sector, on the other hand, had the worst results for 2005 as 50% of those surveyed were not certified, and in 2010 transport was worst with 66.67% of those surveyed not certified. In this last case, it may be due to the fact that companies are more concerned with the standard EN13816, which is specific to the sector.

Planned maintenance work is four to 12 times more efficient than unplanned work [86]; therefore, the percentage of hours spent on corrective maintenance gives an idea of the efficiency of a sector in maintenance activity. From these results, it can be seen that the mode of all sectors in 2010 has levels of corrective maintenance higher than 15%; this the highest level permitted by the best practice benchmark [87]. Diverse firms are the sector with the largest increase in utility (decrease in corrective action) in the percentage of corrective maintenance (127.01%), followed by Energy (95.03%) and Building materials (45.85%). Transport, Automobile and auxiliary industry and Electromechanical constructions, on the other hand, have seen the largest decrease in utility, with −91.73%, −84.76 and −74.58%, respectively. For example, in 2005 in the Automobile and auxiliary industry 100% of those surveyed applied 50% corrective maintenance, while in 2010, 100% of those surveyed applied more than 75%. This increase in corrective action means more and more serious breakdowns, higher spare part and labour costs, as well as having a negative effect on the availability, safety and quality of the plant.

The experience of the head of maintenance is vital to improvements in the state of maintenance. Building materials experienced an increase in utility from 2005 to 2010 of 800.68%. This is because, while in 2005 100% of those surveyed in the sector had less than three years of experience, in 2010, 60% of those surveyed had more than eight years of experience. Other sectors with a noteworthy increase in utility are Transport (198.62%), Automobile and auxiliary industry (56.72%), Buildings (41.61%), Electromechanical constructions (39.64%) and Chemical industry (28.24%). The Food sector, however, had a variation in utility of −56.51%.

It is seen that the level of qualification of heads of maintenance was, in general, higher in 2005 than in 2010. In 2005 in the Transport sector all those surveyed were industrial engineers, while in 2010 33.33% of the heads of maintenance had no qualifications. It should be noted that the Automobile and auxiliary industry has a much lower utility than the other sectors in this criterion in 2005 and 2010 (0.0362 and 0.0241) as, surprisingly, the heads of maintenance have no university qualifications;

rather, they are industrial technicians and in some cases they have no qualifications at all. These results are much worse than those in Carnero [88] for small businesses in Spain where only 20% of heads of maintenance has a university qualification.

A key aspect in maintenance is the level of satisfaction with a CMMS. In this matter, the Energy sector is the one with the biggest improvement, with an increase in utility in this criterion (C_{61}) of 105.65%. This is clear because only 9.09% of those surveyed were very satisfied with the CMMS in 2005, while in 2010 40% of them were. Other sectors with an increase in utility in this criterion are the Automobile and auxiliary industry and Food with 43.19% and 20.13%, respectively. Buildings, on the other hand, was the sector with the biggest drop in utility in this criterion (−25.9%).

In the criterion outsourcing, there are small increases in utility in the Food and Transport sectors, while the other sectors show decreases in utility with Buildings having the highest value of −34.56%. Thus, companies in general have not worked on improving the outsourcing of maintenance.

5.2. Companies with 201 to 500 Employees

Figure 5 shows the results for companies with between 201 and 500 employees in 2005 and 2010. The 2005 results show that the Building sector was in first place with a utility of 0.6476, followed by Iron and steel and mining (0.6354) and the Chemical industry (0.6316).

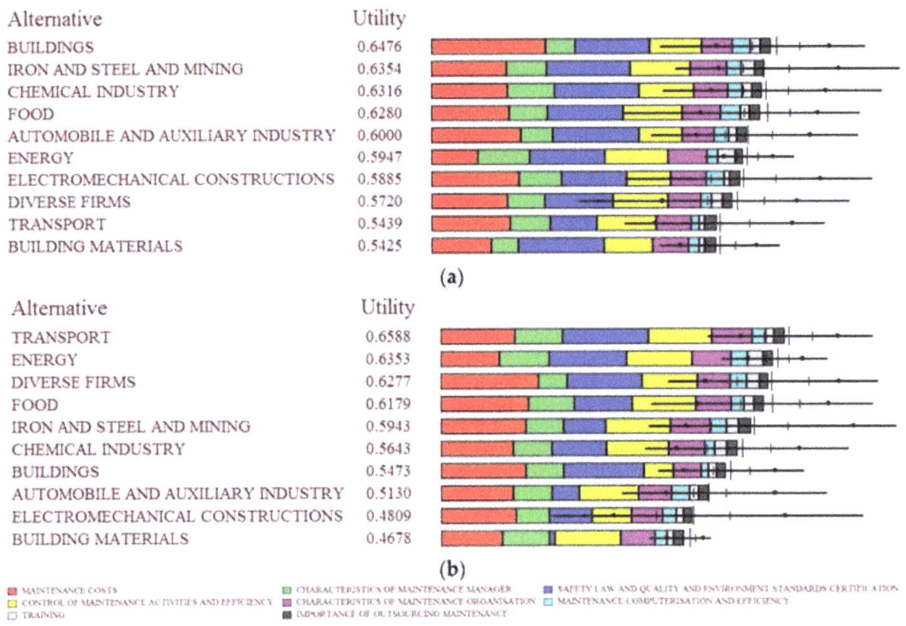

Figure 5. Classification of the state of maintenance by industrial sectors: (**a**) Companies with 201–500 workers in 2005; (**b**) Companies with 201–500 workers in 2010.

The Transport sector was in first place in 2010, followed by Energy and Diverse firms. The Chemical industry and Building sectors considerably reduce the level of maintenance, taking 6th and 7th place, respectively. Building materials, which was in last place in 2005 with a utility of 0.5425, retained that position in 2010, although the utility fell to 0.4678.

The Transport, Diverse firms and Energy sectors underwent improvements in maintenance (increased utility) from 2005 to 2010, which can be quantified as 21.13%, 9.73% and 6.82%, respectively. The remaining sectors see a drop in the level of maintenance, which is largest in Electromechanical constructions with −18.28%, Buildings with −15.49% and Automobile and auxiliary industry with −14.50%.

Despite the economic crisis in Spain that began in 2008, not all sectors have worked to reduce their total maintenance costs. The Energy sector had an improvement in utility in 2010 with respect to 2005 of 400.93%; despite this improvement, it is the sector with the lowest utility, that is, the worst behaviour in costs both in 2005 and 2010. Also, 80% of those surveyed had total maintenance costs above €9,000,000 in 2005, whereas 100% had costs between €4,501,000 and €9,000,000 in 2010. Nevertheless, the results would appear to show that they are aware of the need to continue working on this criterion. Other sectors that have improved in overall maintenance costs are Diverse firms (105.85%), Iron and steel and mining (43.72%), Building materials (40.54%) and Food (33.64%). Transport has seen no changes in costs. Buildings, on the contrary, has seen a decrease in utility of costs of 41.76%; this may be due to the changes in the laws on sustainable buildings in Spain, which state the minimum requirements for energy efficiency, but whose facilities may require more sophisticated and therefore more expensive maintenance. Despite this, it is one of the sectors with the lowest maintenance costs. Other sectors that have worsened with respect to costs are the Automobile and auxiliary industry, Chemical industry and Electromechanical constructions. Change in the behaviour of the Chemical industry in small companies is very similar to companies with from 201 to 500 employees, with decreases in the utility of total maintenance costs of approximately 20%; this could be because the Chemical industry sector is made up of increasingly complex plants, with an ever higher level of automation operating in extreme conditions, requiring high availabilities and where the regulations on safety and the environment require very precise and delicate maintenance activities. The results for this sector in Spain, however, are very varied; while most sectors are concentrated in relatively few measurement levels of the descriptor, the Chemical industry has companies with all the measurement levels.

The Buildings, Iron and Steel and mining and Transport sectors had a utility of 1.000 in 2005 and 2010 with respect to certification to standard ISO 9000. The Energy sector, on the other hand (the same as in companies with fewer than 200 workers), had the lowest utility in 2005 since 40% of those surveyed were not certified; in 2010, the utility was still low (0.5), with only 50% of companies certified to ISO 9000. Building materials, with excellent results in 2005 (utility 1.0) since 100% of those surveyed were certified, had none of them certified to ISO 9000 in 2010.

The experience of the head of maintenance is vital to improvement in the state of maintenance. Building materials shows a decrease in utility from 2005 to 2010 of 477.03% (similar results are found in companies with up to 200 workers). This is because, while in 2005 100% of those surveyed in this sector had from four to seven years of experience as head of maintenance, in 2010, 100% of those surveyed had more than 20 years of experience. Other sectors with significant improvement in utility in 2010 were Buildings (78.02%) and Food (15.81%). The other sectors saw a drop in utility, which was highest in Energy and Diverse firms with values of −60%.

It can be seen that the level of training of heads of maintenance was higher, in general, in 2010 than in 2005 (the opposite of the situation in companies with up to 200 workers). The highest utility (1.000) in 2005 and 2010 was in the Energy sector, as 100% of those surveyed had industrial engineers as heads of maintenance. Transport saw substantial improvement in utility from 2005 (0.0000) to 2010 (0.6095); this was because in 2005 the heads of maintenance surveyed had no qualifications, while in 2010 50% of them were industrial engineers and the other 50% were technical industrial engineers.

With respect to the criterion of satisfaction with a CMMS, the Energy sector had the highest improvement (the same was the case with companies of up to 200 workers), giving an increase in utility of 299.62% in 2010 over 2005. This is because in 2005, 40% of those surveyed were not satisfied with the CMMS, while in 2010 there were no dissatisfied companies. Other sectors with increased utility are Transport, Diverse Firms, Building materials, Iron and Steel and mining and Automobile and auxiliary industry with 139.77%, 144.79%, 49.97%, 23.90% and 16.68%, respectively.

Buildings, on the other hand, was the sector with the highest drop in utility for this criterion (−75.09%), the same as in companies with up to 200 workers.

With regard to outsourcing of maintenance, the sector with the highest utility was Transport in 2005 and Iron and steel and mining in 2010. There was an increase in outsourcing in most sectors (Buildings, Electromechanical constructions, Transport, Automobile and auxiliary industry, Diverse firms and Food). The Iron and Steel and mining, Energy, and Chemical industry sectors had the highest increases in utility, with 29.03%, 16.97% and 5.53%, respectively.

With respect to the criterion of training; the Energy sector had excellent utilities (0.8905 in 2005 and 0.8310 in 2010). Electromechanical constructions, Building materials, and Transport, on the other hand, had very low utility in both 2005 and in 2010. There were, however, improvements in utility in the Transport sector (34.35%), and in the Automobile and auxiliary industry, Diverse firms, Chemical industry and Electromechanical constructions. Iron and steel and mining had the largest decrease in utility (−10.10%).

5.3. Companies with over 500 Employees

Figure 6 shows the results for companies with over 500 employees in 2005 and 2010. In 2005, Building materials had better maintenance, with a utility of 0.6359, followed by the Automobile and auxiliary industry (0.6021) and Transport (0.5960). The Energy sector had the worst results with a utility of 0.4482, but it can be seen that the uncertainty in the results is very high; this is due to the fact that no company from this sector filled out the questionnaire, so the results were obtained using constant probability distributions. However, in 2010, five companies filled out the survey, giving as a result the best level of maintenance with a utility of 0.6272.

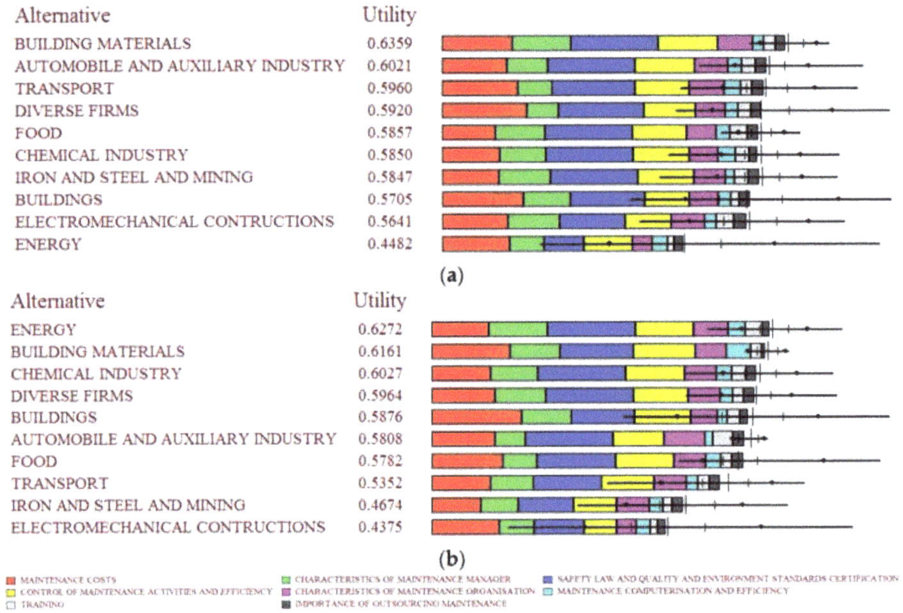

Figure 6. Classification of the state of maintenance by industrial sectors: (**a**) Companies with more than 500 workers in 2005; (**b**) companies with more than 500 workers in 2010.

Like in companies with up to 200 workers and with from 201 to 500 employees, the level of maintenance improved in a few sectors. Energy improved 39.94%. The Chemical industry had an improvement in maintenance of 3.03%, Buildings of 3.00% and Diverse Firms of 0.74%.

However, the Electromechanical constructions, Iron and steel and mining and Transport sectors showed variations in the level of maintenance of −22.44%, −20.06% and −10.20%, respectively.

In general, the results are slightly poorer, as these companies have their own organisational structure for overseeing compliance with Spanish law on Health and Safety at work. This significantly affects maintenance since maintenance workers are more exposed to noise, vibrations, different kinds of radiation, dangerous substances, vapours and gases, heat in summer, cold in winter and high humidity than those in other occupations [89]. In fact, 14–17% of accidents in Spain are related to maintenance activities (the only EU countries with a higher percentage are Belgium with 20% and Finland with 18–19%) [90]. It should also be remembered that bad maintenance can cause accidents in the workplace. The sectors with the highest utility (1.000) in 2005 and 2010 were the Automobile and auxiliary industry, Building materials, Chemical industry, Diverse firms and Food. Energy also had the highest utility in 2010 (in 2005 there were no companies surveyed). The Electromechanical constructions sector had bad results in both 2005 and 2010, with a variation in utility of −14.92%.

In general, the utilities of the criterion total maintenance costs are worse in large companies than in other sizes of company; these results do not agree with Komonen (2002) [23], who says that maintenance costs may triple as the size of a plant decreases; this could be caused by the fact that maintenance in Spain is considered a necessary evil instead of an area that can be optimised and can provide competitive advantage, and so by applying economies of scale, costs may be reduced. Energy was the sector with worst utility in 2010, a result that was repeated in companies with up to 200 workers. Food saw the most important reduction in maintenance costs, as 100% of those surveyed in 2005 had maintenance costs above nine million euros, while in 2010 50% had maintenance costs below 4.5 million euros. Other sectors that reduced their maintenance costs were Electromechanical constructions, Building materials, Chemical industry and Buildings with utility increases of 127.28%, 100%, 75.17% and 32.31%, respectively.

With regard to the percentage of corrective maintenance performed, the Building materials sector had the highest increase in utility (reduction in corrective maintenance) with 2444.53%, followed by Electromechanical constructions (248.71%) and Building (154.86%). Automobile and auxiliary industry, Transport (the same as in companies with up to 200 employees) and Diverse firms, on the other hand, are the sectors that had the biggest drop in this criterion (increase in corrective actions) with −88.13%, −42.62 and −25.61%, respectively.

In experience of heads of maintenance, Building materials had a significant increase in utility in the other company sizes, but did not show the same trend in large companies, where it decreased by −78.09%. It appears that this sector, one of those that has suffered the most in Spain in the recession, has replaced its heads of maintenance with more qualified professionals, as shown by the fact that 100% of heads of maintenance in 2010 were industrial engineers. The same applies to the Automobile and auxiliary industry. The sectors with significant increases in utility in 2010 were Transport, Diverse firms, Food, Energy, Chemical industry and Buildings.

As with companies with from 201 to 500 workers, in general a slight increase in utility can be seen in 2010 with respect to 2005 in the level of training of the heads of maintenance. As well as the Building materials and Automobile and auxiliary industry sectors, already referred to, the Energy (202.94%), Electromechanical constructions (108.92%), Diverse firms (67.65%) and Buildings (19.69%) sectors also underwent improvements in utility. The Food sector, however, showed a decrease in utility of 82.67%, as 25% of those surveyed do not have a university qualification.

Outsourcing has increased in utility in the Chemical industry (25.97%), Iron and steel and mining (11.30%), Food (10.17%) and Diverse firms (2.04%); as with other company sizes, the increase in utility is small compared with other criteria. Building materials had the largest drop in utility in this criterion (−63.61%), probably because of the serious recession in the sector, leading to a reduction in subcontracted services.

Satisfaction in the introduction of a CMMS saw an increase in utility of 279.51% in Building materials, 239.61% in Iron and steel and mining and 46.60% in the Food sector. The Automobile and auxiliary industry had the greatest decrease in utility (−100%) as 100% of those surveyed in 2010 state that the introduction has not achieved the expected results.

With respect to the criterion of qualifications, the Energy sector had the greatest increase in utility (119.49%) in 2010 with respect to 2005. The Automobile and auxiliary industry, Buildings, Transport and Chemical industry sectors also showed improvements with increases in utility of 46.67%, 31.52%, 17.20% and 6.31%, respectively. Electromechanical constructions, on the other hand, had the largest decrease in utility (−56.67%).

6. Discussion

The model presented in this article integrates fuzzy AHP and a Multi-Attribute Utility Theory to assess the state of maintenance in a country by industrial sectors and study its development both overall and by criteria and subcriteria. Furthermore, this model will facilitate the application of benchmarking by comparing the practice of the best sector for a criterion with other sectors. All of this could contribute to the application of tools like benchmarking in the area of maintenance, where the quality of contributions is lower than other areas, and especially in Spain, where the application of benchmarking is much poorer than in other countries.

It should also be noted that the multicriteria model described uses all the results obtained on maintenance in each sector of activity, instead of the mean values generally applied, favouring a more accurate assessment. The use of fuzzy AHP allows weightings to be obtained for the most suitable criteria and subcriteria, as the decision makers have doubts and uncertainties in their judgements, because the importance of the criteria can change slightly over time or by sector analysed.

In general, the level of maintenance is higher in small businesses than in medium and large businesses, where the utility values are lower; this was true both in 2005 and 2010. In large companies, Energy, Chemical industry, Buildings and Diverse Firms had improvements in utility, while this improvement is seen in the Building materials, Iron and Steel and mining and Diverse firms sectors in small enterprises and in Transport, Diverse firms and Energy in companies with from 201 to 500 employees.

Instead of improving maintenance in 2010 with respect to 2005, there is a reversal in some sizes of company. The economic crisis suffered in Spain since 2008 has, then, negatively affected the level of maintenance applied, instead of being considered an area that could provide cost reduction and improvements in productivity and quality to organisations. In general, behaviour with respect to maintenance is seen to be more similar in companies with up to 200 workers and with 201 to 500 workers than in large companies. As the size of a company increases, higher utility values are seen in both the 2005 and 2010 results.

By sectors, in the case of large companies it should be noted that Building materials had the highest utility in 2005 and the second highest in 2010. It should also be noted that the Energy sector had the highest utility in 2010.

Unlike the situation with large companies, in companies with from 201 to 500 employees, Building materials was the sector with the worst utility in 2005 and 2010, a position it also occupied in the case of small businesses in 2005. Thus, it seems that this sector, for small and medium enterprises, is where maintenance has suffered most as a result of the recession. Transport also has a very poor level of maintenance in small businesses and a variable situation in medium enterprises, where it occupied the penultimate position in 2005 and the top position in 2010. In large companies, it was in third place in 2005 and in last place in 2010. All this is worrying since maintenance applied in this sector can have an important influence on equipment anomalies that might affect a large number of users.

Electromechanical constructions is a sector where utility decreases as the size of the company increases. In 2010, in companies with up to 200 workers and large companies, it had the smallest utility and in companies with from 201 to 500 workers, only Building materials had lower utility.

The Automobile and auxiliary industry, a sector that has pioneered the application of maintenance policies like TPM, showed a worsening of behaviour in 2010. Thus, while in companies with up to

200 workers it had the highest utility in 2005 and the fourth place in 2010, in companies with from 201 and 500 employees it went from fifth place in 2005 to eighth place in 2010. In large companies it went from second place in 2005 to sixth place in 2010. So, although it is a sector with high dedication to maintenance, it suffered a significant decrease with respect to the pre-2005 levels.

In future work the intention is to update the model with the behaviour recorded in 2015 or 2020, if new national surveys are carried out. There is also an intention to ascertain whether other countries, particularly EU countries, have suffered the same decline in maintenance as Spain as a result of the economic crisis. There is a further idea of applying the Fuzzy TOPSIS technique instead of FAHP to the model to test the results obtained.

Acknowledgments: This research was supported by the Junta de Comunidades de Castilla-La Mancha and the European Regional Development Fund under Grant number PPII-2014-013-P.

Conflicts of Interest: The author declares no conflict of interest.

References

1. Alsyouf, I. The role of maintenance in improving companies' productivity and profitability. *Int. J. Prod. Econ.* **2007**, *105*, 70–78. [CrossRef]
2. Mazidi, P.; Tohidi, Y.; Sanz-Bobi, M.A. Strategic Maintenance Scheduling of an Offshore Wind Farm in a Deregulated Power System. *Energies* **2017**, *10*, 313. [CrossRef]
3. Raknes, N.T.; Ødeskaug, K.; Stålhane, M.; Hvattum, L.M. Scheduling of Maintenance Tasks and Routing of a Joint Vessel Fleet for Multiple Offshore Wind Farms. *J. Mar. Sci. Eng.* **2017**, *5*, 11. [CrossRef]
4. Carnero, M.C.; Gómez, A. A Multicriteria Model for Optimization of Maintenance in Thermal Energy Production Systems in Hospitals: A Case Study in a Spanish Hospital. *Sustainability* **2017**, *9*, 493. [CrossRef]
5. Bazrafshan, M.; Hajjari, T. Diagnosing maintenance system problems: Theory and a case study. *Qual. Reliab. Eng. Int.* **2012**, *28*, 594–603. [CrossRef]
6. Raouf, A. On evaluating maintenance performance. *Int. J. Qual. Reliab. Manag.* **1993**, *10*, 33–36. [CrossRef]
7. De Groote, P. Maintenance performance analysis: A practical approach. *J. Qual. Maint. Eng.* **1995**, *1*, 4–24. [CrossRef]
8. Martorell, S.; Sanchez, A.; Muñoz, A.; Pitarch, J.L.; Serradell, V.; Roldan, J. The use of maintenance indicators to evaluate the effects of maintenance programs on Npp performance and safety. *Reliab. Eng. Syst. Safe* **1999**, *65*, 85–94. [CrossRef]
9. Löfsten, H. Measuring maintenance performance-in search for a maintenance productivity index. *Int. J. Prod. Econ.* **2000**, *63*, 47–58. [CrossRef]
10. Chan, K.T.; Lee, R.H.K.; Burnett, J. Maintenance performance: A case study of hospitality engineering systems. *Facilities* **2001**, *19*, 494–503. [CrossRef]
11. Kumar, U. Development and implementation of maintenance performance measurement system: Issues and Challenges. In Proceedings of the 1st World Congress on Engineering Asset Management (WCEAM), Gold Coast, Australia, 11–14 July 2006.
12. Parida, A.; Kumar, U. Maintenance performance measurement (MPM): Issues and challenges. *J. Qual. Maint. Eng.* **2006**, *12*, 239–251. [CrossRef]
13. Muchiri, P.; Pintelon, L. Performance measurement using overall equipment effectiveness (OEE): Literature review and practical application discussion. *Int. J. Prod. Res.* **2008**, *46*, 3517–3535. [CrossRef]
14. Tsarouhas, P.H. Evaluation of overall equipment effectiveness in the beverage industry: A case study. *Int. J. Prod. Res.* **2013**, *51*, 515–523. [CrossRef]
15. Kumar, U.; Galar, D.; Parida, A.; Stenström, C.; Berges, L. Maintenance performance metrics: A state-of-the-art review. *J. Qual. Maint. Eng.* **2013**, *19*, 233–277. [CrossRef]
16. Pekkola, S.; Saunila, M.; Ukko, J.; Rantala, T. The role of performance measurement in developing industrial services. *J. Qual. Maint. Eng.* **2016**, *22*, 264–276. [CrossRef]
17. Shohet, I.M.; Nobili, L. Application of key performance indicators for maintenance management of clinics facilities. *Int. J. Strateg. Prop. Manag.* **2017**, *21*, 58–71. [CrossRef]
18. Spanish Maintenance Association (SMA). *The Maintenance in SPAIN*; SMA: Barcelona, Spain, 2010. (In Spanish)

19. Conde, R. El benchmarking en la industria química. In Proceedings of the Technical Meeting about Benchmarking in Industrial Maintenance, Barcelona, Spain, 21–23 May 2007.
20. Álvarez, G. Conclusiones de la encuesta AEM sobre el mantenimiento en la industria de proceso. In Proceedings of the Technical Meeting about Benchmarking in Industrial Maintenance, Barcelona, Spain, 21–23 May 2007.
21. Paredes, P. Conclusiones de la encuesta AEM sobre el mantenimiento en la industria manufacturer. In Proceedings of the Technical Meeting about Benchmarking in Industrial Maintenance, Barcelona, Spain, 21–23 May 2007.
22. González, F.J. El valor del benchmarking en mantenimiento. In Proceedings of the Technical Meeting about Benchmarking in Industrial Maintenance, Barcelona, Spain, 21–23 May 2007.
23. Komonen, K. A cost model of industrial maintenance for profitability analysis and benchmarking. *Int. J. Prod. Econ.* **2002**, *79*, 15–31. [CrossRef]
24. Dwight, R. Frameworks for Measuring the Performance of the Maintenance System in a Capital Intensive Organization. Ph.D. Thesis, University of Wollongong Thesis Collection, Wollongong, Australia, 1999.
25. Madu, C.N. Competing through maintenance strategies. *J. Qual. Reliab. Manag.* **2000**, *17*, 937–948. [CrossRef]
26. Ahrén, T.; Parida, A. Maintenance performance indicators (MPI's) for benchmarking the railway infrastructure: A case study. *Benchmarking* **2009**, *16*, 247–258. [CrossRef]
27. Simoes, J.M.; Gomes, C.F.; Yasin, M.M. A literature review of maintenance performance measurement. A conceptual framework and directions for future research. *J. Qual. Maint. Eng.* **2011**, *17*, 116–137. [CrossRef]
28. Macchi, M.; Fumagalli, L. A maintenance maturity assessment method for the manufacturing industry. *J. Qual. Maint. Eng.* **2013**, *19*, 295–315. [CrossRef]
29. Nachtmann, H.; Collins, T.; Chimka, J.R.; Tong, J. Development of a balanced scorecard for flight line maintenance activities. *J. Qual. Maint. Eng.* **2015**, *21*, 436–455. [CrossRef]
30. Van Horenbeek, A.; Pintelon, L. Development of a maintenance performance measurement framework-using the analytic network process (ANP) for maintenance performance indicator selection. *Omega Int. J. Manag.* **2014**, *42*, 33–46. [CrossRef]
31. Muchiri, A.K.; Ikua, B.W.; Muchiri, P.N.; Irungu, P.K. Development of a theoretical framework for evaluating maintenance practices. *Int. J. Syst. Assur. Eng. Manag.* **2017**, *8*, 198–207. [CrossRef]
32. Carnero, M.C. Multicriteria model for maintenance benchmarking. *J. Manuf. Syst.* **2014**, *33*, 303–321. [CrossRef]
33. Stefanovic, M.; Nestic, S.; Djordjevic, A.; Djurovic, D.; Macuciz, I.; Tadic, D.; Gacic, M. An assessment of maintenance performance indicators using the fuzzy sets approach and genetic algorithms. *Proc. Inst. Mech. Eng. B J. Eng. Manuf.* **2015**, *231*, 15–27. [CrossRef]
34. Kubler, S.; Robert, J.; Derigent, W.; Voisin, A.; Le Traon, Y. A state-of the-art survey & testbed of fuzzy AHP (FAHP) applications. *Expert Syst. Appl.* **2016**, *65*, 398–422.
35. Mardani, A.; Zavadskas, E.K.; Khalifah, Z.; Zakuan, N.; Jusoh, A.; Nor, K.M.; Khoshnoudi, M. A review of multi-criteria decision-making applications to solve energy management problems: Two decades from 1995 to 2015. *Renew. Sustain. Energy Rev.* **2017**, *71*, 216–256. [CrossRef]
36. Zamarrón-Mieza, I.; Yepes, V.; Moreno-Jiménez, J.M. A systematic review of application of multi-criteria decision analysis for aging-dam management. *J. Clean. Prod.* **2017**, *147*, 217–230. [CrossRef]
37. Ashour, O.M.; Okudan, G.E. Fuzzy AHP and utility theory based patient sorting in emergency departments. *Int. J. Collab. Enterp.* **2010**, *1*, 332–358. [CrossRef]
38. Ashour, O.M.; Okudan, G.E. Patient priorization in Emergency Departments: Decision Making under Uncertainty. In *Decision Making in Service Industries: A Practical Approach*; Faulin, J., Juan, A.A., Grasman, S.E., Fry, M.J., Eds.; Taylor & Francis Group (CRC Press): Boca Raton, FL, USA, 2013; pp. 175–204. ISBN 9781138073685.
39. Ashour, O.M.; Okudan, G.E. A simulation analysis of the impact of FAHP-MAUT triage algorithm on the Emergency Department performance measures. *Expert Syst. Appl.* **2013**, *40*, 177–187. [CrossRef]
40. Johal, L.K.; Sandhu, A.S. Developing an efficient utility theory based VHO algorithm to boost user satisfaction in HETNETs. *Indian J. Sci. Technol.* **2016**, *9*, 928–969.
41. Goyal, R.K.; Kaushal, S.; Sangaiah, A.K. The utility based non-linear fuzzy AHP optimization model for network selection in heterogeneous wireless networks. *Appl. Soft Comput.* **2017**, in press. [CrossRef]

42. Carnero, M.C. A decision support system for maintenance benchmarking in big buildings. *Eur. J. Ind. Eng.* **2014**, *8*, 388–420. [CrossRef]
43. Wang, Y.M.; Luo, Y.; Hua, Z. On the extent analysis method for fuzzy AHP and its applications. *Eur. J. Oper. Res.* **2008**, *186*, 735–747. [CrossRef]
44. Buckley, J.J. Fuzzy hierarchical analysis. *Fuzzy Sets Syst.* **1985**, *17*, 233–247. [CrossRef]
45. Jenatabadi, H.S.; Babashamsi, P.; Yusoff, N.I. The Combination of a Fuzzy Analytical Hierarchy Process and the Taguchi Method to Evaluate the Malaysian Users' Willingness to Pay for Public Transportation. *Symmetry* **2016**, *8*, 90. [CrossRef]
46. Bozbura, F.T.; Beskese, A.; Kahraman, C. Prioritization of human capital measurement indicators using fuzzy AHP. *Expert Syst. Appl.* **2007**, *32*, 1100–1112. [CrossRef]
47. Keeney, R.L.; Raiffa, H. *Decisions with Multiple Objectives: Preferences and Value Tradeoffs*; Wiley: New York, NY, USA, 1976.
48. Ikhwan, M.A.H.; Burney, F.A. Maintenance in Saudi Industry. *Int. J. Oper. Prod. Manag.* **1994**, *14*, 70–80. [CrossRef]
49. Assaf, S.A.; Hadidi, L.A.; Hassanain, M.A.; Rezq, M.F. Performance evaluation and benchmarking for maintenance decision making units at petrochemical corporation using a DEA model. *Int. J. Adv. Manuf. Technol.* **2015**, *76*, 1957–1967. [CrossRef]
50. Jonsson, P. The status of maintenance management in Swedish manufacturing firms. *J. Qual. Maint. Eng.* **1997**, *3*, 233–258. [CrossRef]
51. Alsyouf, I. Maintenance practices in Swedish industries: Survey results. *Int. J. Prod. Econ.* **2009**, *121*, 212–223. [CrossRef]
52. Luxhøj, J.T.; Riis, J.O.; Thorsteinsson, U. Trends and Perspectives in Industrial Maintenance Management. *J. Manuf. Syst.* **1997**, *16*, 437–453. [CrossRef]
53. Holgeid, K.K.; Krogstie, J.; Sjøberg, D.I.K. A study of development and maintenance in Norway: Assessing the efficiency of information systems support using functional maintenance. *Inf. Softw. Technol.* **2000**, *42*, 687–700. [CrossRef]
54. Reiman, T.; Oedewald, P. Measuring maintenance culture and maintenance core task with CULTURE-questionnaire-a case study in the power industry. *Saf. Sci.* **2004**, *42*, 859–889. [CrossRef]
55. Pinjala, S.K.; Pintelon, L.; Vereecke, A. An empirical investigation on the relationship between business and maintenance strategies. *Int. J. Prod. Econ.* **2006**, *104*, 214–229. [CrossRef]
56. Muchiri, P.N.; Pintelon, L.; Martin, H.; De Meyer, A.M. Empirical analysis of maintenance performance measurement in Belgian industries. *Int. J. Prod. Res.* **2010**, *48*, 5905–5924. [CrossRef]
57. Cholasuke, C.; Bhardwa, R.; Antony, J. The status of maintenance management in UK manufacturing organizations: Results from a pilot survey. *J. Qual. Maint. Eng.* **2004**, *10*, 5–15. [CrossRef]
58. Connaughton, G.E. The state of the Art of maintenance in North America. In Proceedings of the International Maintenance Congress (Euromaintenance), Göteborg, Sweden, 7–10 March 2000.
59. Reliabilityweb, R.C.M. Benchmarking Report. Available online: http://www.reliabilityweb.com (accessed on 30 December 2015).
60. Wireman, T. *Bechmarking Best Practices in Maintenance Management*; Industrial Press Inc.: New York, NY, USA, 2004.
61. Wireman, T. *Bechmarking Best Practices for Maintenance, Reliability and Asset Management. Updated for ISO 55000*; Industrial Press Inc.: Norwalk, CT, USA, 2015.
62. Tse, P.W. Maintenance practices in Hong Kong and the use of the intelligent scheduler. *J. Qual. Maint. Eng.* **2002**, *8*, 369–380. [CrossRef]
63. Reis, A.C.B.; Costa, A.P.C.S.; Teixeira de Almeida, A. Planning and competitiveness in maintenance management. An exploratory study in manufacturing companies. *J. Qual. Maint. Eng.* **2009**, *15*, 259–270. [CrossRef]
64. Dowlatshahi, S. The role of industrial maintenance in the maquiladora industry: An empirical analysis. *Int. J. Prod. Econ.* **2008**, *114*, 298–307. [CrossRef]
65. Modgil, S.; Sharma, S. Total productive maintenance, total quality management and operational performance: An empirical study of Indian pharmaceutical industry. *J. Qual. Maint. Eng.* **2016**, *22*, 353–377. [CrossRef]
66. Muchiri, A.K.; Ikua, B.W.; Muchiri, P.N.; Irungu, P.K.; Kibicho, K. An evaluation of maintenance practices in Kenya: Preliminary results. *Int. J. Syst. Assur. Eng. Manag.* **2017**, 1–18. [CrossRef]

67. Jelušič, P.; Žlender, B. Discrete Optimization with Fuzzy Constraints. *Symmetry* **2017**, *9*, 87. [CrossRef]
68. Isaai, M.T.; Kanani, A.; Tootoonchi, M.; Afzali, H.R. Intelligent timetable evaluation using fuzzy AHP. *Expert Syst. Appl.* **2011**, *38*, 3718–3723. [CrossRef]
69. Chang, D.Y. Applications of the extent analysis method on fuzzy AHP. *Eur. J. Oper. Res.* **1996**, *95*, 649–655. [CrossRef]
70. Zadeh, L.A. Fuzzy sets. *Inf. Control* **1965**, *8*, 338–353. [CrossRef]
71. Kaufmann, A.; Gupta, M.M. *Fuzzy Mathematical Models in Engineering and Management Science*; North Holland: Amsterdam, The Netherlands, 1988.
72. Büyüközkan, G.; Kahraman, C.; Ruan, D. A fuzzy multicriteria decision approach for software development strategy selection. *Int. J. Gen. Syst.* **2004**, *33*, 259–280. [CrossRef]
73. Buckley, J.J.; Feuring, T.; Hayashi, Y. Fuzzy hierarchical analysis revisited. *Eur. J. Oper. Res.* **2001**, *129*, 48–84. [CrossRef]
74. Opricovic, S.; Tzeng, G.H. Defuzzification within a multicriteria decision model. *Int. J. Uncertain. Fuzziness* **2003**, *11*, 635–652. [CrossRef]
75. Chang, T.H.; Wang, T.C. Using the fuzzy multi-criteria decision making approach for measuring the possibility of successful knowledge management. *Inf. Sci.* **2009**, *179*, 355–370. [CrossRef]
76. Saaty, T.L. *The Analytic Hierarchy Process*; McGraw Hill: New York, NY, USA, 1980.
77. Forman, E.; Selly, M.A. *Decision by Objectives*; World Scientific: London, UK, 2001.
78. Durán, O. Computer-aided maintenance management systems selection based on a fuzzy AHP approach. *Adv. Eng. Softw.* **2011**, *42*, 821–829. [CrossRef]
79. Spanish Maintenance Association (SMA). *The Maintenance in Spain*; SMA: Barcelona, Spain, 2005. (In Spanish)
80. Keeney, R.L. *Value-Focused Thinking: A Path to Creative Decision Making*; Harvard: Cambridge, MA, USA, 1996.
81. Bana e Costa, C.A.; Correa, E.; De Corte, J.M.; Vansnick, J.C. Facilitating bid evaluation in public call for tenders: A socio-technical approach. *Omega Int. J. Manag.* **2002**, *30*, 227–242. [CrossRef]
82. Bana e Costa, C.A.; Ensslin, L.; Costa, A.P. Structuring the Process of Choosing Rice Varieties at the South of Brazil. In *Multicriteria Analysis for Land-Use Management*; Beinat, E., Nijkamp, P., Eds.; Springer: Amsterdam, The Netherlands, 1998.
83. Zhu, K.J.; Jing, Y.; Chang, D.Y. A discussion on Extent Analysis Method and applications of fuzzy AHP. *Eur. J. Oper. Res.* **1999**, *116*, 450–456. [CrossRef]
84. Meixner, O. Fuzzy AHP Group Decision Analysis and its Application for the Evaluation of Energy Sources. In Proceedings of the 10th International Symposium on the Analytic Hierarchy/Network Process, Pittsburgh, PA, USA, 29 July–1 August 2009.
85. Descals, A.M.; Contrí, G.B.; Saura, I.G.; Ruiz-Molina, M.E.; Vallet-Bellmunt, T. *La Distribución de Cerámica y Materiales de Construcción en España*; University of Valencia: Valencia, Spain, 2006.
86. Casto, P. Defining Best Practice for Maintenance Overtime. In Proceedings of the MARCON Conference, Knoxville, TN, USA, 15–18 February 2010.
87. Mitchell, J. *Physical Asset Management Handbook*; Chapter 2: Metrics/Measures of Performance; Clarion Technical Publishers: Houston, TX, USA, 2002.
88. Carnero, M.C. Condition Based Maintenance in small industries. In Proceedings of the 2nd International Workshop on Advanced Maintenance Engineering, Services and Technology (IFAC A-MEST), Sevilla, Spain, 22–23 November 2012.
89. National Institute of Workplace Safety and Hygiene (INSHT). VI Encuesta Nacional de Condiciones de Trabajo (VI ENCT) en España. 2007. Available online: http://www.oect.es/Observatorio/Contenidos/InformesPropios/Desarrollados/Ficheros/Informe_VI_ENCT.png (accessed on 30 December 2015).
90. European Agency for Safety and Health at Work. Maintenance and OSH-A Statistical Picture. Available online: http://osha.europa.eu/en/publications/literature_reviews (accessed on 30 December 2015).

symmetry

MDPI

Article

A Recourse-Based Type-2 Fuzzy Programming Method for Water Pollution Control under Uncertainty

Jing Liu [1], Yongping Li [1,*], Guohe Huang [2] and Lianrong Chen [1]

1 Department of Environmental Engineering, Xiamen University of Technology, Xiamen 361024, China; wsljing0909@gmail.com (J.L.); iLrchen@hotmail.com (L.C.)
2 Sino-Canada Energy and Environmental Research Center, North China Electric Power University, Beijing 102206, China; guohe.huang@outlook.com
* Correspondence: yongping.li33@gmail.com

Received: 29 September 2017; Accepted: 31 October 2017; Published: 4 November 2017

Abstract: In this study, a recourse-based type-2 fuzzy programming (RTFP) method is developed for supporting water pollution control of basin systems under uncertainty. The RTFP method incorporates type-2 fuzzy programming (TFP) within a two-stage stochastic programming with recourse (TSP) framework to handle uncertainties expressed as type-2 fuzzy sets (i.e., a fuzzy set in which the membership function is also fuzzy) and probability distributions, as well as to reflect the trade-offs between conflicting economic benefits and penalties due to violated policies. The RTFP method is then applied to a real case of water pollution control in the Heshui River Basin (a rural area of China), where chemical oxygen demand (COD), total nitrogen (TN), total phosphorus (TP), and soil loss are selected as major indicators to identify the water pollution control strategies. Solutions of optimal production plans of economic activities under each probabilistic pollutant discharge allowance level and membership grades are obtained. The results are helpful for the authorities in exploring the trade-off between economic objective and pollutant discharge decision-making based on river water pollution control.

Keywords: recourse; River Basin; stochastic; type-2 fuzzy; uncertainty; water pollution control

1. Introduction

The trade-off between water pollution control and economic development is of great concern in many basins since it is essential to local sustainable development [1,2]. It is difficult to keep the economy growing under the utilization of water resources and the deterioration of environmental conditions [3,4]. Meanwhile, it is hard to promote human society improvement if the authorities excessively restrict economic development. Under such a contradictory situation, optimization techniques are proper to detect the economic and environmental impacts of alternative pollution control actions from a system point of view, and thus aid the authorities in formulating and adopting cost-effective water pollution plans and policies. However, water pollution control planning is governed by significant sources of uncertainty associated with different variables, and uncertainty is a non-negligible constituent of such a procedure [5]. There are significant uncertainties in not only how the system might develop, but also in how the system is expected to adjust when many system components are altered (e.g., pollutant discharge amount, cost/benefit coefficient and economic activity scale).

In water pollution control problems, values of associated parameters (e.g., cost/benefit coefficients) are usually determined via tests, experiences and expertises, while these methods may fail in determining accurate values, resulting in the parameters being described by fuzzy membership functions. Such deviations in subjective estimations can lead to fuzziness being inherent in the real-world decision problems (e.g., vagueness and/or impreciseness in the outcomes of a water pollution control sample), neglect of which can cause the solutions of problems deviating greatly from their true values. Fuzzy mathematic programming (FMP), based on fuzzy sets theory, can facilitate the analysis of system associated with uncertainties being derived from vagueness and imprecision [6,7]. FMP is capable of handling decision problems under fuzzy goal and constraints and tackling ambiguous coefficients in the objective function and constraints. Previously, a wide range of FMP methods were developed for water pollution control [8–15]. For example, Liu et al. [16] improved a two-stage fuzzy robust programming model for water pollution control to address fuzzy parameters, which were represented by possibility distributions on the left- and right-hand sides of the constraints. Tavakoli et al. [17] developed an interactive two-stage stochastic fuzzy programming method to handle uncertainties expressed as fuzzy boundary intervals (i.e., the lower- and upper-bounds of intervals are presented as possibility distributions). Ji et al. [18] enhanced an inexact left-hand-side chance-constrained fuzzy multi-objective programming approach to cope with fuzziness in the constraints and objectives. Generally, the conventional fuzzy programming methods could only tackle fuzzy uncertainty with precise membership grades, which may encounter difficulty when the membership grades are also obtained as fuzzy sets.

In many real-world situations, related data such as unit net benefits of economic activities and pollutant discharge allowances are often highly uncertain, which could not be handled by the conventional fuzzy programming methods. When it is challenging to identify the membership grade of a fuzzy set as crisp values (e.g., unit benefit), type-2 fuzzy sets (T2FS) can effectively determine the membership function through defying membership grades of T2FS are fuzzy sets within [0,1] [19–22]. In addition, membership functions cannot express uncertainties featured with randomness. Two-stage stochastic programming with recourse is effective for handling decision-making problems in which an examination of policy levels is desired and the system data is characterized by uncertainty [23]. Therefore, as an extension to the existing approaches, a recourse-based type-2 fuzzy programming (RTFP) method incorporating the concepts of type-2 fuzzy programming (TFP) within a two-stage stochastic programming with recourse (TSP) framework will be developed to address the above deficiencies. Then, the RTFP method will be applied to water pollution control in the Heshui River Basin in China. Results of optimal agriculture, industry, forestry, fishery, and livestock husbandry activities will be generated, which will be used for providing insight into the trade-off among system benefit, water pollution control, and sustainability.

2. Methodology

In water pollution control decision making problems, uncertainties may arise due to subjective estimation. For instance, decision makers may estimate the unit benefit from planting fruit/vegetable being [424.1, 575.6] $\times 10^3$ \$/km^2 with additional information as the possibility of "the most possible unit benefit of 499.8 $\times 10^3$ \$/km^2" is 0.8, and the possibilities of "there is no possibility that the unit benefit is lower than 424.1 $\times 10^3$ \$/km^2 or higher than 575.6 $\times 10^3$ \$/km^2" are 0.2 and 0.3, respectively. Under such a situation, unit benefit from planting fruit/vegetable should be described as type-2 fuzzy sets (T2FS). Thus, type-2 fuzzy programming (TFP) can be adopted to tackle such uncertainties, which can be presented as [24]:

$$\text{Max } f = \widetilde{C}X, \tag{1}$$

subject to:

$$\widetilde{A}X \leq B, \tag{2}$$

$$X \geq 0, \tag{3}$$

where $\widetilde{C} \in \{R\}^{1 \times n}$ and $\widetilde{A} \in \{R\}^{m \times n}$ are vectors of T2FS. A type-2 fuzzy set $\widetilde{C}$ in X is a fuzzy set in which the membership function is also fuzzy (i.e., type-2 membership function). The $\widetilde{C}$ defined on the universe of discourse X is represented as [25]:

$$\widetilde{C} = \{((x, u), \mu_{\widetilde{C}}(x, u)) : \forall x \in X, \forall u \in J_x \subseteq [0, 1]\}, \tag{4}$$

where $0 \leq \mu_{\widetilde{C}}(x, u) \leq 1$ is the type-2 membership function, $J_x \subseteq [0, 1]$ is the primary membership of $x \in X$, which is the domain of the secondary membership function $\mu_{\widetilde{C}}(x)$ so that all $u \in J_x$ are the primary membership grades of the point x. The secondary membership function of a triangular $\widetilde{C} = (c_1, c_2, c_3, \theta_1, \theta_r)$ can be defined as:

$$\mu_{\widetilde{C}}(x) = \begin{cases} (\frac{x-c_1}{c_2-c_1} - \theta_1 \frac{x-c_1}{c_2-c_1}, \frac{x-c_1}{c_2-c_1}, \frac{x-c_1}{c_2-c_1} + \theta_r \frac{x-c_1}{c_2-c_1}), \text{ if } x \in [c_1, \frac{c_1+c_2}{2}] \\ (\frac{x-c_1}{c_2-c_1} - \theta_1 \frac{c_2-x}{c_2-c_1}, \frac{x-c_1}{c_2-c_1}, \frac{x-c_1}{c_2-c_1} + \theta_r \frac{c_2-x}{c_2-c_1}), \text{ if } x \in [\frac{c_1+c_2}{2}, c_2] \\ (\frac{c_3-x}{c_3-c_2} - \theta_1 \frac{x-c_2}{c_3-c_2}, \frac{c_3-x}{c_3-c_2}, \frac{c_3-x}{c_3-c_2} + \theta_r \frac{x-c_2}{c_3-c_2}), \text{ if } x \in [c_1, \frac{c_2+c_3}{2}] \\ (\frac{c_3-x}{c_3-c_2} - \theta_1 \frac{c_3-x}{c_3-c_2}, \frac{c_3-x}{c_3-c_2}, \frac{c_3-x}{c_3-c_2} + \theta_r \frac{c_3-x}{c_3-c_2}), \text{ if } x \in [\frac{c_2+c_3}{2}, c_2] \end{cases}, \tag{5}$$

where c_1, c_2, c_3 are real numbers and θ_1; $\theta_r \in [0, 1]$ are two parameters representing the spreads of primary membership grades of $\widetilde{C}$.

One of the main limitations of the TFP method remains in its difficulty in coping with uncertainties described as probability distributions when the available historical data is sufficient (e.g., pollutant discharge allowances) [26–28]. Such a problem can be formulated as a two-stage stochastic programming with a recourse (TSP) model [23]. Through incorporating the TFP method within the TSP framework, a recourse-based type-2 fuzzy programming (RTFP) method can be formulated as follows:

$$\text{Max } f = \sum_{j=1}^{n_1} \widetilde{c}_j x_j - \sum_{j=1}^{n_2} \sum_{s=1}^{S} p_s \widetilde{e}_j y_{js}, \tag{6}$$

subject to:

$$\sum_{j-1}^{n_1} \widetilde{a}_{rj} x_j \leq \widetilde{b}_r, \; r = 1, 2, \ldots, m_1, \tag{7}$$

$$\sum_{j=1}^{n_1} a_{tj} x_j + \sum_{j=1}^{n_2} a'_{tj} y_{jts} \leq \omega_s, \; t = 1, 2, \ldots, m_2; \; s = 1, 2, \ldots, v, \tag{8}$$

$$x_j \geq 0, \; j = 1, 2, \ldots, n_2, \tag{9}$$

$$y_{js} \geq 0, \; j = 1, 2, \ldots, n_2; \; s = 1, 2, \ldots, v. \tag{10}$$

Decision variables are divided into two subsets: those that must be determined before the realizations of random variables are known, and those (recourse variables) that are determined after the realized random variables are disclosed. x_j is the first-stage decision made before the random variable is observed, ω_s is the random variable with a probability level p_s (i.e., the probability of realization of ω_s, with $p_s > 0$ and $\sum_{s=1}^{S} p_s = 1$) [29], and y_{js} is the second-stage adaptive decision, which depends on the realization of the random variable. $\sum_{j=1}^{n_2} \widetilde{e}_j y_{js}$ denotes the second-stage cost function ($\widetilde{e}_j$ is cost coefficient of y_{js}). Inequality (4e) presents the relationship among x_j, y_{js}, and ω_s.

An extra type reduction is needed to convert the output of T2FS into conventional fuzzy sets so that they can be defuzzified to give crisp outputs. Suppose that $h_1, h_2, \ldots, h_i$ are the value of $\widetilde{c}_{ij}$ (for at least one pair of $i \neq j$, define $h_i \neq h_j$) evaluated by t different experts. The relative distances of h_i are used to approximate the center. Values lying closer to the center are considered more important. Generally, the fuzzification of $\widetilde{c}_{ij}$ can be represented as: (1) calculate the relative distance matrix $D = |d_{ij}|_{t \times t}$, where $d_{ij} = |h_i - h_j|$; (2) calculate the average of relative distances

$\bar{d}_i = \sum_{i=1}^{t} d_{ij}/(t-1)$; (3) introduce a pair-wise comparison p_{ij} ($p_{ij} = \bar{d}_j/\bar{d}_i$), and the pair-wise matrix $P = |p_{ij}|_{t\times t}$; (4) obtain the true-importance degree w_i of h_i ($w_j = 1/\sum_{j=1}^{t} p_{ij}$); (5) assess the mode of m ($m = \sum_{i=1}^{t} w_i h_i$) of the fuzzy number; (6) choose $s = \sum_{i=1}^{t} w_i|h_i - m|$ to approximate the unknown mean deviation σ; (7) acquire the ratio of left spread to the right spread $\eta = (m - h^l)/(h' - m)$, with $h^l = \sum_{i\in A} w_i h_i / \sum_{i\in A} w_i$, $h' = \sum_{i\in B} w_i h_i / \sum_{i\in B} w_i$, $A = \{i|h_i < m,\ i \in I\}$, $B = \{i|h_i > m,\ i \in I\}$ and $I = \{1,\ 2,\ \ldots,\ t\}$; (8) obtain $a = m - 3(1+\eta)\eta\sigma/(1+\eta^2)$ and $b = m + 3(1+\eta)\sigma/(1+\eta^2)$. The conventional fuzzy number $F = (a,\ m,\ b)$ of $\tilde{c}_{ij}$ can thus be acquired. Then, the defuzzification of T2FS can be conducted according to the critical value (CV)-based reduction method [30].

Suppose that $\tilde{c}$ is a triangular type-2 fuzzy variable with secondary possibility distribution function $\mu_{\tilde{c}}(x)$ (which represents a regular fuzzy variable). The method is introducing the *CV*s as representing values of the regular fuzzy variable $\mu_{\tilde{c}}(x)$, i.e., $CV^*[\mu_{\tilde{c}}(x)]$ (optimistic *CV*), $CV_*[\mu_{\tilde{c}}(x)]$ (pessimistic *CV*) and $CV[\mu_{\tilde{c}}(x)]$. Then, the corresponding fuzzy variables are derived using these *CV*s of the secondary possibilities:

$$\tilde{c} = \begin{pmatrix} \alpha_1 & \alpha_2 & \alpha_3 \\ m & 1 & n \end{pmatrix}, \tag{11}$$

where α_1, α_2, and α_3 are primary membership grade of $\tilde{c}$, and m, 1, and n are corresponding secondary membership grades.

$$Pos(\tilde{c} \geq \alpha) = \sup_{r\geq\alpha} \mu_{\tilde{c}}(r) = \begin{cases} 1 \text{ if } 0 \leq \alpha \leq \alpha_2 \\ n \text{ if } \alpha_2 \leq \alpha \leq \alpha_3 \\ 0 \text{ if } \alpha_3 \leq \alpha \leq 1 \end{cases}, \tag{12}$$

$$Nec(\tilde{c} \geq \alpha) = 1 - \sup_{r<\alpha} \mu_{\tilde{c}}(r) = \begin{cases} 1 \text{ if } 0 \leq \alpha \leq \alpha_1 \\ m \text{ if } \alpha_1 \leq \alpha \leq \alpha_2 \\ 0 \text{ if } \alpha_2 \leq \alpha \leq 1 \end{cases}, \tag{13}$$

$$Cr\{\tilde{c} \geq \alpha\} = \frac{1}{2}(Pos(\tilde{c} \geq \alpha) + Nec(\tilde{c} \geq \alpha)) = \begin{cases} 1 & \text{if } 0 \leq \alpha \leq \alpha_1 \\ \frac{1+m}{2} & \text{if } \alpha_1 \leq \alpha \leq \alpha_2 \\ \frac{n}{2} & \text{if } \alpha_2 \leq \alpha \leq \alpha_3 \\ 0 & \text{if } \alpha_3 \leq \alpha \leq 1 \end{cases}. \tag{14}$$

Then, $CV^*[\mu_{\tilde{c}}(x)]$, $CV_*[\mu_{\tilde{c}}(x)]$ and $CV[\mu_{\tilde{c}}(x)]$ are defined as follows:

$$CV^*(\tilde{c}) = \sup_{\alpha\in[0,\ 1]} (\alpha \wedge Pos\{\tilde{c} \geq \alpha\}), \tag{15}$$

$$CV_*(\tilde{c}) = \sup_{\alpha\in[0,\ 1]} (\alpha \wedge Nes\{\tilde{c} \geq \alpha\}), \tag{16}$$

$$CV(\tilde{c}) = \sup_{\alpha\in[0,\ 1]} (\alpha \wedge Cr\{\tilde{c} \geq \alpha\}). \tag{17}$$

Finally, to obtain crisp values, centroid method ($\sum_x x\mu_{\tilde{c}}(x)/\sum_x \mu_{\tilde{c}}(x)$) is used for these reduced conventional fuzzy variables. Generally, the detailed computational processes for solving the RTFP method can be summarized as follows:

Step 1. Formulate the RTFP model.
Step 2. Discrete probability distribution into several values with each corresponds to one probability.
Step 3. Convert the output of T2FS into conventional fuzzy sets.
Step 4. Conduct defuzzification of T2FS according to the critical value (CV)-based reduction method.
Step 5. Run the RTFP model.
Step 6. Obtain the optimal solutions of the objective function (f), first-stage decision variable (x_j), and second-stage decision variable (y_{js}).

3. Case Study

Yongxin, a county of Jiangxi Province, is located in the southeast part of China (as shown in Figure 1). It ranges in longitude from 113.83° E to 114.31° E, and in latitude from 26.78° N to 27.23° N. It occupies a total area of around 2194.57 km^2, and the majority of the county (i.e., approximately 1800 km^2) lies within the middle reaches of the Heshui River Basin [31]. The basin features subtropical monsoon humid climatic conditions (abundant rainfall and sunlight, and long frost-free periods) with an average annual precipitation of 1530.7 mm and an average annual temperature of 18.2 °C. The Heshui River has a total length of 225 km, with 77 km flowing within the borders of Yongxin County from west to east. The county mainly relies on the Heshui River to support its agriculture, industry, livestock husbandry, forestry, and fishery [32].

In the county, farmland, orchard, and woodland occupy15.2%, 0.6%, and 70.1% of the total land, receptively. Grassland accounts for 6.1% of the total area, most of which is in good condition and suitable for feeding a large number of livestock. Land for urban construction, traffic, and water conservancy facilities cover 3.4%, 0.4%, and 0.6% of the total land, respectively. Agriculture is traditionally the primary sector of the County. Paddy soil is the main cultivation soil type, and it is not only suitable for rice planting but can also be used for rapeseeds and other cash crops. Fruit orchards produce a variety of fruits such as pear, peach, orange and plum. Agriculture includes paddy, dry and vegetable/fruit farms. The main types of livestock raised in the County are hogs, cattle and poultry. Pork and beef account for around 75.5 and 13.2% of the total meat production, respectively. The second industrial sector in the county is mainly comprised of mining, manufacturing, construction, transportation and other industries. The area possesses more than 20 types of mineral deposits, and the mineral production was 315.0 thousand tonnes [32,33].

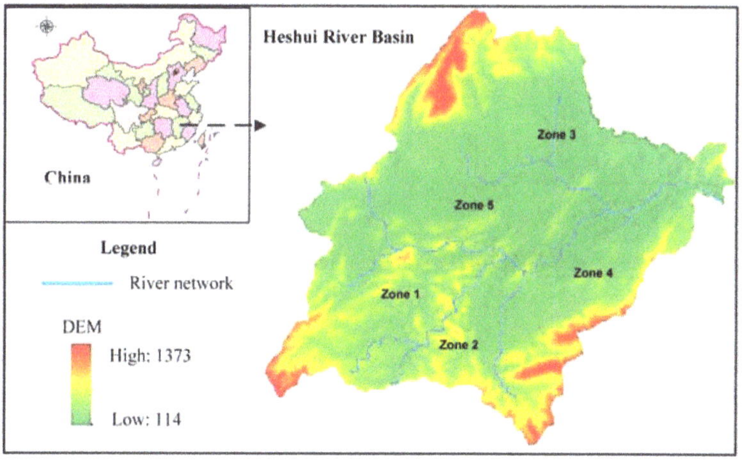

Figure 1. The study area.

In the past twenty years, driven by poverty, authorities have given top priority to the booming economy in the local strategic plans [33]. According to the report of local government in 2016, the total population was approximately 53,139, and the net income per capita was 8520 Chinese yuan (approximately $1290). One major environmental issue in the study area is water pollution, which is mainly caused by excessive pollutant loadings from agriculture, forestry, fishery, and livestock husbandry, and industry. Water pollution problems pose great obstacles to sustainable development in the county; thus, it is essential to optimize economic structure from a systematic point of view. Based on field investigations and related literature, chemical oxygen demand (COD), total nitrogen (TN), total phosphorus (TP), and soil loss have been selected as the pollutant types to control water pollution. Five zones (zones 1 to 5) are chosen to control water pollution. After the first-stage decisions (i.e., planning targets of economic activities) are made, associated pollution, the amount of which is in proportion to the economic production, is discharged into the water body. Based on the calculation of discharged wastewater and the measurement of incoming water quality, the mitigation schemes can be determined as the second-stage decisions in order to meet the environmental standards. If the production targets of them are made too high, the additionally generated pollutants will have to be mitigated in a more expensive way or discharged/drained into the stream (leading to penalties from the government). Conversely, if the production targets are made too low, the system will encounter opportunity losses of economic income, leading to a reduced system benefit. The pollutant discharge allowances are highly uncertain, which are presented as probability distributions. Seven pollutant-discharge allowance levels are generated in association with different probabilities, which are identified as very-low, low, low-medium, medium, medium-high, high, and very-high, respectively. More uncertainties associated with cost/benefit coefficients may come from measurement errors and/or subjective judgment, and they can be expressed as T2FS. Therefore, the developed RTFP method can be adopted to plan water pollution control in the study area. The formulation of the RTFP model is presented as follows:

$$\begin{aligned}
\text{Max } f^{\pm} = \; & \sum_{i=1}^{I_a}\sum_{j=1}^{J} \widetilde{BA}_{ij}\cdot TA_{ij}(+\Delta TA_{ij}u_{ij}) - \sum_{i=1}^{I_a}\sum_{j=1}^{J}\sum_{h=1}^{H}\sum_{k=2}^{K} p_h\cdot PEA_{ik}\cdot \widetilde{DPA}_{ijk}\cdot XA_{ijhk} \\
& + \sum_{i=1}^{I_f}\sum_{j=1}^{J} \widetilde{BF}_{ij}\cdot TF_{ij} - \sum_{i=1}^{I_f}\sum_{j=1}^{J}\sum_{h=1}^{H}\sum_{k=1}^{K-1} p_h\cdot PEF_{ik}\cdot \widetilde{DPF}_{ijk}\cdot XF_{ijhk} \\
& + \sum_{i=1}^{I_l}\sum_{j=1}^{J} \widetilde{BL}_{ij}\cdot TL_{ij} - \sum_{i=1}^{I_l}\sum_{j=1}^{J}\sum_{h=1}^{H} p_h\cdot PEL_i\cdot \widetilde{DPL}_{ij}\cdot XL_{ijh} \\
& \sum_{i=1}^{I_i}\sum_{j=1}^{J} TI_{ij} - \sum_{i=1}^{I_i}\sum_{j=1}^{J}\sum_{h=1}^{H} p_h\cdot PEI_i^{\pm}\cdot \widetilde{DPI}_{ij}\cdot XI_{ijh} \\
& + \sum_{i=1}^{I_w}\sum_{j=1}^{J} \widetilde{BW}_{ij}\cdot TW_{ij} - \sum_{i=1}^{I_i}\sum_{j=1}^{J}\sum_{h=1}^{H} p_h\cdot PEW_i\cdot \widetilde{DPW}_{ij}\cdot XW_{ijh}
\end{aligned} \tag{18}$$

(1) COD discharge constraints:

$$\sum_{i=1}^{I_f}(TF_{ij} - XF_{ijh})\cdot CCO_{ij} \le PCF_{jh}, \quad \forall j,\ h, \tag{19}$$

$$\sum_{i=1}^{I_i}(TL_{ij} - XL_{ijh})\cdot DPL_{ij}^{\pm} \le PCL_{jh}, \quad \forall j,\ h, \tag{20}$$

$$\sum_{i=1}^{I_i}(TI_{ij} - XI_{ijh})\cdot DPI_{ij}^{\pm} \le PCI_{jh}, \quad \forall j,\ h, \tag{21}$$

$$\begin{aligned}
& \sum_{j=1}^{J}\Big[\sum_{i=1}^{I_f}(TF_{ij} - XF_{ijh})\cdot CCO_{ij} + \sum_{i=1}^{I_i}(TL_{ij} - XL_{ijh})\cdot DPL_{ij} \\
& \quad + \sum_{i=1}^{I_i}(TI_{ij} - XI_{ijh})\cdot DPI_{ij}\Big] \le MCL_h, \quad \forall h
\end{aligned}, \tag{22}$$

(2) Phosphorus discharge constraints:

$$\sum_{i=1}^{I_a}(TA_{ij} - XA_{ijh})\cdot(SL_{ij}\cdot AP_{ij} + RA_{ij}\cdot RP_{ij}) \le PAP_{jh}, \quad \forall j,\ h, \tag{23}$$

$$\sum_{i=1}^{I_f}(TF_{ij}-XF_{ijh})\cdot FP_{ij}^{\pm}\leq PFP_{jh},\quad\forall j,\ h, \tag{24}$$

$$\begin{aligned}&\sum_{j=1}^{J}[\sum_{i=1}^{I_a}(TA_{ij}-XA_{ijh})\cdot(SL_{ij}\cdot AP_{ij}+RA_{ij}\cdot RP_{ij})\\&+\sum_{i=1}^{I_f}(TF_{ij}-XF_{ijh})\cdot FP_{ij}]\leq MPL_h,\quad\forall h\end{aligned}, \tag{25}$$

(3) Nitrogen discharge constraints:

$$\sum_{i=1}^{I}(TA_{ij}-XA_{ijh})\cdot(SL_{ij}\cdot AN_{ij}+RA_{ij}RN_{ij})\leq PAN_{jh},\quad\forall j,\ h, \tag{26}$$

$$\sum_{i=1}^{I_f}(TF_{ij}-XF_{ijh})\cdot FN_{ij}\leq PFN_{jh},\quad\forall j,\ h, \tag{27}$$

$$\begin{aligned}&\sum_{i=1}^{I_a}(TA_{ij}-XA_{ijh})\cdot(SL_{ij}\cdot AN_{ij}+RA_{ij}\cdot RN_{ij})\\&+\sum_{i=1}^{I_f}(TF_{ij}-XF_{ijh})\cdot FN_{ij}\leq MNL_h,\quad\forall h\end{aligned}, \tag{28}$$

(4) Soil loss constraints:

$$\sum_{i=1}^{I_i}(TA_{ij}-XA_{ijh})\cdot SL_{ij}\leq PSL_{jh},\quad\forall j,\ h, \tag{29}$$

$$\sum_{i=1}^{I_w}(TW_{ij}-XW_{ijh})\cdot DPW_{ij}\leq PWS_{jh},\quad\forall j,\ h, \tag{30}$$

$$\begin{aligned}&\sum_{j=1}^{J}[\sum_{i=1}^{I_i}(TA_{ij}-XA_{ijh})\cdot SL_{ij}^{\pm}\\&+\sum_{i=1}^{I_w}(TW_{ij}-XW_{ijh})\cdot DPW_{ij}]\leq MSL_h,\quad\forall h\end{aligned}, \tag{31}$$

(5) Water supply balance constraint:

$$\begin{aligned}&\sum_{h=1}^{H}(\sum_{i=1}^{I_i}(TA_{ij}-XA_{ijh})\cdot WA_i+\sum_{i=1}^{I_f}(TF_{ij}-XF_{ijh})\cdot WF_i\\&+\sum_{i=1}^{I_l}(TL_{ij}-XL_{ijh})\cdot WL_i+\sum_{i=1}^{I_i}(TI_{ij}-XI_{ijh})\cdot WI_i\\&+\sum_{i=1}^{I_w}(TW_{ij}-XW_{ijh})\cdot WW_i]\leq MAXW_j,\quad\forall j\end{aligned}, \tag{32}$$

(6) Product demand constraints:

$$TA_{i\,\min}\leq\sum_{j=1}^{J}TA_{ij}\leq TA_{i\,\max},\quad\forall i, \tag{33}$$

$$TF_{i\,\min}\leq\sum_{j=1}^{J}TF_{ij}\leq TF_{i\,\max},\quad\forall i, \tag{34}$$

$$TL_{i\,\min}\leq\sum_{j=1}^{J}TL_{ij}^{-}\leq TL_{i\,\max},\quad\forall i, \tag{35}$$

$$TI_{i\,\min} \leq \sum_{j=1}^{J} TI_{ij}^{-} \leq TI_{i\,\max}, \quad \forall i, \tag{36}$$

$$TW_{i\,\min} \leq \sum_{j=1}^{J} TW_{ij}^{-} \leq TW_{i\,\max}, \quad \forall i, \tag{37}$$

(7) Technical constraints:

$$XA_{ijh},\ XF_{ijh},\ XL_{ijh},\ XI_{ijh},\ XW_{ijh} \geq 0, \quad \forall i,\ j,\ h. \tag{38}$$

The nomenclature of variables and parameters is provided at the end of this ariticle. The imprecise input parameters are investigated according to field surveys, statistical data [31], government reports [34], and related literature [32,33]. Table 1 presents unit benefit of agriculture and livestock husbandry, which are expressed as type-2 fuzzy sets. Table 2 provides pollutant allowances and the associated probabilities of occurrence.

Table 1. Unit benefits of agriculture and livestock husbandry.

Zone	Agriculture		
	Paddy Farm	Dry Farm	Fruit/Vegetable
	Unit Net Benefit (10^3 \$/km^2)		
Zone 1	(201.4, 272.6, 292.3; 0.2, 0.6)	(104.5, 148.4, 163.5; 0.2, 0.6)	(345.3, 457.4, 486.2; 0.2, 0.6)
Zone 2	(174.2, 219.6, 269.6; 0.3, 0.5)	(96.9, 130.3, 140.9; 0.3, 0.5)	(152.9, 205.9, 215.1; 0.3, 0.5)
Zone 3	(169.6, 231.7, 254.4; 0.4, 0.7)	(87.8, 112.1, 127.2; 0.4, 0.7)	(337.7, 472.6, 569.4; 0.4, 0.7)
Zone 4	(187.8, 265.1, 290.8; 0.4, 0.7)	(92.4, 118.1, 124.2; 0.4, 0.7)	(254.5, 343.4, 384.7; 0.4, 0.7)
Zone 5	(184.8, 245.4, 275.7; 0.5, 0.7)	(92.4, 122.7, 133.3; 0.5, 0.7)	(289.3, 390.8, 402.9; 0.5, 0.7)
Zone	**Livestock Husbandry**		
	Pig	**Cattle**	**Poultry**
	Unit Net Benefit (\$/head)		
Zone 1	(152.5, 166.3, 170.1; 0.4, 0.7)	(785.4, 892.4, 998.4; 0.4, 0.7)	(5.3, 6.2, 7.1; 0.4, 0.7)
Zone 2	(139.6, 151.2, 156.6; 0.5, 0.8)	(833.9, 933.6, 974.2; 0.5, 0.8)	(6.2, 6.8, 7.4; 0.5, 0.8)
Zone 3	(139.6, 159.6, 161.2, 0.4, 0.7)	(732.9, 831.1, 868.1; 0.4, 0.7)	(4.7, 6.9, 7.1; 0.4, 0.7)
Zone 4	(154.7, 172.2, 175.4; 0.3, 0.8)	(662.1, 734.9, 760.7; 0.3, 0.8)	(4.9, 6.7, 7.2; 0.3, 0.8)
Zone 5	(158.5, 168.7, 185.2; 0.4, 0.8)	(904.8, 984.2, 999.2; 0.4, 0.8)	(6.2, 7.1, 7.8; 0.4, 0.8)

Table 2. Pollutant allowances under different levels.

Pollutant	Level	Probability	Zone				
			Zone 1	Zone 2	Zone 3	Zone 4	Zone 5
COD (10^3 kg)	Very-low	0.05	2453.5	496.4	3768.4	3284.2	2307.5
	Low	0.10	2469.8	499.3	3780.8	3298.5	2213.3
	Low-medium	0.20	2498.3	500.3	3792.4	3390.7	2309.8
	Medium	0.30	2523.5	502.5	3897.9	3469.3	2468.3
	Medium-high	0.20	2760.7	515.6	4506.2	3607.8	2647.8
	High	0.10	2984.4	535.5	4512.5	3812.9	2851.2
	Very-high	0.05	3286.2	578.7	4714.7	4214.2	3154.9
TN (10^3 kg)	Very-low	0.05	166.5	193.2	235.7	286.3	223.6
	Low	0.10	216.4	297.3	307.3	369.4	374.3
	Low-medium	0.20	266.5	401.4	379.5	451.8	526.8
	Medium	0.30	324.3	506.7	458.8	561.2	679.2
	Medium-high	0.20	381.4	614.4	544.3	683.3	836.2
	High	0.10	457.2	747.5	653.9	815.5	994.1
	Very-high	0.05	545.5	883.9	756.3	935.8	1168.7

Table 2. *Cont.*

Pollutant	Level	Probability	Zone				
			Zone 1	Zone 2	Zone 3	Zone 4	Zone 5
TP (10^3 kg)	Very-low	0.05	24.2	31.1	35.6	43.2	39.9
	Low	0.10	34.4	52.0	50.0	59.5	70.5
	Low-medium	0.20	44.5	73.5	64.5	76.6	101.7
	Medium	0.30	55.0	94.6	81.0	93.6	132.2
	Medium-high	0.20	65.1	118.1	100.2	113.9	165.4
	High	0.10	79.5	142.6	114.8	134.3	200.3
	Very-high	0.05	93.2	166.1	130.4	152.6	231.6
Soil loss (10^3 t)	Very-low	0.05	97.3	92.5	133.4	168.5	61.7
	Low	0.10	92.4	93.6	134.2	169.5	63.0
	Low-medium	0.20	99.6	94.7	135.3	170.7	64.2
	Medium	0.30	100.3	98.9	140.6	176.0	71.5
	Medium-high	0.20	118.0	133.1	155.7	186.1	82.7
	High	0.10	133.1	140.4	165.9	201.3	89.9
	Very-high	0.05	133.2	140.5	171.6	274.4	91.9

Note: COD, Carbon oxygen demand; TN, Total nitrogen; TP, Total phosphorus.

4. Results Analysis and Discussion

Figure 2 represents the benefits under various pollutant discharge allowances and the system benefit, indicating that a higher allowance would correspond to a higher benefit. For instance, when the allowable pollutant discharge changes from very low to very high, the total benefit (i.e., sum of benefit of each activity) would be raise from \$172.1 $\times$ 10^6 to \$195.7 $\times$ 10^6. Decisions at a higher allowable pollutant discharge would lead to a higher system benefit, but the reliability in fulfilling the environmental requirements would decrease; on the contrary, decisions at a lower allowable pollutant discharge would lead to a decreasing of risk for violating the pollutant discharge constraints, but with a lower system benefit. It demonstrates a trade-off between environmental requirement violation risk and benefit due to the uncertainties existing in various system components. In practice, when the plan aims to a higher system benefit, the environmental requirements may not be adequately satisfied; contrarily, planning with a lower system benefit may guarantee that the requirements be met. Additionally, the benefit of agriculture activity would take the largest proportion in total benefit and would increase slightly with the raising of pollutant discharge allowances. Moreover, the benefit of fishery would be stable at a low level, approximately occupying 2.1% of the total benefit.

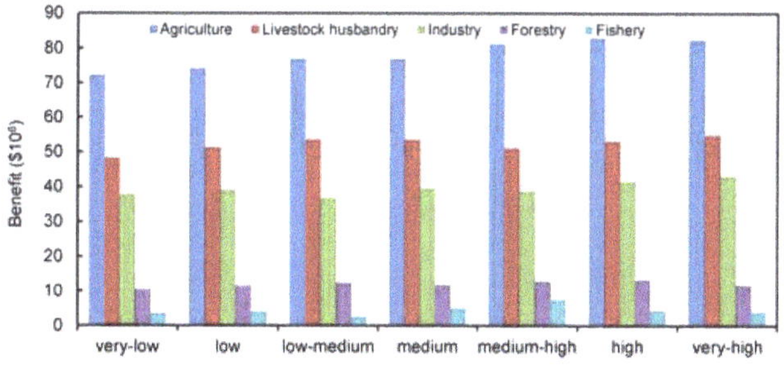

Figure 2. Benefit of each activity under each pollutant discharge allowance.

The water pollution control problem can simplified as a recourse-based fuzzy programming (RFP) problem by transforming the membership grades of T2FS into deterministic values. System benefit obtained from RFP (\$180.8 $\times$ 10^6) would be lower than that from RTFP (\$191.1 $\times$ 10^6). This is due to the fact that cost/benefit coefficients are handled by RFP, resulting in higher loss of uncertain information than that handled by RTFP. Figure 3 presents the optimal target and standard production scale of each activity under RFP and RTFP. Target of each activity discharge pollutant exceeds standards can be calculated through multiplying excess scale of economic activity (i.e., target—standard) by pollutant discharge rate. Results indicate that the excess planning scale of agricultural activity would be high. This may be attributed to their high crop yields and great selling prices. It is also depicted that the excess feeding size of livestock husbandry activity would also maintain high levels due to its high annual incomes. The excess outputs of industrial activity would also be significant since industry is promoted by the authority to push up the local income. Excess fishery and forestry activities would be low due to their limited planning lands. Furthermore, excess economic activities corresponding to RFP would be higher than that corresponding to RTFP. For instance, target of agricultural activity would be 180 ha in zone 1, while the standard production scale under RTFP and RFP would, respectively, be 156.4 ha and 148.5 ha corresponding to very-low level. Thus, excess agricultural activity would be 23.6 ha and 31.5 ha, respectively. It is revealed that the varied uncertain information would affect the water pollution control plans. Any simplifications may result in unreliable or misleading plans.

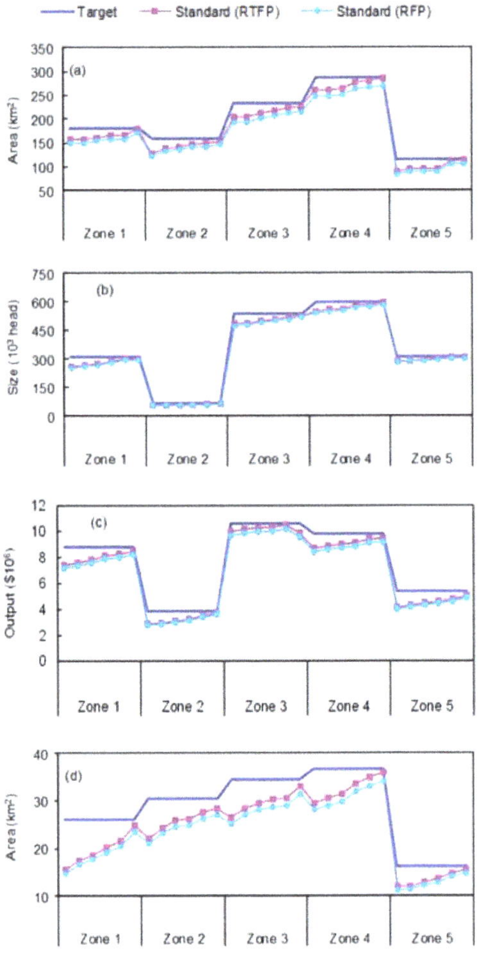

(a)

Figure 3. *Cont.*

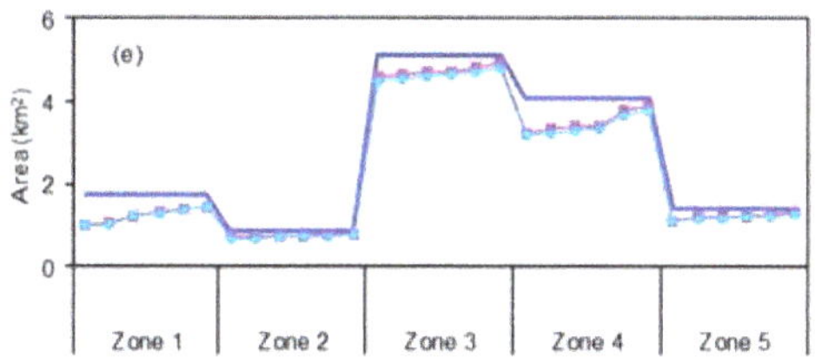

(b)

Figure 3. Optimal target and standard production scale of each activity. (**a**) Agriculture; (**b**) Livestock husbandry; (**c**) Industry; (**d**) Forestry; (**e**) Fishery.

Figure 4 displays the excess pollutant discharges under different pollutant discharge allowance levels corresponding to RFP and RTFP. It is indicated that amounts of excess pollutant discharges would reduce with the increased levels. For instance, under RTFP, the amounts of excess soil loss, COD, TP, and TN discharges would be 52.3×10^3, 152.1, 22.4, and 99.2 t under low pollutant discharge allowance level; in comparison, under high pollutant discharge allowance level, they would respectively decrease to 1.3×10^3, 8.2, 7.2, 5.4 t. In general, results discover that a more restrictive pollution control would result in a higher excess pollutant discharge while a looser pollution control would bring on a lower excess pollutant discharge.

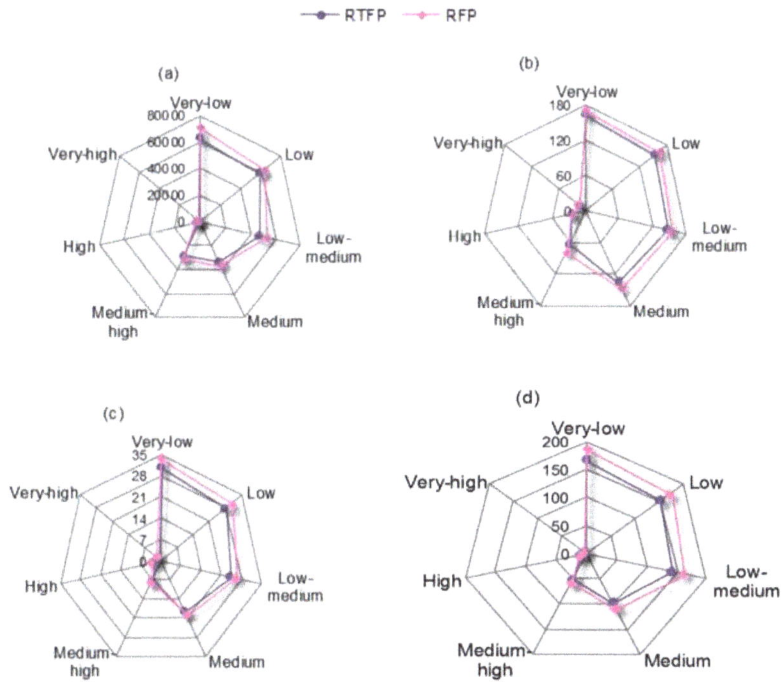

Figure 4. Excess pollutant discharge under different pollutant allowances (unit: t). (**a**) soil loss; (**b**) COD discharge; (**c**) TP discharge; (**d**) TN discharge.

Figure 5 displays proportion of excess pollutant discharges under medium-high level. It is indicated that the excess soil loss, TN, and TP discharges of agriculture would account for about 97%,

88%, and 87% of the of the total discharge amounts, respectively. The high excess pollutants discharges are mainly related to its high targeted planning scale and discharge rates. The excess COD discharged from fishery would be high (accounting for 58% of the total excess discharge) due to its high COD discharge rate. In summary, the trade-off between agricultural income and pollution control (i.e., soil loss, TP, and TN discharges) would be of great concern for the local authority; fishery would be ceased due to its low benefit and high pollutant discharges.

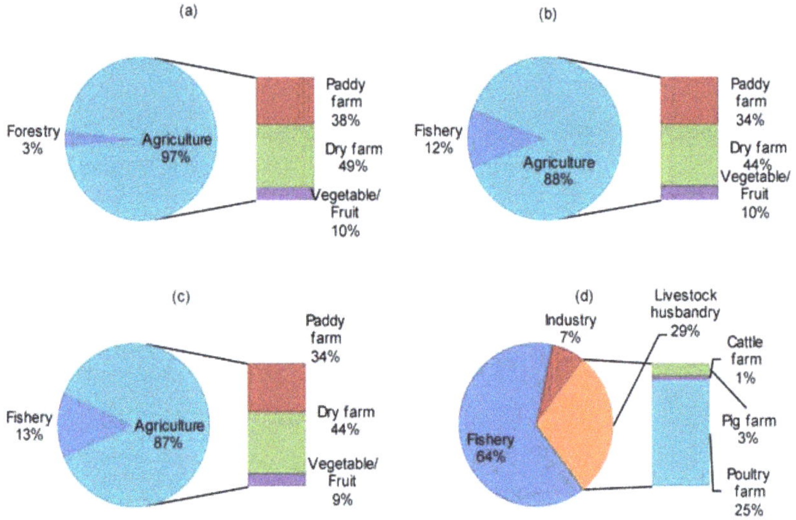

Figure 5. Proportion of pollutant discharges under medium-high level. (**a**) soil loss; (**b**) TN discharge; (**c**) TP discharge; (**d**) COD discharge.

Generally, compared with RTFP, RFP that tackles uncertainty in a single fuzzy set is not satisfactory to sufficiently reflect enough uncertain information in the decision-making process. Such a simplification may lead to unreliable or even misleading solutions. For instance, due to loss of the additional design degree of freedom, which is useful in water pollution control systems where input data are highly uncertain, economic coefficients (unit cost and unit benefit) treated by RFP are higher than that treated by RTFP. The higher economic coefficients lead to a lower system benefit, which may present unreliable decision support for local authorities. RTFP can tackle more complex uncertainty in terms of T2FS due to the extra degree of freedom. Moreover, solutions of economic activities scale and excess pollutant discharge also display that RTFP is more capable of balancing economic development and pollutant discharge control in the study area. Thus, RTFP is more enhanced in uncertainty reflection, as well as water pollution control.

5. Conclusions

In this study, a recourse-based type-2 fuzzy programming (RTFP) approach is developed for water pollution control planning. RTFP has incorporated the techniques of type-2 fuzzy programming and two-stage stochastic programming with recourse (TSP) within a general framework. RTFP can handle uncertainties presented as type-2 fuzzy sets (i.e., a fuzzy set in which the membership function is also fuzzy) and probability distributions. RTFP could provide benefit assessment for random pollutant discharge allowance levels through the two-stage framework. The results are helpful for the authorities in exploring the trade-off between economic objective and pollutant discharge decision-making based on river water pollution control.

The developed RTFP method has been demonstrated through its application to water pollution control of Heshui River Basin. Different levels that pollutant discharge allowances are assumed to be random and benefit/cost coefficients are specified as type-2 fuzzy sets have been investigated. Solutions for production scale of agriculture, livestock husbandry, industry, forestry, and fishery are generated. Several findings can be concluded: (i) results reveal that uncertainties in pollutant discharge allowance level have significant effects on the city's future economic structure; (ii) constrained with low pollutant discharge allowance, the excess pollutant discharge would be high; (iii) the increased restriction of discharge would stimulate the basin taming the pace of economic growth; (iv) results also disclose that the benefit of agriculture activity would take the largest proportion in total benefit and would increase slightly with the raising of pollutant discharge allowances; and (v) system benefit is powerfully impressed by the pollutant discharge allowance level.

It is the first attempt to apply the RTFP method to water pollution control planning, and results indicate that (1) RTFP cannot only handle uncertainty expressed as T2FS, but also effectively cope with uncertainty described as probability distribution; (2) TFCP can help authorities to make trade-offs among environmental violation risk and system benefit. Nevertheless, there are also potential extensions of the proposed method. For example, RTFP has difficulty in addressing uncertainties that cannot be expressed as type-2 fuzzy sets or probability distributions; interval is a proper type to present such a kind of uncertainty. Furthermore, construction of T2FS membership function is usually based on experts' subjective evaluation, which may also lead to high uncertainty. At present, there is no appropriate evaluation method except experts' decisions. It is worthwhile to search for more effective methods to improve the method of constructing membership functions for T2FS, which can be helpful to reduce uncertainty.

Acknowledgments: This research was supported by the National Key Research Development Program of China (2016YFC0502803 and 2016YFA0601502), the Beijing Natural Science Foundation (L160011), and the 111 Project (B14008). The authors are grateful to the editors and the anonymous reviewers for their insightful comments and suggestions.

Author Contributions: Yongping Li and Jing Liu conceived and designed the experiments; Yongping Li performed the experiments; Guohe Huang analyzed the data; Lianrong Chen contributed reagents/materials/analysis tools; Jing Liu wrote the paper

Conflicts of Interest: The authors declare no conflict of interest.

Nomenclatures

i	index for economic activities; for agricultural activities, $i = 1, 2, \ldots, I_a$; for fishery activities $i = 1, 2, \ldots, I_f$ (e.g., fish and prawn farming); for livestock husbandry activities $i = 1, 2, \ldots, I_l$; for industrial activities $i = 1, 2, \ldots, I_i$ (e.g., manufacturing, mining, architecture, transportation and others); for forestry activities $i = 1, 2, \ldots, I_w$
j	index for zones; $j = 1, 2, \ldots, J$
k	index for pollutants; $k = 1, 2, \ldots, K$ (e.g., COD discharge, TN loss, TP loss, and soil loss)
h	allowable pollutant discharge level; $h = 1, 2, \ldots, H$
p_h	probability of occurrence allowable pollutant discharge level h (%)
$\widetilde{BA}_{ij}$	unit benefit from agricultural activity i in zone j (RMB¥/km^2)
TA_{ij}	land area target for agricultural activity i in zone j (km^2)
PEA_{ik}	reduction of net benefit from agricultural activity i for excess discharge of pollutant k (RMB¥/kg when $k = 2, 3$; RMB¥/tonne when $k = 4$)
DPA_{ijk}	discharge rate of pollutant k from agricultural activity i in zone j (kg/km^2 when $k = 2, 3$; tonne /km^2 when $k = 4$)
XA_{ijhk}	decision variables representing amount by which the target of agricultural activity i discharge pollutant k exceeds standards in zone j when level is h (km^2)
$\widetilde{BF}_{ij}$	unit benefit from fishery activity i in zone j (RMB¥/km^2)
TF_{ij}	land area target for fishery farming activity i in zone j (km^2)

PEF_{ik}	reduction of net benefit from fishery activity *i* for excess discharge of pollutant *k* (RMB¥/kg)
DPF_{ijk}	discharge rate of pollutant *k* from fishery activity *i* in zone *j* (kg/km^2)
XF_{ijhk}	decision variables representing amount by which target of fishery activity *i* discharge pollutant *k* exceeds standards in zone *j* when level is *h* (km^2)
$\widetilde{B}L_{ij}^{\pm}$	unit benefit from livestock husbandry activity *i* in zone *j* (RMB¥/head)
TL_{ij}	target for livestock husbandry activity *i* in zone *j* (head)
PEL_i	reduction of net benefit from livestock husbandry activity *i* for excess discharge of pollutant (i.e., COD) (RMB¥/kg)
DPL_{ij}	discharge rate of pollutant (i.e., COD) from livestock husbandry activity *i* in zone *j* (kg/head)
XL_{ijh}	decision variables representing amount by which target of livestock husbandry activity *i* discharge pollutant (i.e., COD) exceeds standards in zone *j* when level is *h* (head)
TI_{ij}	output target for industrial activity *i* in zone *j* (RMB¥)
PEI_i	reduction of net benefit from industrial activity *i* for excess discharge of pollutant (i.e., COD) (RMB¥/kg)
DPI_{ij}	discharge rate of pollutant (i.e., COD) from industrial activity *i* in zone *j* (kg/RMB¥)
XI_{ijh}	decision variables representing amount by which target of industrial activity *i* discharge pollutant (i.e., COD) exceeds standards in zone *j* when level is *h* (RMB¥)
$\widetilde{B}W_{ij}$	unit benefit from forestry activity *i* in zone *j* (RMB¥/head)
TW_{ij}	land area target for forestry activity *i* in zone *j* (unit)
PEW_i	reduction of net benefit from forestry activity *i* for excess discharge of pollutant (i.e., soil loss) (RMB¥/tonne)
DPW_{ij}	discharge rate of pollutant (i.e., soil loss) from forestry activity *i* in zone *j* (tonne/km^2)
XW_{ijh}	decision variables representing amount by which target of forestry activity *i* discharge pollutant (i.e., soil loss) exceeds standards in zone *j* when level is *h* (unit)
COF_{ij}	COD discharge from fishery farming activity *i* in zone *j* (kg/km^2)
PCF_{jh}	maximum allowable COD discharge for fishery farming activities in zone *j* with probability p_h of occurrence under level *h* (kg)
PCL_{jh}	maximum allowable COD discharge for livestock husbandry activities in zone *j* with probability p_h of occurrence under level *h* (kg)
PCI_{jh}	maximum allowable COD discharge for industrial activity *i* in zone *j* with probability p_h of occurrence under level *h* (kg)
MCL_h	maximum allowable COD discharge from economic activities with probability p_h of occurrence under level *h* (kg)
SL_{ij}	soil loss from agricultural activity *i* in zone *j* (tonne/km^2)
AP_{ij}	phosphorous content of soil corresponding to agricultural activity *i* in zone *j* (kg/tonne)
RA_{ij}	runoff from agricultural activity *i* in zone *j* (kg/km^2)
RP_{ij}	dissolved phosphorous content of runoff corresponding to agricultural activity *i* in zone *j* (%)
PAP_{jh}	maximum allowable phosphorous loss from agricultural activities in zone *j* with probability p_h of occurrence under level *h* (kg)
FP_{ij}	dissolved phosphorous loss from fishery farming activity *i* in zone *j* (kg/km^2)
PFP_{jh}	maximum allowable phosphorous loss from fishery farming activities in zone *j* with probability p_h of occurrence under level *h* (kg)
MPL_h	maximum allowable phosphorous loss from economic activities with probability p_h of occurrence under level *h* (kg)
AN_{ij}	nitrogen content of soil corresponding to agricultural activity *i* in zone *j* (kg/tonne)
RN_{ij}	dissolved nitrogen content of runoff corresponding to agricultural activity *i* in zone *j* (%)
PAN_{jh}	maximum allowable nitrogen loss from agricultural activities in zone *j* with probability p_h of occurrence under level *h* (kg)
FN_{ij}	dissolved nitrogen loss from fishery activity *i* in zone *j* (kg/km^2)
PFN_{jh}	maximum allowable nitrogen loss from fishery farming activities in zone *j* with probability p_h of occurrence under level *h* (kg)

MSL_{jh}	maximum allowable soil loss from economic activities with probability p_h of occurrence under level h (tonne)
MNL_h	maximum allowable nitrogen loss from economic activities with probability p_h of occurrence under level h (kg)
$\widetilde{BF}_{ij}$	unit benefit from fishery activity i in zone j (RMB¥/km^2)
PSL_{jh}	maximum allowable soil loss from agricultural activities in zone j with probability p_h of occurrence under level h (tonne)
PWS_{jh}	maximum allowable soil loss from forestry activities in zone j with probability p_h of occurrence under level h (tonne)
WA_i	water demand for agricultural activity i (m^3/km^2)
WF_i	water demand for fishery activity i (m^3/km^2)
WL_i	water demand for livestock husbandry activity i (m^3/head)
WI_i	water demand for industrial activity i (m^3/RMB¥)
WW_i	water demand for forestry activity i (m^3/km^2)
$MAXW_j$	maximum allowable water resources supply amount in zone j (m^3)
$TA_{i\ min}$	minimum demand for agricultural activity i (km^2)
$TA_{i\ max}$	maximum demand for agricultural activity i (km^2)
$TF_{i\ min}$	minimum demand for fishery activity i (km^2)
$TF_{i\ max}$	maximum demand for fishery activity i (km^2)
$TL_{i\ min}$	minimum demand for livestock husbandry activity i (head)
$TL_{i\ max}$	maximum demand for livestock husbandry activity i (head)
$TW_{i\ min}$	minimum demand for industrial activity i (RMB¥)
$TW_{i\ max}$	maximum demand for industrial activity i (RMB¥)

References

1. Pastori, M.; Udías, A.; Bouraoui, F.; Bidoglio, G. A multi-objective approach to evaluate the economic and environmental impacts of alternative water and nutrient management strategies in Africa. *J. Environ. Inform.* **2017**, *29*. [CrossRef]
2. Steenbergen, R.D.J.M.; Gelder van, P.H.A.J.M.; Miraglia, S.; Vrouwenvelder, A.C.W.M. Safety, Reliability and Risk Analysis: Beyond the Horizon. In *Proceedings of the 22nd Annual Conference on European Safety and Reliability (ESREL), Wroclaw, Poland, 14–18 September 2014*; CRC Press-Taylor & Francis Group: Boca Raton, FL, USA, 2014; pp. 1115–1120.
3. David, F.; Sandra, P. On the complementary nature of CGC-MS, CGC-FTIR, and CGC-AED for water pollution control. *J. Sep. Sci.* **2015**, *14*, 554–557. [CrossRef]
4. Pietrucha-Urbanik, K.; Tchorzewska-Cieslak, B. *Water Supply System Operation Regarding Consumer Safety Using Kohonen Neural Network*; Steenbergen, R.D.J.M., van Gelder, P.H.A.J.M., Miraglia, S., Vrouwenvelder, A.C.W.M.T., Eds.; CRC Press: Boca Raton, FL, USA, 2013.
5. Li, Y.P.; Zhang, N.; Huang, G.H.; Liu, J. Coupling fuzzy-chance constrained program with minimax regret analysis for water quality management. *Stoch. Environ. Res. Risk Assess.* **2014**, *28*, 1769–1784. [CrossRef]
6. Jorba, L.; Adillon, R. A generalization of trapezoidal fuzzy numbers based on modal interval theory. *Symmetry* **2017**, *9*, 198. [CrossRef]
7. Hu, B.; Bi, L.Q.; Dai, S.S. The orthogonality between complex fuzzy sets and its application to signal detection. *Symmetry* **2017**, *9*, 175. [CrossRef]
8. Dimitrov, V.; Driankov, D.; Petrov, A. Fuzzy equations and their applications to water pollution control. *IFAC Proc.* **1977**, *10*, 369–371. [CrossRef]
9. Chang, N.B.; Chen, H.W.; Shaw, D.G. Water pollution control in river basin by interactive fuzzy interval multiobjective programming. *J. Environ. Eng.* **1997**, *123*, 1208–1216. [CrossRef]
10. Karmakar, S.; Mujumdar, P.P. Grey fuzzy optimization model for water quality management of a river system. *Adv. Water Resour.* **2006**, *29*, 1088–1105. [CrossRef]
11. Singh, A.P.; Ghosh, S.K.; Sharma, P. Water quality management of a stretch of river Yamuna: An interactive fuzzy multi-objective approach. *Water Resour. Manag.* **2007**, *21*, 515–532. [CrossRef]
12. Sakawa, M. Interactive fuzzy goal programming for multiobjective nonlinear problems and its application to water quality management. *Electron. Commun. Jpn.* **2010**, *68*, 49–56. [CrossRef]

13. Chmielowski, W.Z. *Fuzzy Control in Environmental Engineering*; Springer International Publishing: Basel, Switzerland, 2016.
14. Forio, M.A.E.; Mouton, A.; Lock, K. Fuzzy modelling to identify key drivers of ecological water quality to support decision and policy making. *Environ. Sci. Policy* **2017**, *68*, 58–68. [CrossRef]
15. Maeda, S.; Kuroda, H.; Yoshida, K. A GIS-aided two-phase grey fuzzy optimization model for nonpoint source pollution control in a small watershed. *Paddy Water Environ.* **2016**, 1–14. [CrossRef]
16. Liu, M.; Nie, G.; Hu, M. An interval-parameter fuzzy robust nonlinear programming model for water quality management. *J. Water Resour. Prot.* **2013**, *5*, 12–16. [CrossRef]
17. Tavakoli, A.; Nikoo, M.R.; Kerachian, R. River water quality management considering agricultural return flows: application of a nonlinear two-stage stochastic fuzzy programming. *Environ. Monit. Assess.* **2015**, *187*, 158. [CrossRef] [PubMed]
18. Ji, Y.; Huang, G.; Sun, W. Water quality management in a wetland system using an inexact left-hand-side chance-constrained fuzzy multi-objective approach. *Stoch. Environ. Res. Risk Assess.* **2016**, *30*, 621–633. [CrossRef]
19. Wu, D.R.; Tan, W.W. Computationally Efficient Type-reduction Strategies for a Type-2 Fuzzy Logic Controller. In Proceedings of the 14th IEEE International Conference on Fuzzy System, Reno, NV, USA, 25–25 May 2005; pp. 353–358.
20. Cervantes, L.; Castillo, O. Type-2 fuzzy logic aggregation of multiple fuzzy controllers for airplane flight control. *Inf. Sci.* **2015**, *324*, 247–256. [CrossRef]
21. Castillo, O.; Melin, P. A review on interval type-2 fuzzy logic applications in intelligent control. *Inf. Sci.* **2014**, *279*, 615–631. [CrossRef]
22. Precup, R.E.; Angelov, P.; Costa, B.S.J.; Sayed-Mouchaweh, M. An overview on fault diagnosis and nature-inspired optimal control of industrial process applications. *Comput. Ind.* **2015**, *74*, 75–94. [CrossRef]
23. Li, Y.P.; Huang, G.H.; Li, H.Z. A recourse-based interval fuzzy programming model for point-nonpoint source effluent trading under uncertainty. *J. Am. Water Resour. Assoc.* **2015**, *50*, 1191–1207. [CrossRef]
24. Suo, C.; Li, Y.P.; Wang, C.X. A type-2 fuzzy chance-constrained programming method for planning Shanghai's energy system. *Int. J. Electr. Power Energy Syst.* **2017**, *90*, 37–53. [CrossRef]
25. Alcantud, J.C.R.; García-Sanz, M.D. Evaluations of infinite utility streams: Pareto Efficient and Egalitarian Axiomatics. *Metroeconomica* **2013**, *64*, 432–447. [CrossRef]
26. Precup, R.E.; Sabau, M.C.; Petriu, E.M. Nature-inspired optimal tuning of input membership functions of Takagi-Sugeno-Kang fuzzy models for Anti-lock Braking Systems. *Appl. Soft Comput. J.* **2015**, *27*, 575–589. [CrossRef]
27. Vrkalovic, S.; Teban, T.-A.; Borlea, I.-D. Stable Takagi-Sugeno fuzzy control designed by optimization. *Int. J. Artif. Intell.* **2017**, *15*, 17–29.
28. Chen, Z.; Zhou, S.; Luo, J. A robust ant colony optimization for continuous functions. *Expert Syst. Appl.* **2017**, *81*, 309–320. [CrossRef]
29. Tchórzewska-Cieślak, B.; Pietrucha-Urbanik, K.; Urbanik, M. Analysis of the gas network failure and failure prediction using the Monte Carlo simulation method. *Eksploatacja i Niezawodnosc Maint. Reliab.* **2016**, *18*, 254–259.
30. Qin, R.; Liu, Y.K.; Liu, Z.Q. Methods of critical value reduction for type-2 fuzzy variables and their applications. *J. Comput. Appl. Math.* **2011**, *235*, 1454–1481. [CrossRef]
31. Yongxin Bureau of Statistics Statistical Communiqué on the 2012 Economic and Social Development in Yongxin. Available online: http://old.yongxin.gov.cn/tjj/a/detail-7021.html (accessed on 28 October 2013).
32. Huang, G.H.; Chen, B.; Qin, X.S.; Mance, E. *An Integrated Decision Support System for Developing Rural Eco-Environmental Sustainability in the Mountain-River-Lake Region of Jiangxi Province, China*; Final Report; United Nations Development Program: New York, NY, USA, 2006.
33. Huang, G.H.; Qin, X.S.; Sun, W.; Nie, X.H.; Li, Y.P. An optimisation-based environmental decision support system for sustainable development in a rural area in China. *Civ. Eng. Environ. Syst.* **2009**, *26*, 65–83. [CrossRef]
34. Statistics Bulletin of the National Economic and Social Development in Yongxin County. 2016. Available online: http://www.yongxin.gov.cn/doc/2017/04/25/52136.shtml (accessed on 25 April 2017).

symmetry

MDPI

Article

Selecting Project Delivery Systems Based on Simplified Neutrosophic Linguistic Preference Relations

Sui-Zhi Luo [1], Peng-Fei Cheng [2,*], Jian-Qiang Wang [1] and Yuan-Ji Huang [3,4]

1 School of Business, Central South University, Changsha 410083, China; szlluo@csu.edu.cn (S.-Z.L.); jqwang@csu.edu.cn (J.-Q.W.)
2 School of Business, Hunan University of Science and Technology, Xiangtan 411201, China
3 School of Economics and Management, Hunan University of Science and Engineering, Yongzhou 425199, China; hs326@126.com
4 Scientific Research and Organization Department of Hunan Federation of Social Science Circles, Changsha 410003, China
* Correspondence: 1180033@hnust.edu.cn

Received: 14 July 2017; Accepted: 3 August 2017; Published: 9 August 2017

Abstract: Project delivery system selection is an essential part of project management. In the process of choosing appropriate transaction model, many factors should be under consideration, such as the capability and experience of proprietors, project implementation risk, and so on. How to make their comprehensive evaluations and select the optimal delivery system? This paper proposes a decision-making approach based on an extended linguistic preference structure: simplified neutrosophic linguistic preference relations (SNLPRs). The basic elements in SNLPRs are simplified neutrosophic linguistic numbers (SNLNs). First, several distance measures of SNLNs are introduced. A distance-based consistency index is provided to measure the consistency degree of a simplified neutrosophic linguistic preference relation (SNLPR). When the SNLPR is not acceptably consistent, a consistency-improving automatic iterative algorithm may be used. Afterwards, a decision-making method with SNLPRs is developed. The example of its application in project delivery systems' selection is offered, and a comparison analysis is given in the end as well.

Keywords: project delivery system selection; preference relations; simplified neutrosophic linguistic number; distance-based consistency index; improving consistency

1. Introduction

Construction is not only a carrier of fixed asset investment in a country, but also a channel to adjust products and industrial structures [1,2]. The choice of a project delivery system may be one of the most crucial elements of a project. There are multiple trading models that can be chosen. According to the complexity of projects and the relationships of owners with contractors, the delivery systems can be divided into four categories [3]. The general contract mode, which is the fixed price contract, mainly includes the Design Build (DB), Engineer Procure Construct Turnkey, and Design Build Operate (DBO). The management contract mode, namely the cost plus contract, principally contains the Construction-Management (CM) and Project-Management Contracting. The traditional trading model, called Design Bid Build (DBB), carries out the unit price contract. Others comprise Private Participating Infrastructure, Build Operate Transfer, Private Finance Initiative, and so on.

A lot of aspects, such as projects' characteristics, construction environment, owners' capacity, and market conditions, need to be decided in the system selection process [4]. Nevertheless, the choice of transaction modes is usually based on the subjective consciousness of the decision makers (DMs)

in engineering practice. Only a few scientific and rational decision-making methods related to the selection of project delivery systems have been established. For instance, a fuzzy approach to pick out appropriate transaction systems was presented by Mostafavi and Karamouz [5]. Wang et al. [6] constructed a project delivery system model based on fuzzy sets. After that, the analytical hierarchy process (AHP) for choosing the trade model was provided as well [7,8].

Neutrosophic sets (NSs), as a generalization of intuitionistic fuzzy sets, were originally proposed by Smarandache [9]. They can deal with consistent, hesitant, and inconsistent information at the same time. Figure 1 in Chen and Ye [10] shows the flow chart extended from fuzzy sets to neutrosophic sets (as well as simplified neutrosophic sets, single-valued neutrosophic sets, and interval neutrosophic sets). There are many extensions of NSs [11–15]. Numerous multi-criteria decision-making methods based on NSs and their extensions have been studied. For example, the ELECTRE approaches of NSs and interval neutrosophic sets (INSs) were presented by Peng et al. [16] and Zhang et al. [17], respectively. The extended TODIM [18] and MULTIMOORA [19] methods with NSs were also offered. Peng et al. [20] put forward a likelihood-based QUALIFLEX method of multi-valued neutrosophic numbers. Besides, the VIKOR method in line with INSs was developed by Bausys and Zavadskas [21]. Pouresmaeil et al. [22] proposed an extended TOPSIS and VIKOR approaches based on NSs. Furthermore, other approaches, such as the heronian mean operators [23], correlation coefficients [24,25], the WASPAS method [26–29], and the COPRAS [30,31] method were discussed and applied in diverse areas.

However, no matter which of the methods mentioned above is used, DMs are asked to give their evaluation values directly. It may be not easy for them in some cases. Sometimes, people may be accustomed to make a judgment through comparing each pair of delivery systems, especially when they cannot make direct evaluations for each single model [32]. Hence, the decision-making methods based on a judgment matrix (preference relations) with NSs or other extensions may be valuable and necessary.

In general, there are two main types of preference relations. One expresses quantitative data, such as reciprocal preference relations [33,34], interval fuzzy preference relations [35–37], intuitionistic fuzzy preference relations [38,39] triangular fuzzy preference relations [40,41], hesitant fuzzy preference relations [42,43], and some extensions [44–46]. Another contains qualitative information, like linguistic preference relations (LPRs) [47–51], hesitant fuzzy linguistic preference relations (HFLPRs) [52–54], intuitionistic linguistic preference relations (ILPRs) [55], and so on [56–58].

This paper introduces a new type of preference relations, simplified neutrosophic linguistic preference relations (SNLPRs). The basic element in SNLPRs is a simplified neutrosophic linguistic number (SNLN). The aims of the paper are as follows. On the one hand, a matrix with qualitative information may be more suitable for selecting project delivery systems as many qualitative factors are considered such as owners' ability, the technical difficulty of the project, the uncertainty of the external environment, and so on. On the other hand, the aforementioned preference relations cannot describe DMs' degrees of certainty, hesitation, and negation of their qualitative judgment simultaneously. There is a hypothesis that the membership degree of linguistic values is 1 in LPRs and HFLPRs. ILPRs only express the consistency and inconsistency of linguistic values.

On the basis of linguistic term sets and simplified neutrosophic numbers [59], simplified neutrosophic linguistic numbers (SNLNs) may be one of the most widely used extensions [60–62]. The former stands for the qualitative evaluation values of project systems [63–65], and the latter are the truth-membership, indeterminacy-membership, and false-membership of the qualitative assessment. Most importantly, these degrees are independent of each other. Consequently, considering the fuzziness of human thought and the complexity of reality, expressing preference in terms of SNLNs may be more suitable.

This paper studies a decision-making method with preference relations under simplified neutrosophic linguistic environment. The following are our innovations:

(1) Propose the Hamming distance, Euclidean distance, and Hausdorff distance of two SNLNs. In addition, several relevant properties are discussed.

(2) Present a new concept, SNLPRs. Subsequently, a distance-based consistency index is introduced to measure the consistency degree of SNLPRs.
(3) Develop a consistency-improving algorithm and a ranking method based on aggregation operators. A decision-making approach based on SNLPRs is described as well.
(4) Apply the proposed method to the project transaction model selection process. The practicability and effectiveness are demonstrated in a comparison analysis.

The remains of this paper are arranged as follows. Basic theories about SNLNs and LPRs are introduced in Section 2. Section 3 proposed some distance measures of SNLNs. In Section 4, the consistency-checking and consistency-improving issues of SNLPRs are discussed. Afterwards, there is an example and some analysis in Section 5. At last, some conclusions are drawn.

2. Preliminaries

In this section, some basic concepts and operations of linguistic term sets, SNLNs and LPRs, are reviewed.

A linguistic tem set is a collection of multiple linguistic values, like

$$\overline{S} = \{s_i | i = -u, \ldots, -1, 0, 1, \ldots, u\},$$

where s_i is a possible linguistic value, and the negation operator is $neg(s_i) = s_{-i}$. Furthermore, if and only if $i > j$, then $s_i > s_j$ [66].

Note that the linguistic term set above is discrete. In some cases, the aggregated results may be used, which are not contained in this set. Hence, Xu [66] further defined a continuous term set, like

$$S = \{s_i | i \in [-g, g]\}(g > u)$$

to extend the old one.

The following are some operations of two linguistic terms $s_i, s_j \in S$.

$$s_i \oplus_{Xu} s_j = s_{i+j} \quad (1)$$

$$s_i \oplus_{Xu} s_j = s_j \oplus_{Xu} s_i \quad (2)$$

$$\lambda s_i = s_{\lambda i}, 0 \le \lambda \le 1 \quad (3)$$

Definition 1. *In Reference [67], let* $S = \{s_i | i \in [-g, g]\}$ *be a linguistic term set. The subscript of any element* s_i *can be obtained by the function* $N(s_i) = i$*. The inverse function is* $N^{-1}(i) = s_i$*.*

Definition 2. *In Reference [68], suppose a crisp number* $\vartheta_i \in [0,1]$*. If there is a mapping from* s_i *to* ϑ_i*, then the linguistic scale function* f^* *is denoted as* $f^* : s_i \to \vartheta_i$ $(i = -g, -g+1, \cdots, g-1, g)$*, where* $0 \le \vartheta_{-g} < \vartheta_{-g+1} < \cdots < \vartheta_{g-1} < \vartheta_g$*. And* f^{*-1} *is the inverse function of* f^**.*

The linguistic scale function $f^*(s_i) = \frac{1}{2} + \frac{i}{2g}(i \in [-g, g])$ is used in this paper.

Definition 3. *In References [69,70], let* $S = \{s_i | i \in [-g, g]\}$ *be a linguistic term set.* $\eta = < h_\eta, (T_\eta, I_\eta, F_\eta) >$ *is a SNLN, where* $h_\eta(x) \in S$*, the truth-membership degree* $T_\eta(x) \in [0,1]$*, indeterminacy-membership degree* $I_\eta(x) \in [0,1]$*, and falsity-membership degree* $F_\eta(x) \in [0,1]$*, and* $0 \le T_\eta(x) + I_\eta(x) + F_\eta(x) \le 3$*.*

Definition 4. *In Reference [71],* $A = < h_A, (T_A, I_A, F_A) >$ *and* $B = < h_B, (T_B, I_B, F_B) >$ *are two arbitrary SNLNs, and their operations can be defined as follows:*

(1) *If* $S(a) > S(b)$*, then* $a > b$*;*
(2) *If* $S(a) = S(b)$ *and* $A(a) > A(b)$*, then* $a > b$*;*

(3) *If* $S(a) = S(b)$, $A(a) = A(b)$ *and* $C(a) > C(b)$, *then* $a > b$;
(4) *If* $S(a) = S(b)$, $A(a) = A(b)$ *and* $C(a) = C(b)$, *then* $a = b$.

Definition 5. *In Reference [71], assume* $a_i = (s_{\theta(a_i)}, < T_i, I_i, F_i >)$ $(i = 1, 2, \ldots, n)$ *are a sequence of SNLNs. Then the simplified neutrosophic linguistic arithmetic mean (SNLAM) operator is*

$$SNLAM(a_1, a_2, \ldots, a_n) = \bigoplus_{i=1}^{n} \frac{1}{n} a_i$$
$$= \left\langle f^{*-1}\left(\frac{1}{n}\sum_{i=1}^{n}\left(f^*(s_{\theta(a_i)})\right)\right), \left(\frac{\sum_{i=1}^{n}\left(f^*(s_{\theta(a_i)})T_i\right)}{\sum_{i=1}^{n} f^*(s_{\theta(a_i)})}, \frac{\sum_{i=1}^{n}\left(f^*(s_{\theta(a_i)})I_i\right)}{\sum_{i=1}^{n} f^*(s_{\theta(a_i)})}, \frac{\sum_{i=1}^{n}\left(f^*(s_{\theta(a_i)})F_i\right)}{\sum_{i=1}^{n} f^*(s_{\theta(a_i)})}\right)\right\rangle. \quad (4)$$

Definition 6. *In Reference [71], suppose* $a_i = (s_{\theta(a_i)}, < T_i, I_i, F_i >)$ $(i = 1, 2, \ldots, n)$ *are a sequence of SNLNs. Then the simplified neutrosophic linguistic geometric mean (SNLGM) operator is*

$$SNLGM(a_1, a_2, \ldots, a_n) = \bigotimes_{i=1}^{n} a_i^{\frac{1}{n}} = \left\langle f^{*-1}\left(\prod_{i=1}^{n}\left(f^*(s_{\theta(a_i)})\right)^{\frac{1}{n}}\right), \left(\prod_{i=1}^{n} T_i^{\frac{1}{n}}, 1 - \prod_{i=1}^{n}(1 - I_i)^{\frac{1}{n}}, 1 - \prod_{i=1}^{n}(1 - F_i)^{\frac{1}{n}}\right)\right\rangle. \quad (5)$$

Definition 7. *In Reference [71], let* $a =< h_a, (T_a, I_a, F_a) >$ *be a SNLN. Then the score function is* $SF(a) = \frac{1}{3} f^*(h_a)(T_a + 1 - I_a + 1 - F_a)$, *the accuracy function is* $AF(a) = f^*(h_a)(T_a - F_a)$, *and the certainty function is* $CF(a) = f^*(h_a)T_a$.

Definition 8. *In Reference [71], for two SNLNs* $a =< h_a, (T_a, I_a, F_a) >$ *and* $b =< h_b, (T_b, I_b, F_b) >$, *the comparison method is:*

(1) *If* $S(a) > S(b)$, *then* $a > b$;
(2) *If* $S(a) = S(b)$ *and* $A(a) > A(b)$, *then* $a > b$;
(3) *If* $S(a) = S(b)$, $A(a) = A(b)$ *and* $C(a) > C(b)$, *then* $a > b$;
(4) *If* $S(a) = S(b)$, $A(a) = A(b)$ *and* $C(a) = C(b)$, *then* $a = b$.

Definition 9. *In Reference [72], let* $X = \{x_1, x_2, \ldots, x_n\}$ *be a collection of n alternatives and* $B = (b_{ij})_{n \times n} \subset X \times X$ *be a judgment matrix. If for all* $i, j = 1, 2, \ldots, n$, *there are*

$$b_{ij} \oplus b_{ji} = s_0 \text{ and } b_{ii} = s_0, \quad (6)$$

then $B = (b_{ij})_{n \times n}$ *is a LPR, where* b_{ij} *is the preference degree of the alternative* x_i *over* x_j. *In particular, if* $b_{ij} < s_0$, x_i *is non-preferred to* x_j; x_i *is preferred to* x_j *if* $b_{ij} > s_0$; *if not,* x_j *is equivalent to* x_i.

Definition 10. *In References [72,73], if* $B = (b_{ij})_{n \times n} \subset X \times X$ *is a LPR, and*

$$b_{ij} = b_{ik} \oplus b_{kj}, (i, k, j = 1, 2, \ldots, n) \quad (7)$$

then B is a perfectly consistent LPR.

3. Distance Measures of SNLNs

Ye [70] defined the distance measure between two SNLNs, but this method has some drawbacks. Thus, some distance measures of SNLNs are redefined in this section.

Distance measure is a universal and effective way to calculate the difference between two elements. There are several common distance measures, such as Hamming distance, Euclidean distance, and Hausdorff distance.

Definition 11. *In Reference [70], suppose $\alpha = (T_\alpha, I_\alpha, F_\alpha)$ and $\beta = (T_\beta, I_\beta, F_\beta)$ are two optional SNLNs, and the subscript function is $N(s_i) = i$. $\lambda \geq 0$. The distance between α and β can be defined as below:*

$$d_Y(\alpha,\beta) = (|T_\alpha \cdot N(h_\alpha) - T_\beta \cdot N(h_\beta)|_\lambda + |I_\alpha \cdot N(h_\alpha) - I_\beta \cdot N(h_\beta)|_\lambda + |F_\alpha \cdot N(h_\alpha) - F_\beta \cdot N(h_\beta)|_\lambda)^{\frac{1}{\lambda}}. \quad (8)$$

Specially, when $\lambda = 1$, Equation (8) can be reduced to Hamming distance; when $\lambda = 2$, Equation (8) can be reduced to Euclidean distance.

The limitations of this definition are noticeable. Firstly, the calculation depends on linguistic subscripts directly, and different semantics cannot be distinguished. Secondly, this distance does not satisfy $0 \leq d_Y(\alpha,\beta) \leq 1$ and the property of triangle inequality. Thirdly, the truth-membership, indeterminacy-membership, and false-membership are put on an equal footing in the calculation process. This is intuitively irrational.

To overcome these shortcomings, the following distance measures between two SNLNs are defined.

Definition 12. *For two arbitrary SNLNs $a =< h_a, (T_a, I_a, F_a) >$ and $b =< h_b, (T_b, I_b, F_b) >$, assume the linguistic term set is $S = \{s_i | i \in [-g, g]\}$ and the subscript function is $N(s_i) = i$. Then the Hamming distance $d_H(a,b)$, Euclidean distance $d_E(a,b)$, and Hausdorff distance $d_{Ha}(a,b)$ can be defined as follows:*

$$d_H(a,b) = \frac{1}{6g}[N(|T_a \cdot h_a - T_b \cdot h_b|) + N(|(1-I_a)\cdot h_a - (1-I_b)\cdot h_b|) + N(|(1-F_a)\cdot h_a - (1-F_b)\cdot h_b|)], \quad (9)$$

$$d_E(a,b) = \sqrt{\frac{1}{12g^2}[N(|T_a \cdot h_a - T_b \cdot h_b|)^2 + N(|(1-I_a)\cdot h_a - (1-I_b)\cdot h_b|)^2 + N(|(1-F_a)\cdot h_a - (1-F_b)\cdot h_b|)^2]}, \quad (10)$$

$$d_{Ha}(a,b) = \frac{1}{2g}\max\{N(|T_a \cdot h_a - T_b \cdot h_b|), (|(1-I_a)\cdot h_a - (1-I_b)\cdot h_b|), N(|(1-F_a)\cdot h_a - (1-F_b)\cdot h_b|)\}. \quad (11)$$

Property 1. *Assume Ω is the set of all SNLNs, $a =< h_a, (T_a, I_a, F_a) >$, $b =< h_b, (T_b, I_b, F_b) >$ and $c =< h_c, (T_c, I_c, F_c) >$, $S = \{s_i | i \in [-g, g]\}$, and then the following properties are satisfied:*

(1) $0 \leq d_H(a,b) \leq 1$, $0 \leq d_E(a,b) \leq 1$, *and* $0 \leq d_{Ha}(a,b) \leq 1$, *for* $\forall a, b \in \Omega$;
(2) $d_H(a,b) = d_H(b,a)$, $d_E(a,b) = d_E(b,a)$, *and* $d_{Ha}(a,b) = d_{Ha}(b,a)$, *for* $\forall a, b \in \Omega$;
(3) *If* $a = b$, *then* $d_H(a,b) = 0$, $d_E(a,b) = 0$, *and* $d_{Ha}(a,b) = 0$, *for* $\forall a, b \in \Omega$;
(4) *If* $s_0 \leq h_a \leq h_b \leq h_c$, $T_a \leq T_b \leq T_c$, $I_a \geq I_b \geq I_c$ *and* $F_a \geq F_b \geq F_c$, *then* $d_H(a,b) \leq d_H(a,c)$, $d_H(b,c) \leq d_H(a,c)$, $d_E(a,b) \leq d_E(a,c)$, $d_E(b,c) \leq d_E(a,c)$, $d_{Ha}(a,b) \leq d_{Ha}(a,c)$ *and* $d_{Ha}(b,c) \leq d_{Ha}(a,c)$.

Proof 1.

(1) Because $i \in [-g, g]$ and $0 \leq T_a, T_b \leq 1$, $N(|T_a \cdot h_a - T_b \cdot h_b|) \in [0, 2g]$; Similarly, $0 \leq (1-I_a) \leq 1$, $0 \leq (1-I_b) \leq 1 \Rightarrow N(|(1-I_a)\cdot h_a - (1-I_b)\cdot h_b|) \in [0,2g]$, and $0 \leq (1-F_a) \leq 1$, $0 \leq (1-F_b) \leq 1 \Rightarrow N(|(1-F_a)\cdot h_a - (1-F_b)\cdot h_b|) \in [0,2g]$; thus $0 \leq [N(|T_a \cdot h_a - T_b \cdot h_b|) + N(|(1-I_a)\cdot h_a - (1-I_b)\cdot h_b|) + N(|(1-F_a)\cdot h_a - (1-F_b)\cdot h_b|)] \leq 6g \Rightarrow 0 \leq d_H(a,b) \leq 1$. Likewise, $0 \leq d_E(a,b) \leq 1$ and $0 \leq d_{Ha}(a,b) \leq 1$.
(2) $N(|T_a \cdot h_a - T_b \cdot h_b|) = N(|T_b \cdot h_b - T_a \cdot h_a|)$, $N(|(1-I_a)\cdot h_a - (1-I_b)\cdot h_b|) = N(|(1-I_b)\cdot h_b - (1-I_a)\cdot h_a|)$, and $N(|(1-F_a)\cdot h_a - (1-F_b)\cdot h_b|) = N(|(1-F_b)\cdot h_b - (1-F_a)\cdot h_a|)$, therefore $d_H(a,b) = d_H(b,a)$. Likewise, $d_E(a,b) = d_E(b,a)$ and $d_{Ha}(a,b) = d_{Ha}(b,a)$.
(3) $a = b \Rightarrow N(|T_a \cdot h_a - T_b \cdot h_b|) = 0$, $N(|(1-I_a)\cdot h_a - (1-I_b)\cdot h_b|) = 0$ and $N(|(1-F_a)\cdot h_a - (1-F_b)\cdot h_b|) = 0$, therefore $d_H(a,b) = 0$. Similarly, $d_E(a,b) = 0$ and $d_{Ha}(a,b) = 0$.
(4) Because $s_0 \leq sa_{ij} \leq sb_{ij} \leq sc_{ij}$, $T^a_{ij} \leq T^b_{ij} \leq T^c_{ij}$, and $N(s_i) = i$ is a monotone increasing function, $T^a_{ij} \cdot (sa_{ij}) \leq T^b_{ij} \cdot (sb_{ij}) \leq T^c_{ij} \cdot (sc_{ij}) \Rightarrow |T^a_{ij} \cdot (sa_{ij}) - T^b_{ij} \cdot (sb_{ij})| \leq |T^a_{ij} \cdot (sa_{ij}) - T^c_{ij} \cdot (sc_{ij})|$ and $N(|T^a_{ij} \cdot (sa_{ij}) - T^b_{ij} \cdot (sb_{ij})|) \leq N(|T^a_{ij} \cdot (sa_{ij}) - T^c_{ij} \cdot (sc_{ij})|)$; Likewise, $N(|(1-I^a_{ij}) \cdot (sa_{ij}) - (1-I^b_{ij}) \cdot (sb_{ij})|) \leq N(|(1-I^a_{ij}) \cdot (sa_{ij}) - (1-I^c_{ij}) \cdot (sc_{ij})|)$ and $N(|(1-F^a_{ij}) \cdot (sa_{ij}) - (1-F^b_{ij}) \cdot (sb_{ij})|) \leq N(|(1-F^a_{ij}) \cdot (sa_{ij}) - (1-F^c_{ij}) \cdot (sc_{ij})|)$, so $d_H(a,b) \leq d_H(a,c)$. Similarly, $d_H(b,c) \leq d_H(a,c)$, $d_E(a,b) \leq d_E(a,c)$, $d_E(b,c) \leq d_E(a,c)$, $d_{Ha}(a,b) \leq d_{Ha}(a,c)$ and $d_{Ha}(b,c) \leq d_{Ha}(a,c)$.

Then, the proof is completed. □

Example 1. *$a =< s_1, (0.2, 0.3, 0.6) >$ and $b =< s_2, (0.5, 0.1, 0.4) >$ are two SNLNs, and $g = 4$. Then $d_H(a, b) \approx 0.1125$, $d_E(a, b) \approx 0.1139$ and $d_{Ha}(a, b) \approx 0.1375$.*

4. Decision-Making Method Based on SNLPRs

In this section, the concept of SNLPRs is presented. A decision-making method is proposed after discussing the checking and improving of consistency.

4.1. The Concept of SNLPRs

Definition 13. *Given a group of n alternatives $X = \{x_1, x_2, \ldots, x_n\}$ and a matrix $K = (k_{ij})_{n \times n} \subset X \times X$. If all the elements are presented with SNLNs, $k_{ij} = (s_{ij}, < T_{ij}, I_{ij}, F_{ij} >)$, and satisfy these conditions in the following:*

$$s_{ij} \oplus_{Xu} s_{ji} = s_0 \tag{12}$$

$$T_{ij} = T_{ji}, I_{ij} = I_{ji}, F_{ij} = F_{ji} \tag{13}$$

$$(s_{ii}, < T_{ii}, I_{ii}, F_{ii} >) = (s_0, < 1, 0, 0 >), \ (i, j = 1, 2, \ldots, n), \tag{14}$$

then the matrix K on X can be regarded as a SNLPR, where s_{ij} is the degree of x_i preferred to x_j, and T_{ij}, I_{ij} and F_{ij} represent the truth-membership degree, the indeterminacy-membership degree, and the falsity-membership degree of s_{ij}, respectively.

Specifically, when $T_{ij} = 1$ and $I_{ij} = F_{ij} = 0$ for all $i, j = 1, 2, \ldots, n$, then the SNLPR is reduced to a LPR. Compared to LPRs, SNLPRs contain not only the linguistic values, but also the degrees of accuracy, hesitation, and mistake. The discrete linguistic term set can be extended to be a continuous one and DMs can express their qualitative preference information more flexibly.

From Definition 13, it can be seen that $k_{ij} = (s_{ij}, < T_{ij}, I_{ij}, F_{ij} >)$ is the preferred value of the scheme x_i to x_j, and it could be the same as $k_{ij} = (< s_{ij}, T_{ij} >, < s_{ij}, I_{ij} >, < s_{ij}, F_{ij} >)$, where $< s_{ij}, T_{ij} >$ shows x_i is s_{ij} to x_j with the true possibility T_{ij}; $< s_{ij}, I_{ij} >$ shows x_i is s_{ij} to x_j with the hesitant possibility I_{ij}; $< s_{ij}, F_{ij} >$ shows x_i is s_{ij} to x_j with the false possibility F_{ij}.

As well as LPR, the preference degree of x_j to x_i can be denoted as $< s_{-ij}, T_{ij} >$, $< s_{-ij}, I_{ij} >$ and $< s_{-ij}, F_{ij} >$, individually. That is to say, $k_{ji} = (< s_{-ij}, T_{ij} >, < s_{-ij}, I_{ij} >, < s_{-ij}, F_{ij} >)$ or $k_{ji} = (s_{-ij}, < T_{ij}, I_{ij}, F_{ij} >)$.

Example 2. *There are three alternatives $X = \{x_1, x_2, x_3\}$, and the linguistic term set $S = \{s_i | i \in [-4, 4]\}$ is used, where $s = \{s_{-4} = tremendously\ poorer,\ s_{-3} = much\ poorer, s_{-2} = poorer, s_{-1} = a\ little\ poorer,$ $s_0 = fair, s_1 = a\ little\ better, s_2 = better, s_3 = much\ better,\ s_4 = tremendously\ better\}$. If a decision maker believes the degree of x_1 preferred to x_2 is s_2, but he is not sure that he is absolutely right. According to his professional knowledge and experience in the past, he deems that he is correct with a probability of 40%, but the probability of error is 50%, and the uncertainty is 10%. In that case, his preference can be described using a SNLN, that is, $(s_2, < 0.4, 0.1, 0.5 >)$. In this way, a SNLPR can be obtained as all the alternatives above are compared with each other in a proper sequence. An example is given as follows:*

$$K_1 = \begin{bmatrix} (s_0, < 1, 0, 0 >) & (s_{-1}, < 0.4, 0.2, 0.1 >) & (s_2, < 0.3, 0.1, 0.2 >) \\ (s_1, < 0.4, 0.2, 0.1 >) & (s_0, < 1, 0, 0 >) & (s_{-3}, < 0.2, 0.5, 0.3 >) \\ (s_{-2}, < 0.3, 0.1, 0.2 >) & (s_3, < 0.2, 0.5, 0.3 >) & (s_0, < 1, 0, 0 >) \end{bmatrix}.$$

Definition 14. *Let $X = \{x_1, x_2, \ldots, x_n\}$ be a cluster of n alternatives and a SNLPR be $K = (k_{ij})_{n \times n} \subset X \times X$, where $k_{ij} = (s_{ij}, < T_{ij}, I_{ij}, F_{ij} >)$. Then the matrix $T = (< s_{ij}, T_{ij} >)_{n \times n} \subset X \times X$ is regarded as the true* *linguistic judgment matrix of K, $I = (< s_{ij}, I_{ij} >)_{n\times n} \subset X \times X$ is the hesitant linguistic judgment matrix of K, and $F = (< s_{ij}, F_{ij} >)_{n\times n} \subset X \times X$ is the false linguistic judgment matrix of K, respectively.*

From Definition 14, it can be known that for an arbitrary SNLPR, it is easy to derive its corresponding true linguistic judgment matrix, hesitant linguistic judgment matrix, and false linguistic judgment matrix. Furthermore, these linguistic judgment matrices are all defined based on the continuous linguistic terms.

Example 3. *Suppose a SNLPR is the same as in Example 2. Then, according to Definition 14, its corresponding true linguistic judgment matrix, hesitant linguistic judgment matrix, and false linguistic judgment matrix are* $T_1 = \begin{bmatrix} (< s_0, 1 >) & (< s_{-1}, 0.4 >) & (< s_2, 0.3 >) \\ (< s_1, 0.4 >) & (< s_0, 1 >) & (< s_{-3}, 0.2 >) \\ (< s_{-2}, 0.3 >) & (< s_3, 0.2 >) & (< s_0, 1 >) \end{bmatrix}$, $I_1 = \begin{bmatrix} (< s_0, 0 >) & (< s_{-1}, 0.2 >) & (< s_2, 0.1 >) \\ (< s_1, 0.2 >) & (< s_0, 0 >) & (< s_{-3}, 0.5 >) \\ (< s_{-2}, 0.1 >) & (< s_3, 0.5 >) & (< s_0, 0 >) \end{bmatrix}$ *and* $F_1 = \begin{bmatrix} (< s_0, 0 >) & (< s_{-1}, 0.1 >) & (< s_2, 0.2 >) \\ (< s_1, 0.1 >) & (< s_0, 0 >) & (< s_{-3}, 0.3 >) \\ (< s_{-2}, 0.2 >) & (< s_3, 0.3 >) & (< s_0, 0 >) \end{bmatrix}$.

4.2. Consistency Checking of SNLPRs

The deviation between two SNLPRs is calculated in this subsection, and then a distance-based consistency index is presented as well.

Definition 15. *Assume there are several alternatives $X = \{x_1, x_2, \ldots, x_n\}$. For an arbitrary SLNLPR, if for all $i, j, k = 1, 2, \ldots, n$, there is $(T_{ik} \cdot s_{ik}) \oplus (T_{kj} \cdot s_{kj}) = (T_{ij} \cdot s_{ij})$, then it can be regarded that $T = (T_{ij} \cdot s_{ij})_{n\times n} \subset X \times X$ is consistent; if for all $i, j, k = 1, 2, \ldots, n$, $((1 - I_{ik}) \cdot s_{ik}) \oplus ((1 - I_{kj}) \cdot s_{kj}) = ((1 - I_{ij}) \cdot s_{ij})$, $I = (I_{ij} \cdot s_{ij})_{n\times n} \subset X \times X$ is consistent; Similarly, if for all $i, j, k = 1, 2, \ldots, n$, $((1 - F_{ik}) \cdot s_{ik}) \oplus ((1 - F_{kj}) \cdot s_{kj}) = ((1 - F_{ij}) \cdot s_{ij})$, $F = (< F_{ij} \cdot s_{ij} >)_{n\times n} \subset X \times X$ is consistent.*

Definition 16. *Let $K = (k_{ij})_{n\times n} = (s_{ij}, < T_{ij}, I_{ij}, F_{ij} >)_{n\times n} \subset X \times X$ be a SNLPR. If the following equations are true for all $i, j, k = 1, 2, \ldots, n$:*

$$(T_{ik} \cdot s_{ik}) \oplus (T_{kj} \cdot s_{kj}) = (T_{ij} \cdot s_{ij}) \tag{15}$$

$$((1 - I_{ik}) \cdot s_{ik}) \oplus ((1 - I_{kj}) \cdot s_{kj}) = ((1 - I_{ij}) \cdot s_{ij}) \tag{16}$$

$$((1 - F_{ik}) \cdot s_{ik}) \oplus ((1 - F_{kj}) \cdot s_{kj}) = ((1 - F_{ij}) \cdot s_{ij}), \tag{17}$$

then K is regarded as a consistent SNLPR.

Example 4. *Suppose a SNLPR is the same one as Example 2. Because $(T_{12} \cdot s_{12}) \oplus (T_{23} \cdot s_{23}) = 0.4s_{-1} \oplus 0.2s_{-3} = s_{-1} \neq T_{13} \cdot s_{13}$, based on Definition 16, K_1 is not considered a consistent SNLPR.*

Theorem 1. *Given some alternatives $X = \{x_1, x_2, \ldots, x_n\}$, and the related SNLPR is $K = (k_{ij})_{n\times n} = (s_{ij}, < T_{ij}, I_{ij}, F_{ij} >)_{n\times n} \subset X \times X$. If $T = (T_{ij} \cdot s_{ij})_{n\times n} \subset X \times X$, $I = (I_{ij} \cdot s_{ij})_{n\times n} \subset X \times X$ and $F = (F_{ij} \cdot s_{ij})_{n\times n} \subset X \times X$ all have perfect consistency, then the SNLPR K is consistent, too. On the contrary, when a SNLPR K has complete consistency, then $T = (T_{ij} \cdot s_{ij})_{n\times n} \subset X \times X$, $I = (I_{ij} \cdot s_{ij})_{n\times n} \subset X \times X$ and $F = (F_{ij} \cdot s_{ij})_{n\times n} \subset X \times X$ are all absolutely consistent.*

Proof 2.

(1) Because $T = (T_{ij} \cdot s_{ij})_{n \times n} \subset X \times X$ is consistent, for all $i, j, k = 1, 2, \ldots, n$, there is $(T_{ik} \cdot s_{ik}) \oplus (T_{kj} \cdot s_{kj}) = (T_{ij} \cdot s_{ij})$ based on Definition 15. In the same way, $((1 - I_{ik}) \cdot s_{ik}) \oplus ((1 - I_{kj}) \cdot s_{kj}) = ((1 - I_{ij}) \cdot s_{ij})$ and $((1 - F_{ik}) \cdot s_{ik}) \oplus ((1 - F_{kj}) \cdot s_{kj}) = ((1 - F_{ij}) \cdot s_{ij})$, as $I = (I_{ij} \cdot s_{ij})_{n \times n} \subset X \times X$ and $F = (F_{ij} \cdot s_{ij})_{n \times n} \subset X \times X$ are consistent. That is to say, $(T_{ik} \cdot s_{ik}) \oplus (T_{kj} \cdot s_{kj}) = (T_{ij} \cdot s_{ij})$, $((1 - I_{ik}) \cdot s_{ik}) \oplus ((1 - I_{kj}) \cdot s_{kj}) = ((1 - I_{ij}) \cdot s_{ij})$ and $((1 - F_{ik}) \cdot s_{ik}) \oplus ((1 - F_{kj}) \cdot s_{kj}) = ((1 - F_{ij}) \cdot s_{ij})$ for all $i, j, k = 1, 2, \ldots, n$. On the basis of Definition 16, it can be seen that K is consistent.

(2) Since K has complete consistency, then these equations hold based on Definition 16: $(T_{ik} \cdot s_{ik}) \oplus (T_{kj} \cdot s_{kj}) = (T_{ij} \cdot s_{ij})$, $((1 - I_{ik}) \cdot s_{ik}) \oplus ((1 - I_{kj}) \cdot s_{kj}) = ((1 - I_{ij}) \cdot s_{ij})$ and $((1 - F_{ik}) \cdot s_{ik}) \oplus ((1 - F_{kj}) \cdot s_{kj}) = ((1 - F_{ij}) \cdot s_{ij})$. In the light of Definition 15, $T = (T_{ij} \cdot s_{ij})_{n \times n} \subset X \times X$, $I = (I_{ij} \cdot s_{ij})_{n \times n} \subset X \times X$ and $F = (F_{ij} \cdot s_{ij})_{n \times n} \subset X \times X$ are all consistent as well.

The proof is done now. □

Theorem 2. *Given an arbitrary SNLPR* $K = (k_{ij})_{n \times n} = (s_{ij}, < T_{ij}, I_{ij}, F_{ij} >)_{n \times n} \subset X \times X$, $i, j, k = 1, 2, \ldots n$, *if*

$$T_{ij}^* \cdot s_{ij}^* = \frac{1}{n}\{\oplus_{k=1}^{n}[(T_{ik} \cdot s_{ik}) \oplus (T_{kj} \cdot s_{kj})]\} \tag{18}$$

$$(1 - I_{ij}^*) \cdot s_{ij}^* = \frac{1}{n}\{\oplus_{k=1}^{n}[((1 - I_{ik}) \cdot s_{ik}) \oplus ((1 - I_{kj}) \cdot s_{kj})]\} \tag{19}$$

$$(1 - F_{ij}^*) \cdot s_{ij}^* = \frac{1}{n}\{\oplus_{k=1}^{n}[((1 - F_{ik}) \cdot s_{ik}) \oplus ((1 - F_{kj}) \cdot s_{kj})]\}, \tag{20}$$

then a consistent SNLPR $K^* = (k_{ij}^*)_{n \times n} = (s_{ij}^*, < T_{ij}^*, I_{ij}^*, F_{ij}^* >)_{n \times n}$ *is obtained.*

Proof 3. Since $(T_{ik}^* \cdot s_{ik}^*) \oplus (T_{kj}^* \cdot s_{kj}^*) = \frac{1}{n}\{\oplus_{e=1}^{n}[(T_{ie} \cdot s_{ie}) \oplus (T_{ek} \cdot s_{ek})]\} \oplus \frac{1}{n}\{\oplus_{e=1}^{n}[(T_{ke} \cdot s_{ke}) \oplus (T_{ej} \cdot s_{ej})]\} = \frac{1}{n}\{\oplus_{e=1}^{n}[(T_{ie} \cdot s_{ie}) \oplus (T_{ek} \cdot s_{ek}) \oplus (T_{ke} \cdot s_{ke}) \oplus (T_{ej} \cdot s_{ej})]\} = \frac{1}{n}\{\oplus_{e=1}^{n}[(T_{ie} \cdot s_{ie}) \oplus (T_{ej} \cdot s_{ej}) \oplus (T_{ek} \cdot s_0)]\} = \frac{1}{n}\{\oplus_{e=1}^{n}[(T_{ie} \cdot s_{ie}) \oplus (T_{ej} \cdot s_{ej})]\} = T_{ij}^* \cdot s_{ij}^*$; then $(T_{ik}^* \cdot s_{ik}^*) \oplus (T_{kj}^* \cdot s_{kj}^*) = T_{ij}^* \cdot s_{ij}^*$; Similarly, $((1 - I_{ik}^*) \cdot s_{ik}^*) \oplus ((1 - I_{kj}^*) \cdot s_{kj}^*) = \frac{1}{n}\{\oplus_{e=1}^{n}[((1 - I_{ie}) \cdot s_{ie}) \oplus ((1 - I_{ek}) \cdot s_{ek})]\} \oplus \frac{1}{n}\{\oplus_{e=1}^{n}[((1 - I_{ke}) \cdot s_{ke}) \oplus ((1 - I_{ej}) \cdot s_{ej})]\} = \frac{1}{n}\{\oplus_{e=1}^{n}[((1 - I_{ie}) \cdot s_{ie}) \oplus ((1 - I_{ek}) \cdot s_{ek}) \oplus ((1 - I_{ke}) \cdot s_{ke}) \oplus ((1 - I_{ej}) \cdot s_{ej})]\} = \frac{1}{n}\{\oplus_{e=1}^{n}[((1 - I_{ie}) \cdot s_{ie}) \oplus ((1 - I_{ej}) \cdot s_{ej}) \oplus ((1 - I_{ek}) \cdot s_0)]\} = \frac{1}{n}\{\oplus_{e=1}^{n}[((1 - I_{ie}) \cdot s_{ie}) \oplus ((1 - I_{ej}) \cdot s_{ej})]\} = (1 - I_{ij}^*) \cdot s_{ij}^*$ and $((1 - F_{ik}^*) \cdot s_{ik}^*) \oplus ((1 - F_{kj}^*) \cdot s_{kj}^*) = \frac{1}{n}\{\oplus_{e=1}^{n}[((1 - F_{ie}) \cdot s_{ie}) \oplus ((1 - F_{ek}) \cdot s_{ek})]\} \oplus \frac{1}{n}\{\oplus_{e=1}^{n}[((1 - F_{ke}) \cdot s_{ke}) \oplus ((1 - F_{ej}) \cdot s_{ej})]\} = \frac{1}{n}\{\oplus_{e=1}^{n}[((1 - F_{ie}) \cdot s_{ie}) \oplus ((1 - F_{ek}) \cdot s_{ek}) \oplus ((1 - F_{ke}) \cdot s_{ke}) \oplus ((1 - F_{ej}) \cdot s_{ej})]\} = \frac{1}{n}\{\oplus_{e=1}^{n}[((1 - F_{ie}) \cdot s_{ie}) \oplus ((1 - F_{ej}) \cdot s_{ej}) \oplus ((1 - F_{ek}) \cdot s_0)]\} = \frac{1}{n}\{\oplus_{e=1}^{n}[((1 - F_{ie}) \cdot s_{ie}) \oplus ((1 - F_{ej}) \cdot s_{ej})]\} = (1 - F_{ij}^*) \cdot s_{ij}^*$.

According to Equations (15)–(17), it can be seen that $K^* = (k_{ij}^*)_{n \times n} = (s_{ij}^*, < T_{ij}^*, I_{ij}^*, F_{ij}^* >)_{n \times n}$ is a consistent SNLPR.

This is the end of Proof 3. □

Note that there are only three equations above, but four variables s_{ij}^*, T_{ij}^*, I_{ij}^* and F_{ij}^* are contained. Thus, there may be many possible answers. In order to get a unique solution, the following method is used:

(1) In a general way, assume $s_{ij}^* \geq s_0$, and then $s_0 \leq T_{ij}^* \cdot s_{ij}^* \leq s_{ij}^*$, $s_0 \leq (1 - I_{ij}^*) \cdot s_{ij}^* \leq s_{ij}^*$, and $s_0 \leq (1 - F_{ij}^*) \cdot s_{ij}^* \leq s_{ij}^*$; suppose $\max\{N(T_{ij}^* \cdot s_{ij}^*), N((1 - I_{ij}^*) \cdot s_{ij}^*), N((1 - F_{ij}^*) \cdot s_{ij}^*)\} \in [a - 1, a]$, $a \leq N(s_{ij}^*)$, and then a unique SNLN $(s_{ij}^*, < T_{ij}^*, I_{ij}^*, F_{ij}^* >) = (s_a, \frac{N(T_{ij}^* \cdot s_{ij}^*)}{a}, 1 - \frac{N((1 - I_{ij}^*) \cdot s_{ij}^*)}{a}, 1 - \frac{N((1 - F_{ij}^*) \cdot s_{ij}^*)}{a})$ can be gained.

For instance, if $T_{ij}^* \cdot s_{ij}^* = 0.3$, $(1 - I_{ij}^*) \cdot s_{ij}^* = 1.2$, and $(1 - F_{ij}^*) \cdot s_{ij}^* = 2.7$, there is $\max\{N(T_{ij}^* \cdot s_{ij}^*), N((1 - I_{ij}^*) \cdot s_{ij}^*), N((1 - F_{ij}^*) \cdot s_{ij}^*)\} = \max(0.3, 1.2, 2.7) = 2.7 \in [2, 3]$, so $a = 3$ and $(s_{ij}^*, < T_{ij}^*, I_{ij}^*, F_{ij}^* >) = (s_3, < 0.1, 0.6, 0.1 >)$.

(2) For $s_{ij}^* \leq s_0, s_{ij}^* \leq T_{ij}^* \cdot s_{ij}^* \leq s_0, s_{ij}^* \leq (1 - I_{ij}^*) \cdot s_{ij}^* \leq s_0$ and $s_{ij}^* \leq (1 - F_{ij}^*) \cdot s_{ij}^* \leq s_0$; if $\min\{N(T_{ij}^* \cdot s_{ij}^*), N((1 - I_{ij}^*) \cdot s_{ij}^*), N((1 - F_{ij}^*) \cdot s_{ij}^*)\} \in [a, a+1], a \geq N(s_{ij}^*)$, then a solitary SNLN is $(s_{ij}^*, < T_{ij}^*, I_{ij}^*, F_{ij}^* >) = (s_a, \frac{N(T_{ij}^* \cdot s_{ij}^*)}{a}, 1 - \frac{N((1-I_{ij}^*) \cdot s_{ij}^*)}{a}, 1 - \frac{N((1-F_{ij}^*) \cdot s_{ij}^*)}{a})$.

For example, $T_{ij}^* \cdot s_{ij}^* = -0.3$, $(1 - I_{ij}^*) \cdot s_{ij}^* = -1.2$ and $(1 - F_{ij}^*) \cdot s_{ij}^* = -2.7$. Because $\min\{N(T_{ij}^* \cdot s_{ij}^*), N((1 - I_{ij}^*) \cdot s_{ij}^*), N((1 - F_{ij}^*) \cdot s_{ij}^*)\} = \min(-0.3, -1.2, -2.7) = 2.7 \in [-3, -2]$, then $a = -3$ and $(s_{ij}^*, < T_{ij}^*, I_{ij}^*, F_{ij}^* >) = (s_{-3}, < 0.1, 0.6, 0.1 >)$.

(3) Besides, there are two other situations: one is that the value of $N(T_{ij}^* \cdot s_{ij}^*)$, $N((1 - I_{ij}^*) \cdot s_{ij}^*)$ and $N((1 - F_{ij}^*) \cdot s_{ij}^*)$ are a positive number and two negative numbers; the other one is that there are two negative numbers and a positive number among $N(T_{ij}^* \cdot s_{ij}^*)$, $N((1 - I_{ij}^*) \cdot s_{ij}^*)$ and $N((1 - F_{ij}^*) \cdot s_{ij}^*)$. In these conditions, the final answers may not meet the requirements of $0 \leq T_{ij}^* \leq 1, 0 \leq 1 - I_{ij}^* \leq 1$ or $0 \leq 1 - F_{ij}^* \leq 1$. In other words, the consistent matrix being obtained may not be a SNLPR. But it still does not affect us to measure the consistency degree of the SNLPR, for the reason that the values of $T_{ij}^* \cdot s_{ij}^*$, $(1 - I_{ij}^*) \cdot s_{ij}^*$ and $(1 - F_{ij}^*) \cdot s_{ij}^*$ can be calculated. Thus, the consistency index of the SNLPR can be acquired (more details see Definition 12, Definition 17 and Definition 18).

Example 5. *Suppose the same SNLPR in Example 2 is given. A consistent SNLPR* $K_1^* = \begin{bmatrix} (s_0, < 1, 0, 0 >) & (s_1, < 0.133, 0.433, 0.167 >) & (s_1, < 0.067, 0.567, 0.934 >) \\ (s_{-1}, < 0.133, 0.433, 0.167 >) & (s_0, < 1, 0, 0 >) & (s_{-1}, < 0.067, 0.867, 0.433 >) \\ (s_{-1}, < 0.067, 0.567, 0.934 >) & (s_1, < 0.067, 0.867, 0.433 >) & (s_0, < 1, 0, 0 >) \end{bmatrix}$ *can be obtained in the abovementioned way.*

According to those distance measures of SNLNs in Section 3, several distance measures of SNLPRs are further defined.

Definition 17. *There are two facultative SNLPRs* $A = (a_{ij})_{n \times n}$ *and* $B = (b_{ij})_{n \times n}$, *the linguistic term set is* $S = \{s_i | i \in [-g, g]\}$. *Then Hamming distance* $D_H(A, B)$, *Euclidean distance* $D_E(A, B)$ *and Hausdorff distance* $D_{Ha}(A, B)$ *between A and B are defined as:*

$$D_H(A, B) = \frac{1}{n(n-1)} \sum_{i \neq j}^{n} d_H(a_{ij}, b_{ij}) \tag{21}$$

$$D_E(A, B) = \frac{1}{n(n-1)} \sum_{i \neq j}^{n} d_E(a_{ij}, b_{ij}) \tag{22}$$

$$D_{Ha}(A, B) = \frac{1}{n(n-1)} \sum_{i \neq j}^{n} d_{Ha}(a_{ij}, b_{ij}). \tag{23}$$

Example 6. *Assume* $g = 4$, *and two SNLPR* $A = \begin{bmatrix} (s_0, < 1, 0, 0 >) & (s_1, < 0.2, 0.3, 0.6 >) & (s_3, < 0.5, 0.4, 0.2 >) \\ (s_{-1}, < 0.2, 0.3, 0.6 >) & (s_0, < 1, 0, 0 >) & (s_{-2}, < 0.7, 0.1, 0.6 >) \\ (s_{-3}, < 0.5, 0.4, 0.2 >) & (s_2, < 0.7, 0.1, 0.6 >) & (s_0, < 1, 0, 0 >) \end{bmatrix}$,
$B = \begin{bmatrix} (s_0, < 1, 0, 0 >) & (s_2, < 0.5, 0.1, 0.4 >) & (s_3, < 0.3, 0.6, 0.1 >) \\ (s_{-2}, < 0.5, 0.1, 0.4 >) & (s_0, < 1, 0, 0 >) & (s_{-1}, < 0.2, 0.8, 0.5 >) \\ (s_{-3}, < 0.3, 0.6, 0.1 >) & (s_1, < 0.2, 0.8, 0.5 >) & (s_0, < 1, 0, 0 >) \end{bmatrix}$. *Then Hamming distance* $D_H(A, B) \approx 0.1014$, *Euclidean distance* $D_E(A, B) \approx 0.1083$ *and Hausdorff distance* $D_{Ha}(A, B) \approx 0.1375$.

Theorem 3. *Given two SNLPRs* $A = (a_{ij})_{n \times n}$ *and* $B = (b_{ij})_{n \times n}$, *if* $D_H(A, B)$, $D_E(A, B)$ *and* $D_{Ha}(A, B)$ *can satisfy the following properties:*

(1) $0 \leq D_H(A, B) \leq 1, 0 \leq D_E(A, B) \leq 1$, *and* $0 \leq D_{Ha}(A, B) \leq 1$;
(2) $D_H(A, B) = D_H(B, A)$, $D_E(A, B) = D_E(B, A)$, *and* $D_{Ha}(A, B) = D_{Ha}(B, A)$;
(3) *If* $A = B$, *then* $D_H(A, B) = 0$, $D_E(A, B) = 0$, *and* $D_{Ha}(A, B) = 0$;

(4) *Let* $A = (a_{ij})_{n\times n} = (sa_{ij}, < T_{ij}^a, I_{ij}^a, F_{ij}^a >)_{n\times n}$, $B = (b_{ij})_{n\times n} = (sb_{ij}, < T_{ij}^b, I_{ij}^b, F_{ij}^b >)_{n\times n}$ *and* $C = (c_{ij})_{n\times n} = (sc_{ij}, < T_{ij}^c, I_{ij}^c, F_{ij}^c >)_{n\times n}$ *be three SNLPRs, if* $s_0 \le sa_{ij} \le sb_{ij} \le sc_{ij}$, $T_{ij}^a \le T_{ij}^b \le T_{ij}^c$, $I_{ij}^a \ge I_{ij}^b \ge I_{ij}^c$ *and* $F_{ij}^a \ge F_{ij}^b \ge F_{ij}^c$ *for all* $i, j = 1, 2, \cdots n$, *then* $D_H(A,B) \le D_H(A,C)$, $D_H(B,C) \le D_H(A,C)$, $D_E(A,B) \le D_E(A,C)$, $D_E(B,C) \le D_E(A,C)$, $D_{Ha}(A,B) \le D_{Ha}(A,C)$ *and* $D_{Ha}(B,C) \le D_{Ha}(A,C)$.

Proof 4.

(1) Since $0 \le d_H(a_{ij}, b_{ij}) \le 1 \Rightarrow 0 \le \frac{1}{n(n-1)}\sum_{i\neq j}^n d_H(a_{ij}, b_{ij}) \le 1$, then $0 \le D_H(A,B) \le 1$.

Likewise, $0 \le D_E(A,B) \le 1$ and $0 \le D_{Ha}(A,B) \le 1$.

(2) As $d_H(a_{ij}, b_{ij}) = d_H(b_{ij}, a_{ij}) \Rightarrow \frac{1}{n(n-1)}\sum_{i\neq j}^n d_H(a_{ij}, b_{ij}) = \frac{1}{n(n-1)}\sum_{i\neq j}^n d_H(b_{ij}, a_{ij})$, then $D_H(A,B) = D_H(B,A)$. Similarly, $D_E(A,B) = D_E(B,A)$ and $D_{Ha}(A,B) = D_{Ha}(B,A)$.

(3) Because $A = B$, for all $i, j = 1, 2, \cdots n$, then $a_{ij} = b_{ij} \Rightarrow d_H(a_{ij}, b_{ij}) = 0 \Rightarrow D_H(A,B) = \frac{1}{n(n-1)}\sum_{i\neq j}^n d_H(a_{ij}, b_{ij}) = 0$. In the same way, $D_E(A,B) = 0$ and $D_{Ha}(A,B) = 0$.

(4) As $d_H(a_{ij}, b_{ij}) \le d_H(a_{ij}, c_{ij}) \Rightarrow \frac{1}{n(n-1)}\sum_{i\neq j}^n d_H(a_{ij}, b_{ij}) \le \frac{1}{n(n-1)}\sum_{i\neq j}^n d_H(a_{ij}, c_{ij}) \Rightarrow D_H(a_{ij}, b_{ij}) \le D_H(a_{ij}, c_{ij})$. Similarly, $D_H(b_{ij}, c_{ij}) \le D_H(a_{ij}, c_{ij})$, $D_E(a_{ij}, b_{ij}) \le D_E(a_{ij}, c_{ij})$, $D_E(b_{ij}, c_{ij}) \le D_E(a_{ij}, c_{ij})$, $D_{Ha}(a_{ij}, b_{ij}) \le D_{Ha}(a_{ij}, c_{ij})$ and $D_{Ha}(b_{ij}, c_{ij}) \le D_{Ha}(a_{ij}, c_{ij})$.

This is the end of Proof 4. □

Definition 18. *Let* $K = (k_{ij})_{n\times n}$ *be a SNLPR, and* $K^* = (k_{ij}^*)_{n\times n}$ *be the corresponding consistent SNLPR, the deviation between* K *and* K^* *can be expressed by a consistency index* $C(K)$ *as follows:*

$$CX(K) = 1 - D(K, K^*), \tag{24}$$

where $D(K,K^*)$ *can be replaced by* $D_H(K,K^*)$, $D_E(K,K^*)$, *or* $D_{Ha}(K,K^*)$.

Example 7. *Assume a SNLPR* $K_1 = \begin{bmatrix} (s_0, <1,0,0>) & (s_{-1}, <0.4,0.2,0.1>) & (s_2, <0.3,0.1,0.2>) \\ (s_1, <0.4,0.2,0.1>) & (s_0, <1,0,0>) & (s_{-3}, <0.2,0.5,0.3>) \\ (s_{-2}, <0.3,0.1,0.2>) & (s_3, <0.2,0.5,0.3>) & (s_0, <1,0,0>) \end{bmatrix}$ *is the same with Example 2, and its consistent SNLPR is* $K_1^* = \begin{bmatrix} (s_0, <1,0,0>) & (s_1, <0.133,0.433,0.167>) & (s_1, <0.067,0.567,0.934>) \\ (s_{-1}, <0.133,0.433,0.167>) & (s_0, <1,0,0>) & (s_{-1}, <0.067,0.867,0.433>) \\ (s_{-1}, <0.067,0.567,0.934>) & (s_1, <0.067,0.867,0.433>) & (s_0, <1,0,0>) \end{bmatrix}$ *from Example 5. If* $D_H(K,K^*)$ *is used, then* $CX(K) \approx 0.8569$; *if* $D_E(K,K^*)$ *is used, then* $CX(K) \approx 0.9936$; *if* $D_{Ha}(K,K^*)$ *is used, then* $CX(K) \approx 0.8083$.

Note that since $0 \le D(K,K^*) \le 1$, then $0 \le CX(K) \le 1$. Moreover, the greater the value of $CX(K)$, the more consistent K will be according to Definition 18.

4.3. Improving the Consistency of SNLPRs

Normally, it is difficult for DMs to provide a fully consistent SNLPR. There will be a lot of uncertainty in the decision-making process. For this reason, it is appropriate and necessary to allow the SNLPR presented by DMs satisfy the consistency in some extent. Then, the following is the concept of acceptable consistency.

Definition 19. *Let* CX *be a consistency threshold value. For an arbitrary SNLPR* K, *if the corresponding consistency index is* $CX(K)$, *and*

$$CX(K) > CX, \tag{25}$$

then K *is consistent in some extent. In other words, it has acceptable consistency.*

Zhu and Xu [74] indicated that the consistency index $CX(K)$ obeys a normal distribution, thus providing a method to determine the consistency threshold value CX. This method is used here. When the significance level $\alpha = 0.1$ and the standard deviation $\sigma = 0.2$, the consistency index threshold is shown in Table 1.

Table 1. The consistency threshold value CX.

CX	$n = 3$	$n = 4$	$n = 5$	$n = 6$	$n = 7$	$n = 8$	$n = 9$
$g = 2$	0.8235	0.7576	0.7210	0.6981	0.6824	0.6710	0.6624
$g = 3$	0.8739	0.8269	0.8007	0.7844	0.7731	0.7650	0.7589
$g = 4$	0.9020	0.8653	0.8450	0.8323	0.8235	0.8172	0.8124

Of course, the numbers in Table 1 can be used for reference. DMs can determine the value of thresholds based on their previous experience, preferences, or actual situations as well.

Example 8. *Suppose a SNLPR K_1 is the same as Example 2, and the consistency index $CX(K_1) \approx 0.8569$ from Example 7. If $D_H(K_1, K_1^*)$ is used, for $g = 4$ and $n = 3$, the consistency threshold value can be assigned with $CX = 0.9020$ based on Table 1. $CX(K_1) < CX$, and it demonstrates that K_1 does not have acceptable consistency.*

When the initial SNLPR presented by DMs is not acceptably consistent, a way to improve this SNLPR should be provided. Then, an iterative algorithm (Algorithm 1) is given to achieve acceptable consistency as follows:

Algorithm 1. Consistency-improving process with automatic iteration

Input: The initial SNLPR $K^{(s)} = (k_{ij}^{(s)})_{n\times n} = (s_{ij}^{(s)}, < T_{ij}^{(s)}, I_{ij}^{(s)}, F_{ij}^{(s)} >)_{n\times n}$, and the value of the consistency threshold CX.
Output: The modified SNLPR K_a, and its consistency index $CX(K_a)$.
Step 1: Let $s = 0$ and $ie = 0$. According to Theorem 2, acquire the consistent SNLPR $K^{*(s)} = (k_{ij}^{*(s)})_{n\times n}$ of $K^{(s)}$, where $k_{ij}^{*(s)} = (s_{ij}^{*(s)}, < T_{ij}^{*(s)}, I_{ij}^{*(s)}, F_{ij}^{*(s)} >)$.
Step 2: Choose an applicable distance, and calculate $CX(K^{(s)})$ on the basis of Definition 18.
Step 3: Determine the maximum value of iterative times $ie_{\max} \geq 1$. If $CX(K^{(s)}) > CX$ or $ie > ie_{\max}$, then go to Step 6; otherwise, go to the next step.
Step 4: Confirm the adjusted parameter $\delta \in (0,1)$. Let $T_{ij}^{(s+1)} \cdot s_{ij}^{(s+1)} = \delta(T_{ij}^{(s)} \cdot s_{ij}^{(s)}) \oplus (1-\delta)(T_{ij}^{*(s)} \cdot s_{ij}^{*(s)})$,
$(1 - I_{ij}^{(s+1)}) \cdot s_{ij}^{(s+1)} = \delta((1 - I_{ij}^{(s)}) \cdot s_{ij}^{(s)}) \oplus (1-\delta)((1 - I_{ij}^{*(s)}) \cdot s_{ij}^{*(s)})$
and $(1 - F_{ij}^{(s+1)}) \cdot s_{ij}^{(s+1)} = \delta((1 - F_{ij}^{(s)}) \cdot s_{ij}^{(s)}) \oplus (1-\delta)((1 - F_{ij}^{*(s)}) \cdot s_{ij}^{*(s)})$.
Step 5: Let $ie = ie + 1$ and $s = s + 1$, then $K^{(s)}$ is the adjusted SNFLPR. Return to Step 2.
Step 6: Let $K_a = K^{(s)}$, Output K_a and $CX(K_a)$.

This algorithm above improves the consistency through the iterative process automatically, which is convenient and efficient.

Theorem 4. *Given a SNLPR K, if K does not have acceptable consistency, it will be more consistent using Algorithm 1. That is to say, $C(K^{(s+1)}) < C(K^{(s)})$ is true. Moreover, $\lim_{s\to\infty} C(K^{(s)}) = 0$.*

Proof 5.

(1) From Equation (18), $T_{ij}^{*(s)} \cdot s_{ij}^{*(s)} = \frac{1}{n}\{\oplus_{k=1}^{n}[(T_{ik}^{(s)} \cdot s_{ik}^{(s)}) \oplus (T_{kj}^{(s)} \cdot s_{kj}^{(s)})]\}$, and then $|\mathrm{T}_{\mathrm{ij}}^{(s+1)} s_{ij}^{(s+1)} - \mathrm{T}_{\mathrm{ij}}^{*(s+1)} s_{ij}^{*(s+1)}| = |T_{ij}^{(s+1)} s_{ij}^{(s+1)} - \frac{1}{n}\{\oplus_{k=1}^{n}[(T_{ik}^{(s+1)} \cdot s_{ik}^{(s+1)}) \oplus (T_{kj}^{(s+1)} \cdot s_{kj}^{(s+1)})]\}| =$ $|\delta(T_{ij}^{(s)} \cdot s_{ij}^{(s)}) \oplus (1-\delta)(T_{ij}^{*(s)} \cdot s_{ij}^{*(s)}) - \frac{1}{n}\{\oplus_{k=1}^{n}[(\delta(T_{ik}^{(s)} \cdot s_{ik}^{(s)}) \oplus (1-\delta)(T_{ik}^{*(s)} \cdot s_{ik}^{*(s)}))\oplus(\delta(T_{kj}^{(s)} \cdot$

$s_{kj}^{(s)}) \oplus (1-\delta)(T_{kj}^{*(s)} \cdot s_{kj}^{*(s)}))]\}| \le |\delta(T_{ij}^{(s)} \cdot s_{ij}^{(s)}) \oplus -\frac{1}{n}\{\oplus_{k=1}^{n}[(\delta(T_{ik}^{(s)} \cdot s_{ik}^{(s)}) \oplus (\delta(T_{kj}^{(s)} \cdot s_{kj}^{(s)}))]\}| + |(1-\delta)(T_{ij}^{*(s)} \cdot s_{ij}^{*(s)}) - \frac{1}{n}\{\oplus_{k=1}^{n}[(1-\delta)(T_{ik}^{*(s)} \cdot s_{ik}^{*(s)})) \oplus (1-\delta)(T_{kj}^{*(s)} \cdot s_{kj}^{*(s)}))]\}|$

$= \delta|(T_{ij}^{(s)} \cdot s_{ij}^{(s)}) \oplus -\frac{1}{n}\{\oplus_{k=1}^{n}[(T_{ik}^{(s)} \cdot s_{ik}^{(s)}) \oplus (T_{kj}^{(s)} \cdot s_{kj}^{(s)})]\}| + (1-\delta)|(T_{ij}^{*(s)} \cdot s_{ij}^{*(s)}) - \frac{1}{n}\{\oplus_{k=1}^{n}[(T_{ik}^{*(s)} \cdot s_{ik}^{*(s)}) \oplus (T_{kj}^{*(s)} \cdot s_{kj}^{*(s)})]\}| = \delta|(T_{ij}^{(s)} \cdot s_{ij}^{(s)}) \oplus -T_{ij}^{*(s)} \cdot s_{ij}^{*(s)}| + |(1-\delta)|(T_{ij}^{*(s)} \cdot s_{ij}^{*(s)}) - \frac{1}{n}\{\oplus_{k=1}^{n}[(\frac{1}{n}\{\oplus_{p=1}^{n}[(T_{ip}^{(s)} \cdot s_{ip}^{(s)}) \oplus (T_{pk}^{(s)} \cdot s_{pk}^{(s)})]\}) \oplus (\frac{1}{n}\{\oplus_{p=1}^{n}[(T_{kp}^{(s)} \cdot s_{kp}^{(s)}) \oplus (T_{pj}^{(s)} \cdot s_{pj}^{(s)})]\})]\}|$

$= \delta|(T_{ij}^{(s)} \cdot s_{ij}^{(s)}) \oplus -T_{ij}^{*(s)} \cdot s_{ij}^{*(s)}| + (1-\delta)|(T_{ij}^{*(s)} \cdot s_{ij}^{*(s)}) - (\frac{1}{n}\{\oplus_{p=1}^{n}[(T_{ip}^{(s)} \cdot s_{ip}^{(s)}) \oplus (T_{pj}^{(s)} \cdot s_{pj}^{(s)})]\})|$
$= \delta|(T_{ij}^{(s)} \cdot s_{ij}^{(s)}) \oplus -T_{ij}^{*(s)} \cdot s_{ij}^{*(s)}| + (1-\delta)|(T_{ij}^{*(s)} \cdot s_{ij}^{*(s)}) - (T_{ij}^{*(s)} \cdot s_{ij}^{*(s)})| = \delta|(T_{ij}^{(s)} \cdot s_{ij}^{(s)}) \oplus -T_{ij}^{*(s)} \cdot s_{ij}^{*(s)}|$, so $N(|T_{ij}^{(s+1)}s_{ij}^{(s+1)} - T_{ij}^{*(s+1)}s_{ij}^{*(s+1)}|) \le \delta \cdot N(|(T_{ij}^{(s)} \cdot s_{ij}^{(s)}) \oplus -T_{ij}^{*(s)} \cdot s_{ij}^{*(s)}|)$;

(2) From Equation (19), $(1-I_{ij}^{*}) \cdot s_{ij}^{*} = \frac{1}{n}\{\oplus_{k=1}^{n}[((1-I_{ik}) \cdot s_{ik}) \oplus ((1-I_{kj}) \cdot s_{kj})]\}$, and then $|(1-I_{ij}^{(s+1)}) \cdot s_{ij}^{(s+1)} - (1-I_{ij}^{*(s+1)}) \cdot s_{ij}^{*(s+1)}| = |(1-I_{ij}^{(s+1)}) \cdot s_{ij}^{(s+1)} - \frac{1}{n}\{\oplus_{k=1}^{n}[((1-I_{ik}^{(s+1)}) \cdot s_{ik}^{(s+1)}) \oplus ((1-I_{kj}^{(s+1)}) \cdot s_{kj}^{(s+1)})]\}| = |\delta((1-I_{ij}^{(s)}) \cdot s_{ij}^{(s)}) \oplus (1-\delta)((1-I_{ij}^{*(s)}) \cdot s_{ij}^{*(s)}) - \frac{1}{n}\{\oplus_{k=1}^{n}[(\delta((1-I_{ik}^{(s)}) \cdot s_{ik}^{(s)}) \oplus (1-\delta)((1-I_{ik}^{*(s)}) \cdot s_{ik}^{*(s)})) \oplus (\delta((1-I_{ij}^{(s)}) \cdot s_{kj}^{(s)}) \oplus (1-\delta)((1-I_{kj}^{*(s)}) \cdot s_{kj}^{*(s)}))]\}|$

$\le |\delta((1-I_{ij}^{(s)}) \cdot s_{ij}^{(s)}) - \frac{1}{n}\{\oplus_{k=1}^{n}[(\delta((1-I_{ik}^{(s)}) \cdot s_{ik}^{(s)}) \oplus (\delta((1-I_{kj}^{(s)}) \cdot s_{kj}^{(s)}))]\}| +$

$|(1-\delta)((1-I_{ij}^{*(s)}) \cdot s_{ij}^{*(s)}) - \frac{1}{n}\{\oplus_{k=1}^{n}[(1-\delta)((1-I_{ik}^{*(s)}) \cdot s_{ik}^{*(s)})) \oplus_{k=1}^{n} (1-\delta)((1-I_{kj}^{*(s)}) \cdot s_{kj}^{*(s)}))]\}| = \delta|((1-I_{ij}^{(s)}) \cdot s_{ij}^{(s)}) - \frac{1}{n}\{\oplus_{k=1}^{n}[((1-I_{ik}^{(s)}) \cdot s_{ik}^{(s)}) \oplus ((1-I_{kj}^{(s)}) \cdot s_{kj}^{(s)}))]\}| + (1-\delta)|((1-I_{ij}^{*(s)}) \cdot s_{ij}^{*(s)}) - \frac{1}{n}\{\oplus_{k=1}^{n}[((1-I_{ik}^{*(s)}) \cdot s_{ik}^{*(s)})) \oplus_{k=1}^{n} ((1-I_{kj}^{*(s)}) \cdot s_{kj}^{*(s)}))]\}|$

$= \delta|((1-I_{ij}^{(s)}) \cdot s_{ij}^{(s)}) - (1-I_{ij}^{*(s)}) \cdot s_{ij}^{*(s)}| + (1-\delta)|((1-I_{ij}^{*(s)}) \cdot s_{ij}^{*(s)}) - \frac{1}{n}\{\oplus_{k=1}^{n}[(\frac{1}{n}\{\oplus_{p=1}^{n}[((1-I_{ip}^{(s)}) \cdot s_{ip}^{(s)}) \oplus$

$((1-I_{pk}^{(s)}) \cdot s_{pk}^{(s)})]\}) \oplus_{k=1}^{n} (\frac{1}{n}\{\oplus_{p=1}^{n}[((1-I_{kp}^{(s)}) \cdot s_{kp}^{(s)}) \oplus ((1-I_{pj}^{(s)}) \cdot s_{pj}^{(s)})]\})]\}|$

$= \delta|((1-I_{ij}^{(s)}) \cdot s_{ij}^{(s)}) - (1-I_{ij}^{*(s)}) \cdot s_{ij}^{*(s)}| + (1-\delta)|((1-I_{ij}^{*(s)}) \cdot s_{ij}^{*(s)}) - \frac{1}{n}\{\oplus_{p=1}^{n}[((1-I_{ip}^{(s)}) \cdot s_{ip}^{(s)}) \oplus ((1-I_{pj}^{(s)}) \cdot s_{pj}^{(s)})]\}| = \delta|(1-I_{ij}^{(s)}) \cdot s_{ij}^{(s)}) - (1-I_{ij}^{*(s)}) \cdot s_{ij}^{*(s)}| + (1-\delta)|(1-I_{ij}^{*(s)}) \cdot s_{ij}^{*(s)} - (1-I_{ij}^{*(s)}) \cdot s_{ij}^{*(s)}| = \delta|(1-I_{ij}^{(s)}) \cdot s_{ij}^{(s)} - (1-I_{ij}^{*(s)}) \cdot s_{ij}^{*(s)}|$, so $N((1-I_{ij}^{(s+1)}) \cdot s_{ij}^{(s+1)} - (1-I_{ij}^{*(s+1)}) \cdot s_{ij}^{*(s+1)}|) \le \delta \cdot N(|(1-I_{ij}^{(s)}) \cdot s_{ij}^{(s)} - (1-I_{ij}^{*(s)}) \cdot s_{ij}^{*(s)}|)$;

(3) From Equation (20), $(1-F_{ij}^{*}) \cdot s_{ij}^{*} = \frac{1}{n}\{\oplus_{k=1}^{n}[((1-F_{ik}) \cdot s_{ik}) \oplus ((1-F_{kj}) \cdot s_{kj})]\}$, and then $|(1-F_{ij}^{(s+1)}) \cdot s_{ij}^{(s+1)} - (1-F_{ij}^{*(s+1)}) \cdot s_{ij}^{*(s+1)}| = |(1-F_{ij}^{(s+1)}) \cdot s_{ij}^{(s+1)} - \frac{1}{n}\{\oplus_{k=1}^{n}[((1-F_{ik}^{(s+1)}) \cdot s_{ik}^{(s+1)}) \oplus ((1-F_{kj}^{(s+1)}) \cdot s_{kj}^{(s+1)})]\}| = |\delta((1-F_{ij}^{(s)}) \cdot s_{ij}^{(s)}) \oplus (1-\delta)((1-F_{ij}^{*(s)}) \cdot s_{ij}^{*(s)}) - \frac{1}{n}\{\oplus_{k=1}^{n}[(\delta((1-F_{ik}^{(s)}) \cdot s_{ik}^{(s)}) \oplus (1-\delta)((1-F_{ik}^{*(s)}) \cdot s_{ik}^{*(s)})) \oplus (\delta((1-F_{ij}^{(s)}) \cdot s_{kj}^{(s)}) \oplus (1-\delta)((1-F_{kj}^{*(s)}) \cdot s_{kj}^{*(s)}))]\}|$

$\le |\delta((1-F_{ij}^{(s)}) \cdot s_{ij}^{(s)}) - \frac{1}{n}\{\oplus_{k=1}^{n}[(\delta((1-F_{ik}^{(s)}) \cdot s_{ik}^{(s)}) \oplus (\delta((1-F_{kj}^{(s)}) \cdot s_{kj}^{(s)}))]\}| +$

$|(1-\delta)((1-F_{ij}^{*(s)}) \cdot s_{ij}^{*(s)}) - \frac{1}{n}\{\oplus_{k=1}^{n}[(1-\delta)((1-F_{ik}^{*(s)}) \cdot s_{ik}^{*(s)})) \oplus_{k=1}^{n} (1-\delta)((1-F_{kj}^{*(s)}) \cdot s_{kj}^{*(s)}))]\}|$

$= \delta|((1-F_{ij}^{(s)}) \cdot s_{ij}^{(s)}) - \frac{1}{n}\{\oplus_{k=1}^{n}[((1-F_{ik}^{(s)}) \cdot s_{ik}^{(s)}) \oplus ((1-F_{kj}^{(s)}) \cdot s_{kj}^{(s)}))]\}| + (1-\delta)|((1-F_{ij}^{*(s)}) \cdot s_{ij}^{*(s)}) - \frac{1}{n}\{\oplus_{k=1}^{n}[((1-F_{ik}^{*(s)}) \cdot s_{ik}^{*(s)})) \oplus_{k=1}^{n} ((1-F_{kj}^{*(s)}) \cdot s_{kj}^{*(s)}))]\}|$

$= \delta|((1-F_{ij}^{(s)}) \cdot s_{ij}^{(s)}) - (1-F_{ij}^{*(s)}) \cdot s_{ij}^{*(s)}| + (1-\delta)|((1-F_{ij}^{*(s)}) \cdot s_{ij}^{*(s)}) - \frac{1}{n}\{\oplus_{k=1}^{n}[(\frac{1}{n}\{\oplus_{p=1}^{n}[((1-F_{ip}^{(s)}) \cdot s_{ip}^{(s)}) \oplus$

$((1-F_{pk}^{(s)}) \cdot s_{pk}^{(s)})]\}) \oplus_{k=1}^{n} (\frac{1}{n}\{\oplus_{p=1}^{n}[((1-F_{kp}^{(s)}) \cdot s_{kp}^{(s)}) \oplus ((1-F_{pj}^{(s)}) \cdot s_{pj}^{(s)})]\})]\}|$

$= \delta|((1-F_{ij}^{(s)})\cdot s_{ij}^{(s)}) - (1-F_{ij}^{*(s)})\cdot s_{ij}^{*(s)}| + (1-\delta)|((1-F_{ij}^{*(s)})\cdot s_{ij}^{*(s)}) - \frac{1}{n}\{\oplus_{p=1}^{n}[((1-F_{ip}^{(s)})\cdot s_{ip}^{(s)})\oplus((1-F_{pj}^{(s)})\cdot s_{pj}^{(s)})]\}| = \delta|(1-F_{ij}^{(s)})\cdot s_{ij}^{(s)}) - (1-F_{ij}^{*(s)})\cdot s_{ij}^{*(s)}| + (1-\delta)|(1-F_{ij}^{*(s)})\cdot s_{ij}^{*(s)} - (1-F_{ij}^{*(s)})\cdot s_{ij}^{*(s)}| = \delta|(1-F_{ij}^{(s)})\cdot s_{ij}^{(s)} - (1-F_{ij}^{*(s)})\cdot s_{ij}^{*(s)}|$, so $N((1-F_{ij}^{(s+1)})\cdot s_{ij}^{(s+1)} - (1-F_{ij}^{*(s+1)})\cdot s_{ij}^{*(s+1)}|) \leq \delta\cdot N(|(1-F_{ij}^{(s)})\cdot s_{ij}^{(s)} - (1-F_{ij}^{*(s)})\cdot s_{ij}^{*(s)}|)$; According to (1)–(3), there is $N(|T_{ij}^{(s+1)}s_{ij}^{(s+1)} - T_{ij}^{*(s+1)}s_{ij}^{*(s+1)}|) + N((1-I_{ij}^{(s+1)})\cdot s_{ij}^{(s+1)} - (1-I_{ij}^{*(s+1)})\cdot s_{ij}^{*(s+1)}|) + N((1-F_{ij}^{(s+1)})\cdot s_{ij}^{(s+1)} - (1-F_{ij}^{*(s+1)})\cdot s_{ij}^{*(s+1)}|) \leq \delta\cdot[N(|(T_{ij}^{(s)}\cdot s_{ij}^{(s)})\oplus -T_{ij}^{*(s)}\cdot s_{ij}^{*(s)}|) + N(|(1-I_{ij}^{(s)})\cdot s_{ij}^{(s)} - (1-I_{ij}^{*(s)})\cdot s_{ij}^{*(s)}|) + N(|(1-F_{ij}^{(s)})\cdot s_{ij}^{(s)} - (1-F_{ij}^{*(s)})\cdot s_{ij}^{*(s)}|)]$, then $d_H(k_{ij}^{(s+1)},k_{ij}^{*(s+1)}) \leq \delta d_H(k_{ij}^{(s)},k_{ij}^{*(s)})$, $d_E(k_{ij}^{(s+1)},k_{ij}^{*(s+1)}) \leq \delta d_E(k_{ij}^{(s)},k_{ij}^{*(s)})$, $d_{Ha}(k_{ij}^{(s+1)},k_{ij}^{*(s+1)}) \leq \delta d_{Ha}(k_{ij}^{(s)},k_{ij}^{*(s)})$, so $CI(K^{(s+1)}) = 1 - D(K^{(s+1)},K^{*(s+1)}) = \frac{1}{n(n-1)}\sum_{i\neq j}^{n} d(k_{ij}^{(s+1)},k_{ij}^{*(s+1)}) \geq \frac{1}{n(n-1)}\sum_{i\neq j}^{n}\delta\cdot d(k_{ij}^{(s)},k_{ij}^{*(s)}) \geq CI(K^{(s)})$. In addition, $\lim_{s\to\infty} CI(K^{(s)}) = 1$.

This is the end of Proof 5. □

It can be seen from Theorem 4 that an arbitrary SNLPR that does not have a satisfactory consistency can be adjusted by the above algorithm to an acceptable matrix. The value of the adjustment parameter will have an effect on the process speed and times. DMs or other experts can determine the value of δ based on the actual situation. In general, $\delta = 0.5$ is advised. If the predetermined threshold is not satisfied, the algorithm will be repeated until the maximum number of iterations is reached.

Example 9. *Let a SNLPR K_1 be the same as Example 2. It can be seen that $CX(K_1) \approx 0.8569$, and K_1 does not have acceptable consistency from Example 8. Then Algorithm 1 can be used to improve it.*

Algorithm 1. Consistency-improving process with automatic iteration

Input: The initial SNLPR

$$K^{(0)} = K_1 = \begin{bmatrix} (s_0,<1,0,0>) & (s_{-1},<0.4,0.2,0.1>) & (s_2,<0.3,0.1,0.2>) \\ (s_1,<0.4,0.2,0.1>) & (s_0,<1,0,0>) & (s_{-3},<0.2,0.5,0.3>) \\ (s_{-2},<0.3,0.1,0.2>) & (s_3,<0.2,0.5,0.3>) & (s_0,<1,0,0>) \end{bmatrix}, \text{the}$$

consistency threshold value $CX = 0.9020$, and the maximum value of iterative times $ie_{\max} = 3$.

Output: The modified SNLPR K_a, and its consistency index $CX(K_a)$.

Step 1: As $CX(K^{(0)}) < CX$, go to the next step.

Step 2: Let $T_{ij}^{(1)}\cdot s_{ij}^{(1)} = \frac{1}{2}(T_{ij}^{(0)}\cdot s_{ij}^{(0)})\oplus\frac{1}{2}(T_{ij}^{*(0)}\cdot s_{ij}^{*(0)})$,

$(1-I_{ij}^{(1)})\cdot s_{ij}^{(1)} = \frac{1}{2}((1-I_{ij}^{(0)})\cdot s_{ij}^{(0)})\oplus\frac{1}{2}((1-I_{ij}^{*(0)})\cdot s_{ij}^{*(0)})$

and $(1-F_{ij}^{(1)})\cdot s_{ij}^{(1)} = \frac{1}{2}((1-F_{ij}^{(0)})\cdot s_{ij}^{(0)})\oplus\frac{1}{2}((1-F_{ij}^{*(0)})\cdot s_{ij}^{*(0)})$.

Step 3: Let $ie = 1$,

$$K^{(1)} = (k_{ij}^{(1)})_{n\times n} = \begin{bmatrix} (s_0,<1,0,0>) & (s_{-1},<0.133,0.883,0.867>) & (s_2,<0.167,0.442,0.584>) \\ (s_1,<0.133,0.883,0.867>) & (s_0,<1,0,0>) & (s_{-2},<0.167,0.592,0.334>) \\ (s_{-2},<0.167,0.442,0.584>) & (s_2,<0.167,0.592,0.334>) & (s_0,<1,0,0>) \end{bmatrix}$$

Step 4: The consistent SNLPR

$$K^{*(1)} = \begin{bmatrix} (s_0,<1,0,0>) & (s_1,<0.134,0.434,0.367>) & (s_1,<0.067,0.609,0.934>) \\ (s_{-1},<0.134,0.434,0.367>) & (s_0,<1,0,0>) & (s_{-1},<0.067,0.783,0.434>) \\ (s_{-1},<0.067,0.609,0.934>) & (s_1,<0.067,0.783,0.434>) & (s_0,<1,0,0>) \end{bmatrix}.$$

Step 5: $D_H(K^{(1)},K^{*(1)})$ is used, and $CX(K^{(1)}) \approx 0.9276$ on the basis of Definition 18.

Step 6: As $CX(K^{(1)}) > CX$, go to the next step.

Step 7: Let $K_a = K^{(1)}$, Output

$$K_a = \begin{bmatrix} (s_0,<1,0,0>) & (s_{-1},<0.133,0.883,0.867>) & (s_2,<0.167,0.442,0.584>) \\ (s_1,<0.133,0.883,0.867>) & (s_0,<1,0,0>) & (s_{-2},<0.167,0.592,0.334>) \\ (s_{-2},<0.167,0.442,0.584>) & (s_2,<0.167,0.592,0.334>) & (s_0,<1,0,0>) \end{bmatrix},$$

and $CX(K_a) \approx 0.9276$.

4.4. A Decision-Making Approach with SNLPRs

In this section, a decision-making method based on SNLPRs is presented.

Take a decision-making problem under simplified neutrosophic linguistic environment into consideration. Suppose there are a group of alternatives $X = \{x_1, x_2, \ldots, x_n\}$. The DMs want to get the ranking or select the eligible alternative from them. Then a preference matrix is formed after the linguistic term set $S = \{s_i | i \in [-g, g]\}$ is given. The basic elements in this matrix are SNLNs. Then the method based on SNLPRs is provided as Algorithm 2:

Algorithm 2. Decision-making approach with SNLPRs

Input: The initial SNLPR $K = (k_{ij})_{n\times n} = (s_{ij}, < T_{ij}, I_{ij}, F_{ij} >)_{n\times n}$.
Output: The ranking result and the best alternative x^*.
Step 1: Choose a distance measure and calculate the value of $CX(K)$ according to Equation (24)
Step 2: Determine the threshold value CX. If $CX(K) < CX$, then improve it by Algorithm 1 until it is acceptably consistent.
Step 3: Aggregate each row of preference values in K using the SNLAM or SNLGM operator.
Step 4: Calculate the score function $S(x_i)$ of overall preference degree of each $x_i (i = 1, 2, \ldots, n)$ by Definition 7.
Step 5: Rank the alternatives $x_i (i = 1, 2, \ldots, n)$ on the basis of comparison method in Definition 8, and then output the ranking and the optimal alternative(s) x^*.

Note that it is a common and useful way to use aggregation operators to aggregate preference information, and then get the ranking result according to some comparison rules. However, Hou [75] pointed out that using arithmetic mean aggregation may get a reverse ranking. Therefore, the SNLGM operator to aggregate preference values is better.

5. Application and Comparison

The proposed decision-making method is applied to selecting project delivery models in this section. Some related comparison analyses are presented in the end.

YG Construction Co., Ltd. planned to select two suitable delivery models for a road construction project. The first one is chosen at once, and second one is reserved and accepted if necessary in the future. After a preliminary selection, four satisfactory options, DB, DBB, DBO, and CM, denoted by $\{x_1, x_2, x_3, x_4\}$, respectively, are considered. In order to pick out the right model, the project manager invites two experts to make evaluations together.

According to the properties of this project, the actual environment and the capability of owners, the project manager evaluates four alternatives. He presents his preference information with linguistic values. The linguistic term set used is $S = \{s_i | i \in [-4, 4]\}$, where $s = \{s_{-4} = \textit{much poorer},\ s_{-3} = \textit{a lot poorer},\ s_{-2} = \textit{poorer},\ s_{-1} = \textit{slightly poorer},\ s_0 = \textit{fair},\ s_1 = \textit{slightly better},\ s_2 = \textit{better},\ s_3 = \textit{a lot better},\ s_4 = \textit{much better}\}$. Simultaneously, he gives the corresponding hesitant degrees of each preference value. Then two experts are asked to judge the possibility that the evaluation is inaccurate. In this way, SNLNs may be a good indication of their preference. As an example, the manager holds the view that x_1 is s_{-1} to x_2, but he is not sure of his assessment. He thinks the degree of hesitation is 0.3. Afterwards, there is a probability of 0.9 that s_{-1} is right, and a 0.2 probability of error given by two specialists. Therefore, they can be expressed by a SNLN $k_{12} = (s_{-1}, < 0.9, 0.3, 0.2 >)$.

In the end, all the preference information yields an SNLPR as follows:

$$K = \begin{bmatrix} (s_0, <1,0,0>) & (s_{-1}, <0.9,0.3,0.2>) & (s_1, <0.7,0.5,0.4>) & (s_3, <0.8,0.3,0.1>) \\ (s_1, <0.9,0.3,0.2>) & (s_0, <1,0,0>) & (s_1, <0.5,0.2,0.5>) & (s_{-2}, <0.9,0.2,0.3>) \\ (s_{-1}, <0.7,0.5,0.4>) & (s_{-1}, <0.5,0.2,0.5>) & (s_0, <1,0,0>) & (s_{-3}, <0.6,0.7,0.1>) \\ (s_{-3}, <0.8,0.3,0.1>) & (s_2, <0.9,0.2,0.3>) & (s_3, <0.6,0.7,0.1>) & (s_0, <1,0,0>) \end{bmatrix}.$$

5.1. Illustration

The decision-making method proposed in Section 4.4 is used to rank four options and select two models among them. The following are the specific steps:

Step 1: After discussion, DMs choose $D_H(K, K^*)$, and then calculate $CX(K) \approx 0.8628$ according to Equation (24).

Step 2: Because $g = 4$ and $n = 4$, DMs suggest the threshold value $CX = 0.8653$ from Table 1. And they find $CX(K) < CX$, then use Algorithm 1 to improve it as follows:

Let $K^{(0)} = K = (k_{ij}^{(0)})_{n\times n} = (s_{ij}^{(0)}, < T_{ij}^{(0)}, I_{ij}^{(0)}, F_{ij}^{(0)} >)_{n\times n}$ and $ie = 0$. According to Theorem 2, the consistent SNLPR $K^{*(0)} = (k_{ij}^{*(0)})_{n\times n} = (s_{ij}^{*(0)}, < T_{ij}^{*(0)}, I_{ij}^{*(0)}, F_{ij}^{*(0)} >)_{n\times n} =$

$$\begin{bmatrix} (s_0, <1,0,0>) & (s_1, <0.65,0.5,0.35>) & (s_2, <0.65,0.4875,0.2125>) & (s_1, <0.25,0.625,0.725>) \\ (s_{-1}, <0.65,0.5,0.35>) & (s_0, <1,0,0>) & (s_1, <0.65,0.475,0.075>) & (s_{-1}, <0.4,0.875,0.625>) \\ (s_{-2}, <0.65,0.4875,0.2125>) & (s_{-1}, <0.65,0.475,0.075>) & (s_0, <1,0,0>) & (s_{-2}, <0.525,0.675,0.35>) \\ (s_{-1}, <0.25,0.625,0.725>) & (s_1, <0.4,0.875,0.625>) & (s_2, <0.525,0.675,0.35>) & (s_0, <1,0,0>) \end{bmatrix}.$$

Since $CX(K^{(0)}) = CX(K) \approx 0.8649 < CX$, determine $\delta = \frac{1}{2}$ and $ie_{\max} = 3$. Let $K^{(1)} = K = (k_{ij}^{(1)})_{n\times n} = (s_{ij}^{(1)}, < T_{ij}^{(1)}, I_{ij}^{(1)}, F_{ij}^{(1)} >)_{n\times n} =$

$$\begin{bmatrix} (s_0, <1,0,0>) & (s_{-1}, <0.125,0.9,0.925>) & (s_2, <0.5,0.6188,0.4563>) & (s_2, <0.6625,0.3813,0.2563>) \\ (s_1, <0.125,0.9,0.925>) & (s_0, <1,0,0>) & (s_1, <0.575,0.3375,0.2875>) & (s_{-2}, <0.55,0.5688,0.5563>) \\ (s_{-2}, <0.5,0.6188,0.4563>) & (s_{-1}, <0.575,0.3375,0.2875>) & (s_0, <1,0,0>) & (s_{-2}, <0.7125,0.6125,0>) \\ (s_{-2}, <0.6625,0.3813,0.2563>) & (s_2, <0.55,0.5688,0.5563>) & (s_2, <0.7125,0.6125,0>) & (s_0, <1,0,0>) \end{bmatrix},$$

where $T_{ij}^{(1)} \cdot s_{ij}^{(1)} = \frac{1}{2}(T_{ij}^{(0)} \cdot s_{ij}^{(0)}) \oplus \frac{1}{2}(T_{ij}^{*(0)} \cdot s_{ij}^{*(0)})$, $(1 - I_{ij}^{(1)}) \cdot s_{ij}^{(1)} = \frac{1}{2}((1 - I_{ij}^{(0)}) \cdot s_{ij}^{(0)}) \oplus \frac{1}{2}((1 - I_{ij}^{*(0)}) \cdot s_{ij}^{*(0)})$ and $(1 - F_{ij}^{(1)}) \cdot s_{ij}^{(1)} = \frac{1}{2}((1 - F_{ij}^{(0)}) \cdot s_{ij}^{(0)}) \oplus \frac{1}{2}((1 - F_{ij}^{*(0)}) \cdot s_{ij}^{*(0)})$. As $CX(K^{(1)}) \approx 0.9314 > CX$, output $K^{(1)}$.

Step 3: Aggregate each row of preference values in $K^{(1)} (i \neq j)$ using SNLGM operator. Then the overall preferences are $px_1 \approx (s_{0.7622}, < 0.5966, 0.7132, 0.6882 >)$, $px_2 \approx (s_{-0.3160}, < 0.5570, 0.6943, 0.7127 >)$, $px_3 \approx (s_{-1.7106}, < 0.5039, 0.5392, 0.2710 >)$ and $px_4 \approx (s_{0.1602}, < 0.5102, 0.5307, 0.3090 >)$.

Step 4: Calculate the score function $S(px_i)(i = 1, 2, \ldots, n)$ by Definition 7: $S(px_1) \approx 0.2372$, $S(px_2) \approx 0.1765$, $S(px_3) \approx 0.1616$ and $S(px_4) \approx 0.2896$.

Step 5: Because $S(px_4) > S(px_1) > S(px_2) > S(px_3) \Rightarrow x_4 > x_1 > x_2 > x_3$, and the optimal alternative is $x^* = x_4$, the second alternative is x_1.

5.2. Comparison Analysis

Considering the concept of SNLPRs is newly proposed, several approaches related to other kinds of preference relations are chosen to make a comparison in this subsection.

As the expressions of basic elements in different preference relations are diverse, the first task is to transform the SNLNs in SNLPRs into the corresponding expression. Then, the same problem will be solved. The following are the information conversion process and major steps of the related methods:

(1) Single-valued trapezoidal neutrosophic preference relations (SVTNPRs) [44]

First, SNLNs in SNLPRs can be transformed into single-valued trapezoidal neutrosophic numbers (SVTNNs). A suitable way is changing the linguistic values of SNLNs into trapezoidal fuzzy numbers in SVTNNs and keeping membership degrees. The converted values can be denoted by $A_i = (a_i, b_i, c_i, d_i) = (\max\{\frac{2i+2g-1}{4g+3}, 0\}, \frac{2i+2g}{4g+3}, \frac{2i+2g+1}{4g+3}, \min\{\frac{2i+2g+2}{4g+3}, 1\})$ according to [76]. As an illustration, $(s_1, < 0.7, 0.5, 0.4 >)$ can be regarded as $([0.474, 0.526, 0.579, 0.632], < 0.7, 0.5, 0.4 >)$. Then, the corresponding single-valued trapezoidal neutrosophic matrix is

$$K^{TN} = \begin{bmatrix} ([0.368,0.421,0.474,0.526],<1,0,0>) & ([0.263,0.316,0.368,0.421],<0.9,0.3,0.2>) & ([0.474,0.526,0.579,0.632],<0.7,0.5,0.4>) & ([0.684,0.737,0.790,0.842],<0.8,0.3,0.1>) \\ ([0.474,0.526,0.579,0.632],<0.9,0.3,0.2>) & ([0.368,0.421,0.474,0.526],<1,0,0>) & ([0.474,0.526,0.579,0.632],<0.5,0.2,0.5>) & ([0.158,0.211,0.263,0.316],<0.9,0.2,0.3>) \\ ([0.263,0.316,0.368,0.421],<0.7,0.5,0.4>) & ([0.263,0.316,0.368,0.421],<0.5,0.2,0.5>) & ([0.368,0.421,0.474,0.526],<1,0,0>) & ([0.053,0.105,0.158,0.211],<0.6,0.7,0.1>) \\ ([0.053,0.105,0.158,0.211],<0.8,0.3,0.1>) & ([0.579,0.632,0.684,0.737],<0.9,0.2,0.3>) & ([0.684,0.737,0.790,0.842],<0.6,0.7,0.1>) & ([0.368,0.421,0.474,0.526],<1,0,0>) \end{bmatrix}.$$

Subsequently, using the method in [44] to get the consistent preference matrix, and the ranking is $x_4 > x_1 > x_2 > x_3$.

(2) ILPRs [55]

In the beginning, SNLNs should be converted into intuitionistic linguistic numbers. The linguistic values can remain, $\frac{1}{3}(T_{ij} + 1 - I_{ij} + 1 - F_{ij})$ may be equivalent to u, and $v = 1 - u$. As an example, $(s_{-1}, <0.9, 0.3, 0.2>)$ can be converted into $< s_{-1}, (0.8, 0.2) >$. Then, an intuitionistic linguistic preference relation (ILPR)$K^I = \begin{bmatrix} <s_0,(1,0)> & <s_{-1},(0.8,0.2)> & <s_1,(0.6,0.4)> & <s_3,(0.8,0.2)> \\ <s_1,(0.8,0.2)> & <s_0,(1,0)> & <s_1,(0.6,0.4)> & <s_{-2},(0.8,0.2)> \\ <s_{-1},(0.6,0.4)> & <s_{-1},(0.6,0.4)> & <s_0,(1,0)> & <s_{-3},(0.6,0.4)> \\ <s_{-3},(0.8,0.2)> & <s_2,(0.8,0.2)> & <s_3,(0.6,0.4)> & <s_0,(1,0)> \end{bmatrix}$ is got. As $CI(K^I) \approx 0.8819 < \theta(0.9016)$, let $\beta = 0.8$, and then the consistency index of the adjusted ILPR is $CI(K^{I(1)}) \approx 0.9069 > \theta(0.9016)$. Thus, using the method in [55], the preferred degree matrix $U = \begin{pmatrix} 1/2 & 1 & 1 & 6/7 \\ 0 & 1/2 & 1 & 0 \\ 0 & 0 & 1/2 & 0 \\ 1/7 & 1 & 1 & 1/2 \end{pmatrix}$. Since the preferred degrees $r_1(\frac{47}{14}) > r_4(\frac{37}{14}) > r_2(\frac{3}{2}) > r_3(\frac{1}{2})$, then $x_1 \succ x_4 \succ x_2 \succ x_3$.

(3) HFLPRs [54]

Firstly, SNLNs can be converted to hesitant fuzzy linguistic term sets. For example, $(s_{-0.9}, s_{-0.3}, s_{-0.2})$ can take the place of $(s_{-1}, <0.9, 0.3, 0.2>)$. Hence, the corresponding HFLPR is $K^H = \begin{bmatrix} (s_0) & (s_{-0.9},s_{-0.3},s_{-0.2}) & (s_{0.4},s_{0.5},s_{0.7}) & (s_{0.3},s_{0.9},s_{2.4}) \\ (s_{0.9},s_{0.3},s_{0.2}) & (s_0) & (s_{0.2},s_{0.5},s_{0.5}) & (s_{-1.8},s_{-0.6},s_{-0.4}) \\ (s_{-0.4},s_{-0.5},s_{-0.7}) & (s_{-0.2},s_{-0.5},s_{-0.5}) & (s_0) & (s_{-2.1},s_{-1.8},s_{-0.3}) \\ (s_{-0.3},s_{-0.9},s_{-2.4}) & (s_{1.8},s_{0.6},s_{0.4}) & (s_{2.1},s_{1.8},s_{0.3}) & (s_0) \end{bmatrix}$. The expected 2-tuple linguistic preference relation is $E^{H(1)} = \begin{bmatrix} (s_0,0) & (s_0,-7/15) & (s_1,-7/15) & (s_1,1/5) \\ (s_0,7/15) & (s_0,0) & (s_0,2/5) & (s_{-1},1/15) \\ (s_{-1},7/15) & (s_0,-2/5) & (s_0,0) & (s_{-1},-2/5) \\ (s_{-1},-1/5) & (s_1,-1/15) & (s_1,2/5) & (s_0,0) \end{bmatrix}$. Because $CI(E^{H(1)}) \approx 0.0860 < \overline{\overline{CI}}(0.1347)$, the matrix is acceptably consistent. Then use the aggregation operators and comparison method, and the ranking is $x_1 \succ x_4 \succ x_2 \succ x_3$.

(4) LPRs [77]

At first, the conversion function $k_{ij}^L = \frac{1}{3}(T_{ij} + 1 - I_{ij} + 1 - F_{ij})k_{ij}$ will be used to convert SNLNs in SNLPR K into linguistic variables in a LPR K^L. For instance, $(s_{-1}, <0.9, 0.3, 0.2>)$ can be replaced by $s_{-0.8}$. Then $K^L = \begin{bmatrix} s_0 & s_{-0.8} & s_{0.6} & s_{2.4} \\ s_{0.8} & s_0 & s_{0.6} & s_{-1.6} \\ s_{-0.6} & s_{-0.6} & s_0 & s_{-1.8} \\ s_{-2.4} & s_{1.6} & s_{1.8} & s_0 \end{bmatrix}$. Since $CI(K^L) \approx 0.2667 > \delta_0(0.1347)$, the automatic iterative Algorithm 1 in [77] is used. The consistency index of the adjusted LPR is $CI(K^{L(1)}) \approx 0.1334 < \delta_0(0.1347)$, and then the final ranking is $x_4 > x_1 > x_2 > x_3$.

Subsequently, comparisons with each method in terms of backgrounds, consistency-improving processes, ranking methods, and ranking results are made in Table 2.

Table 2. Comparison of different methods.

Approaches	Backgrounds	Improving Consistency	Ranking Methods	Ranking Orders
Liang et al. [44]	SVTNPRs	Interactive feedback	Arithmetic operator	$x_4 > x_1 > x_2 > x_3$
Meng et al. [55]	ILPRs	Automatic iteration	Preferred degrees	$x_1 \succ x_4 \succ x_2 \succ x_3$
Wu and Xu [54]	HFLPRs	Interactive feedback	Expected values	$x_1 \succ x_4 \succ x_2 \succ x_3$
Jin et al. [77]	LPRs	Automatic iteration	Arithmetic operator	$x_4 > x_1 > x_2 > x_3$
The proposed method	SNLPRs	Automatic iteration	Geometric operator	$x_4 > x_1 > x_2 > x_3$

In the ranking results of Table 2 only the order of x_1 and x_4 varies, which demonstrates the effectiveness of the proposed method. The in-depth comparison analyses are shown as follows:

(1) Comparison with References [44] and [77]: the same ranking results are obtained using the methods in [44,77] and our approach. An interactive feedback is used to improve the consistency in [44]. It may be a little difficult for DMs to do this work, especially when the alternatives are numerous. In addition, the arithmetic operator used may cause a reversal of ranking in some cases. Jin et al. described information with linguistic term sets in Reference [77]. However, all of the membership degrees are missing in LPRs. And the arithmetic operator used in Reference [77] also has the limitation of sorting reversal.
(2) Comparison with Reference [55]: both [55] and our method choose the automatic iteration to improve consistency. The reason for the different ranking results may be that there are only membership and non-membership degrees in ILPRs. The conversion function possibly led to a loss of the original information.
(3) Comparison with Reference [54]: the difference between [54] and our approach is that there is no process of consistency improvement in HFLPRs. Moreover, the truth-membership, indeterminacy-membership, and false-membership of the linguistic values in SNLPRs have identical roles in HFLPRs. This may be another explanation of the different rankings.

According to the analysis above, the strengths of the presented approach are obvious. First of all, the basic elements in SNLPRs, SNLNs, contain three independent membership degrees to describe the consistent, hesitant, and inconsistent information, respectively. It means that the problem of evaluation information being missing is avoided to a greater extent. Thus, the proposed method is more suitable for solving problems in a simplified neutrosophic linguistic environment. Secondly, the consistency-improving process is an automatic iteration algorithm. It saves time and increases convenience for DMs. In addition, as mentioned in Section 4.4, the geometric operator being used may avoid the problem of ranking reversal. It is easy for us to understand and operate. Finally, the flexibility is increased as the linguistic scale function can be changed in different semantic situations.

6. Discussion and Conclusions

Appropriate project delivery systems play an irreplaceable role in promoting the development of the construction industry. The paper provided a decision-making approach with SNLPRs to solve the problem of selecting an optimal system under simplified neutrosophic linguistic circumstances. Several distance measures of SNLNs, which are the basic elements of SNLPRs, were redefined. They can overcome the drawbacks of the definition of Ye [70], so that the differences between two SNLNs can be well distinguished. Moreover, the paper created a distance-based consistency index to check the consistency of SNLPRs. A consistency-improving algorithm was also suggested. The effectiveness and advantages of this method were displayed by an illustration of selecting project delivery systems and the corresponding comparison analysis.

Nevertheless, the proposed method still has some limitations, such as a case of SNLPRs with incomplete assessment information. In order to make the method based on SNLPRs more effectively and widely used in engineering projects, several future works can be planned as follow: (1) Other kinds of consistency, such as the multiplicative consistency of SNLPRs may be presented; (2) a situation where the linguistic term sets are unbalanced [78] may be under consideration; (3) decision-making methods based on incomplete SNLPRs are worth studying.

Acknowledgments: The authors thank the editors and anonymous reviewers for their helpful comments and suggestions. This work was supported by a Key Project of Hunan Social Science Achievement Evaluation Committee (XSP2016040508), the human philosophy social science fund project (15JD21), and the Natural Science Foundation of Hunan Province (2017JJ3181).

Author Contributions: Sui-Zhi Luo, Peng-Fei Cheng, and Jian-Qiang Wang conceived and worked together to achieve this work, Sui-Zhi Luo wrote the paper; Yuan-Ji Huang made contributions to the case study.

Conflicts of Interest: The authors declare no conflict of interest.

References

1. Uhlik, F.T.; Eller, M.D. Alternative delivery approaches for military medical construction projects. *J. Archit. Eng.* **1996**, *5*, 149–155. [CrossRef]
2. Partington, D.; Pellegrinelli, S.; Young, M. Attributes and levels of programme management competence: An interpretive study. *Int. J. Proj. Manag.* **2005**, *23*, 87–95. [CrossRef]
3. Hong, K.E.; Liu, C.; Management, S.O. Research on importance ranking of influencing factors on selecting project delivery system. *J. Civ. Eng. Manag.* **2014**, *3*, 2224–2230.
4. Lahdenperä, P.; Koppinen, T. Financial analysis of road project delivery systems. *J. Financ. Manag. Prop. Constr.* **2009**, *14*, 61–78. [CrossRef]
5. Mostafavi, A.; Karamouz, M. Selecting appropriate project delivery system: Fuzzy approach with risk analysis. *J. Constr. Eng. Manag.* **2010**, *136*, 923–930. [CrossRef]
6. Wang, L.Y.; An, X.W.; Li, H.M. Applying fuzzy set model for selecting project delivery system. In Proceedings of the International Conference on Simulation and Modeling Methodologies, Technologies and Applications, Vienna, Austria, 28–30 August 2014; Volume 60, pp. 1301–1308.
7. Barati, K.; Shiran, G.R.; Sepasgozar, S.M.E. Selecting optimal project delivery system for infrastructural projects using analytic hierarchy process. *Am. J. Civ. Eng. Archit.* **2015**, *3*, 212–217.
8. Mahdi, I.M.; Alreshaid, K. Decision support system for selecting the proper project delivery method using analytical hierarchy process (AHP). *Int. J. Proj. Manag.* **2005**, *23*, 564–572. [CrossRef]
9. Smarandache, F. *A Unifying Field in Logics. Neutrosophy: Neutrosophic Probability, Set and Logic*; American Research Press: Rehoboth, DE, USA, 1999.
10. Chen, J.Q.; Ye, J. Some single-valued neutrophic dombi weighted aggregation operators for multiple attribute decision-making. *Symmetry* **2017**, *9*, 82. [CrossRef]
11. Peng, H.G.; Zhang, H.Y.; Wang, J.Q. Probability multi-valued neutrosophic sets and its application in multi-criteria group decision-making problems. *Neural Comput. Appl.* **2016**, 1–21. [CrossRef]
12. Ye, J. Trapezoidal neutrosophic set and its application to multiple attribute decision-making. *Neural Comput. Appl.* **2015**, *26*, 1157–1166. [CrossRef]
13. Ma, Y.X.; Wang, J.Q.; Wang, J.; Wu, X.H. An interval neutrosophic linguistic multi-criteria group decision-making method and its application in selecting medical treatment options. *Neural Comput. Appl.* **2017**, *28*, 2745–2764. [CrossRef]
14. Li, Y.Y.; Zhang, H.Y.; Wang, J.Q. Linguistic neutrosophic sets and their application in multi-criteria decision-making problem. *Int. J. Uncertain. Quantif.* **2017**, *7*, 135–154. [CrossRef]
15. Tian, Z.P.; Wang, J.; Wang, J.Q.; Zhang, H.Y. An improved MULTIMOORA approach for multi-criteria decision-making based on interdependent inputs of simplified neutrosophic linguistic information. *Neural Comput. Appl.* **2016**, 1–13. [CrossRef]

16. Peng, J.J.; Wang, J.Q.; Zhang, H.Y.; Chen, X.H. An outranking approach for multi-criteria decision-making problems with simplified neutrosophic sets. *Appl. Soft Comput.* **2014**, *25*, 336–346. [CrossRef]
17. Zhang, H.Y.; Wang, J.Q.; Chen, X.H. An outranking approach for multi-criteria decision-making problems with interval-valued neutrosophic sets. *Neural Comput. Appl.* **2016**, *27*, 615–627. [CrossRef]
18. Zhang, M.C.; Liu, P.D.; Shi, L.L. An extended multiple attribute group decision-making TODIM method based on the neutrosophic numbers. *J. Intell. Fuzzy Syst.* **2016**, *30*, 1773–1781. [CrossRef]
19. Stanujkic, D.; Zavadskas, E.K.; Smarandache, F.; Brauers, W.K.M.; Karabasevic, D. A neutrosophic extension of the MULTIMOORA method. *Informatica* **2017**, *28*, 181–192. [CrossRef]
20. Peng, J.J.; Wang, J.Q.; Yang, W.E. A multi-valued neutrosophic qualitative flexible approach based on likelihood for multi-criteria decision-making problems. *Int. J. Syst. Sci.* **2017**, *48*, 425–435. [CrossRef]
21. Bausys, R.; Zavadskas, E.K. Multi-criteria decision making approach by VIKOR under interval neutrosophic set environment. *Econ. Comput. Econ. Cybern. Stud. Res.* **2015**, *49*, 33–48.
22. Pouresmaeil, H.; Shivanian, E.; Khorram, E.; Fathabadi, H.S. An extended method using TOPSIS and VIKOR for multiple attribute decision making with multiple decision makers and single valued neutrosophic numbers. *Adv. Appl. Stat.* **2017**, *50*, 261–292. [CrossRef]
23. Li, Y.H.; Liu, P.D.; Chen, Y.B. Some single valued neutrosophic number heronian mean operators and their application in multiple attribute group decision making. *Informatica* **2016**, *27*, 85–110. [CrossRef]
24. Ye, J. Multiple attribute decision-making method using correlation coefficients of normal neutrosophic Sets. *Symmetry* **2017**, *9*, 80. [CrossRef]
25. Ye, J. Correlation coefficients of interval neutrosophic hesitant fuzzy sets and its application in a multiple attribute decision making method. *Informatica* **2016**, *27*, 179–202. [CrossRef]
26. Zavadskas, E.K.; Baušys, R.; Lazauskas, M. Sustainable assessment of alternative sites for the construction of a waste incineration plant by applying waspas method with single-valued neutrosophic set. *Sustainability* **2015**, *7*, 15923–15936. [CrossRef]
27. Zavadskas, E.K.; Baušys, R.; Stanujkic, D.; Magdalinovic-Kalinovic, M. Selection of lead-zinc flotation circuit design by applying waspas method with single-valued neutrosophic set. *Acta Montan. Slov.* **2016**, *21*, 85–92.
28. Baušys, R.; Juodagalvienė, B. Garage location selection for residential house by WASPAS-SVNS method. *J. Civ. Eng. Manag.* **2017**, *23*, 421–429. [CrossRef]
29. Nie, R.; Wang, J.; Zhang, H. Solving solar-wind power station location problem using an extended weighted aggregated sum product assessment (WASPAS) technique with interval neutrosophic sets. *Symmetry* **2017**, *9*, 106. [CrossRef]
30. Bausys, R.; Zavadskas, E.K.; KAKLAUSKAS, A. Application of neutrosophic set to multi-criteria decision making by COPRAS. *Econ. Comput. Econ. Cybern. Stud. Res.* **2015**, *49*, 91–105.
31. Čereška, A.; Zavadskas, E.K.; Cavallaro, F.; Podvezko, V.; Tetsman, I.; Grinbergienė, I. Sustainable assessment of aerosol pollution decrease applying multiple attribute decision-making methods. *Sustainability* **2016**, *8*, 586. [CrossRef]
32. Herrera-Viedma, E.; Herrera, F.; Chiclana, F.; Luque, M. Some issues on consistency of fuzzy preference relations. *Eur. J. Oper. Res.* **2004**, *154*, 98–109. [CrossRef]
33. Xu, Y.J.; Herrera, F.; Wang, H.M. A distance-based framework to deal with ordinal and additive inconsistencies for fuzzy reciprocal preference relations. *Inf. Sci.* **2016**, *328*, 189–205. [CrossRef]
34. Chiclana, F.; Herrera-Viedma, E.; Alonso, S.; Herrera, F. Cardinal consistency of reciprocal preference relations: A characterization of multiplicative transitivity. *IEEE Trans. Fuzzy Syst.* **2009**, *17*, 14–23. [CrossRef]
35. Meng, F.Y.; Chen, X.H.; Zhang, Y.L. Consistency-based linear programming models for generating the priority vector from interval fuzzy preference relations. *Appl. Soft Comput.* **2016**, *41*, 247–264. [CrossRef]
36. Dong, Y.C.; Li, C.C.; Chiclana, F.; Herrera-Viedma, E. Average-case consistency measurement and analysis of interval-valued reciprocal preference relations. *Knowl. Based Syst.* **2016**, *114*, 108–117. [CrossRef]
37. Wu, J.; Chiclana, F.; Liao, H.C. Isomorphic multiplicative transitivity for intuitionistic and interval-valued fuzzy preference relations and its application in deriving their priority vectors. *IEEE Trans. Fuzzy Syst.* **2016**. [CrossRef]
38. Yang, Q.; Zhang, Z.S.; You, X.S.; Chen, T. Evaluation and classification of overseas talents in China based on the BWM for intuitionistic relations. *Symmetry* **2016**, *8*, 137. [CrossRef]

39. Xu, G.L.; Wan, S.P.; Wang, F.; Dong, J.Y.; Zeng, Y.F. Mathematical programming methods for consistency and consensus in group decision making with intuitionistic fuzzy preference relations. *Knowl. Based Syst.* **2016**, *98*, 30–43. [CrossRef]
40. Wang, Z.J.; Tong, X. Consistency analysis and group decision making based on triangular fuzzy additive reciprocal preference relations. *Inf. Sci.* **2016**, *361–362*, 29–47. [CrossRef]
41. Liu, F.; Pedrycz, W.; Wang, Z.X.; Zhang, W.G. An axiomatic approach to approximation-consistency of triangular fuzzy reciprocal preference relations. *Fuzzy Sets Syst.* **2017**, *322*, 1–18. [CrossRef]
42. Zhu, B.; Xu, Z.S.; Xu, J.P. Deriving a ranking from hesitant fuzzy preference relations under group decision making. *IEEE Trans. Cybern.* **2014**, *44*, 1328. [CrossRef] [PubMed]
43. Xu, Y.J.; Cabrerizo, F.J.; Herrera-Viedma, E. A consensus model for hesitant fuzzy preference relations and its application in water allocation management. *Appl. Soft Comput.* **2017**, *58*, 265–284. [CrossRef]
44. Liang, R.X.; Wang, J.; Zhang, H.Y. A multi-criteria decision-making method based on single-valued trapezoidal neutrosophic preference relations with complete weight information. *Neural Comput. Appl.* **2017**, 1–16. [CrossRef]
45. Yang, Y.; Hu, J.H.; An, Q.X.; Chen, X.H. Group decision making with multiplicative triangular hesitant fuzzy preference relations and cooperative games method. *Int. J. Uncertain. Quantif.* **2017**, *7*, 271–284. [CrossRef]
46. Wang, C.; Wang, J. A multi-criteria decision-making method based on triangular intuitionistic fuzzy preference information. *Intell. Autom. Soft Comput.* **2016**, *22*, 473–482. [CrossRef]
47. Herrera, F.; Herrera-Viedma, E. Choice functions and mechanisms for linguistic preference relations. *Eur. J. Oper. Res.* **2000**, *120*, 144–161. [CrossRef]
48. Nie, R.; Wang, J.; Li, L. A shareholder voting method for proxy advisory firm selection based on 2-tuple linguistic picture preference relation. *App. Soft Comput.* **2017**, *60*, 520–539. [CrossRef]
49. Xu, Y.J.; Wei, C.P.; Sun, H. Distance-based nonlinear programming models to identify and adjust inconsistencies for linguistic preference relations. *Soft Comput.* **2017**, 1–17. [CrossRef]
50. Zhang, G.Q.; Dong, Y.C.; Xu, Y.F. Consistency and consensus measures for linguistic preference relations based on distribution assessments. *Inf. Fusion* **2014**, *17*, 46–55. [CrossRef]
51. Zhang, X.; Wang, J. Consensus-based framework to MCGDM under multi-granular uncertain linguistic environment. *J. Intell. Fuzzy Syst.* **2017**, *33*, 1263–1274. [CrossRef]
52. Wang, J.; Wang, J.Q.; Tian, Z.; Zhao, D. A multi-hesitant fuzzy linguistic multi-criteria decision-making approach for logistics outsourcing with incomplete information. *Int. Trans. Oper. Res.* **2017**. [CrossRef]
53. Wang, H.; Xu, Z.S. Some consistency measures of extended hesitant fuzzy linguistic preference relations. *Inf. Sci.* **2015**, *297*, 316–331. [CrossRef]
54. Wu, Z.B.; Xu, J.P. Managing consistency and consensus in group decision making with hesitant fuzzy linguistic preference relations. *Omega* **2016**, *65*, 28–40. [CrossRef]
55. Meng, F.Y.; Tang, J.; An, Q.X.; Chen, X.H. Decision making with intuitionistic linguistic preference relations. *Int. Trans. Oper. Res.* **2017**. [CrossRef]
56. Nie, R.; Wang, J.; Li, L. 2-tuple linguistic intuitionistic preference relation and its application in sustainable location planning voting system. *J. Intell. Fuzzy Syst.* **2017**, *33*, 885–899. [CrossRef]
57. Zhang, Y.; Ma, H.X.; Liu, B.H.; Liu, J. Group decision making with 2-tuple intuitionistic fuzzy linguistic preference relations. *Soft Comput.* **2012**, *16*, 1439–1446. [CrossRef]
58. Zhang, Y.X.; Xu, Z.S.; Wang, H.; Liao, H.C. Consistency-based risk assessment with probabilistic linguistic preference relation. *Appl. Soft Comput.* **2016**, *49*, 817–833. [CrossRef]
59. Peng, J.; Wang, J.; Yang, L.; Qian, J. A novel multi-criteria group decision-making approach using simplified neutrosophic information. *Int. J. Uncertain. Quantif.* **2017**. [CrossRef]
60. Wang, J.Q.; Yang, Y.; Li, L. Multi-criteria decision-making method based on single valued neutrosophic linguistic Maclaurin symmetric mean operators. *Neural Comput. Appl.* **2016**, 1–19. [CrossRef]
61. Wu, X.; Wang, J. Cross-entropy measures of multi-valued neutrosophic sets and its application in selecting middle-level manager. *Int. J. Uncertain. Quantif.* **2017**, *7*, 155–176. [CrossRef]
62. Tian, Z.P.; Wang, J.; Zhang, H.Y.; Wang, J.Q. Multi-criteria decision-making based on generalized prioritized aggregation operators under simplified neutrosophic uncertain linguistic environment. *Int. J. Mach. Learn. Cybern.* **2016**, 1–17. [CrossRef]
63. Yu, S.; Zhang, H.; Wang, J. Hesitant fuzzy linguistic maclaurin symmetric mean operators and their applications to multi-criteria decision-making problem. *Int. J. Intell. Syst.* **2017**. [CrossRef]

64. Wang, J.; Wang, J.Q.; Zhang, H.Y.; Chen, X.H. Distance-based multi-criteria group decision-making approaches with multi-hesitant fuzzy linguistic information. *Int. J. Inf. Technol. Decis. Mak.* **2017**, *16*, 1069–1099. [CrossRef]
65. Zhang, H.Y.; Peng, H.G.; Wang, J.; Wang, J.Q. An extended outranking approach for multi-criteria decision-making problems with linguistic intuitionistic fuzzy numbers. *Appl. Soft Comput.* **2017**, *59*, 462–474. [CrossRef]
66. Xu, Z.S. Deviation measures of linguistic preference relations in group decision making. *Omega* **2005**, *33*, 249–254. [CrossRef]
67. Zadeh, L.A. The concept of a linguistic variable and its application to approximate reasoning. *Inf. Sci.* **1975**, *8*, 199–249. [CrossRef]
68. Wang, J.Q.; Wu, J.T.; Wang, J.; Zhang, H.Y.; Chen, X.H. Interval-valued hesitant fuzzy linguistic sets and their applications in multi-criteria decision-making problems. *Inf. Sci.* **2014**, *288*, 55–72. [CrossRef]
69. Tian, Z.P.; Wang, J.; Zhang, H.Y.; Chen, X.H.; Wang, J.Q. Simplified neutrosophic linguistic normalized weighted Bonferroni mean operator and its application to multi-criteria decision-making problems. *FILOMAT* **2016**, *30*, 3339–3360. [CrossRef]
70. Ye, J. An extended TOPSIS method for multiple attribute group decision making based on single valued neutrosophic linguistic numbers. *J. Intell. Fuzzy Syst.* **2015**, *28*, 247–255.
71. Tian, Z.P.; Wang, J.; Wang, J.Q.; Zhang, H.Y. Simplified neutrosophic linguistic multi-criteria group decision-making approach for green product development. *Group Decis. Negot.* **2017**, *26*, 597–627. [CrossRef]
72. Zhou, L.G.; Zhou, Y.Y.; Liu, X.; Chen, H.Y. Some ILOWA operators and their applications to group decision making with additive linguistic preference relations. *J. Intell. Fuzzy Syst.* **2015**, *29*, 831–843. [CrossRef]
73. Dong, Y.C.; Xu, Y.; Li, H. On consistency measures of linguistic preference relations. *Eur. J. Oper. Res.* **2008**, *189*, 430–444. [CrossRef]
74. Zhu, B.; Xu, Z.S. Consistency measures for hesitant fuzzy linguistic preference relations. *IEEE Trans. Fuzzy Syst.* **2014**, *22*, 35–45. [CrossRef]
75. Hou, F.J. A hierarchical decision model based on pairwise comparisons. *Fundam. Inform.* **2016**, *144*, 333–348. [CrossRef]
76. Zhang, Z.; Guo, C.H. A method for multi-granularity uncertain linguistic group decision making with incomplete weight information. *Knowl. Based Syst.* **2012**, *26*, 111–119. [CrossRef]
77. Jin, F.F.; Ni, Z.W.; Chen, H.Y.; Li, Y.P. Approaches to decision making with linguistic preference relations based on additive consistency. *Appl. Soft Comput.* **2016**, *49*, 71–80. [CrossRef]
78. Cabrerizo, F.J.; Al-Hmouz, R.; Morfeq, A.; Balamash, A.S.; Martínez, M.A.; Herrera-Viedma, E. Soft consensus measures in group decision making using unbalanced fuzzy linguistic information. *Soft Comput.* **2017**, *21*, 3037–3050. [CrossRef]

symmetry

MDPI

Article

Managing Non-Homogeneous Information and Experts' Psychological Behavior in Group Emergency Decision Making

Liang Wang [1,2], Álvaro Labella [1], Rosa M. Rodríguez [3,*], Ying-Ming Wang [2] and Luis Martínez [1]

[1] Department of Computer Science, University of Jaén, 23071 Jaén, Spain; wangliangg322@hotmail.com (L.W.); alabella@ujaen.es (Á.L.); martin@ujaen.es (L.M.)
[2] Decision Sciences Institute, Fuzhou University, Fuzhou 350116, China; msymwang@hotmail.com
[3] Department of Computer Science and A.I., University of Granada, 18071 Granada, Spain
* Correspondence: rosam.rodriguez@decsai.ugr.es

Received: 25 September 2017; Accepted: 13 October 2017; Published: 18 October 2017

Abstract: After an emergency event (EE) happens, emergency decision making (EDM) is a common and effective way to deal with the emergency situation, which plays an important role in mitigating its level of harm. In the real world, it is a big challenge for an individual emergency manager (EM) to make a proper and comprehensive decision for coping with an EE. Consequently, many practical EDM problems drive group emergency decision making (GEDM) problems whose main limitations are related to the lack of flexibility in knowledge elicitation, disagreements in the group and the consideration of experts' psychological behavior in the decision process. Hence, this paper proposes a novel GEDM approach that allows more flexibility for preference elicitation under uncertainty, provides a consensus process to avoid disagreements and considers experts' psychological behavior by using the fuzzy TODIM method based on prospect theory. Eventually, a group decision support system (GDSS) is developed to support the whole GEDM process defined in the proposed method demonstrating its novelty, validity and feasibility.

Keywords: group emergency decision making; non-homogeneous information; psychological behavior; group decision support system

1. Introduction

Emergencies are defined as events that suddenly take place, causing or having the possibility of provoking intense death and injury, property loss, ecological damage and social hazards. In recent years, various emergency events, such as earthquakes, floods, hurricanes, terrorist attacks, etc., have exerted severely negative impacts on human life and socio-economic development. When an emergency event (EE) occurs, Emergency Decision Making (EDM) is typically characterized by at least uncertainty, time pressure, and lack of information, resulting in potentially serious consequences [1]. Since EDM plays a crucial role in alleviating the losses of properties and lives caused by EEs, it has received increasing attention from both government and academia because of the frequent occurrence of EEs, becoming a very active and important research field in recent years [1–5].

When an EE occurs, it is hard to collect the information related to the event and predict its evolution particularly in the early stage because of the inadequate and uncertain information. Consequently, it is too complex for just one emergency manager (EM) to make comprehensive judgments under emergency situations. Therefore, EDM requires multiple experts from diverse professional backgrounds (such as hydrological, geological, meteorological, sociological, demographic, etc.) to help the EM make a decision. This leads to Group EDM (GEDM) problems. Figure 1 shows a graphical general scheme for GEDM problems, in which experts play a role of think tank in supporting the EM who is in charge of the EE.

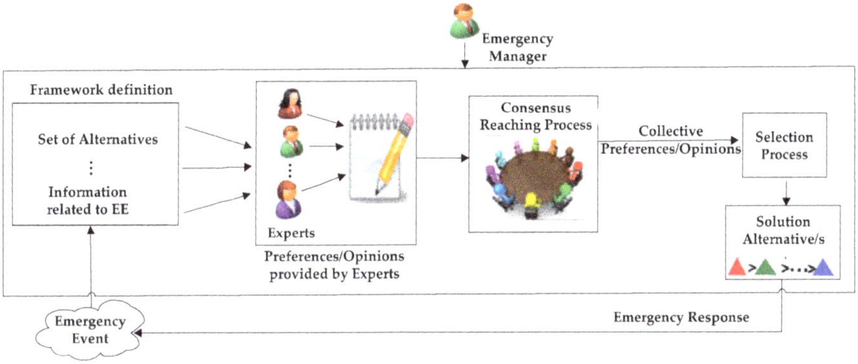

Figure 1. The general scheme of GEDM process.

In the real world, it is common that experts with different background and knowledge might have different attitudes or opinions over different alternatives concerning different criteria. Moreover, criteria defined in a GEDM problem might have different nature, qualitative or quantitative. Therefore, experts might hesitate and express their opinions or assessments by using different types of information according to their knowledge and criteria nature. The complexity of GEDM problems could imply not only the use of a non-homogeneous context in which multiple information types can be utilized by experts to elicit their knowledge and expertise, but also the modeling of uncertain assessments including hesitancy. However, current EDM approaches deal with the information using only one expression domain: numerical values [4], interval values [3] or linguistic information [6].

Traditionally, group decision making (GDM) approaches have shown that a solution can be obtained under disagreement among experts [7,8], however several experts may not accept the decision made because they might consider that their individual opinions have not been taken into account sufficiently [9,10]. Such a situation could be very serious in GEDM driving either to deadlock in the decision or in a harmful decision. Hence, it seems necessary and reasonable to achieve a consensus among all experts involved in the GEDM problem before making the decision. The Consensus Reaching Process (CRP) is a way to integrate group wisdom into one and then reach an agreement among all experts in the GEDM problem. There are already different approaches [1,4,11] focused on how to reach as much agreement as possible among all experts participating in the problem. However, they have strict expression domains [1,11]; or time cost [4,5]. However, time is extremely valuable, because it means lives and chances, thus emergency responses cannot afford a time-consuming consensus model.

Different behavioral experiments [12–14] show that human beings are usually bounded rationally in decision-making processes under risk and uncertainty. Therefore, psychological behavior plays a crucial role in the decision processes. Nevertheless, as far as we know, experts' psychological behavior is neglected in current GEDM [1,4,5,11,15] approaches.

According to the previous limitations presented in current GEDM methods, the aim of this paper is to propose a new GEDM method that overcomes them. Such a method is able:

1. To allow more flexibility for eliciting information by dealing with non-homogeneous information including hesitancy.
2. To include a consensus model with low time cost to achieve an agreement among experts involved in the GEDM problem.
3. To take into account experts' psychological behavior by means of the fuzzy TODIM method [16–18] based on prospect theory [14].

Furthermore, the proposed method is implemented into a Group Decision Support System (GDSS) named GENESIS (Group EmergeNcy dEcision SupportIng System) based on FLINTSTONES (Fuzzy LINguisTic DeciSion Tools eNhacemEnt Suite) [19,20] that supports the whole GEDM process effectively and in a timely way, as shown in an illustrative example.

The remainder of this paper is organized as follows: Section 2 revises briefly different concepts about non-homogeneous information, CRPs and the fuzzy TODIM, which will be used in our proposal together with some related works. Section 3 presents the new GEDM method that integrates the novelties pointed out previously. Section 4 introduces the structure and components of the GDSS, GENESIS, and shows an example to illustrate the feasibility and validity of the proposed method. A sensitive analysis is also presented to study the robustness of the proposal. Section 5 presents some conclusions and future works.

2. Preliminaries

In this section, some basic concepts about non-homogeneous information and CRPs are revised in short in order that readers can understand easily the proposed GEDM model. It also reviews the fuzzy TODIM method that is used in the selection process of the proposal to obtain the ranking of alternatives considering experts' psychological behavior. Eventually, some related works to illustrate the importance of this research are reviewed.

2.1. Non-Homogeneous Information in Decision Making

Nowadays, real-world decision-making problems are more diversified and complex because of rapid socio-economic development, such as EDM problems [2,3], GEDM problems [1,15], and Intelligent GEDM problems [11]. Those problems are usually defined under uncertainty because of inadequate and uncertain information. The complexity of these problems implies multiple experts with different backgrounds and knowledge participating in the decision process.

To model the uncertainty and non-homogeneous information, such as numerical values, interval values and linguistic terms elicited by experts, several approaches have been discussed in current GDM approaches. Some of them [21–25] make the computations using directly the non-homogeneous information [26] and others unify the information into one domain [24,27], being the most common one the linguistic information. Recently, the inclusion of hesitancy is becoming more important [28,29].

The concept of hesitant fuzzy linguistic term sets (HFLTS) [30] has been introduced to model experts' hesitation in qualitative settings and it has been applied in decision making problems obtaining successful results. It is defined as follows.

Definition 1 [30]. *Let* $S = \{s_0, s_1, \ldots, s_g\}$ *be a linguistic term set, a HFLTS* H_S, *is defined as an ordered finite subset of consecutive linguistic terms of* S:

$$H_S = \{s_i, s_{i+1}, \ldots, s_j\}, s_k \in S,\ k \in \{i, \ldots, j\}$$

Nevertheless, when experts provide their opinions and they feel hesitation among several linguistic terms, they do not use multiple linguistic terms, but linguistic expressions close to the natural language used by human beings. Hence, Rodríguez et al. [30] proposed the use of context-free grammars G_H to build complex linguistic expressions more flexible and richer than single linguistic terms [29,30]. The expressions produced by the context-free grammar G_H, may be either a single linguistic term $s_i \in S$, or comparative linguistic expressions S_{ll} (see [29,30] for further detail).

In our proposal, the non-homogeneous information including experts' hesitancy will be transformed into a unified fuzzy domain to facilitate the computations (see Section 3.3).

2.2. Consensus Reaching Processes

GDM problems are usually solved by a selection process that obtains the best alternative as a solution to the problem. However, sometimes the goal of the problem is not to obtain the best solution, but an accepted one for all involved experts in the problem. In such a situation, it seems necessary to apply a CRP. Consensus can be defined as [9] "a state of mutual agreement among members of a group in which the decision made satisfies all of them". Therefore, a consensus process requires that experts modify their opinions making them closer to each other and this way to obtain a collective opinion that is satisfactory for all of them [10,31–34].

In GEDM process, experts play a role of think tank in supporting EM to make a decision, recently several proposals [1,4,5,11,15,34] integrate CRP into GEDM to deal with experts' opinions in order to achieve an agreement among all experts involved and make a right decision. However, these approaches deal just with numerical values [1,5,25] and are not suitable for other types of information, additionally, they have a high time cost [4,5] because of the supervised feedback mechanism that should be avoided in GEDM problems.

Due to these reasons and the type of information used in our proposal, a fuzzy linear programming-based consensus model [34] with low time cost will be utilized to achieve consensus in our proposal. Before introducing the fuzzy linear programming model, it is necessary to revise the definition of the distance between fuzzy numbers, which will be used.

Definition 2 [34]. *Let* $A = (a_1, a_2, a_3, a_4)$ *and* $B = (b_1, b_2, b_3, b_4)$ *be two trapezoidal fuzzy numbers. The distance between* A *and* B *can be obtained as follows, the measure of* d_p *can also be called as* l_p *metric:*

$$d_p(A,B) = \left(\sum_{i=1}^{4} (|a_i - b_i|)^p \right)^{1/p} \tag{1}$$

where p *is an integer* ≥ 1. *Let* U *be the universe of discourse and* $u = \max(U) - \min(U)$. *The similarity between* A *and* B *can be defined as [34,35]:*

$$S_p(A,B) = 1 - \frac{1}{4u^p} (d_p(A,B))^p \tag{2}$$

The dissimilarity is defined as $c - S_p(A,B)$, where c is a constant >1. The selection of c will influence in the final result of the aggregation.

Let $\widetilde{A}_h = (a_{h1}, a_{h2}, a_{h3}, a_{h4})$ be the h-th expert's individual opinion and $\widetilde{O}$ be the overall opinion obtained by aggregating experts' individual opinions.

The fuzzy linear programming model is [34]:

$$\begin{cases} \min \sum_{h=1}^{\overline{K}} (w^h)^{\alpha} (c - S_p(\widetilde{A}_h, \widetilde{O})) \\ s.t.\ d_p(\widetilde{A}_h, \widetilde{O}) \leq \varepsilon_h,\ h = 1, 2, \ldots, \overline{K} \end{cases} \tag{3}$$

where α is an integer ≥ 1, w^h denotes the h-th experts' importance. ε_h denotes a threshold that means the maximum change that the h-th expert can make. $d_p(\widetilde{A}_h, \widetilde{O})$ denotes the distance between $\widetilde{A}_h$ and $\widetilde{O}$, which can be obtained according to Equation (1).

2.3. Fuzzy TODIM Method

Some studies [12–14] have shown that human beings are bounded rationally especially in risk and uncertain decision processes and their psychological behavior is very important in the decision process. Therefore, it seems necessary to consider experts' psychological behavior in GEDM problem.

TODIM method was proposed by Gomes and Lima [36,37]; it is a popular multi-criteria decision making (MCDM) method based on prospect theory [13] considering humans psychological behavior. It has been widely applied to solve different decision problems [38,39]. To cope with complex problems and uncertain information in the real world, the TODIM method has been extended to deal with fuzzy MCDM problems [16,17].

In our proposal, we will use fuzzy TODIM method [16–18] based on prospect theory [14] because of its advantage and capability of capturing the experts' psychological behavior under fuzzy environment.

The fuzzy TODIM was introduced in [18] and briefly summarized below:

Let $P = \{p_1, p_2, \ldots, p_m\}$ be a set of alternatives, $C = \{c_1, c_2, \ldots, c_n\}$ be a set of criteria and $w_c = (w_{c_1}, w_{c_2}, \ldots, w_{c_n})$ be a weighting vector for criteria, where w_{c_j} denotes the weight of criterion c_j. Let $A = (a_{ij})_{m\times n}$ be a fuzzy decision matrix, where $a_{ij} = (a_{ij}^1, a_{ij}^2, a_{ij}^3, a_{ij}^4)$ denotes the rating of the alternative p_i with respect to criterion c_j.

Step 1: To normalize the fuzzy decision matrix $A = (a_{ij})_{m\times n}$ into the correspondent normalized fuzzy decision matrix $G = (g_{ij})_{m\times n}$, according to the cost and benefit criteria.

Step 2: To determine the reference criterion c_r and calculate the relative weight w_{jr} of criterion c_j $(j = 1, 2, \ldots, n)$, i.e.,

$$w_{jr} = w_{c_j} / w_r \tag{4}$$

where $w_r = \max\left\{ w_{c_j} \middle| j = 1, 2, \ldots, n \right\}$.

Step 3: To calculate the dominance degree, $\Phi_j(p_i, p_k)$, of alternative p_i, $(i = 1, 2, \ldots, m)$ over the remaining alternatives p_k $(k = 1, 2, \ldots, m)$ concerning criterion c_j $(j = 1, 2, \ldots, n)$, i.e.,

$$\Phi_j(p_i, p_k) = \begin{cases} \sqrt{w_{jr} / (\sum_{j=1}^{n} w_{jr}) d(g_{ij}, g_{kj})}, \ \mathbb{F}(g_{ij}) - \mathbb{F}(g_{kj}) \geq 0 \\ -\frac{1}{\theta} \sqrt{(\sum_{j=1}^{n} w_{jr}) / w_{jr} d(g_{ij}, g_{kj})}, \ \mathbb{F}(g_{ij}) - \mathbb{F}(g_{kj}) < 0 \end{cases} \tag{5}$$

where θ is the attenuation factor of the losses, $\theta > 0$. $d(g_{ij}, g_{kj})$ denotes the distance between two fuzzy numbers g_{ij} and g_{kj} and $\mathbb{F}(*)$ is a defuzzification function [18].

Step 4: To calculate the dominance degree, $\delta(p_i, p_k)$, of alternative p_i, $(i = 1, 2, \ldots, m)$ over the remaining alternatives p_k $(k = 1, 2, \ldots, m)$, i.e.,

$$\delta(p_i, p_k) = \sum_{j=1}^{n} \Phi_j(p_i, p_k) \tag{6}$$

Step 5: To calculate the overall dominance degree, $\eta(p_i)$, of alternative p_i, $(i = 1, 2, \ldots, m)$, i.e.,

$$\eta(p_i) = \frac{\sum_{k=1}^{m} \delta(p_i, p_k) - \min_i\{\sum_{k=1}^{m} \delta(p_i, p_k)\}}{\max_i\{\sum_{k=1}^{m} \delta(p_i, p_k)\} - \min_i\{\sum_{k=1}^{m} \delta(p_i, p_k)\}} \tag{7}$$

Step 6: According to the overall dominance degrees of each alternative, the corresponding ranking can be determined such that the bigger $\eta(p_i)$, the better alternative p_i.

2.4. Related Works

In order to show the importance of GEDM in the real world, this subsection reviews several important studies in the literature that are related to our research [1,4–6,40].

These studies have approached GEDM problems from different aspects. For example, Wang et al. [40] proposed a group emergency decision method based on prospect theory by using interval values. Xu et al. [4] proposed a consensus model for multi-criteria large group emergency decision making considering non-cooperative behaviors and minority opinions, wherein numerical value is employed to represent experts' assessments. Ju et al. [6] presented a model to evaluate emergency response capacity by using 2-tuple fuzzy linguistic information. Xu et al. [5] proposed a conflict-eliminating approach for GEDM problem. Levy and Taji [1] utilized a group analytic network process to construct a group decision support system to support hazard planning and emergency management under incomplete information.

So far, there is not any proposal in previous GEDM approaches [1,4,5,11,40] that considers the non-homogeneous information together with the experts' hesitation due to uncertain information. In addition, those GEDM approaches [1,4,5,11] dealing with the consensus process; just make use of it with strict expression domains or high time cost. However, time is extremely valuable in EDM process, which means life and opportunity. Furthermore, experts' psychological behavior is neglected in current GEDM approaches [1,4,5] that plays an important role in the GEDM process under risk and uncertainty.

As pointed out in Introduction, our proposed method aims to overcome such limitations and shows the relevance of this research.

3. Managing Non-Homogeneous Information and Experts' Psychological Behavior in GEDM

This section introduces a new GEDM method to overcome the limitations pointed out in the Introduction regarding the current GEDM methods. This proposal is able: (i) to manage non-homogeneous information, including hesitant information (ii) to achieve consensus with low time cost, (iii) to take into account the experts' psychological behavior in the GEDM process.

Our proposal extends the general scheme of a GEDM process shown in Figure 1 by adding two new phases to deal with non-homogeneous information and calculate the criteria weights, and modifying another two phases (CRP and selection process), they are highlighted in Figure 2 by using dashed lines.

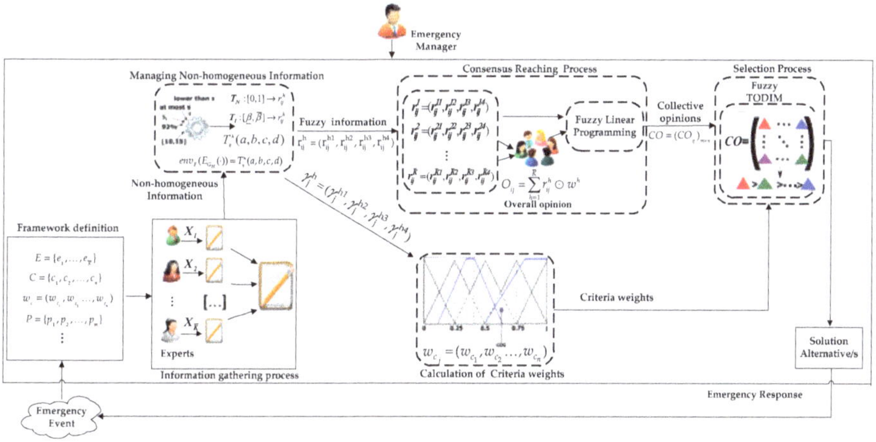

Figure 2. Scheme of proposed GEDM method.

It consists of six main phases:

1. Definition framework. The main features, terminology and expression domains utilized in the proposed GEDM problem are defined.
2. Information gathering process. Opinions or assessments over different alternatives concerning different criteria and importance of criteria provided by experts using multiple types of information are gathered.

3. Managing non-homogeneous information. The non-homogeneous information gathered is unified into a fuzzy domain to deal with the decision computations.
4. Consensus reaching process. A fuzzy linear programming-based consensus model [34] is utilized to deal with fuzzy information and achieve an agreement among all the experts involved in the GEDM problem.
5. Calculation of criteria weights. Criteria weights are calculated by using experts' opinions.
6. Selection process-fuzzy TODIM method. Fuzzy TODIM method is applied to manage experts' psychological behavior in GEDM processes and obtain the ranking of alternatives.

According to the ranking of alternatives, the EM can select the best or more suitable alternative to cope with the EE. These phases are further detailed in the following subsections.

3.1. Definition Framework

The framework for GEDM problem is established by defining its main features and terminology.

- $P = \{p_1, p_2, \ldots, p_m\}$: the set of emergency alternatives, where p_i is the i-th emergency alternative, $i = 1, 2, \ldots, m$.
- $C = \{c_1, c_2, \ldots, c_n\}$: the set of criteria/attributes, where c_j denotes the j-th criterion/attribute, $j = 1, 2, \ldots, n$.
- $w_c = (w_{c_1}, w_{c_2} \ldots, w_{c_n})$: the weighting vector for the criteria, where w_{c_j} denotes the criterion weight of the j-th criterion/attribute, satisfying $\sum_{j=1}^{n} w_{c_j} = 1$, $w_{c_j} \in [0,1]$ $j = 1, 2, \ldots, n$.
- $E = \{e_1, \ldots, e_{\overline{K}}\}$: the set of experts, where e_h denotes the h-th expert, $h = 1, 2, \ldots, \overline{K}$.
- $X_h = (x_{ij}^h)_{m \times n}$: the information matrix provided by the h-th expert, where x_{ij}^h represents the assessments/opinions provided by the h-th expert over the i-th alternative concerning the j-th criterion, $h = 1, 2, \ldots, \overline{K}$, $i = 1, 2, \ldots, m$, $j = 1, 2, \ldots, n$ (see Remark 1).
- $w_h = (w_1^h, w_2^h \ldots, w_n^h)$: the assessment vector of criteria importance provided by the expert e_h, where w_j^h represents the importance provided by the h-th expert on the importance of criterion c_j, $h = 1, 2, \ldots, \overline{K}$, $j = 1, 2, \ldots, n$ (see Remark 2).
- r_{ij}^h: denotes the experts' assessments, x_{ij}^h, unified in a fuzzy domain, $h = 1, 2, \ldots, \overline{K}$, $i = 1, 2, \ldots, m$, $j = 1, 2, \ldots, n$.
- γ_j^h: denotes the experts' opinions regarding the criteria importance, w_j^h, unified in a fuzzy domain, $h = 1, 2, \ldots, \overline{K}$, $j = 1, 2, \ldots, n$.

Remark 1. *In our method, experts can provide their opinions/assessments by utilizing multiple expression domains (numerical values (N), interval values (I), linguistic terms (S) and comparative linguistic expressions (S_{ll})) according to their background, degree of knowledge, hesitancy and criteria nature.*

$$x_{ij}^h \in \begin{cases} N \in R \\ I \in [\xi^L, \xi^U] \\ S = \{s_0, s_1, \ldots, s_g\} \\ S_{ll} \end{cases} \tag{8}$$

Remark 2. *In GEDM problems, the criteria need to be weighted. However, due to the complexity of EEs, it is not easy to collect the related information about the criteria, especially at the early stage of EE. In such situation, a possible way is to calculate the criteria weights from experts' knowledge and experience. In this proposal, experts can express their opinions about the criteria importance by utilizing either S_{ll} or S, because S_{ll} and S are* *more flexible and similar to the natural language utilized by human beings in real-world EE situations, and they are suitable for GEDM problems defined in uncertain contexts.*

$$w_j^h \in \begin{cases} S = \{s_0, s_1, \ldots, s_g\} \\ S_{II} \end{cases} \tag{9}$$

3.2. Information Gathering Process

Once the framework of GEDM problem is defined, experts can provide their judgments over the emergency alternatives p_i concerning each criterion c_j and the importance over different criteria (see Tables 1 and 2) by using the expression domains defined previously.

Table 1. Assessments over alternative p_i concerning criterion c_j.

Experts	Assessments
e_1	$\{x_{ij}^1, \ldots, x_{mn}^1\}$
e_2	$\{x_{ij}^2, \ldots, x_{mn}^2\}$
……	……
$e_{\overline{K}}$	$\{x_{ij}^{\overline{K}}, \ldots, x_{mn}^{\overline{K}}\}$

Table 2. Importance over criteria c_j.

Experts	Assessments
e_1	$\{w_1^1, \ldots, w_n^1\}$
e_2	$\{w_1^2, \ldots, w_n^2\}$
……	……
$e_{\overline{K}}$	$\{w_1^{\overline{K}}, \ldots, w_n^{\overline{K}}\}$

For example, the information on alternatives with respect to criteria provided by expert e_1 can be expressed as:

$$X_1 = \begin{array}{c} \\ p_1 \\ p_2 \\ \vdots \\ p_m \end{array} \begin{array}{c} \begin{array}{cccc} c_1 & c_2 & \cdots & c_n \end{array} \\ \begin{bmatrix} x_{11}^1 & x_{12}^1 & \cdots & x_{1n}^1 \\ x_{21}^1 & x_{22}^1 & \cdots & x_{2n}^1 \\ \vdots & \vdots & \cdots & \vdots \\ x_{m1}^1 & x_{m2}^1 & \cdots & x_{mn}^1 \end{bmatrix} \end{array}$$

where $x_{ij}^1 \in \begin{cases} N \in R \\ I \in [\xi^L, \xi^U] \\ S = \{s_0, s_1, \ldots, s_g\} \\ S_{II} \end{cases}, i = 1, 2, \ldots, m, j = 1, 2, \ldots, n.$

The information on the importance of criterion c_j provided by expert e_1 can be expressed as:

$$w_1 = \begin{array}{c} \begin{array}{cccc} c_1 & c_2 & \cdots & c_n \end{array} \\ \begin{bmatrix} w_1^1 & w_2^1 & \cdots & w_n^1 \end{bmatrix} \end{array}$$

where $w_j^1 \in \begin{cases} S = \{s_0, s_1, \ldots, s_g\} \\ S_{II} \end{cases}, j = 1, 2, \ldots, n.$

3.3. Managing Non-Homogeneous Information

As it was stated in Section 2.1, our proposal deals with non-homogeneous information including hesitant information. Therefore, the expression domains used by experts to provide their assessments in this proposal are the following ones:

- *Numerical value.* Assessments represented as numerical values N belonging to a specific numerical scale R, i.e., $N \in R$.
- *Interval value.* Assessments represented as interval values I, belonging to a specific domain $[\xi^L, \xi^U]$, i.e., $I \in [\xi^L, \xi^U]$.
- *Linguistic terms.* Assessments represented as linguistic terms $s_k \in S = \{s_0, s_1, \ldots, s_g\}$, $k \in \{0, \ldots, g\}$, with granularity $g+1$.
- *Comparative linguistic expressions.* Assessments represented as comparative linguistic expressions S_{ll} generated by a context-free grammar G_H [29,30].

In order to make computations with non-homogeneous information elicited by experts, it is necessary to conduct the different types of information into a unique expression domain. Most approaches unify the non-homogeneous information into linguistic information [23,24]. Nevertheless, in order to keep the uncertainty provided by experts involved in a GEDM problem, we unify the information into a fuzzy domain r_{ij}^h, by introducing some transformation functions (see Figure 3).

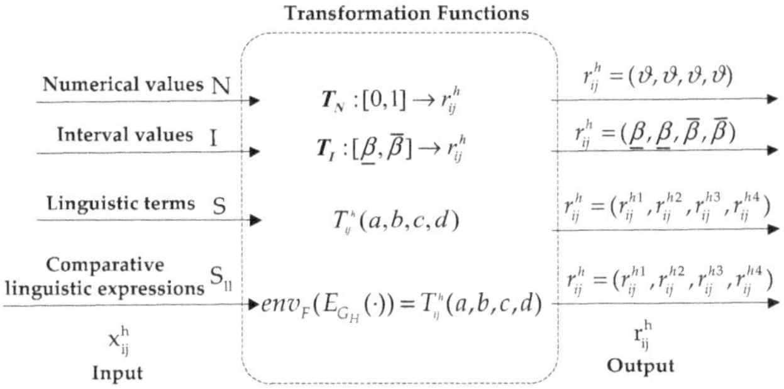

Figure 3. Unification process for non-homogeneous information.

The following transformation functions are defined to unify the information into a fuzzy domain.

1. For numerical values N, they are first normalized into the interval $[0,1]$ and then a transformation function T_N is utilized to transform them into trapezoidal fuzzy numbers. Let R be the domain of the numerical values, N_{ij}^h be the numerical value provided by the h-th expert over the i-th alternative concerning the j-th criterion, N_{ij}^h is normalized into the interval $[0,1]$, as follows:

$$\vartheta = \frac{N_{ij}^h}{N^*}$$

where $\vartheta \in [0,1]$, $N^* = \max_{h=1,2,\ldots,K} \{N_{ij}^h\}$, $i = 1,2,\ldots,m$, $j = 1,2,\ldots,n$.

Definition 3. *A numerical value is transformed into a trapezoidal fuzzy number by utilizing a transformation function* T_N:

$$T_N : [0,1] \to r_{ij}^h \tag{10}$$

$$T_N(\vartheta) = r_{ij}^h = (\vartheta, \vartheta, \vartheta, \vartheta)$$

2. The interval values I are first normalized into $[0,1]$ and then a transformation function T_I is utilized to transform them into trapezoidal fuzzy numbers. Let $[\xi^L, \xi^U]$ be the domain of the interval values, let $[d^L, d^U]_{ij}^h$ be the interval values provided by the h-th expert over the i-th alternative concerning the j-th criterion, where $[d^L, d^U]_{ij}^h \in [\xi^L, \xi^U]$. The interval values $[d^L, d^U]_{ij}^h$ are normalized into $[\underline{\beta}, \overline{\beta}]$ as follows:

$$\underline{\beta} = \frac{d^L - \xi^L}{\xi^U - \xi^L} \text{ and } \overline{\beta} = \frac{d^U - \xi^L}{\xi^U - \xi^L} \quad (11)$$

The transformation function T_I is defined as follows.

Definition 4. *An interval value is transformed into a trapezoidal fuzzy number by utilizing a transformation function* T_I:

$$T_I : [\underline{\beta}, \overline{\beta}] \rightarrow r_{ij}^h \quad (12)$$

$$T_I(\underline{\beta}, \overline{\beta}) = r_{ij}^h = (\underline{\beta}, \underline{\beta}, \overline{\beta}, \overline{\beta})$$

where $\underline{\beta}, \overline{\beta} \in [0,1]$ *and* $\underline{\beta} \leq \overline{\beta}$.

3. The linguistic terms $s_k \in S = \{s_0, s_1, \ldots, s_g\}$, are represented by trapezoidal fuzzy numbers. Therefore, the expert e_h provides his/her opinions over the i-th alternative concerning the j-th criterion as a linguistic term s_k that is represented by a trapezoidal fuzzy number $r_{ij}^h = (r_{ij}^{h1}, r_{ij}^{h2}, r_{ij}^{h3}, r_{ij}^{h4})$.
4. The comparative linguistic expressions, $x_{ij}^h \in S_{ll}$, are transformed into HFLTS by $E_{G_H}(\cdot)$ and its fuzzy envelop $env_F(\cdot)$ obtained by [41],

$$env_F(E_{G_H}(x_{ij}^h)) = T_{ij}^h(a, b, c, d) = r_{ij}^h \quad (13)$$

E_{G_H} is a function that transforms the linguistic expressions obtained by using G_H, into HFLTS [30]. $T_{ij}^h(a, b, c, d)$ is a trapezoidal fuzzy membership function corresponding to the trapezoidal fuzzy number $r_{ij}^h = (r_{ij}^{h1}, r_{ij}^{h2}, r_{ij}^{h3}, r_{ij}^{h4})$.

3.4. Consensus Reaching Process

As stated in Section 2.2, a fuzzy linear programming-based consensus model [34] is used in our proposal to achieve an agreement among all the experts involved in the problem. This model is able to deal with fuzzy information and update experts' opinions automatically without a supervised feedback mechanism [33], which is adequate for GEDM problems defined in fuzzy environment (see Figure 4).

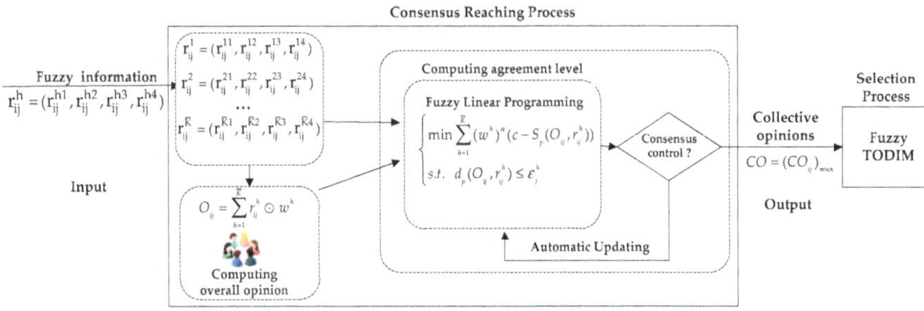

Figure 4. The process of fuzzy linear programming-based consensus model.

The fuzzy linear programming-based consensus model is given by,

$$\begin{cases} \min \sum\limits_{h=1}^{\overline{K}} (w^h)^\alpha (c - S_p(O_{ij}, r_{ij}^h)) \\ s.t.\ d_p(O_{ij}, r_{ij}^h) \le \varepsilon_j^h,\ h = 1,2,...,\overline{K};\ j = 1,2,\ldots,n,\ i = 1,2,\ldots,m. \end{cases} \tag{14}$$

According to Figure 4, the input information is represented in a fuzzy domain, which is obtained from the previous phase. It consists of three steps that are further detailed as follows:

1. *Computing overall opinion*. As introduced in Section 2.2, before applying fuzzy linear programming model, the overall opinions, O_{ij}, are obtained by aggregating the individual expert opinions, r_{ij}^h. Let O_{ij} be the overall opinion over the i-th alternative concerning the j-th criterion. It can be obtained as follows:

$$O_{ij} = \sum_{h=1}^{\overline{K}} r_{ij}^h \odot w^h,\ h = 1,2,\ldots,\overline{K},\ j = 1,2,\ldots,n,\ i = 1,2,\ldots,m. \tag{15}$$

where $\odot$ is an aggregation operator. For example, suppose that $r_{12}^1 = (0.17, 0.34, 0.5, 0.67)$, $r_{12}^2 = (0, 0.17, 0.34, 0.5)$ and $(w^1, w^2) = (0.6, 0.4)$, then O_{12} could be computed by a weighted average operator:

$$\begin{aligned} O_{12} &= 0.6 \odot (0.17, 0.34, 0.5, 0.67) + 0.4 \odot (0, 0.17, 0.34, 0.5) \\ &= (0.102, 0.272, 0.436, 0.602) \end{aligned}$$

2. *Computing agreement level*. In this step, there are two processes:

   (i) *Computing the distance and similarity*. The distance, $d_p(O_{ij}, r_{ij}^h)$, between the overall opinion, O_{ij}, and the individual opinion, r_{ij}^h, and its similarity, $S_p(O_{ij}, r_{ij}^h)$, can be computed according to Equations (1) and (2) respectively.

   (ii) *Determining the threshold values*. The threshold value, ε_j^h, is an important factor in the fuzzy linear programming model, which means the maximum change that the expert e_h can make concerning the j-th criterion. There are different ways to determine the threshold value ε_j^h [34,35]. In this paper, ε_j^h will be calculated by the h-th experts' familiarity degree concerning the j-th criterion using a linguistic term set $S = \{s_0, s_1, \ldots, s_g\}$, because the linguistic terms are flexible and able to deal with uncertain and vague information. The more familiar the expert is with the criterion, the less change he/she will make. Therefore, a negative operator is applied to the familiarity degree to obtain the threshold, which is defined as follows:

Definition 5. *Let* $S = \{s_0, s_1, \ldots, s_g\}$ *be a linguistic term set, a negative operator:*

$$Neg(s_k) = \widetilde{s}_q, \text{ such that } q = g - k,\ k = \{0, \ldots, g\}. \tag{16}$$

where $g + 1$ *is the cardinality of* S.

Thus, the ε_j^h can be computed by using the center of gravity (COG) method [42], i.e., $\varepsilon_j^h = COG(\widetilde{s}_q)$ (see Equation (18)).

3. *Control consensus.* When all constraints meet the conditions in Equation (14), it means that the consensus has been reached, and the final overall opinion, O_{ij}, is the aggregated collective opinion denoted as $CO = (CO_{ij})_{m \times n}$ Which will be used as input in the selection process.

3.5. Calculation of Criteria Weights

In this phase, the weights of criteria, w_{c_j}, are calculated by utilizing the experts' assessments provided over the criteria importance which were unified into a fuzzy domain. Figure 5 shows the process of computing criteria weights.

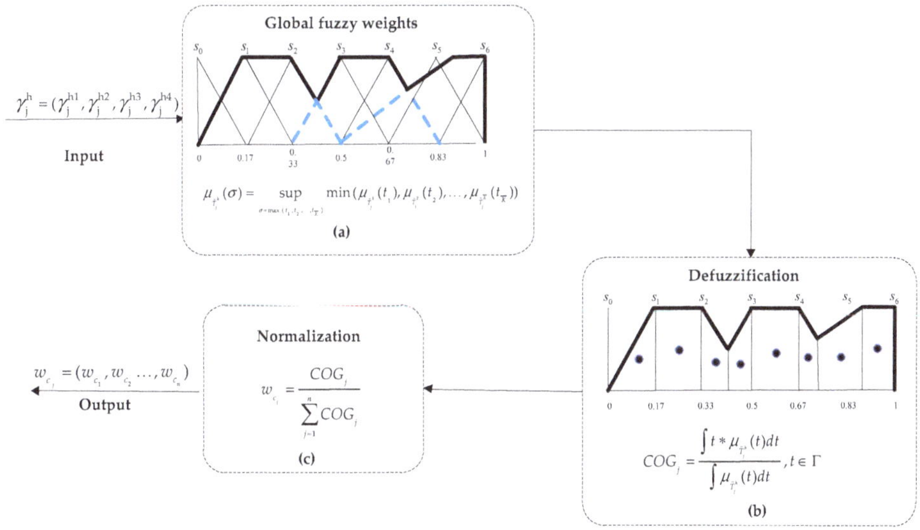

Figure 5. Computing criteria weights.

Three steps are comprised:

1. *Global fuzzy weights.* The fuzzy weights obtained for the criterion c_j are aggregated by using a max-min composition [43,44]:

$$\mu_{\widetilde{T}_j^h}(\sigma) = \sup_{\sigma = \max(t_1, t_2, \ldots, t_{\overline{K}})} \min\left(\mu_{\widetilde{T}_j^1}(t_1), \mu_{\widetilde{T}_j^2}(t_2), \ldots, \mu_{\widetilde{T}_j^{\overline{K}}}(t_{\overline{K}})\right), t_h \in \Gamma, h \in \{1, 2, \ldots, \overline{K}\} \tag{17}$$

where $\widetilde{T}_j^h$ is the fuzzy membership function of w_j^h, $j = 1, 2, \ldots, n$, and Γ is the universe of discourse.

Suppose that three experts provide their opinions w_1^1, w_1^2 and w_1^3 concerning the criterion c_1, the corresponding fuzzy membership functions are $\widetilde{T}_1^1$, $\widetilde{T}_1^2$ and $\widetilde{T}_1^3$ respectively. According to Equation (17), $\mu_{\widetilde{T}_j^h}(\sigma)$ is the area under the bold black line shown in Figure 5a.

2. *Defuzzification*. The COG method [42] is utilized to calculate the weighting value of the global fuzzy weights:

$$COG_j = \frac{\int t * \mu_{\widetilde{T}_j^h}(t)dt}{\int \mu_{\widetilde{T}_j^h}(t)dt}, t \in \Gamma \tag{18}$$

where Γ is the universe of discourse.

For criterion c_1, Equation (18) means that the center of gravity for each small trapezoid (see Figure 5b) is computed and the COG_1 can be obtained by the arithmetic mean of the sum of center of gravity of all small trapezoids.

3. *Normalization*. When COG_j of all criteria are obtained, the criteria weights w_{c_j} are calculated by using the following equation:

$$w_{c_j} = \frac{COG_j}{\sum_{j=1}^{n} COG_j} \tag{19}$$

where $\sum\limits_{j=1}^{n} w_{c_j} = 1, w_{c_j} \in [0,1]\ j = 1,2,\ldots,n$.

3.6. Selection Process—Fuzzy TODIM Method

As it was pointed out in Introduction, the experts' psychological behavior are neglected in current GEDM approaches. However, our proposal takes into account experts' psychological behavior by means of fuzzy TODIM based on prospect theory dealing with the problem defined in a fuzzy environment.

Once the criteria weights w_{c_j} and the aggregated collective opinions $CO = (CO_{ij})_{m\times n}$ are obtained, the fuzzy TODIM method is applied to obtain a ranking of alternatives and select the best one. To do so, the fuzzy TODIM method introduced in Section 2.3 is used. The step 1 is not necessary to do it, because the collective opinion matrix $CO = (CO_{ij})_{m\times n}$, is already normalized and the step 3 has been modified to adapted it to GEDM problem as it is shown below:

Step 3: To calculate the dominance degree, $\Phi_j(p_i, p_k)$, of alternative p_i $(i = 1,2,\ldots,m)$ over the remaining alternatives p_k $(k = 1,2,\ldots,m)$ concerning criterion c_j $(j = 1,2,\ldots,n)$, i.e.,

$$\Phi_j(p_i,p_k) = \begin{cases} \sqrt{d(CO_{ij},CO_{kj})w_{jr}/(\sum_{j=1}^{n} w_{jr})}, & \widetilde{m}(CO_{ij}) - \widetilde{m}(CO_{kj}) \geq 0 \\ -\frac{1}{\theta}\sqrt{d(CO_{ij},CO_{kj})(\sum_{j=1}^{n} w_{jr})/w_{jr}}, & \widetilde{m}(CO_{ij}) - \widetilde{m}(CO_{kj}) < 0 \end{cases} \tag{20}$$

CO_{ij} denotes the trapezoidal fuzzy number $CO_{ij} = (CO_{ij}^1, CO_{ij}^2, CO_{ij}^3, CO_{ij}^4)$ that represents the information about the i-th alternative concerning the j-th criterion. $\widetilde{m}(CO_{ij})$ and $\widetilde{m}(CO_{kj})$ denotes the defuzzified value of the fuzzy number CO_{ij} and CO_{kj}, respectively, where $\widetilde{m}(CO_{ij}) = \frac{CO_{ij}^1 + 2CO_{ij}^2 + 2CO_{ij}^3 + CO_{ij}^4}{6}$ [42]. $d(CO_{ij}, CO_{kj})$ denotes the gains or losses of the alternative p_i over p_k concerning the criterion c_j, where $d(CO_{ij}, CO_{kj}) = \sqrt{\sum_{\ell=1}^{4}(CO_{ij}^{\ell} - CO_{kj}^{\ell})^2}$ [45].

For benefit criteria, $d(CO_{ij}, CO_{kj})$ denotes the gains with $\widetilde{m}(CO_{ij}) - \widetilde{m}(CO_{kj}) \geq 0$ or losses with $\widetilde{m}(CO_{ij}) - \widetilde{m}(CO_{kj}) < 0$, respectively. $\Phi_j(p_i, p_k)$ can be expressed as:

$$\Phi_j(p_i,p_k) = \begin{cases} \sqrt{d(CO_{ij},CO_{kj})w_{jr}/(\sum_{j=1}^{n} w_{jr})}, & \widetilde{m}(CO_{ij}) - \widetilde{m}(CO_{kj}) \geq 0 \\ -\frac{1}{\theta}\sqrt{d(CO_{ij},CO_{kj})(\sum_{j=1}^{n} w_{jr})/w_{jr}}, & \widetilde{m}(CO_{ij}) - \widetilde{m}(CO_{kj}) < 0 \end{cases} \tag{21}$$

For cost criteria, $d(CO_{ij}, CO_{kj})$ denotes the gains with $\widetilde{m}(CO_{ij}) - \widetilde{m}(CO_{kj}) \leq 0$ or losses with $\widetilde{m}(CO_{ij}) - \widetilde{m}(CO_{kj}) > 0$, respectively, $\Phi_j(p_i, p_k)$ can be expressed as:

$$\Phi_j(p_i, p_k) = \begin{cases} \sqrt{d(CO_{ij}, CO_{kj}) w_{jr} / (\sum_{j=1}^{n} w_{jr})}, \ \widetilde{m}(CO_{ij}) - \widetilde{m}(CO_{kj}) \leq 0 \\ -\frac{1}{\theta} \sqrt{d(CO_{ij}, CO_{kj}) (\sum_{j=1}^{n} w_{jr}) / w_{jr}}, \ \widetilde{m}(CO_{ij}) - \widetilde{m}(CO_{kj}) > 0 \end{cases} \quad (22)$$

Finally, the ranking of alternatives can be determined according to their overall dominance degree.

4. Group Decision Support System for GEDM Based on GENESIS: Case Study

EEs are always characterized by complexity, risk and uncertainty, and a delayed or wrong decision may result in extremely serious consequences. Thus, it is necessary to make a decision in short time, taking into account the opinions of multiple experts involved in the problem.

In order to deal properly with real-world GEDM problems and make timely and effective decisions, we have implemented a GDSS named GENESIS to support the proposed GEDM method. This section introduces the structure and components of GENESIS (see Figure 6); and shows a case study to illustrate the applicability and robustness of the proposed method by using GENESIS.

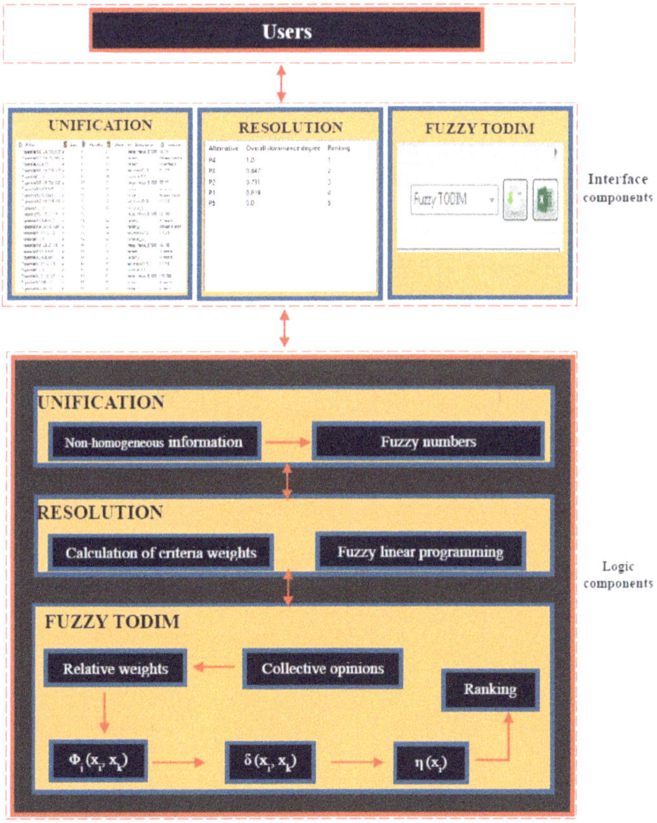

Figure 6. Structure of GENESIS.

4.1. GENESIS: (Group EmergeNcy dEcision SupportIng System)

Since our proposal deals with non-homogeneous and fuzzy information, in order to facilitate the transformation of non-homogeneous information and the decision process of the proposed method in a simple and fast manner, GENESIS has been implemented to use different components and specific functions based on FLINTSTONES [19,20] developed by using Eclipse Rich Client Platform (Eclipse RCP), which is a component-based application [46], a platform that builds and deploys rich client applications.

GENESIS consists of six components (see Figure 6):

(1) Two components taken from FLINTSTONES are adapted to define different transformation functions to unify non-homogeneous information into a fuzzy domain and show its user interface respectively.
(2) Two new components are defined for the resolution processes and show their interface to compute the criteria weights and obtain the consensus opinion based on fuzzy linear programming-based consensus model.
(3) Two new components are introduced to carry out the steps defined in the fuzzy TODIM method such as the computation of the relative weights, dominance degree etc., and show its user interface.

4.2. Case Study

In order to demonstrate the applicability of the proposed GEDM method, this section presents an example adapted from a big explosion of Tianjin Port that occurred in the north of China (Background Information Source. http://www.safehoo.com/Case/Case/Blow/201602/428723.shtml).

The blasts took place at a warehouse at the port that contained hazardous and flammable chemicals, including calcium carbide, sodium cyanide, potassium nitrate, ammonium nitrate and sodium nitrate, etc.

In this problem, we assume that six experts are invited to participate in the EDM process to support the EM to make the final decision. In order to solve this GEDM problem, we have used the proposed method by means of GENESIS.

4.2.1. Framework Definition

When the explosion occurred, the local government organized people located within two kilometers of the explosion area, evacuated them to safety areas and sent short messages to inform people in potentially dangerous areas to prepare for evacuation and keep distances from the dangerous area. Five emergency alternatives $\{p_1, p_2, \ldots, p_5\}$ were put forward taking into account five criteria $\{c_1, c_2, \ldots, c_5\}$, which are described in Tables 3 and 4, respectively.

For the criteria importance, the linguistic term set is S_1 = {*absolutely low importance (ali), very low importance (vli), low importance (li), medium importance (mi), high importance (hi), very high importance (vhi), absolutely high importance (ahi)*}. (see Figure 7 "syntax for S_1")

For criteria C_2 and C_3, the experts provide their opinions using linguistic term sets S_2 = {*none (n), very low seriously (vls), low seriously (ls), medium (m), high seriously (hs), very high seriously (vhs), absolutely seriously (as)*} and S_3 = {*none (n), very low (vl), low (l), medium (m), high (h), very high (vh), absolutely high (ah)*} (see Figure 7 "syntax for C_2" and "syntax for C_3"), respectively.

Table 3. Description of alternatives.

Alternative	Description
Evacuate people (p_1)	Evacuate and inform people, and at same time, assign 9 fire squadrons and 35 fire engines to deal with the emergency event.
Increase help and report (p_2)	Increase to 23 fire squadrons, 93 fire engines and more than 600 fire fighters for participating in dealing with the emergency event; at the same time, the local government report the latest news to the masses in order to avoid causing panic and riot.
Rescue military (p_3)	Local government asks the Chinese professional emergency rescue military for emergency rescue. More than 300 soldiers with professional equipment join the rescue action.
Joint rescue (p_4)	Fire squadrons and the military work together dealing with the problems, at the same time, local government asks neighbor cities for fire police to provide support.
Block boundary of explosion areas (p_5)	Block the boundary of the explosion areas; let the material in the explosion areas burn down.

Table 4. Description of criteria.

Criteria	Expression Domain	Description
People affected (C_1)	Interval values	It means that alternative p_i can protect the number of people from the effects caused by EE in [0,1000].
Negative effect on the environment (C_2)	Linguistic	It is evaluated by experts on linguistic expressions.
Social impacts (C_3)	Linguistic	It means the impacts on social development or people's daily life etc. that are evaluated by experts on linguistic expressions.
Property loss (C_4)	Interval values	It means that the alternative p_i can protect the direct and indirect property losses that are caused by the EE in [0,10]. (in billion RMB).
Cost of alternative (C_5)	Numerical values	The numerical values are 0 and 1. 0 means that expert e_h does not care about the cost; 1 means that he/she cares about it.

Note: assume that above criteria are independent.

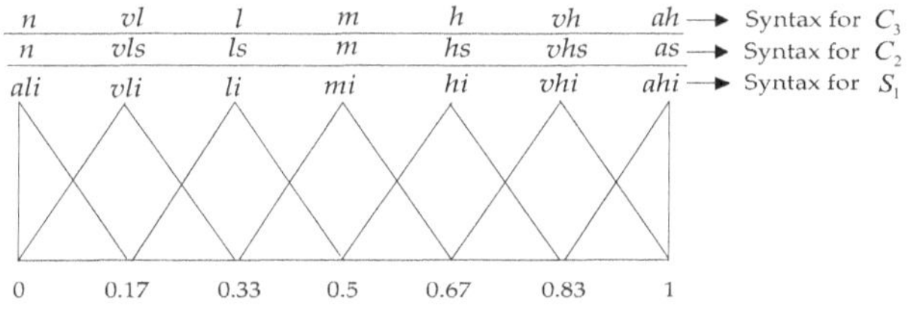

Figure 7. Linguistic term set for S_1, C_2 and C_3.

4.2.2. Information Gathering Process

The assessments provided by experts over the alternatives concerning criteria, and their opinions regarding the criteria importance and the familiarity degree for each criterion are shown in Tables 5–7 respectively. This phase is supported by GENESIS to facilitate the information gathering process (see Figure 8).

Table 5. Assessments provided by all experts on different alternatives concerning each criterion.

Expert	Alternative	Criteria				
		C_1	C_2	C_3	C_4	C_5
		Interval Values [0,1000]	**Linguistic**	**Linguistic**	**Interval Values [0,10]**	**Numerical Values (0,1)**
e_1	P_1	[20,25]	*ls*	*l*	[0.2,0.3]	1
	P_2	[30,35]	*ls*	*m*	[0.2,0.35]	1
	P_3	[50,80]	*m*	*h*	[0.5,0.8]	1
	P_4	[100,150]	*hs*	*bt m* and *h*	[1.0,2.0]	1
	P_5	[60,70]	*vhs*	*vh*	[0.1,0.2]	1
e_2	P_1	[30,50]	*vls*	*At* most *vl*	[0.25,0.4]	1
	P_2	[40,50]	*vls*	*vl*	[0.3,0.5]	1
	P_3	[100,150]	*ls*	*m*	[0.6,1.5]	1
	P_4	[150,250]	*m*	*l*	[2.0,2.5]	1
	P_5	[80,100]	*hs*	*vh*	[0.1,0.25]	1
e_3	P_1	[20,30]	*vls*	*l*	[0.1,0.15]	1
	P_2	[30,60]	*ls*	*l*	[0.15,0.25]	1
	P_3	[60,100]	*bt ls* and *m*	*h*	[0.2,0.3]	1
	P_4	[200,300]	*ls*	*m*	[1.5,2.5]	1
	P_5	[50,80]	*hs*	*bt h* and *vh*	[0.2,0.25]	1
e_4	P_1	[25,40]	*vls*	*vl*	[0.2,0.25]	1
	P_2	[30,45]	*vls*	At most *l*	[0.4,0.5]	1
	P_3	[80,150]	*ls*	*m*	[0.6,1.0]	1
	P_4	[200,250]	*bt ls* and *m*	*l*	[1.5,3.0]	1
	P_5	[50,70]	*vhs*	*vh*	[0.3,0.6]	1
e_5	P_1	[20,30]	*vls*	*l*	[0.25,0.3]	1
	P_2	[30,40]	*ls*	*vl*	[0.3,0.4]	1
	P_3	[50,80]	At most *m*	*m*	[0.5,1.0]	1
	P_4	[150,300]	*vls*	*l*	[2.0,2.5]	1
	P_5	[40,70]	*bt hs* and *vhs*	*vh*	[0.35,0.5]	1
e_6	P_1	[30,40]	*ls*	*vl*	[0.2,0.3]	1
	P_2	[20,50]	*vls*	*vl*	[0.5,0.6]	1
	P_3	[40,70]	*ls*	*l*	[0.4,0.6]	1
	P_4	[200,300]	*m*	*bt vl* and *l*	[2.5,3.5]	1
	P_5	[50,60]	*hs*	*h*	[0.3,0.5]	1

Table 6. The importance of each criterion provided by each expert.

Experts	Criteria				
	C_1	C_2	C_3	C_4	C_5
e_1	*vhi*	*hi*	*hi*	*li*	*mi*
e_2	*bt hi and vhi*	*hi*	*hi*	*mi*	*li*
e_3	*hi*	*mi*	*hi*	*li*	*vli*
e_4	*vhi*	*mi*	*mi*	*li*	*vli*
e_5	*hi*	*mi*	*hi*	*mi*	*li*
e_6	At least *hi*	*hi*	*hi*	*mi*	*li*

Note: "bt" means between in Tables 5 and 6.

Table 7. The familiarity degree provided by all experts for each criterion.

Experts	Criteria				
	C_1	C_2	C_3	C_4	C_5
e_1	*vs*	*s*	*vs*	*m*	*m*
e_2	*s*	*m*	*s*	*vs*	*m*
e_3	*m*	*vs*	*vs*	*m*	*s*
e_4	*vs*	*m*	*s*	*s*	*m*
e_5	*m*	*vs*	*vs*	*s*	*u*
e_6	*s*	*s*	*s*	*m*	*m*

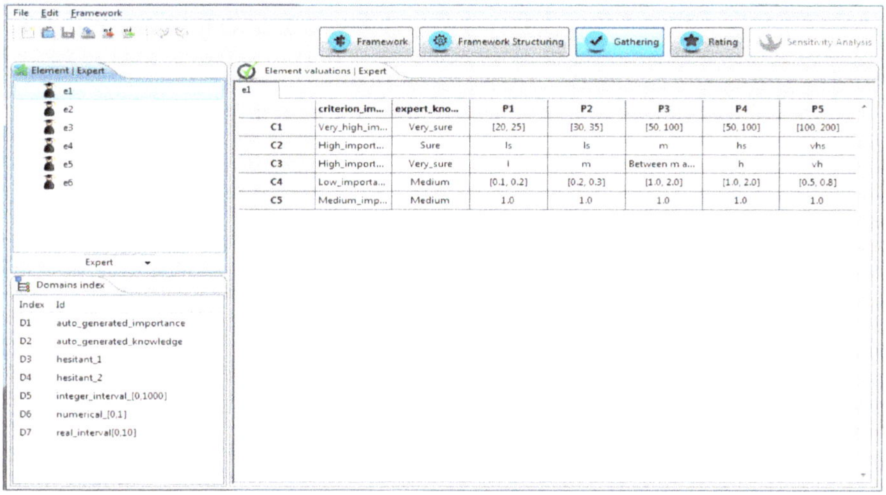

Figure 8. Gathered information in GENESIS.

4.2.3. Managing Non-Homogeneous Information

All experts' assessments are transformed into trapezoidal fuzzy numbers by utilizing the transformation functions defined in Section 3.2. Therefore, GENESIS makes all the necessary computations to unify the non-homogeneous information into a fuzzy domain in a simple and fast way. Figure 9 shows the interface of such a process.

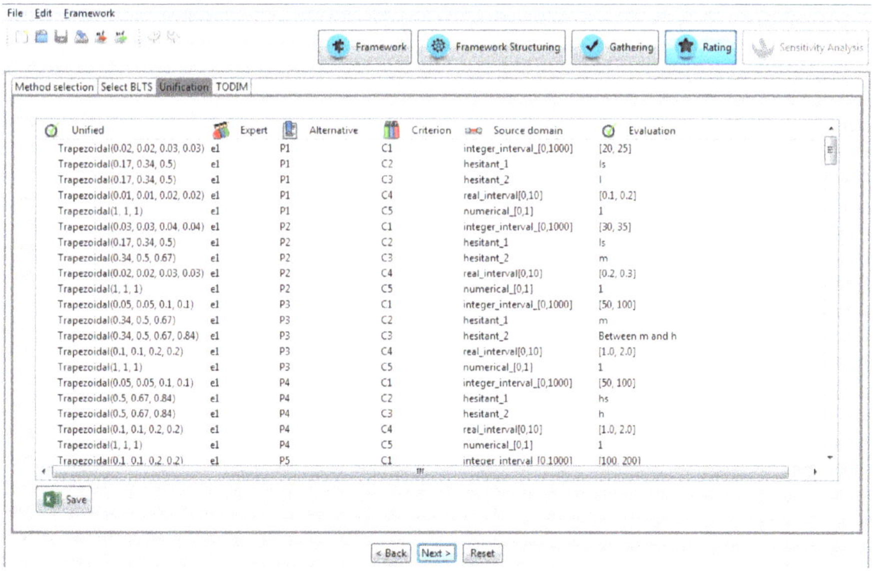

Figure 9. Unification results of non-homogeneous information.

4.2.4. Consensus Reaching Process

The fuzzy linear programming-based consensus model is utilized to achieve the consensus among all experts involved in the GEDM problem and obtain the collective opinion that will be used in the selection process. Before applying the CRP, the threshold values in Equation (14) should be determined.

Let S_4 = {s_0: *none (n)*, s_1: *very unsure (vu)*, s_2: *unsure (u)*, s_3: *medium (m)*, s_4: *sure (s)*, s_5: *very sure (vs)*, s_6: *absolutely sure (as)*} be the linguistic term set (see Figure 10) used by experts to express their familiarity degree for each criterion.

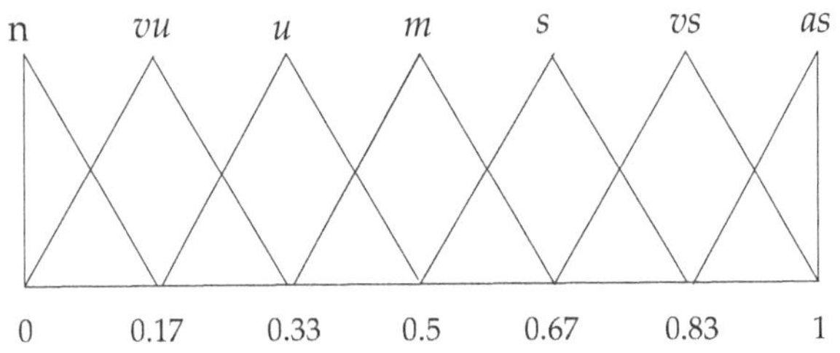

Figure 10. Linguistic term set S_4.

Expert e_h provides his/her familiarity degree for the criterion c_j by using a linguistic term $s_k \in S_4$. According to Equation (16), $\tilde{s}_q = s_{6-k}$, then, the COG of $\tilde{s}_q$ is regarded as the threshold value for the expert e_h about the criterion c_j, shown in Table 8. Table 7 is the familiarity degree provided by all experts for each criterion.

Table 8. Threshold values for $\tilde{s}_q$ transformed by negative operator.

$\tilde{s}_q$	Threshold Value
$\tilde{s}_0$	0
$\tilde{s}_1$	0.17
$\tilde{s}_2$	0.33
$\tilde{s}_3$	0.5
$\tilde{s}_4$	0.67
$\tilde{s}_5$	0.83
$\tilde{s}_6$	1

For example, expert e_1 provides his/her familiarity degree for the criterion c_1, $s_5 = vs$, then according to Equation (16), $\tilde{s}_1 = vu$, and, the COG of $\tilde{s}_1$ is 0.17, i.e., $\varepsilon_1^1 = COG(vu) = 0.17$, it means that the maximum change that expert e_1 can make is 0.17 for the criterion c_1.

In this GEDM problem, experts' weights w^h have the same importance. The parameters p, α and c used in Equation (13) are set, $p = 2$, $\alpha = 2$ and $c = 1.5$ respectively [36].

When all constraints meet the conditions in Equation (14), the aggregated collective opinion, $CO = (CO_{ij})_{5\times 5}$, is obtained.

$$CO = \begin{bmatrix} (0.02,0.02,0.04,0.04) & (0.11,0.22,0.22,0.41) & (0.11,0.22,0.22,0.49) & (0.02,0.02,0.03,0.03) & (1,1,1,1) \\ (0.03,0.03,0.08,0.08) & (0.14,0.19,0.19,0.60) & (0.11,0.17,0.17,0.44) & (0.04,0.04,0.06,0.06) & (1,1,1,1) \\ (0.07,0.07,0.28,0.28) & (0.18,0.30,0.32,0.53) & (0.21,0.37,0.37,0.79) & (0.06,0.06,0.25,0.25) & (1,1,1,1) \\ (0.17,0.17,0.50,0.50) & (0.20,0.38,0.38,0.69) & (0.15,0.30,0.33,0.62) & (0.23,0.23,0.40,0.40) & (1,1,1,1) \\ (0.05,0.05,0.09,0.09) & (0.53,0.70,0.70,0.90) & (0.58,0.73,0.73,0.95) & (0.02,0.02,0.05,0.05) & (1,1,1,1) \end{bmatrix}$$

4.2.5. Calculation of Criteria Weights

Using Table 6, the criteria weights are calculated by GENESIS (see Figure 11).

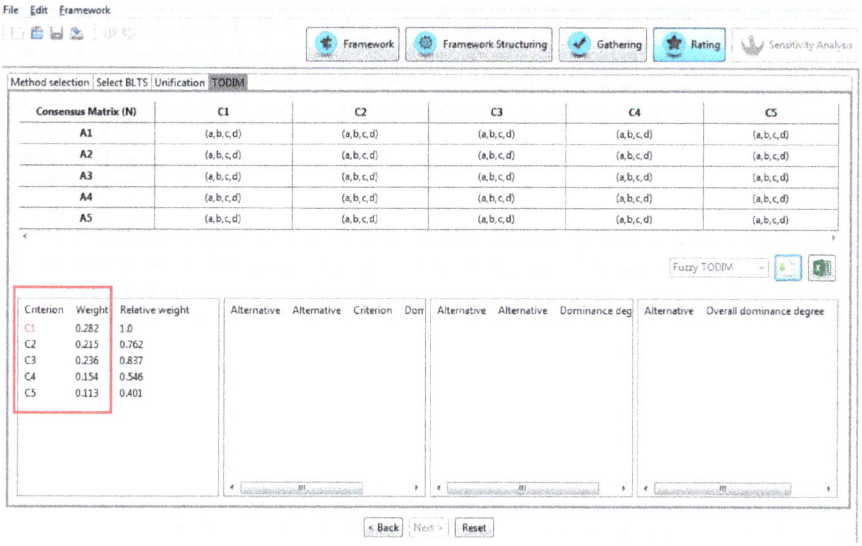

Figure 11. Criteria weights w_{c_j} obtained by GENESIS.

4.2.6. Selection Process-Fuzzy TODIM Method

Once the criteria weights w_{c_j} and the aggregated collective opinion $CO = (CO_{ij})_{m \times n}$ are obtained, fuzzy TODIM method is applied to calculate the overall dominance degree for each alternative and then the ranking of the alternative is obtained. Figure 12 shows the results obtained for each step of the fuzzy TODIM method.

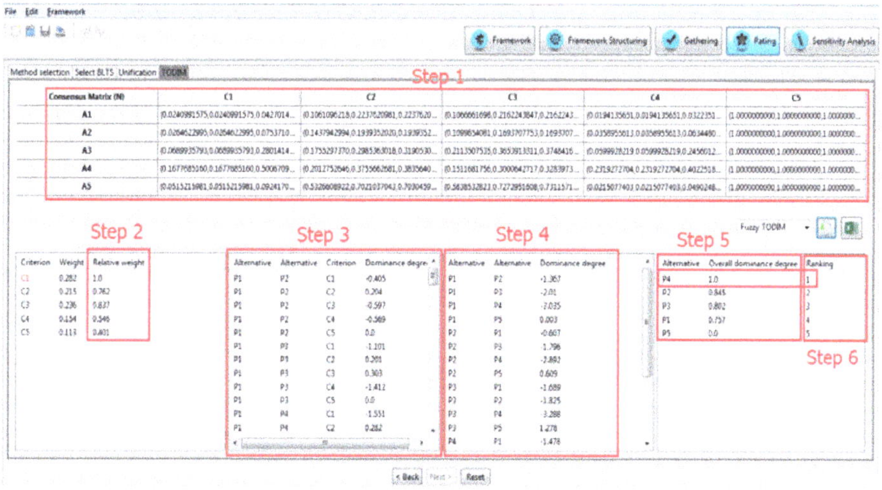

Figure 12. The results of different steps based on fuzzy TODIM by using GENESIS.

The ranking of alternatives is obtained according to the overall dominance degree for each alternative:

$$p_4 \succ p_2 \succ p_3 \succ p_1 \succ p_5$$

Finally, the EM can select p_4, "joint rescue" as the best alternative for the emergency response.

4.2.7. Sensitivity Analysis

To illustrate the feasibility and validity of the proposed method, sensitivity analysis is carried out in a similar way to other TODIM-based proposals in literature [38].

In this case, two aspects of sensitivity analysis are conducted: (i) the analysis about the weight evolution of the most important criterion and (ii) the evolution of attenuation factor θ.

For the weight evolution of the most important criterion, in this case study, it is C_1. First, let the weight of criterion C_1 be equal to the second most important criterion, i.e., $C_1 = 0.236$, then changing the weight of C_1 from 0.236 to 1. The reason for doing this is that the most important criterion is always the same and never changes, hence the relative weights are always calculated according to the same criterion. Applying these changes, the ranking of alternatives does not change.

The attenuation factor θ evolution, is changed from 1 to 15. When these alterations are carried out, there is no any change in the ranking of alternatives.

From the sensitivity analysis, it is easy to see that the ranking of alternatives is consistent with each other. It shows the feasibility, validity and the robustness of the proposed method.

5. Conclusions and Future Works

The non-homogeneous information including experts' hesitancy is not available in current GEDM approaches. To fill such a gap, this paper has taken into account the non-homogeneous information including experts' hesitancy, which extends the scope of non-homogeneous information defined in previous approaches. In order to make computations with non-homogeneous information defined in our proposal, different transformation functions have been presented to unify it into fuzzy numbers. A fuzzy linear programming-based consensus model with a new way for determining the threshold values has been applied to obtain the collective opinion, which is suitable for dealing with the fuzzy information. Experts' psychological behavior is very important in decision processes under risk and uncertainty; however, it is neglected in current GEDM approaches. To address such an important issue, fuzzy TODIM method has been utilized in our proposal due to its advantage of capturing human beings psychological behavior. Furthermore, a case study has been provided to illustrate the feasibility and validity of the proposed method by using GENESIS supporting the whole decision process.

Future research could be the use of computer science and Internet technology for supporting the EDM based on big data, which will lead to more reliable decisions. Furthermore, game theory [47,48] can be applied to deal with the emergency problems under uncertainty.

Acknowledgments: This work was partly supported by the Young Doctoral Dissertation Project of Social Science Planning Project of Fujian Province (Project No. FJ2016C202), National Natural Science Foundation of China (Project Nos. 71371053, 61773123), Spanish National Research Project (Project No. TIN2015-66524-P), and Spanish Ministry of Economy and Finance Postdoctoral Fellow (IJCI-2015-23715) and ERDF.

Author Contributions: All authors have contributed equally to this paper.

Conflicts of Interest: The authors declare no conflict of interest.

References

1. Levy, J.K.; Taji, K. Group decision support for hazards planning and emergency management: A group analytic network process (GANP) approach. *Math. Comput. Model. Int. J.* **2007**, *46*, 906–917. [CrossRef]
2. Liu, Y.; Fan, Z.P.; Zhang, Y. Risk decision analysis in emergency response: A method based on cumulative prospect theory. *Comput. Oper. Res.* **2014**, *42*, 75–82. [CrossRef]

3. Wang, L.; Zhang, Z.X.; Wang, Y.M. A prospect theory-based interval dynamic reference point method for emergency decision making. *Expert Syst. Appl.* **2015**, *42*, 9379–9388. [CrossRef]
4. Xu, X.H.; Du, Z.J.; Chen, X.H. Consensus model for multi-criteria large-group emergency decision making considering non-cooperative behaviors and minority opinions. *Decis. Support Syst.* **2015**, *79*, 150–160. [CrossRef]
5. Xu, Y.; Zhang, W.; Wang, H. A conflict-eliminating approach for emergency group decision of unconventional incidents. *Knowl. Based Syst.* **2015**, *83*, 92–104. [CrossRef]
6. Ju, Y.; Wang, A.; Liu, X. Evaluating emergency response capacity by fuzzy AHP and 2-tuple fuzzy linguistic approach. *Expert Syst. Appl.* **2012**, *39*, 6972–6981. [CrossRef]
7. Herrera, F.; Herrera-Viedma, E.; Verdegay, J.L. A sequential selection process in group decision making with a linguistic assessment approach. *Inf. Sci.* **1995**, *85*, 223–239. [CrossRef]
8. Roubens, M. Fuzzy sets and decision analysis. *Fuzzy Sets Syst.* **1997**, *90*, 199–206. [CrossRef]
9. Butler, C.L.; Rothstein, A. *On Conflict and Consensus: A Handbook on Formal Consensus Decision Making*, 2nd ed.; Citeseer: Food Not Bombs Publishing: Portland, ME, USA, 2007; pp. 21–22.
10. Saint, S.; Lawson, J.R. *Rules for Reaching Consensus: A Modern Approach to Decision Making*; Jossey-Bass: San Francisco, CA, USA, 1994; pp. 27–49.
11. Ping, F.; Chong, W.; Jie, T. Unconventional emergency management based on intelligent group decision-making methodology. *Adv. Inf. Sci. Serv. Sci.* **2012**, *4*, 208–216.
12. Camerer, C. Bounded rationality in individual decision making. *Exp. Econ.* **1998**, *1*, 163–183. [CrossRef]
13. Tversky, A.; Kahneman, D. Advances in prospect theory: Cumulative representation of uncertainty. *J. Risk Uncertain.* **1992**, *5*, 297–323. [CrossRef]
14. Kahneman, D.; Tversky, A. Prospect theory: An analysis of decision under risk. *Econometrica* **1979**, *47*, 263–291. [CrossRef]
15. Yu, L.; Lai, K.K. A distance-based group decision-making methodology for multi-person multi-criteria emergency decision support. *Decis. Support Syst.* **2011**, *51*, 307–315. [CrossRef]
16. Wei, C.; Ren, Z.; Rodriguez, R.M. A hesitant fuzzy linguistic TODIM method based on a score function. *Int. J. Comput. Intell. Syst.* **2015**, *8*, 701–712. [CrossRef]
17. Tosun, Ö.; Akyüz, G. A fuzzy TODIM approach for the supplier selection problem. *Int. J. Comput. Intell. Syst.* **2015**, *8*, 317–329. [CrossRef]
18. Krohling, R.A.; Souza, T.T. Combining prospect theory and fuzzy numbers to multi-criteria decision making. *Expert Syst. Appl.* **2012**, *39*, 11487–11493. [CrossRef]
19. Estrella, F.J.; Espinilla, M.; Herrera, F.; Martínez, L. Flintstones: A fuzzy linguistic decision tools enhancement suite based on the 2-tuple linguistic model and extensions. *Inf. Sci.* **2014**, *280*, 152–170. [CrossRef]
20. Martínez, L.; Rodriguez, R.M.; Herrera, F. Flintstones: A Fuzzy Linguistic Decision Tools Enhancement Suite. In *The 2-Tuple Linguistic Model*; Springer: Cham, Switzerland, 2015; pp. 145–168.
21. Palomares, I.; Rodríguez, R.M.; Martínez, L. An attitude-driven web consensus support system for heterogeneous group decision making. *Expert Syst. Appl.* **2013**, *40*, 139–149. [CrossRef]
22. Cabrerizo, F.J.; Herrera-Viedma, E.; Pedrycz, W. A method based on PSO and granular computing of linguistic information to solve group decision making problems defined in heterogeneous contexts. *Eur. J. Oper. Res.* **2013**, *230*, 624–633. [CrossRef]
23. Herrera, F.; Martínez, L.; Sánchez, P.J. Managing non-homogeneous information in group decision making. *Eur. J. Oper. Res.* **2005**, *166*, 115–132. [CrossRef]
24. Peng, D.H.; Gao, C.Y.; Zhai, L.L. Multi-criteria group decision making with heterogeneous information based on ideal points concept. *Int. J. Comput. Intell. Syst.* **2013**, *6*, 616–625. [CrossRef]
25. Li, D.F.; Huang, Z.G.; Chen, G.H. A systematic approach to heterogeneous multiattribute group decision making. *Comput. Ind. Eng.* **2010**, *59*, 561–572. [CrossRef]
26. Zhang, G.; Lu, J. An integrated group decision-making method dealing with fuzzy preferences for alternatives and individual judgments for selection criteria. *Group Decis. Negot.* **2003**, *12*, 501–515. [CrossRef]
27. Estrella, F.J.; Cevik Onar, S.; Rodríguez, R.M.; Basar, O.; Luis, M.; Cengiz, K. Selecting firms in university technoparks: A hesitant linguistic fuzzy TOPSIS model for heterogeneous contexts. *J. Intell. Fuzzy Syst.* **2017**, *2*, 1155–1172. [CrossRef]
28. Rodríguez, R.M.; Labella, A.; Martínez, L. An overview on fuzzy modelling of complex linguistic preferences in decision making. *Int. J. Comput. Intell. Syst.* **2016**, *9*, 81–94. [CrossRef]

29. Rodríguez, R.M.; Martínez, L.; Herrera, F. Hesitant fuzzy linguistic term sets for decision making. *IEEE Trans. Fuzzy Syst.* **2012**, *20*, 109–119. [CrossRef]
30. Rodríguez, R.M.; Martínez, L.; Herrera, F. A group decision making model dealing with comparative linguistic expressions based on hesitant fuzzy linguistic term sets. *Inf. Sci.* **2013**, *241*, 28–42. [CrossRef]
31. Quesada, F.J.; Palomares, I.; Martínez, L. Managing experts behavior in large-scale consensus reaching processes with uninorm aggregation operators. *Appl. Soft Comput.* **2015**, *35*, 873–887. [CrossRef]
32. Dong, Y.; Luo, N.; Liang, H. Consensus building in multiperson decision making with heterogeneous preference representation structures: A perspective based on prospect theory. *Appl. Soft Comput.* **2015**, *35*, 898–910. [CrossRef]
33. Palomares, I.; Estrella, F.J.; Martínez, L.; Herrera, F. Consensus under a fuzzy context: Taxonomy, analysis framework afryca and experimental case of study. *Inf. Fusion* **2014**, *20*, 252–271. [CrossRef]
34. Liu, J.; Chan, F.T.; Li, Y.; Zhang, Y.; Deng, Y. A new optimal consensus method with minimum cost in fuzzy group decision. *Knowl. Based Syst.* **2012**, *35*, 357–360. [CrossRef]
35. Lee, H.S. Optimal consensus of fuzzy opinions under group decision making environment. *Fuzzy Sets Syst.* **2002**, *132*, 303–315. [CrossRef]
36. Gomes, L.; Lima, M. From modeling individual preferences to multicriteria ranking of discrete alternatives: A look at prospect theory and the additive difference model. *Found. Comput. Decis. Sci.* **1992**, *17*, 171–184.
37. Gomes, L.; Lima, M. TODIM: Basics and application to multicriteria ranking of projects with environmental impacts. *Found. Comput. Decis. Sci.* **1992**, *16*, 113–127.
38. Gomes, L. An application of the TODIM method to the multicriteria rental evaluation of residential properties. *Eur. J. Oper. Res.* **2009**, *193*, 204–211. [CrossRef]
39. Pereira, J.; Gomes, L.F.A.M.; Paredes, F. Robustness analysis in a TODIM-based multicriteria evaluation model of rental properties. *Technol. Econ. Dev. Ecol.* **2013**, *19*, 176–190. [CrossRef]
40. Wang, L.; Wang, Y.M.; Martinez, L. A group decision method based on prospect theory for emergency situations. *Inf. Sci.* **2017**, *418–419*, 119–135. [CrossRef]
41. Liu, H.; Rodríguez, R.M. A fuzzy envelope for hesitant fuzzy linguistic term set and its application to multicriteria decision making. *Inf. Sci.* **2014**, *258*, 220–238. [CrossRef]
42. Chen, S.J.; Chen, S.M. Fuzzy risk analysis based on similarity measures of generalized fuzzy numbers. *IEEE Trans. Fuzzy Syst.* **2003**, *11*, 45–56. [CrossRef]
43. Zadeh, L.A. Similarity relations and fuzzy orderings. *Inf. Sci.* **1971**, *3*, 177–200. [CrossRef]
44. Zadeh, L.A. Outline of a new approach to the analysis of complex systems and decision processes. *IEEE Trans. Syst. Man Cybern.* **1973**, *SMC-3*, 28–44. [CrossRef]
45. Mahdavi, I.; Mahdavi-Amiri, N.; Heidarzade, A.; Nourifara, R. Designing a model of fuzzy TOPSIS in multiple criteria decision making. *Appl. Math. Comput.* **2008**, *206*, 607–617. [CrossRef]
46. Eclipse Foundation. Available online: http://www.eclipse.org (accessed on 18 April 2016).
47. Zheng, X.; Cheng, Y. Modeling cooperative and competitive behaviors in emergency evacuation: A game-theoretical approach. *Comput. Math. Appl.* **2011**, *62*, 4627–4634. [CrossRef]
48. Arena, P.; Fazzino, S.; Fortuna, L.; Maniscalco, P. Game theory and non-liner dynamic: The parrondo paradox case study. *Chaos Solitons Fract.* **2003**, *17*, 545–555. [CrossRef]

symmetry

MDPI

Article

Fishmeal Supplier Evaluation and Selection for Aquaculture Enterprise Sustainability with a Fuzzy MCDM Approach

Tsung-Hsien Wu [1], Chia-Hsin Chen [1], Ning Mao [1] and Shih-Tong Lu [2,*]

[1] Department of Business Administration, Nanjing University, Nanjing 210093, China; dg1002065@smail.nju.edu.cn (T.-H.W.); dg1402041@smail.nju.edu.cn (C.-H.C.); maong@nju.edu.cn (N.M.)

[2] Department of Logistics and Shipping Management, Kainan University, Taoyuan 33857, Taiwan

* Correspondence: stonelu8604@gmail.com

Received: 29 October 2017; Accepted: 14 November 2017; Published: 21 November 2017

Abstract: In the aquaculture industry, feed that is of poor quality or nutritionally imbalanced can cause problems including low weight, poor growth, poor palatability, and increased mortality, all of which can induce a decrease in aquaculture production. Fishmeal is considered a better source of protein and its addition as an ingredient in the aquafeed makes aquatic animals grow fast and healthy. This means that fishmeal is the most important feed ingredient in aquafeed for the aquaculture industry. For the aquaculture industry in Taiwan, about 144,000 ton/USD $203,245,000 of fishmeal was imported, mostly from Peru, in 2016. Therefore, the evaluation and selection of fishmeal suppliers is a very important part of the decision-making process for a Taiwanese aquaculture enterprise. This study constructed a multiple criteria decision-making evaluation model for the selection of fishmeal suppliers using the VlseKriterijumska Optimizacija I Kompromisno Resenje (VIKOR) approach based on the weights obtained with the entropy method in a fuzzy decision-making environment. This hybrid approach could effectively and conveniently measure the comprehensive performance of the main Peruvian fishmeal suppliers for practical applications. In addition, the results and processes described herein function as a good reference for an aquaculture enterprise in making decisions when purchasing fishmeal.

Keywords: aquaculture; fishmeal supplier selection; entropy; VlseKriterijumska Optimizacija I Kompromisno Resenje (VIKOR); fuzzy logic

1. Introduction

On 17 September 2014, the Member States of the United Nations announced the Sustainable Development Goals (SDGs) as part of the 2030 Agenda for sustainable development. The 2030 Agenda set aims for the contribution and conduct of fisheries and aquaculture towards food security and nutrition, and the use of natural resources to ensure sustainable development in economic, social, and environmental terms. According to statistics from the Food and Agriculture Organization (FAO) of the United Nations, aquaculture provided only 7% of fish for human consumption in 1974, but this share has since increased to 26% in 1994, and 44% in 2014. Aquaculture has seen an impressive growth in the supply of farmed fish, which overtook that of wild-caught fish for human consumption in 2014 [1]. This makes the aquaculture industry an important source of aquatic food. Taiwan is one of the top 25 countries in this industry with a total production of 340,600 tons and ranked 19th in 2014 [1].

Aquafeed is a very significant factor for production in the aquaculture industry and it accounts for about 40–60% of the cultivation cost. In addition, the feed quality and its nutrient content greatly affects the growth of aquatic animals as poor-quality feed or nutrient imbalances can cause low weight, poor growth, feed inefficiency, and increase the mortality rate. Therefore, it is very important to choose the best source of feed for sustainable development in aquaculture production. Fishmeal is considered the most nutritious, digestible source of protein for farmed-fish feed. Fishmeal is added to the aquafeed to ensure that the aquatic animals grow fast and healthy, and can improve the quality of the related aquaculture products. Effective screening of the source of this important raw material is necessary to maintain the quality of related products and establish inherent goodwill in the industry. Therefore, an aquaculture enterprise cannot consider price alone as the major consideration for procurement. Meeting the required quality, supplying the appropriate quantity, timely delivery, and long-term partnerships are all factors that should be considered for the evaluation and selection of fishmeal vendors. Taiwan's aquaculture industry imported about 144,000 ton/USD $203,245,000 of fishmeal from Peru, India, Thailand, and Vietnam in 2016. Peru is not only the world's leading exporter of fishmeal, but also the largest supplier for Taiwan's aquaculture industry. However, there are many suppliers of fishmeal in Peru and their supply capacity, product quality, delivery term, and cooperative attitude all vary. Therefore, setting up practical evaluation criteria and a method for the evaluation and selection of fishmeal suppliers from Peru would be helpful for Taiwan's aquaculture enterprises. An example of raw material supplier selection is also given to demonstrate the proposed solution to this kind of problem.

The problems of supplier evaluation and selection have received considerable attention in academic study and in practice. Numerous multiple criteria decision making (MCDM) approaches have been proposed to tackle the problem such as the analytic hierarchy process (AHP), analytic network process (ANP), mathematical programming, technique for order preference by similarity to ideal solution (TOPSIS), preference ranking organization method for enrichment of evaluations (PROMETHEE), VlseKriterijumska Optimizacija I Kompromisno Resenje (VIKOR), and hybrid or extended fuzzy approaches (see Table 1) [2–32]. Amongst these methodologies, many approaches have used criteria weightings that have been determined by subjective evaluations by decision makers or experts for pairwise comparison or direct rating methods. Furthermore, for the VIKOR method, Opricovic and Tzeng [33] have described the advantages of its theory, and Opricovic and Tzeng [34] have compared it with other outranking methods, both of which illustrate the benefits of VIKOR. Therefore, to reduce uncertainties arising from subjective factors, this work adopted the entropy method to objectively determine the criteria weights. Then, based on the entropy weightings, the VIKOR approach was applied to process the performance rating of the alternatives. In addition, to capture and handle the human appraisal of ambiguity, uncertainty, and subjectivity, linguistic variables in the fuzzy sets were integrated into the supplier evaluation and selection process. This hybrid fuzzy MCDM technique was applied to evaluate and select fishmeal suppliers from Peru for Taiwan's aquaculture enterprise. The VIKOR approach was used based on weightings obtained with the entropy method in a fuzzy decision-making environment. The advantage of this approach was that we only needed to evaluate the merits of the alternatives based on linguistic variables under each criterion. These linguistic variables were converted into scores, which were then utilized to calculate the fuzzy entropy weights to help clarify the importance of the criteria. These weights were then applied with the fuzzy VIKOR approach to derive a comprehensive performance evaluation for the complex supplier selection problem. Thus, the overall scores for each supplier in each criterion can be obtained, and the selection decision made accordingly. This method is more effective and convenient in practical applications and provides better decision-making quality. This paper also discusses an empirical case study to demonstrate how an aquaculture enterprise can implement this solution. The results and processes provide a good reference to assist an aquaculture enterprise in Taiwan in the making of fishmeal purchasing decisions.

The remainder of this paper is organized as follows. In Section 2, the criteria for raw material supplier selection are identified. In Section 3, the research methodology including fuzzy entropy and fuzzy VIKOR is introduced. Section 4 includes the numerical case study that uses a Taiwan aquaculture enterprise as an example, thus demonstrating the process of fishmeal supplier evaluation and selection from the proposed model, the procedure, and method. The results of the empirical research are also analyzed. In Section 5, some conclusions are offered.

Table 1. Related research for supplier selection.

Authors	Approaches	Field of Empirical Study
Tam and Tummala, 2001	AHP	Telecommunications company
Dulmin and Mininno, 2003	PROMETHEE-GAIA	Public road and rail transportation
Kumar et al., 2004	Fuzzy integer goal programming	Auto-parts company
Chen et al., 2006	Fuzzy TOPSIS	High-tech company
Kumar et al., 2006	Fuzzy programming	Auto-parts company
Gencer and Gürpinar, 2007	ANP	Electronic firm
Xia and Wu, 2007	AHP	CPU supplier
Sanayei et al., 2008	MAUT and LP	Automobile manufacturer
Yang et al., 2008	Fuzzy AHP	High-tech industries
Amin and Razmi, 2009	Fuzzy set theory	Internet service provider
Boran et al., 2009	Fuzzy TOPSIS	Automotive company
Wang et al., 2009	Fuzzy AHP and TOPSIS	Lithium-ion battery protection IC
Lin et al., 2010	ANP	Semiconductor industry
Chamodrakas et al., 2010	Fuzzy AHP and programming	Electronic marketplaces
Sanayei et al., 2010	Fuzzy VIKOR	Automobile part manufacturing
Shemshadi, et al., 2011	Fuzzy VKOR and Entropy	Petrochemical factory
Jahan et al., 2011	VIKOR	Health care
Amin et al., 2011	fuzzy SWOT and LP	Auto parts company
Kilincci and Onal, 2011	Fuzzy AHP	Washing machine company
Feng et al., 2011,	Multi-objective 0–1 programming	CSA company
Zhao and Yu, 2011	Information entropy	Petroleum enterprises
Vahdani et al., 2012	Locally linear neuro-fuzzy	Cosmetics industry
Chatterjee and Chakraborty, 2012	PROMETHEE II and Gray relation	Rotating machine part
Hsua et al., 2012	DANP with VIKOR	Decoration corporation
Chen and Chao, 2012	AHP and CFPR	Electronic company
Peng and Xiao, 2013	PROMETHEE and ANP	Bush materials
Zhao and Guo, 2014	fuzzy-entropy and fuzzy-TOPSIS	Thermal power equipment
Kuo et al., 2015	DANP with VIKOR	Green Electronics Company
Chung et al., 2016	ANP and IPA	Bicycle manufacturer
Yazdani et al., 2017	DEMATEL	Food-based production company
Wan et al., 2017	ANP and ELECTRE II	Auto manufacture company

2. Criteria for Fishmeal Supplier Selection

Supplier selection is the process of finding appropriate suppliers who are able to provide the buyer with the right quality products and/or services at an acceptable price and delivery time, and in the required quantities. This is one of the most critical activities in establishing an effective supply chain. Obviously, supplier selection is a multiple criteria decision making (MCDM) problem affected by several conflicting factors such as price, quality, delivery, and so on.

Historically, several methodologies have been developed for evaluating, selecting, and monitoring potential suppliers that take into account factors such as quality, logistics, and cost. Dickson [35], in one of the well-known studies on supplier selection, identified 23 important evaluation criteria for supplier selection. Barbarosoglu and Yazgac [36] helped a company find the proper supplier by adopting Dickson's criteria to evaluate supplier performance. In recent years, a number of researchers have begun to identify some of the relevant criteria. Ng [37] constructed a simple and effective supplier evaluation model to deal with problems of supplier selection with supply variety, quality, delivery, and price as the evaluation criteria. Shemshadi et al. [17], Chen et al. [5], Boran et al. [12], and Yang et al. [10] identified product quality, effort to establish cooperation, the supplier's technical level, delay on delivery, price/cost, profitability of supplier, relationship closeness, technological capability, conformance quality, conflict resolution, delivery performance, supplier profile, and risk as factors for determining the best supplier. Chen and Kumar [38] established an evaluation model to obtain the best supplier with the result to be given to a company as a strategy reference. They proposed the following five criteria: overall cost of the product, quality of the product, service performance of supplier, supplier's profile, and risk factor. By consolidating several studies, Sanayei et al. [16] proposed five categories: product quality, on-time delivery, price/cost, supplier's technological level, and flexibility. Shyur and Shih [39] introduced evaluation indictors including on-time delivery, product quality, price/cost, facility and technology, responsiveness to customer needs, professionalism of salespeople, and quality of relationship into the supplier evaluation process. Ávila et al. [40] defined product cost, financial stability, synergy potential, logistics cost, payment flexibility, after sales service cost, and production capacity as supplier evaluation criteria. There have been a significant number of studies discussing supplier selection and a wide range of mathematical methods have been used to provide solutions for supplier selection, as shown in Table 1.

In this study, based on the principles espoused in [35] and in consultation with the management team of a typical aquaculture company in Taiwan, we listed 22 factors that are often used for evaluating fishmeal suppliers in Taiwan's aquaculture enterprise. To evaluate the importance of the factors from an expert viewpoint, questionnaires with responses given in the seven-point Likert scale (from one to seven) were used collect expert opinions, with preferences of very unimportant, essentially unimportant, weakly unimportant, fair, weakly important, essentially important, and very important. Seventeen experts with many years of work experience in Taiwan's aquaculture industry were invited to evaluate the importance of these 22 factors. The demographic information of these 17 respondents is summarized in Table 2. The Cronbach's α was 0.918, which represents excellent internal consistency reliability. The Kaiser-Meyer-Olkin (KMO) measuring sampling adequacy provides an index (between 0 and 1) of the proportion of variance among the variables that might be common variance. A high KMO indicates that sampling is adequate, indicating the existence of a statistically acceptable factor solution representing relationships between the parameters. In our study, the Kaiser-Meyer-Olkin (KMO) value was found to be 0.689, which was better than the suggested value of 0.6 [41]. The factors, as well as the importance of the factors, are summarized in Table 3. The importance values of the 22 factors fell in a range between 5.000 and 6.647. When the importance of factors was identified, it was unrealistic to consider all of the factors simultaneously given the limited time and resources. To improve the evaluation and selection process, 10 major factors were determined as the evaluation criteria given higher priority after discussion with the management team: "Stability of product quality"; "Stability of supply capability"; "Reasonableness of quoted price"; "Financial capability and condition"; "Flexibility in changing shipment schedule"; "Potential cooperation in the future"; "Operating control of pre-delivery"; "Satisfaction with claims for damages"; "Exactness for presenting documents to the bank"; and "Control capability of on-time delivery", as shown in Table 4.

Table 2. Demographic information of the experts who evaluated the criteria.

Demographic Information		Frequency
Gender	Male	3
	Female	14
Age	30–35	10
	35–40	3
	40–55	4
Working experience	Under 5	4
	5–10	6
	Above 10	7
Education level	College	3
	Bachelor	10
	Master	4
Occupation	Purchasing manager	2
	Purchasing specialist	15

Table 3. Importance of the 22 factors.

Factors	Importance	Ranking
1. The ratio of supply quantity to total purchase quantity	5.118	18
2. Stability of supply capability	6.235	2
3. Stability of product quality	6.647	1
4. Completeness of product packaging	5.294	15
5. Operating control of products before delivery	5.529	7
6. Control capability for on-time delivery	5.824	4
7. Records of claim for damages or complaints	5.000	22
8. Satisfaction with handling claims for damages	5.471	8
9. Efficiency of handling claims for damages	5.059	19
10. Facilities and equipment of production plant	5.059	19
11. Financial capability and condition	5.471	8
12. Efficiency of communication	5.176	17
13. Exactness in presenting documents to the bank	5.471	8
14. Operating control of shipping documents	5.353	13
15. Brand awareness	5.235	16
16. Reasonableness of quoted price	5.882	3
17. Reasonableness of shipping freight quotes	5.412	11
18. Flexibility and coordination of order modification	5.412	11
19. Flexibility for changing shipment schedule	5.706	5
20. Service attitudes of operational staff	5.353	13
21. Closeness of previous business relationship	5.059	19
22. Possibility of establishing long-term cooperation	5.588	6

Table 4. Fishmeal supplier selection evaluation criteria.

	Evaluation Criteria	Importance
C_1	Stability of product quality	6.647
C_2	Stability of supply capability	6.235
C_3	Reasonableness of quoted price	5.882
C_4	Control capability of on-time delivery	5.824
C_5	Flexibility for changing shipment schedule	5.706
C_6	Possibility of establishing long-term cooperation	5.588
C_7	Operating control of product before delivery	5.529
C_8	Satisfaction with handling claims for damages	5.471
C_9	Exactness for presenting documents to the bank	5.471
C_{10}	Financial capability and conditions	5.471

3. The Proposed Method

In the process of decision-making, decision-makers often make subjective judgments based on their own knowledge and experience in ambiguous or vague statements, such as good, poor, important, not important, and so on, given in linguistic terms. To deal with the ambiguity and subjectivity of human judgment, linguistic variables have been introduced with these judgments expressed by a membership function within a closed interval of [0, 1] as in fuzzy set theory [42]. Bellman and Zadeh [43] proposed a methodology for decision-making in a fuzzy environment to resolve the lack of precision in assigning the degree of importance of evaluation criteria and the ratings of alternatives based on the evaluation criteria. In this section, we introduce the concepts and processes used to define the linguistic variables, to calculate the entropy weights, and the VIKOR procedure.

3.1. Linguistic Variables and Fuzzy Numbers

A linguistic term or linguistic variable is one whose value is given by words or sentences expressed in a natural language. In this study, we used these kinds of expression in linguistic terms to evaluate the performance of selected alternatives regarding each criterion: "Very poor", "Poor", "Medium poor", "Fair", "Medium good", "Good", and "Very good", with respect to a trapezoidal fuzzy number (TFN) as proposed by [16,44]. A TFN is a fuzzy set $\widetilde{A}$ on X if its membership function is a mapping $\mu_{\widetilde{A}}(x) : X \to [0, 1]$. The membership function of a fuzzy number $\widetilde{A}$ can be described as follows:

$$\mu_{\widetilde{A}}(x) = \begin{cases} 0, & x \leq a_1 \ \ or \ \ x \geq a_4 \\ (x - a_1)/(a_2 - a_1) \ , & a_1 \leq x \leq a_2 \\ (a_4 - x)/(a_4 - a_3) \ , & a_3 \leq x \leq a_4 \\ 1, & a_2 \leq x \leq a_3 \end{cases} \tag{1}$$

The trapezoidal fuzzy number can be denoted by $\widetilde{A} = (a_1, a_2, a_3, a_4)$, where $\{(a_1, a_2, a_3, a_4)|a_1, a_2, a_3, a_4 \in R; a_1 \leq a_2 \leq a_3 \leq a_4\}$ which denotes the smallest possible, the most promising, and the largest possible values, respectively, as shown in Figure 1. Table 5 and Figure 2 show the corresponding TFN for each linguistic variable.

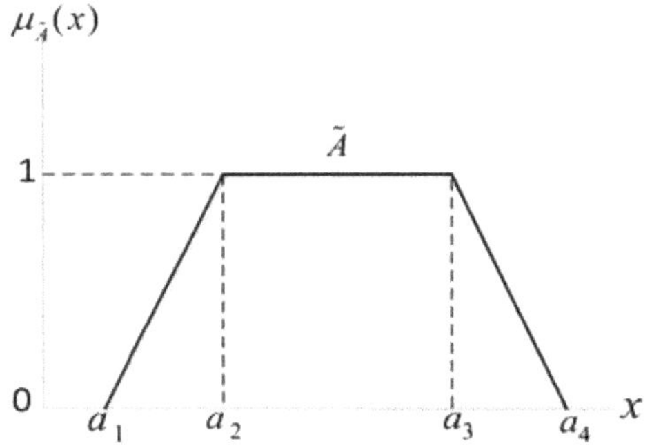

Figure 1. Trapezoidal fuzzy number $\widetilde{A}$.

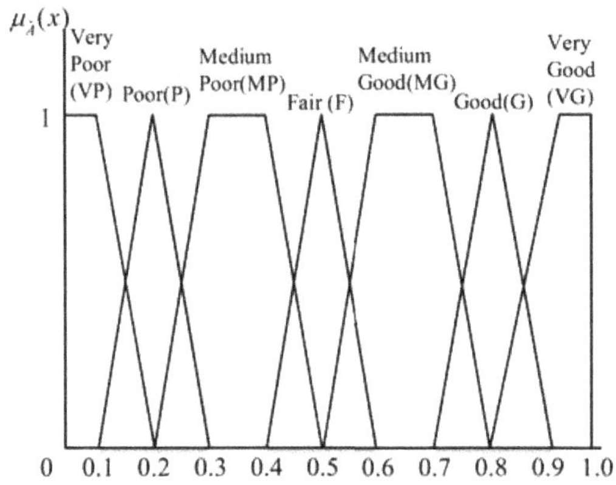

Figure 2. Linguistic variables for the fuzzy rates of alternatives.

The algebraic operations for the two TFNs (addition, subtraction, multiplication, division and reciprocity) applied in this study were based on the arithmetic of special fuzzy numbers as introduced by [44].

After the evaluation process in the fuzzy environment, the results are still in the fuzzy number format. Therefore, it is necessary to further conduct defuzzification to transform the fuzzy numbers to crisp numbers. Based on the center of area (COA) method, TFN $\widetilde{A} = (a_1, a_2, a_3, a_4)$ was defuzzified to a crisp value ($C_{\widetilde{A}}$) as the centroid value of TFN $\widetilde{A}$, as follows:

$$C_{\widetilde{A}} = \frac{\int_{a_1}^{a_4} x\mu_{\widetilde{A}}(x)dx}{\int_{a_1}^{a_4} \mu_{\widetilde{A}}(x)dx} = \frac{\int_{a_1}^{a_2} x(\frac{x-a_1}{a_2-a_1})dx+\int_{a_2}^{a_3} xdx+\int_{a_3}^{a_4} x(\frac{a_4-x}{a_4-a_3})dx}{\int_{a_1}^{a_2} \frac{x-a_1}{a_2-a_1}dx+\int_{a_2}^{a_3} 1dx+\int_{a_3}^{a_4} \frac{a_4-x}{a_4-a_3}dx} = \frac{a_3^2+a_4^2+a_3a_4-a_1^2-a_2^2-a_1a_2}{3(a_3+a_4-a_1-a_2)} \tag{2}$$

Table 5. Fuzzy linguistic assessment variables.

Linguistic Variables	Trapezoidal Fuzzy Number (TFN)
Very poor, VP	(0.0, 0.0, 0.1, 0.2)
Poor, P	(0.1, 0.2, 0.2, 0.3)
Medium poor, MP	(0.2, 0.3, 0.4, 0.5)
Fair, F	(0.4, 0.5, 0.5, 0.6)
Medium good, MG	(0.5, 0.6, 0.7, 0.8)
Good, G	(0.7, 0.8, 0.8, 0.9)
Very good, VG	(0.8, 0.9, 1.0, 1.0)

3.2. Group Decision Making

A good decision-making process is not only comprised of arbitrary decisions made by individuals, but also requires a combination of professional judgments. In this study, for the sustainable management of an enterprise, we developed a method for the evaluation and selection process of fishmeal suppliers to obtain the appropriate and correct results. The group multiple criteria decision making (GMCDM) method includes the following elements: (1) m possible suppliers, $A = \{A_1, A_2, \cdots, A_m\}$; (2) n evaluation criteria, $C = \{C_1, C_2, \cdots, C_n\}$; and (3) k decision-makers, $D = \{D_1, D_2, \cdots, D_k\}$. The performance evaluation matrix of the supplier $A_i(i = 1, 2, \cdots, m)$ with respect to criteria $Cj(j = 1, 2, \cdots, n)$ by decision-maker D_k using the fuzzy linguistic assessment variables can be constructed as follows:

$$\widetilde{E}^k = \left[\widetilde{e}_{ij}^k\right]_{m\times n} = \begin{matrix} & \begin{matrix} C_1 & \cdots & C_n \end{matrix} \\ \begin{matrix} A_1 \\ \vdots \\ A_m \end{matrix} & \begin{pmatrix} \widetilde{e}_{11}^k & \cdots & \widetilde{e}_{1n}^k \\ \vdots & \ddots & \vdots \\ \widetilde{e}_{m1}^k & \cdots & \widetilde{e}_{mn}^k \end{pmatrix}_{m\times n} \end{matrix} \quad i = 1, 2, \cdots, m; \quad j = 1, 2, \cdots, n; \quad k = 1, 2, \cdots r \tag{3}$$

where $\widetilde{e}_{ij}^k = (e_{1ij}^k, e_{2ij}^k, e_{3ij}^k, e_{4ij}^k)$.

Therefore, for the k decision makers conducting the group evaluation process, the aggregated fuzzy performance rating of alternatives with respect to each criterion as an integrated fuzzy decision matrix $\widetilde{E}$ can be calculated as:

$$\widetilde{E} = [\widetilde{e}_{ij}]_{m\times n} = \begin{bmatrix} \widetilde{e}_{11} & \widetilde{e}_{12} & \cdots & \widetilde{e}_{1n} \\ \widetilde{e}_{21} & \widetilde{e}_{22} & \cdots & \widetilde{e}_{2n} \\ \vdots & \vdots & \ddots & \vdots \\ \widetilde{e}_{m1} & \widetilde{e}_{m2} & \cdots & \widetilde{e}_{mn} \end{bmatrix} = [(e_{1ij}, e_{2ij}, e_{3ij}, e_{4ij})]_{m\times n}$$
$$= \begin{bmatrix} (e_{111}, e_{211}, e_{311}, e_{411}) & (e_{112}, e_{212}, e_{312}, e_{412}) & \cdots & (e_{11n}, e_{21n}, e_{31n}, e_{41n}) \\ (e_{121}, e_{221}, e_{321}, e_{421}) & (e_{122}, e_{222}, e_{322}, e_{422}) & \cdots & (e_{12n}, e_{22n}, e_{32n}, e_{42n}) \\ \vdots & \vdots & \ddots & \vdots \\ (e_{1m1}, e_{2m1}, e_{3m1}, e_{4m1}) & (e_{1m2}, e_{2m2}, e_{3m2}, e_{4m2}) & \cdots & (e_{1mn}, e_{2mn}, e_{3mn}, e_{4mn}) \end{bmatrix},$$

where

$$e_{1ij} = \min_k\{e_{1ij}^k\}; \quad e_{2ij} = \frac{1}{r}\sum_{k=1}^{r} e_{2ij}^k; \quad e_{3ij} = \frac{1}{r}\sum_{k=1}^{r} e_{3ij}^k; \quad e_{4ij} = \max_k\{e_{4ij}^k\} \tag{4}$$

3.3. Determination of Criteria Weightings

Shannon [45] introduced the concept of entropy into information theory, which is used to measure information and uncertainty, and to characterize and signal uncertainty for the information sources. If the entropy of an evaluation criterion is smaller, the amount of information provided by the criterion is greater, and the greater the role in the comprehensive evaluation process, the higher the weight. The entropy weight method mainly uses the uncertainty represented by the entropy value as determined by information theory to calculate the decision information that can be transmitted by each evaluation criterion, then obtains the relative weights between the criteria. The relative weight calculated by the entropy weight method is obtained by using the evaluation information for each alternative under each evaluation criterion where there are no subjective factors. In other words, this is an objective weight. For this study, the processes of computing entropy weights were as follows:

(1) According to the established fuzzy decision evaluation matrix $\widetilde{\boldsymbol{E}}$, the fuzzy decision evaluation matrix was defuzzified to a crisp value matrix by

$$\widetilde{\boldsymbol{E}} = [\widetilde{e}_{ij}]_{m\times n} \to \boldsymbol{F} = [f_{ij}]_{m\times n} \tag{5}$$

where $[f_{ij}]_{m\times n} = \begin{array}{c} \\ A_1 \\ \vdots \\ A_m \end{array} \begin{array}{c} \begin{array}{ccc} C_1 & \cdots & C_n \end{array} \\ \begin{pmatrix} f_{11} & \cdots & f_{1n} \\ \vdots & \ddots & \vdots \\ f_{m1} & \cdots & f_{mn} \end{pmatrix}_{m\times n} \end{array} \quad i = 1,2,\cdots,m; \quad j = 1,2,\cdots,n.$

(2) Normalize the evaluation matrix:

$$\boldsymbol{R} = [r_{ij}]_{n\times m} = \begin{array}{c} \\ A_1 \\ \vdots \\ A_m \end{array} \begin{array}{c} \begin{array}{ccc} C_1 & \cdots & C_n \end{array} \\ \begin{pmatrix} r_{11} & \cdots & r_{1n} \\ \vdots & \ddots & \vdots \\ r_{m1} & \cdots & r_{mn} \end{pmatrix}_{m\times n} \end{array} \quad r_{ij} = \frac{f_{ij}}{\sum\limits_{i=1}^{m} f_{ij}}, \quad i = 1,2,\cdots,m; j = 1,2,\cdots,n \tag{6}$$

(3) Calculate the Shannon entropy value of each evaluation criterion:

$$H_j = \left(-\frac{1}{\ln m}\right)\sum_{i=1}^{m} r_{ij}\ln r_{ij}, \quad i = 1,2,\ldots,m; j = 1,2,\ldots n \tag{7}$$

(4) The entropy weights of each evaluation criterion w_j were found as follows:

$$w_j = \frac{1-H_j}{\sum\limits_{j=1}^{n}(1-H_j)}, \quad i = 1,2,\ldots,m; j = 1,2,\ldots n; \quad \boldsymbol{W} = (w_1, w_2, \ldots, w_n), \sum_{j=1}^{n} w_j = 1 \tag{8}$$

3.4. Evaluation and Selection of Alternatives

The VIKOR method proposed by [46] is one of the optimal compromise solution methods used in multiple criteria decision making. The basic concept is to define the positive ideal solution and the negative ideal solution. The so-called positive ideal solution refers to the best alternatives with respect to each evaluation criterion, while the negative ideal solution consists of the worst alternatives for each evaluation criterion. The alternatives are then prioritized by comparing the evaluation values of each alternative with their closeness to the positive ideal solution. To calculate the closeness of the alternatives to the positive ideal solution, the values of the evaluation criteria must be aggregated. In VIKOR, the aggregating function was developed from the L_p-*metric* through a compromise programming method [47] that focuses on ranking and selecting from a set of alternatives to determine a compromise solution that provides the maximum group utility for the majority, and a minimum of individual regret for the opponent, which can help the decision makers reach a final decision. The processes of applying VIKOR for alternative selections are as follows:

(1) Determine the evaluation values of the best and the worst alternatives/suppliers for each criterion j: f_j^+ and f_j^-

$$f_j^+ = \max_i f_{ij}; \quad f_j^- = \min_i f_{ij} \qquad i = 1, 2, \cdots, m; \quad j = 1, 2, \cdots, n \tag{9}$$

(2) Compute the weighted distance ratio to the best value for every alternative/supplier with respect to each criterion: S_i and R_i

$$S_i = \sum_j^n w_j \left(\frac{f_j^+ - f_{ij}}{f_j^+ - f_j^-} \right), \tag{10}$$

$$R_i = \max_j \left(w_j \left(\frac{f_j^+ - f_{ij}}{f_j^+ - f_j^-} \right) \right). \tag{11}$$

(3) Compute the values Q_i as follows:

S_i is the weighted summation of the distance to the best evaluation value of alternative i with respect to all criteria; R_i is calculated by the maximum weighted distance to the best evaluation value of alternative i with respect to the jth criterion; and S_i refers to the overall benefits of the ith alternative where the smaller the value, the larger the benefits. That is, $S^+ = \min S_i$, and $S^- = \max S_i$. R_i refers to the individual regret of the ith alternative where the smaller the value, the smaller the individual regret of the opponent. That is, $R^+ = \min R_i$, and $R^- = \max R_i$. Thus, the index Q_i is based on the consideration of both the group utility and individual regret of the opponent

$$Q_i = v \left(\frac{S_i - S^+}{S^- - S^+} \right) + (1 - v) \left(\frac{R_i - R^+}{R^- - R^+} \right) \tag{12}$$

where v is introduced as a weight for the strategy of maximum group utility, whereas $1 - v$ is the weight of the individual regret of the opponent.

(4) Rank the alternatives by sorting the values S_i, R_i and Q_i in ascending order.

(5) Propose as a comprise solution, alternative $A^{(1)}$, which is best ranked by the measure Q (minimum), if the following two conditions are satisfied:

C1 Acceptable advantage:

$$Q\left(A^{(2)}\right) - Q\left(A^{(1)}\right) \geq DQ = \frac{1}{m - 1} \tag{13}$$

where $A^{(2)}$ is the alternative in the second position in the ranking list bounded by Q and DQ.

C2 Acceptable stability in decision-making:

The alternative $A^{(1)}$ must also be the best ranked by S and/or R. This compromise solution is stable within a decision-making process, which could be the strategy of maximum group utility (when $v > 0.5$ is needed), or "by consensus" $v \approx 0.5$, or "with veto" ($v < 0.5$). Here, v is the weight of the decision-making strategy of the maximum group utility. If one of the conditions is not satisfied, then a set of compromise solutions is proposed, which consists of

(a) Alternatives $A^{(1)}$ and $A^{(2)}$ only if condition C2 is not satisfied;

(b) Alternatives $A^{(1)}, A^{(2)}, \cdots, A^{(m)}$, if condition C1 is not satisfied. $A^{(m)}$ is determined by the relation $Q(A^{(m)}) - Q(A^{(1)}) < DQ$ for maximum m (the positions of these alternatives are "in closeness").

4. Case Study

For aquaculture enterprises, the most important ingredient in the aquafeed is fishmeal. There was a slight decline in global fishmeal production and trade in 2017, but global production is still concentrated among a few top producers. Peru accounts for one-fifth of global production and remains the world's largest producer and exporter of fishmeal, accounting for nearly one-third of the global trade. There are two periods of time where fishing is allowed in the northern and central oceanic areas near Peru. The first period begins around April and runs to July, and the second fishing period starts from November and goes to January of the following year. The government of Peru realizes the importance of protecting its natural oceanic resources and has decided to reduce their fishing quota; therefore, fishermen must fish on the basis of published quotas, which has directly contributed to competitive tension in the supply side of the fisheries, causing a sharp fall in fishmeal production. Due to a decrease in supply due to the fishing quota, the fishmeal market is becoming more competitive. As a result, many small fishmeal factories in Peru have merged into larger main suppliers. About 80% of fishmeal production is now centralized.

To illustrate the proposed method, we considered an example where the managerial board of an aquaculture enterprise in Taiwan has to procure fishmeal for their aquatic stock. There are four main fishmeal companies in Peru considered as possible suppliers (see Table 6), and 10 important factors (as identified in Section 2) for evaluating these companies. To hedge risks, a committee of thirteen experts (decision-makers) with many years of work experience in the aquaculture industry of Taiwan was formed to select the most suitable fishmeal companies. Profiles of these experts are shown in Table 7.

Table 6. Profiles of candidate fishmeal suppliers.

Condition	A_1	A_2	A_3	A_4
Status	Listed	Non-Listed	Non-Listed	Non-Listed
Incorporation Date	25 July 1994	1 August 1945	13 January 1986	5 February 2006
Total Employees	2073	3502	7444	1495
Plants	5	16	9	7
Products	Fishmeal Fish oil	Fishmeal Fish oil	Fishmeal Fish oil Canned food Frozen food	Fishmeal Fish oil Frozen fish

Table 7. Profiles of experts evaluating the alternatives.

Demographic Information		Frequency
Gender	Male	3
	Female	10
Age	30–35	6
	35–40	3
	40–55	4
Working experience	Under 5	4
	5–10	6
	Above 10	3
Education level	College	3
	Bachelor	8
	Master	2
Occupation	Purchasing manager	2
	Purchasing specialist	11

The proposed model applied for fishmeal supplier selection for a firm operating in the field of aquaculture was comprised of the following steps:

Step 1: Using linguistic variable, thirteen decision makers were asked to rate the candidates with respect to each criterion (see Table 5). The ratings of the four suppliers by the decision makers under the various criteria are shown in Table 8.

Step 2: The linguistic evaluations shown in Table 9 was converted into trapezoidal fuzzy numbers. Then, the aggregated fuzzy rating of alternatives as calculated by Equation (4) to construct the fuzzy decision matrix, as shown in Table 10. Then, the data in Table 11 were defuzzified by Equation (2) and normalized by Equation (6). Table 11 shows these processes for calculating the weight of each criterion.

Step 3: With the normalized values in Table 11, the entropy method was applied to determine the weight of each criterion by Equations (7) and (8). The crisp values for the decision matrix and the weight of each criterion were computed as shown in the bottom part of Table 11.

Step 4: Equation (9) was used to determine the best and the worst values of each criterion for the rating of all suppliers from upper part of Table 11, and the results are shown in the upper part of Table 12.

Step 5: The values of S_i, R_i, and Q_i were calculated by Equations (10)–(12) for the four candidate suppliers, as shown in Table 13.

Step 6: The suppliers were ranked by S, R, and Q in decreasing order as shown in Table 9.

Step 7: As seen in Table 9, supplier A_2 was ranked as the best by Q, but condition C1 was not satisfied $Q(A^{(2)}) - Q(A^{(1)}) < \frac{1}{4-1}$. Therefore, A_2 and A_1 were both appropriate choices.

Table 8. Linguistic evaluation of suppliers with respect to criteria by the decision-makers.

Expert/Criteria		C_1	C_2	C_3	C_4	C_5	C_6	C_7	C_8	C_9	C_{10}
D_1	A_1	G	MG	G	G	G	G	F	F	MG	G
	A_2	G	G	G	G	G	VG	MG	MG	MG	MG
	A_3	VG	MG	MG	MG	MG	MG	F	MG	MG	MG
	A_4	G	G	MG	MG	MG	VG	G	MG	MG	G
D_2	A_1	G	VG	G	G	G	VG	G	G	G	G
	A_2	MG	G	G	G	G	VG	MG	MG	MG	MG
	A_3	G	G	G	G	G	G	MG	MG	MG	MG
	A_4	G	G	MG	MG	MG	G	MG	F	F	G
D_3	A_1	G	G	VG	G	MG	VG	F	MG	MG	MG
	A_2	MG	VG	VG	G	G	VG	F	MG	MG	MG
	A_3	VG	MG	MG	MG	MG	G	F	MG	MG	MG
	A_4	VG	VG	G	G	G	VG	G	MG	MG	VG
D_4	A_1	G	G	VG	MG	MG	VG	F	MG	MG	MG
	A_2	MG	VG	VG	G	G	VG	F	MG	MG	MG
	A_3	VG	MG	MG	MG	MG	G	F	MG	MG	MG
	A_4	VG	VG	G	G	G	VG	G	MG	MG	VG
D_5	A_1	G	G	VG	MG	MG	VG	F	MG	MG	MG
	A_2	G	VG	VG	G	G	VG	F	MG	MG	MG
	A_3	VG	MG	MG	MG	MG	G	F	MG	MG	MG
	A_4	G	G	G	G	G	VG	G	MG	MG	G
D_6	A_1	G	G	VG	G	MG	VG	F	MG	F	MG
	A_2	MG	VG	VG	G	G	VG	MG	MG	MG	MG
	A_3	VG	MG	MG	MG	MG	G	F	F	MG	MG
	A_4	VG	VG	G	G	G	VG	G	MG	MG	G

Table 8. *Cont.*

Expert/Criteria		C_1	C_2	C_3	C_4	C_5	C_6	C_7	C_8	C_9	C_{10}
D_7	A_1	G	G	VG	G	MG	VG	F	MG	MG	MG
	A_2	G	VG	VG	G	G	VG	MG	MG	MG	G
	A_3	VG	MG	MG	MG	MG	G	F	F	F	MG
	A_4	VG	VG	G	G	G	VG	G	MG	MG	G
D_8	A_1	G	G	VG	G	MG	VG	F	MG	MG	G
	A_2	MG	VG	VG	G	G	VG	MG	MG	MG	VG
	A_3	VG	MG	MG	MG	MG	G	F	F	F	MG
	A_4	VG	VG	G	MG	G	VG	G	MG	MG	G
D_9	A_1	MG	G	MG	MG	G	G	MG	F	MG	MG
	A_2	MG	G	MG	MG	G	G	G	F	MG	MG
	A_3	MG	MG	MG	F	MG	G	F	MP	F	MG
	A_4	MG	G	MG	MG	G	G	MG	F	MG	MG
D_{10}	A_1	MG	G	MG	MG	MG	G	MG	MG	MG	G
	A_2	MG	G	MG	MG	G	G	MG	F	MG	MG
	A_3	VG	VG	MG	MG	MG	G	MG	F	F	MG
	A_4	MG	MG	MG	F	F	F	F	F	F	F
D_{11}	A_1	G	G	G	G	MG	VG	F	MG	MG	MG
	A_2	MG	VG	VG	G	G	G	F	MG	MG	MG
	A_3	G	G	MG	MG	MG	G	MG	F	F	MG
	A_4	F	F	F	F	F	F	F	F	F	F
D_{12}	A_1	MG	G	MG	MG	G	G	MG	F	MG	MG
	A_2	MG	G	MG	MG	G	G	MG	F	MG	MG
	A_3	MG	G	MG	MG	G	G	MG	F	MG	MG
	A_4	MG	G	MG	MG	G	G	MG	F	MG	MG
D_{13}	A_1	G	G	VG	G	MG	VG	F	MG	MG	MG
	A_2	MG	VG	VG	G	G	VG	MG	MG	MG	MG
	A_3	VG	MG	MG	MG	MG	G	F	MG	MG	MG
	A_4	VG	G	G	G	G	VG	G	MG	MG	G

Table 9. Fuzzy numbers of supplier evaluations with respect to the criteria.

Expert		C_1	C_2	C_3	C_4	C_5
D_1	A_1	(0.7, 0.8, 0.8, 0.9)	(0.5, 0.6, 0.7, 0.8)	(0.7, 0.8, 0.8, 0.9)	(0.7, 0.8, 0.8, 0.9)	(0.7, 0.8, 0.8, 0.9)
	A_2	(0.7, 0.8, 0.8, 0.9)	(0.7, 0.8, 0.8, 0.9)	(0.7, 0.8, 0.8, 0.9)	(0.7, 0.8, 0.8, 0.9)	(0.7, 0.8, 0.8, 0.9)
	A_3	(0.8, 0.9, 1.0, 1.0)	(0.5, 0.6, 0.7, 0.8)	(0.5, 0.6, 0.7, 0.8)	(0.5, 0.6, 0.7, 0.8)	(0.5, 0.6, 0.7, 0.8)
	A_4	(0.7, 0.8, 0.8, 0.9)	(0.7, 0.8, 0.8, 0.9)	(0.5, 0.6, 0.7, 0.8)	(0.5, 0.6, 0.7, 0.8)	(0.5, 0.6, 0.7, 0.8)
D_2	A_1	(0.7, 0.8, 0.8, 0.9)	(0.8, 0.9, 1.0, 1.0)	(0.7, 0.8, 0.8, 0.9)	(0.7, 0.8, 0.8, 0.9)	(0.7, 0.8, 0.8, 0.9)
	A_2	(0.5, 0.6, 0.7, 0.8)	(0.7, 0.8, 0.8, 0.9)	(0.7, 0.8, 0.8, 0.9)	(0.7, 0.8, 0.8, 0.9)	(0.7, 0.8, 0.8, 0.9)
	A_3	(0.7, 0.8, 0.8, 0.9)	(0.7, 0.8, 0.8, 0.9)	(0.7, 0.8, 0.8, 0.9)	(0.7, 0.8, 0.8, 0.9)	(0.7, 0.8, 0.8, 0.9)
	A_4	(0.7, 0.8, 0.8, 0.9)	(0.7, 0.8, 0.8, 0.9)	(0.5, 0.6, 0.7, 0.8)	(0.5, 0.6, 0.7, 0.8)	(0.5, 0.6, 0.7, 0.8)
D_3	A_1	(0.7, 0.8, 0.8, 0.9)	(0.7, 0.8, 0.8, 0.9)	(0.8, 0.9, 1.0, 1.0)	(0.7, 0.8, 0.8, 0.9)	(0.5, 0.6, 0.7, 0.8)
	A_2	(0.5, 0.6, 0.7, 0.8)	(0.8, 0.9, 1.0, 1.0)	(0.8, 0.9, 1.0, 1.0)	(0.7, 0.8, 0.8, 0.9)	(0.7, 0.8, 0.8, 0.9)
	A_3	(0.8, 0.9, 1.0, 1.0)	(0.5, 0.6, 0.7, 0.8)	(0.5, 0.6, 0.7, 0.8)	(0.5, 0.6, 0.7, 0.8)	(0.5, 0.6, 0.7, 0.8)
	A_4	(0.8, 0.9, 1.0, 1.0)	(0.8, 0.9, 1.0, 1.0)	(0.7, 0.8, 0.8, 0.9)	(0.7, 0.8, 0.8, 0.9)	(0.7, 0.8, 0.8, 0.9)
D_4	A_1	(0.7, 0.8, 0.8, 0.9)	(0.7, 0.8, 0.8, 0.9)	(0.8, 0.9, 1.0, 1.0)	(0.5, 0.6, 0.7, 0.8)	(0.5, 0.6, 0.7, 0.8)
	A_2	(0.5, 0.6, 0.7, 0.8)	(0.8, 0.9, 1.0, 1.0)	(0.8, 0.9, 1.0, 1.0)	(0.7, 0.8, 0.8, 0.9)	(0.7, 0.8, 0.8, 0.9)
	A_3	(0.8, 0.9, 1.0, 1.0)	(0.5, 0.6, 0.7, 0.8)	(0.5, 0.6, 0.7, 0.8)	(0.5, 0.6, 0.7, 0.8)	(0.5, 0.6, 0.7, 0.8)
	A_4	(0.8, 0.9, 1.0, 1.0)	(0.8, 0.9, 1.0, 1.0)	(0.7, 0.8, 0.8, 0.9)	(0.7, 0.8, 0.8, 0.9)	(0.7, 0.8, 0.8, 0.9)
D_5	A_1	(0.7, 0.8, 0.8, 0.9)	(0.7, 0.8, 0.8, 0.9)	(0.8, 0.9, 1.0, 1.0)	(0.5, 0.6, 0.7, 0.8)	(0.5, 0.6, 0.7, 0.8)
	A_2	(0.7, 0.8, 0.8, 0.9)	(0.8, 0.9, 1.0, 1.0)	(0.8, 0.9, 1.0, 1.0)	(0.7, 0.8, 0.8, 0.9)	(0.7, 0.8, 0.8, 0.9)
	A_3	(0.8, 0.9, 1.0, 1.0)	(0.5, 0.6, 0.7, 0.8)	(0.5, 0.6, 0.7, 0.8)	(0.5, 0.6, 0.7, 0.8)	(0.5, 0.6, 0.7, 0.8)
	A_4	(0.7, 0.8, 0.8, 0.9)	(0.7, 0.8, 0.8, 0.9)	(0.7, 0.8, 0.8, 0.9)	(0.7, 0.8, 0.8, 0.9)	(0.7, 0.8, 0.8, 0.9)

Table 9. *Cont.*

Expert		C_1	C_2	C_3	C_4	C_5
D_6	A_1	(0.7, 0.8, 0.8, 0.9)	(0.7, 0.8, 0.8, 0.9)	(0.8, 0.9, 1.0, 1.0)	(0.7, 0.8, 0.8, 0.9)	(0.5, 0.6, 0.7, 0.8)
	A_2	(0.5, 0.6, 0.7, 0.8)	(0.8, 0.9, 1.0, 1.0)	(0.8, 0.9, 1.0, 1.0)	(0.7, 0.8, 0.8, 0.9)	(0.7, 0.8, 0.8, 0.9)
	A_3	(0.8, 0.9, 1.0, 1.0)	(0.5, 0.6, 0.7, 0.8)	(0.5, 0.6, 0.7, 0.8)	(0.5, 0.6, 0.7, 0.8)	(0.5, 0.6, 0.7, 0.8)
	A_4	(0.8, 0.9, 1.0, 1.0)	(0.8, 0.9, 1.0, 1.0)	(0.7, 0.8, 0.8, 0.9)	(0.7, 0.8, 0.8, 0.9)	(0.7, 0.8, 0.8, 0.9)
D_7	A_1	(0.7, 0.8, 0.8, 0.9)	(0.7, 0.8, 0.8, 0.9)	(0.8, 0.9, 1.0, 1.0)	(0.7, 0.8, 0.8, 0.9)	(0.5, 0.6, 0.7, 0.8)
	A_2	(0.7, 0.8, 0.8, 0.9)	(0.8, 0.9, 1.0, 1.0)	(0.8, 0.9, 1.0, 1.0)	(0.7, 0.8, 0.8, 0.9)	(0.7, 0.8, 0.8, 0.9)
	A_3	(0.8, 0.9, 1.0, 1.0)	(0.5, 0.6, 0.7, 0.8)	(0.5, 0.6, 0.7, 0.8)	(0.5, 0.6, 0.7, 0.8)	(0.5, 0.6, 0.7, 0.8)
	A_4	(0.8, 0.9, 1.0, 1.0)	(0.8, 0.9, 1.0, 1.0)	(0.7, 0.8, 0.8, 0.9)	(0.7, 0.8, 0.8, 0.9)	(0.7, 0.8, 0.8, 0.9)
D_8	A_1	(0.7, 0.8, 0.8, 0.9)	(0.7, 0.8, 0.8, 0.9)	(0.8, 0.9, 1.0, 1.0)	(0.7, 0.8, 0.8, 0.9)	(0.5, 0.6, 0.7, 0.8)
	A_2	(0.5, 0.6, 0.7, 0.8)	(0.8, 0.9, 1.0, 1.0)	(0.8, 0.9, 1.0, 1.0)	(0.7, 0.8, 0.8, 0.9)	(0.7, 0.8, 0.8, 0.9)
	A_3	(0.8, 0.9, 1.0, 1.0)	(0.5, 0.6, 0.7, 0.8)	(0.5, 0.6, 0.7, 0.8)	(0.5, 0.6, 0.7, 0.8)	(0.5, 0.6, 0.7, 0.8)
	A_4	(0.8, 0.9, 1.0, 1.0)	(0.8, 0.9, 1.0, 1.0)	(0.7, 0.8, 0.8, 0.9)	(0.5, 0.6, 0.7, 0.8)	(0.7, 0.8, 0.8, 0.9)
D_9	A_1	(0.5, 0.6, 0.7, 0.8)	(0.7, 0.8, 0.8, 0.9)	(0.5, 0.6, 0.7, 0.8)	(0.5, 0.6, 0.7, 0.8)	(0.7, 0.8, 0.8, 0.9)
	A_2	(0.5, 0.6, 0.7, 0.8)	(0.7, 0.8, 0.8, 0.9)	(0.5, 0.6, 0.7, 0.8)	(0.5, 0.6, 0.7, 0.8)	(0.7, 0.8, 0.8, 0.9)
	A_3	(0.5, 0.6, 0.7, 0.8)	(0.5, 0.6, 0.7, 0.8)	(0.5, 0.6, 0.7, 0.8)	(0.4, 0.5, 0.5, 0.6)	(0.5, 0.6, 0.7, 0.8)
	A_4	(0.5, 0.6, 0.7, 0.8)	(0.7, 0.8, 0.8, 0.9)	(0.5, 0.6, 0.7, 0.8)	(0.5, 0.6, 0.7, 0.8)	(0.7, 0.8, 0.8, 0.9)
D_{10}	A_1	(0.5, 0.6, 0.7, 0.8)	(0.7, 0.8, 0.8, 0.9)	(0.5, 0.6, 0.7, 0.8)	(0.5, 0.6, 0.7, 0.8)	(0.5, 0.6, 0.7, 0.8)
	A_2	(0.5, 0.6, 0.7, 0.8)	(0.7, 0.8, 0.8, 0.9)	(0.5, 0.6, 0.7, 0.8)	(0.5, 0.6, 0.7, 0.8)	(0.7, 0.8, 0.8, 0.9)
	A_3	(0.8, 0.9, 1.0, 1.0)	(0.8, 0.9, 1.0, 1.0)	(0.5, 0.6, 0.7, 0.8)	(0.5, 0.6, 0.7, 0.8)	(0.5, 0.6, 0.7, 0.8)
	A_4	(0.5, 0.6, 0.7, 0.8)	(0.5, 0.6, 0.7, 0.8)	(0.5, 0.6, 0.7, 0.8)	(0.4, 0.5, 0.5, 0.6)	(0.4, 0.5, 0.5, 0.6)
D_{11}	A_1	(0.7, 0.8, 0.8, 0.9)	(0.7, 0.8, 0.8, 0.9)	(0.7, 0.8, 0.8, 0.9)	(0.7, 0.8, 0.8, 0.9)	(0.5, 0.6, 0.7, 0.8)
	A_2	(0.5, 0.6, 0.7, 0.8)	(0.8, 0.9, 1.0, 1.0)	(0.8, 0.9, 1.0, 1.0)	(0.7, 0.8, 0.8, 0.9)	(0.7, 0.8, 0.8, 0.9)
	A_3	(0.7, 0.8, 0.8, 0.9)	(0.7, 0.8, 0.8, 0.9)	(0.5, 0.6, 0.7, 0.8)	(0.5, 0.6, 0.7, 0.8)	(0.5, 0.6, 0.7, 0.8)
	A_4	(0.4, 0.5, 0.5, 0.6)	(0.4, 0.5, 0.5, 0.6)	(0.4, 0.5, 0.5, 0.6)	(0.4, 0.5, 0.5, 0.6)	(0.4, 0.5, 0.5, 0.6)
D_{12}	A_1	(0.5, 0.6, 0.7, 0.8)	(0.7, 0.8, 0.8, 0.9)	(0.5, 0.6, 0.7, 0.8)	(0.5, 0.6, 0.7, 0.8)	(0.7, 0.8, 0.8, 0.9)
	A_2	(0.5, 0.6, 0.7, 0.8)	(0.7, 0.8, 0.8, 0.9)	(0.5, 0.6, 0.7, 0.8)	(0.5, 0.6, 0.7, 0.8)	(0.7, 0.8, 0.8, 0.9)
	A_3	(0.5, 0.6, 0.7, 0.8)	(0.7, 0.8, 0.8, 0.9)	(0.5, 0.6, 0.7, 0.8)	(0.5, 0.6, 0.7, 0.8)	(0.7, 0.8, 0.8, 0.9)
	A_4	(0.5, 0.6, 0.7, 0.8)	(0.7, 0.8, 0.8, 0.9)	(0.5, 0.6, 0.7, 0.8)	(0.5, 0.6, 0.7, 0.8)	(0.7, 0.8, 0.8, 0.9)
D_{13}	A_1	(0.7, 0.8, 0.8, 0.9)	(0.7, 0.8, 0.8, 0.9)	(0.8, 0.9, 1.0, 1.0)	(0.7, 0.8, 0.8, 0.9)	(0.5, 0.6, 0.7, 0.8)
	A_2	(0.5, 0.6, 0.7, 0.8)	(0.8, 0.9, 1.0, 1.0)	(0.8, 0.9, 1.0, 1.0)	(0.7, 0.8, 0.8, 0.9)	(0.7, 0.8, 0.8, 0.9)
	A_3	(0.8, 0.9, 1.0, 1.0)	(0.5, 0.6, 0.7, 0.8)	(0.5, 0.6, 0.7, 0.8)	(0.5, 0.6, 0.7, 0.8)	(0.5, 0.6, 0.7, 0.8)
	A_4	(0.8, 0.9, 1.0, 1.0)	(0.7, 0.8, 0.8, 0.9)	(0.7, 0.8, 0.8, 0.9)	(0.7, 0.8, 0.8, 0.9)	(0.7, 0.8, 0.8, 0.9)
Expert		C_6	C_7	C_8	C_9	C_{10}
D_1	A_1	(0.7, 0.8, 0.8, 0.9)	(0.4, 0.5, 0.5, 0.6)	(0.4, 0.5, 0.5, 0.6)	(0.5, 0.6, 0.7, 0.8)	(0.7, 0.8, 0.8, 0.9)
	A_2	(0.8, 0.9, 1.0, 1.0)	(0.5, 0.6, 0.7, 0.8)	(0.5, 0.6, 0.7, 0.8)	(0.5, 0.6, 0.7, 0.8)	(0.5, 0.6, 0.7, 0.8)
	A_3	(0.5, 0.6, 0.7, 0.8)	(0.4, 0.5, 0.5, 0.6)	(0.5, 0.6, 0.7, 0.8)	(0.5, 0.6, 0.7, 0.8)	(0.5, 0.6, 0.7, 0.8)
	A_4	(0.8, 0.9, 1.0, 1.0)	(0.7, 0.8, 0.8, 0.9)	(0.5, 0.6, 0.7, 0.8)	(0.5, 0.6, 0.7, 0.8)	(0.7, 0.8, 0.8, 0.9)
D_2	A_1	(0.8, 0.9, 1.0, 1.0)	(0.7, 0.8, 0.8, 0.9)	(0.7, 0.8, 0.8, 0.9)	(0.7, 0.8, 0.8, 0.9)	(0.7, 0.8, 0.8, 0.9)
	A_2	(0.8, 0.9, 1.0, 1.0)	(0.5, 0.6, 0.7, 0.8)	(0.5, 0.6, 0.7, 0.8)	(0.5, 0.6, 0.7, 0.8)	(0.5, 0.6, 0.7, 0.8)
	A_3	(0.7, 0.8, 0.8, 0.9)	(0.5, 0.6, 0.7, 0.8)	(0.5, 0.6, 0.7, 0.8)	(0.5, 0.6, 0.7, 0.8)	(0.5, 0.6, 0.7, 0.8)
	A_4	(0.7, 0.8, 0.8, 0.9)	(0.5, 0.6, 0.7, 0.8)	(0.4, 0.5, 0.5, 0.6)	(0.4, 0.5, 0.5, 0.6)	(0.7, 0.8, 0.8, 0.9)
D_3	A_1	(0.8, 0.9, 1.0, 1.0)	(0.4, 0.5, 0.5, 0.6)	(0.5, 0.6, 0.7, 0.8)	(0.5, 0.6, 0.7, 0.8)	(0.5, 0.6, 0.7, 0.8)
	A_2	(0.8, 0.9, 1.0, 1.0)	(0.4, 0.5, 0.5, 0.6)	(0.5, 0.6, 0.7, 0.8)	(0.5, 0.6, 0.7, 0.8)	(0.5, 0.6, 0.7, 0.8)
	A_3	(0.7, 0.8, 0.8, 0.9)	(0.4, 0.5, 0.5, 0.6)	(0.5, 0.6, 0.7, 0.8)	(0.5, 0.6, 0.7, 0.8)	(0.5, 0.6, 0.7, 0.8)
	A_4	(0.8, 0.9, 1.0, 1.0)	(0.7, 0.8, 0.8, 0.9)	(0.5, 0.6, 0.7, 0.8)	(0.5, 0.6, 0.7, 0.8)	(0.8, 0.9, 1.0, 1.0)
D_4	A_1	(0.8, 0.9, 1.0, 1.0)	(0.4, 0.5, 0.5, 0.6)	(0.5, 0.6, 0.7, 0.8)	(0.5, 0.6, 0.7, 0.8)	(0.5, 0.6, 0.7, 0.8)
	A_2	(0.8, 0.9, 1.0, 1.0)	(0.4, 0.5, 0.5, 0.6)	(0.5, 0.6, 0.7, 0.8)	(0.5, 0.6, 0.7, 0.8)	(0.5, 0.6, 0.7, 0.8)
	A_3	(0.7, 0.8, 0.8, 0.9)	(0.4, 0.5, 0.5, 0.6)	(0.5, 0.6, 0.7, 0.8)	(0.5, 0.6, 0.7, 0.8)	(0.5, 0.6, 0.7, 0.8)
	A_4	(0.8, 0.9, 1.0, 1.0)	(0.7, 0.8, 0.8, 0.9)	(0.5, 0.6, 0.7, 0.8)	(0.5, 0.6, 0.7, 0.8)	(0.8, 0.9, 1.0, 1.0)
D_5	A_1	(0.8, 0.9, 1.0, 1.0)	(0.4, 0.5, 0.5, 0.6)	(0.5, 0.6, 0.7, 0.8)	(0.5, 0.6, 0.7, 0.8)	(0.5, 0.6, 0.7, 0.8)
	A_2	(0.8, 0.9, 1.0, 1.0)	(0.4, 0.5, 0.5, 0.6)	(0.5, 0.6, 0.7, 0.8)	(0.5, 0.6, 0.7, 0.8)	(0.5, 0.6, 0.7, 0.8)
	A_3	(0.7, 0.8, 0.8, 0.9)	(0.4, 0.5, 0.5, 0.6)	(0.5, 0.6, 0.7, 0.8)	(0.5, 0.6, 0.7, 0.8)	(0.5, 0.6, 0.7, 0.8)
	A_4	(0.8, 0.9, 1.0, 1.0)	(0.7, 0.8, 0.8, 0.9)	(0.5, 0.6, 0.7, 0.8)	(0.5, 0.6, 0.7, 0.8)	(0.7, 0.8, 0.8, 0.9)

Table 9. *Cont.*

Expert		C_6	C_7	C_8	C_9	C_{10}
D_6	A_1	(0.8, 0.9, 1.0, 1.0)	(0.4, 0.5, 0.5, 0.6)	(0.5, 0.6, 0.7, 0.8)	(0.4, 0.5, 0.5, 0.6)	(0.5, 0.6, 0.7, 0.8)
	A_2	(0.8, 0.9, 1.0, 1.0)	(0.5, 0.6, 0.7, 0.8)	(0.5, 0.6, 0.7, 0.8)	(0.5, 0.6, 0.7, 0.8)	(0.5, 0.6, 0.7, 0.8)
	A_3	(0.7, 0.8, 0.8, 0.9)	(0.4, 0.5, 0.5, 0.6)	(0.4, 0.5, 0.5, 0.6)	(0.5, 0.6, 0.7, 0.8)	(0.5, 0.6, 0.7, 0.8)
	A_4	(0.8, 0.9, 1.0, 1.0)	(0.7, 0.8, 0.8, 0.9)	(0.5, 0.6, 0.7, 0.8)	(0.5, 0.6, 0.7, 0.8)	(0.7, 0.8, 0.8, 0.9)
D_7	A_1	(0.8, 0.9, 1.0, 1.0)	(0.4, 0.5, 0.5, 0.6)	(0.5, 0.6, 0.7, 0.8)	(0.5, 0.6, 0.7, 0.8)	(0.5, 0.6, 0.7, 0.8)
	A_2	(0.8, 0.9, 1.0, 1.0)	(0.5, 0.6, 0.7, 0.8)	(0.5, 0.6, 0.7, 0.8)	(0.5, 0.6, 0.7, 0.8)	(0.7, 0.8, 0.8, 0.9)
	A_3	(0.7, 0.8, 0.8, 0.9)	(0.4, 0.5, 0.5, 0.6)	(0.4, 0.5, 0.5, 0.6)	(0.4, 0.5, 0.5, 0.6)	(0.5, 0.6, 0.7, 0.8)
	A_4	(0.8, 0.9, 1.0, 1.0)	(0.7, 0.8, 0.8, 0.9)	(0.5, 0.6, 0.7, 0.8)	(0.5, 0.6, 0.7, 0.8)	(0.7, 0.8, 0.8, 0.9)
D_8	A_1	(0.8, 0.9, 1.0, 1.0)	(0.4, 0.5, 0.5, 0.6)	(0.5, 0.6, 0.7, 0.8)	(0.5, 0.6, 0.7, 0.8)	(0.7, 0.8, 0.8, 0.9)
	A_2	(0.8, 0.9, 1.0, 1.0)	(0.5, 0.6, 0.7, 0.8)	(0.5, 0.6, 0.7, 0.8)	(0.5, 0.6, 0.7, 0.8)	(0.8, 0.9, 1.0, 1.0)
	A_3	(0.7, 0.8, 0.8, 0.9)	(0.4, 0.5, 0.5, 0.6)	(0.4, 0.5, 0.5, 0.6)	(0.4, 0.5, 0.5, 0.6)	(0.5, 0.6, 0.7, 0.8)
	A_4	(0.8, 0.9, 1.0, 1.0)	(0.7, 0.8, 0.8, 0.9)	(0.5, 0.6, 0.7, 0.8)	(0.5, 0.6, 0.7, 0.8)	(0.7, 0.8, 0.8, 0.9)
D_9	A_1	(0.7, 0.8, 0.8, 0.9)	(0.5, 0.6, 0.7, 0.8)	(0.4, 0.5, 0.5, 0.6)	(0.5, 0.6, 0.7, 0.8)	(0.5, 0.6, 0.7, 0.8)
	A_2	(0.7, 0.8, 0.8, 0.9)	(0.7, 0.8, 0.8, 0.9)	(0.4, 0.5, 0.5, 0.6)	(0.5, 0.6, 0.7, 0.8)	(0.5, 0.6, 0.7, 0.8)
	A_3	(0.7, 0.8, 0.8, 0.9)	(0.4, 0.5, 0.5, 0.6)	(0.2, 0.3, 0.4, 0.5)	(0.4, 0.5, 0.5, 0.6)	(0.5, 0.6, 0.7, 0.8)
	A_4	(0.7, 0.8, 0.8, 0.9)	(0.5, 0.6, 0.7, 0.8)	(0.4, 0.5, 0.5, 0.6)	(0.5, 0.6, 0.7, 0.8)	(0.5, 0.6, 0.7, 0.8)
D_{10}	A_1	(0.7, 0.8, 0.8, 0.9)	(0.5, 0.6, 0.7, 0.8)	(0.5, 0.6, 0.7, 0.8)	(0.5, 0.6, 0.7, 0.8)	(0.7, 0.8, 0.8, 0.9)
	A_2	(0.7, 0.8, 0.8, 0.9)	(0.5, 0.6, 0.7, 0.8)	(0.4, 0.5, 0.5, 0.6)	(0.5, 0.6, 0.7, 0.8)	(0.5, 0.6, 0.7, 0.8)
	A_3	(0.7, 0.8, 0.8, 0.9)	(0.5, 0.6, 0.7, 0.8)	(0.4, 0.5, 0.5, 0.6)	(0.4, 0.5, 0.5, 0.6)	(0.5, 0.6, 0.7, 0.8)
	A_4	(0.4, 0.5, 0.5, 0.6)	(0.4, 0.5, 0.5, 0.6)	(0.4, 0.5, 0.5, 0.6)	(0.4, 0.5, 0.5, 0.6)	(0.4, 0.5, 0.5, 0.6)
D_{11}	A_1	(0.8, 0.9, 1.0, 1.0)	(0.4, 0.5, 0.5, 0.6)	(0.5, 0.6, 0.7, 0.8)	(0.5, 0.6, 0.7, 0.8)	(0.5, 0.6, 0.7, 0.8)
	A_2	(0.7, 0.8, 0.8, 0.9)	(0.4, 0.5, 0.5, 0.6)	(0.5, 0.6, 0.7, 0.8)	(0.5, 0.6, 0.7, 0.8)	(0.5, 0.6, 0.7, 0.8)
	A_3	(0.7, 0.8, 0.8, 0.9)	(0.5, 0.6, 0.7, 0.8)	(0.4, 0.5, 0.5, 0.6)	(0.4, 0.5, 0.5, 0.6)	(0.5, 0.6, 0.7, 0.8)
	A_4	(0.4, 0.5, 0.5, 0.6)	(0.4, 0.5, 0.5, 0.6)	(0.4, 0.5, 0.5, 0.6)	(0.4, 0.5, 0.5, 0.6)	(0.4, 0.5, 0.5, 0.6)
D_{12}	A_1	(0.7, 0.8, 0.8, 0.9)	(0.5, 0.6, 0.7, 0.8)	(0.4, 0.5, 0.5, 0.6)	(0.5, 0.6, 0.7, 0.8)	(0.5, 0.6, 0.7, 0.8)
	A_2	(0.7, 0.8, 0.8, 0.9)	(0.5, 0.6, 0.7, 0.8)	(0.4, 0.5, 0.5, 0.6)	(0.5, 0.6, 0.7, 0.8)	(0.5, 0.6, 0.7, 0.8)
	A_3	(0.7, 0.8, 0.8, 0.9)	(0.5, 0.6, 0.7, 0.8)	(0.4, 0.5, 0.5, 0.6)	(0.5, 0.6, 0.7, 0.8)	(0.5, 0.6, 0.7, 0.8)
	A_4	(0.7, 0.8, 0.8, 0.9)	(0.5, 0.6, 0.7, 0.8)	(0.4, 0.5, 0.5, 0.6)	(0.5, 0.6, 0.7, 0.8)	(0.5, 0.6, 0.7, 0.8)
D_{13}	A_1	(0.8, 0.9, 1.0, 1.0)	(0.4, 0.5, 0.5, 0.6)	(0.5, 0.6, 0.7, 0.8)	(0.5, 0.6, 0.7, 0.8)	(0.5, 0.6, 0.7, 0.8)
	A_2	(0.8, 0.9, 1.0, 1.0)	(0.5, 0.6, 0.7, 0.8)	(0.5, 0.6, 0.7, 0.8)	(0.5, 0.6, 0.7, 0.8)	(0.5, 0.6, 0.7, 0.8)
	A_3	(0.7, 0.8, 0.8, 0.9)	(0.4, 0.5, 0.5, 0.6)	(0.5, 0.6, 0.7, 0.8)	(0.5, 0.6, 0.7, 0.8)	(0.5, 0.6, 0.7, 0.8)
	A_4	(0.8, 0.9, 1.0, 1.0)	(0.7, 0.8, 0.8, 0.9)	(0.5, 0.6, 0.7, 0.8)	(0.5, 0.6, 0.7, 0.8)	(0.7, 0.8, 0.8, 0.9)

Table 10. Aggregated Fuzzy numbers of supplier evaluations with respect to the criteria.

	C_1	C_2	C_3	C_4	C_5
A_1	(0.5, 0.754, 0.777, 0.9)	(0.5, 0.792, 0.808, 1.0)	(0.5, 0.808, 0.885, 1.0)	(0.5, 0.723, 0.762, 0.9)	(0.5, 0.662, 0.731, 0.9)
A_2	(0.5,0.646,0.723,0.9)	(0.7, 0.862, 0.923, 1.0)	(0.5, 0.815, 0.900, 1.0)	(0.5, 0.754, 0.777, 0.9)	(0.7, 0.800, 0.800, 0.9)
A_3	(0.5,0.838,0.923,1.0)	(0.5, 0.669, 0.746, 1.0)	(0.5, 0.615, 0.708, 0.9)	(0.4, 0.608, 0.692, 0.9)	(0.5, 0.631, 0.715, 0.9)
A_4	(0.4,0.777,0.846,1.0)	(0.4, 0.800, 0.846, 1.0)	(0.4, 0.700, 0.738, 0.9)	(0.4, 0.677, 0.715, 0.9)	(0.4, 0.723, 0.738, 0.9)
	C_6	C_7	C_8	C_9	C_{10}
A_1	(0.7, 0.869, 0.938, 1.0)	(0.4, 0.546, 0.569, 0.9)	(0.4, 0.592, 0.662, 0.9)	(0.4, 0.608, 0.692, 0.9)	(0.5, 0.662, 0.731, 0.9)
A_2	(0.7, 0.869, 0.938, 1.0)	(0.4, 0.585, 0.646, 0.9)	(0.4, 0.577, 0.654, 0.8)	(0.5, 0.600, 0.700, 0.8)	(0.5, 0.638, 0.731, 1.0)
A_3	(0.5, 0.785, 0.792, 0.9)	(0.4, 0.531, 0.562, 0.8)	(0.2, 0.531, 0.585, 0.8)	(0.4, 0.562, 0.623, 0.8)	(0.5, 0.600, 0.700, 0.8)
A_4	(0.4, 0.815, 0.877, 1.0)	(0.4, 0.708, 0.731, 0.9)	(0.4, 0.562, 0.623, 0.8)	(0.4, 0.577, 0.654, 0.8)	(0.4, 0.738, 0.769, 1.0)

Table 11. Aggregated values and weights of supplier evaluations.

	C_1	C_2	C_3	C_4	C_5	C_6	C_7	C_8	C_9	C_{10}
					Defuzzified					
A_1	0.723	0.767	0.786	0.715	0.699	0.871	0.618	0.641	0.650	0.699
A_2	0.694	0.867	0.791	0.723	0.800	0.871	0.637	0.606	0.650	0.725
A_3	0.800	0.734	0.685	0.650	0.689	0.730	0.581	0.521	0.597	0.650
A_4	0.741	0.744	0.675	0.666	0.678	0.753	0.674	0.597	0.606	0.719
					Normalized					
A_1	0.244	0.247	0.268	0.260	0.244	0.270	0.246	0.271	0.260	0.250
A_2	0.235	0.278	0.269	0.262	0.279	0.270	0.254	0.256	0.260	0.260
A_3	0.270	0.236	0.233	0.236	0.241	0.226	0.231	0.220	0.239	0.233
A_4	0.251	0.239	0.230	0.242	0.236	0.233	0.269	0.252	0.242	0.257
					Entropy weights					
H_j	0.9990	0.9984	0.9980	0.9993	0.9984	0.9976	0.9990	0.9980	0.9994	0.9994
$1-H_j$	0.0010	0.0016	0.0020	0.0007	0.0016	0.0024	0.0010	0.0020	0.0006	0.0006
w_j	7.2%	11.8%	14.7%	5.4%	12.0%	17.5%	7.7%	14.9%	4.0%	4.8%

Table 12. The best and the worst values for each criterion and the *S* value of the suppliers.

	C_1	C_2	C_3	C_4	C_5	C_6	C_7	C_8	C_9	C_{10}
f^+	0.800	0.867	0.791	0.723	0.800	0.871	0.674	0.641	0.650	0.725
f^-	0.694	0.734	0.675	0.650	0.678	0.730	0.581	0.521	0.597	0.650
					S value					
A_1	0.052	0.089	0.006	0.006	0.100	0.000	0.046	0.000	0.000	0.017
A_2	0.072	0.000	0.000	0.000	0.000	0.000	0.030	0.044	0.000	0.000
A_3	0.000	0.118	0.134	0.054	0.109	0.175	0.077	0.149	0.040	0.048
A_4	0.040	0.110	0.147	0.042	0.120	0.146	0.000	0.055	0.034	0.004

Table 13. The values and rankings of *S*, *R*, and *Q* of each fishmeal supplier.

	S_i	Ranking	R_i	Ranking	Q_i	Ranking
A_1	0.3154	2	0.0998	2	0.247	2
A_2	0.1460	1	0.0720	1	0.000	1
A_3	0.9038	4	0.1745	4	1.000	4
A_4	0.6968	3	0.1469	3	0.729	3

These results showed that the difference between the Q value of A_2 and A_1 was not satisfied with Equation (13), therefore, two candidate suppliers, A_2 and A_1, are both appropriate choices. A_1 represents the largest fishmeal supplier in Peru, and A_2 is the second largest, and they both owned and operated 16 and 5 fishmeal plants, respectively, in 2011. About 80% of fishmeal production is produced by main 7 suppliers in Peru. In 2011, the largest fishmeal supplier (A_1) produced approximately 350,000 tons of fishmeal (27% of the total exported production), while A_2 produced approximately 200,000 tons (15.4% of total exported production). Obviously, plant size and capacity are of concern.

Regarding the criteria weights, an entropy method was applied to obtain objective weights from the supplier evaluation results. This was different from other methods like AHP, where weights are based on the subjective opinions given by experts. The entropy method results showed that the top three most important criteria were: (1) the possibility of establishing long-term cooperation, C_6 (0.17); (2) reasonableness of the quoted price, C_3 (0.15); and (3) satisfaction with claims for damages C_8 (0.15). The implication of these results is that aquaculture enterprises are concerned about a reduction in the quantities of fishmeal they can purchase due to a decrease in natural ocean resources. Therefore, their desire is to maintain long-term relationships with their supplier to ensure the quantity of supply. This not only affects the amount of aquaculture production, but also the sustainability of those operations. In addition, the reasonableness of the quoted price is also of concern. If the quoted price is too high, it will not attract purchasers to make procurement decisions and will hurt the profits of aquaculture. As seen from the weighting priorities, the related quality criterion C_1 (Stability in product quality) ranked 7th with a weighting of only 7.2%. This showed that controlling the supply of fishmeal resources was more important to the aquaculture industry than the quality requirements. Fishmeal is a special raw material and market demand is greater than supply, therefore making the selection requirements different than usual.

5. Conclusions

In this study, four Peruvian main fishmeal suppliers were evaluated by thirteen experts, which is a typical supplier selection problem often encountered in practice. Fuzzy set theory was an appropriate tool for dealing with this kind of problem. In real decision-making processes, the decision-maker is often unwilling or unable to express their preferences precisely in numerical values, so evaluations are very often expressed in linguistic terms. In this paper, an extension of the VIKOR method with entropy weighting measures in a fuzzy environment was proposed to deal with the qualitative criteria for suitable supplier selections.

From a management perspective, this study dealt with a very practical issue for the aquaculture industry in the selection of fishmeal suppliers, given that fishmeal is a very important raw material. According to the description of the interviewers, the supplier selection processes in this industry are based on personal experience or interpersonal relationships, and lack a scientific or systematic model on which to base these decisions. Given this situation, the management of important raw material suppliers has become less systematic, and does not effectively assess changes in the existing supplier's performance, which results in those suppliers with poor performance being more difficult to manage. Therefore, this study provides a management or evaluation tool for the industry in the event that a supplier is required to improve their performance. At the same time, to implement a concise and efficient questionnaire survey for these practitioners, a suitable research approach must be provided. The proposed method used in this study is expected to be able to obtain relevant information to effectively measure the weights of the evaluation criteria and the performance of the candidate suppliers through a simple questionnaire survey. That is, this study adopted the VIKOR approach based on entropy weights in a fuzzy decision-making environment. Not only can the entropy method reduce uncertainties arising from subjective factors, but also the advantage of this hybrid approach is that the merits of the alternatives can be evaluated with one questionnaire. This can greatly reduce the number of interviews with these fairly busy practitioners, which makes the application of this approach in the practical industry more effective and convenient. Thus, this framework for supplier selection in aquaculture should be helpful in making some progress in the management of the industry.

In addition, the use of fuzzy theory in this study to represent the fuzziness of human decision-making provides judgment linguistic variables that correspond to trapezoidal fuzzy numbers. For future work, if considering the dynamic and interactive group decision-making process, reference can be made to the model proposed by [48]. Alternatively, if future studies wish to consider the interactive consensus analysis of group decision making, it can refer to the integrated linguistic operator weighted average (ILOWA) approach introduced by [49] to obtain more detailed observations and discussion.

It is also worth considering, however, that, when selecting the best supplier, an awareness that all raw materials come from marine resources that can only be provided sustainably under sustainable fishing should be considered. The problem discussed here was based on how aquaculture enterprises make decisions to select the appropriate suppliers. However, if marine resources decrease, aquaculture enterprises will face a lack of raw materials to produce the relevant products, so income might not be enough to operate sustainably. In 2014, the contribution of the aquaculture sector to the supply of fish for human consumption overtook that of wild-caught fish for the first time. The importance of aquaculture in the future is clearly evident, and the best source of protein in the feed is provided by fishmeal. The question of how to provide high quality protein substitutes without relying on wild-caught fish is another topic worthy of discussion.

Acknowledgments: The authors are thankful to the anonymous referees for their valuable comments. The work was supported by the National Natural Science Foundation of China, Project 71372031.

Author Contributions: Tsung-Hsien Wu and Chia-Hsin Chen conceived of the work and wrote part of the manuscript; Ning Mao commented the work; Shih-Tong Lu analyzed the data and wrote part of the manuscript.

Conflicts of Interest: The authors declare no conflict of interest.

References

1. Food and Agriculture Organization (FAO). Contributing to food security and nutrition for all. In *The State of World Fisheries and Aquaculture*; Fisheries and Aquaculture Department, Food and Agriculture Organization of the United Nations: Rome, Italy, 2016.
2. Tam, M.C.Y.; Tummala, V.M. An Application of the AHP in vendor selection of a telecommunications system. *Omega* **2001**, *29*, 171–182. [CrossRef]
3. Dulmin, R.; Mininno, V. Supplier selection using a multi-criteria decision aid method. *J. Purch. Suppl. Manag.* **2003**, *9*, 177–187. [CrossRef]
4. Kumar, M.; Vrat, P.; Shankar, R. A fuzzy goal programming approach for vendor selection problem in a supply chain. *Comput. Ind. Eng.* **2004**, *46*, 69–85. [CrossRef]
5. Chen, C.T.; Lin, C.; Huang, S.F. A fuzzy approach for supplier evaluation and selection in supply chain management. *Int. J. Prod. Econ.* **2006**, *102*, 289–301. [CrossRef]
6. Kumar, M.; Vrat, P.; Shankar, R. A fuzzy programming approach for vendor selection problem in a supply chain. *Int. J. Prod. Econ.* **2006**, *101*, 273–285. [CrossRef]
7. Gencer, C.; Gürpinar, D. Analytic network process in supplier selection: A case study in an electronic firm. *Appl. Math. Model.* **2007**, *31*, 2475–2486. [CrossRef]
8. Xia, W.; Wu, Z. Supplier selection with multiple criteria in volume discount environments. *Omega* **2007**, *35*, 494–504. [CrossRef]
9. Sanayei, A.; Mousavi, S.F.; Abdi, M.R.; Mohaghar, A. An integrated group decision-making process for supplier selection and order allocation using multi-attribute utility theory and linear programming. *J. Frankl. Inst.* **2008**, *345*, 731–747. [CrossRef]
10. Yang, J.L.; Chiu, H.N.; Tzeng, G.H.; Yeh, R.H. Vendor selection by integrated fuzzy MCDM techniques with independent and interdependent relationships. *Inform. Sci.* **2008**, *178*, 4166–4183. [CrossRef]
11. Amin, S.H.; Razmi, J. An integrated fuzzy model for supplier management: A case study of ISP selection and evaluation. *Expert Syst. Appl.* **2009**, *36*, 8639–8648. [CrossRef]
12. Boran, F.E.; Genc, S.; Kurt, M.; Akay, D. A multi-criteria intuitionistic fuzzy group decision making for supplier selection with TOPSIS method. *Expert Syst. Appl.* **2009**, *36*, 11363–11368. [CrossRef]
13. Wang, J.W.; Cheng, C.H.; Huang, K.C. Fuzzy hierarchical TOPSIS for supplier selection. *Appl. Soft Comput.* **2009**, *9*, 377–386. [CrossRef]
14. Lin, Y.T.; Lin, C.L.; Ya, H.C.; Tzeng, G.H. A novel hybrid MCDM approach for outsourcing vendor selection A case study for a semiconductor company in Taiwan. *Expert Syst. Appl.* **2010**, *37*, 4796–4804. [CrossRef]
15. Chamodrakas, I.; Batis, D.; Martakos, D. Supplier selection in electronic marketplaces using satisficing and fuzzy AHP. *Expert Syst. Appl.* **2010**, *37*, 490–498. [CrossRef]
16. Sanayei, A.; Mousavi, S.F.; Yazdankhah, A. Group decision making process for supplier selection with VIKOR under fuzzy environment. *Expert Syst. Appl.* **2010**, *37*, 24–30. [CrossRef]
17. Shemshadi, A.; Shirazi, H.; Toreihi, M.; Tarokh, M.J. A fuzzy VIKOR method for supplier selection based on entropy measure for objective weighting. *Expert Syst. Appl.* **2011**, *38*, 12160–12167. [CrossRef]
18. Jahan, A.; Mustapha, F.; Ismail, M.Y.; Sapuan, S.M.; Bahraminasab, M. A comprehensive VIKOR method for material selection. *Mater. Des.* **2011**, *32*, 1215–1221. [CrossRef]
19. Amin, S.H.; Razmi, J.; Zhang, G.Q. Supplier selection and order allocation based on fuzzy SWOT analysis and fuzzy linear programming. *Expert Syst. Appl.* **2011**, *38*, 334–342. [CrossRef]

20. Kilincci, O.; Onal, S.A. Fuzzy AHP approach for supplier selection in a washing machine company. *Expert Syst. Appl.* **2011**, *38*, 9656–9664. [CrossRef]
21. Feng, B.; Fan, Z.P.; Li, Y. A decision method for supplier selection in multi-service outsourcing. *Int. J. Prod. Econ.* **2011**, *132*, 240–250. [CrossRef]
22. Zhao, K.; Yu, X. A case based reasoning approach on supplier selection in petroleum enterprises. *Expert Syst. Appl.* **2011**, *38*, 6839–6847. [CrossRef]
23. Vahdani, B.; Iranmanesh, S.H.; Mousavi, S.M.; Abdollahzade, M. A locally linear neuro-fuzzy model for supplier selection in cosmetics industry. *Appl. Math. Model.* **2012**, *36*, 4714–4727. [CrossRef]
24. Chatterjee, P.; Chakraborty, S. Material selection using preferential ranking methods. *Mater. Des.* **2012**, *35*, 384–393. [CrossRef]
25. Hsua, C.H.; Wang, F.K.; Tzeng, G.H. The best vendor selection for conducting the recycled material based on a hybrid MCDM model combining DANP with VIKOR. *Res. Conserv. Recycl.* **2012**, *66*, 95–111. [CrossRef]
26. Chen, Y.H.; Chao, R.J. Supplier selection using consistent fuzzy preference relations. *Expert Syst. Appl.* **2012**, *39*, 3233–3240. [CrossRef]
27. Peng, A.H.; Xiao, X.M. Material selection using PROMETHEE combined with analytic network process under hybrid environment. *Mater. Des.* **2013**, *47*, 643–652. [CrossRef]
28. Zhao, H.; Guo, S. Selecting Green Supplier of Thermal Power Equipment by Using a Hybrid MCDM Method for Sustainability. *Sustainability* **2014**, *6*, 217–235. [CrossRef]
29. Kuo, T.C.; Hsu, C.W.; Li, J.Y. Developing a Green Supplier Selection Model by Using the DANP with VIKOR. *Sustainability* **2015**, *7*, 1661–1689. [CrossRef]
30. Chung, C.C.; Chao, L.C.; Lou, S.J. The Establishment of a Green Supplier Selection and Guidance Mechanism with the ANP and IPA. *Sustainability* **2016**, *8*, 259. [CrossRef]
31. Yazdani, M.; Chatterjee, P.; Zavadskas, E.K.; Zolfani, S.H. Integrated QFD-MCDM framework for green supplier selection. *J. Clean. Prod.* **2017**, *142 Pt 4*, 3728–3740. [CrossRef]
32. Wan, S.P.; Xua, G.-L.; Dong, J.Y. Supplier selection using ANP and ELECTRE II in interval 2-tuple linguistic environment. *Inform. Sci.* **2017**, *385–386*, 19–38. [CrossRef]
33. Opricovic, S.; Tzeng, G.H. Compromise solution by MCDM methods: A comparative analysis of VIKOR and TOPSIS. *Eur. J. Oper. Res.* **2004**, *156*, 445–455. [CrossRef]
34. Opricovic, S.; Tzeng, G.H. Extended VIKOR method in comparison with outranking methods. *Eur. J. Oper. Res.* **2007**, *178*, 514–529. [CrossRef]
35. Dickson, G.W. An analysis of vendor selection systems and decisions. *J. Purch.* **1966**, *2*, 5–17. [CrossRef]
36. Barbarosoglu, G.; Yazgac, T. An application of the analytic hierarchy process to the supplier selection problem. *Prod. Inventory Manag. J.* **1997**, *38*, 14–21.
37. Ng, W.L. An efficient and simple model for multiple criteria supplier selection problem. *Eur. J. Oper. Res.* **2008**, *186*, 1059–1067. [CrossRef]
38. Chan, F.T.S.; Kumar, N. Global supplier development considering risk factors using fuzzy extended AHP-based approach. *Omega* **2007**, *35*, 417–431. [CrossRef]
39. Shyur, J.H.; Shih, H.S. A hybrid MCDM model for strategic vendor selection. *Math. Comput. Model.* **2006**, *44*, 749–761. [CrossRef]
40. Ávila, P.; Mota, A.; Pires, A.; Bastos, J.; Putnik, G.; Teixeira, J. Supplier's Selection Model based on an Empirical Study. *Proced. Technol.* **2012**, *5*, 625–634. [CrossRef]
41. Shieh, J.I.; Wu, H.H.; Huang, K.K. A DEMATEL method in identifying key success factors of hospital service quality. *Knowl. Based Syst.* **2010**, *23*, 277–282. [CrossRef]
42. Zadeh, L.A. Fuzzy sets. *Inform. Control* **1965**, *8*, 338–353. [CrossRef]
43. Bellman, R.E.; Zadeh, L.A. Decision-Making in a Fuzzy Environment. *Manag. Sci.* **1970**, *17*, 141–164. [CrossRef]
44. Chen, S.J.; Hwang, C.L. *Fuzzy Multiple Attribute Decision Making: Methods and Applications*; Springer: New York, NY, USA, 1992.
45. Shannon, C.E. A mathematical theory of communication. *Bell Syst. Tech. J.* **1948**, *27*, 379–423. [CrossRef]
46. Opricovic, S.; Tzeng, G.H. Multicriteria planning of post-earthquake sustainable reconstruction. *Comput.-Aided Civ. Infrastruct. Eng.* **2002**, *17*, 211–220. [CrossRef]
47. Zeleny, M. *Multiple Criteria Decision Making*; McGraw-Hill: New York, NY, USA, 1982.

48. Bucolo, M.; Fortuna, L.; La Rosa, M. Complex Dynamics through Fuzzy Chains. *IEEE Trans. Fuzzy Syst.* **2004**, *12*, 289–295. [CrossRef]
49. Rodger, J.A.; George, J.A. Triple bottom line accounting for optimizing natural gas sustainability: A statistical linear programming fuzzy ILOWA optimized sustainment model approach to reducing supply chain global cybersecurity vulnerability through information and communications technology. *J. Clean. Prod.* **2017**, *142*, 1931–1949.

symmetry

MDPI

Article

Fuzzy Logic-Based Model That Incorporates Personality Traits for Heterogeneous Pedestrians

Zhuxin Xue [1,2], Qing Dong [1], Xiangtao Fan [1,3,*], Qingwen Jin [1,2], Hongdeng Jian [1] and Jian Liu [1,3]

1 Key Laboratory of Digital Earth Science, Institute of Remote Sensing and Digital Earth, Chinese Academy of Sciences, Beijing 100094, China; xuezx@radi.ac.cn (Z.X.); dongqing@radi.ac.cn (Q.D.); jinqw@radi.ac.cn (Q.J.); jianhd@radi.ac.cn (H.J.); liujian@radi.ac.cn (J.L.)

2 University of Chinese Academy of Sciences, Beijing 100049, China

3 Hainan Key Laboratory of Earth Observation, Sanya 572029, China

* Correspondence: fanxt@radi.ac.cn

Received: 15 September 2017; Accepted: 13 October 2017; Published: 20 October 2017

Abstract: Most models designed to simulate pedestrian dynamical behavior are based on the assumption that human decision-making can be described using precise values. This study proposes a new pedestrian model that incorporates fuzzy logic theory into a multi-agent system to address cognitive behavior that introduces uncertainty and imprecision during decision-making. We present a concept of decision preferences to represent the intrinsic control factors of decision-making. To realize the different decision preferences of heterogeneous pedestrians, the Five-Factor (OCEAN) personality model is introduced to model the psychological characteristics of individuals. Then, a fuzzy logic-based approach is adopted for mapping the relationships between the personality traits and the decision preferences. Finally, we have developed an application using our model to simulate pedestrian dynamical behavior in several normal or non-panic scenarios, including a single-exit room, a hallway with obstacles, and a narrowing passage. The effectiveness of the proposed model is validated with a user study. The results show that the proposed model can generate more reasonable and heterogeneous behavior in the simulation and indicate that individual personality has a noticeable effect on pedestrian dynamical behavior.

Keywords: pedestrian dynamical behavior; crowd simulation; fuzzy logic; personality trait; multi-agent

1. Introduction

Simulating crowd behavior as an interdisciplinary research field has attracted the keen interest of researchers and managers from various domains, including safety engineering, robotics, computer animation, and social psychology, and in recent years, it has been extensively studied and applied in these fields. For example, crowd simulation technology can be used to predict pedestrian flow [1–3] and recognize abnormal or normal behavior in safety engineering applications [4–6], and it has been implemented for autonomous navigation in robotics [7,8] and enhancing the reality of computer animation [9–11]. The modelling of pedestrian dynamics is a common key issue among these applications.

Over the past three decades, various pedestrian models have been proposed to simulate realistic pedestrian dynamical behavior and understand the potential laws underlying complex crowd phenomena. In general, most of these models can be classified into two main types: macroscopic and microscopic. The former considers pedestrians as a fluid, and it usually uses Navier–Stokes or Boltzmann equations to represent the movement variations in density and speed [1,12]. The latter, which can analyse individual behavior and interactions among members of a crowd, describes a pedestrian as an object driven by rule [13] or force [2]. Compared with macroscopic models that focus on the trend of crowd movement, microscopic models can generate fine-grain simulation results to reflect individual issues and diversities. As a result, over the past few years, many researchers have paid more attention to microscopic modelling, such as the social force model [4], cellular automata model [3], and multi-agent model [14]. However, most microscopic models have been developed based on the assumption that human perception and decision-making in real time can be described using precise values, and they consider that the information obtained from environments can be accurately quantified in a pedestrian perception process and assume that the corresponding behavior can be predicted with certainty in the next decision process. In fact, this assumption does not conform with the nature of human behavior because imprecise concepts are attributes of human cognitive abilities. Therefore, a mathematic formulation with precise values cannot easily describe and predict realistic pedestrian dynamical behavior.

Inspired by soft computing theory, we attempt to incorporate a fuzzy logic system into a multi-agent simulation model to solve the above problem. The theory of fuzzy sets proposed by Zadeh [15] provides a useful modelling tool for many applications when ambiguity is present. For example, the traffic signal alternatives at a midblock crosswalk are controlled using fuzzy logic methods in [16,17], and a collision avoidance system for autonomous vehicles using a fuzzy steering controller is proposed in [8]. The perceptions and decisions of a pedestrian are usually represented by natural language, such as when the pedestrian attempts to pass through the exit at a fast speed rather than at 2 m/s. These processes can be formulated by a set of verbal variables and linguistic rules using a fuzzy logic system. Because of the advantages of accessible input perception-based information and easily steerable output, fuzzy logic-based methods have achieved great progress in the modelling of pedestrian dynamics. To the best of our knowledge, most studies using fuzzy logic have focused on the human ability to perceive their surrounding environments [18–24] and ignored decision-making processes that are vague and imprecise. Compared with previous works, we propose an approach to emulate pedestrian cognitive ability from a new perspective that uses the fuzzy logic system to model subjective decisions. Moreover, a concept of decision preferences is presented to represent the intrinsic control factors underlying decision-making.

Pedestrian dynamical behavior varies from one individual to another when they are confronted with similar situations in the real world. Thus, the modelling of heterogeneous behavior plays an important role in the simulation of a natural and realistic crowd. Among the many factors that promote variations in pedestrian behavior, such as physiological and psychological characteristics, personality has a significant impact on the subjective decision of the pedestrian. Personality is a pattern of behavioral, temperamental, emotional, and mental traits for an individual. People with various personalities have salient behavior characteristics. To generate the heterogeneity of the pedestrian cognitive ability, we introduce the well-known OCEAN (Openness, Conscientiousness, Extraversion, Agreeableness, and Neuroticism) personality model [25] to realize the different decision preferences of heterogeneous pedestrians in this paper.

The main contribution of this study is the proposal of a novel pedestrian model that incorporates fuzzy logic theory into a multi-agent system to address cognitive behavior that introduces uncertainty and imprecision in decision-making. For simulating the heterogeneous subjective decisions of pedestrians, the OCEAN personality model is introduced to generate the different decision preferences that represent the intrinsic control factors of decision-making. The experiments show that the application developed using our model can simulate more reasonable and heterogeneous pedestrian dynamical behavior in several normal or non-panic situations, including a single-exit room, a hallway with obstacles, and a narrowing passage. Furthermore, the results reveal the impact of personality traits on pedestrian dynamical behavior.

The remainder of this paper is organized as follows: Section 2 provides an overview of the related work. Section 3 describes the details of our model that incorporates fuzzy logic into a multi-agent system and uses personality traits for heterogeneous pedestrians. Section 4 presents the simulations and validation of the proposed model. Finally, Section 5 concludes the paper with a summary and outlook.

2. Related Work

Over the years, various microscopic models have been presented to simulate pedestrian dynamical behavior under abnormal and normal situations. These models are usually classified into three types: cellular automata, social force, and multi-agent. The cellular automata model [3,26,27] describes pedestrian flow through a discrete arrangement of space into grids of equal cells that have two states: occupied or unoccupied. According to a set of simple transition rules, the state of cells can be updated to indicate the movement of pedestrians at each time step. However, the pedestrians in this model move in discrete time and space; therefore, the simulation result is highly dependent on the discretization level of time and space, and it is difficult to ignore pedestrian body size and step time to accurately simulate pedestrian dynamical behaviors. The social force model [2,4] inspired by Newton's second law describes the pedestrians' behavior with a mathematical equation of the interaction forces that consist of socio-psychological and physical forces. This model can easily reproduce certain self-organization phenomena in a real evacuation, such as arching, clogging, and the "faster-is-slower effect". However, the computational complexity of the model rapidly increases with the crowd number, and heterogeneous pedestrian behavior is difficult to replicate. The multi-agent model [6,14,28] proposes a computational methodology in which all individuals are modelled as autonomous agents that are capable of interacting with each other. Agents are autonomous software entities with the perception and social ability to perform goal-directed knowledge processing. The key advantage of a multi-agent system is that it can model the dynamics of real-life complex systems. Therefore, these systems are particularly suitable for simulating the cognitive process and behavior of pedestrians. Based on the above discussion, we focus on the multi-agent model for pedestrian modelling.

To improve the believability of crowd simulations, simulating the uncertainty and imprecision of human behavior is an important aspect in many scenarios, such as pedestrian steering and emergency evacuation. A fuzzy logic approach has certain advantages over other approaches, such as its ability to use perceptual information and human experience and knowledge, and to emulate human thought processes. Therefore, this approach represents a natural and suitable tool for modelling pedestrian dynamic behavior. Nasir et al. [18] introduced a fuzzy logic framework to predict the impact of perceived attractive and repulsive stimuli, within the pedestrian's field of view, on movement direction during normal situations. Zhu et al. [19] integrated fuzzy logic with the social force model and reproduced the dynamical features of pedestrian evacuation. Li et al. [20] presented a fuzzy logic-based approach for crowd simulation that extracts fuzzy rules from the realistic videos that can be considered a parameterized behavior model. Dell'Orco et al. [21] proposed a microscopic model of crowd evacuation defined on a continuous space and used a fuzzy logic technique to reproduce human reasoning. Nasir et al. [22] proposed a genetic fuzzy system to model and simulate a pedestrian's steering behavior in a built environment. Fu et al. [23] proposed a fuzzy theory-based behavioral model to investigate evacuation dynamics in a cellular space. Zhou et al. [24] proposed a fuzzy logic approach for simulating pedestrian dynamical behavior, and it integrates the intermediate results of local obstacle-avoiding behavior, regional path-searching behavior and global goal-seek behavior with mutable weighting factors. Although the above studies have achieved positive results for pedestrian dynamics using a fuzzy logic-based approach, the human ability to perceive external information is considered fuzzy, and the subjective differences among crowds is ignored because all individuals are treated as homogeneous. Therefore, the differences in decision-making for heterogeneous individuals are difficult to simulate in these studies.

In order to simulate the different decision-making of heterogeneous pedestrians, researchers have also proposed some approaches using fuzzy logic theory. Akasaka et al. [29] presented a pedestrian navigation system using fuzzy measures and integrals for selecting a route based on users' own preference for routes. Teknomo et al. [30] integrated multi-states of pedestrian situation and pedestrian group behavior into the multi-agent pedestrian simulation through the concept of fuzzy inter-personal spacing. These studies directly simulated pedestrian navigation under some certain conditions from the perspective of subjective selection, but neglected the intrinsic factors influencing pedestrian behavior. In this paper, we attempt to model individual intrinsic characteristics for realizing the different decision preferences of heterogeneous pedestrians. The behavior of a pedestrian can be mainly influenced by physiological and psychological characteristics. For the aspect of physiology, Zheng et al. [9] chose four basic physiological characteristics—gender, age, health, and body shape—to generate heterogeneous crowd behavior. In terms of psychological characteristics, Durupinar et al. [10] integrated the OCEAN personality model into HiDAC (High-Density Autonomous) [31] to simulate crowd behavior by mapping between personality traits and observed behavior types. Guy et al. [11] presented an approach to simulating a heterogeneous crowd using the PEN (Psychoticism, Extraversion, and Neuroticism) personality model based on the RVO (Reciprocal Velocity Obstacle) library [7]. These studies showed that the modelling of pedestrian characteristics plays an important role in simulating realistic and heterogeneous pedestrian dynamical behavior. However, they are based on the assumption that human decision-making can be described using precise values. To solve the problem of uncertainty and imprecision during decision-making, we adopt a fuzzy logic-based approach to map the relationships between the personality traits and the decision preferences.

In this paper, we propose a new pedestrian model that incorporates fuzzy logic theory into a multi-agent system. Our model can emulate pedestrian cognitive ability that introduces uncertainty and imprecision during decision-making. Inspired by previous works, we choose the OCEAN personality model to simulate the individual characteristics for generating the different decision preferences of heterogeneous pedestrians. To describe the fuzzy attributes, a fuzzy inference system is adopted for mapping the relationships between the personality traits and the decision preferences.

3. Model Description

In this section, we first introduce an overview of our proposed model for simulating microscopic pedestrian behavior based on multi-agent systems. Next, we describe the well-known Five-Factor personality model, which is also known as the OCEAN model and used in this study to drive the heterogeneous behavior of individuals within a crowd. Then, we present a concept of decision preferences and define the meanings of these parameters. Finally, a fuzzy logic-based approach is presented to represent the relationship between the personality traits and the decision preferences of agents for addressing uncertainty and imprecision during decision-making.

3.1. Overview of the Proposed Model

Human behavior is a complex phenomenon that is difficult to capture via computers performing mathematical equations. In addition, crowd behavior is not a simple collection of individual behaviors in the crowd but also includes the interactions between people. To generate heterogeneous crowd behavior and understand the complicated motion features of pedestrians, we adopt a framework based on a multi-agent system to model and simulate pedestrian movement. A multi-agent system is an extension of agent technology where a group of loosely connected autonomous agents act in an environment to achieve a common goal [32]. In this framework, each human individual is modelled as an autonomous agent who interacts with a virtual environment and other agents. We believe that such a framework is particularly suitable for simulating individual cognitive processes and behavior in pedestrian dynamics.

In our proposed simulation framework, the virtual environment mainly includes agents, obstacles, and open spaces. The agent consists of three basic components: perception module, decision-making module, and action module.

- **Perception module:** This module of an autonomous agent is included to perceive the surrounding information from the virtual environment, and it has an important role as the information portal of the agent that interacts with the external environment. The perceived information of an agent mainly includes (1) the location of other agents and obstacles; (2) the distance from itself to other agents and obstacles; (3) the speed of other agents; and (4) the range of obstacles. Because of the limited range of the visual field, each agent has a limited sensing capability.

- **Decision-making module:** This module is designed to represent the cognitive and reasoning processes associated with the movement of a pedestrian. The perceived information is used as input in this module for an agent making decisions on steering behavior. The publicly available RVO2 library is used to implement the reasoning process, and the decision preferences are defined using the following five parameters: (1) *NeighborDist*; (2) *MaxNeighbors*; (3) *TimeHorizon*; (4) *Radius*; and (5) *PrefVelocity*. A detailed description of these parameters is presented in Section 3.3. The multi-agent system consists of heterogeneous agents, or agents with different decision-making capabilities.
- **Action module:** In this module, the actions of an agent include the agent's ability to walk, run, turn, and stop, which are the basic locomotion capacities of a pedestrian in a real-life environment. In this article, we use two main variables—the speed and direction of movement—to express the basic actions in the steering activity. Actions driven by decision-making can vary considerably among agents presented with the same perceived information.

To simulate the uncertainty and imprecision of pedestrian dynamical behavior, we incorporate a fuzzy logic system into a multi-agent simulation framework to model the decision-making process of an agent. Compared with previous works, we focus on the different decision-making abilities of heterogeneous agents in this paper. For the sake of simplicity, we use the decision preferences as fuzzy variables to describe qualitative cognitive behaviors in pedestrian decision-making. According to the fuzzy relationship between personality traits and human behavior [33], we introduce the OCEAN personality model to represent heterogeneous pedestrians with various characteristics, such as openness, conscientiousness, extraversion, agreeableness, and neuroticism. In addition, the mapping between personality factors and decision preferences is determined by a fuzzy inference system.

The overall framework of our proposed model is shown in Figure 1. In the following sections, we describe the personality model, decision parameters, and a fuzzy logic-based approach in more detail.

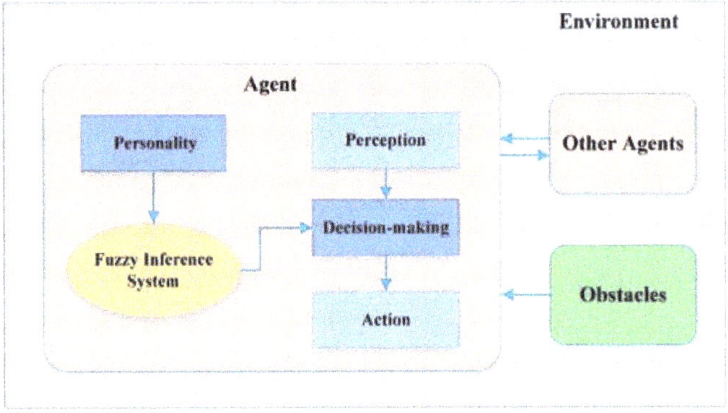

Figure 1. Framework of the proposed model based on a multi-agent system.

3.2. Modelling of Personality

Research in social science and human psychology defines personality as representing the pattern of a person's thoughts, feelings, and behaviors, which distinguish one person from another and persist over time and situations [34]. To exhibit a natural and realistic simulation effect, personalities should be considered an important aspect of modelling heterogeneous pedestrian behavior because pedestrians with different personalities may react differently to the same situation. In addition, understanding how varying personality factors affect the walking behaviors of pedestrians is important for many applications.

Many mature models have been proposed by psychologists to describe the spectrum of personalities, such as Eysenck's three-factor model [35] and the Five-Factor model [25]. In this paper, we choose the famous Five-Factor personality model, which is also known as the OCEAN model, to characterize the personality traits of autonomous agents. This model categorizes personality into five orthogonal factors based on a factor analysis of user studies in which participants use common language adjectives to describe the behavior of people. The five factors have been defined as Openness, Conscientiousness, Extraversion, Agreeableness, and Neuroticism [36]. These factors are explained as follows:

- **Openness** reflects the degree of curiosity and creativity and preferences for novelty and variety.
- **Conscientiousness** describes the level of organization and care exhibited in collective activities.
- **Extraversion** is related to the degree of energy, sociability, and outgoingness.
- **Agreeableness** is a tendency to exhibit compassion and cooperation rather than suspicion and antagonism towards others.
- **Neuroticism** is the tendency to experience unpleasant emotions easily, such as anger, anxiety, depression, or vulnerability, and is the opposite of emotional stability.

To satisfy the goal of generating realistic and heterogeneous crowd behavior, we must model all agents with either salient characteristics or non-characteristics. Each factor of the OCEAN model is bipolar and consists of several traits [37]. We divide each factor into three types, namely, negative, neutral, and positive types, and describe them via descriptive adjectives. The details of the personality factors in our study are given in Table 1, where −, =, and + mean negative, neutral and positive.

Table 1. Personality factors—Hierarchical types—Descriptions of characteristic.

Factor	Type	Characteristic
Openness	−	cautious, narrow, conservative
	=	occasional curiosity, moderate creativity
	+	curious, inventive, explorer
Conscientiousness	−	careless, rude, changeable
	=	spontaneous, reasonable order
	+	persistent, organized, dependable
Extraversion	−	shy, withdrawn, introvert
	=	neutral, self-conscious
	+	social, energetic, outgoing
Agreeableness	−	competitive, negative, harsh
	=	somewhat gentle, moderately tolerant
	+	cooperative, compassionate, friendly
Neuroticism	−	calm, secure, confident
	=	occasional anxiety, basically stable
	+	sensitive, fearful, nervous

This model describes the personality characteristics using a set of five factors rather than separate factors. Therefore, we use a five-dimension vector to model an agent's personality P:

$$P = (V(O),\ V(C),\ V(E),\ V(A),\ V(N)) \tag{1}$$

$$V(i) \in [-100, 100], i \in \{O, C, E, A, N\}$$

where O, C, E, A, and N represent *Openness*, *Conscientiousness*, *Extroversion*, *Agreeableness*, and *Neuroticism*, respectively. The default personality of agents is configured with $P = (0,\ 0,\ 0,\ 0,\ 0)$.

3.3. Decision Preferences

Our main goal is to generate realistic and natural pedestrian dynamic behavior by simulating human cognitive ability with uncertainty and imprecision in movement, and the critical step is determining the method of describing the subjective decision-making of pedestrians. We consider that the parameters representing the influence factors of the decision must satisfy the following requirements:

- The content of these parameters can be associated with the perceptual information obtained from the environment;
- The personality of an individual has a significant impact on the range of these parameters;
- These parameters can directly reflect the differences in decision-making between pedestrians.

To meet the above requirements, we present the concept of decision preferences to represent the intrinsic control factors of decision-making. In this paper, we choose five parameters—*NeighborDist*, *MaxNeighbors*, *TimeHorizon*, *Radius*, and *PrefVelocity*—as the decision preferences based on the RVO2 library. The RVO2 library provides an easy-to-use implementation of the optimal reciprocal collision avoidance (ORCA) formulation [38] for multi-agent simulations. Recently, many multi-agent-based approaches have used these or similar parameters to calculate the locomotion of agents in pedestrian simulation scenarios. These parameters are described below.

- ***NeighborDist*** is the maximal distance of other agents that the agent considers during path-planning. Here, we use this parameter to represent the spatial scope that the agent must consider when making decisions.
- ***MaxNeighbors*** is the maximum number of neighbours affecting the steering behavior of an agent. These neighbours are selected by the rule of nearest in spatial distance. The mutual interactions between agents are an important aspect of the simulation in this study. We use *MaxNeighbors* to describe whether the movement of an agent is susceptible to other agents.
- ***TimeHorizon*** is the minimal amount of time for which the simulated agent velocities are safe with respect to other agents and obstacles. This parameter is used to describe the planning horizon in the path-planning process. Larger *TimeHorizon* values correspond to greater foresight of the agent.
- ***Radius*** is the personal space maintained by the agent to avoid collisions in motion. Agents attempt to preserve this space when other agents are around. In normal circumstances, the agent also tends to wait for available personal space before moving.
- ***PrefVelocity*** is the preferred velocity of the agent that is used if no other agents or obstacles are present. We use this parameter to express the expectation of the pedestrian to achieve the goal. Different individuals may have different velocities when faced with the same situation.

Guy et al. [11] studied the mapping from simulation parameters to the perceived behavior of agents, and they obtained a range of effective parameter values via a data analysis. In this paper, we use their results as empirical data to define the range of the decision preferences. The range and default values of these parameters are shown in Table 2.

Table 2. Range of the decision preferences and their default values.

Parameters	Value			Unit
	Min	Max	Default	
NeighborDist	3	30	15	m
MaxNeighbors	1	100	10	(n/a)
TimeHorizon	1	30	10	s
Radius	0.3	2.0	1.0	m
PrefVelocity	1.2	2.2	1.45	m/s

3.4. *Fuzzy Inference System*

In most decision-making processes, the capacity for addressing uncertainty and imprecision is a key issue and influences the quality of the decisions. The decision preferences reflect the cognitive ability of an individual with subjective differences. Precisely determining the values of these parameters is difficult because imprecise concepts are attributes of human cognitive abilities. For example, pedestrians are frequently described as attempting to move at a 'fast' speed rather than at '2 m/s'. In addition, the OCEAN personality model introduced in Section 3.2 is defined based on linguistic variables. Compared with the traditional methods [10,11], the mapping relationships with uncertainty and imprecision between the personality traits and the decision preferences can be easily determined.

In this paper, we incorporate a fuzzy logic system into the multi-agent simulation framework to handle the imprecise and uncertain issues of cognitive abilities on pedestrian dynamics. The fuzzy inference process includes four parts as follows:

- Under fuzzification, a crisp value of input variables can be transformed into a fuzzy value with membership degree by their membership functions;
- A rule base is the collection of the domain expert knowledge, and it is usually expressed as a set of 'IF-THEN' rules that are used to capture the relationship between inputs and outputs;
- During fuzzy inference, various fuzzy logic operations are used, and all fuzzy rules are triggered and combined to acquire a fuzzy consequence for each output variable;
- Under defuzzification, each fuzzy consequence must be transformed into a crisp value before the results can be used in simulation.

The structure of the proposed fuzzy logic system is shown in Figure 2, and more details are presented in the following section.

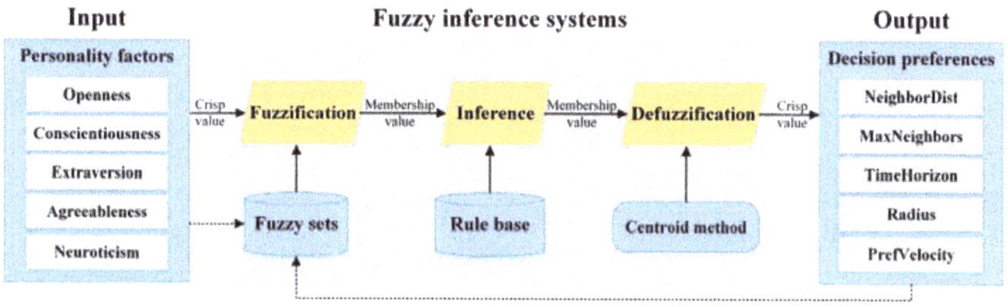

Figure 2. Structure of the proposed fuzzy logic system.

3.4.1. Fuzzy Membership Functions

In a fuzzy logic system, fuzzy sets and their membership functions should be defined for each input and output. Let U represent the universe of discourse, with elements of U denoted by x. A set A is a fuzzy subset of U. The degree to which an element x belongs to set A, which is a real number between 0 and 1, is called the membership value $A(x)$ in the fuzzy set A. The meaning of a fuzzy set A is characterized by a membership function μ_A that maps elements of a universe of discourse U to their corresponding membership values $A(x)$:

$$A(x) = \mu_A(x) \in [0,\ 1],\ x \in U. \tag{2}$$

The fuzzy membership function μ can be represented by a variety of shapes, such as triangles and trapezoids, depending on how the expert relates different domain values to belief values.

In our proposed fuzzy logic-based approach, we consider each of the personality factors and decision preferences as linguistic variables. The inputs are the OCEAN personality factors *Openness*,

Conscientiousness, *Extraversion*, *Agreeableness*, and *Neuroticism*. The outputs are the decision preferences *NeighborDist*, *MaxNeighbors*, *TimeHorizon*, *Radius*, and *PrefVelocity*. According to the description of the personality factors in Section 3.2 and the decision preferences in Section 3.3, we determine fuzzy sets of inputs and outputs as shown in Table 3, and their corresponding fuzzy membership functions are shown in Figure 3.

Table 3. Fuzzy attributes of inputs and outputs.

	Name	Universe of Discourse	Fuzzy Sets	Membership Function
Inputs	Each factor of OCEAN	[−100, 100]	Negative	Trapezoidal
			Neutral	Trapezoidal
			Positive	Trapezoidal
Outputs	NeighborDist	[3, 30]	Near	Trapezoidal
			Moderate	Trapezoidal
			Far	Trapezoidal
	MaxNeighbors	[1, 100]	Small	Trapezoidal
			Medium	Trapezoidal
			Large	Trapezoidal
	TimeHorizon	[1, 30]	Short	Trapezoidal
			Moderate	Triangular
			Long	Trapezoidal
	Radius	[0.3, 2.0]	Small	Trapezoidal
			Medium	Triangular
			Large	Trapezoidal
	PrefVelocity	[1.2, 2.2]	Slow	Triangular
			Moderate	Triangular
			Fast	Trapezoidal

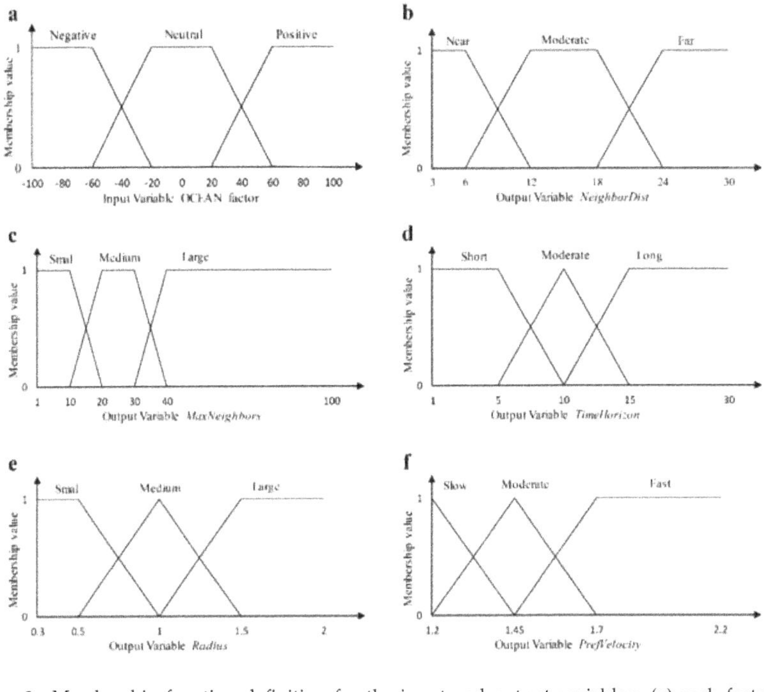

Figure 3. Membership function definition for the input and output variables: (**a**) each factor of the OCEAN personality model; (**b**) *NeighborDist*; (**c**) *MaxNeighbors*; (**d**) *TimeHorizon*; (**e**) *Radius*; (**f**) *PrefVelocity*.

3.4.2. Fuzzy Rules

As the most widely applied reasoning pattern, a set of 'IF-THEN' rules are used by the fuzzy rule-based method to obtain the relationship between inputs and outputs. The antecedent of a fuzzy rule is a logical combination of fuzzy propositions, which is usually in the form of 'x is A'. The consequence of a fuzzy rule is calculated by the degree to which the antecedent is satisfied. In general, fuzzy rules are formulated by the domain expert based on empirical knowledge, and they can be gradually improved with further use. An example of the 'IF-THEN' fuzzy rule is as follows:

IF (*Extraversion* is Positive) AND (*Agreeableness* is Positive) THEN *Radius* is small.

The accuracy of a fuzzy inference system is affected by the number of linguistic fuzzy sets that cover the universe of discourse. However, the size of fuzzy rules grows exponentially with the number of input fuzzy sets of the antecedent. To obtain a trade-off between the computational complexity and accuracy, we select two personality factors as the major impact factors for each decision parameter. In this study, we construct a rule base containing fuzzy rules on the relationship between the OCEAN personality factors and the decision preferences based on previous studies [10,11]. The details are summarized as follows:

- *NeighborDist* is related to *Openness* and *Conscientiousness*. Pedestrians with positive *Openness* are more curious about their surroundings; therefore, they explore a larger scope in the path-planning process. Pedestrians who have a positive *Conscientiousness* trait notice individuals in the distance because they are predictable and self-disciplined.
- *MaxNeighbors* is related to *Openness* and *Neuroticism*. Pedestrians with positive *Openness* are likely to observe the behavior of more people around them. An important facet of the *Neuroticism* factor is sensitivity, with sensitive individuals more easily affected by others.
- *TimeHorizon* is related to *Conscientiousness* and *Agreeableness*. To maintain the orderly pedestrian flow, individuals with positive *Conscientiousness* tend to be prepared for upcoming events in advance. The *Agreeableness* factor describes the cooperative tendency of people. Pedestrians with a stronger cooperative tendency will respond earlier to avoid a collision during path-planning.
- *Radius* is related to *Extraversion* and *Agreeableness*. Individuals with positive *Extraversion* are outgoing and sociable and maintain a small territory in which they feel comfortable. Friendly individuals who have a positive *Agreeableness* trait usually do not react harshly when others are too close.
- *PrefVelocity* is related to *Extraversion* and *Neuroticism*. An individual's energy level is the key factor that determines their preferred velocity. In generally, extroverts tend to be more energetic and thus have a fast *PrefVelocity*, whereas introverts are more lethargic and present opposite characteristics to that of extroverts. Pedestrians with positive *Neuroticism* are prone to be anxious and tense when congestion occurs; therefore, they try to pass at a fast speed.

Because the antecedent has two input variables and each personality factor has three fuzzy sets, a total of 45 (or 5×3^2) 'IF-THEN' rules are established to deduce the decision preferences of pedestrian dynamical behavior. In Table 4, we report several sample fuzzy rules.

Table 4. 'IF-THEN' fuzzy rules for the decision preferences.

Rule Number	IF-THEN Statements	
	Antecedent	**Consequence**
R_1	IF (*O* is Negative) AND (*C* is Negative)	THEN *NeighborDist* is Near
R_2	IF (*O* is Negative) AND (*C* is Neutral)	THEN *NeighborDist* is Near
R_3	IF (*O* is Negative) AND (*C* is Positive)	THEN *NeighborDist* is Moderate
R_4	IF (*O* is Neutral) AND (*C* is Negative)	THEN *NeighborDist* is Moderate
R_5	IF (*O* is Neutral) AND (*C* is Neutral)	THEN *NeighborDist* is Moderate
R_6	IF (*O* is Neutral) AND (*C* is Positive)	THEN *NeighborDist* is Far
R_7	IF (*O* is Positive) AND (*C* is Negative)	THEN *NeighborDist* is Moderate
R_8	IF (*O* is Positive) AND (*C* is Neutral)	THEN *NeighborDist* is Far
R_9	IF (*O* is Positive) AND (*C* is Positive)	THEN *NeighborDist* is Far
...	...	...
R_{44}	IF (*E* is Positive) AND (*N* is Neutral)	THEN *PrefVelocity* is Fast
R_{45}	IF (*E* is Positive) AND (*N* is Positive)	THEN *PrefVelocity* is Fast

3.4.3. Inference Method

Three types of fuzzy inference methods are available: Mamdani, Larsen, and Takagi–Sugeno. In this paper, the Mamdani inference method [39], which includes aggregation, activation, and accumulation steps, is selected to calculate the fuzzy output of each decision parameter based on the Sup-Min composition.

The fuzzy rules proposed in Section 3.4.2 are a multidimensional multiple fuzzy reasoning model. The general format is as follows:

$$\begin{array}{cccccc} A_{11}, & A_{12}, & \cdots, & A_{1n} & \rightarrow & B_1 \\ A_{21}, & A_{22}, & \cdots, & A_{2n} & \rightarrow & B_2 \\ \vdots & \vdots & \cdots & \vdots & & \vdots \\ A_{m1}, & A_{m2}, & \cdots, & A_{mn} & \rightarrow & B_m \\ A_1^*, & A_2^*, & \cdots, & A_n^* & & \\ \hline & & & B^* & & \end{array} \tag{3}$$

where A_{ij} and A_j^* are the fuzzy subsets of U_j; A_{ij} represents the jth input of the ith fuzzy rule in a fuzzy inference model; A_j^* represents the jth input of an actual antecedent; B_i and B^* are the fuzzy subsets of V; B_i represents the output of the ith rule; B^* represents the composite output of an actual antecedent ($i = 1, 2, \cdots, m$; $j = 1, 2, \cdots, n$); m is the number of fuzzy rules for a fuzzy inference model; and n is the number of antecedent inputs of an 'IF-THEN' fuzzy rule. In this study, V and U_j are the universe of discourse of one decision preference and its corresponding personality factor, respectively.

The inference process is written as follows:

$$\begin{aligned} A_1(x) &= \min\{A_{11}(x_1), A_{12}(x_2), \cdots, A_{1n}(x_n)\} \\ A_2(x) &= \min\{A_{21}(x_1), A_{22}(x_2), \cdots, A_{2n}(x_n)\} \\ &\vdots \\ A_m(x) &= \min\{A_{m1}(x_1), A_{m2}(x_2), \cdots, A_{mn}(x_n)\} \\ A^*(x) &= \min\{A_1^*(x_1), A_2^*(x_2), \cdots, A_n^*(x_n)\} \\ B_1^*(y) &= \bigvee_{x \in U}[A^*(x) \wedge A_1(x) \wedge B_1(y)] \\ B_2^*(y) &= \bigvee_{x \in U}[A^*(x) \wedge A_2(x) \wedge B_2(y)] \\ &\vdots \\ B_m^*(y) &= \bigvee_{x \in U}[A^*(x) \wedge A_m(x) \wedge B_m(y)] \\ B^*(y) &= B_1^*(y) \vee B_2^*(y) \vee \cdots \vee B_m^*(y) \end{aligned} \tag{4}$$

where x_j $(j = 1, 2, \cdots, n)$ is the input value of the personality factor, and $B_i^*(y)$ $(i = 1, 2, \cdots, m)$ is the intermediate result of each 'IF-THEN' rule, with m = 9 and n = 2. The operators $\wedge$ and $\vee$ take the minimum and maximum values of the membership functions, respectively; $B^*(y)$ represents a composite fuzzy set of output decision preferences.

3.4.4. Defuzzification Method

Because the consequences of the fuzzy rule set are also fuzzy subsets, these subsets must be transformed into crisp values through the defuzzification process. For defuzzification, several methods are available, including the mean of the maxima, average of the maxima, and centroid. Because of the accuracy of these methods, the most frequently used centroid method is adopted in this paper. The formula is given as follows:

$$y_{final} = \frac{\int_V B^*(y) y dy}{\int_V B^*(y) dy} \quad (5)$$

where y_{final} is a final output of the fuzzy inference system, i.e., each one of the decision preferences.

In order to depict the relationship between inputs of personality factors toward the output of decision preferences, we show the surface plots of input and output of the fuzzy simulation in Figure 4 generated in Matlab.

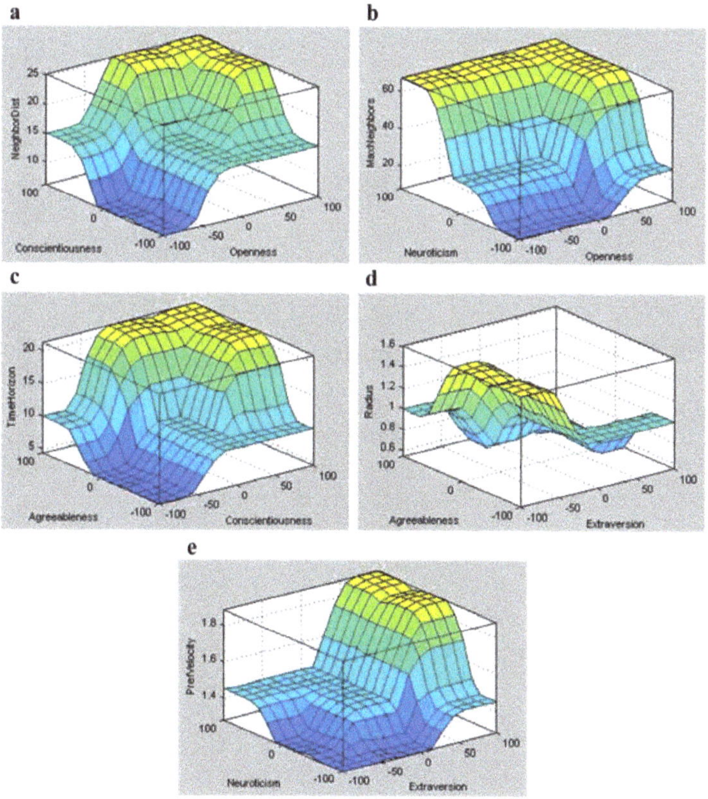

Figure 4. Surface plots of input and output of the fuzzy simulation (**a**) *NeighborDist*, (**b**) *MaxNeighbors*, (**c**) *TimeHorizon*, (**d**) *Radius*, and (**e**) *PrefVelocity*.

4. Simulation and Validation

To test the proposed pedestrian model, we have developed a Visual C++ application based on publicly available Open GL and RVO2 [40] libraries to perform a variety of pedestrian simulation scenarios. All experiments are executed in real time by a PC (Dell OptiPlex 9020) with an Intel Core 3.6 GHz i7-4790 processor, an 8 GB memory and an AMD Radeon R5 240 graphics card with 1 GB memory. In this paper, we focus on the modelling and prediction of heterogeneous pedestrian dynamical behavior under normal or non-panic situations. Therefore, we simulated pedestrians with various personalities in several typical scenarios, including a single-exit room, a hallway with obstacles, and a narrowing passage. Furthermore, the effectiveness of our proposed model is validated by a user study. Although the distribution of personality traits in the population is closer to the normal distribution, we artificially set the distributions in our experiments, in order to obviously demonstrate different behaviors between heterogeneous pedestrians with various personalities.

4.1. Simulation of Heterogeneous Pedestrians

Figure 5 shows the motion trajectories taken by the highlighted agents in a single-exit room scenario. From left to right, the highlighted agents represent the pedestrians with positive (i.e., $V(i) = 80$) *Openness* (blue), *Conscientiousness* (green), *Extraversion* (red), *Agreeableness* (orange), and *Neuroticism* (black) traits. The non-highlighted agents (grey) are given the default personality. The blue agent tends to choose more daring routes and takes a wavy trajectory. The green agent takes a fairly direct trajectory and moves in an orderly way behind the other agents. The red agent moves quickly and often tries to weave through others in the crowd. The orange agent can slightly adjust the direction of movement to avoid collision with the grey agents and thus takes a less direct trajectory. The black agent also moves quickly but takes a tortuous route when congestion occurs near the exit.

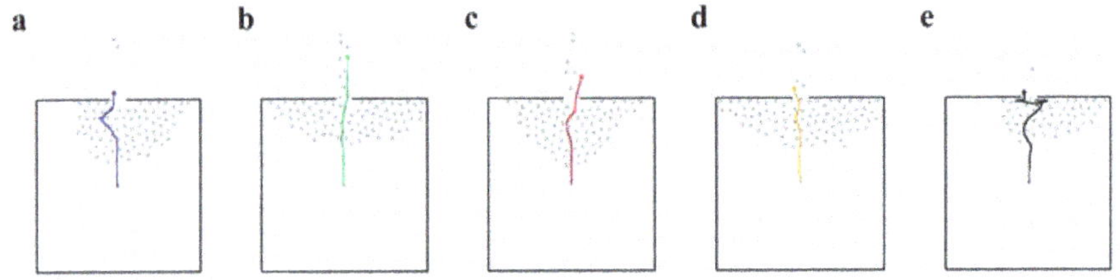

Figure 5. Motion trajectories of agents with default personality factors, although only one factor is positive. From left to right, the traits are (**a**) *Openness*, (**b**) *Conscientiousness*, (**c**) *Extraversion*, (**d**) *Agreeableness*, and (**e**) *Neuroticism*. All trajectories are displayed for an equal length of time.

With the same configuration above, Figure 6 demonstrates the different trajectories of five agents with distinct personality traits in the hallway with obstacles scenario. The agent with a positive *Extraversion* or *Neuroticism* trait can clearly be distinguished from the crowd. The agent with a positive *Extraversion* trait can pass through the crowd at a faster speed than others. The agent with a positive *Neuroticism* trait is easily deflected by the movement of the other agents. As for the remaining types, they take relatively straight trajectories and have no obvious differences under a non-congested situation.

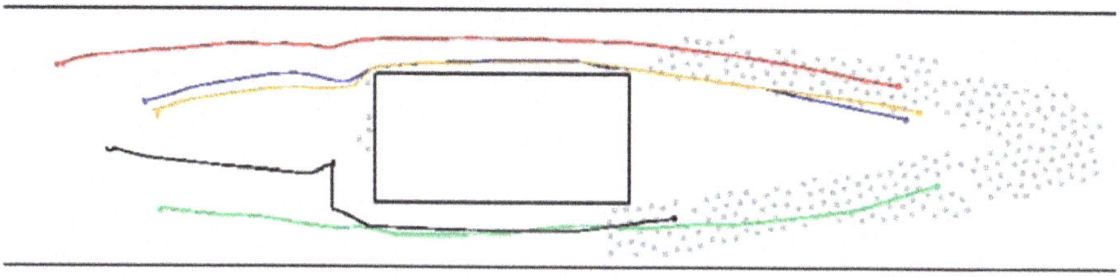

Figure 6. Comparison of motion trajectories among five agents with different personality types in the hallway with obstacles scenario. The agents with positive *Openness* (blue), *Conscientiousness* (green), *Extraversion* (red), *Agreeableness* (orange), and *Neuroticism* (black) traits are highlighted. The non-highlighted agents (grey) have the default personality traits.

In addition to comparing motion trajectories among agents with five personality traits, we have simulated the behavior of agents with a different level of each personality trait. Here, agents are simulated with a positive (i.e., $V(i) = 80$) or negative (i.e., $V(i) = -80$) *Extraversion* trait in the narrowing passage scenario as an example. Figure 7 shows the results at the same time from two simulations in which the red agents are assigned a positive *Extraversion* trait on the left and a negative *Extraversion* trait on the right. Agents with a positive *Extraversion* trait tend to move quicker and closer to others than those with the negative trait. Furthermore, more agents simultaneously pass through the exit on the left. Agents with a positive *Extraversion* trait exit more efficiently in the narrowing passage than those with the negative trait.

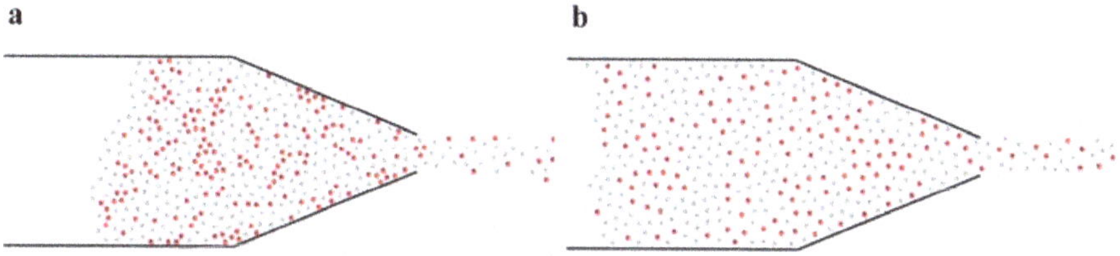

Figure 7. Comparison between (**a**) red agents with positive *Extraversion* and (**b**) red agents with negative *Extraversion* in the narrowing passage scenario for an equal length of time.

Figure 8 shows a comparison between homogeneous and heterogeneous pedestrian simulations in a single-exit room scenario. In the left simulation, all agents have the default personality. In the right simulation, the agents have a variety of personalities that are randomly generated. At the beginning of the two simulations, the agents are randomly distributed and have no significant differences as shown in Figure 8a. When the congestion phenomenon begins to occur (Figure 8b), the heterogeneous agents display various behaviors because of their different personality traits. For example, most of the agents with a positive *Extraversion* trait (red) walk at the front of the crowd and try to quickly exit by weaving through others. We can see from Figure 8c that the results of the homogeneous simulation are symmetrical and artificial when the congestion becomes serious. In contrast, the heterogeneous simulation is more reasonable and natural. In the final stage of the heterogeneous simulation as shown in Figure 8d, the agents with a positive *Conscientiousness* trait (green) are the majority of individuals who have not yet left the room. This result is consistent with the characteristics of self-disciplined pedestrians.

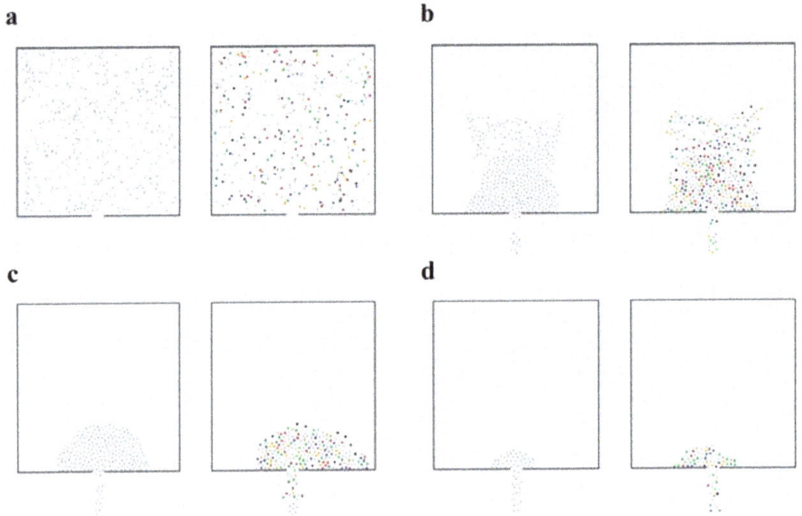

Figure 8. Comparison between homogeneous and heterogeneous pedestrians grouped by 200 individuals in a single-exit room scenario. The left panels show homogeneous pedestrians, and the right panels show heterogeneous pedestrians.

4.2. Effect of Different Personalities During Evacuation

To reveal how holistic crowd behavior is influenced by different personalities, we have simulated a pedestrian evacuation in a single-exit room by changing the percentage of pedestrians with a certain personality trait. Similar to the scenario described above, the scenario is a square room sized 20 m $\times$ 20 m with a 1.5 m wide exit in the middle of the wall. The total number of agents is 200, and their initial locations and directions are given at random in the room. Here, we choose the following five types as an illustrative example: including agents with positive *Openness* (O+), *Conscientiousness* (C+), *Extraversion* (E+), *Agreeableness* (A+), and *Neuroticism* (N+). Except for these agent types, the other agents all have the default personality.

In Figure 9, we plot the evacuation time versus the percentage of a certain type of agent in the crowd. The blue line indicates that the proportion of O+ agents in the crowd has little effect on the total time of evacuation, and the green and orange lines show that with more C+ or A+ agents in the crowd, the evacuation can be more effective. The evacuation time decreases with an increasing proportion of E+ agents in the crowd when the proportion is not very large. Nevertheless, the evacuation time increases if the proportion is larger than approximately 60%. This result is related to the limitation of exit width and the occurrence of congestion, thus reflecting the well-known 'faster-is-slower' phenomenon. The black line shows that a greater proportion of N+ agents leads to a significant increase in evacuation time.

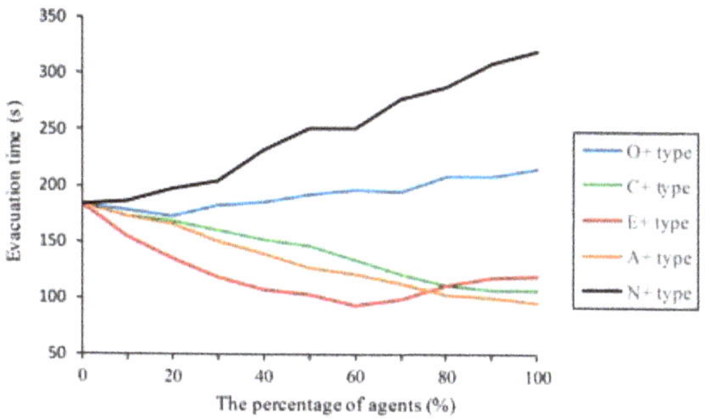

Figure 9. Evacuation time versus the percentage of certain types of agents in the crowd.

4.3. Validation of the Proposed Model

Evaluating the simulation results for pedestrian dynamical behavior influenced by human personality is a challenge. Here, the user study method is used to validate the proposed model because the real data are difficult to obtain and identify. This method has been proven feasible and effective by previous works [5,10,11]. A user study was completed by 72 participants (33 females and 39 males, ages 16 to 51) who had no previous knowledge of the experiment. Moreover, we repeated the same test three times to all participants, using new video examples for each time, to avoid testing them on best-case simulation scenarios only. Before the evaluation, participants were given a brief explanation of each factor of the OCEAN personality model. We designed a user study consisting of two sections (the first contains 5 questions and the second contains 10 questions).

The first section was designed to evaluate how well the different levels of each trait of the OCEAN model could be reproduced by our model. For each question, we created a pair of videos that simulate the movement of agents with a certain trait in the narrowing passage scenario. In each pair of videos, the simulations are generated by the highlighted agents with the positive and negative types of the trait. After watching the videos, the participants were asked to choose which simulation displayed the positive type trait in question. Figure 10 shows that the success rate for distinguishing between different levels of personality traits is high, especially for Neuroticism and Extraversion, which are 95.8% and 94.4% respectively, and the rates for Agreeableness, Conscientiousness and Openness are 81.9%, 79.2% and 73.6%, respectively. To test the statistical significance of these results, we use a two-tailed binomial test to calculate the statistical p-values. The results indicate that the participants could correctly distinguish the positive and negative traits at a statistically significant rate ($p < 0.05$).

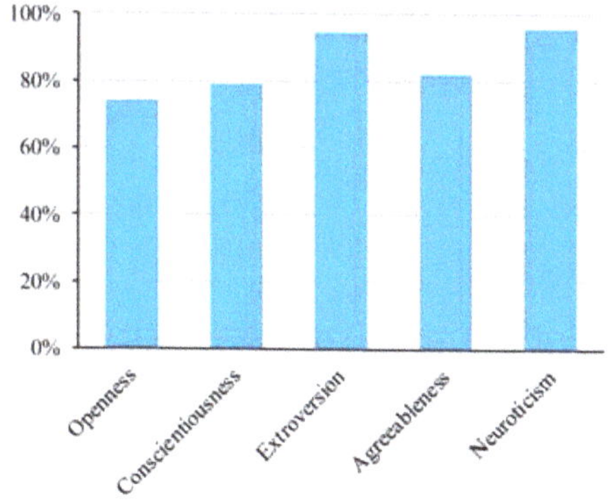

Figure 10. Success rate of distinguishing between the different levels of personality traits.

The second section was intended to reflect the degree to which the five personality traits could be distinguished in the simulation by our model. All five positive traits have been compared with each other; thus, 10 videos presenting evacuation behavior in a single-exit room scenario were created. In each video, two highlighted groups of agents were assigned different personality traits (e.g., O+ and C+). In a corresponding question, the participants were asked to select which group was O+ and which was C+. The results are summarized in Table 5, which shows that most of the OCEAN personality traits could be correctly distinguished by the participants at a statistically significant rate ($p < 0.05$). Among these traits, the E+ and N+ traits were easily distinguished from the other traits. However, the participants had difficulty distinguishing between the C+ and A+ traits because of their similar characteristics in movement.

Table 5. Accuracy and *p*-values for distinguishing traits.

Distinguishing Traits	Accuracy	*p*-Value
O+ and C+	72.2%	2×10^{-4}
O+ and E+	78.8%	2×10^{-6}
O+ and A+	76.4%	8×10^{-6}
O+ and N+	87.5%	4×10^{-11}
C+ and E+	91.7%	8×10^{-14}
C+ and A+	66.7%	6×10^{-3}
C+ and N+	95.8%	2×10^{-15}
E+ and A+	79.2%	7×10^{-7}
E+ and N+	84.7%	2×10^{-9}
A+ and N+	95.8%	2×10^{-15}

5. Conclusions

In this paper, we propose a new pedestrian model that incorporates fuzzy logic theory into a multi-agent system to address cognitive behavior that introduces uncertainty and imprecision during decision-making. This model can describe the imprecise subjective decisions of a pedestrian while steering. Subjective decisions vary for pedestrians confronted with similar situations. To simulate heterogeneous pedestrians, the OCEAN personality model is introduced to model the psychological characteristics of a pedestrian, and it can generate the different decision preferences that represent the intrinsic control factors of decision-making. The fuzzy relationships between personality traits and decision preferences are determined by a fuzzy inference system. Finally, a variety of simulations and validation experiments are implemented in our developed application. The experimental results show that the proposed model can exhibit more reasonable and heterogeneous behavior in various scenarios and improve the credibility of the simulation; thus, it can be used to analyse how different personalities influence the crowd phenomena.

In the future, we intend to consider the aspect of physiological characteristics when simulating the heterogeneous pedestrian behaviors because pedestrian decision-making in movement is always subject to physiological conditions. Accordingly, the simulation effect of pedestrian dynamical behavior can be improved by combining psychological characteristics with physiological characteristics. Moreover, we intend to study pedestrian behavior in abnormal or unusual situations by adding an emotion model for the agents because emergent behavior can be profoundly influenced by instantaneous emotions.

Acknowledgments: This work was partly supported by the National Key Research and Development Program of China (Grant nos. 2016YFB0501502 and 2016YFB0501503), the Hainan Provincial Department of Science and Technology (Grant no. ZDKJ2016021), the Natural Science Foundation of Hainan (Grant no. 20154171) and the 135 Plan Project of Chinese Academy of Sciences (Grant no. Y6SG0200CX).

Author Contributions: Zhuxin Xue, Qing Dong, Xiangtao Fan, and Qingwen Jin conceived and worked together to achieve this work. Hongdeng Jian and Jian Liu performed the experiments. Zhuxin Xue wrote the paper.

Conflicts of Interest: The authors declare no conflict of interest.

References

1. Treuille, A.; Cooper, S.; Popovic, Z. Continuum crowds. *ACM Trans. Graph.* **2006**, *25*, 1160–1168. [CrossRef]
2. Helbing, D.; Molnár, P. Social force model for pedestrian dynamics. *Phys. Rev. E* **1995**, *51*, 4282–4286. [CrossRef]
3. Burstedde, C.; Klauck, K.; Schadschneider, A.; Zittartz, J. Simulation of pedestrian dynamics using a two-dimensional cellular automaton. *Phys. A Stat. Mech. Appl.* **2001**, *295*, 507–525. [CrossRef]
4. Helbing, D.; Farkas, I.; Vicsek, T. Simulating dynamical features of escape panic. *Nature* **2000**, *407*, 487–490. [CrossRef] [PubMed]
5. Cai, L.; Yang, Z.; Simon, X.; Qu, H. Modelling and simulating of risk behaviours in virtual environments based on multi-agent and fuzzy logic. *Int. J. Adv. Robot. Syst.* **2013**, *10*, 387. [CrossRef]
6. Brambilla, M.; Cattelani, L. Mobility analysis inside buildings using distrimobs simulator: A case study. *Build. Environ.* **2009**, *44*, 595–604. [CrossRef]
7. Berg, J.V.D.; Lin, M.; Manocha, D. Reciprocal velocity obstacles for real-time multi-agent navigation. In Proceedings of the IEEE International Conference on Robotics & Automation, Pasadena, CA, USA, 19–23 May 2008; pp. 1928–1935.
8. Llorca, D.F.; Milanes, V.; Alonso, I.P.; Gavilan, M.; Daza, I.G.; Perez, J.; Sotelo, M.Á. Autonomous pedestrian collision avoidance using a fuzzy steering controller. *IEEE Trans. Intell. Transp. Syst.* **2011**, *12*, 390–401. [CrossRef]
9. Zheng, L.; Qin, D.; Cheng, Y.; Wang, L.; Li, L. Simulating heterogeneous crowds from a physiological perspective. *Neurocomputing* **2016**, *172*, 180–188. [CrossRef]
10. Durupinar, F.; Pelechano, N.; Allbeck, J.M.; Gudukbay, U.; Badler, N.I. How the ocean personality model affects the perception of crowds. *IEEE Comput. Graph. Appl.* **2011**, *31*, 22–31. [CrossRef] [PubMed]
11. Guy, S.J.; Kim, S.; Lin, M.C.; Manocha, D. Simulating heterogeneous crowd behaviors using personality trait theory. In Proceedings of the International Conference on Computer Graphics and Virtual Reality, Vancouver, BC, Canada, 5–7 August 2011.
12. Hughes, R.L. A continuum theory for the flow of pedestrians. *Transp. Res. B Methodol.* **2002**, *36*, 507–535. [CrossRef]
13. Reynolds, C.W. Flocks, herds and schools: A distributed behavioral model. *ACM SIGGRAPH Comput. Graph.* **1987**, *21*, 25–34. [CrossRef]
14. Pan, X.; Han, C.S.; Dauber, K.; Law, K.H. A multi-agent based framework for the simulation of human and social behaviors during emergency evacuations. *AI Soc.* **2007**, *22*, 113–132. [CrossRef]
15. Zadeh, L.A. Fuzzy sets. *Inf. Control* **1965**, *8*, 338–353. [CrossRef]
16. Lu, G.; Noyce, D.A. Pedestrian crosswalks at midblock locations: Fuzzy logic solution to existing signal operations. *Transp. Res. Rec.* **2009**, *45*, 63–78. [CrossRef]
17. Niittymaki, J.; Kikuchi, S. Application of fuzzy logic to the control of a pedestrian crossing signal. *Transp. Res. Rec.* **1998**, *1651*, 30–38. [CrossRef]
18. Nasir, M.; Nahavandi, S.; Creighton, D. Fuzzy simulation of pedestrian walking path considering local environmental stimuli. In Proceedings of the 2012 IEEE International Conference on Fuzzy Systems, Brisbane, Australia, 10–15 June 2012; pp. 1–6.
19. Zhu, B.; Liu, T.; Tang, Y. Research on pedestrian evacuation simulation based on fuzzy logic. In Proceedings of the 2008 9th International Conference on Computer-Aided Industrial Design and Conceptual Design, Kunming, China, 22–25 November 2008; pp. 1024–1029.
20. Li, M.; Li, S.; Liang, J. A fuzzy logic based approach for crowd simulation. In *Advances in Electronic Commerce, Web Application and Communication, Volume 2 (Advances in Intelligent and Soft Computing)*; Jin, D., Lin, S., Eds.; Springer: Berlin/Heidelberg, Germany, 2012; pp. 29–35.
21. Dell'Orco, M.; Marinelli, M.; Ottomanelli, M. Simulation of crowd dynamics in panic situations using a fuzzy logic-based behavioural model. In *Computer-Based Modelling and Optimization in Transportation*; de Sousa, J.F., Rossi, R., Eds.; Springer: Cham, Switzerland, 2014; pp. 237–250.
22. Nasir, M.; Lim, C.P.; Nahavandi, S.; Creighton, D. A genetic fuzzy system to model pedestrian walking path in a built environment. *Simul. Modell. Pract. Theor.* **2014**, *45*, 18–34. [CrossRef]
23. Fu, L.; Song, W.; Lo, S. A fuzzy-theory-based behavioral model for studying pedestrian evacuation from a single-exit room. *Phys. Lett. A* **2016**, *380*, 2619–2627. [CrossRef]

24. Zhou, M.; Dong, H.; Wang, F.Y.; Wang, Q.; Yang, X. Modeling and simulation of pedestrian dynamical behavior based on a fuzzy logic approach. *Inf. Sci.* **2016**, *360*, 112–130. [CrossRef]
25. McCrae, R.R.; Costa, P.T. Validation of the five-factor model of personality across instruments and observers. *J. Personal. Soc. Psychol.* **1987**, *52*, 81–90. [CrossRef]
26. Zheng, X.; Li, W.; Guan, C. Simulation of evacuation processes in a square with a partition wall using a cellular automaton model for pedestrian dynamics. *Phys. A Stat. Mech. Appl.* **2010**, *389*, 2177–2188. [CrossRef]
27. Blue, V.J.; Adler, J.L. Cellular automata microsimulation for modeling bi-directional pedestrian walkways. *Transp. Res. B Methodol.* **2001**, *35*, 293–312. [CrossRef]
28. Dai, J.; Li, X.; Liu, L. Simulation of pedestrian counter flow through bottlenecks by using an agent-based model. *Phys. A Stat. Mech. Appl.* **2013**, *392*, 2202–2211. [CrossRef]
29. Akasaka, Y.; Onisawa, T. Individualized pedestrian navigation using fuzzy measures and integrals. In Proceedings of the 2005 IEEE International Conference on Systems, Man and Cybernetics, Waikoloa, HI, USA, 12 October 2005; pp. 1461–1466.
30. Teknomo, K.; Gerilla, G.P. Fuzzy perceptional spacing for intelligent multi agent pedestrian simulation. In Proceedings of the International Symposium of Lowland Technology, Saga, Japan, 14–16 September 2006.
31. Pelechano, N.; Allbeck, J.M.; Badler, N.I. Controlling individual agents in high-density crowd simulation. In Proceedings of the 2007 ACM SIGGRAPH/Eurographics Symposium on Computer Animation, San Diego, CA, USA, 3–4 August 2007; pp. 99–108.
32. Balaji, P.G.; Srinivasan, D. An introduction to multi-agent systems. In *Innovations in Multi-Agent Systems and Applications—1*; Srinivasan, D., Jain, L.C., Eds.; Springer: Berlin/Heidelberg, Germany, 2010; pp. 1–27.
33. Ghasemaghaee, N.; Oren, T.I. In Towards fuzzy agents with dynamic personality for human behavior simulation. In Proceedings of the Summer Computer Simulation Conference, Montreal, PQ, Canada, 20–24 July 2003.
34. Phares, E.J. *Introduction to Psychology*, 3rd ed.; Harper Collins Publishers: New York, NY, USA, 1991.
35. Eysenck, H.J.; Eysenck, M.W. *Personality and Individual Differences: A Natural Science Approach*; Plenum Press: New York, NY, USA, 1985.
36. McCrae, R.R.; John, O.P. An introduction to the five-factor model and its applications. *J. Personal.* **1992**, *60*, 175–215. [CrossRef]
37. Goldberg, L.R. An alternative "description of personality": The big-five factor structure. *J. Personal. Soc. Psychol.* **1990**, *59*, 1216–1229. [CrossRef]
38. Berg, J.V.D.; Guy, S.J.; Lin, M.; Manocha, D. Reciprocal n-body collision avoidance. *Springer Tracts Adv. Robot.* **2011**, *70*, 3–19.
39. Mamdani, E.H. Application of fuzzy algorithms for control of simple dynamic plant. *Proc. Inst. Electr. Eng.* **1974**, *121*, 1585–1588. [CrossRef]
40. Berg, J.V.D.; Guy, S.J.; Snape, J.; Lin, M.; Manocha, D. RVO2 Library: Reciprocal Collision Avoidance for Real-Time Multi-Agent Simulation. Available online: http://gamma.cs.unc.edu/RVO2 (accessed on 4 October 2017).

MDPI
St. Alban-Anlage 66
4052 Basel, Switzerland
Tel. +41 61 683 77 34
Fax +41 61 302 89 18
http://www.mdpi.com

Symmetry Editorial Office
E-mail: symmetry@mdpi.com
http://www.mdpi.com/journal/symmetry

www.ingramcontent.com/pod-product-compliance
Lightning Source LLC
Chambersburg PA
CBHW051706210326
41597CB00032B/5392
* 9 7 8 3 0 3 8 4 2 8 8 7 9 *